Microscale Acoustofluidics

Microscale Acoustofluidics

Edited by

Thomas Laurell
Lund University, Lund, Sweden
Email: thomas.laurell@bme.lth.se

Andreas Lenshof
Lund University, Lund, Sweden
Email: andreas.lenshof@bme.lth.se

Print ISBN: 978-1-84973-671-8
PDF eISBN: 978-1-84973-706-7

A catalogue record for this book is available from the British Library

© The Royal Society of Chemistry 2015

All rights reserved

Apart from fair dealing for the purposes of research for non-commercial purposes or for private study, criticism or review, as permitted under the Copyright, Designs and Patents Act 1988 and the Copyright and Related Rights Regulations 2003, this publication may not be reproduced, stored or transmitted, in any form or by any means, without the prior permission in writing of The Royal Society of Chemistry or the copyright owner, or in the case of reproduction in accordance with the terms of licences issued by the Copyright Licensing Agency in the UK, or in accordance with the terms of the licences issued by the appropriate Reproduction Rights Organization outside the UK. Enquiries concerning reproduction outside the terms stated here should be sent to The Royal Society of Chemistry at the address printed on this page.

The RSC is not responsible for individual opinions expressed in this work.

The authors have sought to locate owners of all reproduced material not in their own possession and trust that no copyrights have been inadvertently infringed.

Published by The Royal Society of Chemistry,
Thomas Graham House, Science Park, Milton Road,
Cambridge CB4 0WF, UK

Registered Charity Number 207890

Visit our website at www.rsc.org/books

Printed and bound by CPI Group (UK) Ltd, Croydon, CR0 4YY

Preface

Over the last fifteen years acoustophoresis, *i.e.* the manipulation of particles by means of acoustic forces, in microfluidic structures has taken giant leaps forward. Much insight and improved understanding of these complex systems have been gained in all aspects, ranging from basic theory to modeling, design and fabrication. As a consequence, the number of applications in which the possibility of non-contact handling of cells and particles in a gentle way is also increasing rapidly.

The origin of this book began at a summer school on acoustofluidics in Udine, Italy in 2010. The information presented was overwhelming, and voices were raised to try to compile this wealth of specialized knowledge in a comprehensive manner for the broader scientific community.

This eventually became a series of tutorial review papers published in the journal *Lab on a Chip* between 2011 and 2013, in which 23 separate papers covered different aspects of acoustofluidics, starting with *Lab Chip*, 2011, **11**, 3579. This tutorial series serves as the foundation for this book. Since then, updates of recent developments and additional chapters have been written to provide a more comprehensive, broader and up to date overview of the field of microscale acoustofluidics.

It is our hope that the presented material can serve as a valuable source of information in your scientific or educational work.

Andreas Lenshof and Thomas Laurell
Lund

Contents

Chapter 1	**Governing Equations in Microfluidics** *Henrik Bruus*	**1**
	1.1 Introduction	1
	1.2 The Basic Continuum Fields	2
	1.3 Mathematical Notation	3
	1.4 Governing Equations	5
	1.4.1 The Continuity Equation	5
	1.4.2 The Navier–Stokes Equation	7
	1.4.3 The Heat-transfer Equation	9
	1.5 Flow Solutions	11
	1.5.1 Hydrostatic Pressure	11
	1.5.2 The Reynolds Number and the Stokes Equation	12
	1.5.3 Poiseuille Flow	13
	1.5.4 Flow Rate	16
	1.6 Equivalent Circuit Modeling	16
	1.6.1 Hydraulic Resistance	17
	1.6.2 Hydraulic Compliance	18
	1.6.3 Hydraulic Inductance	20
	1.7 Scaling Laws in Microfluidics	20
	1.7.1 Flow Rate	21
	1.7.2 Characteristic Dimensionless Numbers	22
	1.7.3 Entrance Length	24
	1.7.4 Inertial Time Scale	25
	References	27

Microscale Acoustofluidics
Edited by Thomas Laurell and Andreas Lenshof
© The Royal Society of Chemistry 2015
Published by the Royal Society of Chemistry, www.rsc.org

Chapter 2	**Perturbation Theory and Ultrasound Resonances**	29
	Henrik Bruus	

	2.1 Introduction	29
	2.2 First-order Perturbation Theory, the Acoustic Wave Equation	30
	2.3 Second-order Perturbation Theory and Acoustic Streaming	32
	2.4 Basics of Acoustic Resonances in Viscous Liquids	35
	2.5 Acoustic Eigenmodes	38
	2.5.1 Boundary Conditions	38
	2.5.2 An Ideal Water-filled Rectangular Channel	39
	2.5.3 Viscous and Radiative Losses in the Rectangular Channel	40
	2.6 Geometrical Effects and FEM Simulations	42
	2.6.1 Finite Element Method (FEM) Simulations	42
	2.6.2 Symmetry Breaking in Acoustic Resonances	43
	References	45

Chapter 3	**Continuum Mechanics for Ultrasonic Particle Manipulation**	46
	Jürg Dual and Thomas Schwarz	

	3.1 Introduction	46
	3.2 Linear Elastodynamics	47
	3.3 Deformations of Structures	51
	3.4 Fluid Structure Interaction at the Device Level	56
	3.4.1 Acoustic Radiation from a Vibrating Surface	56
	3.4.2 Acoustic Radiation from a Plate Vibrating Harmonically	58
	3.4.3 Vibrations of Devices for Particle Manipulation	59
	3.5 Example of a Mechanical Model of an Ultrasonic Cavity used for Particle Manipulation	61
	3.6 Conclusions	63
	References	64

Chapter 4	**Acoustic Radiation Force on Small Particles**	65
	Henrik Bruus	

	4.1 Introduction	65
	4.2 The Acoustic Radiation Force	66
	4.3 Scattering Theory	67
	4.3.1 Scattering and the Radiation Force	68
	4.3.2 The Near-field Potential	70
	4.3.3 The Monopole Coefficient f_1	71

Contents ix

	4.3.4 The Dipole Coefficient f_2	71
	4.3.5 The Resulting Radiation Force for a Standing Plane Wave	72
4.4	Acoustophoretic Particle Tracks	74
4.5	Energy Density as a Function of the Applied Piezo Voltage	77
4.6	Viscous Corrections to the Radiation Force	78
	References	79

Chapter 5 Piezoelectricity and Application to the Excitation of Acoustic Fields for Ultrasonic Particle Manipulation — 81
Jürg Dual and Dirk Möller

5.1	Introduction	81
5.2	Basic Equations	82
5.3	Vibration of a Free Piezoelectric Element Excited by an Applied Electrical Voltage	85
5.4	Piezoelectric Transducers Used to Excite Mechanical Vibrations in a Structure	87
5.5	FEM Model Example of an Ultrasonic Cavity used for Particle Manipulation	95
5.6	Conclusions	98
	References	98

Chapter 6 Building Microfluidic Acoustic Resonators — 100
Andreas Lenshof, Mikael Evander, Thomas Laurell, and Johan Nilsson

6.1	Introduction	100
6.2	Choice of Material	103
6.3	Design Configurations	106
	6.3.1 Layered Resonators	106
	6.3.2 Transversal Resonators	107
	6.3.3 SAW Devices	108
	6.3.4 Flow Splitter Design	109
	6.3.5 1D and 2D Acoustic Focusing in Continuous Flow	110
	6.3.6 Acoustic Traps	111
	6.3.7 Capillaries	114
6.4	Actuation	116
	6.4.1 Transducer Coupling	116
	6.4.2 Coupled Resonance Modes	118
	6.4.3 Electrode and Transducer Modifications	119
	6.4.4 Focused Transducers	121
6.5	Conclusions	123
	References	124

Chapter 7 Modelling and Applications of Planar Resonant Devices for Acoustic Particle Manipulation 127
Peter Glynne-Jones, Rosemary J. Boltryk, and Martyn Hill

   7.1 Introduction 127
   7.2 Acoustic Radiation Forces 127
   7.3 Generating the Required Acoustic Field 128
   7.4 Modelling of Planar Resonators 128
   7.5 Resonator Configurations 131
      7.5.1 Half-Wave Devices 133
      7.5.2 Quarter-Wave Resonators 137
      7.5.3 Thin-Reflector Resonators 138
   7.6 Position Control in Resonators 138
   7.7 Applications 141
      7.7.1 Filtration, Washing and Separation 141
      7.7.2 Sensing and Detection 142
      7.7.3 Cell-Interaction Studies and Sonoporation 143
   7.8 Conclusions 144
   References 144

Chapter 8 Applications in Continuous Flow Acoustophoresis 148
Andreas Lenshof, Per Augustsson, and Thomas Laurell

   8.1 Introduction 148
   8.2 Concentration 150
   8.3 Clarification 152
   8.4 Acoustic Particle Sorting and Cytometry Applications 154
   8.5 Medium Exchange of Cells and Microparticles 157
      8.5.1 Transport Mechanisms 158
      8.5.2 Acoustic Forces in Stratified Liquids 163
      8.5.3 Devices for Buffer Exchange 163
   8.6 Separation by Acoustophysical Properties 167
      8.6.1 Free Flow Acoustophoresis (FFA) 168
      8.6.2 FFA Devices 169
      8.6.3 Pre-Alignment and Fractionation 172
      8.6.4 Bi-Directional Fractionation 174
      8.6.5 Altering the Acoustic Properties of the Medium 176
   8.7 Cells Bound to Beads 176
      8.7.1 Positive Contrast Particles 176
      8.7.2 Negative Contrast Particles 177
   8.8 Frequency Switching 177
   8.9 Acoustic Radiation Forces in Stratified Liquids 179

8.10	Considerations	181
	8.10.1 Variations of the Acoustic Field	181
	8.10.2 Temperature Aspects	182
	8.10.3 Cell Analysis Following Acoustofluidic Handling	182
	8.10.4 Measuring Separation Performance	183
	8.10.5 Size Limitations	184
	References	185

Chapter 9 Applications in Acoustic Trapping — 189
Mikael Evander and Johan Nilsson

9.1	Introduction	189
9.2	Theory	190
9.3	Methods for Acoustic Trapping	193
9.4	Applications in Acoustic Trapping	196
	9.4.1 Particle Studies	196
	9.4.2 Bioassays on Trapped Microparticles	199
9.5	Cell Population Studies	201
	9.5.1 Enrichment/Washing of Cells	204
	9.5.2 Size Sorting and Separations	208
9.6	Conclusion	208
	References	208

Chapter 10 Ultrasonic Microrobotics in Cavities: Devices and Numerical Simulation — 212
Jürg Dual, Philipp Hahn, Andreas Lamprecht, Ivo Leibacher, Dirk Möller, Thomas Schwarz, and Jingtao Wang

10.1	Introduction	212
10.2	Particle Manipulation in Cavities	214
	10.2.1 Pattern Formation in Microchambers	214
	10.2.2 Rotation of Micro-particles	220
	10.2.3 Particle Transport	223
	10.2.4 Particle Positioning Combined with a Microgripper	227
10.3	Numerical Techniques for the Time-averaged Acoustic Effects	229
	10.3.1 Perturbation Method	231
	10.3.2 Solving the N–S Equation by FVM	233
	10.3.3 Numerical Examples	234
	10.3.4 Summary of the Numerical Examples	239
10.4	Summary and Conclusions	239
	References	239

Chapter 11 Acoustic Manipulation Combined with Other Force Fields   242
Peter Glynne-Jones and Martyn Hill

    11.1 Introduction   242
    11.2 Gravitational Forces   243
    11.3 Hydrodynamic Forces   245
    11.4 Forces Induced by Electrical Fields   248
    11.5 Magnetic Forces   250
    11.6 Optical Forces   252
    11.7 Conclusions   253
    References   254

Chapter 12 Analysis of Acoustic Streaming by Perturbation Methods   256
Satwindar Singh Sadhal

    12.1 Introduction   256
        12.1.1 Oscillatory Flows   259
        12.1.2 Incompressible Flow Approximation   263
    12.2 One- and Two-Dimensional Cases   266
        12.2.1 The Quartz Wind   266
        12.2.2 Rayleigh Streaming   268
        12.2.3 Stokes Drift   270
        12.2.4 Streaming between Two Plates   271
    12.3 Interaction of Solid Particles and Drops with Ultrasound   276
        12.3.1 Acoustic Levitators   278
    12.4 Solid Particles and Drops Placed Between Nodes   279
        12.4.1 Solution   282
        12.4.2 Discussion   286
        12.4.3 Streaming around a Solid Sphere Placed between Nodes   289
        12.4.4 Solid Particle/Liquid Drop at the Velocity Antinode   290
        12.4.5 Solid Sphere at the Velocity Node   294
        12.4.6 Discussion   296
    12.5 Bubbles in Acoustic Fields   296
        12.5.1 Transverse Oscillations with Fixed Volume   297
        12.5.2 Radial and Transverse Oscillations of Bubbles   299
        12.5.3 Semi-cylindrical Bubble   302
    12.6 Concluding Remarks   305
    Acknowledgements   308
    Nomenclature   308
    Roman Symbols   308
    Greek Letters   309

Contents xiii

	Subscripts, Superscripts and Accents	309
	References	310

Chapter 13 Applications of Acoustic Streaming 312
Roy Green, Mathias Ohlin, and Martin Wiklund

13.1 Introduction 312
13.2 A Qualitative Description of Acoustic Streaming Phenomena 313
  13.2.1 Inner and Outer Boundary Layer Acoustic Streaming 314
  13.2.2 Eckart Streaming 316
  13.2.3 Cavitation Microstreaming 317
13.3 Microfluidic Applications of Acoustic Streaming 318
  13.3.1 Applications of Rayleigh Streaming 318
  13.3.2 Applications of Eckart Streaming 326
  13.3.3 Applications of Cavitation Microstreaming 327
  13.3.4 Surface Acoustic Wave Induced Streaming Applications 331
13.4 Conclusion 333
References 333

Chapter 14 Theory of Surface Acoustic Wave Devices for Particle Manipulation 337
Michael Gedge and Martyn Hill

14.1 Introduction 337
14.2 Rayleigh Waves 338
14.3 Leaky Rayleigh Waves 344
14.4 Scholte Waves 347
14.5 Interface Waves 347
14.6 Stoneley Waves 349
14.7 Anisotropic Media and Piezoelectric Considerations 349
14.8 Generation of Surface Acoustic Waves 350
14.9 Conclusions 352
References 352

Chapter 15 Lab-on-a-chip Technologies Enabled by Surface Acoustic Waves 354
Xiaoyun Ding, Peng Li, Sz-Chin Steven Lin, Zackary S. Stratton, Nitesh Nama, Feng Guo, Daniel Slotcavage, Xiaole Mao, Jinjie Shi, Francesco Costanzo, Thomas Franke, Achim Wixforth, and Tony Jun Huang

15.1 Introduction 354

15.2	Travelling Surface Acoustic Wave (TSAW) Microfluidics	357
	15.2.1 Theory Involved with TSAW	357
	15.2.2 Microfluidic Technologies Enabled by TSAWs	358
	15.2.3 Microfluidic Technologies Enabled by Phononic Crystal-assisted TSAWs	370
15.3	Standing Surface Acoustic Wave (SSAW) Microfluidics	375
	15.3.1 Theory Involved with SSAW	375
	15.3.2 Microfluidic Technologies Enabled by SSAWs	377
15.4	Conclusions and Perspectives	389
Acknowledgements		391
References		392

Chapter 16 Surface Acoustic Wave Based Microfluidics and Droplet Applications 399
Thomas Franke, Thomas Frommelt, Lothar Schmid, Susanne Braunmüller, Tony Jun Huang, and Achim Wixforth

16.1	Introduction	399
16.2	Acoustic Streaming Effects	400
16.3	Acoustically Induced Mixing	403
16.4	Acoustic Droplet Actuation	405
16.5	PDMS Microfluidics	407
	16.5.1 SAW Excitation on a Piezosubstrate and Acoustic Coupling to Standard PDMS Devices	408
	16.5.2 Pumping in Closed PDMS Channels	410
	16.5.3 Droplet Based Fluidics	412
16.6	Conclusions	417
References		417

Chapter 17 Ultrasound-Enhanced Immunoassays and Particle Sensors 420
Martin Wiklund, Stefan Radel, and Jeremy Hawkes

17.1	Introduction	420
17.2	Ultrasound-Enhanced Bead-Based Immunoassays	421
	17.2.1 Agglutination Assays	422
	17.2.2 Fluorescence Assays	425
17.3	Ultrasound-Enhanced Particle Sensors	427
	17.3.1 The Distinctive Pattern of Clumps in Contact with a Surface	428
	17.3.2 The Need for a Cell Attracting Wall in Microsystems	428

		17.3.3	Device Design	430
		17.3.4	The Cell Attractor Wall	430
		17.3.5	Immuno-Based Selective Cell Capture and Detection by Light	431
		17.3.6	The Next Stage of Developments	434
	17.4	Ultrasound-Enhanced Vibrational Spectroscopy		435
		17.4.1	Agglomeration of Crystals for Raman Spectroscopy	436
		17.4.2	Enhancement of Stopped Flow FT–IR ATR Spectroscopy	439
		17.4.3	ATR Probe for Inline Bioprocess Monitoring	443
	17.5	Conclusions		448
	References			448

Chapter 18 Multi-Wavelength Resonators, Applications and Considerations   452
Jeremy J. Hawkes and Stefan Radel

	18.1	Introduction		452
	18.2	Acoustic Filters		455
		18.2.1	Ultrasonically Enhanced Sedimentation	456
		18.2.2	Enhanced Sedimentation	459
		18.2.3	Influences on Separation Efficiency (UES)	460
		18.2.4	Flow Splitters in the h-Shape Separator	461
	18.3	Resonators		462
		18.3.1	Construction	462
		18.3.2	Tools for Development	465
		18.3.3	Models and Measurements	465
		18.3.4	Gel Technique	468
		18.3.5	Control of Acoustic Signal	470
		18.3.6	Dimensions	471
		18.3.7	Thickness Limits and Scale Considerations	472
		18.3.8	Acoustic Contrast	473
	18.4	Flow Changes Produced by Scale-up		474
		18.4.1	Non-Turbulent Flow Required	474
		18.4.2	Calculating the Onset of Turbulence	475
		18.4.3	Effect of Scale-up on the Initiation of Turbulence	476
		18.4.4	Heating	476
		18.4.5	Entrance Condition	479
		18.4.6	Flow Expansion and Contraction without Disruption or Dead Volumes	480
	18.5	New Design Approaches		482
		18.5.1	Heating Mitigation	482
		18.5.2	Creating a Standing Wave: Selection of a Particular Wall Mode Is Not Always Required	482

		18.5.3 Modelling Resonant Parts	483
		18.5.4 Modelling the Chamber Wall	484
		18.5.5 Experimental Confirmation	486
	18.6	Appendix	487
	Nomenclature		487
	Greek Letters		488
	References		488
Chapter 19	**Microscopy for Acoustofluidic Micro-Devices**		**493**
	Martin Wiklund, Hjalmar Brismar, and Björn Önfelt		
	19.1	Introduction	493
	19.2	Basic Principles of Optical Microscopy	494
		19.2.1 Illumination System	497
	19.3	Modes of Optical Microscopy	506
		19.3.1 Bright-Field Microscopy	507
		19.3.2 Fluorescence Microscopy	509
	19.4	Implementation of Optical Microscopy in an Acoustofluidic Micro-Device	512
		19.4.1 Design Criteria	512
		19.4.2 Applications of Microscopy to Acoustofluidics	515
	19.5	Conclusions	516
	Acknowledgements		517
	References		517
Chapter 20	**Experimental Characterization of Ultrasonic Particle Manipulation Devices**		**520**
	Jürg Dual, Philipp Hahn, Ivo Leibacher, Dirk Möller, and Thomas Schwarz		
	20.1	Introduction	520
	20.2	Laser Interferometry	521
		20.2.1 Obtainable Resolution in Interferometry	523
	20.3	Frequency Analysis, Admittance Curves and Modal Analysis	524
	20.4	Schlieren Imaging of 2D Pressure Fields in Cavities	527
	20.5	Measurements	529
		20.5.1 Characterization of a Transducer	529
		20.5.2 Characterization of Manipulation Devices	539
	20.6	Conclusions	542
	References		543

Chapter 21	**Biocompatibility and Cell Viability in Acoustofluidic Resonators**	**545**
	Martin Wiklund	
	21.1 Introduction	545
	21.2 Physical Mechanisms of Ultrasound Causing a Bioeffect	547
	21.2.1 Thermal Effects	547
	21.2.2 Cavitation-Based Effects	550
	21.2.3 Effects of Acoustic Radiation Forces	553
	21.2.4 Effects of Acoustic Streaming	554
	21.2.5 Effects Not Caused by Ultrasound	555
	21.3 Observed Bioeffects on Cells in Ultrasonic Standing Wave Manipulation Devices	555
	21.4 Methods for Measuring the Impact of Ultrasound on Cell Viability	558
	21.5 Conclusions	561
	Acknowledgements	562
	References	562
Subject Index		**566**

CHAPTER 1

Governing Equations in Microfluidics

HENRIK BRUUS

Department of Physics, Technical University of Denmark, Lyngby, Denmark
E-mail: bruus@fysik.dtu.dk

1.1 Introduction

Microfluidics deals with the flow of fluids and suspensions in channels of sub-millimetre-sized cross-sections under the influence of external forces.[1-6] Here, viscosity dominates over inertia, ensuring the absence of turbulence and the appearance of regular and predictable laminar flow streams, which implies an exceptional spatial and temporal control of solutes. The combination of laminar flow streams and precise control of external forces acting on particles in solution, has resulted in particle handling methods useful for analytical chemistry and bioanalysis, based on different physical mechanisms including inertia,[7] electrokinetics,[8] dielectrophoretics,[9] magnetophoretics,[10] as well as mechanical contact forces.[11]

Acoustofluidics, *i.e.* ultrasound-based external forcing of microparticles in microfluidics, has attracted particular attention because it allows gentle, label-free separation based on purely mechanical properties: size, shape, density, and compressibility. The early acoustophoretic microparticle filters[12-16] were soon developed into the first successful on-chip acoustophoretic separation devices.[17,18] Many different biotechnical applications of acoustophoresis have subsequently emerged including cell trapping,[19-21] plasmapheresis,[22] forensic analysis,[23] food analysis,[24] cell sorting using surface acoustic waves,[25] cell synchronization,[26] and cell differentiation.[27]

Furthermore, substantial advancements in understanding the fundamental physics of microsystems acoustophoresis have been achieved through full-chip imaging of acoustic resonances,[28] surface acoustic wave generation of standing waves,[29] multi-resonance chips,[30] advanced frequency control,[31,32] on-chip integration with magnetic separators,[33] acoustics-assisted microgrippers,[34] *in situ* force calibration,[35] automated systems,[36] and full 3D characterization of acoustophoresis.[37]

In this chapter, adapted from Bruus,[38,39] we study the governing equations in microfluidics formulated in terms of the classical continuum field description of velocity $\boldsymbol{v}$, pressure p, and density ρ. We also present some of the basic flow solutions, equivalent circuit modeling useful for predicting the flow rates in networks of microfluidic channels, and scaling laws for various microfluidic phenomena.

1.2 The Basic Continuum Fields

In the following we use the so-called Eulerian picture of the continuum fields, where we observe how the fields evolve in time at each fixed spatial position $\boldsymbol{r}$. Consequently, the position $\boldsymbol{r}$ and the time t are independent variables. The Eulerian picture is illustrated by the velocity field in Figure 1.1. In general, the value of any field variable $F(\boldsymbol{r}, t)$ is defined as the average value $\langle F_{\text{mol}}(\boldsymbol{r}', t) \rangle$ of the corresponding molecular quantity for all the molecules contained in some liquid particle of volume $\Delta V(\boldsymbol{r})$ around $\boldsymbol{r}$ at time t

$$F(\boldsymbol{r}, t) = \langle F_{\text{mol}}(\boldsymbol{r}', t) \rangle_{\boldsymbol{r}' \in \Delta V(\boldsymbol{r})}. \tag{1.1}$$

If we, for brevity, let m_i and $\boldsymbol{v}_i$ be the mass and the velocity of molecule i, respectively, and furthermore let $i \in \Delta V$ stand for all molecules i present inside the volume $\Delta V(\boldsymbol{r})$ at time t, then the definition of the density $\rho(\boldsymbol{r}, t)$ and the velocity field $\boldsymbol{v}(\boldsymbol{r}, t)$ can be written as

$$\rho(\boldsymbol{r}, t) \equiv \frac{1}{\Delta V} \sum_{i \in \Delta V} m_i, \tag{1.2a}$$

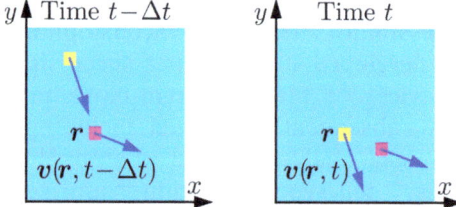

Figure 1.1 The Eulerian picture: the spatial coordinates r do not follow the flow. Instead, the velocity field $\boldsymbol{v}$ at the fixed point $\boldsymbol{r}$ is defined by the molecules in the purple region at time $t - \Delta t$, and by those in the yellow region at time t. Reproduced with permission from ref. 38.

$$v(\mathbf{r},t) \equiv \frac{1}{\rho(\mathbf{r},t)\Delta V} \sum_{i \in \Delta V} m_i \mathbf{v}_i. \tag{1.2b}$$

Here, we have introduced the symbol "≡" to mean "equal-to-by-definition". Notice how the velocity is defined through the more fundamental concept of momentum.

The dependent field variables in microfluidics can be scalars (such as density ρ, viscosity η, pressure p, temperature T, and free energy Φ), vectors (such as velocity $\mathbf{v}$, current density $\mathbf{J}$, pressure gradient ∇p, force density $\mathbf{f}$, and electric fields $\mathbf{E}$), and tensors (such as stress tensor $\boldsymbol{\sigma}$ and velocity gradient $\nabla \mathbf{v}$).

1.3 Mathematical Notation

The mathematical treatment of microfluidic problems is complicated due to the presence of several scalar, vector and tensor fields and the non-linear partial differential equations that govern them. To facilitate the treatment, some simplifying notation is called for, and here we follow Bruus.[4] First, a suitable co-ordinate system must be chosen. We shall mainly work with Cartesian co-ordinates (x, y, z) with corresponding basis vectors $\mathbf{e}_x$, $\mathbf{e}_y$, and $\mathbf{e}_z$ of unity length and that are mutually orthogonal. The position vector $\mathbf{r} = (r_x, r_y, r_z) = (x, y, z)$ is written as

$$\mathbf{r} = r_x \mathbf{e}_x + r_y \mathbf{e}_y + r_z \mathbf{e}_z = x \mathbf{e}_x + y \mathbf{e}_y + z \mathbf{e}_z. \tag{1.3}$$

Any vector $\mathbf{v}$ can be written in terms of its components v_i (where for Cartesian co-ordinates $i = x, y, z$) as

$$\mathbf{v} = \sum_{i=x,y,z} v_i \mathbf{e}_i \equiv v_i \mathbf{e}_i. \tag{1.4}$$

In the last equality we have introduced the Einstein summation convention: a repeated index implies a summation over that index (unless noted otherwise). Other examples of this compact, so-called index notation, are the vector scalar product $\mathbf{u} \cdot \mathbf{v}$, the length v of a vector $\mathbf{v}$, and the ith component of a vector $\mathbf{u}$ given as a matrix M multiplied by a vector $\mathbf{v}$,

$$\mathbf{v} \cdot \mathbf{u} = v_i u_i, \tag{1.5a}$$

$$v = |\mathbf{v}| = \sqrt{v^2} = \sqrt{\mathbf{v} \cdot \mathbf{v}} = \sqrt{v_i v_i}, \tag{1.5b}$$

$$u_i = M_{ij} v_j. \tag{1.5c}$$

For the partial derivatives of a given function $F(\mathbf{r}, t)$, we use the symbols ∂_i, with $i = x, y, z$, and ∂_t

$$\partial_x F \equiv \frac{\partial F}{\partial x}, \quad \text{and} \quad \partial_t F \equiv \frac{\partial F}{\partial t}. \tag{1.6}$$

The vector differential operator nabla ∇ contains the three partial space derivatives ∂_i. It plays an important role in differential calculus, and it is defined by

$$\nabla \equiv \boldsymbol{e}_x \partial_x + \boldsymbol{e}_y \partial_y + \boldsymbol{e}_z \partial_z = \boldsymbol{e}_i \partial_i = \partial_x^2 + \partial_y^2 + \partial_z^2. \tag{1.7}$$

The Laplace operator, which appears in numerous partial differential equations in theoretical physics, is just the scalar product of the nabla operator with itself,

$$\nabla \cdot \nabla \equiv \nabla^2 \equiv \partial_i \partial_i. \tag{1.8}$$

In terms of ∇, the total time derivative of a quantity $F(\boldsymbol{r}(t), t)$ flowing along with the fluid can be written as

$$\frac{dF}{dt} = \partial_t F + (\partial_t r_i)\partial_i F = \partial_t F + (\boldsymbol{v} \cdot \nabla)F. \tag{1.9}$$

Since ∇ is a differential operator, the order of the factors does matter in a scalar product containing it. So, whereas $\boldsymbol{v} \cdot \nabla$ in the previous equation is a scalar differential operator, the product $\nabla \cdot \boldsymbol{v}$ with the reversed order of the factors is a scalar function. The latter appears so often in mathematical physics that it has acquired its own name, namely the divergence of the vector field,

$$\nabla \cdot \boldsymbol{v} \equiv \partial_x v_x + \partial_y v_y + \partial_z v_z = \partial_i v_i. \tag{1.10}$$

Concerning integrals, we denote the 3D integral measure by $d\boldsymbol{r}$, so that in Cartesian coordinates we have $d\boldsymbol{r} = dx\,dy\,dz$. We also consider definite integrals as operators acting on integrands, thus we keep the integral sign and the associated integral measure together to the left of the integrand. As an example, the integral in spherical coordinates (r, θ, ϕ) over a spherical body with radius a of the scalar function $S(\boldsymbol{r})$ is written as

$$\int_{\text{sphere}} dx\,dy\,dz\, S(x,y,z) = \int_{\text{sphere}} d\boldsymbol{r}\, S(\boldsymbol{r}) = \int_0^a r^2 dr \int_0^\pi \sin\theta d\theta \int_0^{2\pi} d\phi\, S(r, \theta, \phi). \tag{1.11}$$

When working with vectors and tensors it is advantageous to use the following two special symbols: the Kronecker delta δ_{ij} and the Levi-Civita symbol ε_{ijk},

$$\delta_{ij} = \begin{cases} 1, & \text{for } i = j, \\ 0, & \text{for } i \neq j, \end{cases} \tag{1.12a}$$

$$\varepsilon_{ijk} = \begin{cases} +1, & \text{if } (ijk) \text{ is an even permutation of } (xyz), \\ -1, & \text{if } (ijk) \text{ is an odd permutation of } (xyz), \\ 0, & \text{otherwise.} \end{cases} \tag{1.12b}$$

In the index notation, the Levi-Cevita symbol appears directly in the definition of the ith component of the cross-product $\boldsymbol{u} \times \boldsymbol{v}$ of two vectors $\boldsymbol{u}$ and $\boldsymbol{v}$, and of the rotation $\nabla \times \boldsymbol{v}$,

$$(\boldsymbol{u} \times \boldsymbol{v})_i \equiv \varepsilon_{ijk}\, u_j u_k, \quad \text{and} \quad (\nabla \times \boldsymbol{v})_i \equiv \varepsilon_{ijk} \partial_j v_k. \tag{1.13}$$

When evaluating the rotation of a rotation, the "pairing-minus-antipairing" theorem is useful,

$$\varepsilon_{ijk} \varepsilon_{ilm} = \delta_{jl}\delta_{km} - \delta_{jm}\delta_{kl}. \tag{1.14}$$

As a last mathematical subject, we mention Gauss's theorem, which we shall employ repeatedly below. For a given vector field $V(r)$, it relates the volume integral in a given region Ω of the divergence $\nabla \cdot V$ to the integral over the surface $\partial\Omega$ of the flux $V \cdot n\, da$ through an area element da with the surface normal n,

$$\int_\Omega d\boldsymbol{r}\, \nabla \cdot V = \oint_{\partial\Omega} da\, \boldsymbol{n} \cdot V \quad \text{or} \quad \int_\Omega d\boldsymbol{r}\, \partial_j V_j = \oint_{\partial\Omega} da\, n_j V_j \tag{1.15}$$

By definition, the surface normal n of a closed surface is an outward-pointing unit vector perpendicular to the surface, hence the name divergence. For a tensor T_{jk} of rank 2 the theorem states $\int_\Omega d\boldsymbol{r}\, \partial_j T_{jk} = \oint_{\partial\Omega} da\, n_j T_{jk}$ with a straightforward generalization to higher ranks.

1.4 Governing Equations

1.4.1 The Continuity Equation

The first governing equation to be derived is the continuity equation, which expresses the conservation of mass. We consider a compressible fluid, *i.e.* the density ρ may vary as a function of space and time, and a fixed, arbitrarily shaped region Ω in the fluid as sketched in Figure 1.2. The total mass $M(\Omega, t)$ inside Ω can be expressed as a volume integral over the density ρ,

$$M(\Omega, t) = \int_\Omega d\boldsymbol{r}\, \rho(r, t). \tag{1.16}$$

Since mass can neither appear nor disappear spontaneously in non-relativistic mechanics, $M(\Omega, t)$ can only vary due to a mass flux through the surface $\partial\Omega$ of the region Ω. The mass current density J is defined as the mass density ρ times the advection velocity v, *i.e.* the mass flow per oriented area per time (kg m^{-2} s^{-1}),

$$J(r, t) = \rho(r, t) v(r, t). \tag{1.17}$$

As the region Ω is fixed, $\partial_t M(\Omega, t)$ can be calculated either by differentiating the volume integral eqn (1.16),

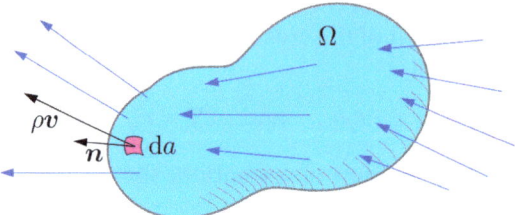

Figure 1.2 A sketch of the mass current density field ρv (long arrows) flowing through an arbitrarily shaped, but fixed region Ω (blue). Any infinitesimal area da (purple) is associated with an outward-pointing unit vector $\boldsymbol{n}$ (short arrow) perpendicular to the local surface. The current flowing out through the area da is given by da times the projection $\rho \boldsymbol{v} \cdot \boldsymbol{n}$ of the current density on the surface unit vector. Reproduced with permission from ref. 38.

$$\partial_t M(\Omega, t) = \partial_t \int_\Omega d\boldsymbol{r} \rho(\boldsymbol{r}, t) = \int_\Omega d\boldsymbol{r} \, \partial_t \rho(\boldsymbol{r}, t), \tag{1.18}$$

or as a surface integral over $\partial\Omega$ of the mass current density using eqn (1.17) and Figure 1.2,

$$\partial_t M(\Omega, t) = \int_{\partial\Omega} da(-\boldsymbol{n}) \cdot \boldsymbol{J} = -\int_{\partial\Omega} da \, \boldsymbol{n} \cdot (\rho \boldsymbol{v}) = -\int_\Omega d\boldsymbol{r} \, \boldsymbol{\nabla} \cdot [\rho(\boldsymbol{r},t)\boldsymbol{v}(\boldsymbol{r},t)]. \tag{1.19}$$

The last expression is obtained by applying Gauss's theorem eqn (1.15) to the vector field $\boldsymbol{V} \equiv \rho \boldsymbol{v}$. We have used $-\boldsymbol{n}$ because this is the direction of *entering* the region. It follows immediately from eqn (1.18) and (1.19) that

$$\int_\Omega d\boldsymbol{r} \, \partial_t \rho(\boldsymbol{r}, t) = -\int_\Omega d\boldsymbol{r} \, \boldsymbol{\nabla} \cdot [\rho(\boldsymbol{r},t)\boldsymbol{v}(\boldsymbol{r},t)]. \tag{1.20}$$

This result is true for any choice of region Ω. However, this is only possible if the integrands are identical. Thus we have derived the continuity equation,

$$\partial_t \rho = -\boldsymbol{\nabla} \cdot (\rho \boldsymbol{v}) \quad \text{or} \quad \partial_t \rho = -\partial_j(\rho v_j). \tag{1.21}$$

In many cases, especially in microfluidics, where the flow velocity is much smaller than the velocity of speed of sound (pressure waves), the fluid appears incompressible. This means that any volume change $\Delta V = \oint_{\partial\Omega} da(\boldsymbol{n} \cdot \boldsymbol{v}) \Delta t$ in the time interval Δt for an arbitrary region Ω must be zero, or by Gauss's theorem $\int_\Omega d\boldsymbol{r} \boldsymbol{\nabla} \cdot \boldsymbol{v} = 0$, and the continuity equation simplifies to.

$$\boldsymbol{\nabla} \cdot \boldsymbol{v} = 0 \quad \text{or} \quad \partial_i v_i = 0. \tag{1.22}$$

In Chapter 2, compressibility $\rho = \rho(p)$ will be discussed further in connection with acoustics.

Governing Equations in Microfluidics   7

1.4.2 The Navier–Stokes Equation

The second governing equation, the Navier–Stokes equation, is the equation of motion for the Eulerian velocity field directly related to momentum conservation and the momentum density ρv. It is derived using an approach similar to that which led us to the continuity equation. We consider the ith component $P_i(\Omega, t)$ of the total momentum of the fluid inside an arbitrarily shaped, but fixed, region Ω. In analogy with the mass eqn (1.18), the rate of change of the momentum is given by

$$\partial_t P_i = \partial_t \int_\Omega \mathrm{d}\mathbf{r}\, \rho(\mathbf{r},t) v_i(\mathbf{r},t) = \int_\Omega \mathrm{d}\mathbf{r}\, [(\partial_t \rho) v_i + \rho \partial_t v_i]. \tag{1.23}$$

In contrast to the mass inside Ω, which according to eqn (1.19) can only change by advection through the surface $\partial\Omega$, the momentum $P_i(\Omega, t)$ can change both by advection and by the action of forces given by Newton's second law. These forces can be divided into body forces that act on the interior of Ω, e.g. gravitational and electrical forces, and contact forces that act on the surface $\partial\Omega$ of Ω, e.g. pressure and viscosity forces. Thus, the rate of change of the ith component of the momentum can be written as

$$\partial_t P_i = \partial_t P_i^{\mathrm{body}} + \partial_t P_i^{\mathrm{adv}} + \partial_t P_i^{\mathrm{pres}} + \partial_t P_i^{\mathrm{visc}}. \tag{1.24}$$

A body force $\mathbf{f}^{\mathrm{body}}$ is an external force that acts throughout the entire body of the fluid. The change in the momentum of Ω due to $\mathbf{f}^{\mathrm{body}}$, e.g. gravity in terms of the density ρ and the acceleration of gravity g, is.

$$\partial_t P_i^{\mathrm{body}} = \int_\Omega \mathrm{d}\mathbf{r}\, f_i^{\mathrm{body}} = \int_\Omega \mathrm{d}\mathbf{r}\, (\rho \mathbf{g})_i = \int_\Omega \mathrm{d}\mathbf{r}\, \rho g_i. \tag{1.25}$$

For the advection of momentum ρv into Ω, we note that it is described in terms of the tensor $(\rho v) v$, just as advection of density ρ is described by the vector $(\rho) v$. Considering the ith momentum component, we see that the flux of momentum into Ω through the infinitesimal area $\mathrm{d}a$ is given by $(\rho v_i) v \cdot (-\mathbf{n}) \mathrm{d}a$. Thus the total change $\partial_t P_i^{\mathrm{adv}}(\Omega, t)$ of momentum in Ω due to advection is

$$\partial_t P_i^{\mathrm{adv}} = \oint_{\partial\Omega} \mathrm{d}a (-\mathbf{n}) \cdot (\rho v_i \mathbf{v}) = -\oint_{\partial\Omega} \mathrm{d}a\, n_j \rho v_i v_j \tag{1.26}$$

The rate of momentum change due to pressure is the sum of the pressure force $p(-\mathbf{n})\mathrm{d}a$ from the surroundings on each infinitesimal area $\mathrm{d}a$ of the surface $\partial\Omega$. Thus, for the ith component of the momentum along the unit vector $\mathbf{e}_i$ we obtain

$$\partial_t P_i^{\mathrm{pres}} = \oint_{\partial\Omega} \mathrm{d}a (-\mathbf{n}) \cdot (p \mathbf{e}_i) = -\oint_{\partial\Omega} \mathrm{d}a\, n_j p \delta_{ij} \tag{1.27}$$

In the last equation we use $\mathbf{n} \cdot \mathbf{e}_i = n_j \delta_{ij}$, whereby $\mathbf{n}$ can be ascribed the same free index j differing from the momentum component index i as in eqn (1.26).

The momentum in Ω is also changed by viscous friction at the surface $\partial\Omega$ from the surrounding fluid. The frictional force $d\mathbf{F}$ on a surface element da with the normal vector $\mathbf{n}$ must be characterized by a tensor rank of two as two vectors are involved. This tensor is denoted as the viscous stress tensor σ'_{ij}, and it expresses the ith component of the friction force per area acting on a surface element oriented with its surface normal parallel to the jth unit vector $\mathbf{e}_j$. So we have $dF_i = \sigma'_{ij} n_j \, da$, which leads directly to the change in the momentum of Ω due to the viscous forces at the surface $\partial\Omega$,

$$\partial_t P_i^{\text{visc}}(\Omega, t) = \oint_{\partial\Omega} dF_i = \oint_{\partial\Omega} da \, n_j \sigma'_{ij} \tag{1.28}$$

The internal friction is only non-zero when fluid particles move relative to each other, hence the viscous stress tensor σ'_{ij} depends only on the spatial derivatives of the velocity. For the small velocity gradients encountered in microfluidics we can safely assume that only first-order derivatives are relevant. Thus, σ'_{ij} must thus depend linearly on the velocity gradients $\partial_i v_j$. Further analysis shows that it must be symmetric, and one way of writing the tensor of rank two satisfying these conditions is[4]

$$\sigma'_{ij} = \eta(\partial_j v_i + \partial_i v_j) + (\beta - 1)\eta(\partial_k v_k)\delta_{ij}, \tag{1.29}$$

where the first term relates to the dynamic shear viscosity η of an incompressible fluid, and the second term appears when compressibility-induced dilatational viscosity (proportional to $\nabla \cdot \mathbf{v}$) cannot be neglected. The value of η is determined experimentally, and for water we have

$$\eta_{\text{water}}(20\,^\circ\text{C}) = 1.002 \times 10^{-3} \text{ Pa s} = 1.002 \text{ mPa s}. \tag{1.30}$$

It varies strongly with temperature from $\eta_{\text{water}}(0\,^\circ\text{C}) = 1.787$ mPa s to $\eta_{\text{water}}(100\,^\circ\text{C}) = 0.282$ mPa s. The coefficient β is related to the ratio between the two types of viscosity; for simple monatomic gases it is known to be $\beta = 1/3$, while for water at 25 °C it is determined experimentally[40] to be $\beta = 1/3 + (2.4 \text{ mPa s})/\eta \approx 3.0$.

Combining the results from eqn (1.23) to (1.29) yields

$$\int_\Omega d\mathbf{r}[(\partial_t \rho)v_i + \rho\partial_t v_i] = \oint_{\partial\Omega} da \, n_j \left[-\rho v_i v_j - p\delta_{ij} + \sigma'_{ij}\right] + \int_\Omega d\mathbf{r} \, \rho g_i \tag{1.31}$$

Utilizing Gauss's theorem, the surface integral involving n_j can be rewritten as a volume integral involving ∂_j. Since the resulting volume integral equation is valid for any region Ω, the integrands must be identical. After some rewriting and using eqn (1.21), we finally arrive at the general equation of motion for the Eulerian velocity field of a viscous fluid,

$$\rho \partial_t v_i + \rho v_j \partial_j v_i = -\partial_i p + \partial_j \sigma'_{ij} + \rho g_i. \tag{1.32}$$

The left-hand side can be interpreted as inertial force densities, density times the sum of the local and the advective acceleration, while the right-hand side is the sum of intrinsic or applied force densities. Normally, for the

Governing Equations in Microfluidics

so-called Newtonian fluids at a given temperature, the viscosity coefficients η and β can be taken as constants, and eqn (1.32) reduces to the celebrated Navier–Stokes equation for compressible fluids,

$$\rho[\partial_t v_i + v_j \partial_j v_i] = -\partial_i p + \eta \partial_j^2 v_i + \beta \eta \partial_i (\partial_j v_j) + \rho g_i, \quad (1.33a)$$

$$\rho[\partial_t \boldsymbol{v} + (\boldsymbol{v} \cdot \boldsymbol{\nabla})\boldsymbol{v}] = -\boldsymbol{\nabla} p + \eta \nabla^2 \boldsymbol{v} + \beta \eta \boldsymbol{\nabla}(\boldsymbol{\nabla} \cdot \boldsymbol{v}) + \rho \boldsymbol{g}. \quad (1.33b)$$

For incompressible fluids ($\boldsymbol{\nabla} \cdot \boldsymbol{v} = 0$) it becomes

$$\rho[\partial_t \boldsymbol{v} + (\boldsymbol{v} \cdot \boldsymbol{\nabla})\boldsymbol{v}] = -\boldsymbol{\nabla} p + \eta \nabla^2 \boldsymbol{v} + \rho \boldsymbol{g}. \quad (1.34)$$

1.4.3 The Heat-transfer Equation

The third and last governing equation to be established is the heat-transfer equation, building on the energy density flux and conservation of energy. The thermodynamic quantities for fluids are usually taken per unit mass, directly relating them to the molecules of the fluid. Thus, we work with the internal energy ε per unit mass, the entropy s per unit mass, and the volume $1/\rho$ per unit mass instead of the energy E, the entropy S, and the volume ς of the fluid. The first law of thermodynamics relates internal energy $d\varepsilon$, heat $T\,ds$, and pressure work $-p\,d(1/\rho)$, and per unit mass it becomes

$$d\varepsilon = T\,ds - p\,d\left(\frac{1}{\rho}\right) = T\,ds + \frac{p}{\rho^2}\,d\rho. \quad (1.35)$$

The densities of the quantities involved are obtained by multiplying them by the mass density ρ, e.g. the energy density is written as $\rho \varepsilon$.

In analogy with the mass and momentum densities above, we consider the rate of change $\partial_t E(\Omega, t)$ of the energy (power conversion) of the fluid inside some fixed region Ω. As the energy density is given by the sum of the kinetic energy density $\frac{1}{2}\rho v^2$ and the internal energy density $\rho \varepsilon$, we obtain

$$\partial_t E = \partial_t \int_\Omega d\boldsymbol{r} \left[\frac{1}{2}\rho v^2 + \rho \varepsilon\right] = \int_\Omega d\boldsymbol{r}\, \partial_t \left[\frac{1}{2}\rho v^2 + \rho \varepsilon\right]. \quad (1.36)$$

Here, E can change by energy advection through the surface $\partial \Omega$ of Ω, by work done by pressure and friction forces from the surroundings acting on the surface $\partial \Omega$, and by heat conduction due to thermal gradients at the surface. For simplicity, we disregard heat sources and sinks that could be present inside Ω, so $\partial_t E$ can be written as

$$\partial_t E = \partial_t E^{\text{adv}} + \partial_t E^{\text{pres}} + \partial_t E^{\text{visc}} + \partial_t E^{\text{cond}}. \quad (1.37)$$

The terms $\partial_t \rho$ and $\partial_t v_j$ appearing at the right-hand side of eqn (1.36) are rewritten using the continuity eqn (1.21) and the equation of motion, eqn (1.32), whereby

$$\partial_t E = \int_\Omega d\mathbf{r}\left[-\left(\frac{1}{2}v^2+\varepsilon\right)\partial_j(\rho v_j) - \rho v_k \partial_k\left(\frac{1}{2}v^2\right) - v_j\partial_j p + v_j\partial_k\sigma'_{jk} + \rho\partial_t\varepsilon\right]. \tag{1.38}$$

The last term $\rho\partial_t\varepsilon$ can be further rewritten by using the first law of thermodynamics, eqn (1.35), thereby bringing the entropy s into play as $\rho\partial_t\varepsilon = \rho T \partial_t s + (p/\rho)\partial_t\rho = \rho T \partial_t s - (p/\rho)\partial_j(\rho v_j)$. Likewise, the third term is rewritten as $-v_j\partial_j p = -\rho v_j\partial_j(\varepsilon + p/\rho) + \rho T v_j\partial_j s$.

Similar to eqn (1.19) and (1.26), the advection of energy into the region is easily expressed in terms of the energy flux density $\mathbf{J}_\varepsilon = (\frac{1}{2}\rho v^2 + \rho\varepsilon)\mathbf{v}$,

$$\partial_t E^{adv} = \oint_{\partial\Omega} da(-\mathbf{n})\cdot\mathbf{J}_\varepsilon = -\oint_{\partial\Omega} da\, n_j v_j\left[\frac{1}{2}\rho v^2 + \rho\varepsilon\right] \tag{1.39}$$

The power transferred into the region Ω through the work done by the stress forces due to pressure and viscosity at the surface is given by the product of the stress force vector $\sigma_{jk}\, n_j\, da$ and the velocity of the fluid v_k,

$$\partial_t E^{pres} + \partial_t E^{visc} = \oint_{\partial\Omega} da\, n_j\left[-p\delta_{jk} + \sigma'_{jk}\right]v_k \tag{1.40}$$

Thermal conduction occurs in any medium given a spatially varying temperature field $T(\mathbf{r})$. The heat flux density $\mathbf{J}_{heat}$, which is the heat-transfer per area per time given in units of J m^{-2} s^{-1} or W m^{-2}, can therefore be expanded in derivatives of the temperature. For small temperature variations only the first derivative ∇T is significant, and we arrive at Fourier's law of heat conduction for an isotropic medium,

$$\mathbf{J}_{heat} = -\kappa\nabla T, \tag{1.41}$$

where κ is the thermal conductivity of the fluid. For water, $\kappa_{water} = 0.597$ Wm^{-1} K^{-1} at 20 °C. The rate of change of energy by conduction is readily found through the heat flux density and Fourier's law,

$$\partial_t E^{cond}(\Omega, t) = \oint_{\partial\Omega} da(-\mathbf{n})\cdot\mathbf{J}_{heat} = \oint_{\partial\Omega} da\, n_j(\kappa\partial_j T). \tag{1.42}$$

The heat-transfer equation now follows from eqn (1.37) using eqn (1.38), (1.39) and (1.40), and (1.42). As before, we use Gauss's theorem to convert the surface integrals into volume integrals, and then equate the integrands to obtain

$$\rho T[\partial_t s + v_j\partial_j s] = \sigma'_{jk}\partial_k v_j + \partial_j[\kappa\partial_j T], \tag{1.43a}$$

$$\rho T[\partial_t s + (\mathbf{v}\cdot\nabla)s] = \boldsymbol{\sigma}' : \nabla\mathbf{v} + \nabla\cdot(\kappa\nabla T). \tag{1.43b}$$

The left-hand side is ρT times the total time derivative of the entropy per unit mass, hence expressing the total gain in heat density per unit time, while the right-hand side represents the sources for heat gain: viscous friction and thermal conduction.

In microfluidics, the fluid velocities are generally much smaller than the speed of sound in the fluid. Consequently, pressure variations are minute, leading to the constant pressure approximation, for which $ds = c_p dT$, where c_p is the specific heat at constant pressure. In this case the heat-transfer equation reduces to

$$\rho c_p [\partial_t T + (v_j \partial_j) T] = \partial_j (\kappa \partial_j T) + \sigma'_{jk} \partial_k v_j. \tag{1.44}$$

For a fluid at rest ($v_j = 0$) with a constant thermal conductivity κ, the equation reduces to the Fourier equation,

$$\partial_t T = D_{th} \nabla^2 T, \tag{1.45}$$

which introduces the thermal diffusivity $D_{th} = \kappa/(\rho c_p)$. For water, $D_{th} = 1.43 \times 10^{-7}$ m^2 s^{-1} at 20 °C. From eqn (1.43b), we see that the dissipation $\partial_t E^{visc}$ due to viscosity can be written as

$$\partial_t E^{visc} = \int_\Omega d\mathbf{r}\, \boldsymbol{\sigma}' : \boldsymbol{\nabla} \mathbf{v} = \eta \int_\Omega d\mathbf{r} (\partial_i v_j + \partial_j v_i) \partial_i v_j = \frac{1}{2} \eta \int_\Omega d\mathbf{r} (\partial_i v_j + \partial_j v_i)^2. \tag{1.46}$$

1.5 Flow Solutions

Mathematically, the richness and beauty of hydrodynamic phenomena are spawned by the nonlinear term $\rho(\mathbf{v} \cdot \boldsymbol{\nabla})\mathbf{v}$ in the Navier–Stokes equation. However, this term is also responsible for complicating the mathematical treatment of the equation; the solutions of the equation have never been fully characterized.

In the following, we study two different cases where this complicating non-linear term can be neglected. The first case is the limit of low flow velocities, a limit highly relevant for microfluidic systems. Here, the non-linear term can be neglected as it is much smaller than the viscous term, and we enter the linear regime of the so called Stokes flow or creeping flow. The second case is when translation symmetry of long, straight channels implies the exact vanishing of the non-linear term. In particular we study the steady-state pressure-driven Poiseuille flow.

1.5.1 Hydrostatic Pressure

The simplest case is a fluid in mechanical equilibrium. Such a fluid is at rest relative to the walls of the vessel containing it, and the velocity field is therefore trivially zero everywhere, $\mathbf{v} = \mathbf{0}$. If we let the z-axis point upwards, the gravitational acceleration is $\mathbf{g} = -g\mathbf{e}_z$, and the Navier–Stokes eqn (1.33b) becomes

$$0 = -\boldsymbol{\nabla} p_{hs} - \rho g \mathbf{e}_z, \tag{1.47}$$

where the subscript "hs" refers to hydrostatic. For an incompressible fluid, say water, the density ρ is constant and p_{hs} is easily found to be

$$p_{\text{hs}}(z) = p^* - \rho g z, \tag{1.48}$$

where p^* is the pressure at the arbitrarily defined zero level $z = 0$. In many microfluidic applications this is the only manifestation of gravity. It is therefore customary to write the total pressure as $p_{\text{tot}} = p + p_{\text{hs}}$, such that in the Navier–Stokes equation the gravitational body force is canceled by the gradient of hydrostatic pressure. The resulting Navier–Stokes equation thus contains the auxiliary pressure p and no gravitational body force.

The hydrostatic pressure p_{hs} provides an easy way of generating pressure differences in liquids: the pressure at the bottom of a liquid column of height ΔH is higher by $\Delta p = \rho g \Delta H$ than the pressure at the top. For water, $\rho g \approx 10^4$ Pa m^{-1}, so a vertical water column of height 10 cm creates $\Delta p = 1$ kPa, while it takes a height of 10 m to create 10^5 Pa $= 1$bar ≈ 1 atm.

1.5.2 The Reynolds Number and the Stokes Equation

For small flow velocities the non-linear term is expected to be negligible. To determine the condition for this, we non-dimensionalize the Navier–Stokes equation: all physical variables are expressed in units of characteristic scales. For a system characterized by only one length scale L_0 and one velocity scale V_0, the expression of co-ordinates and velocity in terms of dimensionless co-ordinates and velocity is

$$\boldsymbol{r} = L_0 \tilde{\boldsymbol{r}}, \quad \text{and} \quad \boldsymbol{v} = V_0 \tilde{\boldsymbol{v}} \tag{1.49a}$$

where the tilde indicates a non-dimensionalized quantity, *i.e.* a pure number. Once L_0 and V_0 have been fixed, the scales T_0 and P_0 for time and pressure follow,

$$t = \frac{L_0}{V_0}\tilde{t} = T_0 \tilde{t}, \quad \text{and} \quad p = \frac{\eta V_0}{L_0}\tilde{p} = P_0 \tilde{p} \tag{1.49b}$$

Viscosity dominates in microfluidics, so we choose $P_0 = \eta V_0 / L_0$ and not as ρV_0^2. By insertion of eqn (1.49a) and (1.49b) into the Navier–Stokes eqn (1.33b) excluding the body-forces as well as the compressibility term, and using the straightforward scaling of the derivatives, $\partial_t = \left(\frac{1}{T_0}\right)\tilde{\partial}_t$ and $\boldsymbol{\nabla} = \left(\frac{1}{L_0}\right)\tilde{\boldsymbol{\nabla}}$, we obtain

$$\rho\left[\frac{V_0}{T_0}\tilde{\partial}_t\tilde{\boldsymbol{v}} + \frac{V_0^2}{L_0}(\tilde{\boldsymbol{v}}\cdot\tilde{\boldsymbol{\nabla}})\tilde{\boldsymbol{v}}\right] = -\frac{P_0}{L_0}\tilde{\boldsymbol{\nabla}}\tilde{p} + \frac{\eta V_0}{L_0^2}\tilde{\boldsymbol{\nabla}}^2\tilde{\boldsymbol{v}} \tag{1.50}$$

Dividing by $\eta V_0/L_0^2$ this becomes

$$Re\left[\tilde{\partial}_t \tilde{v} + (\tilde{v}\cdot\tilde{\nabla})\tilde{v}\right] = -\tilde{\nabla}\tilde{p} + \tilde{\nabla}^2 \tilde{v} \qquad (1.51a)$$

$$Re = \frac{\rho V_0 L_0}{\eta}, \qquad (1.51b)$$

where the dimensionless parameter Re, the so-called Reynolds number, has been introduced.

For $Re \ll 1$ the left-hand side of eqn (1.51a) can be neglected. For water in microfluidics, typical values are $\eta/\rho = 10^{-6}$ m^2 s^{-1}, $L_0 \approx 10^{-4}$ m, and $V_0 \approx 10^{-3}$ m s^{-1}, which leads to a small Reynolds number, $Re \approx 0.1$.

Returning to physical variables in the limit of a low Reynolds number, the non-linear Navier–Stokes equation is reduced to the linear Stokes equation.

$$0 = -\nabla p + \eta \nabla^2 v. \qquad (1.52)$$

If the time dependence is controlled by some external time scale different from the intrinsic scale $T_0 = L_0/V_0$, the time derivative is not necessarily negligible, and we must employ the time-dependent, linear Stokes equation,

$$\rho \partial_t v = -\nabla p + \eta \nabla^2 v. \qquad (1.53)$$

For zero pressure gradient, and introducing the kinematic viscosity ν, it reduces to a diffusion equation

$$\partial_t v = \nu \nabla^2 v, \quad \text{with} \quad \nu = \frac{\eta}{\rho}, \qquad (1.54)$$

where ν acts as the diffusivity of momentum in the fluid.

An important example of Stokes flow relates to particle solutions. Let an external force move a spherical particle of radius a with velocity v_p through a fluid that itself moves with velocity v far from the particle. Given the no-slip boundary condition on the particle surface $\partial\Omega_p$, the velocity and pressure field of the fluid can be found from eqn (1.52), and from that follows the viscous Stokes drag F_{drag} on the particle,[4]

$$F_{\text{drag}} = \int_{\partial\Omega_p} da(-pn + \sigma'\cdot n) = 6\pi\eta a(v - v_p). \qquad (1.55)$$

Changing the particle concentration from zero to a few volume percent does not affect the flow profile significantly, while the viscosity may increase by a few percent.

1.5.3 Poiseuille Flow

Our prime example of a solution to the Navier–Stokes equation in the dynamic case, is the pressure-driven, steady-state flow in a straight channel, also known as Poiseuille flow, see Figure 1.3. This class of flow is of major

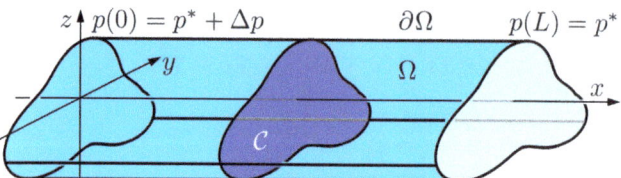

Figure 1.3 Poiseuille flow of liquid through a straight channel Ω (blue), where the flow is subject to the no-slip boundary condition on the surface $\partial\Omega$. The channel is translational invariant in the x direction, and it has an arbitrarily shaped cross-section X (dark blue) in the yz-plane. The pressure at $x = 0$ (left) is Δp higher than at $x = L$ (right). Reproduced with permission from ref. 38.

importance for the basic understanding of liquid handling in microfluidic lab-on-a-chip systems.

In a Poiseuille flow, the fluid is driven through a long, straight, and rigid channel of length L by imposing a pressure difference Δp between the two ends of the channel. The channel is placed horizontally along the x-axis, so along the z-axis gravity is balanced by the hydrostatic pressure. Furthermore, the cross-section of the channel is constant along the x-axis, so the liquid in the channel is only affected by the force from the pressure drop along the x-axis. The velocity field is therefore assumed to have only an x-component, and this component depends only on the transverse co-ordinates y and z, such that $\boldsymbol{v} = v_x(y, z)\,\boldsymbol{e}_x$. This implies $(\boldsymbol{v}\cdot\nabla)\boldsymbol{v} = (v_x\partial_x)\,v_x(y, z)\boldsymbol{e}_x = \boldsymbol{0}$, and the non-linear Navier–Stokes equation becomes the linear Stokes equation.

For the velocity field, we employ the so-called no-slip boundary condition: on all points on the solid surface $\partial\Omega$, the velocity of the fluid equals that of the wall $\boldsymbol{v}_{\text{wall}}$ (often equal to zero),

$$\boldsymbol{v}(\boldsymbol{r}) = \boldsymbol{v}_{\text{wall}}, \quad \text{for} \quad \boldsymbol{r} \in \partial\Omega \text{ (no-slip).} \tag{1.56}$$

The microscopic origin of this condition is the assumption of complete momentum relaxation between the molecules of the wall and the outermost molecules of the fluid that collide with the wall. The momentum is relaxed on a length scale, which approximately is the molecular mean free path in the fluid, which for liquids means one intermolecular distance ($\simeq 0.3$ nm).

The final form of the steady-state Navier–Stokes equation for the Poiseuille flow thus becomes

$$\boldsymbol{v}(\boldsymbol{r}) = v_x(y, z)\boldsymbol{e}_x, \tag{1.57a}$$

$$0 = -\nabla p + \eta \nabla^2[v_x(y, z)\boldsymbol{e}_x]. \tag{1.57b}$$

Since the y and z components of the velocity field are zero, it follows that both $\partial_y p$ and $\partial_z p$ are zero, and consequently the pressure field depends only on x, $p(\boldsymbol{r}) = p(x)$. Using this result, the x component of the Navier–Stokes eqn (1.57b) becomes

$$\eta\left[\partial_y^2 + \partial_z^2\right]v_x(y, z) = \partial_x p(x). \tag{1.58}$$

Governing Equations in Microfluidics

Here, it is seen that the left-hand side is a function of y and z, while the right-hand side is a function of x. The only possible solution is therefore that the two sides of the Navier–Stokes equation equal the same constant. However, a constant pressure gradient $\partial_x p(x)$ implies that the pressure must be a linear function of x, and using the boundary conditions for the pressure we obtain

$$p(\mathbf{r}) = \frac{\Delta p}{L}(L - x) + p^*. \tag{1.59}$$

With this we arrive at the second-order partial differential equation for $v_x(y, z)$ in the cross-section C obeying no-slip boundary conditions at the solid walls $\partial \Omega$,

$$\left[\partial_y^2 + \partial_z^2\right] v_x(y, z) = -\frac{\Delta p}{\eta L}, \quad \text{for} \quad (y, z) \in C \tag{1.60a}$$

$$v_x(y, z) = 0, \quad \text{for} \quad (y, z) \in \partial \Omega. \tag{1.60b}$$

The resulting velocity field can be determined analytically for a limited number of cross-section shapes, as done by Bruus.[4] Here, we present three of these solutions: a circular cross-section of radius a; the space between two horizontal infinite parallel plates placed at $z = 0$ and $z = h$; and a rectangular cross-section $-\frac{1}{2}w < y < \frac{1}{2}w$ and $0 < z < h$, see Figure 1.4(a),

$$v_x(y, z) = \frac{\Delta p}{4 \eta L}(a^2 - y^2 - z^2), \quad \text{circular channel,} \tag{1.61a}$$

$$v_x(y, z) = \frac{\Delta p}{2 \eta L}(h - z)z, \quad \text{parallel-plate channel,} \tag{1.61b}$$

$$v_x(y, z) = \frac{4 h^2 \Delta p}{\pi^3 \eta L} \sum_{n \text{ odd}}^{\infty} \frac{1}{n^3} \left[1 - \frac{\cosh\left(n\pi \frac{y}{h}\right)}{\cosh\left(n\pi \frac{w}{2h}\right)}\right] \sin\left(n\pi \frac{z}{h}\right), \tag{1.61c}$$

rectangular channel $h < w$.

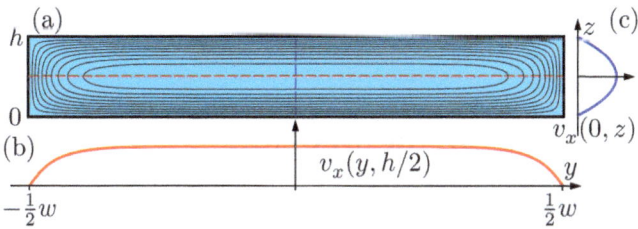

Figure 1.4 (a) Contour lines for the velocity field $v_x(y, z)$ for a Poiseuille-flow in a rectangular channel. The contour lines are shown in steps of 10% of the maximal value $v_x(0, h/2)$. (b) A plot (red) of $v_x(y, h/2)$ along the long centerline. (c) A plot (blue) of $v_x(0, z)$ along the short centerline. Reproduced with permission from ref. 38.

Direct substitution into eqn (1.60) shows that these solutions are correct. Note how the flow profile in the rectangular case is independent of y along a large part $w - h$ of the long width, Figure 1.4(b), but parabolic across the short height, Figure 1.4(c).

1.5.4 Flow Rate

Once the velocity field is determined, it is possible to calculate the so-called volumetric flow rate Q, which is defined as the fluid volume discharged by the channel per unit time. In the case of the geometry of Figure 1.3 we have

$$Q = \int_C dy\, dz\, v_x(y, z) = A v_{\text{avr}}, \qquad (1.62)$$

where $v_{\text{avr}} = (1/A) \int_C dy\, dz\, v_x(y, z)$ is the average velocity and A is the cross-section area. The flow rate for three selected Poiseuille flows are

$$Q = \frac{\pi a^4}{8\eta L} \Delta p, \quad \text{circular channel}, \qquad (1.63a)$$

$$Q = \frac{h^3 w}{12\eta L} \Delta p, \quad \text{parallel-plate channel}, \qquad (1.63b)$$

$$Q \approx \left[1 - 0.630\frac{h}{w}\right] \frac{h^3 w}{12\eta L} \Delta p, \quad \text{rectangular channel with } h < w. \qquad (1.63c)$$

For the Poiseuille flow in a rectangular channel of height h and width w, an analytical solution in closed form has not yet been found. However, the error of the approximative result (1.63c) is just 13% for the worst case (a square with $h = w$), while already for an aspect ratio of a half (a rectangle with $h = w/2$), it has decreased to 0.2%.

The SI unit of flow rate is m^3 s^{-1}, but in microfluidics, volume is often measured in µL = mm^3 and time in minutes, so the following conversion factors are useful.

$$1\frac{\mu L}{s} = 10^{-9} \frac{m^3}{s}, \quad 1\frac{\mu L}{\min} = 1.67 \times 10^{-11} \frac{m^3}{s}. \qquad (1.64)$$

1.6 Equivalent Circuit Modeling

Often it suffices to find the flow rate through a given microfluidic network rather than the detailed flow field. To this end, the method of equivalent circuit modeling may be used. This type of modeling is only strictly valid for steady Stokes flow in the limit of zero velocity, however, it is a good approximation for $Re \lesssim 1$ in long, narrow channels, for which entrance effects are negligible, and for changes propagating slower than the speed of sound.

Governing Equations in Microfluidics

1.6.1 Hydraulic Resistance

Above we have found that a constant pressure drop Δp results in a constant flow rate Q. This result can be summarized in the Hagen–Poiseuille law

$$\Delta p = R_{\text{hyd}} Q, \quad \text{or} \quad R_{\text{hyd}} = \frac{\Delta p}{Q}, \tag{1.65}$$

where we have introduced the proportionality factor R_{hyd} known as the hydraulic resistance; a central concept in characterizing and designing microfluidic channels in lab-on-a-chip systems, see the list in Table 1.1. The SI units used in the Hagen–Poiseuille law are

$$[Q] = \frac{\text{m}^3}{\text{s}}, \quad [\Delta p] = \text{Pa} = \frac{\text{kg}}{\text{m s}^2}, \quad [R_{\text{hyd}}] = \frac{\text{Pa s}}{\text{m}^3} = \frac{\text{kg}}{\text{m}^4 \text{s}}. \tag{1.66}$$

Table 1.1 The hydraulic resistance for water flowing through straight channels with different cross-sectional shapes. The numerical values are calculated using $\eta = 1$ mPa s, $L = 1$ m, $a = 100$ µm, $b = 33$ µm, $h = 100$ µm, and $w = 300$ µm. Note that the areas are different.

Shape and parameters	R_{hyd} expression	R_{hyd} $\left[10^{12} \dfrac{Pa\ s}{m^3}\right]$
circle (a)	$\dfrac{8}{\pi} \eta L \dfrac{1}{a^4}$	25
ellipse (a, b)	$\dfrac{4}{\pi} \eta L \dfrac{1 + (b/a)^2}{(b/a)^3} \dfrac{1}{a^4}$	393
triangle (a)	$\dfrac{320}{\sqrt{3}} \eta L \dfrac{1}{a^4}$	1850
two parallel plates (h, w)	$12 \eta L \dfrac{1}{h^3 w}$	40
rectangle (h, w)	$\dfrac{12 \eta L}{1 - 0.63(h/w)} \dfrac{1}{h^3 w}$	51
square (h)	$28.4\, \eta L \dfrac{1}{h^4}$	284
parabola (h, w)	$\dfrac{105}{4} \eta L \dfrac{1}{h^3 w}$	88
perimeter $\mathcal{P}$, area $\mathcal{A}$	$\approx 2 \eta L \dfrac{\mathcal{P}^2}{\mathcal{A}^3}$	—

The Hagen–Poiseuille law is analogous to Ohm's law, $\Delta V = RI$, for a wire carrying the electrical current I and having the electrical resistance R and the potential drop ΔV. In hydraulic systems volume is moved, Q is volume per time, while in electric systems charge is moved, I is charge per time. Likewise, Δp is energy per volume as ΔV is energy per charge. Hydraulic power is $Q\Delta P$ (vol/time × energy/vol) while electric power is $I\Delta V$ (charge/time × energy/charge). For low Reynolds numbers, fluid flow is described by the linear Stokes equation, and to a good approximation the hydraulic resistances obey the same rules for series and parallel coupling as the electric resistances in linear circuit theory. Thus for two hydraulic resistances R_1 and R_2 we have

$$R_{\text{hyd}}^{\text{series}} = R_1 + R_2 \quad \text{and} \quad R_{\text{hyd}}^{\text{parallel}} = \left(\frac{1}{R_1} + \frac{1}{R_2}\right)^{-1}. \tag{1.67}$$

For a generic fluidic network, Kirchhoff's laws apply:

(a) The sum of mass flow rates entering/leaving any node in the circuit is zero.
(b) The sum of all pressure differences in any closed loop of the circuit is zero. (1.68)

1.6.2 Hydraulic Compliance

The analogy between hydraulic and electric systems can be taken one step further. When the pressure increases by Δp in a liquid inside an elastic channel, the volume available to the liquid increases by ΔW. This is analogous to the charging of a capacitor where an increase in voltage by ΔV increases the charge on the capacitor by $\Delta q = C\Delta V$. The electric capacitance is given by $C = \partial q/\partial V$, and we are led to introduce hydraulic capacitance C_{hyd}, also known as compliance, given by.

$$C_{\text{hyd}} \equiv \frac{dW}{dp}, \quad \text{with} \quad [C_{\text{hyd}}] = m^3 \, \text{Pa}^{-1}. \tag{1.69}$$

As an example of compliance we consider a simple model of a soft-walled channel filled with an incompressible liquid as sketched in Figure 1.5(a). If the pressure increases inside the channel, the latter will expand. The compliance C_{hyd} of the channel is a given constant related to the geometry and the material properties of the channel walls. As a simplification we model the channel as consisting of two subchannels with hydraulic resistances R_1 and R_2, respectively, connected in series. The pressure p_c at the point, where the two subchannels connect, determines the expansion of the whole channel. The equivalent model is seen in Figure 1.5(b). We let the pressure at the inlet be p^* for time $t < 0$ and $p^* + \Delta p$ for time $t > 0$. The flow rate at the inlet and the outlet are given by the Hagen–Poiseuille law $Q_1 = (p^* + \Delta p - p_c)/R_1$ and $Q_2 = (p_c - p^*)/R_2$, respectively, while the rate of volume expansion inside the chamber is given by $Q_c = \partial_t W = C_{\text{hyd}} \, \partial_t p_c$. Since

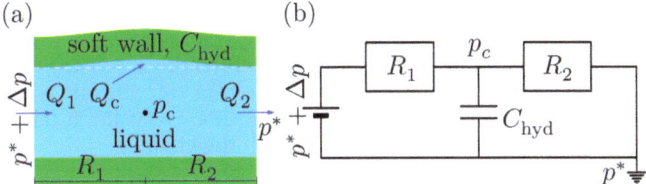

Figure 1.5 (a) Compliance in a soft-walled channel (green) filled with liquid (light blue). The liquid center pressure is denoted by p_c, while R_1 and R_2 denote the hydraulic resistances of the first and second part of the channel, respectively. Mass conservation yields $Q_1 = Q_2 + Q_c$. (b) The equivalent circuit diagram corresponding to the soft-walled channel of panel (a), where R_1 and R_2 are the hydraulic resistances of each part of the channel, while C_{hyd} is the compliance of the soft wall. Reproduced with permission from ref. 38.

the liquid is assumed to be incompressible, conservation of mass leads to $Q_1 = Q_2 + Q_c$, and we arrive at a differential equation for the liquid center pressure p_c,

$$\partial_t p_c = -\left(\frac{1}{\tau_1} + \frac{1}{\tau_2}\right) p_c + \left(\frac{1}{\tau_1} + \frac{1}{\tau_2}\right) p^* + \frac{1}{\tau_1} \Delta p, \qquad (1.70)$$

where $\tau_1 = R_1 C_{hyd}$ and $\tau_2 = R_2 C_{hyd}$ are the hydraulic RC times. The solution,

$$p_c(t) = p^* + \left(1 - e^{-[\tau_1^{-1} + \tau_2^{-1}]t}\right) \frac{\tau_2}{\tau_1 + \tau_2} \Delta p, \qquad (1.71)$$

is analogous to the voltage across a capacitor being charged through a voltage divider.

Often the external tubing may lead to long transient times in the external system due to the RC-time.

$$\tau_{RC} = R_{hyd} C_{hyd} \qquad (1.72)$$

arising from the hydraulic resistance R_{hyd} and compliance C_{hyd}. For an elastic tube of fixed length L, inner radius a, wall thickness d, and thickness ratio $\delta = d/a$, C_{hyd} can be derived from the basic theory of elasticity,[41]

$$C_{hyd}^{tube} = 2\pi(1+\bar{\nu}) \frac{(1-2\bar{\nu}) + (1+\delta)^2}{(1+\delta)^2 - 1} \frac{a^2 L}{Y}. \qquad (1.73)$$

Here, $\bar{\nu}$ is the Poisson ratio, and Y is Young's modulus. Specific examples of transient RC times are listed in Table 1.2. If a gas bubble is trapped inside a microfluidic channel, an extra compliance C_{hyd}^{bubble} is added to the system,

$$C_{hyd}^{bubble} = \frac{dW}{dp} = -\frac{dW^{bubble}}{dp} = \frac{4\pi}{3} \frac{a^3}{p^*}. \qquad (1.74)$$

Here, we have assumed isothermal compression of the gas bubble around ambient pressure p^*. For isentropic compression, the hydraulic compliance is multiplied by the adiabatic factor $\gamma \approx 1.4$. A large bubble with $a = 1$ mm

Table 1.2 The hydraulic resistance R_{hyd}, compliance C_{hyd}, and inductance L_{hyd} as well as the RC and RL times (τ_{RC} and τ_{RL}) for water ($\eta = 1$ mPa s, $\rho = 998$ kg m^{-3}) in a straight circular channel (length $L = 1$ m, radius $a = 0.1$ mm, thickness ratio $\delta = 1$) made of either soft silicone rubber ($Y = 2.1$ MPa, $\bar{\nu} = 0.49$) or hard teflon ($Y = 0.5$ GPa, $\bar{\nu} = 0.45$). It takes 0.7 ms for a pressure wave to propagate through the channel.

Material	R_{hyd} $\left[10^{12} \dfrac{\text{Pa s}}{\text{m}^3}\right]$	C_{hyd} $\left[10^{-15} \dfrac{\text{m}^3}{\text{Pa}}\right]$	L_{hyd} $\left[10^9 \dfrac{\text{Pa s}^2}{\text{m}^3}\right]$	τ_{RC} [ms]	τ_{RL} [ms]
Silicone	25	60	32	1500	1.3
Teflon	25	0.25	32	6.2	1.3

leads to $C_{hyd}^{bubble} = 42 \times 10^{-15}$ m^3/Pa comparable to the silicone tube in Table 1.2.

1.6.3 Hydraulic Inductance

The last analogy is inductance. The self-inductance L_{el}, or the electric inertia, is the ability of an electric system to maintain a given current I. A rate of change $\partial_t I$ in the current induces a potential drop $\Delta V = L_{el} \partial_t I$. Hydraulic inductance L_{hyd} therefore relates to maintaining an existing volume current $Q = A v_{avr}$. Since the rate of change $\partial_t Q = A \partial_t v_{avr}$ involves acceleration, L_{hyd} must correspond to inertia. Consider the flow through a channel of length L and cross-section area A. If the force F driving the flow arises from a pressure drop Δp, we find from Newton's second law that $\Delta p = F/A = (\rho L A) \partial_t v_{avr}/A$. This leads to $\Delta p = (\rho L/A) \partial_t Q$, from which we can read off the hydraulic inductance L_{hyd} as

$$L_{hyd} = \frac{\rho L}{A}, \quad \text{with} \quad [L_{hyd}] = \frac{\text{Pa s}^2}{\text{m}^3}. \tag{1.75}$$

In analogy with the RC-time there is also a transient RL-time given by $\tau_{RL} = L_{hyd}/R_{hyd}$. For a tube it is

$$\tau_{RL}^{tube} = \frac{L_{hyd}}{R_{hyd}} = \frac{\rho a^2}{8\eta} = \frac{a^2}{8\nu}, \tag{1.76}$$

where in the last equality we have used the kinematic viscosity ν from eqn (1.54). Specific examples of RL times are listed in Table 1.2. For a more complete example of circuit modeling in a microfluidic system with pulsatile flow including compliance and air bubbles see Vedel et al.[42]

1.7 Scaling Laws in Microfluidics

In physics, scaling laws are statements about how dependent variables scales with the independent variables, e.g. the flow rate Q in a microfluidic system scales like the pressure drop Δp to the power one, written as $Q \propto \Delta p$, and the

Governing Equations in Microfluidics

speed of sound c_0 in a liquid scales like the density ρ_0 of the liquid to the power minus one-half, written as $c_0 \propto (\rho_0)^{-\frac{1}{2}}$. Scaling laws are helpful in gaining basic physical understanding of a given system, and they are intimately related to dimensional analysis and to the concept of characteristic dimensionless groups or numbers, such as the Reynolds number.

In the following subsections, adapted from Bruus,[39] we shall see examples of scaling laws relevant for microfluidics and also identify important dimensionless groups, often derived by dimensional analysis combined with physical insight. We shall see how the scaling laws are useful for understanding, designing, and analyzing microfluidic devices.

1.7.1 Flow Rate

As a first example of scaling laws and dimensional analysis, we take the Poiseuille flow analyzed in Section 1.5.3 and sketched in Figure 1.6. Let us ask the question of how the flow rate Q scales with the physical parameters of the problem: pressure drop Δp, channel length L, cross-sectional dimensions (radius a for a circle, and width w and height h for parallel plates), as well as density ρ_0 and viscosity η of the liquid. From Section 1.5.4, we know the full answer to be

$$Q = \frac{\pi a^4}{8\eta L}\Delta p, \quad \text{circular channel}, \tag{1.77a}$$

$$Q = \frac{h^3 w}{12\eta L}\Delta p, \quad \text{parallel-plate channel}, \tag{1.77b}$$

but how can we obtain this from dimensional analysis?

First we list the SI units (marked by square brackets) of the involved quantities: $[Q] = \text{m}^3\,\text{s}^{-1}$, $[\Delta p] = \text{Pa}$, $[\eta] = \text{Pa s}$, $[L] = [h] = [w] = [a] = \text{m}$. We leave out ρ_0, as density-related inertial effects are negligible in microfluidics systems.[38] Then we look at the driving force, here the pressure drop Δp, and note that in the linear regime, the flow resistance ensures that flow rate scales

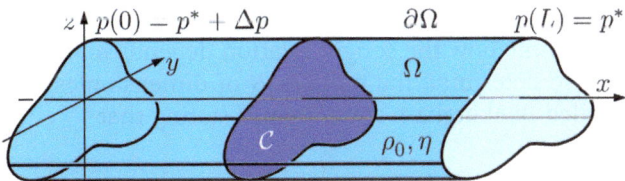

Figure 1.6 Poiseuille flow of a liquid with density ρ_0 and viscosity η through a straight channel Ω (blue), where the flow is subject to the no-slip boundary condition on the surface $\partial\Omega$. The channel is translational invariant in the x direction, and it has an arbitrarily shaped cross-section X (dark blue) in the yz-plane. The pressure at $x = 0$ (left) is Δp higher than at $x = L$ (right). Reproduced with permission from ref. 39.

like $Q \propto \Delta p$. Since the left-hand side contains s^{-1}, we must divide the right-hand side by η, the only remaining quantity containing the unit of time, $Q \propto \Delta p/\eta$. As a result, the right-hand side needs to be multiplied by some length ℓ to the power 3 to balance the units, $Q \propto \ell^3 \Delta p/\eta$. But which physical lengths enters ℓ^3 and to which power? Here we must go beyond the dimensional analysis and bring in some physical insight. Clearly, the longer the channel, the smaller a flow rate for a given pressure drop Δp, so for linear response we get $Q \propto \ell^4 \Delta p/(\eta L)$, where the length scale ℓ^4 now only involves the cross-sectional geometry. Since $Q \propto vA$, where v is the average flow velocity and A the area of the cross section, we have

$$Q \propto \frac{\ell^2 A}{\eta L} \Delta p, \qquad (1.78)$$

where ℓ is now solely related to the scaling of v with the cross-sectional geometry. From the general nature of viscous flows, we know that ℓ must be associated with the smallest distance from the center of the channel to the side walls, as this determines the shear stress $\eta v/\ell$ dominating the flow resistance. The scaling law eqn (1.78) is how far we can come from a purely dimensional analysis of the Poiseuille flow for a given constant pressure Δp. A circular tube of radius a will have $\ell = a$ and $A = \pi a^2$, while a parallel-plate channel with $h < w$ will have $\ell = h/2$ and $A = wh$, whereby we nearly recover the exact result, eqn (1.77a) and (1.77b), except for the numerical factors 1/8 and 1/3, respectively. Without solving any differential equations, we can infer the expected proportionality $Q \propto \Delta p/(\eta L)$, and the less obvious, but experimentally very important, dependence of Q on the smallest length scale of the cross-section, $Q \propto a^4$ (circle) and $Q \propto h^3$ (parallel plate). A linear downscaling of a channel cross section by a factor of 2 thus reduces the flow rate by a factor of 16 for the circular channel and by 8 for a parallel-plate channel.

1.7.2 Characteristic Dimensionless Numbers

The above example shows the problems that occur when dimensionless groups or numbers appear, e.g. the length ratio ℓ/L. Dimensional analysis alone cannot determine to which power it enters the final expression. It is therefore important firstly to identify the dimensionless numbers, and secondly to determine how the sought relation depends on them.

A formal way to determine the number N of dimensionless groups is by using the so-called Buckingham π-theorem,[43] stating that

$$N \equiv D - F, \qquad (1.79)$$

where D is the number of dependent variables in the given problem, and Φ is the number of independent, fundamental dimensions (3 in mechanics: length L, time T, and mass M). Let us apply the theorem for the Poiseuille flow in a circular tube described by $D = 6$ dependent variables, a, v, Δp, η, ρ_0, and L. According to Buckingham's theorem, this results in $N = 6 - 3 = 3$ dimensionless numbers.

Governing Equations in Microfluidics

Normally, it is straightforward to determine the dimensionless groups, but there exists a formal method to determine them if needed.[39,43] The first step is to resolve the physical dimension of the D dependent variables $d_k \sim L^{\alpha_{1k}} T^{\alpha_{2k}} M^{\alpha_{3k}}$ ($k = 1, 2, \ldots, D$) in powers of the Φ fundamental dimensions (with index $j = 1, 2, \ldots \Phi$), such as in $[\Delta p] = \text{Pa} = \text{kg}\,(\text{s}^{-2}\,\text{m}^{-1}) \sim L^{-1}\,T^{-2}\,M^1$. Then we construct the $F \times D$ matrix $\mathbf{P}$ with elements $(\mathbf{P})_{jk} = \alpha_{jk}$, which in our case becomes

$$\mathbf{P} = \begin{pmatrix} & a & v & \rho_0 & \eta & \Delta p & L \\ L & 1 & 1 & -3 & -1 & -1 & 1 \\ T & 0 & -1 & 0 & -1 & -2 & 0 \\ M & 0 & 0 & 1 & 1 & 1 & 0 \end{pmatrix}, \quad (1.80)$$

where for clarity the row and column labels are shown. The dimensionless groups or numbers can now be determined as the N null vectors $\boldsymbol{\pi}_i$ (here $\boldsymbol{\pi}_1$, $\boldsymbol{\pi}_2$ and $\boldsymbol{\pi}_3$) of the matrix $\mathbf{P}$,

$$\mathbf{P} \cdot \boldsymbol{\pi}_i = 0, \quad \text{for } i = 1, 2, \ldots, N, \quad (1.81)$$

because the components of these vectors contains the powers β_{jk} of the dependent variables d_k leading to the power zero of fundamental jth dimension, $[d_1^{\beta_{j1}} d_2^{\beta_{j2}} \ldots d_D^{\beta_{jD}}] = 0$. Three possible dimensionless groups for the matrix in eqn (1.80) are

$$\boldsymbol{\pi}_1 = \begin{pmatrix} 1 \\ 1 \\ 1 \\ -1 \\ 0 \\ 0 \end{pmatrix}, \quad \boldsymbol{\pi}_2 = \begin{pmatrix} 1 \\ 0 \\ 0 \\ 0 \\ 0 \\ -1 \end{pmatrix}, \quad \boldsymbol{\pi}_3 = \begin{pmatrix} -1 \\ 1 \\ 0 \\ 1 \\ -1 \\ 0 \end{pmatrix}, \quad (1.82)$$

corresponding to the dimensionless groups or numbers

$$\pi_1 \sim N_1 = \frac{a v \rho_0}{\eta}, \quad \pi_2 \sim N_2 = \frac{a}{L}, \quad \pi_3 \sim N_3 = \frac{\eta v}{a \Delta p}. \quad (1.83)$$

The first of these numbers is the Reynolds number Re,[38] the second is the aspect ratio of the channel, and the third is ratio of the shear-induced stress $\eta(v/a)$ and the driving pressure Δp. The values of these respective numbers reveal information about the physical state of the system. It is only if the Reynolds number is small, $Re = N_1 \lesssim 1$, that it makes sense to exclude the density as a relevant parameter in the dimensional analysis in Section 1.6.1. Likewise, it is only if the channel aspect ratio is small, $N_2 \lesssim 1$, that we can determine ℓ^2 in eqn (1.78). Consequently, the identification of the dimensionless numbers is an important part of dimensional analysis.

An important caveat regarding the dimensionless numbers is that they are not uniquely determined. Any linear combination of the null vectors $\boldsymbol{\pi}_i$ results in a null vector. The following set is therefore also a possible solution,

$$\boldsymbol{\pi}_1^* = \boldsymbol{\pi}_1, \quad \boldsymbol{\pi}_2^* = -\boldsymbol{\pi}_2, \quad \boldsymbol{\pi}_3^* = \boldsymbol{\pi}_1 + \boldsymbol{\pi}_3, \quad (1.84)$$

corresponding to the dimensionless groups or numbers

$$N_1^* = \frac{av\rho_0}{\eta}, \quad N_2^* = \frac{L}{a}, \quad N_3^* = \frac{\rho_0 v^2}{\Delta p}. \tag{1.85}$$

Here $N_1^* = N_1$ is the Reynolds number Re as before, $N_2^* = 1/N_2$ is the inverse aspect ratio of the channel, while $N_3^* = N_1 N_3$ is the ratio of the kinetic pressure to the driving pressure. For very low flow velocities, $N_3^* \ll N_3$ and the shear stress dominates over the kinetic pressure. For higher velocities the situation reverses, as is of course evident from the relationship $N_3^*/N_3 = Re$.

1.7.3 Entrance Length

In acoustofluidics, one is often in a situation where several flow streams are brought together at the entrance of the region (here a long straight channel) of the microfluidic chip where the active acousto-activated particle separation takes place. The question naturally arises at which distance L down-stream from the entrance point is the laminar flow a fully developed steady-state Poiseuille flow profile.

For very low flow velocities, it is expected that the entrance length L scales like the smallest distance ℓ from the center of the channel to the side walls in the cross-sectional geometry (a for the circle and $h/2$ for the parallel-plate with $h < w$), $L \propto \ell$. This must be so, as ℓ in this case is the only length scale in the problem. However, as the velocity is increased and the Reynolds number exceeds unity, another (kinematic) length scale appears, namely $L \propto Re \, \ell$.

As in Section 1.7.1, more physical insight is needed to obtain a more precise statement. Let us consider a simplified case, where instead of several flow streams meeting at the inlet of the channel, we just assume that the velocity profile differs from the fully developed Poiseuille profile and is, say, the constant v across the inlet. Due to mass conservation in the long straight channel we are analyzing, the average velocity remains v. In a reference frame where the channel walls are fixed, the steady-state flow profile has a spatial development, it simply changes its shape as a function of axial position x. In a reference system moving with the average, constant flow speed v, the development of the flow profile is temporal and not spatial.

This situation can be analyzed using the Navier–Stokes equation in the form of the velocity diffusion equation eqn (1.54), $\partial_t v = \nu \nabla^2 v$, where $\nu = \eta/\rho_0$ is the momentum diffusivity or kinematic viscosity, which for water is $\nu = 10^{-6}$ m^2 s^{-1}. Diffusion length (here the shortest, transverse channel dimension ℓ) and time τ are described by the exponential factor $\exp[-\ell^2/(2\nu\tau)]$ in one spatial dimension as studied in the following section. If we define the full development of the flow profile as when the developing profile is within 5% of the steady-state value, then $\exp[-\ell^2/(2\nu\tau)] \approx 0.05$ or $\ell^2 \approx 6\nu\tau$. The entrance length L in the fixed frame of reference can therefore be

Governing Equations in Microfluidics

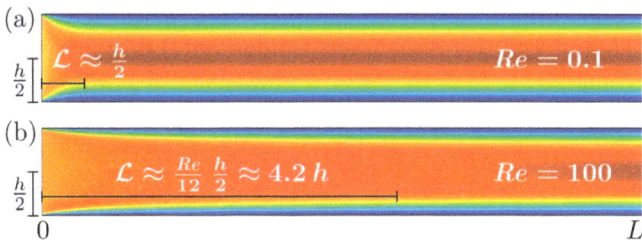

Figure 1.7 Color plots of the axial velocity from zero (dark blue) to v_{max} (dark red) in a parallel-plate channel (side view) of length L and height h. The velocity field on the inlet to the left is set to be a constant. (a) In the low Reynolds number limit ($Re = 0.1$), the entrance length over which a full Poiseuille flow profile is established is given by $\frac{1}{2}h$. (b) In the medium Reynolds number limit ($Re = 100$), the entrance length is given by $\frac{Re}{24}h$. Reproduced with permission from ref. 39.

estimated as $L \approx v\tau \approx \ell^2 v/(6\nu) = \frac{1}{12} Re\, l$, where we use the usual definition of the Reynolds number $Re = (2\ell)v\rho_0/\eta$. In summary, our scaling law for the entrance length in straight-channel Poiseuille flow becomes

$$L \approx \max\left\{\ell, \frac{Re}{12}\ell\right\}. \quad (1.86)$$

Figure 1.7 shows a numerical simulation of the entrance effects for the axial velocity field in a straight parallel plate channel of height h. The inlet velocity profile is the constant v in the axial direction at $x = 0$. In this case, $\ell = h/2$, and in panel (a) it is seen how $L \approx h/2$ for the low Reynolds number $Re = 0.1$, while $L \approx (Re/12)h/2 = 0.042 Re\, h = 4.2h$ for the medium Reynolds number $Re = 100$ in accordance with eqn (1.86). We note that for typical microchannels used in acoustophoresis with $h = 0.2$ mm, the Reynolds number $Re = 12$ for which the scaling changes from one behavior to another in eqn (1.86) corresponds to the relatively high average velocity $v \approx 6$ cm s^{-1}. We therefore expect that the entrance length in acoustofluidics in most cases is given by half the channel height, $L \approx \ell$, for $h < w$.

1.7.4 Inertial Time Scale

In spite of the dominance of viscosity in microfluidics systems, it can nevertheless be of interest to determine the time scale of inertial effects of the liquid flow. In Section 1.6.3, this was briefly treated in terms of hydraulic inductance. As an example we study here the inertial-related time τ_{iner} it takes a liquid with density ρ_0, viscosity η, and kinematic viscosity $\nu = \eta/\rho_0$, to reach rest after an instantaneous removal of the pressure difference driving a steady-state Poiseuille flow, through a circular microchannel of radius a and length L. Here "instantaneous" of course refers to the small acoustic time scale $\tau_{acou} = L/c_0$ of pressure (or sound) wave propagation along the channel given the speed of sound c_0

of the liquid. For this problem we have $D = 6$ dependent variables (a, L, τ_{iner}, c_0, η, ρ_0) and $F = 3$ independent fundamental dimensions (L, T, M). We can thus construct $N = D - F = 3$ dimensional groups or numbers. One possibility is

$$N_1 = \frac{a}{L}, \quad N_2 = \frac{\tau_{acou}}{\tau_{iner}}, \quad N_3 = \frac{a^2}{\nu \tau_{iner}}. \quad (1.87)$$

The first of these is the aspect ratio of the channel, the second is the ratio of the axial propagation time of pressure waves relative to the stopping time, while the last is the ratio of the momentum diffusion time relative to the stopping time. In the case of a long, narrow channel ($N_1 \ll 1$), and a propagation time of sound much shorter than the stopping time ($N_2 \ll 1$), the liquid flow stops due to the sidewards diffusion of momentum over the distance a with the momentum diffusivity ν. Dimensional analysis thus leads to $\tau_{iner} \approx a^2/\nu$, while standard diffusion theory for two spatial dimensions provides another factor of 1/4. We therefore arrive at the scaling law for the inertial stopping time,

$$\tau_{iner} \approx \frac{a^2}{4\nu}. \quad (1.88)$$

To provide a numerical example, we take the channel dimensions to be $a = 0.1$ mm and $L = 10$ mm, and the liquid to be water at room temperature, i.e. $\nu = 10^{-6}$ m^2 s^{-1} and $c_0 = 1500$ m s^{-1}. The pressure propagation time becomes $L/c_0 \approx 10$ µs and the inertial stopping time is $\tau_{iner} \approx 3$ ms, and the conditions $N_1 \ll 1$ and $N_2 \ll 1$ for the analysis are thus fulfilled. We note that $\tau_{iner} \approx 3$ ms for $a = 0.1$ mm combined with the a^2 scaling law implies that for channels 10 times wider, the inertial time scale approaches 1 s, while 10 times more narrow leads to inertial time scales approaching the pressure propagation time scale. Figure 1.8 shows in detail how the axial velocity

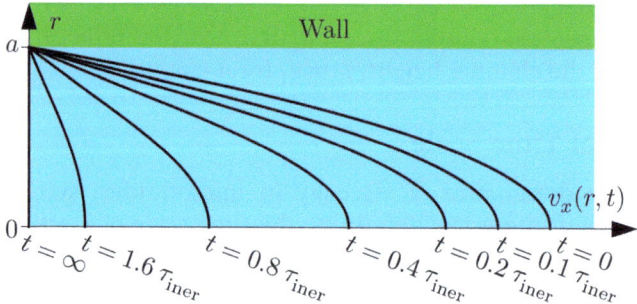

Figure 1.8 The evolution in time from the initial Poiseuille paraboloid at $t = 0$ to a zero velocity profile at $t = \infty$ of the velocity profile $v_x(r, t)$ in a cylindrical channel of radius a for a decelerating Poiseuille flow due to the abrupt disappearance of the driving pressure Δp at $t = 0$. The time is expressed in units of the inertial time scale $\tau_{iner} = a^2/(4\nu)$ of eqn (1.88). Reproduced with permission from ref. 39.

profile $v_x(r, t)$ develops in time from the initial removal of the driving pressure at $t = 0$ in units of the characteristic stopping time τ_{iner} of eqn (1.88); for details of the calculation see the textbooks by Bruus[4] and Batchelor.[44] The figure shows how the velocity profile indeed does decay on a time scale given by τ_{iner}.

References

1. P. Tabeling, *Introduction to Microfluidics*, Oxford University Press, Oxford, 2005.
2. G. Karniadakis, A. Beskok and N. Aluru, *Microflows and Nanoflows: Fundamentals and Simulation*, Springer Science + Business Media, Inc., New York NY, 2005.
3. N.-T. Nguyen and S. T. Wereley, *Fundamentals and Applications of Microfluidics*, Artech House, Norwood MA, 2nd edn., 2006.
4. H. Bruus, *Theoretical Microfluidics*, Oxford University Press, Oxford, 2008.
5. J. Berthier and P. Silberzan, *Microfluidics for Biotechnology*, Artech House, Norwood MA, 2nd edn., 2010.
6. B. Kirby, *Micro- and Nanoscale Fluid Mechanics: Transport in Microfluidic Devices*, Cambridge University Press, New York, 2010.
7. A. J. Mach and D. Di Carlo, *Biotechnol. Bioeng.*, 2010, **7**, 302–311.
8. S. Jacobson, R. Hergenroder, L. Koutny and J. Ramsey, *Anal. Chem.*, 1994, **66**, 1114–1118.
9. P. R. C. Gascoyne, J. Noshari, T. J. Anderson and F. F. Becker, *Electrophoresis*, 2009, **30**, 1388–1398.
10. M. A. M. Gijs, *Microfluid. Nanofluid.*, 2004, **1**, 22–40.
11. L. Huang, E. Cox, R. Austin and J. Sturm, *Science*, 2004, **304**, 987–990.
12. D. Johnson and D. Feke, *Sep. Technol.*, 1995, **5**, 251–258.
13. K. Yasuda, S. Umemura and K. Takeda, *Jpn. J. Appl. Phys.*, 1995, **34**, 2715–2720.
14. J. J. Hawkes and W. T. Coakley, *Sens. Actuators, B*, 2001, **75**, 213–222.
15. J. J. Hawkes, W. T. Coakley, M. Groschl, E. Benes, S. Armstrong, P. J. Tasker and H. Nowotny, *J. Acoust. Soc. Am.*, 2002, **111**, 1259–1266.
16. N. R. Harris, M. Hill, S. Beeby, Y. Shen, N. M. White, J. J. Hawkes and W. T. Coakley, *Sens. Actuators, B*, 2003, **95**, 425–434.
17. F. Petersson, A. Nilsson, C. Holm, H. Jönsson and T. Laurell, *Analyst*, 2004, **129**, 938–943.
18. F. Petersson, A. Nilsson, C. Holm, H. Jönsson and T. Laurell, *Lab Chip*, 2005, **5**, 20–22.
19. J. Hultström, O. Manneberg, K. Dopf, H. M. Hertz, H. Brismar and M. Wiklund, *Ultrasound Med. Biol.*, 2007, **33**, 145–151.
20. M. Evander, L. Johansson, T. Lilliehorn, J. Piskur, M. Lindvall, S. Johansson, M. Almqvist, T. Laurell and J. Nilsson, *Anal. Chem.*, 2007, **79**, 2984–2991.
21. J. Svennebring, O. Manneberg, P. Skafte-Pedersen, H. Bruus and M. Wiklund, *Biotechnol. Bioeng.*, 2009, **103**, 323–328.

22. A. Lenshof, A. Ahmad-Tajudin, K. Jaras, A.-M. Sward-Nilsson, L. Aberg, G. Marko-Varga, J. Malm, H. Lilja and T. Laurell, *Anal. Chem.*, 2009, **81**, 6030–6037.
23. J. V. Norris, M. Evander, K. M. Horsman-Hall, J. Nilsson, T. Laurell and J. P. Landers, *Anal. Chem.*, 2009, **81**, 6089–6095.
24. C. Grenvall, P. Augustsson, J. R. Folkenberg and T. Laurell, *Anal. Chem.*, 2009, **81**, 6195–6200.
25. T. Franke, S. Braunmueller, L. Schmid, A. Wixforth and D. A. Weitz, *Lab Chip*, 2010, **10**, 789–794.
26. P. Thevoz, J. D. Adams, H. Shea, H. Bruus and H. T. Soh, *Anal. Chem.*, 2010, **82**, 3094–3098.
27. P. Augustsson, R. Barnkob, C. Grenvall, T. Deierborg, P. Brundin, H. Bruus and T. Laurell, Measuring the acoustophoretic contrast factor of living cells in microchannels, in ed. S. Verporte, H. Andersson, J. Emneus and N. Pamme, *Proc. 14th MicroTAS, 3–7 October 2010, Groningen, The Netherlands, CBMS*, 2010, pp. 1337–1339.
28. S. M. Hagsäter, T. G. Jensen, H. Bruus and J. P. Kutter, *Lab Chip*, 2007, **7**, 1336–1344.
29. J. Shi, H. Huang, Z. Stratton, Y. Huang and T. J. Huang, *Lab Chip*, 2009, **9**, 3354–3359.
30. O. Manneberg, S. M. Hagsäter, J. Svennebring, H. M. Hertz, J. P. Kutter, H. Bruus and M. Wiklund, *Ultrasonics*, 2009, **49**, 112–119.
31. O. Manneberg, B. Vanherberghen, B. Onfelt and M. Wiklund, *Lab Chip*, 2009, **9**, 833–837.
32. P. Glynne-Jones, R. J. Boltryk, N. R. Harris, A. W. J. Cranny and M. Hill, *Ultrasonics*, 2010, **50**, 68–75.
33. J. D. Adams, P. Thevoz, H. Bruus and H. T. Soh, *Appl. Phys. Lett.*, 2009, **95**, 254103–254111.
34. S. Oberti, D. Moeller, A. Neild, J. Dual, F. Beyeler, B. J. Nelson and S. Gutmann, *Ultrasonics*, 2010, **50**, 247–257.
35. R. Barnkob, P. Augustsson, T. Laurell and H. Bruus, *Lab Chip*, 2010, **10**, 563–570.
36. P. Augustsson, R. Barnkob, S. T. Wereley, H. Bruus and T. Laurell, *Lab Chip*, 2011, **11**, 4152–4164.
37. P. B. Muller, M. Rossi, A. G. Marín, R. Barnkob, P. Augustsson, T. Laurell, C. J. Kähler and H. Bruus, *Phys. Rev. E*, 2013, **88**, 023006.
38. H. Bruus, *Lab Chip*, 2011, **11**, 3742–3751.
39. H. Bruus, *Lab Chip*, 2012, **12**, 1578–1586.
40. A. S. Dukhin and P. J. Goetz, *J. Chem. Phys.*, 2009, 130.
41. L. D. Landau and E. M. Lifshitz, *Theory of Elasticity. Course of Theoretical Physics*, Pergamon Press, Oxford, 3rd edn., vol. 7, 1986.
42. S. Vedel, L. H. Olesen and H. Bruus, *J. Micromech. Microeng.*, 2010, **20**, 035026.
43. E. Buckingham, *Phys. Rev.*, 1914, **4**, 345–376.
44. G. K. Batchelor, *An Introduction to Fluid Dynamics*, Cambridge University Press, Cambridge, 1967.

CHAPTER 2

Perturbation Theory and Ultrasound Resonances

HENRIK BRUUS

Department of Physics, Technical University of Denmark, Lyngby, Denmark
E-mail: bruus@fysik.dtu.dk

2.1 Introduction

Ultrasound acoustics in the low MHz-range is well suited for applications within microfluidics. When frequencies $f \gtrsim 1.5$ MHz are combined with the speed of sound in water at room temperature, $c_{wa} \approx 1.5 \times 10^3$ m s^{-1}, the corresponding wavelengths $\lambda_{wa} \lesssim 1$ mm may fit into the submillimetre-sized channels and cavities in microfluidic systems and result in the formation of standing pressure waves known as resonance modes. For several reasons it is often advantageous to operate an acoustofluidic device at these resonance modes: they are usually both stable and reproducible, their spatial patterns are controlled by the geometry of the microfluidic channel, and at resonance, a maximum of acoustic power is delivered from the transducer to where it is needed in the system in the form of acoustic radiation force on suspended particles[1-3] or acoustic streaming of the solvent.[4-7] For microchips, the ultrasound is typically generated as bulk or surface acoustic waves driven by an ac voltage applied to externally mounted piezo-ceramic transducers[8,9] or to interdigitated metal electrodes deposited internally on a piezo-electric substrate,[21,10] respectively.

In this chapter, adapted from Bruus,[11] we derive the linear wave equation for the acoustic field in a fluid using regular, first-order perturbation theory. A more complete treatment of this theory can be found in textbooks by

Lighthill,[12] Pierce,[13] and Landau and Lifshitz.[14] Here, as in the previous chapter, we shall use the notation of the textbook by Bruus.[22] The derivation is based on a combination of the thermodynamic equation of state expressing pressure p in terms of density ρ, the kinematic continuity equation for ρ, and the dynamic Navier–Stokes equation for the velocity $\boldsymbol{v}$ (see more details in Chapter 1),

$$p = p(\rho), \tag{2.1a}$$

$$\partial_t \rho = -\boldsymbol{\nabla} \cdot (\rho \boldsymbol{v}), \tag{2.1b}$$

$$\rho \partial_t \boldsymbol{v} = -\boldsymbol{\nabla} p - \rho(\boldsymbol{v} \cdot \boldsymbol{\nabla})\boldsymbol{v} + \nu \nabla^2 \boldsymbol{v} + \beta \eta \boldsymbol{\nabla}(\boldsymbol{\nabla} \cdot \boldsymbol{v}). \tag{2.1c}$$

For water, the values of the viscosity η and the viscosity ratio β are given in Table 2.1.

Throughout this chapter, we disregard for simplicity all external fields, *e.g.* gravity and electromagnetism, and study only the adiabatic case to avoid involving the heat transfer equation. Even so, the set eqn (2.1) of coupled non-linear, partial differential equations is notoriously difficult to solve analytically, but good and useful approximate solutions can be found by perturbation theory.

2.2 First-order Perturbation Theory, the Acoustic Wave Equation

Consider a quiescent liquid, which before the presence of any acoustic wave has constant density ρ_0 and pressure p_0. Let an acoustic wave constitute tiny perturbations (subscript 1) in the fields of density ρ, pressure p, and velocity $\boldsymbol{v}$,

$$\rho = \rho_0 + \rho_1, \quad p = p_0 + c_0^2 \rho_1, \quad \text{and} \quad \boldsymbol{v} = \boldsymbol{v}_1. \tag{2.2}$$

In the (isentropic) expansion of the equation of state $p(\rho) = p_0 + (\partial p/\partial \rho)_s \rho_1$, the derivative has the dimension of a velocity squared, written as c_0^2. Below we find that c_0 can be identified with the (isentropic) speed of sound in the liquid. Inserting eqn (2.2) into eqn (2.1b) and (2.1c), and neglecting products of first-order terms, leads to the first-order continuity and Navier–Stokes equation,

$$\partial_t \rho_1 = -\rho_0 \boldsymbol{\nabla} \cdot \boldsymbol{v}_1, \tag{2.3a}$$

$$\rho_0 \partial_t \boldsymbol{v}_1 = -c_0^2 \boldsymbol{\nabla} \rho_1 + \eta \nabla^2 \boldsymbol{v}_1 + \beta \eta \boldsymbol{\nabla}(\boldsymbol{\nabla} \cdot \boldsymbol{v}_1). \tag{2.3b}$$

A single equation for ρ_1 is obtained by taking the time derivative of eqn (2.3a) followed by insertion of eqn (2.3b) in the resulting expression.

$$\partial_t^2 \rho_1 = -\boldsymbol{\nabla} \cdot (\rho_0 \partial_t \boldsymbol{v}_1) = c_0^2 \nabla^2 \rho_1 - (1+\beta)\eta \nabla^2 (\boldsymbol{\nabla} \cdot \boldsymbol{v}_1) = c_0^2 \left[1 + \frac{(1+\beta)\eta}{\rho_0 c_0^2} \partial_t \right] \nabla^2 \rho_1. \tag{2.4}$$

In microfluidics, where $v_0 \lesssim 0.1$ m s^{-1}, eqn (2.3) and (2.4) are approximately correct to order $v_0/c_0 \lesssim 10^{-4}$, see Landau and Lifshitz.[14]

To make further analytical progress, we assume harmonic time dependence of all fields,

$$\rho_1(\mathbf{r},t) = \rho_1(\mathbf{r})e^{-i\omega t}, \quad p_1(\mathbf{r},t) = c_0^2 \rho_1(\mathbf{r})e^{-i\omega t}, \quad \text{and} \quad \mathbf{v}_1(\mathbf{r},t) = \mathbf{v}_1(\mathbf{r})e^{-i\omega t}, \tag{2.5}$$

where $\omega = 2\pi f$ is the angular frequency and f the frequency of the acoustic field. The harmonic time dependence is expressed by the complex phase $e^{-i\omega t}$ to ease the mathematical treatment. The physical fields are obtained simply by taking the real part. With this, each time derivative ∂_t in eqn (2.4) gives a factor $-i\omega$, and the equation for the pressure becomes

$$\nabla^2 p_1 = -k^2 p_1, \tag{2.6a}$$

$$k = (1+i\gamma)k_0 = (1+i\gamma)\frac{\omega}{c_0}, \quad \text{with} \quad \gamma = \frac{(1+\beta)\eta\omega}{2\rho_0 c_0^2} \approx 10^{-5}, \tag{2.6b}$$

where we have used $p_1 = c_0^2 \rho_1$ as well as introduced the complex-valued wavenumber k, the wavenumber $k_0 = \omega/c_0$, and the viscous damping factor γ (for values see Table 2.1). Eqn (2.6a) is the Helmholtz equation for a damped wave with wavenumber k and angular frequency ω. As $\gamma \ll 1$, we can neglect the viscosity in the bulk part of the acoustic wave, and going back to the explicitly time-dependent eqn (2.4) using $\omega \to i\partial_t$, we obtain the wave equation

$$\nabla^2 p_1 = \frac{1}{c_0^2}\partial_t^2 p_1, \quad \text{for} \quad \eta = 0. \tag{2.7}$$

The solutions in 1D to this standard wave equation have the form $p_1 = p_1(x \pm c_0 t)$ showing that a pressure perturbation $p_1(x)$ at $t = 0$ propagates a distance $\mp c_0 t$ in time t, and thus that c_0 indeed can be interpreted as the speed of sound. In the inviscid limit, it furthermore follows by inserting eqn (2.2) into eqn (2.3b) that $\mathbf{v}_1 = \mathbf{v}(\mathbf{r})\,e^{-i\omega t}$ is a gradient of a potential ϕ_1,

$$\mathbf{v}_1 - -i\frac{1}{\rho_0\omega}\nabla p_1 = \nabla\phi_1, \quad \text{where} \quad \phi_1 = \frac{-i}{\rho_0\omega}p_1 \quad \text{for} \quad \eta - 0. \tag{2.8a}$$

Table 2.1 Parameters at 25 °C used for acoustic modeling. For water $\eta = 0.89$ mPas, $\beta = 3.0$, and (at $f = 1$ MHz) $\gamma = 6 \times 10^{-6}$.

Material	Long. speed of sound	Density	Young's modulus	Poisson's ratio
Water	$c_{wa} = 1497$ m s^{-1}	$\rho_{wa} = 998$ kg m^{-3}	—	—
Polystyrene	$c_{ps} = 2350$ m s^{-1}	$\rho_{ps} = 1050$ kg m^{-3}	—	—
Silicon	$c_{si} = 8490$ m s^{-1}	$\rho_{si} = 2331$ kg m^{-3}	$Y_{si} = 164$ GPa	$\bar{\nu}_{si} = 0.10$
Pyrex	$c_{py} = 5661$ m s^{-1}	$\rho_{py} = 2230$ kg m^{-3}	$Y_{py} = 63$ GPa	$\bar{\nu}_{py} = 0.22$

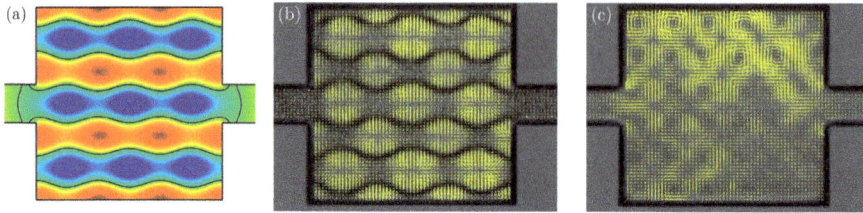

Figure 2.1 An acoustic first-order pressure field and second-order acoustic streaming and radiation forces at the 2.17 MHz-acoustic resonance in a square silicon/glass chamber of area 2 mm × 2 mm and depth 0.2 mm. (a) Color plot of the pressure field p_1 (red positive, green zero, blue negative) calculated by numerical simulation in the 2D chamber model of the 2.17 MHz pressure eigenmode, see Section 2.4. Nodal lines are shown in black. (b) Top-view gray-scale photograph of 5 µm-diameter polystyrene beads undergoing acoustophoretic motion in the water-filled chamber due to the acoustic radiation force. The particle velocity (overlayed yellow arrows) 1 ms after the onset of the 2.17 MHz ultrasound wave were measured by micro-PIV. After 1 s the particles have accumulated at the pressure nodal lines (black wavy lines). (c) Experiments on 1 µm-diameter polystyrene beads under the same conditions as in the previous panel. In this case the acoustic radiation force is much weaker than the Stokes drag from the acoustic streaming motion of the water, so in the resulting 6 × 6 flow roll pattern, no particle accumulation at the pressure nodal lines is observed. Reproduced with permission from ref. 11, see also ref. 23.

Thus both the density ρ_1 and velocity v_1 can be calculated from the pressure p_1, which itself is determined by the Helmholtz equation and the boundary conditions.

An example of a calculated and measured first-order pressure field p_1 is shown in the color-scale and gray-scale plots of Figure 2.1(a) and (b), respectively. Before investigating further the first-order acoustic fields in Sections 2.4 and 2.5, we briefly treat the second-order fields.

2.3 Second-order Perturbation Theory and Acoustic Streaming

A number of acoustic wave effects is well-described by the above first-order fields. However, the first-order theory is insufficient to explain phenomena that are slowly evolving compared to the fast µs time scale of the acoustic oscillation period, such as the steady acoustic streaming flow rolls shown in Figure 2.1(c). Slowly evolving processes are governed by time-averaged fields, and since the time average of the harmonically varying first-order fields is zero, it is necessary to continue the first-order expansion in eqn (2.2) to second order,

$$p = p_0 + p_1 + p_2, \tag{2.9a}$$

Perturbation Theory and Ultrasound Resonances

$$\rho = \rho_0 + \rho_1 + \rho_2, \tag{2.9b}$$

$$\boldsymbol{v} = 0 + \boldsymbol{v}_1 + \boldsymbol{v}_2. \tag{2.9c}$$

Here, all zero-and first-order terms are assumed to be known. The expansions (2.9) are inserted into eqn (2.1), and upon collecting all second-order terms, we find that the second-order fields p_2, ρ_2, and $\boldsymbol{v}_2$ fulfill the following second-order equation of state, continuity equation, and Navier–Stokes equation,

$$p_2 = c_0^2 \rho_2 + \frac{1}{2}(\partial_\rho c_0^2)_0 \rho_1^2, \tag{2.10a}$$

$$\partial_t \rho_2 = -\rho_0 \boldsymbol{\nabla} \cdot \boldsymbol{v}_2 - \boldsymbol{\nabla} \cdot (\rho_1 \boldsymbol{v}_1), \tag{2.10b}$$

$$\rho_0 \partial_t \boldsymbol{v}_2 = -\boldsymbol{\nabla} p_2 + \eta \nabla^2 \boldsymbol{v}_2 + \beta \eta \boldsymbol{\nabla}(\boldsymbol{\nabla} \cdot \boldsymbol{v}_2) - \rho_1 \partial_t \boldsymbol{v}_1 - \rho_0 (\boldsymbol{v}_1 \cdot \boldsymbol{\nabla}) \boldsymbol{v}_1. \tag{2.10c}$$

Normally, the second-order fields would be negligible compared to the first-order fields. However, if the latter have a harmonic time dependence as in eqn (2.5), they do not contribute to any time-averaged effect since $\langle \cos(\omega t) \rangle = 0$. Here, the symbol $\langle X \rangle$ denotes the time average over a full oscillation period τ of a quantity $X(t)$,

$$\langle X \rangle \equiv \frac{1}{\tau} \int_0^\tau dt \, X(t). \tag{2.11}$$

In contrast, the time average of a product of two first-order terms both proportional to $\cos(\omega t)$ is non-zero, as $\langle \cos^2(\omega t) \rangle = \frac{1}{2}$. Assuming harmonic time dependence, the time average of the second-order continuity and Navier–Stokes eqn (2.10b) and (2.10c), becomes

$$\rho_0 \boldsymbol{\nabla} \cdot \langle \boldsymbol{v}_2 \rangle = -\boldsymbol{\nabla} \cdot \langle \rho_1 \boldsymbol{v}_1 \rangle, \tag{2.12a}$$

$$\eta \nabla^2 \langle \boldsymbol{v}_2 \rangle + \beta \eta \boldsymbol{\nabla}(\boldsymbol{\nabla} \cdot \langle \boldsymbol{v}_2 \rangle) - \boldsymbol{\nabla} \langle p_2 \rangle = \langle \rho_1 \partial_t \boldsymbol{v}_1 \rangle + \rho_0 \langle (\boldsymbol{v}_1 \cdot \boldsymbol{\nabla}) \boldsymbol{v}_1 \rangle. \tag{2.12b}$$

Clearly, the time-averaged, second-order fields will in general be non-zero, due to the time-averaged products of first-order terms acting as source terms on the right-hand side of the governing equations. Physically, the non-zero velocity $\langle \boldsymbol{v}_2 \rangle$ is the so-called acoustic streaming, see Figure 2.1(c), where the bulk fluid is moving steadily in time due to viscous stresses generated in the fluid near the walls, when the oscillatory velocity in the acoustic wave is forced by the no-slip boundary condition to be zero at the walls.[4–7] The non-zero pressure $\langle p_2 \rangle$ gives rise to the acoustic radiation force coming from scattering of acoustic waves on the particles and causing acoustophoretic motion of the particles.[1–3]

We postpone the treatment of the radiation force to Chapter 4, and here deal with the acoustic streaming velocity $\langle \boldsymbol{v}_2 \rangle$ using scaling arguments. The order of magnitude v_{str} of the streaming velocity follows from eqn (2.12b), where it is seen that the driving force on the right-hand side is proportional to v_a^2, the square of the amplitude of the first-order field v_1. The only other velocity in the problem is the speed of sound c_0, so

$$v_{\text{str}} = |\langle \mathbf{v}_2 \rangle| = \Psi \frac{v_a^2}{c_0}, \qquad (2.13)$$

where Ψ is a geometry-dependent factor, which for a plane rigid wall is $\Psi = 3/8$.[15] As we saw in eqn (2.6b), viscous effects are negligible in the bulk, where the viscous parameter γ is minute, $\gamma \approx 10^{-5}$. However, due to the no-slip boundary condition at rigid walls, the first-order field v_1 changes over a short distance δ from its value v_a in the bulk to zero at the wall. This means that in eqn (2.3b), a term like $\eta \nabla^2 \mathbf{v}_1$ has the same order of magnitude as $\rho_0 \partial_t \mathbf{v}_1$. In fact, these two terms constitute a diffusion equation for momentum, $\rho_0 \partial_t \mathbf{v}_1 = \eta \nabla^2 \mathbf{v}_1$, as eqn (1.54), which for the given harmonic time dependence can be written as $\nabla^2 \mathbf{v}_1 = -\mathrm{i}(\omega/\nu)\mathbf{v}_1$. From this, it follows that the characteristic distance δ over which the velocity field adapts to the no-slip boundary condition is given by[4,14,16]

$$\delta = \sqrt{\frac{2\nu}{\omega}} \approx 0.6 \ \mu\mathrm{m}, \qquad (2.14)$$

where the value is computed for 1-MHz ultrasound in water at room temperature. The viscosity-dependent distance δ is also known as the width of the acoustic boundary layer, a layer which surrounds any rigid wall. Interestingly, neither viscosity nor the viscous boundary-layer width enter the scaling law, eqn (2.13), for the magnitude of the streaming velocity v_{str}.

Figure 2.1 shows an example from Hagsäter *et al.*[23] of these effects observed by measuring the motion of polystyrene tracer beads in a water-filled, square silicon/glass chamber of side length 2 mm and depth 0.2 mm under the influence of an ultrasound field at the resonance frequency 2.17 MHz applied through an externally attached piezo transducer. Panel (a) is a color plot of the pressure field p_1 at a given time calculated using the theory of Sections 2.5 and 2.6.2. Red represents positive, blue negative, and green zero pressure. Half a cycle later, the sign of the pressure field has changed, but the green pressure nodal lines remain fixed in space appearing as six horizontal sinusoidal waves undergoing three full oscillations across the chamber. In panel (b) is shown the result of measuring on 5 μm-diameter tracer beads. Initially, the beads are distributed homogeneously in the chamber. After the onset of the ultrasound field, the particles move away from the high pressure oscillations under the influence of the acoustic radiation force. The first 1 ms of this motion is recorded by micro-particle image velocimetry (micro-PIV), and the obtained velocity of the particles is represented by the yellow arrows. After 1 s, all tracer particles have moved to the nodal lines, where they appear as the black wavy line on the gray-scale top-view photograph of the device. In panel (c), the same experiment is repeated with 1 μm-diameter tracer beads. For these smaller particles, the acoustic radiation force is weak, and the particle

2.4 Basics of Acoustic Resonances in Viscous Liquids

As mentioned in the introduction, acoustic resonance modes are particularly useful to establish for acoustic handling of cells and particles in microfluidic systems. To illustrate the fundamental properties of such resonance modes, we study the simple 1D setup sketched in Figure 2.2(a). Two planar walls are placed parallel to the yz-plane at $x = -L$ and $x = L$, respectively, and the gap is filled with water. To mimic the action of the piezo transducer, the walls are forced to oscillate in anti-phase at a frequency $f \approx 1$ MHz and with an amplitude $\ell \approx 0.1$ nm. As a simplification we neglect the actual tiny displacement of the walls and instead model the oscillation by the velocity boundary condition sketched in Figure 2.2(a).

$$v_1(-L, t) = -\omega \ell\, e^{-i\omega t}, \quad v_1(+L, t) = +\omega \ell\, e^{-i\omega t}. \tag{2.15}$$

Starting from rest, the resonance builds up until the power injected at the moving walls equals the heat dissipation due to viscosity in the liquid, and a steady state at constant temperature is reached. The standing 1D wave $v_1 = f(x)\, e^{-i\omega t} e_x$ has $\nabla \times v_1 = 0$, so $v_1 = \nabla \phi_1$ and $\partial_j v_i = \partial_i v_j$. To find the viscid velocity potential ϕ_1, we note that $\partial_j \partial_j v_i = \partial_j \partial_i v_j = \partial_i \partial_j v_j$, i.e. $\nabla^2 v_1 = \nabla(\nabla \cdot v_1)$,

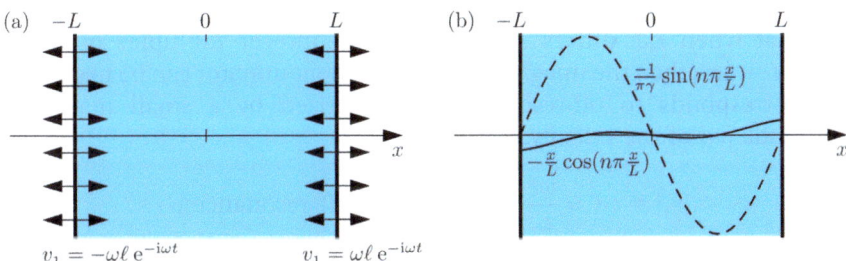

Figure 2.2 (a) A liquid slab (blue) between two parallel planar walls (thick lines) that oscillates harmonically in counter-phase (double arrows). As the amplitude is minute, $\ell \ll L$, the wall positions are considered fixed, while the first-order velocity $v_1(t)$ at the walls is changing harmonically, $v_1(t) = \pm \omega\, \ell\, e^{-i\omega t}$. (b) Sketch of the two terms in the resonant velocity field v_1, eqn (2.24b). The small component (full line) proportional to $(x/L)\cos(\pi x/L)$ obeys the oscillatory boundary condition with amplitude $\pm \omega\, \ell$. The large resonant component (dashed line) proportional to $(1/\pi\gamma)\sin(\pi x/L)$ is an eigenmode obeying the hard-wall condition with zero velocity amplitude. Reproduced with permission from ref. 11.

and from eqn (2.3a) we have $\nabla \cdot v_1 = i\omega p_1/(\rho_0 c_0^2)$. Inserting these two expressions together with $v_1 = \nabla \phi$ into eqn (2.3b), we find,

$$\phi_1(r,t) = \frac{-i}{\omega\rho_0(1+i\gamma)^2} p_1(r,t), \quad \text{for} \quad \nabla \times v_1 = 0. \tag{2.16}$$

Because $\phi_1 \propto p_1$, the wave eqn (2.6a) also holds for ϕ_1, and we can therefore write the solution for ϕ_1 as a superposition of a pair of counter-propagating plane waves with the complex-valued wave number $k = (1+i\gamma)k_0$, with $k_0 = \omega/c_0$, and unknown coefficients ϕ_+ and ϕ_-,

$$\phi_1(x,t) = [\phi_+ e^{ikx} + \phi_- e^{-ikx}] e^{-i\omega t}. \tag{2.17}$$

The corresponding first-order velocity is

$$v_1(x,t) = \partial_x \phi_1(x,t) = ik[\phi_+ e^{ikx} + \phi_- e^{-ikx}] e^{-i\omega t}. \tag{2.18}$$

The antisymmetric boundary condition on v_1 in eqn (2.15) combined with eqn (2.18) leads to $\phi_+ = \phi_-$, as well as an explicit expression for the coefficients,

$$\phi_+ = \phi_- = \frac{-\omega\ell}{2k\sin(kL)}. \tag{2.19}$$

From this, we can obtain the following expression for v_1,

$$v_1(x,t) = \omega\ell \frac{\sin(kx)}{\sin(kL)} e^{-i\omega t} \approx \omega\ell \frac{\sin(k_0 x) + i\gamma k_0 x\cos(k_0 x)}{\sin(k_0 L) + i\gamma k_0 L\cos(k_0 L)} e^{-i\omega t}, \tag{2.20}$$

where we have used $\gamma k_0 L \ll 1$ to make Taylor expansions in kL around $k_0 L$. We note that when $k_0 L$ differs sufficiently from integer multiples of π, i.e. $\gamma \ll |k_0 L - n\pi|$, then the imaginary part of the denominator can be neglected. This corresponds to off-resonance characterized by a small maximum magnitude $|v_1|$ of the velocity,

$$|v_1| \approx \omega\ell = \frac{\omega\ell}{c_0} c_0 \approx 10^{-7} c_0, \quad \text{(off resonance)}, \tag{2.21}$$

where the value is calculated by assuming $\omega \approx 10^7$ rad s^{-1}, $\ell \approx 0.1$ nm and $c_0 \approx 10^3$ m s^{-1}.

Perhaps more interesting are the acoustic resonances, where the acoustic field acquires particularly large amplitudes and thus contains a relatively large amount of stored energy, see Figure 2.2(b). Theoretically, the resonances are identified by the minima in the denominators of v_1 in eqn (2.20), i.e. by $\sin(k_0 L) = 0$ or $k_0 L = n\pi$, $n = 1, 2, 3, \ldots$,

$$k_0 = k_n \equiv n\frac{\pi}{L}, \quad n = 1, 2, 3, \ldots \quad \text{(resonance condition)}. \tag{2.22}$$

Perturbation Theory and Ultrasound Resonances

In practice, the resonance is achieved by tuning the frequency ω to the resonance frequency ω_n given by.

$$\omega_n \equiv c_0 k_n = n\frac{\pi c_0}{L}, \quad n = 1, 2, 3, \ldots. \quad (2.23)$$

At the nth resonance $\sin(k_n L) = 0$ and $\cos(k_n L) = (-1)^n$, so the acoustic fields become.

$$\phi_1(x,t) \approx (-1)^n c_0 \ell \, e^{-i\omega_n t} \left[\frac{i}{n\pi\gamma} \cos\left(n\pi\frac{x}{L}\right) + \frac{x}{L}\sin\left(n\pi\frac{x}{L}\right) \right], \quad (2.24a)$$

$$v_1(x,t) \approx (-1)^n w\ell \, e^{-i\omega_n t} \left[\frac{-i}{n\pi\gamma} \sin\left(n\pi\frac{x}{L}\right) + \frac{x}{L}\sin\left(n\pi\frac{x}{L}\right) \right]. \quad (2.24b)$$

From these expressions it follows that each of the fields acquires a resonant component with an amplitude that is a factor of $1/(n\pi\gamma) \approx (1/n) \times 10^5$ larger than the non-resonant component,

$$|v_1(x,t)| \approx \frac{1}{n\pi\gamma}\frac{\omega_n \ell}{c_0} c_0 \approx \frac{1}{n}10^{-2} c_0, \quad (n\text{th resonance}). \quad (2.25)$$

Despite the huge multiplication factor, $1/(\pi\gamma) \approx 10^4$, the perturbation approach remains valid as the density fluctuation is small, $|\rho_1|/\rho_0 = |v_1|c_0 \approx \frac{1}{n}10^{-2} = 1$.

From eqn (2.24b), we see that the term actually obeying the velocity boundary condition $v_1(\pm L) = \pm \omega\ell$ is 10^5 times smaller than the other term, which obeys the hard-wall condition $v_1(\pm L, t) = 0$ and thus in fact is an eigenmode of the system. Consequently, when coupling into a system with a frequency near an eigenmode frequency, the corresponding eigenmode gets excited with a huge amplitude, approximately a factor $1/\gamma$ larger than the coupling amplitude, independent of the actual boundary condition, see Figure 2.2(b).

As the energy density E_{ac} for an harmonically oscillating system is twice the time-averaged kinetic energy density, eqn (2.20) implies $E_{\text{ac}} = \frac{1}{2L}\int_{-L}^{L} dx \frac{1}{2}\rho_0 |v_1(x)|^2 = \frac{1}{4}\rho_0\omega^2\ell^2/|\sin(kL)|^2$. By Taylor expanding this expression in $kL = (1+i\gamma)\frac{L}{c_0}\omega$ around the resonance $k_n L = n\pi$, we find a Lorentzian line shape as a function of frequency.

$$E_{\text{ac}}(\omega) \frac{\frac{1}{4}\rho_0 \omega^2 \ell^2}{\left|\frac{L}{c_0}(\omega - \omega_n) + i\gamma n\pi\right|^2} = \frac{\rho_0 \omega^2 \ell^2}{4n^2\pi^2} \frac{\omega_n^2}{(\omega - \omega_n)^2 + \gamma^2 \omega_n^2}, \quad \text{for } \omega \approx \omega_n. \quad (2.26)$$

At resonance, $E_{\text{ac}}(\omega_n) = \frac{1}{(n\pi\gamma)^2} \times \frac{1}{4}\rho_0\omega_n^2\ell^2$, which is much larger than the off-resonance energy density of about $\rho_0\omega^2\ell^2$. At $\omega = \omega_n$, the energy density $E_{\text{ac}}(\omega)$ is at its maximum, while it decreases to half of this value when changing the frequency to $\omega = \omega_n + \gamma\omega_n$. Therefore, the full width $\Delta\omega$ at half maximum of the resonance peak is seen to be $\Delta\omega = 2\gamma\omega_n$, and by definition the quality factor Q of the resonance therefore becomes.

$$Q = \frac{\omega_n}{\Delta\omega} = \frac{1}{2\gamma} \approx 10^5. \qquad (2.27)$$

We emphasize that this result overestimates the actual Q-factor, because it builds on the simplifying assumption that viscous dissipation in the bulk liquid is the only source to loss of acoustic energy. In a real device, acoustic energy is also lost as viscous friction in the acoustic boundary layer near the walls and as sound waves emitted into the chip holder and fluidic connectors as well as into the surrounding air.[18] This is discussed further in Section 2.5.3.

2.5 Acoustic Eigenmodes

The above result indicates that we can gain insight into the nature of acoustic resonances in externally driven systems by analyzing the eigenmodes $p_n = p_n(\mathbf{r})e^{-i\omega_n t}$ of the equivalent isolated liquid-filled chamber. If more specific details are needed, it is necessary to calculate the acoustic response of the entire system driven by the piezo transducer, *i.e.* combining the wave equation of the liquid with the elastic equations of the surrounding solids. More complete descriptions are discussed by Dual *et al.*[19]

2.5.1 Boundary Conditions

In the simple chamber model of the liquid domain, the acoustic pressure eigenmodes $p_n = p_n(\mathbf{r})e^{-i\omega_n t}$ in the liquid are determined by the Helmholtz eqn (2.6a) with appropriate boundary conditions. In the following, we use either the soft-wall, the hard-wall, or the lossy-wall boundary condition.

The soft-wall condition is used when the medium interfacing with the liquid cannot sustain any appreciable pressure. Examples of this are a free liquid/air interface or, as in ref. 20, liquid surrounded by a thin glass wall and then air. The mathematical form of the boundary condition is.

$$p_1 = 0, \quad \text{soft-wall boundary condition.} \qquad (2.28a)$$

The hard-wall condition applies when the liquid is interfacing with an infinitely hard wall not yielding to the velocity v_1 of the liquid. Consequently, the normal velocity of the liquid at the wall is zero, which by eqn (2.8a) leads to a zero normal gradient of the pressure.

$$\mathbf{n} \cdot \nabla p_1 = 0, \quad \text{hard-wall boundary condition.} \qquad (2.28b)$$

The lossy-wall condition is an approximative description of partial radiative acoustic losses from the liquid to the surrounding medium. Consider a planer configuration with the liquid placed in the half-space $x < 0$ next to an ideally absorbing medium (subscript m) at $x > 0$. A radiative loss can be described by a wave in the medium having only the right-ward propagating component: $\phi_m = Ae^{-ik_m x}$. From eqn (2.8) follows $v_m = ik_m\phi_m$ and $p_m = i\omega\rho_m\phi_m$, and thus $p_m = \rho_m c_m v_m$ at the interface. Combining this with continuity in pressure $(p_1 = p_m)$ and velocity $(v_1 = v_m)$ across the interface, and with $\mathbf{n}\cdot\nabla p_1 = i\omega\rho_0 \mathbf{n}\cdot\mathbf{v}_1$ from eqn (2.8), we arrive at.

$$\mathbf{n}\cdot\nabla p_1 = i\frac{\omega\rho_0}{\rho_m c_m} p_1, \quad \text{lossy-wall boundary condition,} \quad (2.28c)$$

where the planar case is generalized to any curved surface. The product $\rho_m c_m$ is known as the specific acoustic impedance of the medium. The soft-wall condition, eqn (2.28a), is recovered from eqn (2.28c) in the limit of zero impedance of the surrounding medium, while the hard-wall condition, eqn (2.28b), corresponds to infinitely large impedance. If the liquid and the medium have the same impedance, acoustic waves pass the interface without suffering any scattering.

Finally, we note that since $(\rho_{si}c_{si})/(\rho_{wa}c_{wa}) = 13.2$ and $(\rho_{py}c_{py})/(\rho_{wa}c_{wa}) = 8.4$ (using the longitudinal sound velocities of Table 2.1), it may be a reasonable first approximation to use the hard-wall boundary condition for water channels in silicon/glass chips.

2.5.2 An Ideal Water-filled Rectangular Channel

Next, we study the acoustic modes for a rectangular water-filled channel placed along the coordinate axes with its opposite corners at $(0, 0, 0)$ and (ℓ, w, h) surrounded by either hard or soft walls. As can be verified easily by direct substitution, the eigenmodes of this ideal rectangular channel are.

$$p_1(x,y,z) = p_a\sin(k_x x)\sin(k_y y)\sin(k_z z), \quad \text{with} \quad k_j = n_j\frac{\pi}{L_j}, \quad \text{soft wall,} \quad (2.29a)$$

$$p_1(x,y,z) = p_a\cos(k_x x)\cos(k_y y)\cos(k_z z), \quad \text{with} \quad k_j = n_j\frac{\pi}{L_j}, \quad \text{hard wall,} \quad (2.29b)$$

where p_a is the pressure amplitude, where $(L_x, L_y, L_z) = (\ell, w, h)$, and where $n_j = (0)\, 1, 2, 3, \ldots$ is the number of half wavelengths ($n_j > 0$ for the sine waves). The corresponding three-index resonance frequencies $f_{n_x,n_y,n_z} = \omega_{n_x,n_y,n_z}/(2\pi)$ are

$$f_{n_x,n_y,n_z} = \frac{c_{wa}}{2}\sqrt{\frac{n_x^2}{\ell^2} + \frac{n_y^2}{w^2} + \frac{n_z^2}{h^2}}, \quad \text{with} \quad n_x, n_y, n_z = (0,)1, 2, 3, 4, \ldots \quad (2.30)$$

Examples of these analytically determined eigenmodes are shown in Figure 2.3. Note the relatively low frequency of (d) and (e) having $n_z = 0$ along the smallest dimension in contrast to $n_z = 1$ of the other four eigenmodes. In (f) one half-length is squeezed in along the z-direction ($n_z = 1$) and the frequency increases significantly. In fact, as (a) and (f) have the same indices

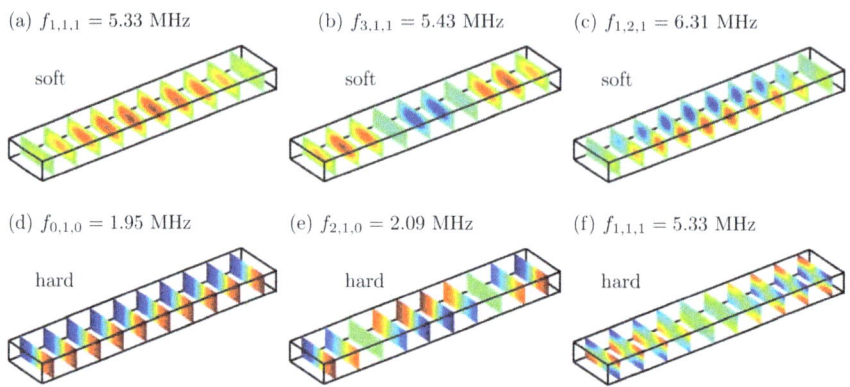

Figure 2.3 Color slice plots (red positive, green zero, blue negative) in the inviscid limit of some eigenmodes of the pressure field p_1 in a rectangular, single, water-filled microchannel of length $\ell = 2$ mm, width $w = 0.38$ mm, and height $h = 0.15$ mm. (a)–(c) Soft-wall boundary conditions $p_1 = 0$ at the surface, *i.e.* a zero-density wall surrounds the channel. (d)–(f) Hard-wall boundary condition $\boldsymbol{n}\cdot\nabla p_1 = 0$, *i.e.* the surrounding wall is of infinite density. Reproduced with permission from ref. 11.

they also have the same frequency, namely $f_{1,1,1}$ despite their different boundary conditions. It turns out that the anti-symmetric resonance (d) having a perfect nodal plane in the vertical center plane is an ideal configuration for acoustophoretic separation.

2.5.3 Viscous and Radiative Losses in the Rectangular Channel

When energy dissipates from the acoustic field near resonance, the resonance peak in the form of acoustic energy density plotted *versus* frequency acquires a finite full width at half maximum, $\Delta\omega = \omega_n/Q$, where Q is the quality factor of the resonance at frequency ω_n. If the only source of energy dissipation is the viscous damping in the liquid, then it follows from eqn (2.6b) and (2.27) that the Q-factor is of the order $Q \approx 10^5$ for water. However, it is found experimentally that typical Q-factors are in the range $100 < Q < 1000$, see, for example, Figure 2.4 and a paper by Groschl,[18] mainly due to dissipation in the acoustic boundary layer and the radiative losses to the sample holder, the surrounding air, and inlet/outlet tubes.

The loss $\partial_t E^{\text{visc}}$ due to viscous dissipation in the acoustic boundary layer of a rectangular channel of length ℓ, width w and height h can be estimated using eqn (1.46). For the half-wavelength resonance $f_{0,1,0}$, the boundary layers at the top and bottom walls each have the volume $\ell w \delta$ and inside them, the velocity gradient is approximately v_a/δ. Thus, $\partial_t E^{\text{visc}} \approx 2\ell w\delta \tfrac{1}{2}\eta(v_a/\delta)^2 = \tfrac{1}{2}\ell w\delta\rho_0 v_a^2\omega$. Expressing the Q-factor as the ratio of the total energy $E_{\text{ac}}\ell wh$, eqn (2.26), stored in the resonator volume ℓwh over the time-averaged dissipated energy per oscillation $\langle\partial_t E^{\text{visc}}\rangle/\omega$, we obtain.

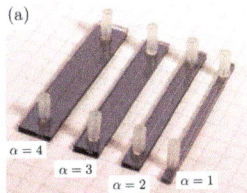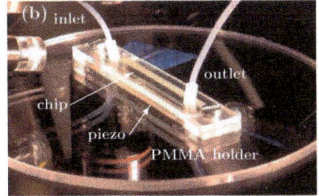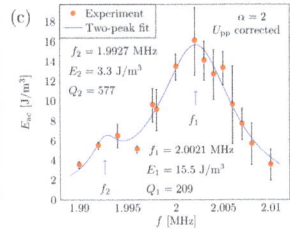

Figure 2.4 Experimental determination of the Q factor of an acoustic mode. (a) The silicon/glass chips contain straight channels of length $\ell = 40$ mm, width $w = 377$ μm, and height $h = 157$ μm. The channels are etched down into the silicon chip of thickness 350 μm, and they are covered by a pyrex glass lid of thickness 1.13 mm. The chips are 50 mm long and have widths of 2.5 mm ($\alpha = 1$), 4.7 mm ($\alpha = 2$), 6.8 mm ($\alpha = 3$), and 9.0 mm ($\alpha = 4$). (b) A photograph of the experimental setup showing how a chip is coupled mechanically to the piezo actuator, the PMMA holder, and the inlet/outlet tubes. (c) Plot of the measured acoustic energy density E_{ac} versus frequency f (red dots) for the $\alpha = 2$ chip. The two observed acoustic resonance peaks are fitted well by the sum of two Lorentzian line shapes (eqn (2.26), blue line). The resonance frequencies are $f_1 = 2.0021$ MHz and $f_1 = 1.9927$ MHz, while the Q factors are $Q_1 = 209$ and $Q_2 = 577$. Reproduced with permission from ref. 11, see also ref. 25.

$$Q^{\mathrm{visc}} \approx \frac{E_{\mathrm{ac}}\ell w h}{\langle \partial_t E^{\mathrm{visc}}\rangle/\omega} = \frac{\frac{1}{4}\rho_0 v_a^2 \ell w h}{\frac{1}{4}\ell w \delta \rho_0 v_a^2} = \frac{h}{\delta}. \quad (2.31)$$

For the device in Figure 2.4, we find $Q^{\mathrm{visc}} \approx 160/0.4 = 400$.

In the simple chamber model, the contribution to the measured Q-factor from radiative losses due to absorbing walls, can be modeled using the lossy-wall boundary condition (2.28c) with the acoustic impedance $\rho_m c_m$ as a fitting parameter. For the rectangular channel, this can be argued as follows: consider the transverse half-wavelength standing wave between the walls at $y = \pm L/2$ given by $p_1(y) = A\sin[\pi(1 + i\tilde{\gamma})y/L]$, where the unknown $\tilde{\gamma}$ is the damping factor given by the radiative loss. Note that this form is similar to eqn (2.20), but with a different interpretation of $\tilde{\gamma}$. In this case, the lossy-wall boundary conditions become

$$(1 + i\tilde{\gamma})\cos\left[\frac{\pi}{2} + i\tilde{\gamma}\frac{\pi}{2}\right] = i\frac{\rho_0 c_0}{\rho_m c_m}\sin\left[\frac{\pi}{2} + i\tilde{\gamma}\frac{\pi}{2}\right]. \quad (2.32)$$

Assuming $\tilde{\gamma} \ll 1$, a Taylor expansion leads to

$$\tilde{\gamma} \approx -\frac{2\rho_0 c_0}{\pi \rho_m c_m}, \quad (2.33)$$

so that the damping factor $\tilde{\gamma}$ due to radiation loss can be thought of as a ratio between the impedances of water and the surrounding (ideally absorbing) medium.

Radiative losses to ideally absorbing walls can lower the Q-factor to as much as $Q^{\text{rad}} \approx 10$. However, the real surroundings (sample holder, inlet/outlet tubes, and the air) are not completely absorbing, which is why the measured Q-factors are in the range.

$$Q = \left[\frac{1}{Q^{\text{rad}}} + \frac{1}{Q^{\text{visc}}}\right]^{-1} \approx 100 - 1000. \tag{2.34}$$

2.6 Geometrical Effects and FEM Simulations

For geometrical shapes more irregular than rectangular it is in general not possible to find analytical solutions to the acoustic wave equation and one must resort to numerical simulations. Below we sketch the basic principle of the finite element method (FEM), one of the most widely used numerical methods.

With numerical methods at hand, it is possible to study the effects of specific geometrical shapes on the resonance modes in the liquid domain. In fact, we have already seen an example of this in Figure 2.1(a). According to eqn (2.29b), the corresponding pressure resonance of a perfect hard-walled square cavity of side length L in the xy-plane should have the form $p_1(x, y) = p_a \cos(6\pi x/L) \cos(6\pi y/L)$, which is symmetric along the diagonal $x = y$. However, the figure clearly shows that the presence of the inlet and outlet channels along the x-direction breaks this symmetry.

2.6.1 Finite Element Method (FEM) Simulations

A general and versatile method to solve the acoustic wave equations (and other continuum field equations) is the so-called weak form suitable for implementation in the finite element method (FEM) used by the COMSOL Multiphysics software (www.comsol.com). In FEM analysis, the governing equations are not satisfied in each and every point of the computational domain Ω, but only on average in each of the small regions defined by a mesh.

As a main example, we study the Helmholtz equation $\partial_k \partial_k p_1 = -\frac{\omega^2}{c_0^2} p_1$ for acoustic waves, eqn (2.6a). To discretize this equation, the domain Ω is divided by a mesh into a large number M of small cells m. A corresponding number of scalar test functions $\tilde{p}_m$ are introduced, each being different from zero only in their respective cell m. The given field $p_1(r)$ is represented by a linear combination of test functions with coeffcients P_m.

$$p_1(r) = \sum_{m=1}^{M} P_m \tilde{p}_m(r) \tag{2.35}$$

In the weak form, a given differential equation in the continuous space is transformed into M equations, one for each coeffcient P_m, by multiplying it by each of the test functions, integrating of the domain, and demand that all these integrals should be zero.

$$\int_\Omega d\mathbf{r}\, \tilde{p}_m \left[\partial_k \partial_k p_1 + \frac{\omega^2}{c_0^2} p_1\right] = 0, \quad m = 1, 2, \ldots, M. \tag{2.36}$$

Perturbation Theory and Ultrasound Resonances

As a result the integrand is zero on average, and the equation $\partial_k \partial_k p_1 + \frac{\omega^2}{c_0^2} p_1 = 0$ is fulfilled approximatively. By partial integration only first derivatives are left in the integrand, and a specific surface integral appears, through which the boundary conditions can be applied

$$\int_{\partial\Omega} da\, \tilde{p}_m n_k \partial_k p_1 + \int_{\Omega} d\mathbf{r} \left[\partial_k \tilde{p}_m (-\partial_k p_1) + \tilde{p}_m \frac{\omega^2}{c_0^2} p_1 \right] = 0, \qquad (2.37)$$

where $\mathbf{n}$ is the outward-pointing surface normal of $\partial\Omega$. The Neumann boundary condition $\mathbf{n} \cdot \nabla p_1 = N(\mathbf{r})$ is therefore imposed by

$$\int_{\partial\Omega} da\, \tilde{p}_m N(\mathbf{r}) + \int_{\Omega} d\mathbf{r} \left[\partial_k \tilde{p}_m (-\partial_k p_1) + \tilde{p}_m \frac{\omega^2}{c_0^2} p_1 \right] = 0. \qquad (2.38)$$

For the hard-wall condition, eqn (2.28b), $N(\mathbf{r}) = 0$ and the surface integral vanishes.

The Dirichlet condition $p_1 = D(\mathbf{r})$ is more tricky to impose on the boundary. In this case the normal derivative on the boundary is free to vary such that the imposed Dirichlet boundary condition indeed is fulfilled. This case therefore requires the introduction of an auxiliary field $f(\mathbf{r})$ on the boundary together with a number J of associated test functions $\tilde{f}_j(\mathbf{r}), j = 1, 2, \ldots, J$, the so-called Lagrange multipliers, each defined along the edges of the outer edges of the J mesh cells at the surface $\partial\Omega$

$$\int_{\partial\Omega} da [\tilde{p}_m f + \tilde{f}_j (D(\mathbf{r}) - p_1)] + \int_{\Omega} d\mathbf{r} \left[\partial_k \tilde{p}_m (-\partial_k p_1) + \tilde{p}_m \frac{\omega^2}{c_0^2} p_1 \right] = 0. \qquad (2.39)$$

As the Lagrange multiplier test functions $\tilde{f}_j$ only exist on the boundary, the demand for zero valued integrals forces the coeffcient $(D(\mathbf{r}) - p_1)$ to be zero for any converged solution of the problem. On the other hand, the coeffcient $f(\mathbf{r})$ of the pressure test function $\tilde{p}_m$ is a dependent variable, which through ∇p_1 couples to the terms of the volume integral. Since this coeffcient is also the normal derivative of the pressure, we find that it has been determined as $\mathbf{n} \cdot \nabla p_1 = f$ for a converged solution.

Using this method we can calculate the acoustic resonance modes for any shape of acoustic cavities with any boundary condition. The soft-wall condition eqn (2.28a) is obtained by setting $D(\mathbf{r}) - 0$, while the lossy-wall condition eqn (2.28c) requires $D(\mathbf{r}) = -i\frac{\rho_m c_m}{\omega \rho_0} \mathbf{n} \cdot \nabla p_1$.

2.6.2 Symmetry Breaking in Acoustic Resonances

As an example of the effects of non-trivial geometrical shapes on the acoustic resonances, we take the study of symmetry breaking in a water-filled silicon/glass microchamber from the paper by Hagsäter et al.,[23] see Figure 2.5. The 2 mm by 2 mm square chamber has a depth of 0.2 mm, and two inlet/outlet channels of width 0.4 mm and unequal lengths (12.5 mm and 10.5 mm) are attached on opposite sides as sketched in Figure 2.5(a). Experimentally, two closely lying resonances of nearly identical resonance patterns are observed at 2.06 MHz and 2.08 MHz, see Figure 2.5(b) and (c). The pattern of the

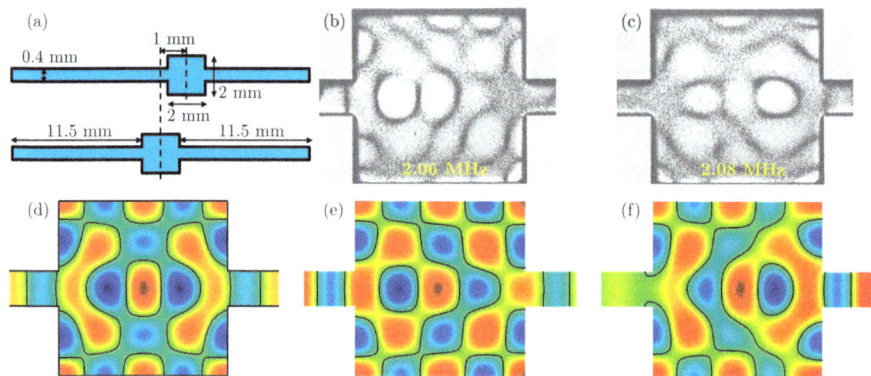

Figure 2.5 (a) Sketches of a 2 mm square microfluidic chamber placed symmetrically and asymmetrically (shifted 1 mm to the right) relative to 0.4 mm wide and 11.5 mm long inlet/outlet channels (not drawn to scale). (b) Photograph of the position of 5 μm-diameter tracer beads (black bands) in a silicon/pyrex chip having the asymmetric geometry of panel (a) after 1 s of motion in the 2.06 MHz ultrasound resonance mode. (c) The same as the previous panel, but for the 2.08 MHz resonance mode. (d) Color plot of the pressure field p_1 (blue negative, green zero, red positive) of the resonance near the 2.1 MHz in the symmetric geometry with hard-wall boundary conditions found by numerical simulation using COMSOL. Notice the left–right symmetry of p_1. (e) Same as the previous panel, except for the asymmetric device. Notice the slightly larger amplitude of p_1 left of the symmetry line as in panel (b). (f) Same as the previous panel, but for the resonance mode with a frequency 0.028 MHz higher. Notice the slightly larger amplitude of p_1 right of the symmetry line as in panel (c). Reproduced with permission from ref. 11, see also ref. 23.

lower-frequency resonance is shifted slightly to the left, towards the longer and thus less confining channel, while the opposite is the case with the higher-frequency resonance.

A similar trend is seen in the numerical simulation presented in the color plots of the pressure field p_1 in Figure 2.5(d)–(f) obtained by solving the 2D Helmholtz eqn (2.6a) with the hard-wall condition (2.28b). Figure 2.5(d) shows the result for the symmetric case of equal-length (11.5 mm) inlet/outlet channels: a left-right symmetric pressure field. To quantitatively explain the non-symmetric pressure modes observed experimentally, the pressure resonances are also calculated in the asymmetric case of inlet/outlet channels of unequal lengths (12.5 mm and 10.5 mm). Two nearly identical resonance patterns are found near the frequency of the symmetric case in Figure 2.5(d), one mode with a frequency 0.028 MHz lower than the other similar to the observed frequency difference. In agreement with the experimental observations, the lower-frequency mode is seen to be shifted towards the left inlet channel, which is longer and thus supports a resonance at a slightly lower frequency, while the other is shifted towards the short, right inlet channel thus giving rise to a slight increase in resonance frequency.

The simple acoustic model, where the chip surrounding the water channel is merely modeled as a hard wall, thus offers some qualitative insight into the behavior of acoustic resonance modes.

References

1. L. V. King, *Proc. R. Soc. London, Ser. A,* 1934, **147**, 212–240.
2. K. Yosioka and Y. Kawasima, *Acustica,* 1955, **5**, 167–173.
3. L. P. Gorkov, *Phys.- Dokl.,* 1962, **6**, 773–775.
4. L. Rayleigh, *Philos. Trans. R. Soc. London,* 1884, **175**, 1–21.
5. H. Schlichting, *Phys. Z,* 1932, **33**, 327–335.
6. C. Eckart, *Phys. Rev.,* 1948, 68–76.
7. J. Lighthill, *J. Sound Vibr.,* 1978, **61**, 391–418.
8. A. Nilsson, F. Petersson, H. Jönsson and T. Laurell, *Lab Chip,* 2004, **4**, 131–135.
9. O. Manneberg, J. Svennebring, H. M. Hertz and M. Wiklund, *J. Micromech. Microeng.,* 2008, **18**, 095025.
10. Z. Guttenberg, H. Muller, H. Habermuller, A. Geisbauer, J. Pipper, J. Felbel, M. Kielpinski, J. Scriba and A. Wixforth, *Lab Chip,* 2005, **5**, 308–317.
11. H. Bruus, *Lab Chip,* 2012, **12**, 20–28.
12. J. Lighthill, *Waves in Fluids,* Cambridge University Press, 2002.
13. A. D. Pierce, *Acoustics,* Acoustical Society of America, Woodbury, 1991.
14. L. D. Landau and E. M. Lifshitz, *Fluid Mechanics, Course of Theoretical Physics,* Pergamon Press, Oxford, 2nd edn, 1993, vol. 6).
15. P. B. Muller, R. Barnkob, M. J. H. Jensen and H. Bruus, *Lab Chip,* 2012, **12**, 4617–4627.
16. S. Sadhal, *Lab Chip,* 2012, **12**, 2292–2300.
17. R. Barnkob, P. Augustsson, T. Laurell and H. Bruus, *Phys. Rev. E,* 2012, **86**, 056307.
18. M. Groschl, *Acustica,* 1998, **84**, 432–447.
19. J. Dual and T. Schwarz, *Lab Chip,* 2012, **12**, 244–252.
20. B. Hammarstrom, M. Evander, H. Barbeau, M. Bruzelius, J. Larsson, T. Laurell and J. Nillsson, *Lab Chip,* 2010, **10**, 2251–2257.
21. J. Shi, H. Huang, Z. Stratton, Y. Huang and T. J. Huang, *Lab Chip,* 2009, **9**, 3354–3359.
22. H. Bruus, *Theoretical Microfluidics,* Oxford University Press, Oxford, 2008.
23. S. M. Hagsäter, T. G. Jensen, H. Bruus and J. P. Kutter, *Lab Chip,* 2007, 7, 1336–1344.
24. P. B. Muller, M. Rossi, A. G. Marín, R. Barnkob, P. Augustsson, T. Laurell, C. J. Kähler and H. Bruus, *Phys. Rev. E,* 2013, **88**, 023006.
25. R. Barnkob, P. Augustsson, T. Laurell and H. Bruus, *Lab Chip,* 2010, **10**, 563–570.

CHAPTER 3

Continuum Mechanics for Ultrasonic Particle Manipulation

JÜRG DUAL* AND THOMAS SCHWARZ

Institute of Mechanical Systems, Department of Mechanical and Process Engineering, ETH Zentrum, CH-8092 Zurich, Switzerland
*E-mail: dual@imes.mavt.ethz.ch

3.1 Introduction

Ultrasonic manipulation is performed in a fluid volume. This volume is either partially (manipulation in droplets[1] or flow-through channels[2–6]) or fully bounded by solid bodies.[7] Because the impedance of these solid boundaries is finite, a pressure field in the fluid will necessarily give rise to motion and deformations of the solid surrounding it. Therefore the modelling of these surroundings is important in order to understand the system as a whole. In particular, resonances and damping of the various modes encountered and used in the fluid are influenced by the neighbouring materials including the container, the transducers used to excite the acoustic modes and possible glue layers that attach the transducer to the container. The harmonic motion of the structural elements at resonance, given the scale of typical devices, will have MHz frequencies as has the pressure variation in the fluid. The resulting strains are normally small, such that the solid motion can be treated in the framework of linear elasticity. This chapter will start with the basic theory in three dimensions. Then the motion of structures is introduced, in particular plates, which are the most relevant elements used in particle manipulation systems. Then the solid is coupled to the fluid. Practical systems need to be solved using the finite element method (FEM), which is able to take care of complicated shapes. The chapter ends with an example of an ultrasonic cavity as used for ultrasonic manipulation.

3.2 Linear Elastodynamics

In contrast to the description of the behavior of a fluid, a solid has a defined reference configuration. Referring to Figure 3.1, a solid body is moving from its reference configuration λ_0 to the current configuration λ. The position of a material point P with respect to a coordinate system with base vectors e_i ($i = 1,2,3$) is X in the reference configuration and x in the current configuration. The displacement vector $u(x_i,t)$, which is dependent on the position x and time t is defined as

$$u = x - X = u_i e_i \tag{3.1}$$

The summation convention is used for repeated indices. For the limiting case of small deformations and rotations, the basic equations of linear elasticity describe the relevant quantities. The strain tensor γ_{ij}, which describes the local deformation and the stress tensor σ_{ij}, which describes the local state of stress. Both γ_{ij} and σ_{ij} are symmetric second rank tensors. The symmetry of the stress tensor expresses the local conservation equation for the angular momentum. The linear constitutive law connects the two and is therefore in general a tensor of rank 4. A detailed explanation can be found in ref. 8.

For a given displacement field $u_i(x_j,t)$, the strain tensor γ_{ij} is computed using the kinematical relations eqn (3.2)

$$\gamma_{ij} = \frac{1}{2}(u_{i,j} + u_{j,i}), \tag{3.2}$$

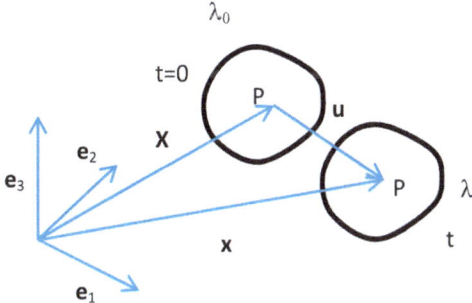

Figure 3.1 A solid body moving from its reference configuration λ_0 to the current configuration λ. Reproduced from Dual et al.[22]

where "$u_{i,j}$" means taking the derivative with respect to x_j. For small deformations and rotations $u_{i,j} \ll 1$ is assumed.

$$\gamma_{11} = u_{1,1} \tag{3.3}$$

is the longitudinal strain in the x_1 direction. This corresponds to the normalized length change of a material line element that extends in the direction of e_1.

$$\gamma_{12} = \frac{1}{2}(u_{1,2} + u_{2,1}) \tag{3.4}$$

is the shear strain in the 1-2 plane, corresponding to half of the change in angle between two material line elements extending in the directions of e_1 and e_2, respectively. By taking the symmetric part of the deformation gradient $u_{i,j}$ as a measure for the deformation, the local rigid body rotation of an infinitesimal volume element is excluded.

Using the summation convention the volume strain is given by

$$\gamma_{ii} = \gamma_{11} + \gamma_{22} + \gamma_{33}. \tag{3.5}$$

With the components of the stress tensor σ_{ij} we can formulate the local linear momentum equation, where a superposed dot corresponds to the time derivative:

In direction x_1:

$$\sigma_{11,1} + \sigma_{12,2} + \sigma_{13,3} = \rho \ddot{u}_1, \tag{3.6}$$

or in indicial notation for the direction x_i:

$$\sigma_{ij,j} = \rho \ddot{u}_i. \tag{3.7}$$

Here, ρ is the density. Body forces are neglected, because of the linearity of the theory and because the focus here is on the high frequency ultrasonic motion.

An isotropic linearly elastic material is fully described by two material constants E, G or λ and μ, where E, G are Young's and shear modulus, respectively, and λ and μ are the Lamé constants, respectively.

$$\sigma_{ij} = \lambda \gamma_{kk} \delta_{ij} + 2\mu \gamma_{ij}. \tag{3.8}$$

δ_{ij} is the unity tensor (Kronecker symbol).

$$\delta_{ij} = \begin{cases} 1 & \text{if } i = j \\ 0 & \text{if } i \neq j \end{cases} \tag{3.9}$$

λ and μ are related to E and G by

$$\mu = G = \frac{E}{2(1+\nu)},$$
$$\lambda = \frac{\nu E}{(1+\nu)(1-2\nu)} = 2G \frac{\nu}{1-2\nu}. \tag{3.10}$$

ν is Poisson's ratio.

Using E and ν, the inverse of eqn (3.8) can be easily obtained, e.g. for γ_{11}

$$\gamma_{11} = \frac{1}{E}(\sigma_{11} - \nu(\sigma_{22} + \sigma_{33})). \tag{3.11}$$

From eqn (3.11) and its permutations, the physical meaning of E and ν becomes obvious. A stress σ_{11} will result in a tensile strain $\gamma_{11} = \sigma_{11}/E$ and lateral contraction strains $\gamma_{22} = \gamma_{33} = -\nu\gamma_{11}$.

Combining eqn (3.2), (3.7) and (3.8), we obtain

$$\sigma_{ij,j} = (\lambda + \mu)u_{k,ki} + \mu u_{i,kk} = \rho\ddot{u}_i. \tag{3.12}$$

These are three coupled differential equations that need to be solved together with initial conditions $g_i(x_j)$ and $h_i(x_j)$ for displacement and velocity, respectively.

$$\begin{aligned} u_i(x_j, 0) &= g_i(x_j), \\ \dot{u}_i(x_j, 0) &= h_i(x_j). \end{aligned} \tag{3.13}$$

For the boundary conditions, the displacement $\boldsymbol{u}$ and the stress vector $\boldsymbol{t} = \boldsymbol{\sigma} \cdot \boldsymbol{n}$ are relevant, where $\boldsymbol{n}$ is the outward unit normal vector of the boundary surface of the solid. Either the stress vector or the displacement can be prescribed on the boundary in normal and tangential directions. For interface conditions, the associated field variables are not prescribed, but are determined from solving the coupled equations. Conditions for both displacement and stress vector then need to be imposed. As an example, for the interface of an inviscid, quiescent fluid with the solid structure, we have

$$\begin{aligned} u_{nS} &= u_{nF}, \\ \boldsymbol{t}_{nS} &= -p\boldsymbol{n}, \\ \boldsymbol{t}_{tS} &= 0, \end{aligned} \tag{3.14}$$

where the indices n and t relate to the normal and tangential directions, respectively, F and S relate to the fluid and solid, respectively, and p is the pressure in the fluid. The first of eqn (3.14) is sometimes formulated as a velocity or acceleration boundary condition. Note that because of zero viscosity, no boundary condition applies for the tangential motion of solid and fluid.

There exists a unique solution to these equations, if all the eqn (3.12)–(3.14) are set up properly.[9]

Eqn (3.12)–(3.14) cannot be solved in a straightforward manner. They can, however, be simplified using the Helmholtz decomposition[10]

$$u_i = \varphi_{,i} + \varepsilon_{ijk}\Psi_{k,j}, \tag{3.15}$$

where a scalar displacement potential φ and a vector potential $\boldsymbol{\Psi}$ with zero divergence have been introduced as well as the permutation tensor ε_{ijk}. ε_{ijk} is +1 when i,j,k are even permutations of 1,2,3, −1 for odd permutations and

0 otherwise. When eqn (3.15) is inserted into eqn (3.12), it is seen that two types of waves exist, which satisfy the classical wave equations

$$\varphi_{,ii} = \frac{1}{c_1^2}\ddot{\varphi},$$
$$\Psi_{k,ii} = \frac{1}{c_2^2}\ddot{\Psi}_k,$$
(3.16)

with wave speeds

$$c_1^2 = \frac{\lambda + 2\mu}{\rho},$$
$$c_2^2 = \frac{\mu}{\rho},$$
(3.17)

where c_1 and c_2 correspond to the wave speeds of the primary and secondary waves, respectively. The terms come from geophysics and result from the fact that $c_1 > c_2$ and hence the primary wave arrives first. They are often abbreviated as P- and S-waves, respectively. For plane waves in isotropic materials, P-waves are longitudinal, while S-waves are transverse.

Typical values are $c_1 = 6300$ m s^{-1} and $c_2 = 3140$ m s^{-1} for aluminium and $c_1 = 2650$ m s^{-1} and $c_2 = 1080$ m s^{-1} for PMMA, respectively. For single crystal silicon (SCSi), the situation is more complicated as it is anisotropic. Therefore the wave speeds are dependent on the direction of propagation and the constitutive law must be replaced by[11]

$$\sigma_{ij} = c_{ijkl}\gamma_{kl}$$
$$\gamma_{ij} = s_{ijkl}\sigma_{kl}$$
(3.18)

The 4th order tensors c_{ijkl} and s_{ijkl} are the stiffness and the compliance tensor of the material, respectively. They are symmetric with respect to the indices i and j, as well as k and l, due to the symmetry of σ and γ. They are also symmetrical with respect to index pairs ij and kl, because for elastic materials, a strain energy exists, which is given by

$$U = \frac{1}{2}\sigma_{ij}\gamma_{ij} = \frac{1}{2}c_{ijkl}\gamma_{ij}\gamma_{kl}.$$
(3.19)

In general therefore an anisotropic elastic material is characterized by 21 independent material constants.

For a specific material, the stiffness and compliance tensors reflect the symmetries of the material. SCSi is a crystalline cubic material where all three coordinate axes taken parallel to the crystal axes are equivalent. For easier readability, sometimes a different representation is introduced (Voigt notation), which reflects the symmetries of the quantities. As seen in the first part of eqn (3.20), double indices running from 1 to 3 each are replaced by one single index running from 1 to 6.

$$\begin{pmatrix} \sigma_{11} \\ \sigma_{22} \\ \sigma_{33} \\ \sigma_{23} \\ \sigma_{13} \\ \sigma_{12} \end{pmatrix} = \begin{pmatrix} \sigma_1 \\ \sigma_2 \\ \sigma_3 \\ \sigma_4 \\ \sigma_5 \\ \sigma_6 \end{pmatrix} = \begin{bmatrix} c_{11} & c_{12} & c_{12} & 0 & 0 & 0 \\ c_{12} & c_{11} & c_{12} & 0 & 0 & 0 \\ c_{12} & c_{12} & c_{11} & 0 & 0 & 0 \\ 0 & 0 & 0 & c_{44} & 0 & 0 \\ 0 & 0 & 0 & 0 & c_{44} & 0 \\ 0 & 0 & 0 & 0 & 0 & c_{44} \end{bmatrix} \begin{pmatrix} \gamma_1 \\ \gamma_2 \\ \gamma_3 \\ \gamma_4 \\ \gamma_5 \\ \gamma_6 \end{pmatrix} \quad (3.20)$$

The corresponding components of the stiffness tensor for SCSi in the notation of eqn (3.20) are

$$c_{11} = 165.7 \text{ GPa}, c_{12} = 63.9 \text{ GPa}, c_{44} = 79.6 \text{ GPa.} \quad (3.21)$$

While the analytical treatment of wave propagation in such materials is quite involved and beyond the scope of the present tutorial, for a numerical vibrational analysis of an anisotropic structure, only the material properties are changed when compared to the isotropic solution.[11]

The two types of waves for the isotropic material interfere and get reflected at boundaries and interfaces to yield the modes of vibrations of the system. It must be noted that in general, a P-wave is reflected as both a P-wave and an S-wave. The same is true for an incident S-wave. At interfaces, a part of the energy is transmitted, another part is reflected depending on material properties and incident angle. The amount of energy that flows through an interface can strongly influence the amount of damping for a particular resonance mode in a micromanipulation device. For normal incidence, the reflected and transmitted energy flows R and T, respectively, divided by the incident energy flow are given by

$$R = \frac{(Z_2 - Z_1)^2}{(Z_2 + Z_1)^2},$$
$$T = \frac{4 Z_1 Z_2}{(Z_2 + Z_1)^2} \quad (3.22)$$

Where $Z_i = \rho_i c_i$ are the characteristic impedances of the two materials. $R + T = 1$ because of conservation of energy and both coefficients are independent of the direction of the energy flow. For a silicon [110] water interface and normal incidence, $R = 0.75$ and $T = 0.25$. Even though a hard wall boundary condition is often assumed in simplified models, which would correspond to $R = 1$, in reality, 25% of the incident energy flow will pass through the interface.

In addition, interface waves exist at free boundaries (*e.g.* Rayleigh Waves) and at interfaces (*e.g.* Stoneley Waves).[10]

3.3 Deformations of Structures

If one or two dimensions of a structure are much smaller than the third, there exist a number of models that allow simple modelling of the behaviour, which take into account the stress free boundary conditions and are based upon certain assumptions regarding the displacement distribution. These assumptions have originally been made by physical insight; they can also be made in a rigorous way using matched asymptotic expansions.[12] These models can be grouped as

- Beam (length $\gg$ cross-sectional dimension) with longitudinal, torsional and bending motion,
- Plate (flat structure, thickness $h \ll$ other dimensions) with longitudinal, bending, shearing and twisting motion,
- Shell: as plate, except that the middle surface has a curvature.

In view of the applications to micromanipulation where often a cover is used above and or below a chamber or channel, we focus here on bending vibrations of thin plates as an example for structural models.

In the modelling of plates, it is customary to refer all quantities to the middle surface and the coordinate system has its origin and axes located at this middle surface in the reference configuration. For simplicity we consider a displacement $w(x_1,t) = u_3(x_1,x_3 = 0,t)$ of the middle surface in the x_3 direction independent of x_2 (Figure 3.2). The bending motion is analyzed in the context of the Kirchhoff plate theory,[10] which is the equivalent to the Euler–Bernoulli beam theory.[8] The displacement assumption is that plane cross-sections orthogonal to the middle surface in the reference configuration remain plane and orthogonal to the middle surface during deformation.

For a homogeneous, isotropic, linearly elastic plate of unit width, small strains, small deformation and wavelengths that are large with respect to the thickness h of the plate, the following equations are valid:

Displacement assumption

$$u_1 = -w_{,1}x_3. \tag{3.23}$$

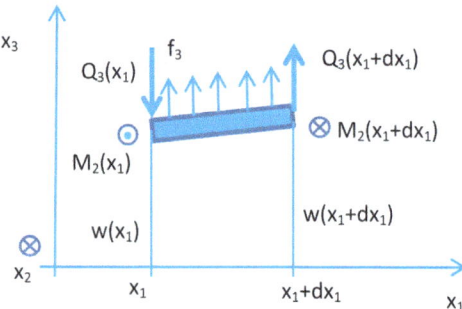

Figure 3.2 Infinitesimal element of a plate (length dx_1 and unit width) with stress resultants shear force Q_3 and bending moment M_2. Reproduced from Dual *et al.*[22]

Constitutive law relating bending moment and curvature

$$M_2 = \int_{-\frac{h}{2}}^{+\frac{h}{2}} \sigma_{11} x_3 \, dx_3 = -\frac{E}{(1-\nu^2)} \frac{h^3}{12} w_{,11}. \tag{3.24}$$

Angular momentum equation

$$-Q_3 + M_{2,1} = 0 \tag{3.25}$$

Linear momentum equation

$$f_3 + Q_{3,1} = \rho h \ddot{w}. \tag{3.26}$$

Here, M_2 is the bending moment, Q_3 the shear force, both per unit width and f_3 the load per unit area. Please note that shear deformation has been neglected in the kinematic eqn (3.23) and rotatory inertia has been neglected in the angular momentum eqn (3.25). This is consistent with the assumption that the wavelength is much larger than the thickness.

$E/(1-\nu^2)$ is called the longitudinal plate modulus, which results from the assumptions of plane strain ($\gamma_{22} = 0$) and negligible transverse normal stress ($\sigma_{33} = 0$).

Combining eqn (3.24) and (3.26), we obtain

$$\frac{E}{(1-\nu^2)} \frac{h^3}{12} w_{,1111} + \rho h \ddot{w} = f_3. \tag{3.27}$$

In order to solve eqn (3.27), we consider harmonic solutions with angular frequency ω of the type

$$w = w_0 e^{i(\omega t - k x_1)}, \quad k = \frac{\omega}{c} = \frac{2\pi}{\lambda}, \tag{3.28}$$

where k is the wavenumber and c the phase speed. If w satisfies eqn (3.27), for $f_3 = 0$, we obtain

$$c^2 = \sqrt{\frac{Eh^2}{12\rho(1-\nu^2)} \omega} = c_3 \, j_2 \omega, \tag{3.29}$$

where c_3

$$c_3 = \sqrt{\frac{E}{\rho(1-\nu^2)}} \tag{3.30}$$

is the longitudinal wave speed in the plate and j_2

$$j_2 = \sqrt{\frac{h^2}{12}} \tag{3.31}$$

is the radius of inertia of the cross-section.

The phase speed c is a function of frequency, therefore bending waves are dispersive. This dispersion is called geometric dispersion and is a function of the ratio of plate thickness h to wavelength λ. Using $\omega = ck$, we obtain:

$$c = c_3\, j_2 k,$$
$$\omega = c_3\, j_2 k^2. \tag{3.32}$$

The dispersion relation is plotted in Figure 3.3 for the Kirchhoff plate theory and the Mindlin plate theory.[10] The deviations between the two theories are due to rotatory inertia and transverse shear deformation, which must be taken into account for about $j_2 k > 0.1$.

Note that in general for plate vibrations with $w(x_1, x_2, t)$, the differential equation is.

$$D(w_{,1111} + 2 w_{,1122} + w_{,2222}) + m\ddot{w} = f_3, \tag{3.33}$$

where

$$D = \frac{E h^3}{12(1 - \nu^2)}, \tag{3.34}$$
$$m = \rho h.$$

D is the plate bending stiffness and m is the mass per unit area of the plate.

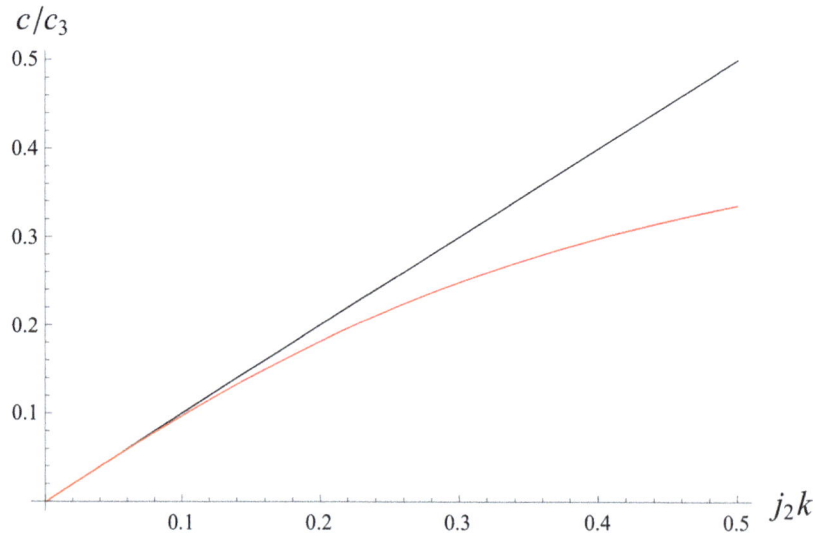

Figure 3.3 Normalized dispersion curve (phase speed c vs. wave number k) for bending waves in a plate according to the Kirchhoff plate theory (black) and according to the Mindlin plate theory (red).[10] Reproduced from Dual et al.[22]

As an example let us consider vibrations of a plate that are independent of x_2, with pinned–pinned boundary conditions (*i.e.* zero displacement and zero bending moment) at $x_1 = 0$ and $x_1 = L$. We assume

$$w(x_1, t) = f(x_1)e^{i\omega t}. \qquad (3.35)$$

After insertion into eqn (3.33), we get

$$f_{,1111} - k^4 f = 0, \qquad (3.36)$$

with the solution

$$f = a_1 \sin(kx_1) + a_2 \cos(kx_1) + a_3 \sinh(kx_1) + a_4 \cosh(kx_1). \qquad (3.37)$$

a_1 to a_4 are obtained from the application of the boundary conditions

$$f(0) = f(L) = f_{,11}(0) = f_{,11}(L) = 0. \qquad (3.38)$$

Because the differential equation is of fourth order, we need four boundary conditions. This yields a 4×4 homogeneous system of equations for the unknowns a_1 to a_4, the determinant of which must vanish for a nontrivial solution to exist, resulting in

$$\begin{aligned} \sin(kL) &= 0, \\ k_n L &= n\pi. \end{aligned} \qquad (3.39)$$

For the corresponding eigenfrequencies f_n one obtains using eqn (3.32),

$$f_n = c_3 j_2 \frac{n^2}{2} \frac{\pi}{L^2}. \qquad (3.40)$$

For other boundary conditions, the solution is more complicated, *e.g.* for clamped–clamped (zero displacement and zero rotation at both ends), we have[10]

$$k_n L = 4.73, 7.85, 10.99, 14.14, \ldots \qquad (3.41)$$

Note that the resonance frequencies are proportional to $1/L^2$ and increase with n^2.

Material damping can easily be incorporated into the equations, if we restrict ourselves to harmonic motion. According to the theory of linear viscoelastic materials,[13] and for small damping, the plate material is described by a complex Young's modulus E^*.

$$E^* = E_0(1 + i\varphi) \qquad (3.42)$$

Here $\varphi \ll 1$ is the loss tangent, which characterizes the intrinsic damping. E_0 is the elastic part. Similarly complex values can be introduced into the Lamé constants of eqn (3.8).

For metals, φ is about 0.01–0.0001 and E^* is nearly independent of frequency. For plastic materials, damping is much larger, depends on temperature and a strong frequency dependence might be present. For a given device, the intrinsic damping might strongly affect the pressure amplitudes that can be achieved in an ultrasonic manipulation device. Glue layers are of particular relevance in this context.

3.4 Fluid Structure Interaction at the Device Level

When a fluid is in contact with a solid, fluid structure interaction takes place. Both the field equations in the solid and in the fluid must be satisfied, as well as the boundary conditions. In addition, at the interface, certain conditions must be satisfied, depending on how the fluid is modelled.

For an inviscid fluid, the interface conditions have been given in eqn (3.14). At the interface, the normal displacements must be equal, the normal stress of the solid must be equal to the negative of the pressure in the fluid, and the shear stress in the solid must vanish, if the fluid's viscosity is neglected.

If the cavity becomes very small, the viscosity might become important, which changes the boundary conditions. Then the tangential part of the stress vector must be set equal to the viscous stress in the fluid and the tangential velocities must also be equal. The boundary conditions, eqn (3.14), must therefore be replaced by

$$\begin{aligned} \boldsymbol{u}_F &= \boldsymbol{u}_S, \\ \boldsymbol{t}_F &= \boldsymbol{t}_S. \end{aligned} \quad (3.43)$$

A harmonic tangential motion of the surface will give rise to a viscous boundary layer of thickness δ in the fluid. It is given by[14]

$$\delta = \sqrt{\frac{2\eta}{\rho_F \omega}}, \quad (3.44)$$

where ρ_F and η are density and viscosity of the fluid, respectively. For water at 1 MHz, δ is about 0.5 µm and less than 10^{-3} of the wavelength. Viscous effects at the boundary are usually neglected, unless the cavity is comparable in size to δ. They will cause additional damping however.

In order to understand the physics of interaction between a harmonically vibrating surface and an inviscid fluid, two simple cases are given here.

In the first case, the solid moves without influence from the fluid. In the second case, there is a strong coupling between the two domains.

3.4.1 Acoustic Radiation from a Vibrating Surface

Referring to Figure 3.4, a plane situation is considered, where the surface of a solid half space ($x_3 < 0$) is vibrating harmonically with a given displacement distribution $u_3(x_1,0) = u(x_1)\exp(i\omega t)$. The fluid that occupies the half space $x_3 > 0$ does not influence the motion. We look for the solution in the fluid, which must satisfy.

$$p_{,ii} = \frac{1}{c_F^2}\ddot{p} = -k_F^2 p, \quad (3.45)$$

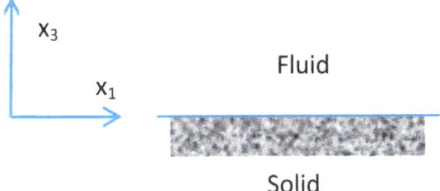

Figure 3.4 Boundary between a fluid and a solid half space. Reproduced from Dual et al.[22]

with the boundary condition at $x_3 = 0$ (eqn (3.14))

$$-\omega^2 \rho_F u_3 + p_{,3} = 0, \qquad (3.46)$$

where c_F and k_F are acoustic wave speed and wave number in the fluid.

(a) Constant displacement $u = u_0 e^{i\omega t}$

We obtain the solution by setting

$$p = p_0 e^{i(\omega t - k_F x_3)}, \qquad (3.47)$$

where we have assumed that there is no energy flowing back from the fluid towards the surface (radiation condition). Eqn (3.47) satisfies eqn (3.45) and the boundary condition yields

$$p_0 = -\frac{\rho_F \omega^2}{i k_F} u_0 = \rho_F c_F v_0, \qquad (3.48)$$

i.e. the pressure amplitude is equal to the characteristic impedance times velocity amplitude as expected by linear acoustic theory. Energy is radiated into the fluid.

(b) Sinusoidal displacement $u = u_0 \sin(k_1 x_1)$, k_1 given

We assume a solution of the form

$$p = p_0 \sin(k_1 x_1) e^{i(\omega t - k_3 x_3)}, \qquad (3.49)$$

which satisfies eqn (3.45) if

$$k_3^2 = k_F^2 - k_1^2. \qquad (3.50)$$

We must discriminate between two cases:

(b1) $k_F > k_1$

i.e. the wavelength $\lambda_1 = 2\pi/k_1$ of the surface motion is larger than the wavelength in the fluid for the particular frequency. k_2 is a real number and we obtain a wave propagating away from the surface. In the limit $k_1 \to 0$, the situation (a) is recovered.

(b2) $k_F < k_1$

i.e. the wavelength λ_1 of the surface motion is smaller than the wavelength in the fluid for the particular frequency. k_3 is a purely imaginary number. When inserted into the assumption eqn (3.49), we obtain

$$p = p_0 \sin k_1 x_1 e^{-\alpha x_3} e^{i\omega t},$$
$$\alpha = \sqrt{k_1^2 - k_F^2}. \tag{3.51}$$

This is an exponentially decaying pressure field, and no acoustic radiation occurs.

The fluid is pumped back and forth between neighbouring peaks and valleys of the surface motion. Therefore, for low frequencies, i.e. $f < c_F/\lambda_1$, no acoustic radiation occurs.

3.4.2 Acoustic Radiation from a Plate Vibrating Harmonically

We now combine the Kirchhoff plate equations in 2D with the acoustic solutions by looking at a situation in which a plate vibrates in contact with a fluid half space $x_3 > 0$ on one side.[16] The plate motion satisfies eqn (3.27), where f_3 is given by the fluid pressure.

The fluid satisfies eqn (3.45). The boundary condition at $x_3 = 0$ is

$$-\rho_F \omega^2 w = -p_{,3}. \tag{3.52}$$

Note that the top surface of the plate is taken as $x_3 = 0$. With this boundary condition, we now consider the full interaction between the sound field and the motion of the plate. This will result in a modified wave speed for the wave in the plate. Assuming a wave traveling in the $+x_1$-direction we look for coupled solutions of the form

$$p = P(x_3) e^{i(\omega t - k_1 x_1)},$$
$$w = w_0 e^{i(\omega t - k_1 x_1)}, \tag{3.53}$$

Here, k_1 is unknown and common to both fields. When inserted into eqn (3.45), we obtain for the fluid pressure:

$$P_{,33} + (k_F^2 - k_1^2)P = 0, \tag{3.54}$$

resulting in

$$P = A e^{i\beta x_3} + B e^{-i\beta x_3},$$
$$\beta^2 = (k_F^2 - k_1^2). \tag{3.55}$$

As before, we have to discriminate two cases, while keeping in mind that k_1 is now not constant, but a function of frequency.

(a) $k_F > k_1$

i.e. the wavelength of the plate motion is larger than the wavelength in the fluid for the particular frequency. We obtain a wave propagating away from the surface.

The radiation condition yields $A = 0$.
The boundary condition eqn (3.52) at the plate yields B and after insertion into eqn (3.27), we obtain a modified dispersion relation

$$Dk_1^4 - \left(\rho h + \frac{i\rho_F}{\sqrt{(k_F^2 - k_1^2)}}\right)\omega^2 = 0. \tag{3.56}$$

The solution k_1 must be obtained numerically. It is complex, representing the fact that the acoustic radiation will introduce an exponential decay of the traveling wave. It is called a leaky bending wave.

(b) $k_F < k_1$

i.e. the wavelength of the plate motion is smaller than the wavelength in the fluid for the particular frequency.

When combining eqn (3.52) and (3.27), we obtain as in section D.1, an exponential decay of the fluid motion into the fluid and a modified dispersion relation.

$$Dk_1^4 - \left(\rho h + \frac{\rho_F}{\sqrt{(k_1^2 - k_F^2)}}\right)\omega^2 = 0. \tag{3.57}$$

In this case, the dispersion relation is only modified in a way where the mass term is increased by fluid being pumped back and forth.

This is the situation that one would like to have, if one wants to make a density sensor.

A special case of this is when $k_F \ll k_1$. We can then neglect k_F in eqn (3.57), neglect $\rho_0\, k_1 h$ when compared to ρ (thin plate) and get

$$k_1^5 = \frac{\omega^2 \rho_F}{D}. \tag{3.58}$$

Physically speaking, all the stiffness is provided by the plate and all the mass is provided by the fluid.

In Figure 3.5, the interaction between a silicon plate (thickness 100 μm) and water is shown by comparing the relative wave numbers.

In a micromanipulation device, the assumption of the semi-infinite fluid domain is not valid. Therefore reflections from the solid adjacent boundary need to be considered.[17]

3.4.3 Vibrations of Devices for Particle Manipulation

For general geometries as they occur in particle manipulation devices, a numerical solution must be found, as there is no analytical solution available. The method used normally is the FEM.[18] In short, in the FEM, the spatial domain is split up into standard subdomains (elements), for which certain displacement assumptions are made. The elements are connected to their neighbors at nodes. The displacements within the elements are expressed in

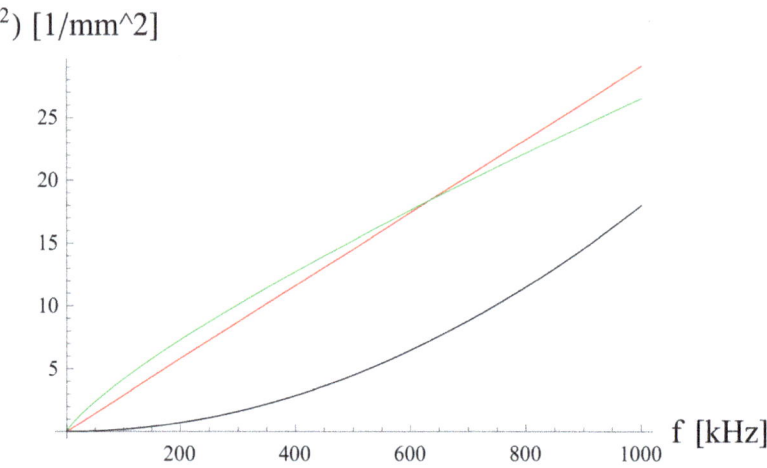

Figure 3.5 Square of the wave number for waves in water (black) and bending waves in a silicon plate of thickness 100 µm, without fluid (red, eqn (3.32)) and interacting with an adjacent fluid (green, eqn (3.58)). Reproduced from Dual et al.[22]

terms of the displacement at the nodes. The elements are then assembled to yield a discretized version of the problem as shown in eqn (3.59).

$$[M]\{\ddot{x}\} + [D]\{\dot{x}\} = [K]\{x\} = \{F\}. \qquad (3.59)$$

Here M is the mass matrix, D the damping matrix, K the stiffness matrix and F the loading vector. x is the vector of the nodal displacements. These equations are then solved to yield a discretized solution, consisting of the nodal vector x.

Eqn (3.59) contains the resonance frequencies and modes of the system. Often D is set to zero and then M and K are diagonalized to yield modes independent of each other corresponding to the modal coordinates y. D is then taken as a diagonal matrix, *i.e.* damping is assigned to the modes individually. Each mode is then considered as a single degree of freedom system (SDOF). A necessary condition for this assumption is that the modes are separated sufficiently, *i.e.* the bandwidth of each mode must be smaller than the spacing between the frequencies of adjacent modes. If a SDOF system is taken as a model for a mode, the quality factor (Q-factor) is a good parameter to quantify the damping of the particular mode. A SDOF system with mass m, stiffness k, damping δ and driving force f is described by the equation

$$m\ddot{y} + \delta\dot{y} + ky = f. \qquad (3.60)$$

The resonance frequency ω_0 and quality factor Q are then defined as

$$\begin{aligned} \omega_0 &= \sqrt{\frac{k}{m}}, \\ Q &= \frac{m\omega_0}{\delta} = \frac{\omega_0}{\Delta\omega}, \end{aligned} \qquad (3.61)$$

where $\Delta\omega$ is the bandwidth. Q can also be seen as the ratio between the dissipated energy per cycle D_{tot} divided by $2\pi U$, where U is the maximum stored energy. In an ultrasonic manipulation system, many damping mechanisms are present; each one will contribute to the resulting Q according to its activity level. As U is fixed, we have

$$\frac{1}{Q_{tot}} = \frac{D_{tot}}{2\pi U_{max}} = \frac{1}{Q_{fluid}} + \frac{1}{Q_{solid}} + \frac{1}{Q_{glue}} + \frac{1}{Q_{support}}. \quad (3.62)$$

$1/Q_i$ corresponds to the respective normalized damping energies D_i. The fluid contribution $1/Q_{fluid}$ relates to viscous damping, both within the bulk of the fluid as well as in the viscous boundary layer. For a solid described with a complex modulus (3.42), the contribution is

$$\frac{1}{Q_{solid}} = \varphi. \quad (3.63)$$

If glue layers are explicitly considered, their damping contributions $1/Q_{glue}$ can be modelled as in eqn (3.63). $1/Q_{support}$ is a simplified term to describe the net energy that is radiated from the system to neighbouring systems by mechanical waves. This is sometimes difficult to model exactly.

Examples of a single-mode analysis of an experimentally determined acoustofluidic resonance have been given in ref. 15, 19 and 20.

The usual steps when applying the FEM are:

- Definition of the geometry.
- Definition of material properties.
- Choosing suitable elements for the spatial discretization. Important in this decision is whether structural elements (beams, plates, shells) or 3D brick elements are used. Also the approximation of the displacement functions within the element is an important consideration.
- Definition of boundary conditions, interface conditions and loading conditions (possibly depending on the element).
- Choosing the mode of analysis (*e.g.* harmonic analysis).
- Computation of the results.
- Analysis of the results and quality check.

3.5 Example of a Mechanical Model of an Ultrasonic Cavity used for Particle Manipulation

Finite element modelling is a useful tool for the simulation of fluid structure interaction. A practical simulation is only possible by including the resonant cavity and the surrounding mechanical structure, which are coupled and therefore influence each other. The modelling helps to understand the response of the system and can be very useful for the design and optimization process. The frequency at which a strong pressure field is built up inside the fluid cavity that is suitable for particle manipulation can be determined.

The 2D simulation presented here was done with COMSOL Multiphysics 4.1 and represents the three-dimensional situation where all quantities are independent of the third dimension. The resonating system model is a typical microfluidic device for ultrasonic manipulation.[20] It consists of the three parts as shown in Figure 3.6(a). The main part consists of silicon, where a cavity for a fluid is etched inside. The system is sealed with a glass part at the top. The silicon dimensions are 5 mm × 0.5 mm. The water filled cavity is 3 mm × 0.2 mm and the glass top has the dimensions 5 mm × 0.5 mm. The piezoelectric transducer, which is normally driving such a system, is omitted here for simplicity reasons. Therefore the excitation is done with a prescribed displacement amplitude of 0.1 nm in the y direction in the middle of the silicon bottom over a length of 0.25 mm (represented by the red part in Figure 3.6(a)). This excitation amplitude leads to pressure amplitudes in the fluid that are comparable with the excitation achieved with a piezoelectric transducer.

The geometry is built in COMSOL using the above dimensions and produces the three domains 1–3 shown in Figure 3.6(a). To every domain, a material with its properties must be assigned. The following material properties have been used: water with a density of 998 kg m^{-3} and a speed of sound of $1481(1 + i/2000)$ ms^{-1}. Damping has been included with complex stiffness parameters for solids and a complex wave speed for the fluid.[21] Silicon is an anisotropic material with stiffness parameters as given in eqn (3.21) and a density of 2330 kg m^{-3}. The damping of the silicon has been neglected as it is small compared to the other materials used. The glass has a Young's modulus of $63(1 + i/400)$ GPa, a Poisson's ratio of $\nu = 0.2$ and a density of 2220 kg m^{-3}.

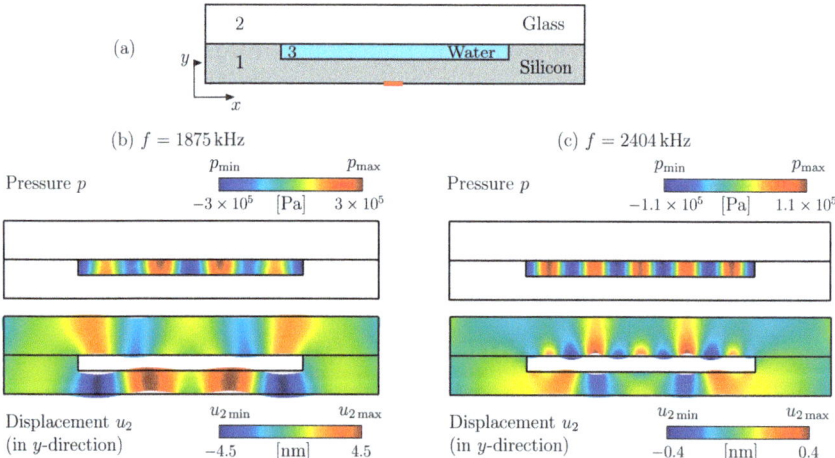

Figure 3.6 (a) Model of a microfluidic device for ultrasonic manipulation. (b) Results for a frequency of 1875 kHz, the pressure field in the fluid and displacement, deformation of the solid structure in the y direction (deformation scaled by a factor of 10^5) are depicted. (c) Pressure and displacement field for a frequency of 2404 kHz.

For the modelling of the water domain, the pressure acoustic physics-module (acpr) has been chosen, where the dependent variable is the acoustic pressure p. The coupling for all cavity boundaries with the surrounding structure is defined with an acceleration a_n (acceleration denoted by (solid/lemm1) in COMSOL) normal to the interface which is coupled to the solid mechanics physics-module. The first boundary condition used here is equivalent to eqn (3.14), where the normal displacement is used instead of the normal acceleration.

The silicon and glass parts are represented with the solid mechanics physics-module (solid). There the dependent variable is the displacement field u with its field components u_1 and u_2. Two linear elastic material models are needed, one with anisotropic properties for the silicon and the other with isotropic properties for the glass. The boundary conditions used were as follows: at all outside boundaries, a free displacement was implemented, with the exception of the boundary for the prescribed displacement amplitude of 0.1 nm in the y direction. For the fluid structure interaction, a boundary load was defined with a so-called "load defined as force per unit area" (acoustic load per unit area denoted by (acpr/pam1) in COMSOL) from the pressure acoustic physics-module. This boundary load is defined by COMSOL and can be understood as $-p\boldsymbol{n}$ as shown in eqn (3.14), which is equal to the normal stress vector t_{nS} in the solid.

The meshing was done automatically with about 13 500 "free quad" elements of equal size for all three domains. The maximum element size depends on the speed of sound of the materials used and the frequency and must be much smaller than the wavelength. A mesh convergence test, where the results with different element sizes are compared, can be used to ascertain that the spatial discretization is sufficient. A frequency response analysis (frequency domain) was performed. In the simulation, a "fully coupled" PARDISO solver was applied as both fields (acoustic, solid) are influencing each other. The simulation was done parametric over a frequency range from 1–3 MHz in 1 kHz steps. As an example of the simulation, two results at frequencies of $f = 1875$ kHz and $f = 2404$ kHz are represented in Figure 3.6(b) and (c), respectively. The pressure field in the fluid and the displacement and deformation of the solid structure in the y direction are presented. The deformation is scaled by a factor of 10^5.

Strong pressure fields are created at these frequencies and the system is in resonance, therefore the excitation is 0.1 nm smaller than the maximum displacement of the solid. These results show nicely the coupling between the two modules. The excitation of the system is done with the displacement of the silicon and is exciting the fluid cavity to resonance. The standing pressure wave inside the fluid cavity is strongly influencing the surrounding structure.

3.6 Conclusions

In this chapter, linear elastic continuum mechanics has been summarized and applied to a typical problem that occurs in ultrasonic manipulation devices. While some physical insight can be gained by studying simple systems, for an analysis of a practical device, numerical tools such as FEM are needed. It has been shown that for systems where the fluid used is a fluid with water like properties, the relatively small impedance difference results in motion for the whole system and not just for the acoustic cavity. When designing a device, normally resonance frequencies are sought, because they have much stronger fields, which helps in the manipulation efficiency. These resonance frequencies and modes are influenced by all the components of the system. For the absolute value of the pressure maxima in the cavity, the damping of the resonance mode is also important, which is normally difficult to model on first principles.

References

1. S. Oberti, A. Neild, R. Quach and J. Dual, *Ultrasonics,* 2009, **49**, 47.
2. W. T. Coakley, D. W. Bardsley, M. A. Grundy, F. Zamani and D. J. Clarke, *J. Chem. Technol. Biotechnol.,* 1989, **44**, 43.
3. N. R. Harris, M. Hill, S. Beeby, Y. Shen, N. M. White, J. J. Hawkes and W. T. Coakley, *Sens. Actuators B Chem.,* 2003, **95**, 425.
4. J. Nilsson, M. Evander, B. Hammarstrom and T. Laurell, *Anal. Chim. Acta,* 2009, **649**, 141.
5. A. Nilsson, F. Petersson, H. Jonsson and T. Laurell, *Lab Chip,* 2004, **4**(2), 131.
6. A. Neild, S. Oberti, F. Beyeler, J. Dual and B. J. Nelson, *J. Micromech. Microeng.,* 2006, **16**, 1562.
7. S. Oberti, A. Neild and J. Dual, *J. Acoust. Soc. Am.,* 2007, **121**, 778.
8. S. P. Timoshenko and J. N. Goodier, *Theory of Elasticity*, MacGraw-Hill, 1970.
9. E. Sternberg and M. E. Gurtin, On the completeness of certain stress functions in the linear theory of elasticity, *ASME, Proc. 4th U.S. Nat. Congr. Appl. Mech. (Univ. California, Berkeley, Calif., 1962)* American Society of Mechanical Engineers, New York, USA, 1962, vol. 2, pp. 793–797.
10. K. F. Graff, *Wave Motion in Elastic Solids*, Ohio State University Press, 1975.
11. V. M. Ristic, *Principles of Acoustic Devices*, Wiley, 1983.
12. M. Sayir, *Ingenieurarchiv,* 1980, **49**, 309.
13. R. M. Christensen, *Theory of Viscoelasticity*, Academic Press, 1971.
14. L. D. Landau and E. M. Lifshitz, *Fluid Mechanics*, Pergamon, 1959.
15. H. Bruus, *Lab Chip,* 2012, **12**, 20.
16. F. Fahy, *Sound and Structural Vibration*, Academic Press, 1987.
17. A. Haake and J. Dual, *Ultrasonics,* 2004, **42**, 75.
18. H.-J. Bathe and E. Wilson, *Numerical Methods in Finite Element Analysis*, Prentice-Hall, 1976.
19. R. Barnkob, P. Augustsson, T. Laurell and H. Bruus, *Lab Chip,* 2010, **10**, 563.
20. A. Neild, S. Oberti and J. Dual, *Sens. Actuators B Chem.,* 2007, **121**, 452.
21. M. Gröschl, *Acustica,* 1998, **84**, 432.
22. J. Dual and T. Scharz, *Lab Chip,* 2012, **12**, 244.

CHAPTER 4

Acoustic Radiation Force on Small Particles

HENRIK BRUUS

Department of Physics, Technical University of Denmark, Lyngby, Denmark
E-mail: bruus@fysik.dtu.dk

4.1 Introduction

When an ultrasound field is imposed on a fluid containing a suspension of particles, the latter will be affected by the so-called acoustic radiation force arising from the scattering of the acoustic waves on the particle. The particle motion resulting from the acoustic radiation force is denoted as acoustophoresis, and it plays a key role in on-chip microparticle handling. The observed acoustophoretic motion is not resolved on the µs time scale of the imposed MHz ultrasound wave, but stems from the radiation force time-averaged over a full oscillation cycle.

The studies of acoustic radiation forces on suspended particles in inviscid fluids have a long history. The analysis of incompressible particles in acoustic fields dates back to the work in 1934 by King,[1] while the forces on compressible particles in plane acoustic waves was calculated in 1955 by Yosioka and Kawasima.[2] Their work was admirably summarized and generalized by Gorkov in 1962 in a short paper,[3] which in turn was extended to include a more detailed derivation as well as the effect of viscosity by Settnes and Bruus.[4] In this chapter, adapted from ref. 5, we discuss the acoustic radiation force following the latter two approaches.

The theory of the radiation force relies on the perturbation expansion of the acoustic fields in the fluid as described in Chapter 2. Restricting the following analysis to inviscid fluids, we obtain the wave equation for the velocity potential $\phi_1 = -ip_1/(\rho_0\omega)$ from eqn (2.7),

$$\nabla^2 \phi_1 = \frac{1}{c_0^2}\partial_t^2 \phi_1 = -\frac{\omega^2}{c_0^2}\phi_1. \qquad (4.1)$$

Together with eqn (2.12b) for the time-averaged, second-order acoustic pressure $\langle p_2 \rangle$, this equation forms the starting point for the scattering theory used below to calculate the acoustic radiation force acting on the particle. In the bulk fluid, where viscosity can be neglected, see eqn (2.6a), $\langle p_2 \rangle$ is given by

$$\nabla \langle p_2 \rangle = -\langle \rho_1 \partial_t \mathbf{v}_1 \rangle - \rho_0 \langle (\mathbf{v}_1 \cdot \nabla)\mathbf{v}_1 \rangle, \quad \text{or} \quad \langle p_2 \rangle = \frac{1}{2\rho_0 c_0^2}\langle p_1^2 \rangle - \frac{1}{2}\rho_0 \langle v_1^2 \rangle. \qquad (4.2)$$

In the latter equality we have used eqns (2.3b) and (2.8a) to obtain $-\langle \rho_1 \partial_t \mathbf{v}_1 \rangle = 1/(2\rho_0 c_0^2)\nabla\langle p_1^2 \rangle$ and $\langle (\mathbf{v}_1 \cdot \nabla)\mathbf{v}_1 \rangle = (1/2)\nabla\langle v_1^2 \rangle$, respectively. We note that the physical, real-valued time average $\langle f\, g \rangle$ of two harmonically varying fields f and g with the complex representation eqn (2.5) is given by the real-part rule

$$\langle f\, g \rangle = \frac{1}{2}\mathrm{Re}[f(\mathbf{r})g^*(\mathbf{r})], \qquad (4.3)$$

where the asterisk denotes complex conjugation.

4.2 The Acoustic Radiation Force

Below we calculate the acoustic radiation force on a compressible, spherical, micrometre-sized particle of radius a suspended in an inviscid fluid in an ultrasound field of wavelength λ. A small particle, i.e. $a \ll \lambda$, of density ρ_p and compressibility κ_p acts as a weak point-scatterer of acoustic waves, which thus can be treated by first-order scattering theory. An incoming wave described by some given velocity potential ϕ_{in}, results in a scattered wave ϕ_{sc} propagating away from the particle. For sufficiently weak incoming and scattered waves, the total first-order acoustic field ϕ_1 is given by the sum of the two as sketched in Figure 4.1(a),

$$\phi_1 = \phi_{\mathrm{in}} + \phi_{\mathrm{sc}}, \qquad (4.4a)$$

$$\mathbf{v}_1 = \nabla\phi_1 = \nabla\phi_{\mathrm{in}} + \nabla\phi_{\mathrm{sc}}, \qquad (4.4b)$$

$$p_1 = i\rho_0\omega\phi_1 = i\,\rho_0\omega\phi_{\mathrm{in}} + i\,\rho_0\omega\,\phi_{\mathrm{sc}}. \qquad (4.4c)$$

Once the first-order scattered field ϕ_{sc} has been determined for the given incoming first-order field ϕ_{in}, the acoustic radiation force $\mathbf{F}^{\mathrm{rad}}$ on the particle can be calculated as the surface integral of the time-averaged second-order

Acoustic Radiation Force on Small Particles

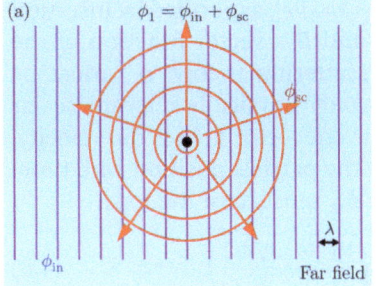
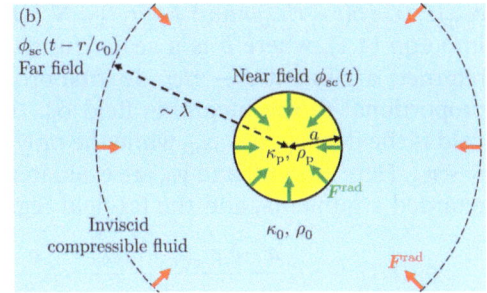

Figure 4.1 (a) Sketch of the far-field region $r \gg \lambda$ of an incoming acoustic wave ϕ_{in} (blue lines) of wavelength λ scattering off a small particle (black dot) with radius $a \ll \lambda$, leading to the outgoing scattered wave ϕ_{sc} (red circles and arrows). The resulting first-order wave is $\phi_1 = \phi_{\text{in}} + \phi_{\text{sc}}$. (b) Sketch of a compressible spherical particle (yellow disk) of radius a, compressibility κ_p, and density ρ_p, surrounded by the compressible inviscid bulk fluid (light blue) of compressibility κ_0 and density ρ_0. The fluid is divided into the near-field region for $r \ll \lambda$, with the instantaneous field $\phi_{\text{sc}}(t)$, and the far-field region with the time-retarded field $\phi_{\text{sc}}(t - r/c_0)$. The radiation force $\boldsymbol{F}^{\text{rad}}$ (red arrows) evaluated at any surface in the far-field region (dashed circle) equals that (green arrows) at the surface of the sphere. Reproduced with permission from ref. 5, see also ref. 4.

pressure $\langle p_2 \rangle$ and momentum flux tensor $\rho_0 \langle v_1 v_1 \rangle$ at a fixed surface just outside the oscillating sphere, see Figure 4.1(b). This follows from the general method of calculating the rate of change of the momentum used in Section 1.4, but is now applied to the inviscid fluid. The expression for $\boldsymbol{F}^{\text{rad}}$ becomes

$$\boldsymbol{F}^{\text{rad}} = -\oint_{\partial \Omega} \mathrm{d}a \{ \langle p_2 \rangle \boldsymbol{n} + \rho_0 \langle (\boldsymbol{n} \cdot \boldsymbol{v}_1) \boldsymbol{v}_1 \rangle \}$$

$$= -\oint_{\partial \Omega} \mathrm{d}a \left\{ \left[\frac{\langle p_1^2 \rangle}{2 \rho_0 c_0^2} - \frac{\rho_0}{2} \langle v_1^2 \rangle \right] \boldsymbol{n} + \rho_0 \langle (\boldsymbol{n} \cdot \boldsymbol{v}_1) \boldsymbol{v}_1 \rangle \right\} \quad (4.5)$$

As there are no body forces in this problem, any fixed surface $\partial \Omega$ encompassing the sphere experiences the same force, and given the result of the following scattering theory analysis, it is advantageous to choose a sphere of radius $r \gg \lambda$ in the far-field region, see the dashed circle in Figure 4.1(b), with its center coinciding with that of the spherical particle.

4.3 Scattering Theory

In scattering theory, the scattered field ϕ_{sc} from a point scatterer at the center of the coordinate system $r = 0$, is represented by a time-retarded multipole expansion. In the far-field region, the monopole and dipole components dominate, $\phi_{\text{sc}} \approx \phi_{\text{mp}} + \phi_{\text{dp}}$, and these two components have the form

$\phi_{\mathrm{mp}}(\mathbf{r},t) = b(t-r/c_0)/r$ and $\phi_{\mathrm{dp}}(\mathbf{r},t) = \nabla \cdot [\mathbf{B}(t-r/c_0)/r]$, as verified by insertion into eqn (4.1), where b is a scalar function and $\mathbf{B}$ a vector function of the retarded argument $t - r/c_0$. In first-order scattering theory, ϕ_{sc} must be proportional to the incoming field ϕ_{in}. The only physically relevant scalar field is the density, $b \sim \rho_{\mathrm{in}}$, while the only relevant vector field is the velocity, $\mathbf{B} \sim \mathbf{v}_{\mathrm{in}}$. Here both ρ_{in} and $\mathbf{v}_{\mathrm{in}}$ are evaluated at the particle position with time-retarded arguments, and the far-field region ϕ_{sc} must have the form

$$\phi_{\mathrm{sc}}(\mathbf{r},t) = -f_1 \frac{a^3}{3\rho_0} \frac{\partial_t \rho_{\mathrm{in}}(t-r/c_0)}{r} - f_2 \frac{a^3}{2} \nabla \cdot \left(\frac{\mathbf{v}_{\mathrm{in}}(t-r/c_0)}{r} \right), \quad r \gg \lambda \quad (4.6)$$

where the particle radius a, the unperturbed density ρ_0, and the time derivative ∂_t are introduced to ensure the correct physical dimension of ϕ_{sc}, namely $\mathrm{m}^2\,\mathrm{s}^{-1}$. The factors 1/3 and 1/2 are inserted for later convenience. The main goal of the calculation is to determine the dimensionless scattering coefficients f_1 and f_2.

In the following, we use a spherical coordinate system with unit vectors $(\mathbf{e}_r, \mathbf{e}_\theta, \mathbf{e}_\phi)$ located at the instantaneous center of the particle. Due to the azimuthal symmetry of the problem, the velocities have no azimuthal component, $\mathbf{v} = v_r \mathbf{e}_r + v_\theta \mathbf{e}_\theta$, and all fields depend only on r and θ. The polar axis $\mathbf{e}_z$ points along the instantaneous direction of the incoming velocity $\mathbf{v}_{\mathrm{in}}$, such that $\mathbf{v}_{\mathrm{in}} = v_{\mathrm{in}} \mathbf{e}_z$. By the azimuthal symmetry of the problem, the particle must also move in that direction, $\mathbf{v}_{\mathrm{p}} = v_{\mathrm{p}} \mathbf{e}_z$,

$$\mathbf{v}_{\mathrm{in}} = v_{\mathrm{in}} \mathbf{e}_z = \cos\theta\, v_{\mathrm{in}} \mathbf{e}_r - \sin\theta\, v_{\mathrm{in}} \mathbf{e}_\theta, \quad (4.7\mathrm{a})$$

$$\mathbf{v}_{\mathrm{p}} = v_{\mathrm{p}} \mathbf{e}_z = \cos\theta\, v_{\mathrm{p}} \mathbf{e}_r - \sin\theta\, v_{\mathrm{p}} \mathbf{e}_\theta. \quad (4.7\mathrm{b})$$

4.3.1 Scattering and the Radiation Force

Before determining the values of the scattering coefficients, we express the radiation force $\mathbf{F}^{\mathrm{rad}}$ in terms of the incoming acoustic wave evaluated at the particle position as well as the coefficients f_1 and f_2 as follows. When inserting the velocity potentials eqn (4.4a) and (4.6) into eqn (4.5) for $\mathbf{F}^{\mathrm{rad}}$, we obtain a sum of terms each proportional to the square of $\phi_1 = \phi_{\mathrm{in}} + \phi_{\mathrm{sc}}$. This results in three types of contributions: (*i*) squares of ϕ_{in} containing no information about the scattering and therefore yielding zero, (*ii*) squares of ϕ_{sc} proportional to the square of the particle volume a^6 and therefore negligible compared to (*iii*) the mixed products $\phi_{\mathrm{in}} \phi_{\mathrm{sc}}$ proportional to particle volume a^3 and therefore the most dominant contribution to the radiation force. Keeping only these mixed terms, which physically can be interpreted as interference between the incoming and the scattered wave, and using the index notation of Section 1.2 (including summation of repeated indices), the ith component of eqn (4.5) becomes

$$F_i^{\mathrm{rad}} = -\oint_{\partial\Omega} \mathrm{d}a\, n_j \left\{ \left[\frac{c_0^2}{\rho_0} \langle \rho_{\mathrm{in}} \rho_{\mathrm{sc}} \rangle - \rho_0 \langle v_k^{\mathrm{in}} v_k^{\mathrm{sc}} \rangle \right] \delta_{ij} + \rho_0 \langle v_i^{\mathrm{in}} v_j^{\mathrm{sc}} \rangle + \rho_0 \langle v_i^{\mathrm{sc}} v_j^{\mathrm{in}} \rangle \right\} \quad (4.8\mathrm{a})$$

$$= -\int_\Omega d\mathbf{r}\; \partial_j\left\{\left[\frac{c_0^2}{\rho_0}\langle\rho_{\text{in}}\rho_{\text{sc}}\rangle - \rho_0\langle v_k^{\text{in}}v_k^{\text{sc}}\rangle\right]\delta_{ij} + \rho_0\langle v_i^{\text{in}}v_j^{\text{sc}}\rangle + \rho_0\langle v_i^{\text{sc}}v_j^{\text{in}}\rangle\right\} \quad (4.8\text{b})$$

$$= -\int_\Omega d\mathbf{r}\;\left\{\frac{c_0^2}{\rho_0}[\langle\rho_{\text{in}}\partial_i\rho_{\text{sc}}\rangle + \langle\rho_{\text{sc}}\partial_i\rho_{\text{in}}\rangle] + \rho_0\left[\langle v_i^{\text{in}}\partial_j v_j^{\text{sc}}\rangle + \langle v_i^{\text{sc}}\partial_j v_j^{\text{in}}\rangle\right]\right\} \quad (4.8\text{c})$$

$$= -\int_\Omega d\mathbf{r}\;\left\{-\langle\rho_{\text{in}}\partial_t v_i^{\text{sc}}\rangle - \langle\rho_{\text{sc}}\partial_t v_i^{\text{in}}\rangle + \rho_0\langle v_i^{\text{in}}\partial_j v_j^{\text{sc}}\rangle - \langle v_i^{\text{sc}}\partial_t\rho_{\text{in}}\rangle\right\} \quad (4.8\text{d})$$

$$= -\int_\Omega d\mathbf{r}\;\left\{\langle v_i^{\text{in}}\partial_t\rho_{\text{sc}}\rangle + \rho_0\langle v_i^{\text{in}}\partial_j v_j^{\text{sc}}\rangle\right\} \quad (4.8\text{e})$$

$$= -\int_\Omega d\mathbf{r}\; \rho_0 \left\langle v_i^{\text{in}}\left(\partial_j\partial_j\phi_{\text{sc}} - \frac{1}{c_0^2}\partial_t^2\phi_{\text{sc}}\right)\right\rangle. \quad (4.8\text{f})$$

Here, we have used $p_1 = c_0^2\rho_1$ in eqn (3.8a), Gauss's theorem in eqn (4.8b), exchange of indices $\partial_i v_k = \partial_i\partial_k\phi = \partial_k\partial_i\phi = \partial_k v_i$ to cancel terms in eqn (4.8c), introduction of time derivatives by the continuity equation $\partial_t\rho_1 = -\rho_0\partial_j v_{1,j}$ and the Navier–Stokes equation $\rho_0\partial_t v_{1,i} = -\partial_i p_1 = -c_0^2\partial_i\rho_1$ in eqn (4.8d), vanishing of time-averages of total time derivatives $\langle\partial_t(\rho v_i)\rangle = 0$ or $\langle\rho\partial_t v_i\rangle = -\langle v_i\partial_t\rho\rangle$ for cancellation and rearrangement in eqn (4.8e), and finally reintroduction of the vector potential ϕ_{sc} in eqn (4.8e).

The d'Alembert wave operator $\partial_j\partial_j - (1/c_0^2)\partial_t^2$ acting on ϕ_{sc} appears in the integrand of eqn (4.8f), and since ϕ_{sc} is a sum of simple monopole and dipole terms, significant simplifications are possible. Just as the Laplace operator acting on the monopole potential $\phi = q/(4\pi\varepsilon_0 r)$ yields the point-charge distribution, $\partial_j^2\phi = -(q/\varepsilon_0)\delta(\mathbf{r})$, in the static case, the d'Alembert operator acting on the retarded-time monopole and dipole expressions (4.6) also yields delta function distributions,

$$\partial_j^2\phi_{\text{sc}} - \frac{1}{c_0^2}\partial_t^2\phi_{\text{sc}} = f_1\frac{4\pi a^3}{3\rho_0}\partial_t\rho_{\text{in}}\delta(\mathbf{r}) + f_2 2\pi a^3 \boldsymbol{\nabla}\cdot[\mathbf{v}_{\text{in}}\delta(\mathbf{r})], \quad r \gg \lambda. \quad (4.9)$$

Now we see the great advantage of working in the far-field limit: the appearance of the two delta functions promises simple integrations. The first term is easily integrated, after substitution of eqn (4.9) into the right-hand side of eqn (4.8f). However, for the second term we need to get rid of the divergence operator acting on the delta function before we can evaluate the integral. This is achieved by the following argument relying Gauss's theorem.

First we note that $\boldsymbol{\nabla}\cdot[f(\mathbf{r})\mathbf{u}(\mathbf{r})] = -f\boldsymbol{\nabla}\cdot\mathbf{u} + \mathbf{u}\cdot\boldsymbol{\nabla}f$ for any scalar function f and vector function $\mathbf{u}$. Therefore, $\oint_{\partial\Omega} da\, \mathbf{n}\cdot(f\mathbf{u}) = \int_\Omega d\mathbf{r}\,\boldsymbol{\nabla}\cdot(f\mathbf{u}) = \int_\Omega d\mathbf{r}(f\boldsymbol{\nabla}\cdot\mathbf{u} + \mathbf{u}\cdot\boldsymbol{\nabla}f)$ by Gauss's theorem, and from this relation, with $f = v_i^{\text{in}}$ and $\mathbf{u} = \mathbf{v}_{\text{in}}\delta(\mathbf{r})$, we obtain the following useful identity $\int_\Omega d\mathbf{r}\, v_i^{\text{in}}\boldsymbol{\nabla}\cdot[\mathbf{v}_{\text{in}}\delta(\mathbf{r})] = -\int_\Omega d\mathbf{r}\,\delta(\mathbf{r})\mathbf{v}_{\text{in}}\cdot\boldsymbol{\nabla}v_i^{\text{in}} + \oint_{\partial\Omega} da\,\mathbf{n}\cdot[\mathbf{v}_{\text{in}}\delta(\mathbf{r})]$. Thus, we can rewrite the right-hand side of eqn (4.8f) into the sum of a volume integral encompassing the delta function singularity, which results in a non-zero contribution, and a surface integral on a surface not containing the delta function singularity, which therefore becomes zero. Consequently, the resulting expression for $\boldsymbol{F}^{\text{rad}}$ becomes

$$\boldsymbol{F}^{\text{rad}} = -\frac{4\pi}{3}a^3\langle f_1 \boldsymbol{v}_{\text{in}}\partial_t\rho_{\text{in}}\rangle + 2\pi a^3\rho_0\langle f_2(\boldsymbol{v}_{\text{in}}\cdot\boldsymbol{\nabla})\boldsymbol{v}_{\text{in}}\rangle \tag{4.10a}$$

$$= \frac{4\pi}{3}a^3\langle f_1 \rho_{\text{in}}\partial_t\boldsymbol{v}_{\text{in}}\rangle + 2\pi a^3\rho_0\langle f_2(\boldsymbol{v}_{\text{in}}\cdot\boldsymbol{\nabla})\boldsymbol{v}_{\text{in}}\rangle \tag{4.10b}$$

$$= -\frac{4\pi}{3\rho_0 c_0^2}a^3\langle f_1 p_{\text{in}}\boldsymbol{\nabla}p_{\text{in}}\rangle + 2\pi a^3\rho_0\langle f_2(\boldsymbol{v}_{\text{in}}\cdot\boldsymbol{\nabla})\boldsymbol{v}_{\text{in}}\rangle \tag{4.10c}$$

$$= -\text{Re}[f_1]\frac{2\pi}{3\rho_0 c_0^2}a^3\boldsymbol{\nabla}\langle p_{\text{in}}^2\rangle + \text{Re}[f_2]\,\pi a^3\rho_0\boldsymbol{\nabla}\langle v_{\text{in}}^2\rangle, \tag{4.10d}$$

with p_{in} and v_{in} evaluated at $r = 0$.

Here, we have integrated over the delta function in eqn (4.10a), applied the previously used rule $\langle \rho_{\text{in}}\partial_t\boldsymbol{v}_{\text{in}}\rangle = -\langle \boldsymbol{v}_{\text{in}}\partial_t\rho_{\text{in}}\rangle$ in eqn (4.10b), inserted $\rho_{\text{in}} = p_{\text{in}}/c_0^2$ and $\partial_t\boldsymbol{v}_{\text{in}} = -\boldsymbol{\nabla}p_{\text{in}}/\rho_0$ in eqn (4.10c), and finally pulled f_1 and f_2 outside the time averages (using eqn (4.3)) together with the nabla operator in eqn (4.10d). The final expression for the radiation force acting on a small particle ($a \ll \lambda$) is,

$$\boldsymbol{F}^{\text{rad}} = \frac{4\pi}{3}a^3\boldsymbol{\nabla}\left[\frac{\text{Re}[f_1]}{2\rho_0 c_0^2}\langle p_{\text{in}}^2\rangle - \frac{3}{4}\text{Re}[f_2]\rho_0\langle v_{\text{in}}^2\rangle\right]. \tag{4.11}$$

Now we need to determine the coefficients f_1 and f_2.

4.3.2 The Near-field Potential

As sketched in Figure 4.1(b), the time-retarded argument $t - r/c_0$ of the acoustic field ϕ_1 can be replaced by the instantaneous argument t in the vicinity of the particle of radius a. The reason is that within one oscillation period $\tau = 2\pi/\omega$, the retardation time over the distances $r \simeq a$ is negligible, $r/c_0 \simeq a/c_0 = \lambda/c_0 = \tau$. Therefore, in the near-field region ϕ_{in}, eqn (4.7a), and the scattering potential, eqn (4.6), with its monopole and dipole term become

$$\phi_{\text{in}}(r,\theta) = v_{\text{in}}r\cos\theta, \tag{4.12a}$$

$$\phi_{\text{sc}}(r,\theta) = \phi_{\text{mp}}(r) + \phi_{\text{dp}}(r,\theta), \quad r = \lambda, \tag{4.12b}$$

$$\phi_{\text{mp}}(r) = -f_1\frac{a^3}{3\rho_0}\partial_t\rho_{\text{in}}\frac{1}{r}, \tag{4.12c}$$

$$\phi_{\text{dp}}(r,\theta) = +f_2\frac{a^3}{2}v_{\text{in}}\frac{\cos\theta}{r^2}, \tag{4.12d}$$

where $\partial_t\rho_{in}$ and v_{in} are evaluated at the position of the particle, $r = 0$. In first-order scattering theory, the monopole and dipole parts of the problem do not mix: f_1 is the coefficient in the monopole scattering potential ϕ_{mp} from a stationary sphere in the incoming density wave ρ_{in}, while f_2 is the coeffcient in the dipole scattering potential ϕ_{dp} from an incompressible sphere moving with velocity v_p in the incoming velocity wave v_{in}.

4.3.3 The Monopole Coefficient f_1

The presence of the particle gives rise to a mass rate $\partial_t m$ of scattered fluid mass given by the first-order, scattered mass flux $\rho_0 \nabla \phi_{mp}$. By integration over the surface of the sphere we obtain

$$\partial_t m = \oint_{\partial\Omega} da\, \boldsymbol{e}_r \cdot (\rho_0 \nabla \phi_{mp}) = f_1 \frac{4\pi}{3} a^3 \partial_t \rho_{in}. \tag{4.13}$$

The factor $1/3$ was introduced in eqn (4.6) to make the particle volume $V_p = (4\pi/3)a^3$ appear here. The rate of scattered fluid mass can also be written in terms of the rate of change of the incoming density $\rho_0 + \rho_{in}$ multiplied by V_p as $\partial_t m = \partial_t[\{\rho_0 + \rho_{in}(t)\}V_p(t)]$. Then using that $\rho_0 \partial_t V_p = \rho_0 \partial_p V_p \partial_t p_{in} = -\rho_0 V_p \kappa_p c_0^2 \partial_t \rho_{in} = -(\kappa_p/\kappa_0)V_p \partial_t \rho_{in}$ in terms of the compressibility $\kappa_p = -(1/V)(\partial V/\partial p)$ and $\kappa_0 = 1/(\rho_0 c_0^2)$ for the particle and the liquid, respectively, we obtain

$$\partial_t m = \left[1 - \frac{\kappa_p}{\kappa_0}\right] V_p \partial_t \rho_{in}. \tag{4.14}$$

Consequently, by equating eqn (4.13) and (4.14), we determine f_1 to be

$$f_1(\tilde{\kappa}) = 1 - \tilde{\kappa}, \quad \text{with} \quad \tilde{\kappa} = \frac{\kappa_p}{\kappa_0}. \tag{4.15}$$

4.3.4 The Dipole Coefficient f_2

The dipole coeffcient f_2 is related to the translational motion of the particle. For an inviscid fluid, the boundary condition at $r = a$ only involves the radial-direction components of the particle velocity v_p and the dipole part of the fluid velocity: $\boldsymbol{e}_r \cdot \boldsymbol{v}_p = \boldsymbol{e}_r \cdot \nabla(\psi_{in} + \phi_{dp})$. The left-hand side is determined by eqn (4.7b) and the right-hand side by eqn (4.12a) and (4.12d), whereby the radial velocity boundary condition becomes

$$v_p = (1 - f_2)\, v_{in}. \tag{4.16}$$

On the other hand, the particle velocity v_p is given by Newton's second law $m_p \partial_t v_p = F^{pres}$ involving the dipole part $p_{in} + p_{dp}$ of the fluid pressure acting on the surface of the sphere,

$$-i\frac{4}{3}\pi\, a^3 \rho_p \omega\, v_p = -2\pi\, a^2 \int_{-1}^{1} d(\cos\theta)(p_{in} + p_{dp})\cos\theta. \tag{4.17}$$

From eqn (4.4c), (4.12a) and (4.12d), we obtain

$$p_{in} + p_{dp} = i\rho_0\omega(\phi_{in} + \phi_{dp}) = i\rho_0\omega a\left[1 + \frac{1}{2}f_2\right]v_{in}\cos\theta, \qquad (4.18)$$

which together with eqn (4.17) leads to

$$\tilde{\rho}v_p = \left[1 + \frac{1}{2}f_2\right]v_{in}, \quad \text{with} \quad \tilde{\rho} = \frac{\rho_p}{\rho_0}. \qquad (4.19)$$

The dipole coefficient f_2 follows from eqn (4.16) and (4.19).

$$f_2(\tilde{\rho}) = \frac{2(\tilde{\rho} - 1)}{2\tilde{\rho} + 1}. \qquad (4.20)$$

4.3.5 The Resulting Radiation Force for a Standing Plane Wave

In summary, the resulting radiation force F^{rad} on a small, spherical particle ($a \ll \lambda$) in an inviscid fluid is a gradient force evaluated at the particle position $r = 0$ and given by

$$\boldsymbol{F}^{rad} = -\frac{4\pi}{3}a^3\boldsymbol{\nabla}\left[f_1\frac{1}{2\rho_0 c_0^2}\langle p_{in}^2\rangle - f_2\frac{3}{4}\rho_0\langle v_{in}^2\rangle\right], \qquad (4.21a)$$

$$f_1(\tilde{\kappa}) = 1 - \tilde{\kappa}, \quad \text{with} \quad \tilde{\kappa} = \frac{\kappa_p}{\kappa_0} \qquad (4.21b)$$

$$f_2(\tilde{\rho}) = \frac{2(\tilde{\rho} - 1)}{2\tilde{\rho} + 1}, \quad \text{with} \quad \tilde{\rho} = \frac{\rho_p}{\rho_0} \qquad (4.21c)$$

Our prime example of the acoustic radiation force is the 1D planar standing $\lambda/2$-wave, $p_1(z) = p_a\cos(kz)$, where $k = 2\pi/\lambda = \omega/c_0$ and $\lambda/2 = w$, w being the channel width, see Figure 4.2. This has been realized in numerous applications in microchannel acoustophoresis, some of which have been described in other chapters of this book. Here, we analyze the radiation force resulting from such a field. The first-order, incoming, acoustic fields are given by

$$\phi_{in}(z,t) = \frac{p_a}{\rho_0\omega}\cos(kz)\cos(\omega t), \qquad (4.22a)$$

$$p_{in}(z,t) = p_a\cos(kz)\sin(\omega t), \qquad (4.22b)$$

$$\boldsymbol{v}_{in}(z,t) = -\frac{p_a}{\rho_0 c_0}\sin(kz)\cos(\omega t)\,\boldsymbol{e}_z, \qquad (4.22c)$$

where we have used the usual real-time representation. With these fields, the time averages needed in eqn (3.21) are simply $\langle\cos^2(\omega t)\rangle = \langle\sin^2(\omega t)\rangle = \frac{1}{2}$,

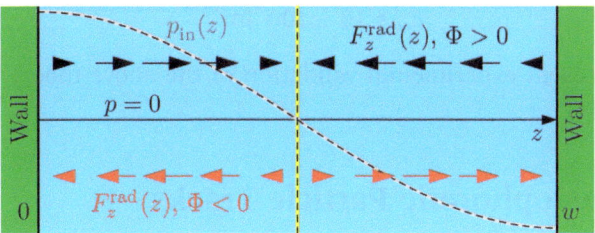

Figure 4.2 A cross-sectional sketch of a straight, hard-walled (green) water-filled channel (blue) of width w with a transverse, standing, ultrasound $\lambda/2$-pressure resonance $p_{in}(z) = p_a \cos(kz)$ (grey dashed line), $k = \pi/w$. Relative to $p_{in}(z)$, the radiation force $F_z^{rad}(z)$ on a small suspended particle is period doubled and phase shifted. For a contrast factor $\Phi > 0$, we have $F^{rad}(z) \propto +\sin(2kz)$ (black arrows), and for $\Phi < 0$, we have $F^{rad}(z) \propto -\sin(2kz)$ (red arrows). Consequently, the resulting particle motion is towards and away from the nodal plane (yellow dashed line), respectively. Reproduced with permission from ref. 5.

and we obtain $\langle p_{in}^2 \rangle = \frac{1}{2}p_a^2 \cos^2(kz)$ and $\rho_0 c_0 \langle v_{in}^2 \rangle = \frac{1}{2}p_a^2 \sin^2(kz)$. Thus we arrive at the following expression for the radiation force

$$F_z^{rad} = -\pi a^3 \frac{p_a^2}{\rho_0 c_0^2} \partial_z \left[\frac{f_1}{3}\cos^2(kz) - \frac{f_2}{2}\sin^2(kz) \right] = 4\pi \Phi(\tilde{\kappa}, \tilde{\rho}) k a^3 E_{ac} \sin(2kz) \quad (4.23a)$$

$$E_{ac} = \frac{p_a^2}{4\rho_0 c_0^2} = \frac{1}{4}\rho_0 v_a^2, \quad (4.23b)$$

$$\Phi(\tilde{\kappa}, \tilde{\rho}) = \frac{1}{3}f_1(\tilde{\kappa}) + \frac{1}{2}f_2(\tilde{\rho}) = \frac{1}{3}\left[\frac{5\tilde{\rho}-2}{2\tilde{\rho}+1} - \tilde{\kappa}\right], \quad (4.23c)$$

where E_{ac} is the acoustic energy density and $\Phi(\tilde{\kappa}, \tilde{\rho})$ is the so-called acoustophoretic contrast factor. The factor $\sin(2kz)$ makes the radiation force period doubled and phase shifted relative to the pressure wave $p_a\cos(kz)$. Note that Φ is positive (negative) if the density-related ratio $(5\tilde{\rho}-2)/(2\tilde{\rho}+1)$ is bigger (smaller) than the compressibility ratio $\tilde{\kappa}$. This sign-difference was used by the Laurell group in its seminal work from 2004 on acoustophoretic separation of red blood cells and lipid particles.[6–8] A sketch of the acoustic radiation force is shown in Figure 4.2.

Most of the parameters needed as input for theoretical calculations can easily be estimated from table values of materials and from the geometry of the given acoustofluidic device. However, the energy density is not so easy to estimate, since the coupling of acoustic energy from the piezo transducer into the fluidic system is hard to predict as discussed in ref. 9 and 10. A typical value for low-voltage ($\lesssim 10$ V) piezo transducers running at a few MHz on silicon/glass chips is[11–14]

$$E_{ac} \approx 10 - 100 \text{ Jm}^{-3}. \tag{4.24}$$

Next, we present experimental validation of the above theory of acoustic radiation force.

4.4 Acoustophoretic Particle Tracks

Basic physical properties of acoustophoresis, such as energy density, local pressure amplitudes, acoustophoretic velocity fields, resonance line shapes, and resonance Q factors, are most easily studied in simple rectangular channels embedded in silicon/glass chips such as those described in ref. 15. Examples of this approach are given in ref. 13 and 14. In both of these papers, the microfluidic chips under study contain a straight channel with one inlet and one outlet. Typical dimensions of the channels are length $\ell = 40$ mm, width $w = 0.38$ mm, and height $h = 0.16$ mm. The particles were liquid suspensions of 5 μm diameter polystyrene microbeads in concentrations ranging from 0.1 g L^{-1} to 0.5 g L^{-1}. The ultrasound frequency is around 2 MHz corresponding to λ around 0.75 mm ensuring the validity of the basic assumption $a \ll \lambda$ of the theory.

In Section 2.5, we have already reviewed the determination in ref. 13 of the acoustic resonance properties, such as the Q-factor and the resonance width. Here we will illustrate the experimental validation of the above theory by briefly reviewing the study of the acoustophoretic particle tracks in ref. 13 and 14.

Assuming the channel is aligned with the x-axis and the ultrasonic standing wave is applied in the transverse z-direction, the path of a microbead moving by acoustophoresis is traced out by the time-dependent co-ordinates $(x(t), z(t))$. A particularly simple analytical expression for the transverse part $z(t)$ of such a path can be obtained from the acoustic radiation force eqn (4.23a) valid for the 1D planar standing $\lambda/2$-wave $p_1(z, t)$, eqn (4.22b). For slowly moving micrometre-sized particles, we can safely neglect inertial effects and determine the transverse path $z(t)$ by balancing the acoustophoretic force F_z^{rad} with the viscous Stokes drag force $F_z^{\text{drag}} = -6\pi\eta a v_p$ from the quiescent liquid. This force balance results in an expression for the position-dependent particle speed v_p,

$$v_p(z) = \frac{2\Phi k a^2 E_{ac}}{3\eta} \sin(2kz). \tag{4.25}$$

Writing $v_p = dz/dt$, the resulting differential equation for the transverse particle path $z(t)$ can be solved analytically by separation in the variables z and t, and using the integral $2\int ds/\sin(2s) = \ln|\tan(s)|$,

$$z(t) = \frac{1}{k}\arctan\left\{\tan[kz(0)]\exp\left[\frac{4\Phi}{3}(ka)^2\frac{E_{ac}}{\eta}t\right]\right\}, \tag{4.26}$$

where $z(0)$ is the transverse position at time $t = 0$.

Acoustic Radiation Force on Small Particles

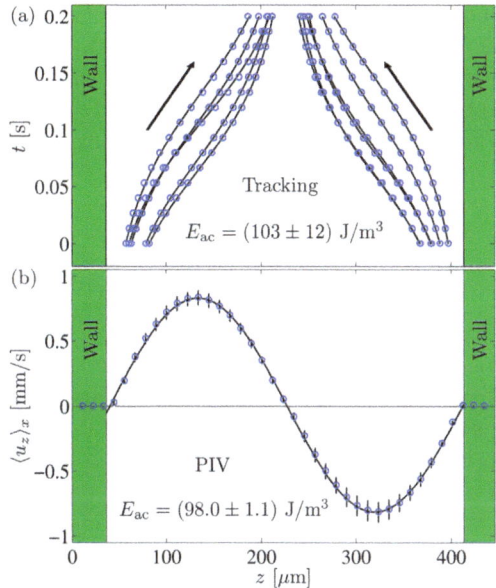

Figure 4.3 Experimental validation of the theory of the acoustic radiation force for 5 μm-diameter polystyrene particles in water ($\tilde{\kappa} = 0.72, \tilde{\rho} = 1.05$, and $\Phi = 0.11$) in a 1D planar standing λ/2-pressure wave. (a) Plot of the transverse path $z(t)$ (blue circles) and corresponding fitting lines (4.26) (black lines) for 16 out of 100 measured particle tracks. For each fit, E_{ac} is the only fitting parameter. The average of all fits resulted in $E_{ac} = (103 \pm 12)$ J m^{-3}. (b) Plot of the average $\langle u_z \rangle_x$ (blue circles) along the channel of the z-component $u_z(x, z)$ of the particle velocity measured by micro-PIV. The data are obtained from the first image-pair in 100 repeated experiments of acoustophoretic focusing, each starting from a homogeneous particle distribution. The acoustophoretic particle velocity v_p of eqn (4.25) is fitted to data with E_{ac} as the only fitting parameter resulting in $E_{ac} = (98.0 \pm 1.1)$ J m^{-3}. Reproduced with permission from ref. 5, see also ref. 14.

Figure 4.3(a) shows the experimental validation of eqn (4.26) from ref. 14. The blue points are data for actual particle paths determined by particle tracking frame by frame from a recorded CCD video of the particle motion. The black lines are fitted to data using eqn (4.26) with only one fitting parameter, the acoustic energy density E_{ac}. The average of the determination of E_{ac} for 100 particle tracks by this method resulted in $E_{ac} = (103 \pm 12)$ J m^{-3}.

Figure 4.3(b) shows micro-particle image velocimetry (micro-PIV) measurements of the transverse acoustophoretic velocity $v_p(z)$. For a microscope field-of-view covering a 0.85 mm-long segment of the 0.38 mm-wide channel, the data are obtained from the first image-pair in 100 repeated experiments of acoustophoretic focusing, each starting from a homogeneous particle distribution. After averaging along the channel length, the data are fitted to the acoustophoretic particle velocity $v_p(z)$ of eqn (4.25) with E_{ac} as the only fitting parameter. The resulting value of $E_{ac} = (98.0 \pm 1.1)$ J m^{-3} is in excellent agreement with, and more precise than, the particle tracking method. These results provide a good validation of the theory.

Inverting the expression, we can also calculate the time t it takes a particle to move from any initial position $z(0)$ to any final position $z(t)$,

$$\begin{aligned} t &= \frac{3\eta}{4\Phi(ka^2)E_{ac}} \ln\left[\frac{\tan[kz(t)]}{\tan[kz(0)]}\right] \\ &= \frac{3}{4\Phi} \frac{c_0^2}{\omega^2 a^2} \frac{\eta}{E_{ac}} \ln\left[\frac{\tan[kz(t)]}{\tan[kz(0)]}\right]. \end{aligned} \quad (4.27)$$

This expression is important for designing acoustofluidic devices to separate particles with acoustophoretic contrast factors Φ with the same sign. In this case, separation must be based on variations in the time $t(w)$ it takes a particle to be focused transversely given the width w of the microfluidic channel. If the axial convection speed of the carrier liquid is v_0, then the distance $\Delta\ell(v_0, w)$ a given particle has to flow along the channel before it has moved the transverse focus distance w can be written as.

$$\Delta\ell(v_0, w) = v_0 t(w) \propto a^{-2}\Phi^{-1} v_0 w^{-2} E_{ac}^{-1}, \quad (4.28)$$

The larger a particle, the shorter it has to be convected before it has been focused. An analysis of the acoustophoretic focus time in terms of the focussing ability is provided in ref. 14.

Figure 4.4 shows measurements by Muller *et al.*[16] of particle tracks in three dimensions: large 5 μm diameter particles and small 0.5 μm diameter particles under the same conditions in a channel with a horizontal half-wave resonance. The large particles are dominated by the acoustic radiation force, and show a clear tendency to simply move directly to the vertical nodal plane

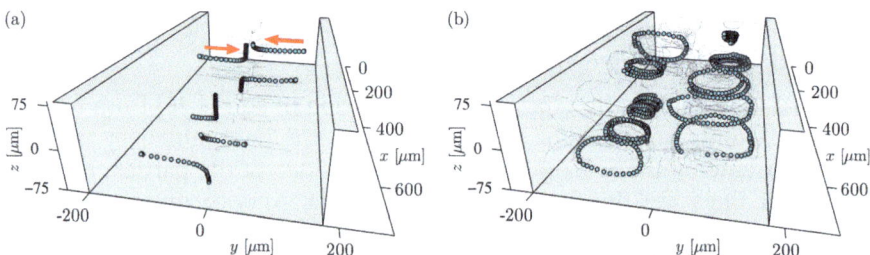

Figure 4.4 Measured particle trajectories (thin lines) obtained using the 3D astigmatism particle-tracking velocimetry technique in a microchannel (grey walls) actuated at the horizontal half-wave resonance. For selected trajectories, the particle positions are represented by dots. (a) Large 5-μm-diameter particles moving (thick arrows) to the vertical center plane $y = 0$ dominated by the radiation force. (b) Small 0.5 μm diameter particles exhibiting circular motion due to acoustic streaming. Reproduced with permission from ref. 16.

in the center of the channel. The small particles are dominated by the drag force from the acoustic streaming rolls and are seen to move in circular paths.

The value of the critical particle diameter $2a_c$ determining the cross-over from radiation force to streaming drag force dominated behavior can be estimated by balancing the two forces, $F^{\text{rad}} = 6\pi\eta a_c v_{\text{str}}$. Using the expressions eqns (2.13) and (4.23a) for v_{str} and F^{rad}, respectively, we arrive at

$$2\pi a_c^3 k\rho_0 v_a^2 \Phi = 6\pi\eta a_c \Psi \frac{v_a^2}{c_0}. \tag{4.29}$$

From this we obtain the expression for the critical particle diameter $2a_c$,

$$2a_c = \sqrt{12\frac{\Psi}{\Phi}\delta} \approx 2\ \mu\text{m}. \tag{4.30a}$$

The value is calculated using $\Psi = 3/8$, valid for a planar wall,[17] and $\Phi = 0.165$, obtained for polystyrene particles in water as used in Figure 4.4.[16] A more detailed theoretical and experimental study of the cross-over as a function of particle size can be found in ref. 18.

4.5 Energy Density as a Function of the Applied Piezo Voltage

In ref. 10, the basic theory of piezoelectric actuation of ultrasonic resonances in water-filled silicon/glass microchannels is presented. It is shown that a linear relation exists between the applied peak-to-peak voltage U_{pp} of the piezo transducer responsible for exciting the ultrasonic resonance and the induced acoustic pressure amplitude p_a. By eqn (4.23b), E_{ac} thus scales with the square of U_{pp}.

$$E_{\text{ac}} \propto p_a^2 \propto U_{\text{pp}}^2. \tag{4.31}$$

This scaling law was tested in ref. 13 by plotting the values of E_{ac} extracted by the above-mentioned particle-tracking method *versus* the applied piezo-transducer voltage U_{pp}. The result is shown in Figure 4.5 for ten values of U_{pp} in the range from 0.4 V to 1.9 V. A power-law fit resulted in the power 2.07, less than 5% from the expected power of 2.

The measurements of the acoustic energy density E_{ac} also allow for a determination of the pressure amplitude p_a. For water at room temperature, we obtain from eqn (4.23b) that

$$p_a = 2\sqrt{\rho_0 c_0^2 E_{\text{ac}}} = 0.094\ \text{MPa}\sqrt{\frac{E_{\text{ac}}}{1\ \text{Jm}^{-3}}}. \tag{4.32}$$

For the energy density $E_{\text{ac}} \approx 100\ \text{J m}^{-3}$, obtained in Figure 4.3, we find $p_a \approx 1$ MPa or 4×10^{-4} times the cohesive energy density 2.6 GPa of water. Equivalently, the density fluctuations are 4×10^{-4} times ρ_0, and thus the acoustic perturbation theory holds even at resonance.

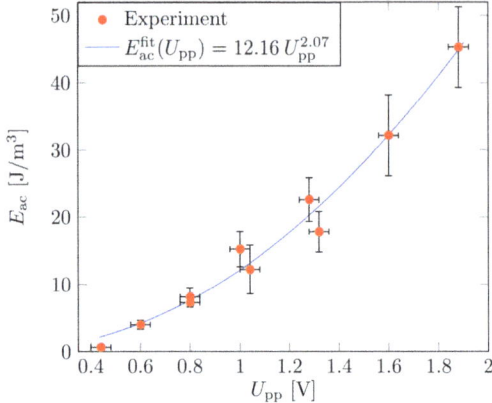

Figure 4.5 Measured acoustic energy density E_{ac} *versus* applied peak-to-peak voltage U_{pp} on the piezo transducer (points) using the particle-tracking method on 5 μm diameter polystyrene particles in water. The power law fit (full line) to the data is close to the expected square law, $E_{ac} \propto U_{pp}^2$. Reproduced with permission from ref. 5, see also ref. 13.

4.6 Viscous Corrections to the Radiation Force

The theory of acoustic radiation force has so far been developed under the assumption of an inviscid fluid. Going back to the perturbation theory reviewed in Section 2.2, this amounts to neglecting the viscous term $\eta \nabla^2 \boldsymbol{v}_1$ relative to $\rho_t \boldsymbol{v}_1$ in the Navier–Stokes equation. Far from any rigid boundaries, this is a good approximation. However, the bulk velocity oscillating at an ultrasound frequency ω must match the no-slip boundary condition at any given rigid wall, and it is well known by momentum diffusion considerations[19,20] that during one oscillation period, the presence of a wall can be felt up to the penetration depth δ, given in terms of the kinematic viscosity $\nu = \eta/\rho_0$ as,

$$\delta = \sqrt{\frac{2\nu}{\omega}} \approx 0.6 \text{ μm}, \quad (4.33)$$

where the value is for 1 MHz ultrasound in water at room temperature. For distances within a few times δ, large velocity gradients may occur in eqn (2.3b), such that $\eta v_1/\delta^2 \gtrsim \rho_0 \partial_t v_1$, and viscosity cannot be neglected. This viscous fluid layer surrounding a given particle is referred to as the acoustic boundary layer. For a particle radius $a \gg \delta$ the boundary layer is of negligible relative size, and the inviscid theory is expected to be a good approximation.

In previous works by Doinikov[21] and by Danilov and Mironov,[22] general theoretical schemes for the radiation force have been developed, but analytical expressions were only provided in the special limits of $a \ll \delta \ll \lambda$ and $\delta \ll a \ll \lambda$. Given the magnitude of δ above, the range of applicability of these published expressions for viscous corrections is severely limited. In recent work by Settnes and Bruus,[4] an analytical expression for the radiation force was derived for any (small) particle size δ, $a \ll \lambda$ using the classic Prandtl–Schlichting boundary-layer theory combined with a stream-function formulation of the acoustic boundary layer.[20] The velocity field in the inviscid bulk is coupled to the motion of the particle through the boundary layer, and not directly as in eqn (4.16) above.

The result of the analysis given in ref. 4 is that the monopole scattering coefficient f_1 is unchanged (the mass scattering and the compressibility are unaffected by viscosity), while f_2 becomes complex-valued,

$$f_2(\tilde{\rho},\tilde{\delta}) = \frac{2[1-\gamma(\tilde{\delta})](\tilde{\rho}-1)}{2\tilde{\rho}+1-3\gamma(\tilde{\delta})}, \text{ with } \gamma(\tilde{\delta}) = -\frac{3}{2}[1+\mathrm{i}(1+\tilde{\delta})]\tilde{\delta}, \text{ and } \tilde{\delta} = \frac{\delta}{a}. \tag{4.34}$$

The viscosity-dependent correction to the final expression (4.21) consists in replacing $f_2(\tilde{\rho})$ by $\mathrm{Re}[f_2(\tilde{\rho},\tilde{\delta})]$. In the inviscid case $\tilde{\delta}=0$, we find $f_2(\tilde{\rho},\tilde{\delta}=0) = f_2(\tilde{\rho})$, as expected, and for neutral-buoyancy particles ($\tilde{\rho}=1$), f_2 is identically zero. As a function of decreasing particle radius a, the value of $f_2(\tilde{\rho},\tilde{\delta})$ saturates asymptotically, and $f_2(\tilde{\rho},\tilde{\delta}\gg 1) = (2/3)(\tilde{\rho}-1)$.

Using tabulated values of the material parameters, it is found that the relative change in the acoustic contrast factor Φ is about 1% or less for 5-μm-diameter, near-neutral-buoyancy polystyrene particles in water, but as much as 50% for pyrex glass particles of the same size.[4]

We have not discussed size effects for particles with radius a comparable to or larger than the acoustic wavelength λ. A good entry to such studies is the theoretical analysis by Hasegawa.[23] Another aspect not touched upon here, and which also needs more studies, is particle–particle interactions. We have only studied the single-particle theory. However, at least two effects play a role as the concentration of the suspended particles is increased. One is hydrodynamic interaction, where one particle feels the Stokes drag from the wake produced by the motion of another particle. A very good and general introduction is given in the textbook by Happel and Brenner,[24] while an explicit example of many-particle effects in microchannel magnetophoresis as a function of concentration is given by Mikkelsen and Bruus.[25] In the coming years, more work may well appear on high particle-concentration acoustophoresis and its application to biomedical samples.

References

1. L. V. King, *Proc. R. Soc. London, Ser. A,* 1934, **147**, 212–240.
2. K. Yosioka and Y. Kawasima, *Acustica,* 1955, **5**, 167–173.
3. L. P. Gorkov, *Phys.-Dokl.,* 1962, **6**, 773–775.
4. M. Settnes and H. Bruus, *Phys. Rev. E,* 2012, **85**, 016327.
5. H. Bruus, *Lab Chip,* 2012, **12**, 1014–1021.

6. H. Jönsson, C. Holm, A. Nilsson, F. Petersson, P. Johnsson and T. Laurell, *Ann. Thoracic Surg.*, 2004, **78**, 1572–1578.
7. F. Petersson, A. Nilsson, C. Holm, H. Jönsson and T. Laurell, *Analyst*, 2004, **129**, 938–943.
8. F. Petersson, A. Nilsson, C. Holm, H. Jönsson and T. Laurell, *Lab Chip*, 2005, **5**, 20–22.
9. J. Dual and T. Schwarz, *Lab Chip*, 2012, **12**, 244–252.
10. J. Dual and D. Möller, *Lab Chip*, 2012, **12**, 506–514.
11. M. Wiklund, P. Spégel, S. Nilsson and H. M. Hertz, *Ultrasonics*, 2003, **41**, 329–333.
12. J. Hultström, O. Manneberg, K. Dopf, H. M. Hertz, H. Brismar and M. Wiklund, *Ultrasound Med. Biol.*, 2007, **33**, 145–151.
13. R. Barnkob, P. Augustsson, T. Laurell and H. Bruus, *Lab Chip*, 2010, **10**, 563–570.
14. P. Augustsson, R. Barnkob, S. T. Wereley, H. Bruus and T. Laurell, *Lab Chip*, 2011, **11**, 4152–4164.
15. A. Lenshof, M. Evander, T. Laurell and J. Nilsson, *Lab Chip*, 2012, **12**, 684–695.
16. P. B. Muller, M. Rossi, A. G. Marín, R. Barnkob, P. Augustsson, T. Laurell, C. J. Kähler and H. Bruus, *Phys. Rev. E*, 2013, **88**, 023006.
17. P. B. Muller, R. Barnkob, M. J. H. Jensen and H. Bruus, *Lab Chip*, 2012, **12**, 4617–4627.
18. R. Barnkob, P. Augustsson, T. Laurell and H. Bruus, *Phys. Rev. E.*, 2012, **86**, 056307.
19. L. Rayleigh, *Philos. Trans. R. Soc. London*, 1884, **175**, 1–21.
20. L. D. Landau and E. M. Lifshitz, *Fluid Mechanics, Course of Theoretical Physics*, Pergamon Press, Oxford, 2nd edn., vol. 6, 1993.
21. A. A. Doinikov, *J. Acoust. Soc. Am.*, 1997, **101**, 722–730.
22. S. D. Danilov and M. A. Mironov, *J. Acoust. Soc. Am.*, 2000, **107**, 143–153.
23. T. Hasegawa, *J. Acoust. Soc. Am.*, 1977, **61**, 1445–1448.
24. J. Happel and H. Brenner, *Low Reynolds Number Hydrodynamics with Special Applications to Particulate Media*, Martinus Nijhoff Publishers, The Hague, 1983.
25. C. Mikkelsen and H. Bruus, *Lab Chip*, 2005, **5**, 1293–1297.

CHAPTER 5

Piezoelectricity and Application to the Excitation of Acoustic Fields for Ultrasonic Particle Manipulation

JÜRG DUAL* AND DIRK MÖLLER

Institute of Mechanical Systems, Department of Mechanical and Process Engineering, ETH Zentrum, CH-8092 Zurich, Switzerland
*E-mail: Dual@imes.mavt.ethz.ch

5.1 Introduction

Exciting and detecting motion in solids and fluids by means of piezoelectric materials, in which an electric signal is converted into a mechanical motion and *vice versa*, has been used extensively for ultrasonic particle manipulation. This method has several advantages:

- With the availability of programmable signal generators, waves of complex shape and frequency content can be produced with a high degree of repeatability, which opens the way to various signal processing techniques.
- By appropriate tailoring of the transducer set-up, different modes of waves and vibrations can be excited and measured selectively.

A necessary condition to make use of the above-mentioned advantages is the availability of materials with sufficiently high piezoelectric constants. This is the case in piezoelectric ceramics. While quartz has a piezoelectric charge constant of about 10^{-12} C/N, the same constant for Pz26, which is a modified lead zirconate titanate (PZT) manufactured by Ferroperm Piezoceramics, amounts to 10^{-10} C/N.[1] The interaction between the piezoelectric transducer and the attached material will be described in the following section. Basics regarding the theory can be found in a book by Ristic.[2]

Piezoelectric materials have been used in ultrasonic particle manipulation from the very beginning. Red blood cells were segregated using planar ultrasound transducers.[3] Standing and travelling waves were used to manipulate cells for a number of applications.[4] The acoustic energy can be focussed by special non-planar transducers.[5,6] Modes can be selectively excited or two-dimensional patterns are formed by segmented transducers.[7,8] As an alternative, miniaturized transducers generate patterns of ultrasonic traps,[9] an approach that will be of increasing importance in view of improved methods to produce batch fabricated systems with piezoelectric elements. Shear transducers are suitable to excite bending vibrations in cover plates,[10] or piezoelectric LiNb crystals might be brought directly into contact with the fluid.[11] They are often used to generate surface acoustic waves by interdigitated electrodes.

A typical transducer as used in ultrasonics is shown in Figure 5.1. It is brought in contact with the structure to be excited using a coupling agent or glue. Classical ultrasonic transducers work in a resonant mode, *i.e.* they have a defined frequency (resonance) at which they work best. Other transducers are used far below their resonance frequency and give a more broad band excitation.

When used in a continuous mode, one has to be careful not to heat up the transducer too much. This is particularly true for ceramic transducers. Heat is generated by the mechanical and electrical damping of the piezoelectric material, which is typically much larger for PZT than for quartz. PZT loses its polarization when heated above its Curie temperature.

5.2 Basic Equations

Piezoelectric materials have an intrinsic polarization density **P**. Depending on the direction of the applied electric field **E**, which is controlled by the electrodes, extension/contraction occurs (for **E** parallel to **P**) or shear (for **E** orthogonal to **P**). In the following, the linear theory of piezoelectric materials, which is suitable for a first order approach to ultrasonic particle

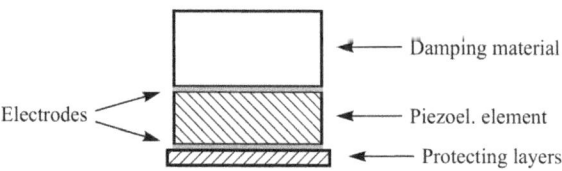

Figure 5.1 Characteristic layout of a mounted piezoelectric element. Reproduced from ref. 16.

manipulation, is described. Therefore, the constitutive law is assumed to be linear. Nonlinear terms in the strain displacement relation are also neglected.

Constitutive relations for piezoelectric materials relate the four quantities (given together with their SI units)

E_i	Electric field	1st order tensor	N C^{-1}
D_i	Electric displacement	1st order tensor	C m^{-2}
γ_{ij}	Mechanical strain	2nd order tensor	—
σ_{ij}	Mechanical stress	2nd order tensor	N m^{-2}

by the material properties

s_{Eijkl}	Mechanical compliance for constant electric field	4th order tensor	m^2 N^{-1}
d_{ijk}	Piezoelectric charge constant	3rd order tensor	N C^{-1} or m V^{-1}
$\varepsilon_{\sigma ij} = \varepsilon_{ij}$	Permittivity at constant mechanical stress	2nd order tensor	C^2 m^{-2} N^{-1}

The indicial notation is used, referring all quantities to an orthonormal base system with coordinates x_i, $i = 1,2,3$. Einstein's summation convention for repeated indices is invoked.

The electric displacement is defined as

$$D_i = \varepsilon_0 E_i + P_i, \tag{5.1}$$

where $\varepsilon_0 = 8.85 \; 10^{-12}$ C^2 (Nm^2)$^{-1}$ is the vacuum permittivity. Eqn (5.1) expresses the fact that the electron cloud shifts with respect to the nucleus under the influence of an external electric field.

Constitutive equations for a piezoelectric material can then be given in the form

$$\gamma_\lambda = s_{E\lambda\mu}\sigma_\mu + d_{k\lambda}E_k,$$
$$D_i = d_{i\mu}\sigma_\mu + \varepsilon_{ik}E_k, \tag{5.2}$$

where for simplicity Greek indices are introduced as in Dual et al.[12] Greek indices take values from 1 to 6, and the correspondence between Greek matrix indices and pairs of Latin tensor indices is given by

ij	11	22	33	23, 32	13, 31	12, 21
λ	1	2	3	4	5	6

It should be noted that piezoelectric ceramics have an axis of symmetry, which is parallel to the direction of poling. The material is transversely isotropic. It is customary to take the 3-direction as the axis of symmetry. The permittivity tensor is then diagonal with $\varepsilon_{11} = \varepsilon_{22}$. The tensor containing the charge constants is zero except for the elements d_{33}, $d_{31} = d_{32}$ and $d_{15} = d_{24}$.

Typical values for the material constants in extension and shear for Pz26 are summarized in Table 5.1.

Other materials have other material symmetries depending on their crystal structure.[2]

Particle manipulation is in most cases done using harmonic time signals. If the excitation voltage has the form $V = \text{Re}(V_0 e^{i\omega t})$, all other quantities will also have the same time-dependence because of the linearity of the equations.

Therefore all quantities will have a time dependence of the form

$$A(\mathbf{r}, t) = A(\mathbf{r})e^{i\omega t}. \tag{5.3}$$

In the following the exponential factor will be omitted, and it is understood that of all the quantities, only the real part will have physical meaning.

The equation of mechanical equilibrium for the case of harmonic loading is then

$$\sigma_{ij,j} + \rho\omega^2 u_i = 0, \tag{5.4}$$

where $\mathbf{u}$, ρ and ω are the mechanical displacement, density and circular frequency, respectively, and $\sigma_{ij,j}$ is the derivative with respect to x_j.

The mechanical strain is defined as

$$\gamma_{ij} = \frac{1}{2}(u_{i,j} + u_{j,i}),$$
$$2\gamma_{ij} = (1 + \delta_{ij})\gamma_\lambda. \tag{5.5}$$

Maxwell's first equation for dielectric materials is given as

$$D_{i,i} = 0. \tag{5.6}$$

The right hand side is zero, because there are no free charges in a dielectric material. Eqns (5.2)–(5.6) are the basic equations that need to be solved together with appropriate boundary conditions for the mechanical and electrical quantities. Normally, the stress vector $s_i = \sigma_{ij} n_j$ and displacement u_i are assumed to be continuous across the interface of transducer and

Table 5.1 Material constants for Pz26 piezoelectric ceramics.[a]

	Units	Extension	Shear in plane orthogonal to P
s_E	10^{-12} m^2 N^{-1}	$s_{E33} = 19.6$	$s_{E55} = 33.2$
s_D	10^{-12} m^2 N^{-1}	$s_{D33} = 10.5$	$s_{D55} = 23.1$
d	10^{-10} C N^{-1}	$d_{33} = 3.28 d_{31} = -1.28$	$d_{15} = 3.27$
ε	10^{-8} C^2 (m^{2}N)	$\varepsilon_{33} = 1.17$	$\varepsilon_{22} = \varepsilon_{11} = 1.06$
g	10^{-3} m^2 C^{-1}	$g_{33} = 28.0 g_{31} = -10.9$	$g_{15} = 38.9$
ρ	10^3 kg m^{-3}	$\rho = 7.7$	

[a]Mechanical compliances $s_{E\lambda\mu}$, $s_{D\lambda\mu}$, charge constant $d_{i\lambda}$, permittivity ε_{ij}, voltage constant $g_{i\lambda}$, density ρ.

structure. If an electrical voltage V is applied across the electrodes a and b of the piezoelectric element, then

$$V = -\int_a^b \mathbf{E} \cdot d\mathbf{r} \qquad (5.7)$$

is the electrical boundary condition. The integral is a curve integral from a to b. The potential is spatially constant on electrodes, as they are conductive. On the other hand, for surfaces with no electrodes

$$\mathbf{D} \cdot \mathbf{n} = 0 \qquad (5.8)$$

is the electrical boundary condition, where $\mathbf{n}$ is the unit normal.

For simplicity, a uniaxial state will be assumed for both electrical and mechanical quantities in the analytical examples of sect. 5.3 and 5.4. All coupling effects with other components of stress, strain, electric displacement and electric field will be neglected and also indices will be dropped for simplicity of writing. For more complicated situations, a finite element analysis is necessary (sect. 5.5).

Eqn (5.2) can be rewritten with σ and D as independent quantities:

$$\gamma = s_D \sigma + gD,$$
$$E = -g\sigma + \frac{D}{\varepsilon}, \qquad (5.9)$$

where $g = d/\varepsilon$ is the piezoelectric voltage constant and $s_D = s_E - d^2/\varepsilon = s_E - gd$ is the mechanical compliance for constant electric displacement.

In the uniaxial case, one obtains from eqn (5.6)

$$D = D_0, \qquad (5.10)$$

where D_0 is constant.

5.3 Vibration of a Free Piezoelectric Element Excited by an Applied Electrical Voltage

Referring to Figure 5.2, the axial motion u of a slender transducer ($L \gg D$) with stress-free boundaries is considered. The stress state is uniaxial and the influence of lateral inertia is neglected in accordance with the theory for a slender bar.

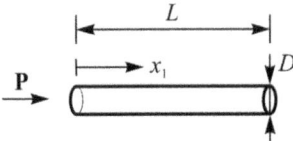

Figure 5.2 Slender transducer ($L \gg D$) with stress-free boundaries and polarization density **P**. Reproduced from ref. 16.

Using eqn (5.4)–(5.10), one obtains the differential equation and boundary conditions

$$u_{,11} + k_D^2 u = 0,$$
$$k_D^2 = s_D \rho \omega^2,$$
$$u_{,1}(0) = gD_0,$$
$$u_{,1}(L) = gD_0,$$
(5.11)

which yield

$$u = \frac{gD_0}{k_D}\left(\sin(k_D x) + \frac{\cos(k_D L) - 1}{\sin(k_D L)}\cos(k_D x)\right),$$
(5.12)

where k_D is the wavenumber and D_0 will be determined later. From eqn (5.5) and (5.9), the corresponding σ and E are computed

$$\sigma = \frac{2gD_0}{s_D}\left(\frac{\sin\left(\frac{1}{2}k_D(L-x)\right)}{\cos\left(\frac{k_D L}{2}\right)}\sin\left(\frac{k_D x}{2}\right)\right),$$

$$E = \frac{D_0}{\varepsilon} - \frac{2g^2 D_0}{s_D}\left(\frac{\sin\left(\frac{1}{2}k_D(L-x)\right)}{\cos\left(\frac{k_D L}{2}\right)}\sin\left(\frac{k_D x}{2}\right)\right).$$
(5.13)

The electric potential difference is given as

$$V = -\int_0^L E dx = -\frac{D_0 L}{\varepsilon} + \left(\frac{D_0 g^2\left(2\tan\left(\frac{1}{2}k_D L\right) - k_D L\right)}{s_D k_D}\right).$$
(5.14)

For a given applied voltage V_0, the resulting D_0 is inserted into eqn (5.12) to yield

$$u(0) = V_0 \frac{ds_D}{2s_E} \frac{\tan\left(\frac{1}{2}k_D L\right)}{\frac{1}{2}k_D L - k^2 \tan\left(\frac{1}{2}k_D L\right)},$$
(5.15)

where $k^2 = gd/s_E$ is the electromechanical coupling coefficient, which describes how well electrical energy is converted into mechanical energy.

If in addition we impose $k_D L \ll 1$, eqn (5.12) to (5.15) can be simplified to yield for the quasistatic or "long wavelength" case:

$$D_0 = -\frac{\varepsilon V}{L}(1 + O((k_D L)^2)),$$

$$u = dV\left(\frac{1}{2} - \frac{x}{L}\right)(1 + O((k_D L)^2)),$$

$$\sigma = 0 + O((k_D L)^2),$$

$$E = -\frac{V}{L}(1 + O((k_D L)^2)),$$
(5.16)

where $O(k_D L)$ denotes terms of order $k_D L$ and smaller. The transducer behaves electrically as a capacitor. Mechanically, it contracts with constant strain, while the center of mass remains unmoved. The total length change ΔL is given by dV.

5.4 Piezoelectric Transducers Used to Excite Mechanical Vibrations in a Structure

In the next step, a situation is considered in which the element of sect. 5.3 excites longitudinal motion in a circular rod (Figure 5.3).

The transfer function between excitation voltage and the resulting motion is determined for the long wavelength case, i.e. the wavelength both in the transducer and in the rod is assumed to be much larger than the diameter of the rod and transducer. Again a uniaxial state of stress results for all elements involved.

To develop the equations the same procedure is used as before. The only difference is the boundary conditions, which are now

$$u_{,1}(0) = gD_0,$$
$$u_{,1}(L) = s_D \sigma_0 + gD_0. \tag{5.17}$$

σ_0 is the stress between transducer and rod and can be expressed in terms of $u_0 = u(L)$ and the mechanical impedance Z_R of the rod.

$$\sigma_0 = i\omega Z_R u_0. \tag{5.18}$$

Dependent on the type of motion, the mechanical impedance is

(a) for a wave propagation problem in an infinite rod

$$u = u_0 e^{-ik_R x_R}, \quad k_R = \frac{\omega}{c_R}, \quad c_R = \sqrt{\frac{E_R}{\rho_R}},$$
$$\sigma_0 = -iE_R k_R u_0, \tag{5.19}$$
$$Z_R = -\sqrt{E_R \rho_R},$$

where E_R and ρ_R are Young's modulus and density of the rod, respectively.

(b) In ultrasonic manipulation devices, standing waves are often used, which correspond to system resonances. Therefore, a resonant vibration of the rod is considered next. For a vibration problem in a rod of length L_R and stress-free boundary at $x_R = L_R$

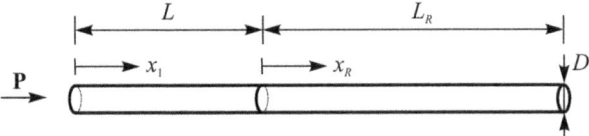

Figure 5.3 Slender transducer ($L \gg D$) with length L and polarization density **P** coupled to a circular rod of equal diameter and length L_R. Reproduced from ref. 16.

$$u = u_0(\cos(k_R x_R) + \tan(k_R L_R)\sin(k_R x_R)),$$
$$\sigma_0 = Ek_R u_0 \tan(k_R L_R), \quad (5.20)$$
$$Z_R = -i\sqrt{E_R \rho_R} \tan(k_R L_R).$$

One should note at this point that the mechanical impedance gets very small for the case of resonance ($k_R L_R = n\pi$), depending on the amount of damping.

For a weakly viscoelastic material in harmonic loading, a complex modulus is introduced[12]

$$E_R = E_0(1 + i\varphi), \quad \varphi \ll 1 \quad (5.21)$$

$$k_R = k_0(1 - i\varphi/2), \quad k_0 = \frac{\omega}{c_0}, \quad c_0 = \sqrt{\frac{E_0}{\rho}}, \quad (5.22)$$

where φ is the loss angle of the rod material. In the case of resonance, where $\text{Im}(Z_R) = 0$, and assuming $k_0 L_R \varphi \ll 1$

$$Z_R = -\sqrt{E_0 \rho_R} \, k_0 L_R \varphi / 2. \quad (5.23)$$

From eqn (5.17) and (5.18) and using the same procedure as in Section 5.3, one obtains

$$D_0 = \frac{V_0 k_D s_D}{g^2 \beta},$$
$$\beta = \sin(k_D L) + \alpha(\cos(k_D L) - 1) - k_D L\left(1 + \frac{s_D}{dg}\right),$$
$$\alpha = \frac{\cos(k_D L) - 1 - i\frac{Z_R}{Z_p}\sin(k_D L)}{\sin(k_D L) + i\frac{Z_R}{Z_p}\cos(k_D L)}, \quad (5.24)$$

where Z_p is the characteristic mechanical impedance of the piezoelectric material

$$Z_p = \sqrt{\frac{\rho}{s_D}}, \quad (5.25)$$

and

$$u = \frac{V_0 s_D}{g\beta}(\sin(k_D x) + \alpha\cos(k_D x)),$$
$$\sigma = \frac{V_0 k_D}{g\beta}(\cos(k_D x) - 1 - \alpha\sin(k_D x)),$$
$$E = -\frac{V_0 k_D}{\beta}\left(\cos(k_D x) - \alpha\sin(k_D x) - \left(1 + \frac{s_D}{dg}\right)\right), \quad (5.26)$$
$$G(\omega) := \frac{u(L)}{V_0}$$

where G denotes the transfer function between applied voltage and displacement at the transducer interface. The electrical impedance Z as seen by the electrical network is computed from eqn (5.24) by

$$Z = \frac{V_0}{i\omega A D_0} = \frac{\beta g^2}{i\omega A k_D s_D}, \qquad (5.27)$$

where A is the electrode area of the transducer. In the limit $k_D L \ll 1$, the above equations reduce to

$$D_0 = -\frac{V_0 \varepsilon}{L \Psi},$$

$$u = -\frac{V_0 d}{\Psi}\left(\frac{x}{L} - \gamma\right),$$

$$\sigma = \frac{V_0 d}{\Psi}\rho\omega^2 L\left(\frac{x^2}{2L^2} - \gamma\frac{x}{L}\right),$$

$$E = -\frac{V_0}{\Psi L}\left(1 - \gamma\frac{dg}{s_D}k_D^2 L x\right), \qquad (5.28)$$

$$\gamma = \frac{\frac{k_D L}{2} + i\frac{Z_R}{Z_p}}{k_D L + i\frac{Z_R}{Z_p}},$$

$$\Psi = 1 - \gamma\frac{dg}{s_D}\frac{(k_D L)^2}{2}.$$

If the transducer is used to excite waves, Z_R and Z_p have the same order of magnitude. Eqn (5.28) can be simplified and the transfer function $G(\omega)$ can be calculated.

$$\Psi = 1,$$

$$\gamma = 1 + \frac{i}{2}\frac{Z_p}{Z_R}k_D L - \frac{1}{2}\left(\frac{Z_p}{Z_R}k_D L\right)^2, \qquad (5.29)$$

and

$$G(\omega): = \frac{u(L)}{V_0} = \frac{d}{2}\frac{\omega\rho L}{\sqrt{E_R \rho_R}}\left(i - \frac{\omega\rho L}{\sqrt{E_R \rho_R}}\right). \qquad (5.30)$$

This expression for the transfer function can also be obtained directly by considering a rigid mass (transducer), the center of which is displaced by an amount $\Delta L/2 = dV_0/2$ relative to the interface with the rod. One obtains

$$G(\omega) = \frac{d}{2}\frac{\omega}{\omega + i\frac{Z_R}{\rho L}}, \qquad (5.31)$$

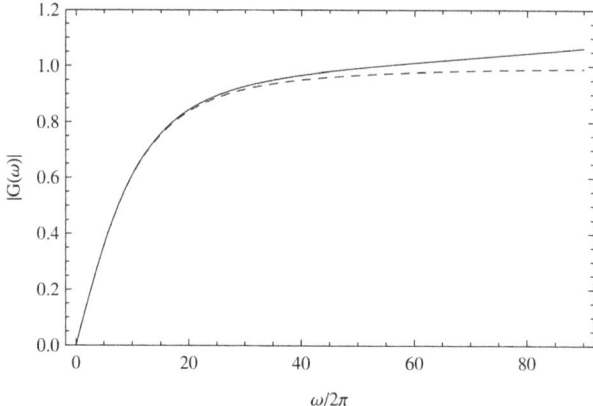

Figure 5.4 Normalized magnitude of the transfer function for the excitation of waves in an infinite rod according to eqn (5.24) to (5.26) (solid) and eqn (5.31) (dashed) for the low frequency range. Transducer: Pz26, $L = 0.004$ m; Rod: lucite with $E_R = 5.29 \cdot 10^9$ N m^{-2}, $\rho_R = 1.2 \cdot 10^3$ kg m^{-3}. Reproduced from ref. 16.

and with eqn (5.19)

$$G(\omega) = \frac{d}{2} \frac{\omega}{\omega + i\omega_C}, \qquad (5.32)$$

where $\omega_C = \dfrac{\sqrt{E_R \rho_R}}{\rho L}$.

This result is equivalent to eqn (5.30) for $\omega \ll \omega_C$ and represents a high-pass behavior with ω_C as the cut-off frequency. The magnitude of the transfer function for a typical configuration of transducer and rod is given in Figure 5.4 for the exact solution according to eqn (5.24)–(5.26) and the rigid mass approximation of eqn (5.31). The magnitude is normalized to yield unity in the high frequency limit of eqn (5.31).

Up to a value of about twice the cut-off frequency of 12.7 kHz, eqn (5.31) is an excellent approximation. In this frequency range, the transducer works off resonance resulting in low electromechanical coupling. For higher frequencies, the effect of the transducer resonance is noticeable as an increase of the displacement amplitude.

A frequency range up to the first transducer resonance frequency is shown in Figure 5.5. The amplitude gets very large for the resonance frequency of the transducer. These effects adversely affect wave propagation experiments, when it is desirable to have excitation that is constant over a wide frequency range. Digital filtering can be used to overcome the problem. On the other hand it is beneficial for cases where a high amplitude is sought at minimal applied voltage, which is often the case for ultrasonic manipulation devices.

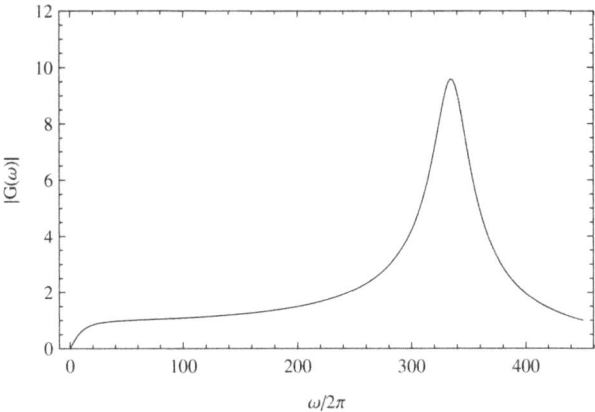

Figure 5.5 Normalized (as in Figure 5.4) magnitude of the transfer function for the excitation of waves in an infinite rod according to eqn (5.24) to (5.26) up to the first transducer resonance. Transducer: Pz26, $L = 0.004$ m; rod: lucite with $E_R = 5.29 \cdot 10^9$ N m^{-2}, $\rho_R = 1.2 \cdot 10^3$ kg m^{-3}. Reproduced from ref. 16.

If the transducer is used to excite resonance vibrations, the situation is different. In eqn (5.23), it was shown that the impedance of the rod Z_R in the vicinity of resonance is proportional to the loss angle φ and can be quite small. If we impose the condition for resonance

$$\mathrm{Re}\left\{\frac{u(L)}{V_0}\right\} = 0, \tag{5.33}$$

assuming that $\varphi \ll 1$, $k_D L \ll 1$ and using the expression from eqn (5.20) for the impedance of the rod, a modified characteristic equation for resonance is obtained

$$\varepsilon^* k_0 L_R + \tan(k_0 L_R) = 0, \tag{5.34}$$

with $\varepsilon^* = \dfrac{\text{mass of the transducer}}{\text{mass of the rod}}$.

This corresponds to the characteristic equation for the resonance of a rod with an attached rigid mass. Using eqn (5.28) and (5.34), it can be shown that

$$\gamma = 1 - i\frac{\rho L}{\rho_R L_R \varphi} = 1 - i\frac{\varepsilon^*}{\varphi}, \tag{5.35}$$

at resonance. For low damping φ, γ becomes very large and mechanical quantities in the transducer are completely dominated by a rigid mass type behavior: u is constant and the stress is linearly distributed.

Eqn (5.28) can only be simplified further for sufficiently high damping. Because dg/s_D is of the order 1, one is allowed to set

$$\Psi = 1 \quad \text{for} \quad \varphi \gg \varepsilon^*(k_D L).$$

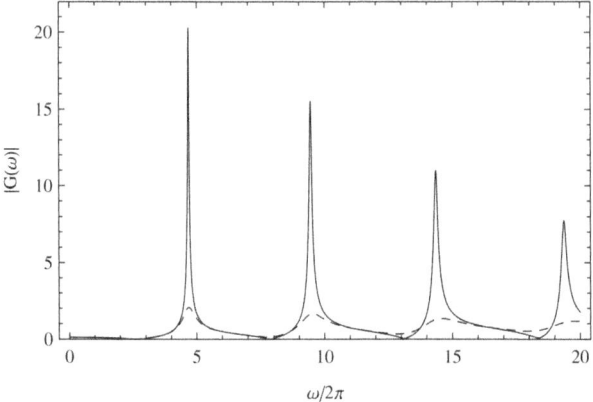

Figure 5.6 Normalized (as in Figure 5.4) magnitude of the transfer function according to eqn (5.24) to (5.26) for excitation of a resonant rod at low frequencies and two values of the damping φ: $\varphi = 0.01$ (solid) and $\varphi = 0.1$ (dashed). Transducer: Pz26, $L = 0.004$ m; rod: lucite with $E_R = 5.29 \cdot 10^9$ N m^{-2}, $\rho_R = 1.2 \cdot 10^3$ kg m^{-3}, $L_R = 0.2$ m. Reproduced from ref. 16.

For this case, the mechanical motion does not influence the electrical circuit, *i.e.* the coupling is low. Again the rigid mass approximation of eqn (5.31) yields the same result. On the other hand, if the damping is too small, very little energy pumped into the system will produce very large displacements, which in turn change the electric displacement. Then, no simplification of eqn (5.28) is possible.

The transfer function for a transducer that excites resonant vibrations is shown in Figure 5.6. It is completely dominated by the longitudinal resonances in the rod and the amount of damping present.

If we extend the frequency range and take a smaller rod length, we obtain Figure 5.7. At low frequencies, the peaks of the rod are visible, then they diminish in magnitude because of damping and increase again, because of the transducer resonance.

When characterizing devices for micromanipulation, very often the electrical impedance Z (eqn (5.27)) or admittance $1/Z$ for the transducer are plotted as shown in Figure 5.8 and Figure 5.9. It is seen that the low frequency peaks almost disappear, because of the low electromechanical coupling far away from the transducer resonance.

If we increase the damping ten times, all the system resonance peaks disappear and only the transducer resonance remains as seen in Figure 5.10.

Piezoelectric elements usually have a much greater impedance than water or plastics. In order to maximize energy transmission into the structure as well as the pressure amplitude in the cavity, impedance matching between the piezoelectric element and water using a suitable layer of matching material is a possibility. These layers usually have a thickness of $\lambda/4$, *i.e.* when

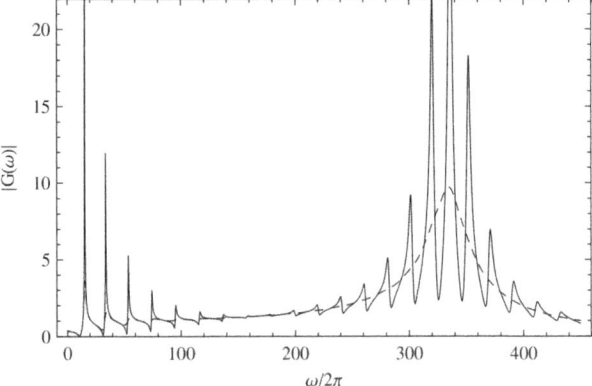

Figure 5.7 Normalized (as in Figure 5.4) magnitude of the transfer function according to eqn (5.24) to (5.26) for excitation of a resonant rod at higher frequencies and two values of the damping φ: $\varphi = 0.01$ (solid) and $\varphi = 0.1$ (dashed). Transducer: Pz26, $L = 0.004$ m; rod: lucite with $E_R = 5.29 \times 10^9$ N m^{-2}, $\rho_R = 1.2 \times 10^3$ kg m^{-3}, $L_R = 0.05$ m. Reproduced from ref. 16.

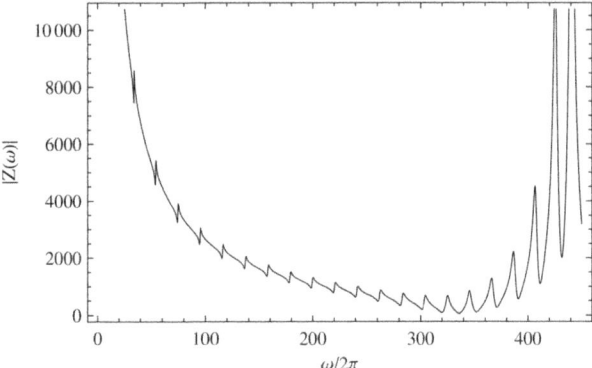

Figure 5.8 Electrical impedance magnitude plot of the transducer according to eqn (5.24) to (5.27) for excitation of a resonant rod at higher frequencies and damping $\varphi = 0.01$. Transducer: Pz26, $L = 0.004$ m; rod: lucite with $E_R = 5.29 \cdot 10^9$ N m^{-2}, $\rho_R = 1.2 \cdot 10^3$ kg m^{-3}, $L_R = 0.05$ m. Reproduced from ref. 16.

designing the layer the frequency must be known already. The specific impedance Z_2 of this layer can be determined with the Collin[13] or Desilets[14] model.

$$Z_2 = \sqrt{Z_1 Z_3} \qquad (5.36)$$

Multilayers might give an even better performance, at the expense of more complicated manufacturing.

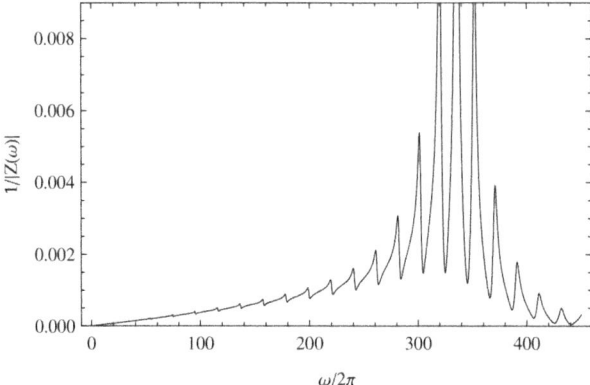

Figure 5.9 Electrical admittance magnitude plot of the transducer according to eqn (5.24) to (5.27) for excitation of a resonant rod at higher frequencies and damping $\varphi = 0.01$ Transducer: Pz26, $L = 0.004$ m; rod: lucite with $E_R = 5.29 \cdot 10^9$ N m^{-2}, $\rho_R = 1.2 \cdot 10^3$ kg m^{-3}, $L_R = 0.05$ m. Reproduced from ref. 16.

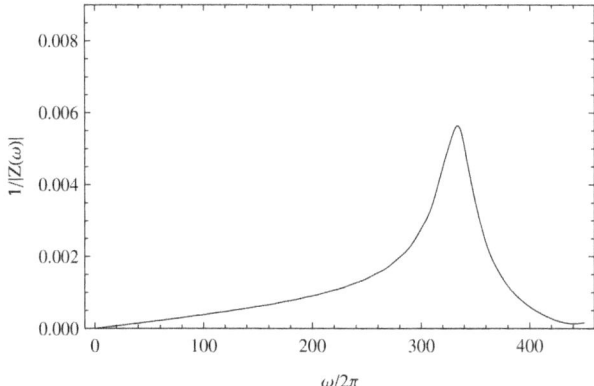

Figure 5.10 Electrical admittance magnitude plot of the transducer according to eqn (5.24) to (5.27) for excitation of a resonant rod at higher frequencies and damping $\varphi = 0.1$ Transducer: Pz26, $L = 0.004$ m; rod: lucite with $E_R = 5.29 \cdot 10^9$ N m^{-2}, $\rho_R = 1.2 \cdot 10^3$ kg m^{-3}, $L_R = 0.05$ m. Reproduced from ref. 16.

This section has shown that even for a simple one-dimensional system there is a complex interplay between the transducer resonances and the system resonances that one wants to excite. Often in a device, these are tailored to coincide approximately in order to obtain maximum efficiency. However, as the transducer resonance depends on the structural impedance, this might not be an easy task. In addition, damping both within the system and within the piezoelectric element plays an important role.

5.5 FEM Model Example of an Ultrasonic Cavity used for Particle Manipulation

A typical ultrasonic standing wave device consists of a fluidic domain, a surrounding mechanical structure and a piezoelectric transducer. For more complex models, *e.g.* when going beyond 1D, a numerical simulation is a very strong tool. For the fluidic domain, the acoustics problem needs to be solved. The solid mechanical parts are described with a linear elastic material model and both the fluidics and the mechanical structure can be coupled with a fluid structure interaction. Such a model can be further extended with a piezoelectric material model.

While a numerical model is useful to find the resonance frequencies of a system, it is also a strong tool to study and understand the response of the system. This can then be used to design or optimize the system further *e.g.* by performing parametric studies with different electrode layouts and different geometrical dimensions.

The resonating system model presented here is a typical microfluidic device for ultrasonic manipulation.[7] It consists of four domains as shown in Figure 5.11. The base is a 5 mm × 0.5 mm piece of silicon with a 3 mm × 0.2 mm cavity filled with water. The cavity is sealed on top with a glass plate with the dimensions 5 mm × 0.5 mm. The transducer plate is attached to the silicon just below the fluidic cavity. It has the same width as the cavity with final dimensions of 3 mm × 0.5 mm. This piezoelectric element is driven in longitudinal mode with a strip of electrode, *i.e.* one part of one electrode is connected to the drive signal, while the rest is grounded. Complicated patterns of electrodes can be produced photolithographically.

The finite element method simulation presented here was performed in 2D and was done with COMSOL Multiphysics 4.2. For the four domains, the following material properties have been used, where damping parameters are included as in eqn (5.21) with complex elastic constants.[15] Silicon is an anisotropic material with a density of 2330 kg m^{-3} and components of the stiffness tensor of $c_{11} = 165.7$ GPa, $c_{12} = 63.9$ GPa and $c_{44} = 79.6$ GPa. Because in silicon the damping is very low, it is neglected here. The glass has a Young's modulus of $63(1 + i/400)$ GPa with complex damping factor, a Poisson's ratio of $\nu = 0.2$ and a density of 2220 kg m^{-3}. Water has a density

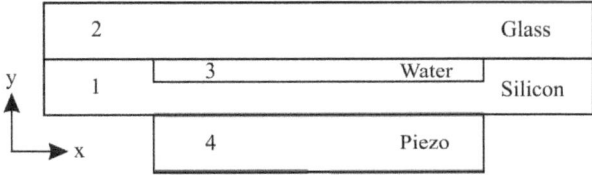

Figure 5.11 Model of a microfluidic device for ultrasonic manipulation with electrode configuration (red = 20 V, blue = ground). Reproduced from ref. 16.

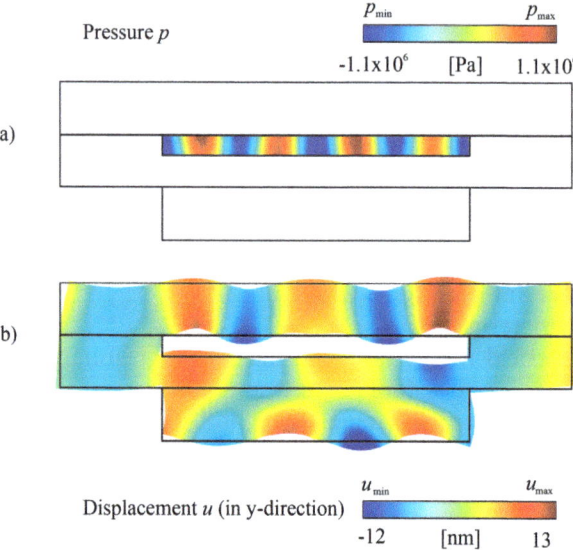

Figure 5.12 (a) Pressure field in a fluid cavity at 1.9 MHz; (b) displacement in the y-direction in color, total displacement outlined. Reproduced from ref. 16.

of 998 kg m^{-3} and a complex speed of sound of $1481(1 + i/2000)$ m s^{-1}. The piezoelectric material is Pz26 with a density given in Table 5.1 and a stiffness matrix with damping $\varphi = 1/180$ with the components $c_{E11} = c_{E22} = 168$ GPa, $c_{E33} = 123$ GPa, $c_{E44} = c_{E55} = 30.1$ GPa, $c_{E66} = 28.8$ GPa, $c_{E12} = c_{E21} = 110$ GPa and $c_{E13} = c_{E23} = c_{E31} = c_{E32} = 99.9$ GPa a coupling matrix with the components $e_{15} = e_{24} = 9.86$ C m^{-2}, $e_{31} = e_{32} = -2.8$ C m^{-2}, $e_{33} = 14.7$ C m^{-2} and a complex relative permittivity $(1 - i \cdot 0.003)$ with the components $\varepsilon_{rS11} = \varepsilon_{rS22} = 828$ and $\varepsilon_{rS33} = 700$.

The fluid structure interaction is modeled with an inward acceleration for the acoustic domain and a boundary load defined as force per unit area for the mechanical structure. The outer mechanical boundaries are free. The electrodes of the piezoelectric element are indicated in Figure 5.11. On the electrode shown in red a sinusoidal voltage with a magnitude of 20 V is applied, while the remaining electrodes shown in blue are set to ground. The other boundaries of the piezoelectric element are set with zero charge ($\mathbf{D} \cdot \mathbf{n} = 0$).

A frequency response analysis (frequency domain) was performed. The dependent variables are the pressure, the displacement field and the voltage. The stationary solver used is a direct MUMPS (MUltifrontal Massively Parallel sparse direct Solver) solver fully coupled. The system has been solved in a parametric study over a frequency range of 1.5–2.5 MHz with a step size of 500 Hz. Two characteristic modes within this range are presented here. Figure 5.12 and Figure 5.13 show a pressure resonant mode at 1.9 MHz and

Piezoelectricity and Application to the Excitation of Acoustic Fields 97

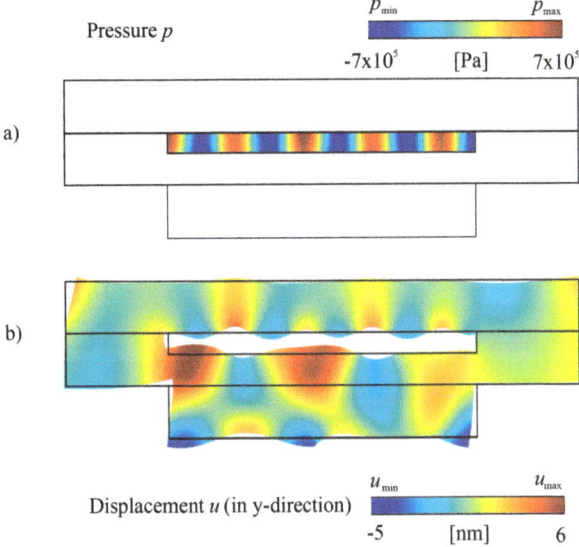

Figure 5.13 (a) Pressure field in a fluid cavity at 2.15 MHz; (b) displacement in the y-direction in color, total displacement outlined. Reproduced from ref. 16.

2.15 MHz respectively. In figure (a), the pressure field is shown and in (b), the displacement in the y-direction is shown as a color graph where its outline shows the scaled up total displacement. Figure 5.12 shows a symmetric pressure field with eight pressure nodal lines, whereas Figure 5.13 shows an unsymmetrical pressure field with nine pressure nodal lines. Alternatively, the symmetrical or antisymmetrical resonance could be suppressed by using an antisymmetrical or symmetrical electrode configuration, respectively. As an example, if the excited red electrode is arranged symmetrically with respect to the symmetry axis of the device as shown in Figure 5.14(a), only symmetrical modes are excited. By using a symmetrical electrode pattern, the symmetric mode at 1.9 MHz is stronger, both in terms of maximum pressure and displacement, when compared to the same mode in Figure 5.12. Periodic arrangements of electrode patterns can be used to selectively excite corresponding wavenumbers.[11]

The quality of the pressure field is exemplary in the two cases presented here and can be reproduced for most higher and lower modes. While the pressure field at resonance is as expected, the corresponding displacement field is more complex and hard to predict without a numerical simulation.

When experimenting with micromanipulation devices, temperature often plays a crucial role. Because all the material properties are functions of temperature, the behavior of the device will change accordingly. Often the most important effect might be the temperature increase in glue layers, which will change the damping of the system and can drastically decrease the pressure amplitude in the cavity.

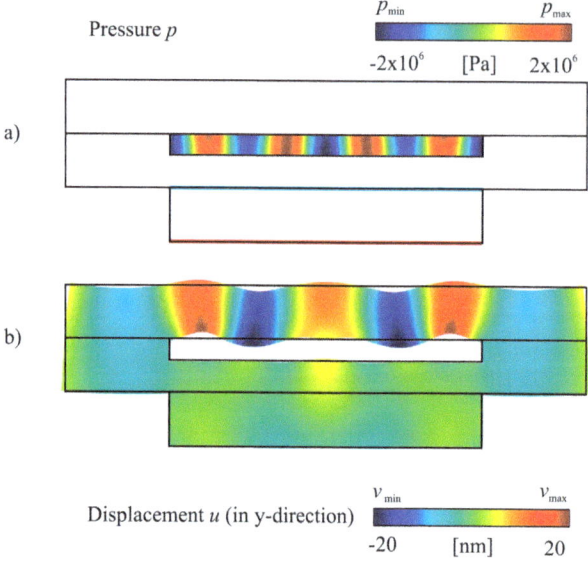

Figure 5.14 (a) Pressure field in a fluid cavity at 1.9 MHz; (b) displacement in the y-direction in color, total displacement outlined. Reproduced from ref. 16.

5.6 Conclusions

The basic equations describing piezoelectric materials for their usage in ultrasonic manipulation devices have been described. A one-dimensional system analysis of a transducer attached to a resonating bar has shown that a complex interplay exists between the resonances of transducer and structure, which is strongly influenced by the damping. This damping on the other hand depends on all the elements of the device, in particular also the glue layers that connect transducer and microfluidic chip. In electrical admittance curves, the resonances will only show up if there is strong electromechanical coupling. That is the system in which resonances must be reasonably close to a transducer resonance. The low coupling of the electrical circuit with the mechanical motion at frequencies far below the resonance frequencies of the piezoelectric transducer on the other hand allows broad band excitation at the expense of pressure amplitude in the device. For an actual device it has been shown that specific modes are selectively excited by special transducer setups, *e.g.* using segmentation.

References

1. Ferroperm Piezoceramics A/S, http://www.ferroperm-piezo.com/.
2. V. M. Ristic, *Principles of Acoustic Devices*, Wiley, 1983.
3. N. V. Baker, *Nature,* 1972, **239**, 398.

4. W. T. Coakley, D. W. Bardsley, M. A. Grundy, F. Zamani and D. J. Clarke, *J. Chem. Technol. Biotechnol.,* 1989, **44**, 43.
5. H. M. Hertz, *J. Appl. Phys.,* 1995, **78**, 4875.
6. O. Manneberg, B. Vanherberghen, J. Svennebring, H. M. Hertz, B. Onfeld and M. Wiklund, *Appl. Phys. Lett.,* 2008, **93**, 063901.
7. A. Neild, S. Oberti and J. Dual, *Sens. Actuators B,* 2007, **121**, 452.
8. A. Neild, S. Oberti and J. Dual, *J. Acoust. Soc. Am.,* 2007, **121**, 778.
9. T. Lilliehorn, U. Simu, M. Nilsson, M. Almqvist, T. Stepinski, T. Laurell, J. Nilsson and S. Johansson, *Ultrasonics,* 2005, **43**, 293.
10. A. Haake and J. Dual, *Ultrasonics,* 2004, **42**, 75.
11. J. Friend and L. Y. Yeo, *Rev. Mod. Phys.,* 2011, **83**, 647.
12. J. Dual and T. Schwarz, *Lab Chip,* 2012, **12**, 244.
13. R. Collin, *Proc. IRE,* 1955, **43**, 179–185.
14. C. Desilets, J. Fraser and G. Kino, *IEEE Trans. Sonics Ultrason.,* 1978, **25**, 115–125.
15. M. Gröschl, *Acustica,* 1998, **84**, 432–447.
16. J. Dual and D. Möller, *Lab Chip,* 2012, **12**, 506.

CHAPTER 6

Building Microfluidic Acoustic Resonators

ANDREAS LENSHOF, MIKAEL EVANDER, THOMAS LAURELL, AND JOHAN NILSSON*

Department of Biomedical Engineering, Lund University, Sweden
*E-mail: johan.nilsson@bme.lth.se

6.1 Introduction

The combination of microfluidics and acoustic standing wave technology has become a viable route to develop integrated systems for non-contact and in-chip manipulation of cells and particles. Acoustic standing wave technology offers means to move and spatially localise cells and particles by utilising acoustic standing wave forces in an acoustic resonator *i.e.* acoustophoresis. When performing acoustophoresis in a continuous flow based system, commonly a mode of separation is sought where particles having different acoustophysical properties can be differentiated. Acoustophoresis can also be utilised to induce retention against flow or aggregation of particles in defined locations, which commonly is performed in so called acoustic traps (or acoustic tweezers). Acoustic trapping allows for, for example, detailed investigations of free floating cell aggregates over an extended period of time. A fundamental requirement to accomplish these modes of operation is that the microfluidic system is designed with well-defined acoustic resonators.

A classical way of designing an acoustic resonator for cell and particle handling comprises a compartment where one wall of the resonator incorporates a piezoceramic transducer glued to a coupling layer of glass or metal and the opposing wall serves as a passive reflector,[1] commonly referred to as

Building Microfluidic Acoustic Resonators

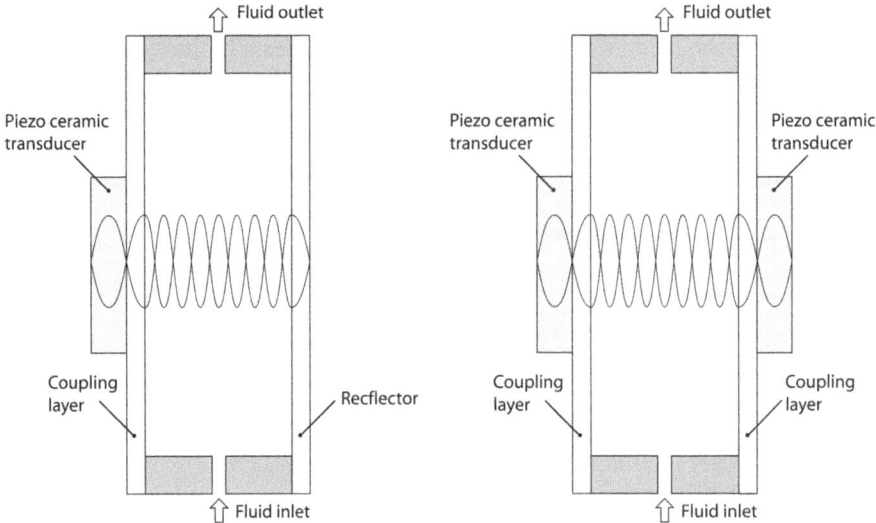

Figure 6.1 Classic configurations of a layered resonator with either a single transducer and a reflector layer or two opposing transducers.

a layered resonator. Optionally the opposing wall can be supplied with a second transducer operating at the same frequency as the first transducer,[2] Figure 6.1. These resonators are typically designed to operate in a multi node resonance mode and of centimetre dimensions or larger.[3,4]

As half wavelength resonators started to appear, as described by Mandralis *et al.*[5] in 1990 and were experimentally presented in 1993,[6] the resonator dimensions were inherently reduced to a millimetre domain or smaller. Early work by Yasuda *et al.* in 1992 described the use of half wavelength resonators in the MHz regime.[7] The patent outlined precise focusing of cells in the standing wave node in the flow channel and further downstream, acoustically enriched fractions of particles were selectively collected in the cell enriched flow segment, Figure 6.2.

The development of half wavelength resonators brought the system fluidics into a mode of low Reynolds numbers and hence a laminar flow domain where integration with microfluidic components became a natural development.[2,8,9] The transition to microsystem integration yields benefits both in terms of precision microfabrication where multiple resonators operating at different frequencies can be realised on a chip. MEMS based manufacturing also enables control for further dimensional reduction of the acoustic resonator, which aligns well with the scaling laws of the primary axial radiation force,[10,11] which in a simplified version described as a one-dimensional resonance mode is expressed in eqn (6.1). The primary acoustic radiation acoustic force is proportional to the employed frequency and hence by reducing the resonator dimension, the frequency is increased in a half wavelength resonator. An increased frequency yields a higher magnitude of the primary acoustic force, which is beneficial for efficient manipulation of particles/cells in continuous flow based systems. It should, however, be noted that reducing the resonator dimension rapidly induces a higher system back pressure and hence impacts the system throughput due to the laminar flow domain. Still, resonator dimensions in the range of a few hundred micrometres yield acoustic forces that enable cell or particle processing at flow rates in the range of 100 uL min^{-1} or higher depending on the system design.

$$F_{Ax} = 4\pi R^3 E k \sin(2kx)\Phi \qquad (6.1)$$

where

$$\Phi = \frac{\rho_p + \frac{2}{3}(\rho_p - \rho_0)}{2\rho_p + \rho_0} - \frac{1}{3}\frac{\rho_0 c_0^2}{\rho_p c_p^2} \qquad (6.2)$$

F_{Ax} = primary axial radiation force
E = acoustic energy density
R = particle radius
x = particle position in the wave propagation direction
ρ_p and ρ_0 = density of particle and fluid respectively
c_p and c_0 = speed of sound in particle material and fluid respectively
$k = 2\pi f/c_0$
Φ = acoustic contrast factor
f = frequency

The combination of acoustic control of particle location in a fluid and the simultaneous transition to a laminar flow domain in half wavelength resonators has opened up acoustic particle handling to become a powerful means to manipulate particles and cells in microfluidic systems.[12,13]

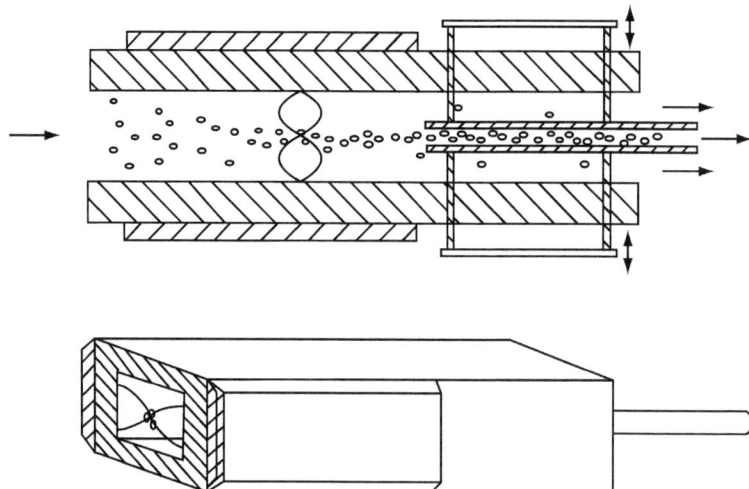

Figure 6.2 Half wavelength acoustic resonator for particle and cell focusing as disclosed by Yasuda et al.[7]

A key feature of this is the ability to allow solely the geometrical dimensions and material properties of the acoustic resonator to govern the fundamental acoustic forces that allow cell or particle manipulation. An important consequence is that the acoustic forces acting on the particles/cells are to a large extent independent of properties such as ionic strength, pH or surface charge, hence making acoustophoresis a generic tool to manipulate biological suspensions, be that blood, plasma, urine, fermentation broths, milk, cell cultures *etc*. Thus the mere design and composition of the resonator in terms of geometrical parameters and material choice becomes a key engineering challenge to accomplish efficient acoustic resonators for particle and cell manipulation. This tutorial will review the literature and outline basic considerations that have to be made when building acoustic resonators for continuous flow-based acoustophoresis or resonators for acoustic cell/particle trapping.

6.2 Choice of Material

The choice of material is very important when designing an acoustic resonator. However, the type of resonator system also influences the material used. There are three main types of acoustophoretic systems: the layered resonator, the transversal resonator and the surface acoustic wave (SAW) resonator.

The layered resonator is a quite complex resonator as the different layers require precise control of thicknesses in order to achieve a resonator with a high Q-value, Figure 6.3. However, as it is the system Q-value that is important, there is actually room for using materials that are often not considered as good resonator materials, such as polymers that are known for their high acoustic attenuation. As the main reflection occurs between the air

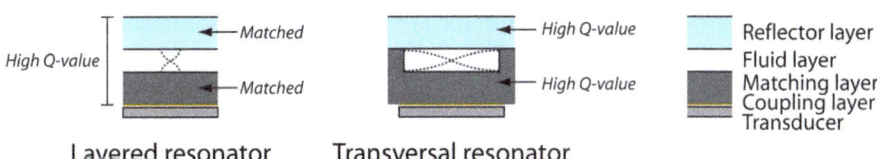

Figure 6.3 The choice of material depends on what type of resonator is to be designed. The layered resonator requires carefully matched reflection and matching layers with regards to the wavelength in order to achieve a system with high Q-value. Although, as it is the system Q-value that is important here, it is possible to use materials which themselves are not acoustically optimal, such as polymers, and some losses could be acceptable as long as the system is well matched. Transversal resonators, on the other hand, rely more on materials with high characteristic acoustic impedance and are less sensitive to matched layers as the whole system resonates as one body. They are thus easier to design, but are limited to the choice of materials which can be utilized. Reproduced from ref. 14 with permission.

backed reflector layer and the transducer, some losses in the supporting layers can be acceptable in order to maximize the energy density in the fluid layer. The attenuation in the polymers can also be an advantage as it results in resonators with larger bandwidths.

The transversal separator, on the other hand, relies on reflections between the channel walls and requires materials of high characteristic acoustic impedance, such as glass, silicon or metals. The high Q-value materials make this resonator type less sensitive to different thicknesses and matched layers as it is the entire bulk that resonates as one body, even though the bonding of the lid to the bulk structure is important to minimize attenuation of the resonance.

The resonance in SAW devices is created differently as they rely on waves propagating into a fluidic compartment *via* a wave guiding substrate. In order not to create interfering resonances, the material enclosing the fluid should be of similar characteristic acoustic impedance as the fluid, making polymers suitable.

A "good" resonator material is thus very device specific. There are some tools that can be useful when designing the acoustic resonator, such as the acoustic impedance. The characteristic acoustic impedance (Z) is comprised of the density (ρ) and speed of sound (c) of the material according to eqn (6.3).

$$Z = \rho c \quad (6.3)$$

The characteristic acoustic impedance is useful for calculating the reflection and transmission coefficient.[15] For normal incidence, the pressure reflection coefficient is:

$$R_p = \frac{Z_2 - Z_1}{Z_1 + Z_2} \quad (6.4)$$

where Z_1 is the characteristic acoustic impedance of the first medium and Z_2 of the second. The transmission coefficient of a wave of normal incidence is subsequently

$$T_p = 1 - R_p \quad (6.5)$$

To avoid acoustic losses due to reflection, the acoustic impedances of two adjacent layers should be carefully matched so that the acoustic energy density in the fluidic layer is maximised. For instance, when designing the matching layer, the characteristic acoustic impedance of the layer should be lower than that of the transducer but higher than the material comprising the fluidic cavity.

There have been numerous different material combinations in acoustophoretic devices over recent years. It should be noted that these materials have very different material properties, as can be seen in Table 6.1. One of the most common material choices is to make the flow channel from silicon.[16,17] The monocrystalline structure of silicon enables precise structures to be fabricated using etching techniques. With the correct mask alignment with regards to the etch planes, it is possible to achieve channels with vertical

Table 6.1 Density, speed of sound and characteristic acoustic impedance for some materials. The parameters are collected from different sources and should be considered as typical values. Different compositions of a polymer from different manufacturers may display varying numbers, so the actual value for the polymer in use must be collected from the supplier or measured before a system is designed.

	Density ($kg\ m^{-3}$)	Speed of sound ($m\ s^{-1}$)	Characteristic acoustic impedance ($10^6\ kg\ m^{-2}\ s^{-1}$)
Silicon	2331	8490	19.79
Pyrex	2230	5647	12.59
Steel – stainless 347	7890	5790	45.68
Aluminium	2700	6420	17.33
Titanium	4506	6070	27.35
Polymethyl methacrylate (PMMA)	1150	2590	2.98
Polycarbonate (PC)	1200	2160	2.59
Polystyrene (PS)	1050	1700	1.79
Polymethylsiloxane (PDMS)	965	1076 (10 : 1)	1.04
		1119 (5 : 1)	1.08
H_2O (25 °C)	997	1497	1.49
PZT transducer	7700	4000	30.8
Air	1	343	0.00034

walls, which are good for standing wave acoustics.[16] A borofloat glass lid to seal the flow channel is preferable as it provides good visual access to the flow channel and the anodic bond process provides a very strong chemical bond that makes the two pieces resonate as one body.

When designing transversal resonators it was initially thought that a resonator channel with a homogeneous width, *i.e.* vertical walls, was a prerequisite for standing wave resonances to occur and that an amorphous material such as glass, which gives a semi-circular cross-section if wet etched, was not suitable. Glass itself has very good properties for standing wave acoustofluidics as it has quite a large difference in density compared to aqueous fluids and reasonably high speed of sound, as well as being chemically inert. Evander *et al.* showed that a resonator consisting of a wet etched flow channel in glass, covered with a fusion bonded glass lid, showed similar performance to silicon channels,[18] even though there was the substantial difference in channel cross-section appearance.

From a material properties point of view, steel is another suitable resonator material. Although not very commonly used, as the microfabrication processes which is used with silicon and glass are not applicable, Hawkes *et al.* reported a layered steel resonator.[19] The resonator was made through wire erosion, although the flow channel itself was defined by a rubber gasket and the height determined by a brass spacer. Steel has high density, good sound propagating properties and also has the advantage of good heat transport capabilities.

Polymers have many desirable properties such as low cost and easy mass fabrication (such as by injection moulding), making them suitable for disposable devices to be used for medical and clinical purposes. Unfortunately, polymers are less optimal from an acoustical point of view. They show poor reflective abilities to water, which is undesired when designing transversal resonators and the relatively high absorbance of acoustic waves makes it more difficult to create standing waves without heavy losses. However, PMMA/SU-8 devices have proven to be useful in layered resonators where the intension is to have the nodal plane less than a ¼ wavelength from the wall.[20] A way to create standing waves without being limited to the restricted reflection abilities of polymers is to use two opposing transducers. This enabled the use of soft polymers such as PDMS since the resonator is defined by the two actuators.[21] In other resonator configurations it would be difficult to set up a standing wave in a cavity of this material as the acoustic impedance of PDMS is close to water. This property is, however, exploited in surface acoustic wave (SAW) devices where PDMS is used as a channel material on top of a piezoelectric substrate.[22,23]

6.3 Design Configurations

The first thing to consider when designing the resonator is the size range of the particles to be manipulated. The most common and practical range is about 1 μm to 20 μm in diameter. For sizes below 1 μm, the primary radiation force is quite weak as it scales with the volume of the particle, and at that level other forces such as viscous drag from acoustic streaming will start to dominate over the primary radiation force. At sizes above 20 μm, gravity will start to affect the particles even more and sedimentation effects may occur, although that is highly dependent on the densities of the particles compared to the fluid medium. The force on the particle is also proportional to the acoustic frequency, see eqn (6.1). Smaller particles that are affected less by the radiation force can still be manipulated by choosing a higher frequency. However, the wavelength decreases with the increasing frequency, meaning that the width of the fluid layer has to decrease accordingly. It must also be decided if the device is to work at the fundamental frequency (*i.e.* a single pressure node) or at a higher harmonic (multiple pressure nodes). Although multiple node resonators can sometimes be beneficial,[24] the most common systems use a half-wavelength resonator. When perfused continuously these devices yield a single particle band located in the center of the fluidic channel. Generally speaking, a frequency range of 1–10 MHz is suitable for the manipulation of particles in the range of 1–20 μm.

6.3.1 Layered Resonators

A layered resonator is a structure composed of different layers which all have a very specific role in the resonator system as the sound wave passes through them, building up the resonance. The different layers can be seen in

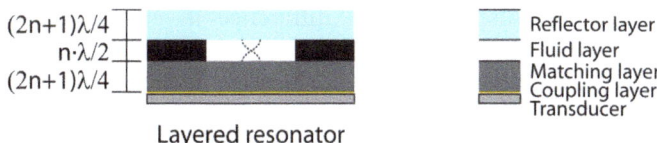

Figure 6.4 The structure of a layered resonator. Reproduced from ref. 14 with permission.

Figure 6.4. The transducer generates the sound and is followed by the coupling layer, which is needed in order to get good acoustic transmission into the system. Next follows the matching layer, sometimes called the transmission layer, which forms the bottom of the resonator chamber and thus also acts as a reflector surface of the standing wave. The fluid layer contains the liquid and cells or beads. At the other end of the system is the reflector layer that is responsible for reflecting the incoming wave back into the fluid layer, giving rise to the standing wave.

The thicknesses of the different layers in the layered resonators are of great importance in order to get as powerful a resonance as possible. Hawkes *et al.* have shown in simulations that a good half wavelength layered resonator should be constructed as follows: a matching layer of a quarter wavelength, fluid layer of half a wavelength and a reflector layer of a quarter wavelength thickness.[8] The quarter wavelength in the matching layer is chosen to maximize the Q-value of the layered resonator. The quarter wavelength in the reflection layer is chosen because of the phase shift of the wave when it reflects to a less dense medium (air). This configuration results in a pressure minimum in the center of the fluidic channel and a pressure maximum at the channel walls, a configuration used in most cell manipulation applications. However, if other locations of the pressure node are desired, such as in the application by Hill[25] where a pressure node is located by one of the walls in a quarter wavelength configuration, other layer parameters are preferable. As this resonance system is not as stable as the half wavelength systems, simulations are needed to predict the layer thicknesses for the special frequency conditions required.[26] The quarter wavelength fluid cavity can be used to drive cells towards a surface that is modified for capturing specific particles.[27]

6.3.2 Transversal Resonators

In contrast to the layered resonator, the transversal resonator is operated such that a standing wave perpendicular to the incident direction of actuation is obtained, Figure 6.5. By exciting the resonator structure at a frequency that matches a half wavelength criterion with respect to the channel width, a transversal mode of operation can be accomplished. Historically, the transversal mode of operation has predominately been reported in resonators made of high Q-value materials, such as glass, silicon or metal. Since the entire microfluidic component is actuated in the transversal mode, the general advice is that the resonator in this case is made of materials with low acoustic losses that display a high difference in characteristic acoustic impedance *versus* the fluid. By matching the dimension of the bulk material, enhanced operating characteristics can be obtained with good acoustic focusing properties.[11] A clear benefit of the transversal resonator configuration is that the acoustic manipulation is performed in plane with the resonator chip, facilitating visual observation of the focusing event. Also, since the actuation is performed in plane with the chip, integration with other microchip based fluidic functions is easily accomplished.

6.3.3 SAW Devices

Surface acoustic wave (SAW) devices utilize surface waves to manipulate particles in a channel, commonly formed in PDMS. The surface waves are generated by one or more interdigitated electrode transducers positioned outside the channel, normally fabricated in PDMS using soft lithography,[28] Figure 6.6. The interdigitated electrode transducers are created by forming

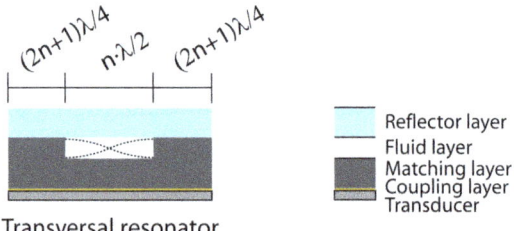

Figure 6.5 The structure of a transversal resonator. Reproduced from ref. 14 with permission.

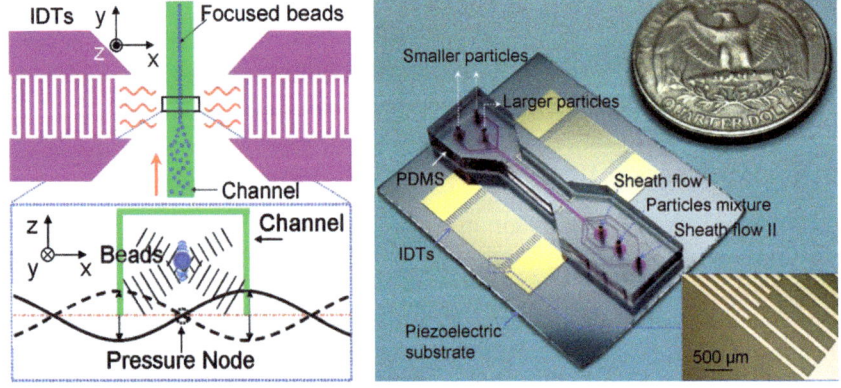

Figure 6.6 Left: basic principle for bead focussing using SSAW.[22] Right: example of an SSAW device with a channel fabricated in PDMS.[23] Reproduced with permission from ref. 22 and.23.

metal strips on a piezo-electric substrate, commonly LiNbO$_3$, by metal evaporation and lift-off. The PDMS channel is fabricated by moulding PDMS on a silicon or SU-8 master. The PDMS channel mould is then bonded against the transducer wafer/plate. The basic principle and an example of a device are shown in Figure 6.6.

In the most straightforward configurations, the inter-electrode distances, λ_{SAW}, are given by the frequency, f, and the surface wave-velocity in the piezo-electric material, v_{SAW}, according to:

$$\lambda_{SAW} = v_{SAW}/f \qquad (6.6)$$

By using opposing transducers, a standing surface acoustic wave (SSAW) is generated in the substrate that couples into the channel and generates the pressure field. Although the PDMS/water interface does not compose a resonant system, the phase relationship between the two opposing fields entering the channel will determine the position of the SSAW in the channel. This opens up the possibility of controlling the positioning of the particles within the channel by individually controlling the phase of the signals to the transducers. Since the transducers are situated at a distance equal to several wavelengths from the channel, accurate temperature control may be an issue due to the temperature dependence of the surface wave velocity in the substrate. A more in-depth discussion on SAW devices for microfluidic acoustic particle manipulation is presented in Chapters 14–16 in this book.

6.3.4 Flow Splitter Design

A flow splitter is an essential component in many acoustophoretic applications, *e.g.* for collecting and maintaining the integrity of separated species. Different flow splitter designs fabricated using anisotropic etching of silicon can be seen in Figure 6.7. The geometric design of the flow splitter is of great importance for the operation.[29] Generally, as with most microfluidic devices, sharp 90° corners should be avoided as gas bubbles are prone to gather there, disturbing the flow.[16] Also, sharp features, such as ridges or flow splitters, can induce considerable acoustic streaming that may interfere with the laminar

Figure 6.7 Left: silicon flow channel with 90° flow splitters etched on a <100>-oriented silicon wafer. Center: silicon flow channel with 45° flow splitters etched on a <100>-oriented silicon wafer. Right: silicon flow channel with straight flow splitters etched on a <110>-oriented silicon wafer. From Laurell *et al.*[29] Reproduced with permission from ref. 29.

flow.[30] However, when working with crystalline materials such as silicon there are certain limitations to which designs are possible, at least if the anisotropic wet etching properties of silicon are to be utilized. The smaller the angle of your flow splitter, the less likely it is that a gas bubble will actually get stuck in your flow channel, which makes the 45° in Figure 6.7(b) preferable over 6.7(a). The splitters in Figure 6.7(a) and (b) are etched on the same wafer but along different crystal orientations. The 90° splitter in Figure 6.7(a) has vertical walls in all channels while the 45° splitter in Figure 6.7(b) has slanting walls in the side channels. The splitter in Figure 6.7(c) is etched in <110> silicon which is more challenging than <100> silicon as the etch planes are not symmetrical and thus limit the downstream fluidic network design possibilities.

6.3.5 1D and 2D Acoustic Focusing in Continuous Flow

One of the largest reasons that a continuous flow acoustophoresis system does not work at optimal performance is the fact that particles are flowing at different flow speeds in the channel because of the parabolic flow profile. Particles close to the walls travel at very low velocity in comparison to the ones near the center of the channel. Additionally, the acoustic primary radiation force is very weak at the walls. These two effects results in particles not being able to focus at the same time as others and in a separation application, the efficiency and recovery of the targeted cells will be poor. One way to counter this problem is to include a pre-focus region where the particles are focused in two dimensions such that all particles end up in the pressure nodes. When the pre-focused particles then enter the main channel, all particles will have the same starting position spatially and thus the one-dimensional acoustic separation will occur purely on the physical properties of the particle and not be influenced by the particles' location in the channel.

The best way of implementing this is to use two different active acoustic regions and two different transducers. This configuration is preferable since the pre-focusing acoustic power is to be maintained at a constant level, while the main acoustic power may be varied depending on the sample and application. In theory, one single transducer may be used for both the pre-focus and main focus step if the channel width is tailored correspondingly, but for the reasons mentioned previously, this may not be advantageous. Augustsson et al.[31] used a pre-focusing channel (150 µm × 300 µm), which corresponds to a double node resonance at 5 MHz, and a main separation channel (150 µm × 375 µm), which resonated with a single node in the center of the channel at 2 MHz, Figure 6.8.

The importance of 2-dimensional pre-focusing before acoustic sorting was further illustrated by Jakobsson et al.[32] in an application that utilized acoustic actuated fluorescence activated sorting of microparticles. Besides the importance of all particles having the same flow vector, as mentioned above, the pre-focusing also positioned the particles where the acoustic radiation force is at its maximum. This is very important in a particle

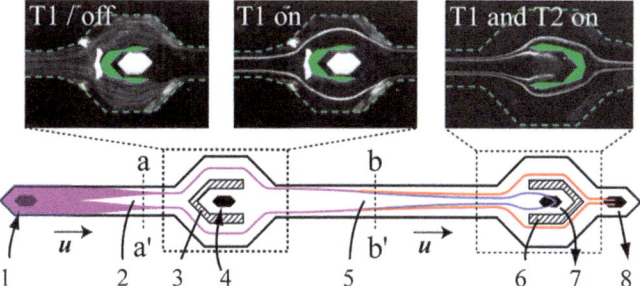

Figure 6.8 The setup used by Augustsson *et al.*[31] where particles were two-dimensionally pre-focused in two nodes. The two particle streams were then split and entered the main separation channel from the sides and subsequently separated. The outtakes clearly show the effect of the pre-focusing transducer (T1) turned "on" and "off". Reproduced with permission from ref. 31. Copyright 2012 American Chemical Society.

switching application as it minimizes the acoustic actuation time for deflecting the targeted particle, and thus the actual performance and throughput of the device as the switching frequency increases.

6.3.6 Acoustic Traps

In a resonator designed for an acoustophoresis system, the axial primary radiation force is commonly utilised to translate objects into the pressure node and the fluid flow subsequently transports the particles to *e.g.* a flow-splitter. In a trapping system, however, the goal is to use the acoustic forces to retain particles against a flow at a fixed position in a channel. Apart from the axial component of the primary radiation force, this setup relies heavily on the lateral component of the primary radiation force to hold the particles in place. The acoustic radiation force is based on both the pressure and the velocity gradient in the standing wave and in order to achieve an efficient trap, large, lateral gradients will be required. The force resulting from the lateral velocity gradient for a dense particle with a radius, R, situated in the pressure node of a standing wave with a constant amplitude gradient that can be expressed as:[33]

$$F_{LAT} = \pi \rho \omega^2 R^2 u_0 u_m \qquad (6.7)$$

ρ denotes the fluid density, ω the angular frequency of the ultrasound, u_0 is the displacement amplitude at the center of the particle and u_m is the difference in displacement amplitude at the edge of the particle compared to the center. As can be seen, the lateral component is not as size dependent as the axial component of the radiation force and the greater the difference in amplitude between the center and the edge of the particle, the larger the trapping force. So in order to design a strong lateral trap, a large, lateral gradient is needed.

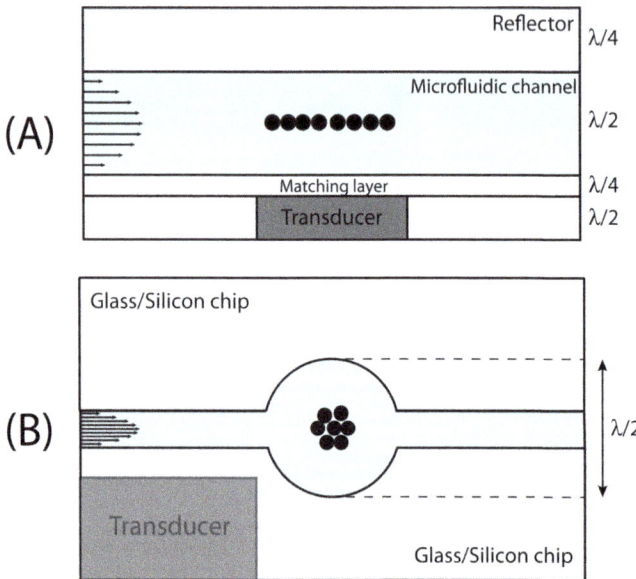

Figure 6.9 (A) shows a side-view drawing of a trapping system using a localized acoustic field from a small transducer. The transducer is coupled to a microfluidic channel through a matching layer and creates a local standing wave that will trap and hold objects. (B) shows a top-view of an alternative approach where a resonance cavity is designed in a sealed and bonded glass/silicon chip. An external transducer actuates the entire chip and a standing wave will be supported only in the resonance cavity where objects can be trapped. Reproduced from ref. 14 with permission.

Another force that is more important in trapping systems than in acoustophoresis systems is the secondary radiation force that is created from the main acoustic field scattering on the objects that are being trapped. The force is strongly dependent on the distance between the particles (d^{-4}) and therefore mainly acts on the particles when they are already focused into the pressure nodal plane by the primary radiation force and brought into close contact (10–100 μm) by the lateral component of the primary radiation force.[33] The secondary force attracts particles to each other and helps to build and stabilize the trapped cluster.

In order to create a strong lateral pressure gradient, trapping systems are usually designed to use either a highly localized acoustic field (Figure 6.9(A)) or create a localized resonance (Figure 6.9(B)).

Since the active trapping area in an acoustic trap using a transducer that creates a localized field near the transducer is essentially decided by the area of the transducer, it is generally easier to trap larger numbers of objects than in the resonance cavity approach. It is also usually easier to design an efficient system as it follows the same design rules as a layered resonator.

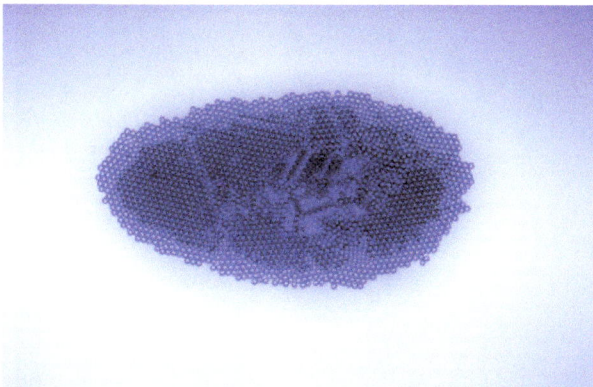

Figure 6.10 12 μm polystyrene particles levitating in a 4 MHz standing wave field above a rectangular transducer.

However, these systems are typically harder to integrate with other lab on a chip systems and more difficult to scale up as they tend to rely on manually assembled transducers. An example of particles trapped over a localized transducer can be seen in Figure 6.10.

The resonance cavity approach is, on the other hand, easy to scale up as it is fabricated using standard photolithography and etching. Although the trapping capacity of an individual cavity may be smaller than that of a localized transducer, it is easy to compensate for this by creating an array of cavities. The main disadvantage here is the difficulty in predicting the behavior of the system as the entire chip is actuated and can support many modes that might interfere with each other.

Examples of trapping systems with localized acoustic fields include designs with miniature transducers coupled directly to a microfluidic channel[34,35] and designs that use matching layers to couple transducers into flow systems.[36–39] Spengler and Coakley originally used a transducer with a rubber gasket and a stainless steel container around it to create a resonance cavity for trapping.[37] The system was later changed to include a stainless steel matching layer and a single inlet and outlet to allow for cell loading.[36] Lilliehorn and Evander et al. used a printed circuit board (PCB) with miniature transducers that compose the bottom of a fluidic channel when a reflector lid with an etched channel was placed on top of the PCB.[34,35] This system was modified by Hammarström et al. to use a capillary as a fluidic channel rather than the etched channel lid, creating a more robust system that used disposable fluidic parts.[39]

The alternative approach, to obtain a localized resonance, is usually achieved by designing a resonance cavity that will support a standing wave when the entire chip is actuated,[40–43] see Figure 6.11. A common approach used is to create a silicon layer with fluidic channels linked to a resonance cavity using DRIE and then anodically bond a glass lid to the chip. Manneberg et al.

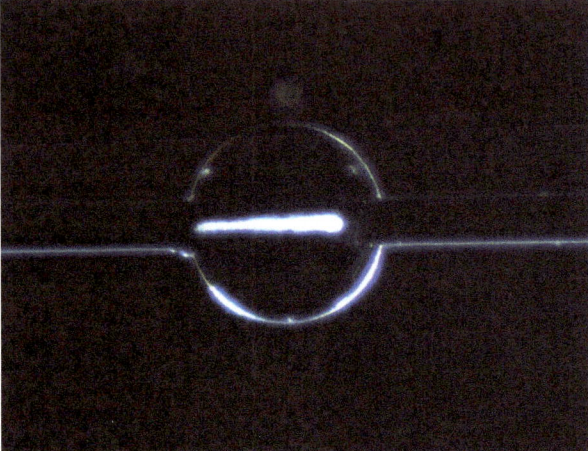

Figure 6.11 A cluster of polyamide particles that are trapped in a resonance cavity and held against a flow. The resonance cavity is etched into an all-glass chip, actuated at 2 MHz. Reproduced from ref. 14 with permission.

created a resonance cavity that allowed for three-dimensional, selective standing waves using external transducers[43] and Vanherberghen *et al.* recently demonstrated a trapping cavity array that was used for microscopy studies of cells.[42]

6.3.7 Capillaries

The use of readily available fluidic components is of special importance when working with very sensitive detection methods or potentially hazardous samples and many times a disposable component is a prerequisite. In this perspective capillaries were successfully used in early approaches both for trapping and acoustophoresis. Capillaries or cuvettes have well defined dimensions, known surface properties (typically glass) and are available without the need of microfabrication or bonding.

Tilley and Coakley did an early experiment in 1987 using a capillary for studies of induced erythrocyte adhesion.[44] The system used a cuvette as a sample reservoir and the sample was drawn into the capillary by capillary forces. The cuvette rested on a 1 MHz transducer that created a standing wave in the direction of the capillary and aggregated the erythrocytes in order for them to be studied by microscopy. Wiklund *et al.* improved the capillary approach further by reducing the capillary dimensions significantly and designing an 8.5 MHz focused transducer and a reflector that allowed for a quartz capillary to pass through the centre of both units,[45] see Figure 6.12. The entire assembly with capillary, transducer and reflector was immersed in a water bath to ensure that the acoustic waves from the focused transducer entered the capillary. This way, a flow-through system was created that could

Building Microfluidic Acoustic Resonators   115

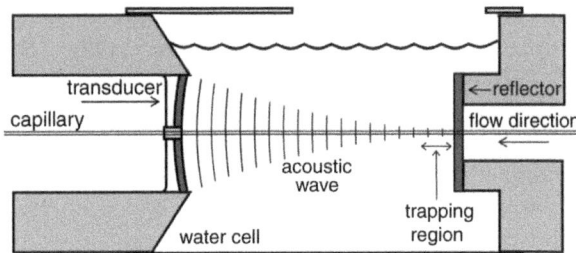

Figure 6.12 A schematic drawing showing how a focused transducer and a continuous flow capillary was used for trapping. Reprinted with permission from ref. 45. Copyright 2001, American Institute of Physics.

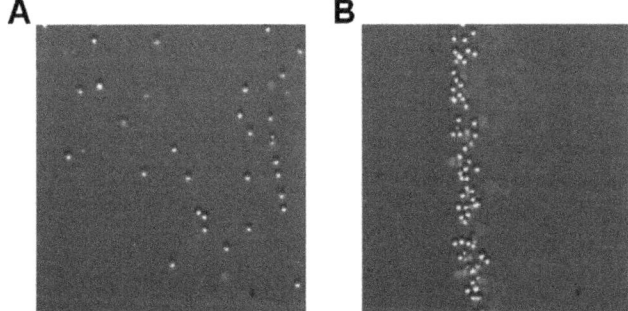

Figure 6.13 (A) shows unfocused 10 μm polystyrene beads flowing through a glass capillary (I.D. 1.9 mm) at a speed of 5 mm s^{-1}. The capillary is actuated using a 417 kHz PZT attached to the outer wall of the capillary and aligns the particles as can be seen in (B). Reprinted with permission from Goddard et al.[48] Copyright 2006, John Wiley and Sons.

trap and hold 3–4 μm particles. The system was later combined with capillary electrophoresis to investigate the possibilities of increasing the limit of detection by selective trapping of particles in an immunoagglutination assay.[46]

Goddard et al. used a capillary to create a continuous flow particle concentrator that was later used to improve the accuracy of a flow cytometer system.[47–49] A glass capillary with an inner diameter of 1.9 mm was actuated at around 460 kHz to focus 10 μm particles into a narrow band that would decrease the spread in the scattered laser signal as well as remove the need of a particle focusing sheath fluid, see Figure 6.13. The transducer was applied using gel to the side of the capillary. To maximize the acoustic energy input into the system they used an identical transducer on the other side of the capillary as a sensor. The system was driven at the frequency that maximized the amplitude of the detected signal at the sensor transducer. By using this system, they could track the resonance and compensate for any drift that might occur.

Hammarström et al. recently presented a PCB-platform with miniature transducers coupled to a rectangular borosilicate capillary through a thin glycerol layer.[39] This way a localized standing wave for trapping could be created in the capillary while still having a disposable microfluidic component that could easily be changed between assays. The system was used for creating a trapping pipette (cf. Figure 6.17(a)) which, when coupled with matrix-assisted laser desorption/ionization mass spectrometry (MALDI-MS), could be used for efficient sample preparation[50] and the study of drug partitioning in minute cell populations.[51] More information on this capillary system can be found in Chapter 9.

To expand the fluidic functionality of capillaries, microfluidic PMMA-interfaces can be coupled to the capillaries. This creates a hybrid device where the acoustics can be performed in the glass capillary and the more advanced fluidic functions can be performed in the polymer chips.[52]

6.4 Actuation

6.4.1 Transducer Coupling

The coupling medium between the transducer and the resonator chip is crucial in deciding the efficiency of energy transfer to the resonator. Effects of the bonding/coupling layer were investigated by Hill et al.[53] using a transducer and impedance transfer model. They concluded that the thin layer of adhesive used to couple the transducer only has a relatively small effect on the modelled electrical characteristics, as long as the layer was thin and "well bonded". Well bonded in this sense means no air voids between the transducer and the chip as they will act as a total reflector and reduce acoustic energy transfer to the chip. Most commonly the transducer is bonded to the resonator chip by, for example, epoxy or cyanoacrylate glues. Hard glues provide a better coupling as compared to, for example, silicone rubber based bonding layers.

If a system in which the transducer can be decoupled from the resonator chip is desired, the use of a hydrogel or glycerol as a coupling layer between the transducer and the chip can be employed,[54] much like current practice in diagnostic ultrasound. Since the bond in this case is viscous, the transducer must be clamped to the chip to retain a constant pressure in the bond and the clamping must be spatially fixed to prevent the sandwich structure from sliding. Coupling of the transducer using a hydrogel is a convenient method when evaluating different resonator set ups and thus allows for rapid reconfiguration of the acoustic system. When using harder glues, *e.g.* cyanoacrylate glues, this can be dissolved/softened in acetone overnight, enabling remounting or change of the transducer on the resonator.

Some publications describe systems where the transducers have been incorporated in the microchannel as one of the resonator cavity walls, delivering ultrasound directly in contact with the fluid.[34,55-57] This alleviates any undesired effects that are linked to a variance in the coupling layer. A side effect, however, is that the acoustic interference patterns from the transducer become more prominent. A benefit is that the lateral pressure gradients that arise in such a system can be utilized in creating strong acoustic trapping systems.[55]

Rather than mounting the transducer directly under the trapping zone, Haake et al. have shown how surface waves can be used to perform two-dimensional particle manipulation.[58] By gluing shear transducers to the sidewalls of a glass plate, it was possible to create standing flexural waves that will couple to the fluid underneath the glass plate and cause particles or cells to aggregate in lines or points.[59] By slowly changing the driving frequency it was also demonstrated that the nodes in the standing wave could be moved and that cells could also be transported in this fashion.

The location of the transducer in relation to the resonance cavity varies throughout the literature. Layered resonators are always constructed with the transducer in line with the fluid and the reflector. For transversal resonators, the most common positioning of the transducer is underneath the channel.[16,60,61] However, the transducer does not have to be located directly under the channel in order to generate a standing wave in the resonator. As long as the transducer delivers acoustic energy into the transversal resonator chip at a frequency that matches the resonator, a standing wave will form. Augustsson et al. demonstrated this—when gluing a small transducer die at the side of the chip, they were still able to actuate a long resonator channel.[62,63] However, as pointed out by Hagsäter et al., the obtained resonance pattern is dependent on the position of the transducer, indicating that different system resonance modes are being actuated in relation to the spatial location of the transducer.[40] Yet, the chip can be actuated from any transducer position although some positions will be more efficient than others in terms of transmitted acoustic energy and hence heat generation. Wiklund et al. introduced a slightly different approach to the transducer coupling, by the use of metal wedges glued as the coupling medium between transducer and chip. The wedges work as refractive elements in coupling the acoustic energy into the resonator structure.[64,65] One particular reason for this design was the need for optical access across the flow channel while still having an actuation mode that translates particles in the plane of the chip, similar to the transversal mode of operation.

As losses in the piezo electric transducer are the primary source of heating to the acoustophoretic system, the coupling layer can be designed to work as a heat sink in order to prevent an increase of temperature in the flow channel and thus a drift in resonance frequency. As an example, at a resonance of 2 MHz, an increase from 25 °C to 30 °C in water results in resonance frequency increase of 0.8% ($\approx$ 17 kHz). This could be sufficient to drive the system off resonance or at least weaken the magnitude of the standing wave considerably. It is therefore important to avoid temperature variations during operation and measures should be taken to counteract them by careful design or heat sinks, or integrated temperature control.

The intrinsic heat generation in the piezo transducer can be used to elevate the operating temperature of the microfluidic system. Evander et al. demonstrated that temperature control of an acoustic trap can be accomplished by calibrating the operating temperature at different actuation voltages in the acoustic trap.[56] By this means, a temperature control of the trap was accomplished such that culturing and proliferation of yeast cells could be performed at 30 °C over an extended period of time. An optional approach was proposed by Wiklund where a heating fan was linked to the micro system, enabling controlled temperature elevation of the system independent of the driving of the piezo actuator.[66] More recently, Augustsson *et al.* demonstrated the profound impact that temperature changes actually have on acoustic resonators and also proposed a Peltier based feedback system to fully control system temperature independent of the driving conditions of the actuator.[67]

Aluminium offers good heat transporting capabilities and also has good acoustic properties. By adding a block of aluminium between the transducer and the chip, an acoustophoretic microchip can be operated at high driving voltages without getting overheated.[24] However, it is important to match the thickness of the aluminium block to a multiple of half a wavelength such that a resonance in the block itself is induced. This ensures that the acoustic energy delivered to the microchip is relatively high yet at moderate actuation amplitudes. An example of this is given below where a silicon chip with eight parallel channels was used together with aluminium coupling blocks of different thicknesses, D, to illustrate the importance of wave length matched distance, Figure 6.14. Each channel width was matched to a half wavelength resonance at an actuation frequency of 2 MHz. As the system was actuated, the microparticles passing through the channels in the chip were focused in the channel centre and at the channel trifurcation outlet the percentage of particles being collected in the central outlet served as a measure of the acoustic focusing efficiency. As seen in Figure 6.15, a focusing chip with aluminium block distances that are unmatched, $(2n + 1)\lambda/4$, shows a significantly lower focusing efficiency as compared to the matched blocks under identical power input conditions.

6.4.2 Coupled Resonance Modes

When designing acoustic resonators that are composed of several building blocks that each display individual acoustic resonances, undesired results may appear at system level. The complete acoustic system with the actuator coupled to the resonator cavity may not display a resonance at the targeted frequency. A very good example of this was presented by Hill *et al.*,[53] where a layered resonator was designed to have a half wavelength resonance at 3 MHz, which was coupled to a piezo transducer bonded to a matching layer with a combined resonance frequency of 3 MHz. The combined resonance of the two parts did, however, not resonate at 3 MHz. Instead a frequency split was observed and two resonances appeared at about 2.9 MHz and 3.1 MHz. By choosing the transducer resonance and the design of the resonator cavity such that they have slightly mismatched resonances, a more stable and robust acoustic system is accomplished. More details on layered resonators can be found in Chapter 7 in this book.

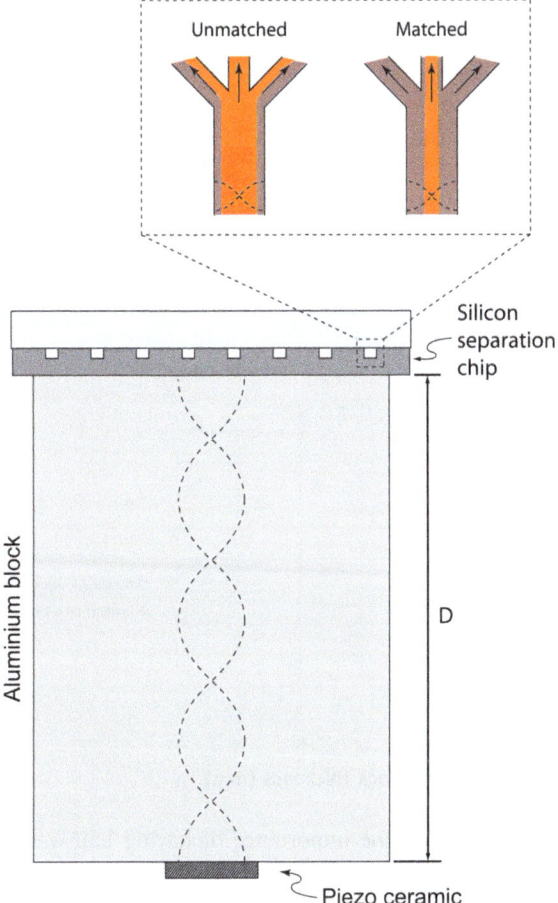

Figure 6.14 A silicon multi-channel chip with an aluminium block in between transducer and separation chip which acts as a heat sink. Reproduced from ref. 14 with.

6.4.3 Electrode and Transducer Modifications

Modifications of the standard piezoelectric transducers, for example, modifying the electrode geometry, can make a big difference in the output. Neild *et al.* showed that by creating "strip electrodes" they were able to increase the number of resonance modes in a transducer and thereby actuate harmonics in their resonator.[68] In the example in Figure 6.16, orthogonal strip electrodes were used to generate two independent pressure fields in a chamber. By superimposing the fields, a pressure distribution could be created in a fluid chamber that positioned cells in a 2D-array format.[69]

120 Chapter 6

The bulk wave transducers are normally used in thickness mode, where the wanted resonance frequency is determined by the thickness of the transducer. However, the transducers constitute Helmholz resonators and several resonance modes may be active at a specific drive frequency. Lateral modes normally have lower fundamental resonance frequencies and harmonics may interfere at the wanted thickness mode frequency. This is especially of interest when automatic frequency tracking systems are used for optimising the drive frequency to circumvent drift, *e.g.* due to temperature changes or changes in acoustic properties of the medium in the channel. The frequency

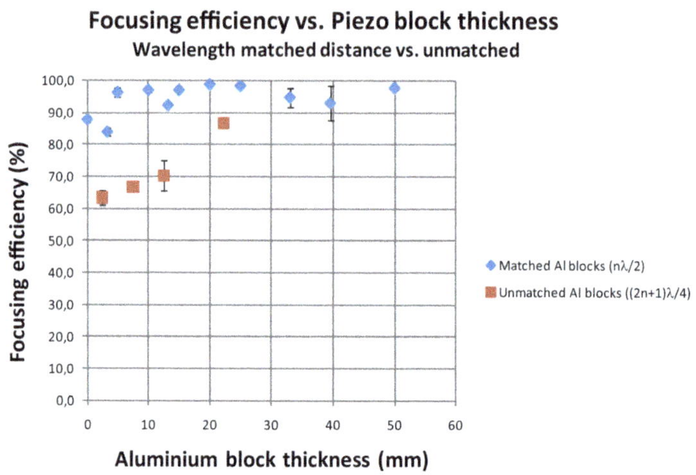

Focusing efficiency vs. Piezo block thickness
Wavelength matched distance vs. unmatched

◆ Matched Al blocks ($n\lambda/2$)
■ Unmatched Al blocks ($(2n+1)\lambda/4$)

Focusing efficiency (%) vs. Aluminium block thickness (mm)

Figure 6.15   Figure showing the importance of having half wavelength matched coupling layers. Unmatched aluminium blocks (quarter wave length matched) reduce the amount of acoustic energy transferred into the channel and affect the separation efficiency negatively. Reproduced from ref. 14 with permission.

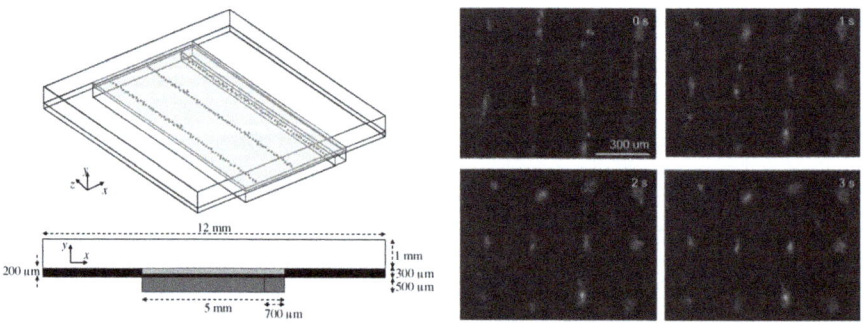

Figure 6.16   Left: 700 µm strip electrode on a 5 mm-wide piezo-electric transducer. Reprinted from Neild *et al.*[68] with permission from Elsevier. Right: 2D-positioning of cells in a fluid chamber actuated by a single transducer with orthogonal strip electrodes.[69] Copyright 2007, John Wiley and Sons.

Building Microfluidic Acoustic Resonators

tracking systems often use the impedance spectrum of the transducers to select the best drive frequency, which may result in a selection of a lateral mode instead of the wanted system resonance. Hammarström *et al.* introduced kerfs in the transducer to suppress the lateral resonance modes to improve the performance of a frequency tracking system (Figure 6.17).[70] It should be noted that introducing kerfs in a transducer also lowers the thickness mode resonance frequency, which means that a slightly thinner transducer with a higher thickness mode resonance frequency should be used as a starting point.[70,71]

Figure 6.18 shows a comparison of the impedance magnitude spectrum of unmounted and mounted solid and kerfed tranducers for a wanted resonance frequency of 4 MHz. The small width wanted for the application in the trapping capillary system results in a strong lateral resonance at 2 MHz for the solid 4 MHz transducer. After mounting, the resonances are attenuated through the increased losses in the PCB but the lateral resonance at 2 MHz is still dominating. The kerfed transducer was fabricated from a 5 MHz transducer element. In contrast to the solid transducer, the kerfed transducer displays a close to ideal 4 MHz resonance impedance spectrum.

6.4.4 Focused Transducers

Focused transducers have mainly been used in acoustic trapping systems where they help in creating the large acoustic gradient that is needed to trap objects. They have been used in two different configurations, either with two focused transducers facing each other or a single focused transducer aimed at a reflector, see Figure 6.19. In the first case, you can either form a standing wave between the transducers[72] or use the pressure gradients to create a potential well between the transducers.[73] In the second case, a hemispheric

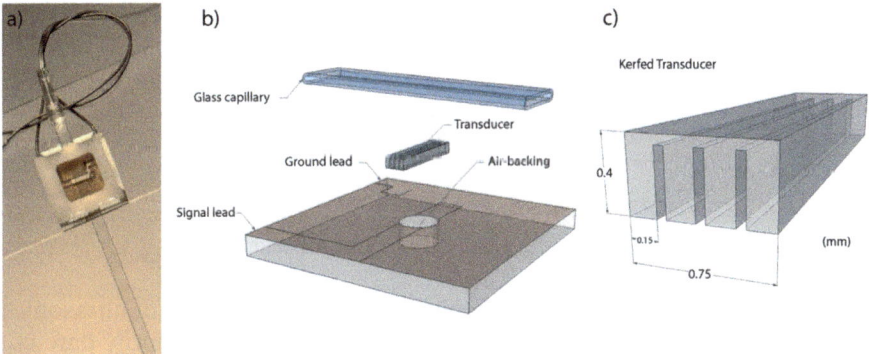

Figure 6.17 Trapping system (a) based on a rectangular cross-section glass capillary and a kerfed transducer (c) to suppress lateral resonance modes of the transducer. The transducer was soldered to a PCB with a through-hole for air-backing (b). Reproduced from ref. 70 with permission.

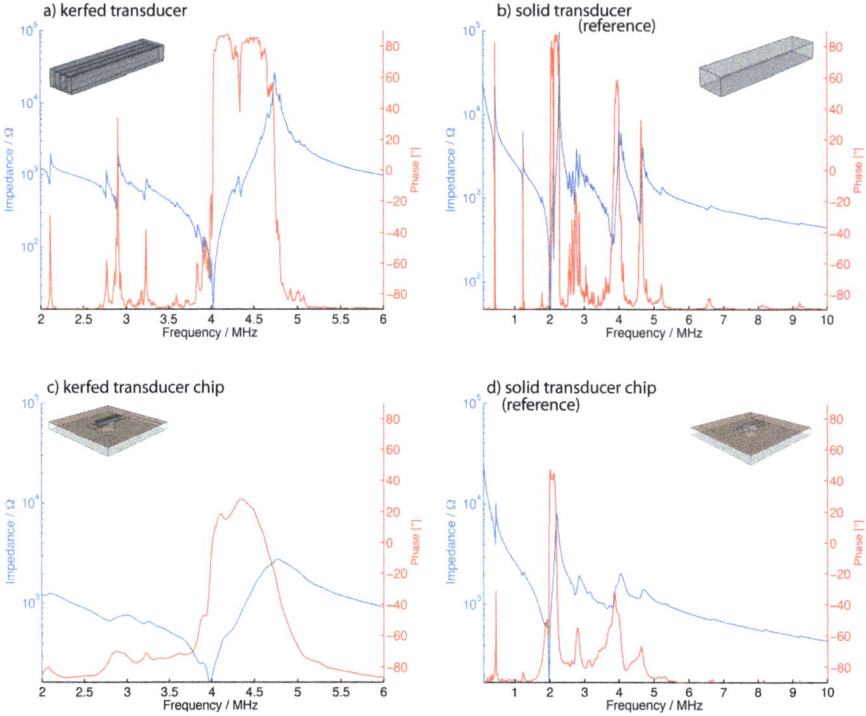

Figure 6.18 Comparison of unmounted and mounted kerfed and solid transducers aimed at 4 MHz thickness mode resonance frequencies. a) a free kerfed transducer element before mounting, b) a free kerfless transducer block, c) a finished transducer chip using a kerfed transducer mounted on a PCB with 1.5 mm air-backing, and d) a mounted kerfless transducer block. The finished transducer chip is dominated by a clear resonance at 4 MHz, as shown by the large impedance dip and the zero phase transition. The kerfless transducer is included for reference only, as this shows the large lateral 2 MHz resonance that is removed by kerfing the transducer. When comparing mounted to free transducers, the attenuation of spurious modes from the 3.2 mm length of the transducer (reoccurring as closely spaced overtones to each fundamental) can also be observed. Reproduced from ref. 70 with permission.

standing wave can be created that allows for object trapping in several nodes.[45,46,74]

Yasuda *et al.* envisioned a device consisting of phased arrays of ultrasonic elements in different configurations in a US patent with a priority date of 1992.[7] The phased arrays could focus the ultrasonic waves at an arbitrary position and create gradients that could be used for focusing, trapping and moving of particles.

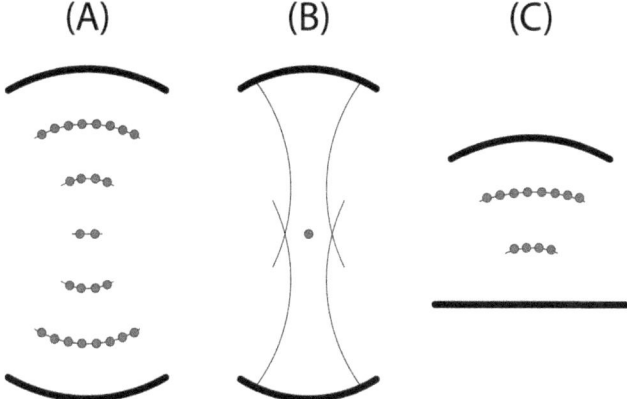

Figure 6.19 Three different approaches for using focused transducers when performing particle trapping. In (A), two focused transducers are aimed at each other and create a standing wave where particles can be trapped. (B) uses the pressure gradient that is created when two transducers are focused with a short distance between the focal points to trap objects. In (C), a hemispheric standing wave is created by having a planar reflector closer than one radius of curvature from the transducer. Reproduced from ref. 14 with permission.

6.5 Conclusions

When designing an acoustic manipulation system, the size range of the particles to be manipulated has to be taken into account. Smaller particles (1–3 µm) will require a higher operating frequency than larger particles to achieve the same radiation force. The frequency and the desired number of pressure nodes will in turn determine the dimensions of the acoustic resonator. Typical frequencies used are 1–10 MHz resulting in resonator dimensions of 750 µm to 75 µm when operated at a half wavelength resonance.

There are several suitable materials for acoustic applications although their choice depends on the resonator design. Transversal resonators are preferably fabricated from materials with high characteristic acoustic impedance such as metals, glass and silicon. The latter two being most commonly used in microfabrication techniques enables precise and reproducible structures in the desired size range. Layered resonators can also be constructed from high acoustic impedance materials although they could benefit from polymers even though polymers attenuate ultrasound. Polymers are also successfully used as microfluidic channels in SSAW devices.

Finally, to ensure a high Q-value in the system, a well bonded chip, matched reflection and matching layers and a thin coupling layer are required.

References

1. O. Doblhoffdier, T. Gaida, H. Katinger, W. Burger, M. Groschl and E. Benes, *Biotechnol. Prog.*, 1994, **10**, 428–432.
2. K. Yasuda, S. Umemura and K. Takeda, *Jpn. J. Appl. Phys., Part 1,* 1995, **34**, 2715–2720.
3. F. Trampler, S. A. Sonderhoff, P. W. S. Pui, D. G. Kilburn and J. M. Piret, *Biotechnology,* 1994, **12**, 281–284.
4. M. Groschl, W. Burger, B. Handl, O. Doblhoff-Dier, T. Gaida and C. Schmatz, *Acustica,* 1998, **84**, 815–822.
5. Z. I. Mandralis, D. L. Feke and R. J. Adler, *Fluid/Part. Sep. J.,* 1990, **3**, 115–121.
6. Z. I. Mandralis and D. L. Feke, *AIChE J.,* 1993, **39**, 197–206.
7. K. Yasuda, S. Umemura, K. Kawabata, K. Takeda, K. Uchida, Y. Harada, M. Kamahori and K. Sasaki, United States Patent, US6216538B1, 1996.
8. J. J. Hawkes, M. Gröschl, E. Benes, H. Nowotny and W. T. Coakley, Special Session PHA -01: Acoustics of Dispersed Particulate Matter, *Proc. Forum Acusticum*, Sevilla, Spain, 2002.
9. D. A. Johnson and D. L. Feke, *Sep. Technol.,* 1995, **5**, 251–258.
10. L. P. Gorkov, *Sov. Phys. Doklady,* 1962, **6**, 773–775.
11. R. Barnkob, P. Augustsson, T. Laurell and H. Bruus, *Lab Chip,* 2010, **10**, 563–570.
12. A. Lenshof and T. Laurell, *Chem. Soc. Rev.,* 2010, **39**, 1203–1217.
13. J. Nilsson, M. Evander, B. Hammarstrom and T. Laurell, *Anal. Chim. Acta,* 2009, **649**, 141–157.
14. A. Lenshof, M. Evander, T. Laurell and J. Nilsson, *Lab Chip,* 2012, **12**, 684–695.
15. J. D. N. Cheeke, *Fundamentals and Applications of Ultrasonic Waves*, CRC Press LLC, Boca Raton, Fl, 2002.
16. A. Nilsson, F. Petersson, H. Jonsson and T. Laurell, *Lab Chip,* 2004, **4**, 131–135.
17. N. R. Harris, M. Hill, S. Beeby, Y. Shen, N. M. White, J. J. Hawkes and W. T. Coakley, *Sens. Actuators, B,* 2003, **95**, 425–434.
18. M. Evander, A. Lenshof, T. Laurell and J. Nilsson, *Anal. Chem.,* 2008, **80**, 5178–5185.
19. J. J. Hawkes and W. T. Coakley, *Sens. Actuators, B,* 2001, **75**, 213–222.
20. I. Gonzalez, L. J. Fernandez, T. E. Gomez, J. Berganzo, J. L. Soto and A. Carrato, *Sens. Actuators, B,* 2010, **144**, 310–317.
21. S. Kapishnikov, V. Kantsler and V. Steinberg, *J. Stat. Mech.: Theory Exp.,* 2006, P01012.
22. J. J. Shi, X. L. Mao, D. Ahmed, A. Colletti and T. J. Huang, *Lab Chip,* 2008, **8**, 221–223.
23. J. J. Shi, H. Huang, Z. Stratton, Y. P. Huang and T. J. Huang, *Lab Chip,* 2009, **9**, 3354–3359.
24. C. Grenvall, P. Augustsson, J. R. Folkenberg and T. Laurell, *Anal. Chem.,* 2009, **81**, 6195–6200.

25. M. Hill, *J. Acoust. Soc. Am.,* 2003, **114**, 2654–2661.
26. R. J. Townsend, M. Hill, N. R. Harris and M. B. McDonnell, *Ultrasonics,* 2008, **48**, 515–520.
27. S. P. Martin, R. J. Townsend, L. A. Kuznetsova, K. A. J. Borthwick, M. Hill, M. B. McDonnell and W. T. Coakley, *Biosens. Bioelectron.,* 2005, **21**, 758–767.
28. Z. C. Wang and J. A. Zhe, *Lab Chip,* 2011, **11**, 1280–1285.
29. T. Laurell, F. Petersson and A. Nilsson, *Chem. Soc. Rev.,* 2007, **36**, 492–506.
30. S. M. Hagsäter, PhD Thesis, Technical Univeristy of Denmark, 2008.
31. P. Augustsson, C. Magnusson, M. Nordin, H. Lilja and T. Laurell, *Anal. Chem.,* 2012, **84**, 7954–7962.
32. O. Jakobsson, C. Grenvall, M. Nordin, M. Evander and T. Laurell, *Lab Chip,* 2014, **14**, 1943–1950.
33. M. Groschl, *Acustica,* 1998, **84**, 432–447.
34. T. Lilliehorn, M. Nilsson, U. Simu, S. Johansson, M. Almqvist, J. Nilsson and T. Laurell, *Sens. Actuators, B,* 2005, **106**, 851–858.
35. M. Evander, L. Johansson, T. Lilliehorn, J. Piskur, M. Lindvall, S. Johansson, M. Almqvist, T. Laurell and J. Nilsson, *Anal. Chem.,* 2007, **79**, 2984–2991.
36. W. T. Coakley, D. Bazou, J. Morgan, G. A. Foster, C. W. Archer, K. Powell, K. A. Borthwick, C. Twomey and J. Bishop, *Colloids Surf., B,* 2004, **34**, 221–230.
37. J. F. Spengler, M. Jekel, K. T. Christensen, R. J. Adrian, J. J. Hawkes and W. T. Coakley, *Bioseparation,* 2000, **9**, 329–341.
38. J. Hultstrom, O. Manneberg, K. Dopf, H. M. Hertz, H. Brismar and M. Wiklund, *Ultrasound Med. Biol.,* 2007, **33**, 145–151.
39. B. Hammarström, M. Evander, H. Barbeau, M. Bruzelius, J. Larsson, T. Laurell and J. Nilsson, *Lab Chip,* 2010, **10**, 2251–2257.
40. S. M. Hagsäter, T. G. Jensen, H. Bruus and J. P. Kutter, *Lab Chip,* 2007, **7**, 1336–1344.
41. J. Svennebring, O. Manneberg, P. Skafte-Pedersen, H. Bruus and M. Wiklund, *Biotechnol. Bioeng.,* 2009, **103**, 323–328.
42. B. Vanherberghen, O. Manneberg, A. Christakou, T. Frisk, M. Ohlin, H. M. Hertz, B. Onfelt and M. Wiklund, *Lab Chip,* 2010, **10**, 2727–2732.
43. O. Manneberg, B. Vanherberghen, J. Svennebring, H. M. Hertz, B. Onfelt and M. Wiklund, *Appl. Phys. Lett.,* 2008, **93**, 063901.
44. D. Tilley, W. T. Coakley, R. K. Gould, S. E. Payne and L. A. Hewison, *Eur. Biophys. J.,* 1987, **14**, 499–507.
45. M. Wiklund, S. Nilsson and H. M. Hertz, *J. Appl. Phys.,* 2001, **90**, 421.
46. M. Wiklund, P. Spégel, S. Nilsson and H. M. Hertz, *Ultrasonics,* 2003, **41**, 329–333.
47. G. Goddard and G. Kaduchak, *J. Acoust. Soc. Am.,* 2005, **117**, 3440.
48. G. Goddard, J. C. Martin, S. W. Graves and G. Kaduchak, *Cytometry, Part A,* 2006, **69A**, 66–74.
49. G. R. Goddard, C. K. Sanders, J. C. Martin, G. Kaduchak and S. W. Graves, *Anal. Chem.,* 2007, **79**, 8740–8746.

50. B. Hammarström, H. Yan, J. Nilsson and S. Ekstrom, *Biomicrofluidics,* 2013, **7**, 11.
51. B. Hammarström, T. Laurell, M. Evander, J. Nilsson and S. Ekström, *The 13th International Conference on Miniaturized Systems for Chemistry and Life Sciences*, *μTAS 2009*, The Chemical and Biological Microsystems Society, Jeju, Korea, 2009.
52. M. Evander and M. Tenje, *J. Micromech. Microeng.,* 2014, **24**, 027003.
53. M. Hill, Y. J. Shen and J. J. Hawkes, *Ultrasonics,* 2002, **40**, 385–392.
54. F. Petersson, A. Nilsson, C. Holm, H. Jonsson and T. Laurell, *Analyst,* 2004, **129**, 938–943.
55. T. Lilliehorn, U. Simu, M. Nilsson, M. Almqvist, T. Stepinski, T. Laurell, J. Nilsson and S. Johansson, *Ultrasonics,* 2005, **43**, 293–303.
56. M. Evander, L. Johansson, T. Lilliehorn, J. Piskur, M. Lindvall, S. Johansson, M. Almqvist, T. Laurell and J. Nilsson, *Anal. Chem.,* 2007, **79**, 2984–2991.
57. J. V. Norris, M. Evander, K. M. Horsman-Hall, J. Nilsson, T. Laurell and J. P. Landers, *Anal. Chem.,* 2009, **81**, 6089–6095.
58. A. Haake and J. Dual, *Ultrasonics,* 2004, **42**, 75–80.
59. A. Haake, A. Neild, G. Radziwill and J. Dual, *Biotechnol. Bioeng.,* 2005, **92**, 8–14.
60. J. D. Adams and H. T. Soh, *Appl. Phys. Lett.*, 2010, **97**, 064103.
61. J. J. Hawkes, R. W. Barber, D. R. Emerson and W. T. Coakley, *Lab Chip,* 2004, **4**, 446–452.
62. P. Augustsson, L. B. Aberg, A. M. K. Sward-Nilsson and T. Laurell, *Microchim. Acta,* 2009, **164**, 269–277.
63. P. Augustsson, J. Persson, S. Ekstrom, M. Ohlin and T. Laurell, *Lab Chip,* 2009, **9**, 810–818.
64. M. Wiklund, C. Gunther, R. Lemor, M. Jager, G. Fuhr and H. M. Hertz, *Lab Chip,* 2006, **6**, 1537–1544.
65. O. Manneberg, J. Svennebring, H. M. Hertz and M. Wiklund, *J. Micromech. Microeng.*, 2008, **18**, 095025.
66. J. Svennebring, O. Manneberg and M. Wiklund, *J. Micromech. Microeng.,* 2007, **17**, 2469–2474.
67. P. Augustsson, R. Barnkob, S. T. Wereley, H. Bruus and T. Laurell, *Lab Chip*, 2011, **11**, 4152–4164.
68. A. Neild, S. Oberti and J. Dual, *Sens. Actuators, B,* 2007, **121**, 452–461.
69. A. Neild, S. Oberti, G. Radziwill and J. Dual, *Biotechnol. Bioeng.,* 2007, **97**, 1335–1339.
70. B. Hammarström, M. Evander, J. Wahlstrom and J. Nilsson, *Lab Chip,* 2014, **14**, 1005–1013.
71. R. S. Cobbold, *Foundations of Biomedical Ultrasound*, Oxford University Press, New York, 2007.
72. H. M. Hertz, *J. Appl. Phys.,* 1995, **78**, 4845–4849.
73. J. Wu, *J. Acoust. Soc. Am.,* 1991, **89**, 2140–2143.
74. M. Wiklund, J. Toivonen, M. Tirri, P. Hänninen and H. M. Hertz, *J. Appl. Phys.,* 2004, **96**, 1242.

CHAPTER 7

Modelling and Applications of Planar Resonant Devices for Acoustic Particle Manipulation

PETER GLYNNE-JONES*, ROSEMARY J. BOLTRYK, AND MARTYN HILL

Engineering Sciences, University of Southampton, Southampton SO17 1BJ, UK
*E-mail: P.Glynne-Jones@soton.ac.uk

7.1 Introduction

There is a need to manipulate micron-scale particles and cells in many areas of physics, analytical chemistry and the biosciences. Techniques such as filtration, centrifugation and sedimentation are well-established in macro-scale applications, but in microfluidic systems the use of other approaches including optical, magnetic, dielectrophoretic and acoustic forces are of interest. Acoustic radiation forces, typically at ultrasonic frequencies in the hundreds of kHz to tens of MHz region, have wavelengths that are well matched to microfluidic channel scales, yet are capable of generating potential wells with significantly larger length scales. The technology is also relatively straightforward to integrate within microfluidic systems.

7.2 Acoustic Radiation Forces

Acoustic radiation forces can be generated on particles by both travelling and standing acoustic fields. Those in standing wave fields (of primary interest in the applications considered here) are generated by the nonlinear interaction between the acoustic field scattered by the particle and the standing wave field itself. The time averaged radiation force $F(r)$ on a small (in comparison with a wavelength) spherical particle of volume V located at r within a stationary acoustic field was shown by Gor'kov[1] to relate to the gradients of the time averaged kinetic and potential energy densities (E_{kin} and E_{pot} respectively) within the field:

$$F(r) = V \nabla \left(\frac{3(\rho_p - \rho_f)}{(2\rho_p + \rho_f)} E_{\text{kin}}(r) - \left(1 - \frac{\beta_p}{\beta_f}\right) E_{\text{pot}}(r) \right) \qquad (7.1)$$

The kinetic energy density gradient (a function of the acoustic velocity field within the standing wave) is weighted by a function of the densities (σ_p and σ_f) of the particle and the surrounding fluid respectively. The potential energy density gradient (a function of the acoustic pressure within the standing wave) is weighted by a function of the compressibilities (β_p and β_f) of the particle and the surrounding fluid. Most particles and cells of interest are denser and less compressible than typical suspending fluid, so there is a force on them that tends to move them to the acoustic pressure node, and the acoustic velocity antinode. In a planar resonator these are co-located.

7.3 Generating the Required Acoustic Field

Many approaches to generating the required acoustic wave field have been reported in the literature. These include the use of near-field effects[2] or focussed ultrasound[3,4] to trap particles and cells, the excitation of lateral standing waves in which the predominant energy gradients run parallel to the face of the excitation transducer,[5,6] and the generation of cylindrical resonances to focus[7] or arrange particles.[8] Particles can be held and moved within standing waves excited by plate waves coupled into the containing fluid[9,10] and there is currently significant interest in the use of surface acoustic waves[11-13] or interface waves[14] to provide acoustic energy within microfluidic channels.

The most straightforward approach to establish a standing wave, however, is to use a planar, layered resonant device which, to a first approximation, establishes a one dimensional resonance pattern along the axis of the outgoing and returning acoustic field.[15-17] A number of investigators have employed a planar design incorporating pairs of opposing transducers to modulate the standing wave field,[18,19] but this chapter focuses on planar systems with a single transducer.

7.4 Modelling of Planar Resonators

For the purposes of obtaining insight into device operation, and to provide high efficiency numerical formulations to enable parameter space exploration for device design, a planar resonator can be approximated as a 1D device. Figure 7.1 shows a typical configuration composed of a sequence of

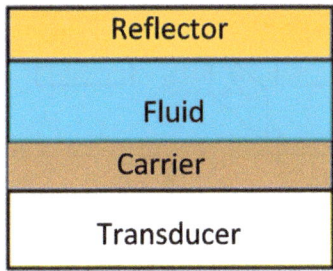

Figure 7.1 Layered components of a planar resonator. Reproduced from ref. 82.

different layers. The piezoelectric transducer couples acoustic energy into a carrier layer. The carrier layer is also sometimes described as a matching layer. In the context of travelling waves, the function of a matching layer is to couple energy more efficiently into subsequent layers, however, in the case of a resonant system the role of this layer is less straightforward, and such tuning can also be accomplished by suitable choice of the other layers (as demonstrated later). In this case the function of the carrier layer may be more structural, or to isolate the transducer from the fluid layer. The fluid layer is the region in which the particle manipulation is to occur, and the reflector layer serves to reflect energy back into the device, and hence maintain strongly resonant operation. At both ends of the device the layers are usually backed by air. The glue layer used to bond the transducer to carrier layers can also have a significant effect on device operation, especially in the quarter-wave devices described later.

The sequence of layers after the transducer can be modelled using an approximation of 1D linear acoustic propagation through the layers, and models have been described in a variety of forms.[17,20,21] Aside from varying degrees of accuracy in the transducer representation (see the next paragraph), approaches such as "transfer impedance", "transfer matrix", and "equivalent circuit models" are all mathematically similar. These models typically consider the system of layers as analogous to a set of electrical transmission lines. For an excellent discussion of one-dimensional modelling for the design of resonators of multiple wavelengths in size, readers are referred to the series of papers by Gröschl.[15,22] A worked example can be found in Bruus et al.[23]

The representation of the transducer driving the layers can be a simple forcing function at the carrier layer boundary. However, this ignores the coupling between the finite impedance transducer and the remainder of the resonator. It is more physically accurate to use a representation such as the Mason model (or the commonly used KLM model). These use an equivalent circuit to model the electro-acoustic response of a thickness mode resonance in a piezoelectric plate (other types of resonance mode can also be represented with alternative model parameters). The Mason model configuration is presented in Figure 7.2; the values of the parameters can be found

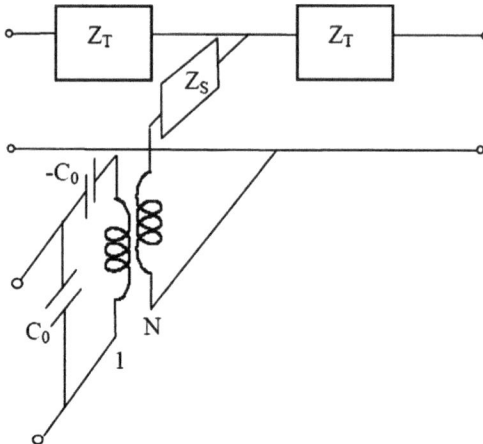

Figure 7.2 Mason model configuration (after Sherrit et al.[24]). The top-left and -right ports represent the acoustic connections to subsequent layers (acoustic pressure represented as a voltage), and the bottom port the electrical connections to the transducer. Reproduced from ref. 82.

elsewhere.[24] The Mason model is derived from the constitutive piezoelectric relations described in part 4 of the paper by Dual et al.[25] The KLM model[26] was developed as an easier-to-use alternative to the Mason model; its layout is similar, but incorporates a transmission line in place of ZT, and reduces the number of components. It has been shown to produce similar results to the Mason model under typical boundary conditions if care is taken to apply complex loss values consistently.[24] However, comparing the response of the two models to an electrical impulse, a time-delay is seen in the output of the KLM model as the impulse propagates through the transmission line. Having deduced the pressure distribution within the fluid layer, eqn (7.1) can be used to predict the radiation force on particles in the resonant field.

An alternative method for predicting device performance is to use a finite element analysis (FEA) package such as ANSYS or COMSOL to represent the various layers, and solve the resulting linear equations.[27–29] These packages are able to model the piezoelectric interactions using the full constitutive relations,[25] and couple the resulting acoustic field into a detailed structural model of the device components. Once again, by including eqn (7.1) above in a post-solution step, it is possible to create an FEA model that predicts radiation forces from the geometry, electrical drive and material properties.

By using suitable symmetry conditions a 1D model can be created; Figure 7.3 compares the transfer impedance model with a KLM transducer representation to an ANSYS FEA model for a typical quarter-wave configuration (presented elsewhere[30]), and shows good agreement. Small inconsistencies are to be expected due to the differences in the ways the models implement the damping and transducer representation.

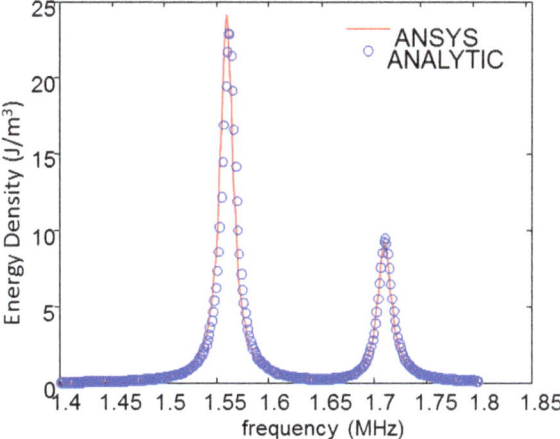

Figure 7.3 FEA model compared to transfer impedance model of a layered resonator for a typical quarter-wave configuration. The graph shows how the energy density in the fluid layer varies with excitation frequency. Reproduced from ref. 82.

The 1D approach is an approximation that ignores the effect of lateral resonances within the various layers, plate waves, surface waves, and edge effects. In practice, significant departures can be seen from the ideal[31] and this can be observed in Figure 7.4, which uses an ANSYS FEA model to predict the field in a silicon microfabricated filtration device[27] (this is an early FEA model that uses a pressure boundary condition rather than a finite element transducer representation). In such devices, the following non-ideal phenomena are typically seen: (a) width-wise variation in node position—the height of the node in the fluid channel often shows a corrugated pattern (as shown in Figure 7.4), (b) width-wise variation in focusing force (particularly a reduction in force near edges), (c) width-wise forces that caused particles to form into bands, and (d) trapping forces capable of holding particles and agglomerates against a flow. In longer devices with particle flow over an acoustic field region, these effects may be less pronounced due to an averaging of the differing acoustic environments. However, the averaging effect is not perfect, particularly for lower flow rates. Frequency sweeping can also reduce these effects by effectively averaging in time the force-fields of similar acoustic modes.[32]

7.5 Resonator Configurations

Any combination of layer thicknesses will show acoustic resonances at certain frequencies; some configurations will be much more strongly resonant than others, and it is these that are discussed here, as the increased energy density resulting from the resonance makes the devices more useful in practice. It should also be noted that devices can be run at frequencies

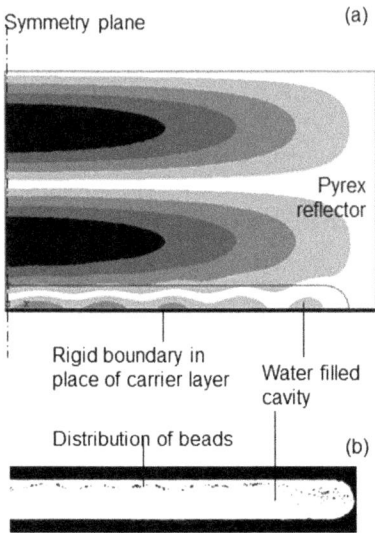

Figure 7.4 (a) Modelled acoustic pressure amplitude distribution within a particle separator shows non-ideal behaviour (arbitrary scale: black = maximum; white = zero) (Townsend et al.[27]). This is a modal analysis with the carrier layer replaced by a rigid boundary condition. The white pressure node can be seen to be corrugated in shape; the acoustic radiation forces (proportional to the spatial gradient of the pressure amplitude squared) can be seen to be variable across the device width and weaker near the right hand edge. (b) Cross sectional image taken from the quarter-wave device described by Glynne-Jones et al.[45] showing the effect of a similar field on a population of 10 μm polystyrene beads as they flow through the device; corrugation and a weaker field near the edge can be seen. Reproduced from ref. 82.

other than their resonant frequency—this is rarely useful in lightly damped devices as the energy density decreases rapidly away from resonance.

Layered resonators can be classified into a number of categories based upon the number of wavelengths and position of nodes within the various layers. The distinctions described here are popularly used in the literature, and are useful as each class displays distinct characteristics. In devices made with low loss materials the distinctions are clear; however, in devices made from more lossy polymeric materials the distinctions are less clear, and since layer thicknesses can vary continuously there are designs that do not fit easily into the classification. Although these devices are usually less energy dense, they have the advantage that the node position is more flexible (*i.e.* shifts in drive frequency can move the nodal position) and less determined by the geometry of the fluid layer.[33] Hawkes[34] explores a range of configurations, some of which are described below, based upon layer thicknesses that are multiples of $\lambda/4$. Figure 7.5 shows typical pressure distributions in the various classes of device.

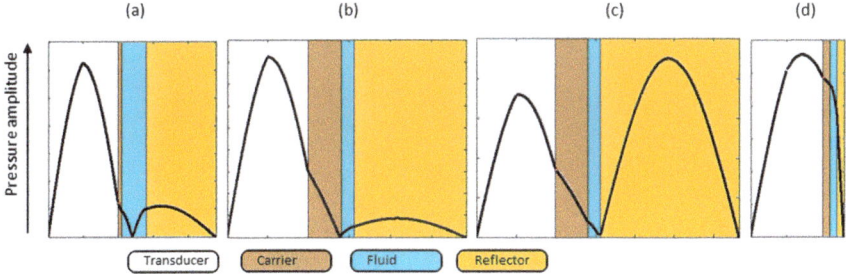

Figure 7.5 Typical pressure amplitude distributions in (a) half-wave, (b) inverted quarter-wave, (c) quarter-wave, (d) thin-reflector resonators. The differing positions of the pressure nodes yield various behaviours. Reproduced from ref. 82.

7.5.1 Half-Wave Devices

Half-wave devices can be defined as resonators in which the fluid channel is close to a half wavelength in thickness. For systems with a reflector that has higher acoustic impedance than the fluid, this leads to a resonance across the thickness of the channel, with a pressure minimum close to the channel centre. This mode is thus useful for focusing particles into a plane away from the chamber walls.

Generating half wavelength modes at high ultrasonic frequencies implies sub-millimetre chamber depths and such chambers were investigated by Hawkes et al.[35] for particle filtration[16] and cell washing.[36]

The transfer impedance model mentioned earlier[17] can be used to explore the parameter space for the optimum selection of layer thicknesses for a planar half-wave design. The following optimisation is based on a design with a PZT4D transducer (speed of sound, $c = 4530$ m s^{-1}, density, $\rho = 7700$ kg m^{-3}, piezo-constant $h_{33} = 2.37 \times 10^9$ V m^{-1}, relative permittivity (clamped) $\varepsilon^S r = 700$), and carrier and reflector layers of glass ($\rho = 2240$ kg m^{-3}, $c = 6000$ m s^{-1}), and water as the fluid ($\rho = 1000$ kg m^{-3}, $c = 1480$ m s^{-1}). The damping was approximated by a Q-factor (the ratio of the stored to dissipated energy) of 100 in all layers (this also accounts for other losses in the structure). Damping was implemented by introducing complex values[24,37] for c, h_{33} and $\varepsilon^S r$. The energy densities quoted later are for a drive voltage of 10 Vpp. While the results are specific to these materials, the optimisation shows some useful design principles that could be extended to other material systems.

The PZT thickness was fixed at 1 mm (see the discussion of scaling later). For each combination of parameters, the frequency with the maximum radiation force was found, as shown in Figure 7.6. In order to visualise the optimisation of the three variables, Figure 7.6 is plotted first for the case when there is no carrier layer, then for a $\lambda/4$ reflector layer. It can be seen that the optimum combination is for a fluid layer of 360 µm, reflector layer 800 µm thick, and zero carrier layer. At the resonant frequency of this

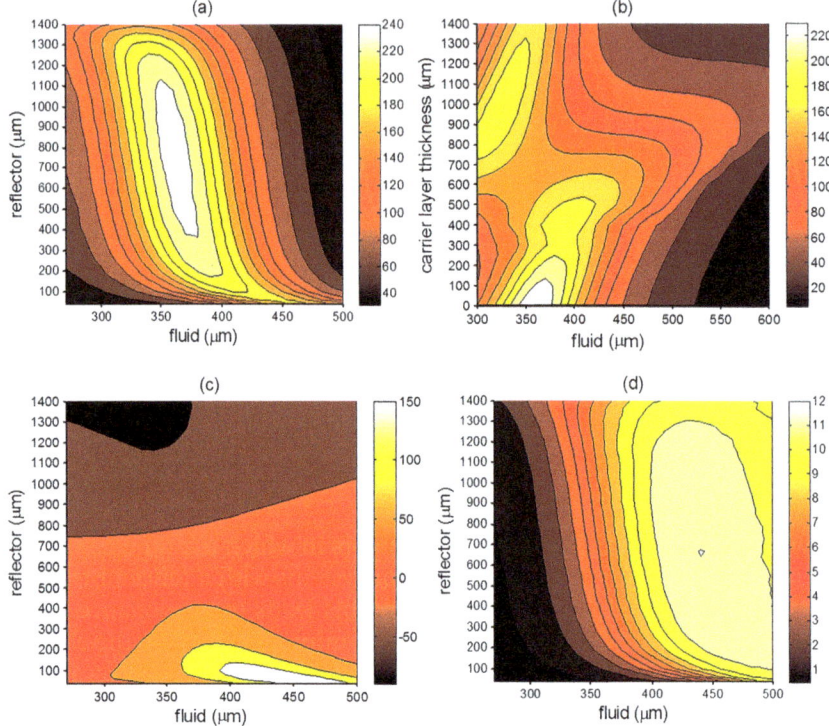

Figure 7.6 Parameter space optimisation for a half-wave resonator. Maximum forces (pN) on a 10 μm polystyrene bead for (a) combinations of fluid and carrier layer thickness with a λ/4 reflector thickness, (b) combinations of fluid and reflector layer thicknesses, (c) force on a 10 μm polystyrene bead at reflector/fluid boundary, (d) performance number for combinations of fluid and reflector layer thicknesses. (b)–(d) are for zero carrier-layer thickness; the transducer thickness is 1 mm. Reproduced from ref. 82.

combination, 1.90 MHz, these can be expressed as fractions of the acoustic wavelength in their corresponding layers: transducer $0.84 \times \lambda_{xdr}/2$; fluid layer $0.93 \times \lambda_f/2$; reflector thickness $1.0 \times \lambda_r/4$.

In a previous paper,[38] the variations of carrier thickness were explored for a fixed fluid layer of λ/2 thickness, leading to the conclusion that a non-zero carrier layer thickness was optimum. This fixed fluid layer case is supported here, but we see that a stronger radiation force is found for a thinner fluid layer and zero carrier layer thickness. The optimum carrier layer thickness is also different to that suggested by Gröschl,[22] which was for an analysis where the operating frequency was constrained to be identical to the resonance frequency of the free transducer.

The optimum found here (the highest force for a given transducer drive voltage) is not very sensitive to variations: reflector thickness can vary by ±60% to cause a drop of only 10% in radiation force; similarly, fluid layer thickness can vary by ±10% for the same drop. When the reflector thickness is close to $\lambda/2$, the radiation force becomes much weaker (we can also question whether the resulting field pattern can really be called a half-wave design), but recovers effective operation even quite close to this.

If we explore this optimum point in more detail, we find that for these layer thicknesses there are two possible resonances, shown by the twin energy density peaks in Figure 7.7(a). The corresponding pressure distributions through the layers for each of these resonances are also plotted; the higher frequency of the pair has a pressure node in the PZT, which (since the radiation force is proportional to the gradient of the potential energy density and hence gradient of pressure amplitude squared, see eqn (7.1)) will lead to a force towards the carrier layer near the bottom of the fluid layer; in contrast, the lower frequency resonance tends to push particles away from the carrier layer. It is found that the lower frequency case is always more energetic in the region plotted in Figure 7.6(a) and (b). In addition to being stronger, this lower frequency resonance is also more likely to be practically useful as it tends to drive (dense) particles away from the transducer/matching layer side of the fluid channel.

The effect of reflector thickness on the radiation force near the fluid/reflector boundary is shown in Figure 7.6(c). Interestingly, we find that we can take advantage of the insensitivity to reflector thickness to choose whether there is a force into the reflector, or away from it (positive force is into the reflector). In many applications we want to ensure there is a force away from the reflector and will choose a reflector thickness between a quarter and a half wavelength (*e.g.* $3\lambda/8$).

These results are all for a PZT thickness of 1 mm. However, planar resonator designs can be scaled linearly, *i.e.* if the PZT thickness is doubled, the corresponding optimum thicknesses for the other layers are also doubled (and the operating frequency halved).

We can summarise these findings in the following half-wave design rules:

(1) Make the carrier layer as thin as possible, or better still, omit it entirely.
(2) Optimum transducer thickness close to $\sim 0.84 \times \lambda/2$ (corresponding to a transducer running close to its own resonant frequency).
(3) Optimum fluid thickness $\sim 0.93 \times \lambda/2$ (less if more damping, see later).
(4) Optimum reflector thickness $\sim \lambda/4$ (very insensitive).
(5) For force away from reflector, choose a reflector thickness greater than $\lambda/4$.
(6) For force away from the carrier layer, choose a lower frequency resonance (of the two possibilities available for any given combination of layer thicknesses).

This analysis is for an optimum radiation force for a given transducer drive voltage. In some applications where self-heating is the limiting factor, rather than drive voltage we may wish to optimise for the performance number as

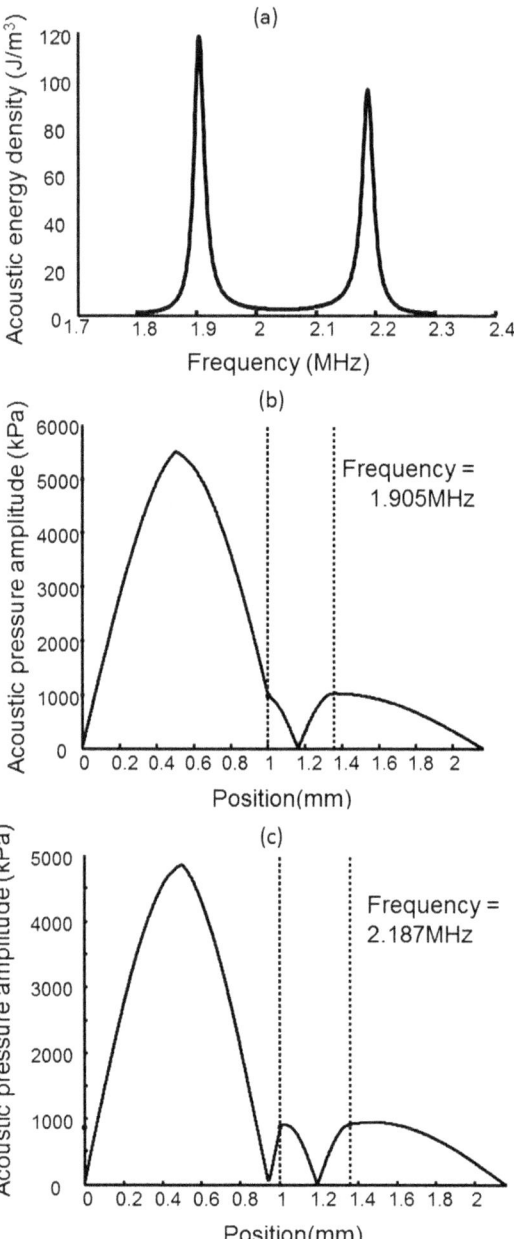

Figure 7.7 Examining the resonances found in the configuration with the optimum combination of layer thicknesses. (a) Two energetic frequencies can be seen corresponding to half-wave resonances with differing force distributions across the fluid layer. (b), (c) The pressure amplitude distributions at these two frequencies. Dotted lines indicate the boundaries between the transducer, fluid and reflector layers (transducer on the left). Reproduced from ref. 82.

defined by Gröschl.[15] This non-dimensional metric is defined as the ratio of energy density in the fluid layer to unit electrical energy per cycle entering the system. We plot in Figure 7.6(d) the performance number over a range of layer thicknesses, and see that the optimum region in this case is much less sensitive to fluid thickness.

The exact positions of the optimum dimensions will depend on material properties, however, the trends above should be valid for materials with a high acoustic impedance. Q-factors in the range 50 to 500 were also explored, and similar trends found—the position of the optimum radiation force or performance number was essentially unchanged. For a given frequency, the force on a particle in the channel is proportional to the acoustic energy density (see eqn (7.1)). For these material parameters, the configuration with maximum energy density was found to always coincide with that of maximum radiation force. We should note that one may also wish to optimise for average force, or force at a particular boundary that may not coincide with this optimum set of layer thicknesses.

While the half-wave resonator has a fluid layer thickness of $\sim\lambda/2$, similar configurations with thickness $\lambda/2$ behave in a similar manner, but with multiple trapping planes (see applications section later).

Further considerations for the design of resonators are described in detail by Gröschl,[22] including the effects of lack of planarity, and acoustic absorption.

7.5.2 Quarter-Wave Resonators

Quarter-wave devices are characterised by a resonant reflector layer that effectively imposes a boundary release pressure condition at the top of the fluid layer. They are designed to bring particles towards the reflector/fluid interface.[39] The term reflector layer can be confusing as this layer functions differently in this case than in a half-wave device; however, for consistency the term has been retained. In Chapter 11, the term "attractor wall" is used to describe the fluid/solid boundary of this layer.

In contrast to half-wave designs, quarter-wave devices are very sensitive to their reflector thickness; this is largely due to the necessity of tuning two resonances: the resonance across the transducer, carrier and fluid layers, and also the resonance in the reflector layer. Townsend et al.[40] show that for a typical design a change in the reflector thickness of 5% is sufficient to cause a drop in radiation force of 50%. In addition, lateral effects (from non-ideal resonances that have a lateral variation), can cause the node location to vary between a position in the reflector and a position within the fluid channel making it difficult to push particles uniformly onto the reflector surface. It is also found that an antinode can exist with the channel causing particles to move towards the carrier layer rather than the reflector depending on the initial position of the particle. The quarter-wave design is so sensitive to layer thicknesses that general design rules of the type described earlier for a half-wave device are hard to form, and a 1D computer-based layer model is required for effective design.

It is also possible to have an inverted quarter-wave resonant mode in which the transducer and carrier layers are in resonance, coupled to a resonance across the fluid and reflector layers such that there is a node close to the carrier/fluid (boundary), pushing particles towards this interface.

7.5.3 Thin-Reflector Resonators

An alternative to the quarter-wave mode described is provided by the "thin reflector mode".[41] This is also designed to bring particles towards the reflector/fluid interface. In this mode, the first order resonance of the whole layered structure is utilised. The reflector, fluid and carrier layers are all thin compared to the wavelength (*e.g.* 0.1λ) such that particles in the fluid channel are attracted towards the pressure node at the reflector/air interface. In contrast to the quarter-wave mode, this mode is relatively insensitive to layer thickness variations, and when suitably designed there are positive forces towards the reflector/fluid interface from all positions within the fluid channel. Since relatively little of the acoustic energy is stored within the carrier or reflector layers, this mode is particularly suitable for implementation in polymers to provide disposable devices.

Figure 7.8 shows layer thickness optimisation for a thin reflector device using the same material properties as described for a half-wave device and a 1 mm thick transducer. Again, it is found (graphs not shown) that having no carrier layer is the most efficient configuration for achieving both maximum average radiation force, and maximum performance number. The average radiation force (spatially averaged over the fluid layer) shows a broad maximum over a range of fluid layers having over 90% of its maximum value over the range $t_f = 0...70$ μm (corresponding to zero to $0.08 \times \lambda_f$).

7.6 Position Control in Resonators

The resonant designs described are limited in the flexibility with which the equilibrium particle position can be manipulated. The nodal position is designed by an appropriate choice of layer thickness, and once created there is little scope for moving this position by, for example, changing frequencies, as radiation force is lost away from the resonant frequency. The node position is essentially fixed by the device geometry. A number of strategies have been demonstrated to overcome this limitation, and are described later. Some of these are more applicable to manipulation in 2D and 3D devices and are discussed in more detail in Chapters 8 and 9.

Kozuka *et al.*[42] demonstrated an array of multi-wavelength focussed transducers acting in air, and showed that particles tended to become positioned above active elements. Demore *et al.*[43] integrated this into a microfluidic environment, and showed that this effect is due to secondary, kinetic energy gradient driven forces.

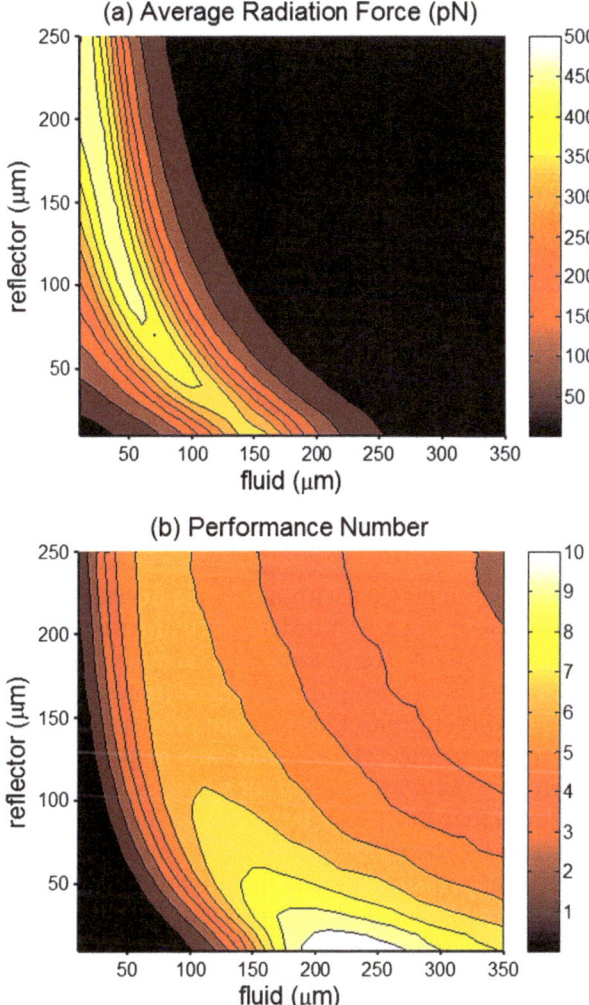

Figure 7.8 Parameter space optimisation for a thin-reflector resonator. (a) Average radiation force in the fluid layer on a 10 μm polystyrene bead. (b) Performance number. In both cases the transducer thickness is 1 mm, and there is no carrier layer. Reproduced from ref. 82.

If high radiation force acoustic fields are not required, then wider bandwidth polymer device layers[33] increase the bandwidth of the device. This allows the resonator to be run over a range of frequencies and the axial nodal positions can then be adjusted simply by changing the driving frequency.

In multi-wavelength devices, sweeping the frequency in a saw tooth pattern enables particles to be moved continuously over a distance of many wavelengths.[9,32,42,44]

An alternative approach has been demonstrated by Glynne-Jones et al.[45] This allows continuous particle positioning in a device that supports both a half-wave and a quarter-wave mode. It was shown that rapidly switching between the two modes at a sufficiently high frequency (e.g. 100 Hz) provides an average force on a particle so as to form a new equilibrium position between the half- and quarter-wave nodal positions. Since this is a stable equilibrium, no feedback control is required to achieve this positioning, and the location of the average node is determined by the duty cycle of the switching waveform. This driving voltage can be written as

$$V(t) = A\sin(w_1 t)S(t) + A\sin(w_2 t)(1 - S(t)) \tag{7.2a}$$

where A represents a voltage amplitude, and $S(t)$ is a unit amplitude switching function with variable mark to space ratio. It would also be possible to achieve this control by applying a waveform that is a sum of the two modal frequencies, and using amplitude weighting to effect control of the nodal position.

$$V(t) = A_1 \sin(w_1 t) + A_2 \sin(w_2 t) \tag{7.2b}$$

The use of multiple frequencies within a single device has also been used in other circumstances.[28] To a first order approximation, the effect on the acoustic radiation force of combining two acoustic fields at different frequencies is linear—the resulting time averaged forces are the sum of those that would have acted if either frequency were present by itself. This is true both for when the frequency components are applied through a single transducer driven by a combined signal, or when two transducers are used (with the advantage that each can be tuned to be resonant at its drive frequency). This can be shown by considering the expression for the time averaged second order pressure as used in derivations of radiation force by, for example, Yosioka and Kawasima[46] or Gor'kov.[1] This is given by

$$\langle p_2 \rangle = \frac{1}{2\rho_0 c_a^2} \langle p_1^2 \rangle - \frac{1}{2}\rho_0 \langle v_1^2 \rangle \tag{7.3}$$

where ρ_0 is the quiescent fluid density, and c_a is the speed of sound in the medium, p_1 and v_1 are the first order acoustic pressure and velocity, and the angled brackets denote time average. If p_1 and v_1 are each a sum of two sinusoids of different frequencies then the time average of p_1^2 or v_1^2 is given by the sum of the time average of the independent components. A similar argument holds for the momentum flux terms that enter the radiation force derivations. It should be noted that this only applies when the time average is taken over many cycles—if the two frequencies are very close, a zero-average beat frequency will be seen as described by Oberti et al.,[47] who also consider the case when two signals of the same frequency are superimposed.

Switching of frequencies has also been applied to the fractionation of particle suspensions. This was demonstrated by Mandralis et al.[48] in combination with a synchronised field-flow fractionation system, whereby particles were cycled away and towards a surface, with larger particles moving further and experiencing regions of higher flow.

Another fractionation method achieved by frequency switching and relying on response time of the particle has been proposed by Harris et al.[49] who used a half-wavelength mode to position particles to a known location in the flow field, then by switching between the second and third modes, particles were fractionated across the depth of the channel. Using continuous flow, different fractions can be extracted using flow splitters. Similar approaches have been demonstrated experimentally in lateral devices.[50,51]

7.7 Applications

An exhaustive description of the potential applications of planar ultrasonic resonators is beyond the scope of this chapter, but this section discusses a number of illustrative applications.

7.7.1 Filtration, Washing and Separation

Amongst the earliest applications for planar resonator ultrasonic particle manipulators was filtration. Some of these were on a relatively large, multiple wavelength scale.[52] The commercial Sonosep™ system was described by Pui et al.[53] as a component for bioreactor systems and production of therapeutic proteins. In their system, a planar resonator was used to aggregate hybridoma cells which then fall more rapidly under gravity, enhancing sedimentation and clarification stages.

However, working on a much smaller system, Hawkes et al.[16] developed a filtration technique based on a half-wave resonator of 250 µm in depth. They demonstrated the filter's ability to generate a particle depleted sample using yeast and latex particles of between 1.5 and 25 µm in diameter. A similar approach was taken by Harris et al.,[54] but in this case the channel was etched into Pyrex which was anodically bonded to a silicon substrate into which inlet and outlet channels had been anisotropically etched. Such microfabricated devices are expensive and complex to produce in small batches so there has been a recent tendency to use lower-cost disposable elements such as capillaries[55] which may then be integrated into a microfluidic system. Using disposable elements removes the need for expensive sterilisation processes between samples and prevents cross-contamination: essential to applications such as medical diagnosis and forensics. The flexibility of disposable or modular systems is also advantageous during the development of devices; microfluidic channels can be contained by large, inexpensive components making handling and modification more straightforward for the experimentalist.

As the physical properties of different populations of cells or particles vary, it is conceivable that they may be fractionated based on their acoustic properties or size, and therefore magnitude of the radiation force. An early illustration of this was reported by Johnson and Feke[56] by introducing a particle stream alongside a 'carrier fluid' stream to ensure the particles were constrained near the channel wall, so that on passing through a half-wavelength, acoustic field particles then fractionated across the channel's width. The same group also demonstrated that it is possible to enhance separation of certain particle combinations by adjusting the properties of the carrying fluid.[57] The use of a 'carrier fluid' or sheath flow within a planar resonator provides the potential to wash cells and particles, by moving them from the initial suspending fluid into the carrier fluid.[36] A limit on this, however, is the radiation force on the fluid/fluid boundary itself, as this can result in a mixing of the two fluid phases.[58]

7.7.2 Sensing and Detection

The use of ultrasound to form a concentrated stream of particles can improve the sensitivity of a sensor. This is possible either by collecting the concentrated stream and removing excess fluid or by moving the plane of particles directly onto a sensor surface. The former approach, essentially a version of some of the filtration approaches described earlier, has been developed for biohazard detection by Townsend *et al.*[40] who use a quarter-wavelength system to concentrate 1 micron sized spores close to the reflector surface with this concentrated region then extract them through one of two outlets and separate them from the remaining clarified region. Quarter-wavelength systems, although typically less efficient than half-wavelength systems, do allow the fluidic system to be simplified; as particles move to the surface, the flow only needs to be split into two streams to separate the concentrate from the clarified flow, as opposed to a half-wavelength where three streams are required to remove the clarified flow from both sides of the concentrate. The fluidic design is a continual challenge for planar flow-through systems; whilst the flow regime in such devices is highly laminar (and therefore predictable, and typically uniform), the acoustic field is less so and likewise the particle trajectories.

Hawkes *et al.*[59] developed a device to push bacterial spores directly onto an antibody coated surface, and demonstrated a significant improvement on the sensitivity and a 200-fold increase in spore capture. In these sensing applications, it is important for the interface to be less than a quarter wavelength from a pressure antinode in order to produce a net force towards the surface. The sensitivity, in particular, of the node location to the reflector thickness was demonstrated by Martin *et al.*[39] This concept has also been combined with sensor elements, for example by Glynne-Jones *et al.*[30] who designed a system incorporating an optical waveguide using a planar resonance to drive beads to the functionalised waveguide surface. On switching the excitation frequency, unbound beads can be driven away leaving only the immobilised beads on the sensor surface, and therefore a means to improve the sensitivity of bead-based bioassays.

Similar concepts have been explored to facilitate the measurement or monitoring of chemical reactions. For example, Radel *et al.*[60] demonstrated the effective use of different excitation frequencies to control the movement of particles onto and away from an infrared spectroscopy probe for online periodic monitoring of chemical reactions, with potential applications in bio-reactors. This multiple-wavelength system, operating in the low MHz range, uses periodic frequency switching to control whether a positive or negative force is seen at the probe surface, enabling particulates to be periodically screened by the probe, then returned to suspension and the probe surface to be cleaned.

Chemical reactions can influence the acoustic properties of particulates, which can be an advantage for the purposes of sensing. This has been demonstrated by Kanazaki *et al.*[61] while monitoring ion-exchange processes. Replacement of counterions causes the water content of ion exchange resins to change, leading to a change in their acoustic properties and the resultant radiation force acting on a resin bead. Kanazaki *et al.* use a planar resonator to levitate a resin bead and the bead's location is determined by the balance between acoustic radiation forces and gravity. Following ion-exchange, the bead is seen to relocate to a new equilibrium position with the displacement indicative of the fluid's ion content.

The relatively large scale of action of ultrasonic manipulation in comparison with other manipulation technologies means that there is an increasing interest in combining ultrasonic approaches with techniques such as dielectrophoresis[62,63] or optical trapping[64] within integrated microfluidic systems.

7.7.3 Cell-Interaction Studies and Sonoporation

Planar resonators are particularly suited to forming 2D monolayers (and discoid aggregates) of cells, with much potential for future cell-interaction studies. Ultrasonic radiation forces provide a method that can produce useful forces on cells without adversely effecting viability,[65–68] an aspect more thoroughly explored later in this series of tutorials. Additionally, due to the expansion of microsystem technology and its synergy with ultrasonic resonators, the suitability of ultrasound devices for biological fluid processing and bio-sensing has attracted increasing interest.

Both Bazou *et al.*[69] and Edwards *et al.*[70] illustrate the formation of 2D aggregates of hepatocyte cells in circular planar resonators. A uniform axial field causes the cells to move into a controlled plane free from the solid substrate, and such that significantly weaker lateral forces agglomerate the cells to form a closely packed monolayer of cells enabling cell–cell interaction and adhesion processes to be studied. This has also been demonstrated with neural cells.[71]

Where the lateral forces are strong enough, it is possible to introduce flow of a suitable medium as shown by Hultström *et al.*[68] They studied the effects of ultrasound exposure on adherent cell proliferation rate in a system that demonstrated the trapping of adherent cells in a 2D agglomeration against fluid flow and isolated from the influence of any surfaces. Previously, Morgan *et al.*[72] used a planar system combined with weaker lateral forces to increase mass transfer of toxicant to hepatocyte cells to expedite toxicological tests. In this work the lateral force also improves formation of spheroidal clusters where cells move out of plane creating 3D constructs, conducive to mimicking *in vivo* hepatocyte cell function. Similarly, the introduction of fluid flow also serves to mimic the action of blood where Bazou *et al.*[73] used this to investigate the development of cancer cells clusters, again for toxicological assessment. An extension of this concept is to immobilise these cell constructs in a gel matrix as shown by Gherardini *et al.*[8] using yeast cells, but more recently by Garvin *et al.*[74] studying vascularisation and Bazou *et al.*[75,76] who used discoids of an hepatocyte cell line immobilised in alginate for rapid creation of *in vitro* tissue models. More detailed examinations of biological applications can be found in Chapter 21.

There has been much research on the use of ultrasound to transfect cells (sonoporation) to permit bio-molecules to cross the cell membrane, for example to effect toxicological analysis or gene therapy. Some of these studies use a standing wave field, specifically with planar resonators. Although the mechanisms are unclear, there is evidence to suggest that these systems have the potential to produce high transfection rates and improve viability over that reported using inertial cavitation.[77-79] Whilst the potential for poration in planar resonators is promising,[80,81] there remains little evidence of the cell physiology, cell dynamics, acoustic environment and poration mechanism in such systems.

7.8 Conclusions

Particle and cell manipulation using ultrasonic standing waves is well matched to lab-on-a-chip technology and of the various techniques available to the researcher for generating an appropriate field, the planar layered resonator remains the most straightforward. Its simplicity lends itself to modeling and optimization using highly efficient one-dimensional models, but the results of these need to be considered along with two- and three-dimensional representations, due to the importance of lateral field variations in many devices.

References

1. L. P. Gor'kov, *Sov. Phys. Doklady*, 1962, **6**, 773–775.
2. T. Lilliehorn, U. Simu, M. Nilsson, M. Almqvist, T. Stepinski, T. Laurell, J. Nilsson and S. Johansson, *Ultrasonics*, 2005, **43**, 293–303.
3. H. M. Hertz, *J. Appl. Phys.*, 1995, **78**, 4845–4849.
4. J. Lee, S.-Y. Teh, A. Lee, H. H. Kim, C. Lee and K. K. Shung, *Ultrasound Med. Biol.*, 2010, **36**, 350–355.
5. P. Augustsson, J. Persson, S. Ekstrom, M. Ohlin and T. Laurell, *Lab Chip*, 2009, **9**, 810–818.
6. F. Petersson, A. Nilsson, C. Holm, H. Jonsson and T. Laurell, *Lab Chip*, 2005, **5**, 20–22.

7. G. Goddard and G. Kaduchak, *J. Acoust. Soc. Am.,* 2005, **117**, 3440–3447.
8. L. Gherardini, C. M. Cousins, J. J. Hawkes, J. Spengler, S. Radel, H. Lawler, B. Devcic-Kuhar and M. Groschl, *Ultrasound Med. Biol.,* 2005, **31**, 261–272.
9. A. Haake, A. Neild, G. Radziwill and J. Dual, *Biotechnol. Bioeng.,* 2005, **92**, 8–14.
10. A. Neild, S. Oberti, F. Beyeler, J. Dual and B. J. Nelson, *J. Micromech. Microeng.,* 2006, **16**, 1562–1570.
11. M. Bok, H. Y. Li, L. Y. Yeo and J. R. Friend, *Biotechnol. Bioeng.,* 2009, **103**, 387–401.
12. J. Shi, X. Mao, D. Ahmed, A. Colletti and T. J. Huang, *Lab Chip,* 2008, **8**, 221–223.
13. C. D. Wood, J. E. Cunningham, R. O'Rorke, C. Walti, E. H. Linfield, A. G. Davies and S. D. Evans, *Appl. Phys. Lett.,* 2009, **94**, 054101.
14. V. Yantchev, J. Enlund, I. Katardjiev and L. Johansson, *J. Micromech. Microeng.,* 2010, **20**, 035031.
15. M. Gröschl, *Acustica,* 1998, **84**, 432–447.
16. J. J. Hawkes and W. T. Coakley, *Sens. Actuators, B,* 2001, **75**, 213–222.
17. M. Hill, Y. Shen and J. J. Hawkes, *Ultrasonics,* 2002, **40**, 385–392.
18. Y. Abe, M. Kawaji and T. Watanabe, *Exp. Therm. Fluid Sci.,* 2002, **26**, 817–826.
19. C. Courtney, C.-K. Ong, B. Drinkwater, P. Wilcox, C. Démoré, S. Cochran, P. Glynne-Jones and M. Hill, *J. Acoust. Soc. Am.,* 2010, **128**, EL195–199.
20. H. Nowotny, E. Benes and M. Schmid, *J. Acoust. Soc. Am.,* 1991, **90**, 1238–1245.
21. P. D. Wilcox, R. S. C. Monkhouse, P. Cawley, M. J. S. Lowe and B. A. Auld, *NDT&E Int.,* 1998, **31**, 51–64.
22. M. Gröschl, *Acustica,* 1998, **84**, 632–642.
23. H. Bruus, *Lab Chip,* 2012, **12**, 20–28.
24. S. Sherrit, S. Leary, B. Dolgin and Y. Bar-Cohen, *Comparison of the Mason and KLM Equivalent Circuits for Piezoelectric Resonators in the Thickness Mode*, Ultrasonics Symposium, 1999, Proceedings IEEE, vol. 2, Nevada USA, 1999.
25. J. Dual and D. Möller, *Lab Chip,* 2012, **12**, 506.
26. R. Krimholt, D. A. Leedom and G. L. Matthaei, *Electron. Lett.,* 1970, **6**, 398–399.
27. R. J. Townsend, M. Hill, N. R. Harris and N. M. White, *Ultrasonics,* 2006, **44**, e467–e471.
28. O. Manneberg, J. Svennebring, H. M. Hertz and M. Wiklund, *J. Micromech. Microeng.,* 2008, **18**, 095025.
29. S. M. Hagsater, T. G. Jensen, H. Bruus and J. P. Kutter, *Lab Chip,* 2007, 7, 1336–1344.
30. P. Glynne-Jones, R. J. Boltryk, M. Hill, F. Zhang, L. Q. Dong, J. S. Wilkinson, T. Melvin, N. R. Harris and T. Brown, *Anal. Sci.,* 2009, **25**, 285–291.
31. S. M. Hagsater, A. Lenshof, P. Skafte-Pedersen, J. P. Kutter, T. Laurell and H. Bruus, *Lab Chip,* 2008, **8**, 1178–1184.

32. O. Manneberg, B. Vanherberghen, B. Onfelt and M. Wiklund, *Lab Chip*, 2009, **9**, 833–837.
33. I. Gonzalez, L. J. Fernandez, T. E. Gomez, J. Berganzo, J. L. Soto and A. Carrato, *Sens. Actuators, B*, 2010, **144**, 310–317.
34. J. J. Hawkes, M. Gröschl, E. Benes, H. Nowotny and W. T. Coakley, *Revista de Acustica*, 2002, **33**.
35. J. J. Hawkes, D. Barrow and W. T. Coakley, *Ultrasonics*, 1998, **36**, 925–931.
36. J. J. Hawkes, R. W. Barber, D. R. Emerson and W. T. Coakley, *Lab Chip*, 2004, **4**, 446–452.
37. S. Sherrit, *IEEE Transactions on Ultrasonics, Ferroelectrics, and Frequency Control*, 2008, **55**(11), 2479–2483.
38. M. Hill, R. J. Townsend and N. R. Harris, *Ultrasonics*, 2008, **48**, 521–528.
39. S. P. Martin, R. J. Townsend, L. A. Kuznetsova, K. A. J. Borthwick, M. Hill, M. B. McDonnell and W. T. Coakley, *Biosens. Bioelectron.*, 2005, **21**, 758–767.
40. R. J. Townsend, M. Hill, N. R. Harris and M. B. McDonnell, *Ultrasonics*, 2008, **48**, 515–520.
41. P. Glynne-Jones, R. J. Boltryk, M. Hill, N. R. Harris and P. Baclet, *J. Acoust. Soc. Am.*, 2009, **126**, EL75–EL79.
42. T. Kozuka, T. Tuziuti, H. Mitome and T. Fukuda, *Jpn. J. Appl. Phys., Part 1*, 1998, **37**, 2974–2978.
43. P. Glynne-Jones, C. E. M. Demore, C. Ye, Y. Qiu, S. Cochran and M. Hill, *IEEE Transactions on Ultrasonics, Ferroelectrics and Frequency Control*, 2012, **59**(6), 1258–1266.
44. T. L. Tolt and D. L. Feke, *Chem. Eng. Sci.*, 1992, **48**, 527–540.
45. P. Glynne-Jones, R. J. Boltryk, N. R. Harris, A. W. J. Cranny and M. Hill, *Ultrasonics*, 2010, **50**, 68–75.
46. K. Yosioka and Y. Kawasima, *Acustica*, 1955, **5**, 167–173.
47. S. Oberti, A. Neild and J. Dual, *J. Acoust. Soc. Am.*, 2007, **121**, 778–785.
48. Z. Mandralis, W. Bolek, W. Burger, E. Benes and D. L. Feke, *Ultrasonics*, 1994, **32**, 113–121.
49. N. Harris, R. Boltryk, P. Glynne-Jones and M. Hill, *Phys. Procedia*, 2010, **3**, 277–281.
50. K. M. Lim and Y. Liu, *Lab Chip*, 2011, **11**, 3167–3173.
51. F. Petersson, *PhD Thesis*, Lund University, 2007.
52. S. Gupta and D. L. Feke, *AIChE J.*, 1998, **44**, 1005–1014.
53. P. W. S. Pui, F. Trampler, S. A. Sonderhoff, M. Groeschl, D. G. Kilburn and J. M. Piret, *Biotechnol. Prog.*, 1995, **11**, 146–152.
54. N. R. Harris, M. Hill, S. P. Beeby, Y. Shen, N. M. White, J. J. Hawkes and W. T. Coakley, *Sens. Actuators, B*, 2003, **95**, 425–434.
55. B. Hammarstrom, M. Evander, H. Barbeau, M. Bruzelius, J. Larsson, T. Laurell and J. Nillsson, *Lab Chip*, 2010, **10**, 2251–2257.
56. D. A. Johnson and D. L. Feke, *Sep. Technol.*, 1995, **5**, 251–258.
57. S. Gupta, D. L. Feke and I. Manaszloczower, *Chem. Eng. Sci.*, 1995, **50**, 3275–3284.

58. L. Johansson, S. Johansson, F. Nikolajeff and S. Thorslund, *Lab Chip*, 2009, **9**, 297–304.
59. J. J. Hawkes, M. J. Long, W. T. Coakley and M. B. McDonnell, *Biosens. Bioelectron.*, 2004, **19**, 1021–1028.
60. S. Radel, M. Brandstetter and B. Lendl, *Ultrasonics*, 2010, **50**, 240–246.
61. T. Kanazaki, S. Hirawa, M. Harada and T. Okada, *Anal. Chem.*, 2010, **82**, 4472–4478.
62. S. K. Ravula, D. W. Branch, C. D. James, R. J. Townsend, M. Hill, G. Kaduchak, M. Ward and I. Brener, *Sens. Actuators, B*, 2008, **130**, 645–652.
63. M. Wiklund, C. Günther, R. Lemor, M. Jäger, G. Fuhr and H. M. Hertz, *Lab Chip*, 2006, **6**, 1537–1544.
64. G. Thalhammer, R. Steiger, M. Meinschad, M. Hill, S. Bernet and M. Ritsch-Marte, *Biomed. Opt. Express*, 2011, **2**, 2859–2870.
65. H. Bohm, P. Anthony, M. R. Davey, L. G. Briarty, J. B. Power, K. C. Lowe, E. Benes and M. Groschl, *Ultrasonics*, 2000, **38**, 629–632.
66. D. Bazou, R. Kearney, F. Mansergh, C. Bourdon, J. Farrar and M. Wride, *Ultrasound Med. Biol.*, 2011, **37**, 321–330.
67. B. Vanherberghen, O. Manneberg, A. Christakou, T. Frisk, M. Ohlin, H. M. Hertz, B. Onfelt and M. Wiklund, *Lab Chip*, 2010, **10**, 2727–2732.
68. J. Hultström, O. Manneberg, K. Dopf, H. M. Hertz, H. Brismar and M. Wiklund, *Ultrasound Med. Biol.*, 2006, **33**, 175–181.
69. D. Bazou, G. P. Dowthwaite, I. M. Khan, C. W. Archer, J. R. Ralphs and W. T. Coakley, *Mol. Membr. Biol.*, 2006, **23**, 195–205.
70. G. O. Edwards, D. Bazou, L. A. Kuznetsova and W. T. Coakley, *Cell Commun. Adhes.*, 2007, **14**, 9–20.
71. D. Bazou, G. A. Foster, J. R. Ralphs and W. T. Coakley, *Mol. Membr. Biol.*, 2005, **22**, 229–240.
72. J. Morgan, J. F. Spengler, L. Kuznetsova, W. T. Coakley, J. Xu and W. M. Purcell, *Toxicol. In Vitro*, 2004, **18**, 115–120.
73. D. Bazou, M. J. Santos-Martinez, C. Medina and M. W. Radomski, *Br. J. Pharmacol.*, 2011, **162**, 1577–1589.
74. K. A. Garvin, D. C. Hocking and D. Dalecki, *Ultrasound Med. Biol.*, 2010, **36**, 1919–1932.
75. D. Bazou, W. T. Coakley, A. J. Hayes and S. K. Jackson, *Toxicol. In Vitro*, 2008, **22**, 1321–1331.
76. D. Bazou, *Cell Biol. Toxicol.*, 2010, **26**, 127–141.
77. Y. H. Lee and C. A. Peng, *Gene Ther.*, 2005, **12**, 625–633.
78. Y. H. Lee and C. A. Peng, *Ultrasound Med. Biol.*, 2007, **33**, 734–742.
79. S. Rodamporn, N. R. Harris, S. P. Beeby, R. J. Boltryk and T. Sanchez-Elsner, *IEEE Trans. Biomed. Eng.*, 2011, **58**, 927–934.
80. D. Carugo, D. N. Ankrett, P. Glynne-Jones, L. Capretto, R. J. Boltryk, X. Zhang, P. A. Townsend and M. Hill, *Biomicrofluidics*, 2011, **5**, 044108.
81. W. Longsine-Parker, H. Wang, C. Koo, J. Kim, B. Kim, A. Jayaraman and A. Han, *Lab Chip*, 2013, **13**(11), 2144–2152.
82. P. Glynne-Jones, R. J. Boltryk and M. Hill, *Lab Chip*, 2012, **12**, 1417–1426.

CHAPTER 8

Applications in Continuous Flow Acoustophoresis

ANDREAS LENSHOF[a], PER AUGUSTSSON[a,b], AND THOMAS LAURELL*[a,c]

[a]Department of Biomedical Engineering, Lund University, Sweden;
[b]Department of Electrical Engineering and Computer Science, Massachusetts Institute of Technology, Cambridge, MA, USA; [c]Department of Biomedical Engineering, Dongguk University, Seoul, Republic of Korea
*E-mail: thomas.laurell@bme.lth.se

8.1 Introduction

Acoustophoresis offers means to integrate a multitude of particle or cell handling unit operations in continuous flow based microfluidic systems. The most evident mode of operation is to utilize the primary acoustic radiation force to translate particulate matter in a non-contact mode across streamlines into an acoustic standing wave pressure node or antinode, thereby deriving a concentrated sample stream. This is analogous to a conventional centrifugation step where either a clear supernatant or the sediment is the desired fraction. The same function can also be used to perform continuous flow based washing or buffer exchanges of particles in suspension, which finds applications in *e.g.* cell handling and microbead-based affinity extraction. Similarly, particles or cells can be transiently exposed to a chemical component by rapidly switching particles across streamlines of different laminated buffers. Acoustophoresis systems can also perform rare event sorting by combining *e.g.* a fluorescence detection unit that triggers an acoustic switching of the rare object to a target outlet. The ability to pre-align cells and particles in a well-defined position in a streamline has enabled a variety of acoustophoresis based cytometry applications both utilizing the precise alignment for in-chip optical interrogation of cells or in Coulter counter configurations.

The fact that the acoustophoretic mobility scales with the square of the radius inherently enables size based separation of particles and has also been realized in the free flow acoustophoresis concept. Since the direction of the primary acoustic radiation force depends on the acoustic contrast factor, acoustophoretic systems can be designed to perform bidirectional separation of two particle types that exhibit positive and negative acoustic contrast factors, respectively. The most common case is the removal of lipid emulsions from particles of cell suspensions. Bidirectional separation can also be accomplished by manipulating the acoustophysical properties of the carrier medium to enhance the difference in acoustic contrast factor between two particle types. In cases where *e.g.* a specific cell type cannot be differentiated against the background population by its acoustic properties, the acoustophoretic mobility of the target cell can then be modified by employing affinity specific beads. The affinity beads bind only to the target cells, and hence the cell–bead complex now displays different acoustophoretic mobility *versus* the background population. The separation can subsequently be performed in a free flow acoustophoresis configuration. This mode of operation is analogous to the laboratory standard magnetic bead based cell separations but now enables a continuous flow mode of operation. The acoustophoresis tool box and several of its application areas are further detailed in this chapter and are followed by a discussion on the limits and challenges of acoustophoresis.

The basic acoustofluidic configuration for the systems described in this chapter is shown in Figure 8.1. A liquid carries cells or microbeads along a microchannel while acoustic radiation forces, perpendicular to the flow,

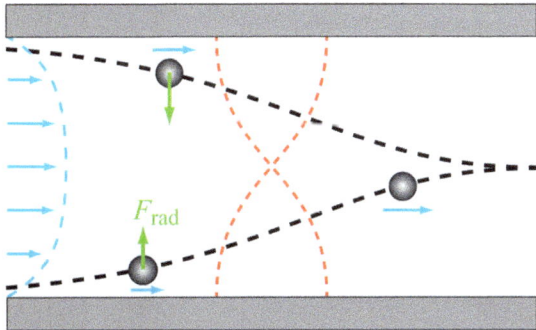

Figure 8.1 Schematic of microchannel acoustophoresis. Particles are flowing (blue arrows) along a microchannel while acoustic radiation forces (green arrows) push them towards the acoustic pressure minimum (dashed red lines) at the center of the channel.

push the cells towards the nearest pressure minima of an acoustic resonance. The trajectories of the particles are dictated by the microchannel's Hagen–Poiseuille flow velocity distribution and the acoustic potential field inside the channel.[1]

8.2 Concentration

In continuous flow mode, the most straightforward acoustophoresis unit operation is to concentrate particles in suspension. This is generally done by focusing the particles into a pressure node, thereby depleting the surrounding medium of particles. The flow is subsequently split into one fraction containing the concentrated particles while the particle-free medium is collected in the other fraction, Figure 8.2. The fundamental requirement in particle concentration or clarification is that the particles must be able to move relative to the liquid in the sound field. The limiting particle size depends on the material of the particle and the frequency of operation, described in greater detail in Chapter 4. For biological particles, the lower limit is approximately 2 µm at a resonance frequency of 2 MHz.

The concentration factor (f_c) of the device is the output cell concentration (C_{out}) divided by the input particle concentration (C_{in}). It depends on the volume flow rate (Q) through the sample inlet and outlet and the recovery (R) of cells in the particle outlet. For a sample containing N cells in a volume V_{in} the concentration factor can be expressed as.

$$f_c = \frac{C_{out}}{C_{in}} = \frac{N \cdot R}{V_{out}} \bigg/ \frac{N}{V_{in}} = \frac{Q_{in}}{Q_{out}} R \qquad (8.1)$$

The channel and the external microfluidic components are normally pre-filled with particle-free liquid before each run. This volume of liquid will end up in the cell outlet and dilute the output suspension. Furthermore, the last fraction of particles to enter the device will never reach the outlet unless the

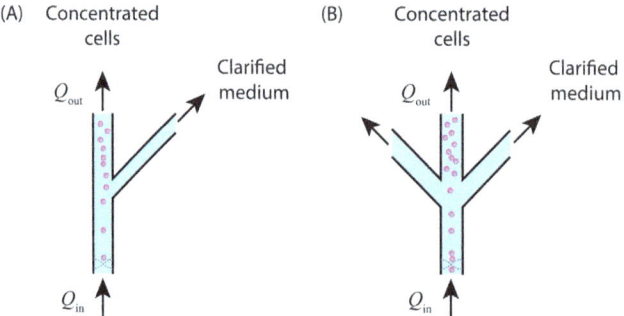

Figure 8.2 Schematics of acoustophoretic devices concentrating a sample. Both work on the principle of focusing the particles to the pressure node of the standing wave while excess particle-free medium is diverted into (A) one or (B) two side branches.

sample is run dry and the particle flow-path becomes air filled. This loss can be estimated by calculating the inner volume of the cell flow-path relative to the volume of the input sample yielding the theoretical maximum achievable recovery.

$$R = 1 - V_{\text{path}}/V_{\text{in}} \tag{8.2}$$

The actual recovery is a function of all the losses in the system, many of which are difficult to calculate or measure individually. Examples are: insufficient acoustic intensity, flow instabilities, too high particle numbers, sedimentation in stagnant flow zones or particles adhering to surfaces.

To achieve a high concentration factor, the flow rate through the particle outlet stream must be low relative to the flow rate of the clarified medium. For a single passage through a device, concentration factors much higher than 10 times become challenging due to the narrow hydrodynamic aperture through which the particles must exit to reach the central outlet. Any small flow imbalances or fluctuations will impact recovery. Also, any offset of the acoustic pressure node relative to the channel center[2,3] will induce a performance loss. For low central outlet flow rates, spurious modes of the acoustic field near the outlet can also cause deflection of the particles when reaching the outlet. To overcome these problems and to concentrate the sample even further, samples can either be rerun through the device or sequential concentration steps can be incorporated in series on the same chip.

One of the early microfluidic continuous cell concentrators was described by Yasuda *et al.* who used a quartz chamber of a half wavelength width. In this case, red blood cells were concentrated into a single band in the flow cell.[4] Using this set-up, it was found that the ultrasonic focusing of the red blood cells did not harm the cells as measured by the release of intracellular components. However, the system did not separate the concentrated fraction from the suspending medium. This was addressed in a follow up work by the same group, introducing a capillary into the chamber through which the concentrated cell fraction was removed.[5]

Cylindrical capillaries have been used as acoustic resonators to concentrate particles in continuous flow mode as described by Goddard *et al.*,[6] where a transducer was glued directly to the capillary and the glass walls of the capillary served both as a transmission layer and a reflector, Figure 8.3. The acoustic capillary focusing was also proposed as an alternative to hydrodynamic focusing in FACS (fluorescence activated cell sorting) applications where the use of acoustic standing waves could replace the sheath flow responsible for focusing the cells or particles into a single stream.[7] This application has recently also been implemented in commercial flow cytometry instrumentation, Attune TM, marketed by Life Technologies.

Harris *et al.* realized a silicon concentrator chip with one inlet and two outlets where concentrated sample exited through one outlet and clarified medium exited through the other.[8,9] It was constructed as a parallel-plate type layered resonator where the particles were concentrated in a nodal plane

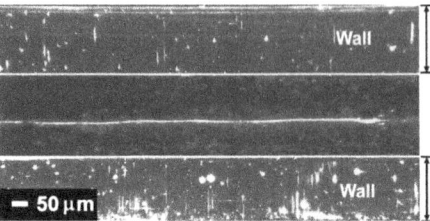

Figure 8.3 Photo of cells focused in a glass capillary. Reprinted with permission from Goddard et al.[6] Copyright 2005, Acoustic Society of America.

parallel to the flow. Nilsson et al.[10] presented the first transverse mode operated acoustophoretic silicon device with two outlets for extraction of clarified medium along the sides of the acoustophoresis channel and one central outlet that collected the focused particles. In the transverse mode, the standing wave is directed perpendicular to the flow and in plane with the chip, which enables visual inspection and hence more easy on-line tuning of acoustophoretic experiments. Acoustic sample concentration has also been implemented in polymers using surface acoustic wave devices. That topic is covered in Chapter 15 of this book and will not be covered further in this chapter.

Nordin et al.[11] designed a system for concentrating dilute cell and particle suspensions using a chip with sequential drains for particle free liquid. It also featured two-dimensional acoustic focusing of the particles that enabled concentration up to two hundred times. It was concluded that the two-dimensional focusing provided a much more controlled separation, as pictured in Figure 8.4, since the particles were all located in the same flow velocity region of the parabolic flow profile. In the one-dimensional case, particles close to the top and bottom of the channel, where the flow velocity approaches zero, will spend longer in the acoustic field and are much more vulnerable to non-uniform resonances and streaming effects which commonly occur in the flow splitter zone. Slow-moving particles may thus be deflected from their original trajectory, resulting in particle losses to the side outlets and hence a lower particle recovery and concentration factor.

8.3 Clarification

Contrary to concentration, acoustic focusing can also be used to remove solid components, producing a cell- or particle-free liquid. The same configuration as in the concentration application can be used, with a flow splitter after a half-wavelength acoustic focusing step.[12,13] However, this basic configuration is only applicable at modest particle concentrations if a reasonable particle clearance is to be obtained. What usually fails at high concentrations is overloading of particles in the central pressure node and a consequent overflow into the clarified medium fraction. To overcome this

Applications in Continuous Flow Acoustophoresis 153

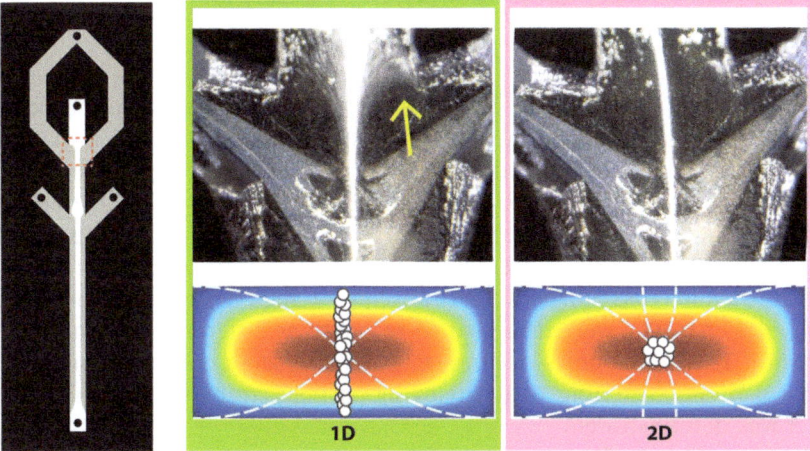

Figure 8.4 The sequential concentration chip developed by Nordin *et al.*[11] The use of one-dimensional standing waves was shown to be less effective as compared to employing two orthogonal standing waves. Particles are exposed to an unwanted acoustic field near the flow splitter. By moving the particles away from the stagnant flow near the top and bottom of the channel, the two-dimensional focusing avoids particles from being deflected from the central outlet. Adapted from ref. 11 with permission.

problem, Lenshof *et al.*[14] reported a device where cells were sequentially removed from the dense stream of cells in the central region of the channel. This enabled the development of a plasmapheresis chip that was capable of handling undiluted whole blood, providing a clarified blood plasma. The focused blood cells were removed through exit holes located at the bottom along the center of the channel where the density of cells was high. The sequential exit holes lowered the cell concentration gradually and at the end of the separation channel, the final fraction of the remaining cells was removed through a flow splitter. Blood plasma of low cell counts was obtained that fulfilled the requirements for clinical diagnostics. The plasmapheresis platform was further developed by Tajudin *et al.*[15] to include an integrated immunoaffinity-capture platform for detection of PSA from whole blood samples.

Particle and cell separation using acoustic standing wave technology has been widely researched where enhanced cell sedimentation is maybe the most common application[16] and is currently also in industrial use *e.g.* in fermentation processes.[17] Sedimentation is accomplished by aggregating particles and cells into the pressure nodal planes of a standing wave. If the nodal planes are aligned with the direction of gravity, the aggregates can be allowed to settle to the bottom of the fluid container. The prerequisite for this is that the acoustic force resulting from any vertical component of the pressure gradient does not exceed that of gravity.

8.4 Acoustic Particle Sorting and Cytometry Applications

The possibility to translate particles across streamlines using acoustic standing wave forces, as outlined in Figure 8.1, enables the principle of valveless switching of particles and cells. By constructing a bifurcation tree of channels, which are actuated by different transducers, it is possible to direct particles to given outlets. An acoustic switch valve device with a single sample inlet and four different addressable outlets was demonstrated by Sundin et al.[18] To avoid cross talk between the excitation regions at the two switching levels, the channels were designed to resonate at different frequencies. By actuating the transducers following a binary sequence as depicted in Figure 8.5, the switching of particles or particle streams to a designated outlet was demonstrated.

The use of multiple resonances for particle stream switching was also presented by Manneberg et al.,[20] where a microchannel had a stepwise narrowed channel width that sequentially supported $3\lambda/2$, $2\lambda/2$ and $\lambda/2$ resonances, Figure 8.6. At the same time, an out of plane resonance was exited at

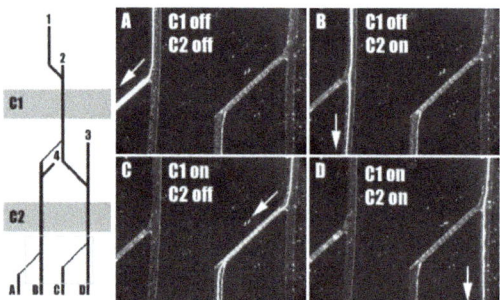

Figure 8.5 Acoustic valving of particles by sequential activation of two transducers in a binary sequence enabled addressing of particles to four outlets (A–D) by the acoustic control setting. Reproduced from ref. 19 with permission.

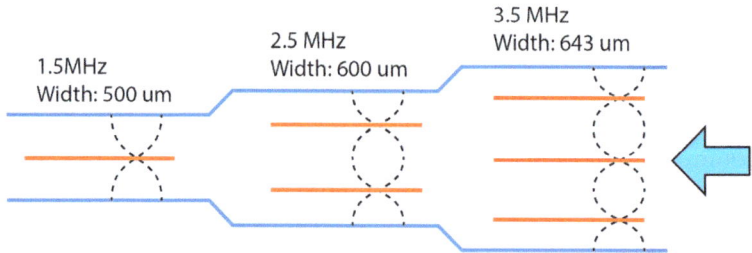

Figure 8.6 Schematic of a multi resonance channel that sequentially along the flow (right to left) merges three different particle streams. Adapted from ref. 20.

a λ/2, levitating the particles at mid height between the channel top and bottom. An alternative particle streamline manipulation method was also implemented in a channel with flow splitters with channels of different dimensions along the flow channel and hence locally different resonance frequencies. The system demonstrated alignment of particles in three streamlines that could be merged according to user specified sequences. The level of resonance cross-talk between neighboring resonances in these systems was also investigated by PIV measurements stating that channel width increments larger than 7% were recommended to avoid significant resonance cross-talk. It should, however, be noted that, the level of cross-talk will also depend on the Q-value of the local channel resonators which is why multi resonator systems need a more in depth analysis.

Manneberg et al. demonstrated in a similar set-up that two-dimensional confinement of particles in the cross-section of a flow channel reduces the particle velocity distribution. In this case, they used two transducers, one for lateral focusing and one for the vertical mode.[21]

Acoustophoretic switching of particles across streamlines has also spurred efforts to develop microfluidic systems that integrate cell and particle sorting on a chip. Chip based cell sorters are analogous to conventional fluorescence activated cell sorters (FACS). The conventional FACS instrumentation uses an external force—an electric field—to deflect cells in droplets at high frequency to a pre-designated collection bin. The modality change from electrostatic force acting on droplets to acoustic standing wave forces acting on targeted cells in a continuous flow truly poses some engineering challenges. The standard way of operating an acoustically driven rare event sorter is to introduce particles in a stream laminated along one side wall. As a rare event is detected, e.g. by fluorescence, the ultrasonic standing wave is actuated and the particle will move to the pressure node and hence be guided into a different flow stream leading to a separate outlet. This principle of sorting was used by Johansson et al., as illustrated in Figure 8.7. The cells pass the interrogation zone A and if sorted they end up in region B and otherwise in zone C.[22] The acoustic FACS was demonstrated to sort cells at a rate of three cells per second. It should be noted that rather than acting on the cells themselves, the acoustic force in this case acted on their surrounding medium, of high density, such that the fluid compartment with particles was translated orthogonal to the flow direction past a flow splitter. Section 8.9 discusses acoustic standing wave forces acting on the interface between fluids of different composition.

In order to increase the sorting rate of an acoustic FACS, it is important that the cells pass the detector and the sorting zone in the same spatial location every time. If the sample enters the detection at an arbitrary position in the parabolic flow profile, the time of arrival to the acoustic switching zone will vary considerably and thus severely limit the system throughput in relation to a theoretical maximum if all cells travelled at the same speed through the system. In conventional FACS systems, this is realized by aligning the cells centrally in the flow stream by a cylindrical sheath flow

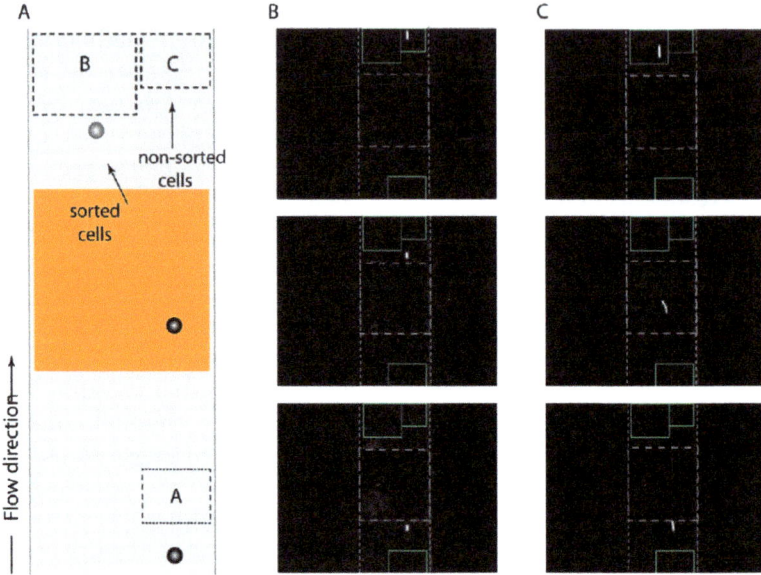

Figure 8.7 Schematic of the cell sorter by Johansson et al.[22] (A) The excitation of the transducer (orange rectangle) is controlled by the optical fluorescent detection of a cell in detection zone A. Sorted cells end up in zone B and non-sorted in zone C. (B) Image where a particle is not actively sorted. (C) The white particle is deflected using an acoustic standing wave and is directed to zone B. Reprinted with permission from ref. 22. Copyright (2009) American Chemical Society.

buffer and ideally the system performance is then defined by the Poisson distributed arrival of particles to the sorting zone. Grenvall et al. solved the need for a uniform particle velocity in a chip based FACS by introducing a pre-alignment region operated at a higher frequency that would not crosstalk with the sorting frequency.[23] The pre-alignment standing wave comprised multiple pressures nodes across the channel width and the particles were pre-aligned in the node closest to the sidewall, Figure 8.8. At the same time, a vertical half wavelength resonance confined the particles to the mid height in the channel giving the particles a uniform flow velocity. The narrow velocity distribution increased the switching frequency and the prototype system reported a switching frequency of 50 Hz. Later, Jakobsson and Grenvall[24] improved this set-up, reporting an acoustic actuation time of 500 μs to be sufficient to deflect particles for recovery in the target outlet. Acoustic switching of particles and cells can also be performed by surface acoustic waves and is further discussed in Chapter 15.

Acoustic focusing of cells lends itself well to applications in cytometry as precise positioning is crucial for high resolution detection. As mentioned earlier, work by Goddard et al.[7] demonstrated sheath-less focusing of cells for flow cytometry applications by setting up an acoustic standing wave in

Applications in Continuous Flow Acoustophoresis

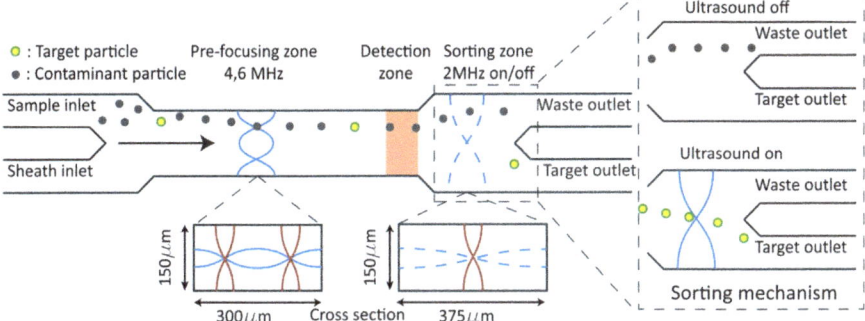

Figure 8.8 Schematic of the acoustic FACS as presented by Jakobsson et al.[24] Particles are laminated against the top side wall in a pre-focusing region (channel width 300 µm and height 150 µm) and are pre-aligned by a λ-resonance at 4.6 MHz, in the first pressure node at λ/4 distance from the wall. At the same time they are focused by a λ/2 resonance to mid height in the channel as they enter the detection zone. Fluorescent particles trigger a burst activation of the 2 MHz transducer which translates particles in the sorting zone to the pressure node in the channel center. The split-flow ratio in the sorting zone is set to direct particles that are in the channel center to the target outlet. Reproduced from Jakobsson et al.[24] with permission.

a cylindrical capillary. The cylindrical geometry inherently yielded focusing in two dimensions. Suthanthiraraj et al.[25] have developed a chip-based cytometry approach to a high throughput domain where a chip with 13 parallel channels, each supporting a 3λ/2 standing wave, provided 39 stable focused particle or cell streams that were analyzed in parallel in an epifluorescence microscope. Measurements on calibration beads in this high throughput mode demonstrated a fluorescence peak intensity coefficient of variation of 24% as compared to 16% for reference measurements with an Accuri C6 FACS instrument.

Acoustic pre-alignment has also been realized for impedance cytometry, where a Coulter counter was incorporated on a glass chip, with a two-dimensional pre-alignment zone in a 420 µm wide and 150 µm deep channel. Sound was actuated by two transducers operated at 2 MHz and 5 MHz respectively, Figure 8.9. The pre-alignment channel, connected to the Coulter counter aperture where impedance measurements were performed, demonstrated chip derived Coulter counter data that approached the performance of conventional table top Coulter counter instruments.[26]

8.5 Medium Exchange of Cells and Microparticles

A routine procedure in any cell lab is the transfer of cells from one suspending medium to another. Typically, this is carried out by centrifugation of the cell suspension prior to removing the supernatant and adding new

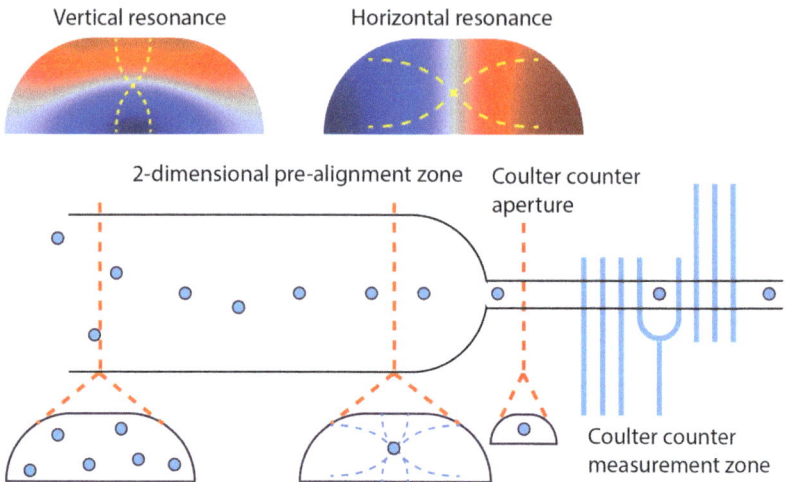

Figure 8.9 Two-dimensional pre-alignment implemented in a glass chip integrated Coulter counter offering a simple fabrication system with single-sided measurement electrodes covered by a glass chip holding the microfluidic channels.

medium. For adherent cells this is routinely carried out when re-plating them at a lower concentration in fresh cell medium for each passage or when dividing the population. For non-adherent cells this is the typical way to shut off a chemical stimulus after exposure to some chemical. In analysis such as flow cytometry several cell washing steps are typically performed in the process of staining cells with fluorescent probes used for categorization of cells, to avoid excess dye molecules in the suspension to interfere with the fluorescence readout.

Microchannel acoustophoresis offers a non-contact, gentle and rapid way of transferring cells or particles across laminar flow streams and is therefore well suited for applications involving medium exchange of cells or particles. Yet another application is short time exposure of cells or microbeads to chemical stimulus by transfer across a sequence of buffer conditions.

8.5.1 Transport Mechanisms

To understand how to design efficient acoustophoresis devices for medium exchange of cells or microbeads, some of the fundamental transport mechanisms in these systems are discussed in the following sections. The fundamental operation is to transfer particles or cells from their initial suspending medium near a wall, to a new carrier liquid occupying the central part of the channel. Minute species from the original suspension, such as unbound dye molecules,[27,28] antibody-carrying phages,[29] cell debris,[30] or peptides,[31] should preferably remain near the channel walls. Figure 8.10 introduces the setting where the particles to be translated into a new carrier

Applications in Continuous Flow Acoustophoresis

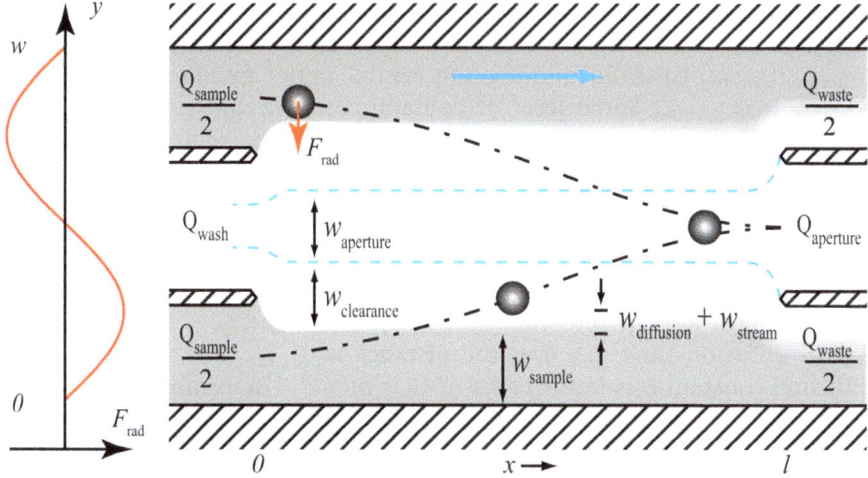

Figure 8.10 Schematic of a generalized acoustophoresis carrier medium exchange channel. Q denotes volume flow rate in the inlets and outlets. Microbeads or cells are focused to the center of the channel by the force (red arrow) exerted on them through the $\lambda/2$ acoustic resonance across the width of the channel. Minute species (gray) remain flow laminated along the channel walls. Each lamina has been assigned a width (w), which can be tuned by adjusting the relative volume flow rates (Q) in the inlets and outlets. Reproduced and modified, with permission, from Augustsson.[32]

medium (white) enter the system in a liquid containing some minute contaminant (gray). Here x and y denote position along the length and width of the channel, respectively, y being the axis along which the sound propagates.

To recover the cells or microbeads in the central outlet while discarding the initial suspending medium, the volume flow rates (Q) in the inlets and outlets should be balanced with the acoustophoretic velocity u_{rad} directed transversely to the flow. This can be achieved by adjusting the acoustic energy density E_{ac} such that the particles reach the center of the flow before arriving at the outlet trifurcation. The center of the channel here corresponds to the part of the flow of width w_{aperture} demarked with a blue line in Figure 8.10, which is determined by the relative flow rates Q_{aperture} and Q_{waste}. According to eqn (4.25), the transverse velocity u_{rad} of a particle is proportional to the square of its radius a. Any minute species of radius $\ll a$ will therefore be efficiently filtered away, regardless of their acoustic properties. We shall see in the following, however, that other effects also come into play.

By adjusting the relative flow rates to the inlets, Q_{sample} and Q_{wash}, and the outlets, Q_{aperture} and Q_{waste}, it is possible to tune the distance $w_{\mathrm{clearance}}$ that a particle must travel sideways in order to reach the central outlet. To maximize $w_{\mathrm{clearance}}$ the central fluid flow Q_{wash} can be increased relative to Q_{sample} while minimizing the central outlet flow rate Q_{aperture}. The washing efficiency and stability of the system can be assumed to depend strongly on $w_{\text{clearance}}$. Such a clearance will only exist under the condition $Q_{\text{wash}}/Q_{\text{sample}} > Q_{\text{aperture}}/Q_{\text{waste}}$. To avoid diluting the central outlet fraction, one can set $Q_{\text{sample}} = Q_{\text{aperture}}$. Some level of contamination in the central outlet is inevitable and can be attributed to any of the following sources.

8.5.1.1 Diffusion

For a channel of length (l), width (w) and height (h), the diffusion of a contaminant introduced through the side inlet can be investigated by comparing the average retention time in the channel $\tau_{\text{pass}} = lwh/Q_{\text{tot}}$, to the typical diffusion time τ_{diff} over the distance $w_{\text{clearance}}$. For small ions the diffusion constant D is in the order of 10^{-9} m^2 s^{-1}. For example, for a clearance $w_{\text{clearance}} = 100$ µm, the passage time in the device should be less than 1 to 10 s to avoid any notable influence of diffusion. The broadening w_{diff} of the sample stream can be expressed as.

$$w_{\text{diff}} \approx \sqrt{D\tau_{\text{pass}}} = \sqrt{D\frac{lwh}{Q_{\text{tot}}}} \tag{8.3}$$

Diffusion can thus be minimized either by increasing the volume flow rate Q_{tot}, or by reducing the channel dimensions.

8.5.1.2 Acoustic Streaming

Lateral acoustic streaming (further details in Chapter 12), often referred to as Rayleigh streaming, can convect liquid in the cross-section transverse to the flow and thereby transfer minute species to the central outlet. Taking a look at Figure 8.11(a), we note that for $z = h/2$ the streaming velocity u_{str} coincides with the acoustophoretic velocity of particles u_{rad} and will cause transfer of contaminant to the central outlet.

Figure 8.11(b) is a schematic of the helical motion of two particles caused by acoustic streaming within the bulk fluid of a rectangular shaped microchannel. Figure 8.11(c) shows an experimental image of the phenomenon as viewed by confocal microscopy where a stream of fluorescently tagged lipoproteins enters the system from the side inlets.[33] The diameters of the vesicles (50 nm < d < 100 nm) are small enough for acoustic streaming to dominate their motion (eqn (4.30a)) while being sufficiently large for diffusion to be minimal.

When washing small particles, the streaming broadening due to acoustic streaming will be large because the acoustic intensity must be higher in order to focus the particles. Higher acoustic intensity leads to faster acoustic streaming and thereby higher levels of contaminant in the central outlet. For particles larger than 10 µm the acoustic streaming velocity is in general negligible relative to the velocity of a cell due to the dominance of the acoustic radiation force.

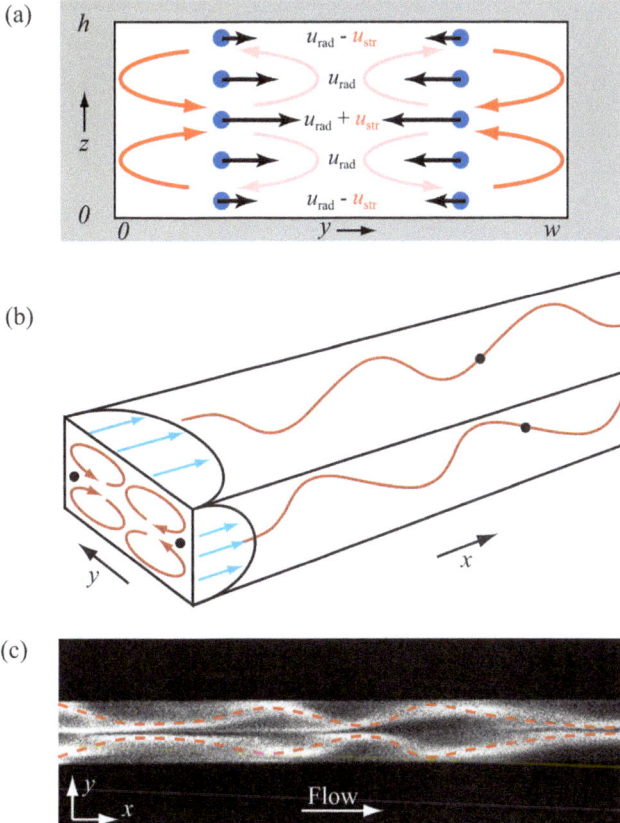

Figure 8.11 (a) Rayleigh streaming in the cross-section orthogonal to the flow induces a swirling motion (red arrows) in the bulk of the liquid. The velocity of a particle (black arrows) is a sum of the acoustic radiation velocity and the acoustic streaming velocity. (b) Schematic of the flow path of streaming dominated particles (*i.e.* particles smaller than the critical diameter). The combination of acoustic streaming and the laminar flow profile lead to a helical motion. (c) Experimental evidence of fluorescently labeled lipoproteins (VLDL) exhibiting streaming motion in an acoustophoresis channel. The image should be viewed as a lengthwise (along the channel) image slice of the distribution of the fluorophore as the vesicles follow the Rayleigh stream. Adapted from Augustsson.[33]

8.5.1.3 Hydrodynamic Interactions between Particle and Fluid

Contaminant can be transported by the drag from the cells or particles as they move at terminal velocity through the liquid. When an object is dragged through a liquid by a force F_{rad}, it will exert that same force on the liquid through friction at the surface, Figure 8.12(a). The volume force exerted on the fluid increases linearly with the volume fraction of suspended particles.

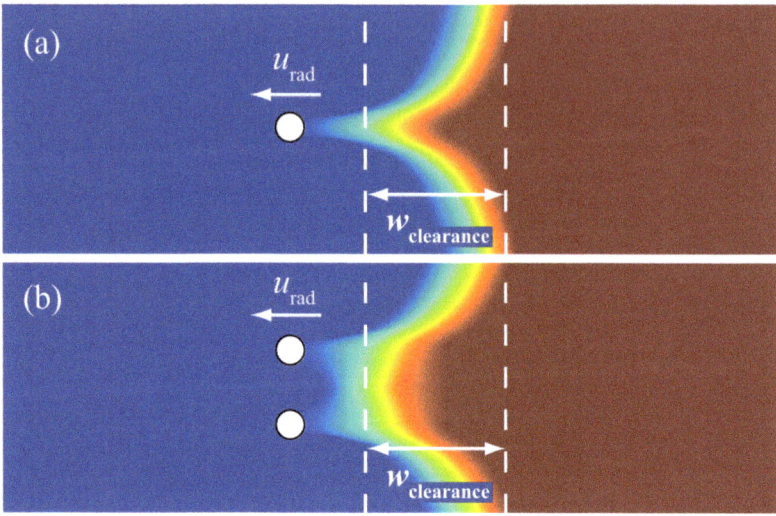

Figure 8.12 (a) Schematic drawing of a sphere moving through an interface between two fluids. The concentration profile of the contaminant (red: high, blue: low) is affected by the drag from the sphere. (b) The combined drag from two closely spaced spheres will convect more contaminant than two largely spaced spheres due to the hydrodynamic interaction. Adapted from Augustsson[32] with permission.

This volume force causes convection of the fluid, and when this velocity becomes substantial relative to the velocity of the individual particles, the particles can be considered to be hydrodynamically coupled, Figure 8.12(b). Modeling of such hydrodynamic coupling has been reported by Mikkelsen et al.[34] in 2005. Numerical calculations for paramagnetic beads ($a = 0.5$ μm) in a magnetic field yielded that the hydrodynamic interaction between spheres was minimal for concentrations up to 10^8 m L^{-1} which correspond to an average inter-particle distance $d \approx 20$ μm. For higher concentrations the interaction increases dramatically, and at 10^{10} m L^{-1} ($d \approx 5$ μm) the particles are fully coupled and will drag along all intermediate fluid. Petersson et al.[28] reports a washing experiment with particle concentrations ranging from 2×10^8 m L^{-1} to 10^9 m L^{-1}, using polyamide beads ($a = 2.5$ μm). From the experiment it was clear that the washing efficiency decreased dramatically for concentrations above 5×10^8 m L^{-1}. Hawkes et al.[27] reports a breakdown in washing efficiency at a concentration of yeast cells of 10^8 m L^{-1}. It is reasonable to assume that hydrodynamic coupling between particles is an important factor in this type of breakdown.

8.5.1.4 Flow Perturbations

Reproducibility and efficiency in acoustophoresis medium exchange rely on stable flow conditions. Consider a situation where the flow in the channel is temporarily disrupted by a short plug of air entering the channel inlet and passing through the main channel, causing complete mixing of the two inlet streams. Before and after this disturbance the separation is ideal, *i.e.* no contaminant is transferred to the central outlet. The impact of such a perturbation can be estimated from the channel dimensions, and the flow rates in the inlets and outlets. The number of molecules in the channel at complete mixing can be expressed as.

$$N_{channel} = \frac{C_{sample} \cdot V_{channel} \cdot Q_{sample}}{Q_{tot}} \qquad (8.4)$$

The central outlet concentration will be the fraction of those molecules that reach the central outlet sample divided by the final volume of the processed sample.

$$C_{outlet} = \frac{N_{channel} \cdot Q_{aperture}}{V_{outlet} \cdot Q_{tot}} \qquad (8.5)$$

For a separation channel of an inner volume $V_{channel} = 2$ µL, flow rates $Q_{sample} = Q_{aperture} = Q_{tot}/4$ and a processed volume $V_{outlet} = 1$ mL, the contamination level in the central outlet due to this single perturbation will be on the order of 100 ppm relative to the input concentration. Refs. 29 and 35 deal with the separation of microbeads from non-specific molecules. In the context of *e.g.* bacteriophage antibody library selections, it can be noted that this level of contamination corresponds to the contamination levels after a manual washing procedure repeated twice. This type of disturbance to the flow is one of the main concerns regarding stability and reproducibility in microfluidic cell and bead handling systems in general and washing applications specifically.

8.5.2 Acoustic Forces in Stratified Liquids

Acoustic radiation forces can induce relocation of the suspension relative to the central inlet cell-free medium. The effect is evident in configurations where the density and sound velocity of the cell suspension is higher than that of the central inlet cell-free medium. While flowing through the channel, the liquids will relocate to a stable configuration where the heavier liquid occupies the central part of the flow along the acoustic pressure node. Such relocation is obviously detrimental for the washing performance of the device. The phenomenon is described in more detail in Section 8.9.

8.5.3 Devices for Buffer Exchange

In 2004, Hawkes *et al.*[27] presented a micro-scale acoustic SPLITT separator for continuous cell washing and mixing. It was demonstrated on a sample of yeast cells in an aqueous suspension containing dissolved fluorescein salt, Figure 8.13(a)–(c). The device consisted of a flat resonance chamber between parallel plates of quartz and steel, positioned 250 µm apart. Adjusting the

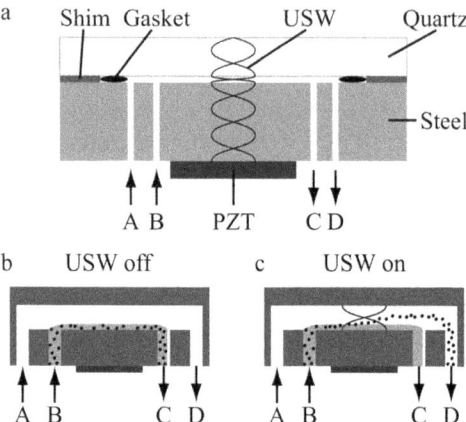

Figure 8.13 (a) A schematic of the device for continuous cell washing presented by Hawkes et al.[27] (b) Yeast cells suspended in fluorescein solution were injected through inlet B while yeast- and fluorescein-free fluid was injected through A. (c) The half wavelength acoustic resonance deflects particles to the mid height of the flow while the original medium remains near the channel floor. At outlets C and D, the flow streams are divided and the cells are separated from the fluorescein dye. Reproduced from ref. 27 with permission.

actuator frequency to ~3 MHz resulted in a half wavelength resonance across the channel so that cells could be transferred from their initial position near one of the walls at the inlet, to the pressure node centered between the plates. This particular channel was 10 mm wide and the volume flow rate was typically 10 mL min^{-1}, which is quite remarkable for a microfluidic system. In principle, by expanding the width of the cavity, the volume throughput can be increased proportionally, while maintaining the same flow velocity and acoustic field amplitude. This resulted in about 90% of the fluorescein being washed away from the yeast cells.

In 2005, Petersson et al.[28] presented a device for acoustophoretic carrier medium exchange of blood cells in a silicon/glass microchannel structure, Figure 8.14. The proposed areas of application were blood component washing and bio-affinity assays. A particle suspension was introduced along the side walls of an acoustophoresis microchannel while the exchange/wash liquid was infused *via* a central inlet. As the particles flowed along the channel the acoustic radiation force pushed them towards the center of the channel into the stream of new wash liquid. At the end of the channel, the flow was split into a central outlet for the particles in the new medium, and two side outlets containing the original suspending medium and particles of low acoustophoretic mobility. The layout of the inlets and outlets is well suited for half wavelength resonators because of the symmetry of the acoustic field with an acoustic pressure node positioned in the center of the flow. While the channel width (w) in these systems matches half a wavelength

Applications in Continuous Flow Acoustophoresis 165

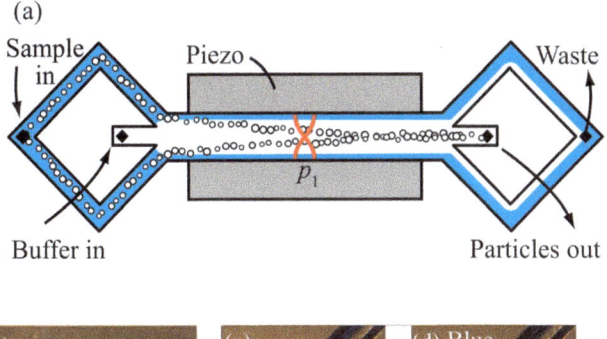

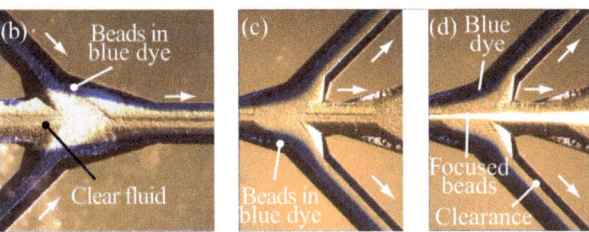

Figure 8.14 The carrier fluid exchange principle introduced by Petersson et al.[28] (a) Particles (white) in contaminated carrier fluid (blue) were introduced through a common side's inlet. Particles were drawn to the center of the channel by the acoustic radiation force while the dye liquid (blue) remained unaffected at the sides. Experimental images of (b) the inlet of the channel, the trifurcation outlet (c) before onset of ultrasound, and (d) during ultrasound actuation. Reproduced from Augustsson et al.[35] with permission.

resonance condition, the channel height (h) is often kept well below half a wavelength to avoid any resonance component along this direction. Wash efficiency was measured as the relative amount of Evans blue dye in the waste (side) outlet. A washing efficiency (contaminant removal) of about 95% was reported at a bead concentration of 1.5% volume. For increased washing efficiency, sequential steps can be integrated on the same chip.[31]

Although high volume throughputs and relatively high recoveries of particulate material were achieved, neither of the above mentioned publications report washing efficiencies sufficiently high to be applicable for bio-affinity assays. This may have been attributed to cross-contamination due to diffusion, acoustically induced streaming or too high particle or cell concentrations.

In 2008, Persson et al.[29] used an acoustophoresis microchannel to select candidates from antibody-displaying bacteriophage libraries (a.k.a. phage display) and thereby outlining a microfluidic based mode of automation of such processes. Microbeads coated with an antigen were incubated with a library of antibody-expressing phages prior to selection. High affinity bacteriophages would thus bind to the microbeads and were separated from non-specific bacteriophages in the chip due to the vast difference in acoustic velocity of the microbeads as compared to the individual phage particles. To achieve sufficient washing efficiency, two serially linked fluid exchange units were incorporated on the same device. Any one out of two antibodies from a small scale model library could be enriched 10^3- to 10^5-fold with respect to the other by coating microbeads with the corresponding antigen. From a full library comprising $\sim 10^9$ different antibodies, it was shown that acoustic washing of microbeads yielded antibodies that were specific to the grass pollen allergen Phlp 5 that had been coated on the microbeads. A single run through the device was found to be comparable to four manual magnetic bead washing steps, in terms of washing efficiency, for non-specific antibodies. The same device was later used for sample decomplexing prior to mass spectrometry readout.[35] A phosphorylated peptide spiked in BSA could be extracted using metal oxide affinity capture beads from the background of highly abundant molecules.

Lenshof *et al.*[30] used an all glass fabricated transverse mode acoustophoresis channel for sample preparation of fluorescently labeled cells prior to FACS analysis. Samples of blood were incubated with fluorescently tagged antibodies prior to osmotic RBC-lysis. 60% of the RBC-debris was removed while recovering 99% of the WBC-fraction. The unbound fluorescent dye label was reduced to 1% of the initial concentration. This wash step is usually done manually by sequential centrifugation steps and by removing the centrifugation steps the analysis time is decreased and the risk of sample-handling errors is reduced.

In 2008, Augustsson *et al.*[36] proposed a method for sequential carrier fluid exchange of cells or microbeads, Figure 8.15. Particles were initially focused to the center of an acoustophoresis prealignment channel, depleting the regions of the flow near the channel walls. At a first intersecting flow junction, a fraction of the main channel flow was extracted on one side of the channel while new carrier fluid was injected on the opposing side. After escaping this flow junction, particles were refocused to the center of the main channel flow. The process of shifting new medium into the main channel was then repeated in a sequence of consecutive flow junctions. The washing performance was found to increase with the number of sequential shifts but it should be pointed out that the improved performance comes at the price of higher complexity of the fluidic network.

An attractive feature of the sequential medium exchange chip is the possibility to perform multiple sequential buffer exchanges,[37,38] where biological and bioanalytical process steps can be performed in a serial fashion, avoiding sample loss and facilitating sample handling. Cell surface bound peptides were extracted in continuous flow by arranging the side inlet channels' liquids as a sequence of buffers of increasing pH. The pH gradient resulted in a pI-dependent dissociation of peptides and proteins from the cell surface in each buffer stream. The corresponding buffer aliquots were collected and analyzed using solid phase extraction and MALDI-TOF mass spectrometry.

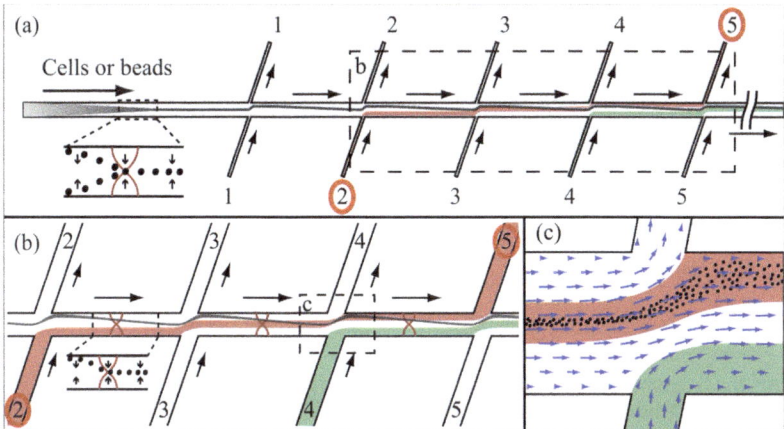

Figure 8.15 The principle of the sequential acoustic carrier fluid exchange channel.[36,38] (a) Cells or microbeads are initially focused to the center of the flow. The carrier fluid is thereafter exchanged in a number of sequential flow junctions allowing the cells to be exposed to a number of different buffer conditions. (b) If the side channel volume flow rate is set to 1/3 of the main channel flow, a buffer solution infused through e.g. side inlet 2 will exit through outlet 5, containing the eluted molecules from the interaction between cells and buffer. (c) At each intersecting flow junction, the particles will follow the fluid flow to the side of the channel and a new buffer will occupy the central part of the channel. Reproduced and modified with permission from ref. 37. Copyright 2008, The Chemical and Biological Microsystems Society.

8.6 Separation by Acoustophysical Properties

Particles and cells can be sorted based on material properties or size using acoustophoresis. The trajectory of a particle when flowing through the acoustic field will be determined by a combination of properties of the particle, the suspending liquid, the flow profile and the sound field. We recall that the acoustic radiation force on a particle increases linearly with its volume and that the direction and magnitude of this motion depend also on the acoustic contrast relative to the suspending liquid. The contrast in turn is a function of the relative density and relative compressibility of the particle with respect to the suspending liquid (eqn (4.23c)). Separation of particles or cells can be performed in two modes of operation. First, if the particles have same-sign acoustic contrast, thus moving in the same direction, they can be separated based on how fast they move in the acoustic field. This method will be referred to as free flow acoustophoresis (FFA). Second, in those more rare cases when the particles have different-sign acoustic contrast a more robust separation scheme can be employed which will be referred to here as bi-directional acoustophoresis.

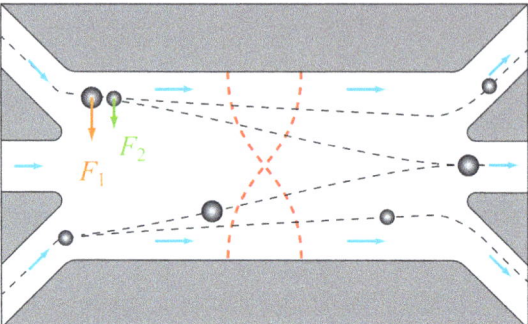

Figure 8.16 Schematic of size based separation of two particles of the same material. The particles enter through the side branches of a trifurcated inlet. The trajectory (black dashed lines) of each particle is determined by the flow (blue arrows) and the magnitude of the acoustic radiation force (F_{rad}, orange and green arrows) on the particle in the oscillating acoustic pressure field (red dashed lines). Larger particles move faster in the acoustic field and exit through the central outlet while smaller particles exit through the side branches of the outlet trifurcation. Reproduced and modified from ref. 32 with permission.

8.6.1 Free Flow Acoustophoresis (FFA)

This section deals with the situation where all particles have positive acoustic contrast and therefore move towards a pressure node located along the center of the channel. Figure 8.16 shows conceptually the separation of small particles from bigger ones given that they have the same acoustic properties. According to eqn (4.25), the acoustophoretic velocity (u_{rad}) of a particle scales with its diameter to the second power. Therefore, a small variation in size translates to a substantial difference in sideways translocation as the particles flow through the separation zone.

For cells, even though the acoustic contrast indeed varies between cell types,[39,40] the acoustophoretic velocity is dominated by the size of the cell in this mode of operation. When suspended in aqueous isotonic medium such as NaCl 9 mg mL^{-1}, PBS or cell culture media, the vast majority of all cells have positive contrast with an exception for adipocytes (fat cells). The special case where particles are separated based on different sign of the acoustic contrast is treated in Section 8.6.4.

The general concept of performing size-based particle separation in continuous micro-scale flows was pioneered by Giddings[41] where an external force is allowed to act on a particle suspension orthogonal to the direction of flow. Particles that experience a stronger force are translated faster across the laminar flow streams and can be collected through a separate outlet, Figure 8.17(a).

Applications in Continuous Flow Acoustophoresis 169

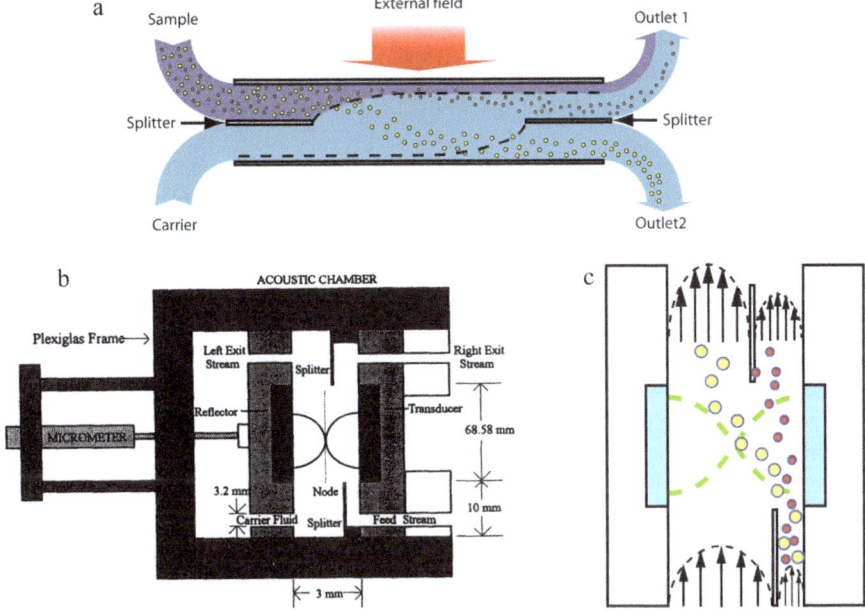

Figure 8.17 (a) The principle of the split-flow thin (SPLITT) fractionation. A particle mixture is flow laminated along one of the walls of the fractionation chamber where it is exposed to an external force field. The larger particles (yellow) experience a higher force than the smaller (red) particles. The force and the flows are tuned such that the particles that are most affected by the force field travel past the flow splitter. Reproduced from ref. 44 with permission. (b) The first acoustic continuous flow SPLITT fractionator, presented in 1995. Reprinted from by Johnson and Feke[42] with permission from Elsevier. (c) Schematic of the half wavelength based resonator for particle size separation as reported by Johnson and Feke.[42] The piezo actuators are indicated in blue and the oscillating acoustic pressure field is indicated by dashed green lines. Reproduced from Lenshof et al.[45] with permission.

8.6.2 FFA Devices

An early implementation of a size-based acoustic standing wave particle separator was reported by Johnson and Feke in 1995.[42] The system had a dual inlet flow with a sample suspension laminated next to a running buffer flow that streamed into a $\lambda/2$ acoustic separation zone with a central pressure node, Figure 8.17(b) and (c). As they flow through the chamber, larger particles translocate sideways towards the pressure node at a velocity proportional to the square of the radius (eqn (4.25)). By supplying the device with a split-flow outlet in line with the original SPLITT design, larger particles (yellow) can be collected in one outlet whereas smaller particles (red) are collected in the other outlet. Later the same group demonstrated the separation of hybridoma cells and *Lactobacillus rhamnosus* cells in a similar set-up. Although not reporting a microfluidic system, as the flow channel was 3.5 mm wide and hence the piezo ceramic actuator was operated at 210 kHz, a proof of concept of acoustophoresis based cell sorting was given.[43]

This concept was later translated into a planar structure as a microfluidic component by Petersson *et al.*[46] This greatly facilitates integration of in-line particle separation with downstream microfluidic unit operations, Figure 8.18(a). Multiple particle or cell fractions can be obtained at the end of the flow channel if the laminar flow is divided into multiple outlets, Figure 8.18(b). This microchip integrated FFA using a $\lambda/2$-resonator was demonstrated for sorting polystyrene particle mixtures (2 µm, 5 µm, 7 µm and 10 µm). The fractionation cut-off was tuned by adjusting the flow rate in each outlet relative to the total flow rate. Separation of diluted whole blood into fractions of erythrocytes, leukocytes and platelets was also reported.[46]

The strategy of using FFA to separate a suspension into two size distributions lends itself well to connecting these devices in series to enable extraction of a population of narrow size distribution. Adams *et al.*[47] demonstrated a tunable band-pass particle size filter that could

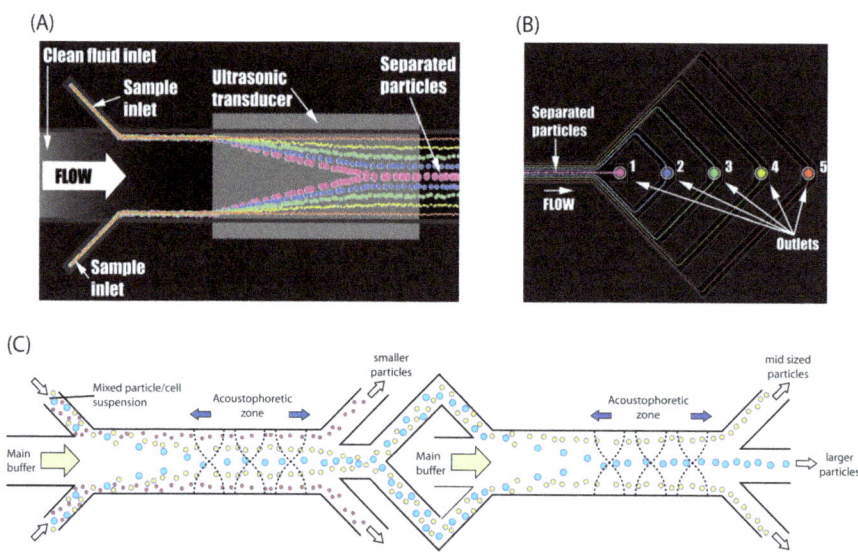

Figure 8.18 (A) A schematic of microchip based free flow acoustophoresis. (B) At the outlet the flow can be branched off to collect particles of different acoustic mobility. (A) and (B) are reproduced from Petersson *et al.*[46] Copyright 2007 American Chemical Society. (C) Schematic of the serially linked dual FFA system as described by Adams, where a "low pass", "band pass" and a "high pass" sizing of microparticles were reported.[47] Reproduced from ref. 45 with permission from the Royal Society of Chemistry.

Applications in Continuous Flow Acoustophoresis 171

discriminate between 3-, 5-, and 10 µm polystyrene particles, Figure 8.18(c). The approach is similar to the multiple outlet FFA described by Petersson et al.,[46] the main difference being that the size cut-off for the low- and high-pass channels can be tuned independently of each other using two separate transducers.

Thevoz et al.[48] showed that FFA can be used to distinguish between cells in different phases of the cell cycle by investigating the breast cancer cell line MDA-MB-231. Cells in the G2/M and S phase, which move faster in the acoustic field due to their relatively large size, were removed *via* the central outlet, while a purified fraction of G1 cells, which are smaller, was collected *via* the side outlets. In a similar device, Yang and Soh isolated viable cancer cells from dead cells.[49]

Dykes et al. reported an FFA system for the removal of excess platelets in peripheral blood progenitor cell (PBPC) apheresis products. The system relied on the acoustophoretic mobility of PBPCs being higher than that of platelets and hence a purified fraction of PBPCs was obtained in the central outlet, Figure 8.19. A mean recovery of WBCs of 98% and a depletion of platelets by 89% was achieved. It was also shown that platelet activation was low and that the viability of the PBPCs was unchanged after acoustophoresis.[50] SAW-based separation of cells has also been reported where platelets and red blood cells were separated based on their different acoustophoretic mobility.[51]

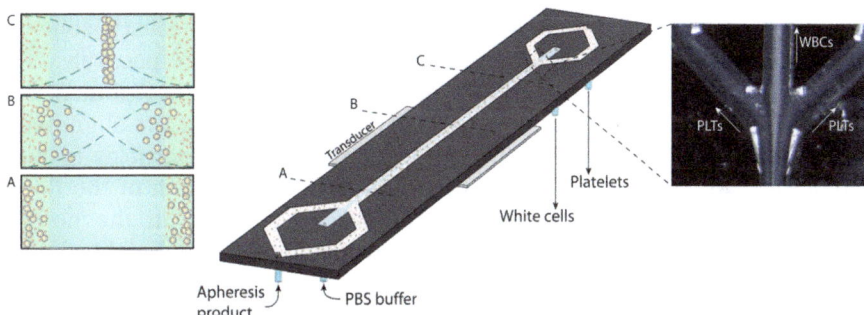

Figure 8.19 Schematic of the chip for platelet removal in an apheresis product. Insets to the left are cross-section schematics of how the cells are affected by the acoustic standing wave in the channel. (A) The cell sample is flow laminated at either side of the channel. (B and C) The larger cells (leukocytes) experience a stronger force than the smaller cells (platelets) and are thus transferred faster to the center of the flow. The inset to the right is an image of the trifurcation outlet during processing of the apheresis product from patients undergoing hematopoietic stem cell treatment. PBPCs and leukocytes are focused to the central outlet while the platelets are exiting *via* the side outlets. Reproduced from Lenshof et al.[45] with permission from the Royal Society of Chemistry.

Adams et al. demonstrated that increased dimensionality in the separation can be accomplished if a combination of magnetic beads and non-magnetic beads are used and both acoustic and magnetic forces are employed.[52] Combined fields are covered in more detail in Chapter 11.

Continuous flow acoustophoresis has also been reported in microfluidic systems that utilize surface acoustic wave-based excitation, where two opposing transducers on a base plate are phase matched to generate a standing wave in the center of a PDMS channel clamped on top of the base plate. Analogous to Figure 8.16, separation of 0.9 μm and 4.2 μm polystyrene beads was demonstrated.[53]

8.6.3 Pre-Alignment and Fractionation

The main limitation in FFA is related to the Hagen–Poiseuille flow velocity distribution inherent with laminar flows. Particles flowing through the channel near the top or bottom of the channel will have longer retention times in the acoustic resonant field than particles flowing at mid height where the flow velocity is higher. Consequently, as shown in Figure 8.20(a), a dispersion of particles in the height dimension at the inlet translates to dispersion in the width dimension at the outlet of the channel. The reason that sorting can be carried out at all despite this dispersion is because the flow rate is inherently low near the walls. The majority of the processed particles will thus reside in the central part of the flow and have therefore similar retention times and sorting conditions.

To achieve high quality, deterministic, sorting within a single passage through the sorting chamber, all particles must enter the acoustic zone at the same point in the height and width dimension, respectively. To accomplish this, an acoustic pre-alignment channel can be incorporated on the chip. Augustsson et al.[54] used a 5 MHz resonance in a pre-alignment channel of width 300 μm and height 150 μm to confine particles to two geometrically equivalent points in the transverse cross-section of the flow. At 5 MHz, the cavity supports a $\lambda/2$-resonance condition in the height dimension focusing particles to a single pressure node at mid height. In the width dimension the same frequency supports a λ-resonance that has two pressure nodes positioned a distance of $\lambda/4$ from the channel center on either side.

The pre-aligned particles thereafter enter a 375 μm wide $\lambda/2$ separation chamber actuated at 2 MHz as was first proposed by Petersson et al.,[46] where cells or particles exit through a trifurcation outlet where the central branch captures objects of high acoustic mobility while low mobility objects exit the through any of the two side branches. Figure 8.20(b) and (c) show how size discrimination of 5- and 7 μm beads is improved when employing pre-alignment FFA. By tuning the piezo actuator voltage to the level of maximum separation, a mix of 50% pure beads could be sorted into two fractions, each of 99% pure 5- and 7 μm beads.

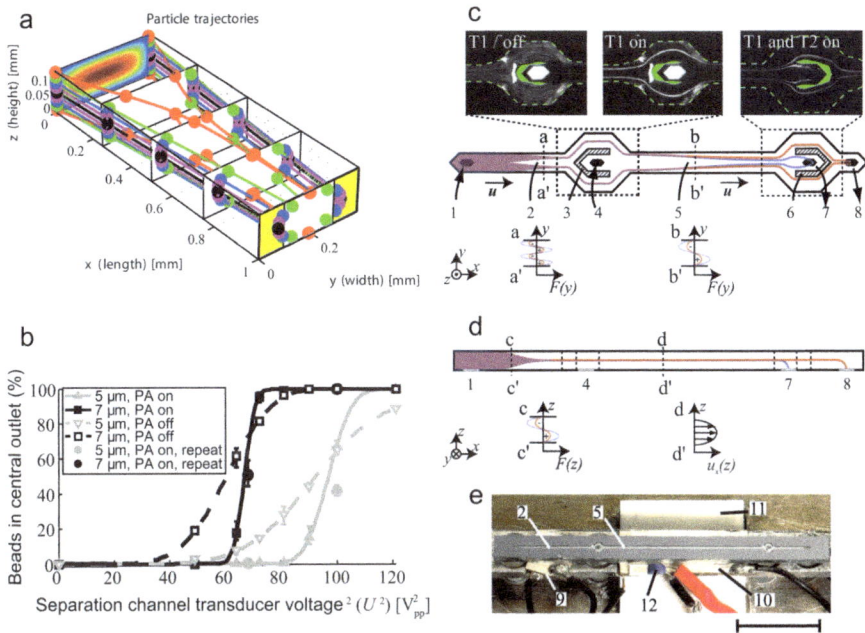

Figure 8.20 (a) Simulated trajectories of identical particles with different starting positions in height. Reproduced from Augustsson[32] with permission. (b) Sorting of 5- and 7 μm beads with and without pre-alignment activated. (c) Top view schematic of the pre-alignment FFA chip. A suspension of cells or particles enters the system through the inlet (1), after which cells are pre-aligned in an acoustophoresis channel (2) by means of an acoustic field (a-a′) in the yz-plane. The two bands of cells are bifurcated (3) to two sides of a central inlet fluid flow (4) and the pre-aligned cells are thereafter flow laminated to proximity of the walls of a separation channel (5), where the trajectories of individual cells are deflected in an acoustic field (b-b′) according to their intrinsic acoustic properties and morphology. At the trifurcation outlet (6), a subgroup of cells can be selectively guided to the central outlet of the chip (7) by tuning the intensity of the second acoustic field while cells of low acoustophoretic mobility will be guided to the side outlet (8). Insets show pre-aligned (T1 on) and non-pre-aligned (T1 off) microbeads at the end of the pre-alignment channel, and 5- and 7 μm beads separated at the central outlet (T1 and T2 on). (d) Side view schematic. Cells/particles are pre-aligned in the vertical direction by means of an acoustic force (c-c′) to minimize the influence of the parabolic flow profile (d-d′) in the channel, which may otherwise affect the trajectories of the cells. (e) A photo showing the positions of the piezoceramic transducers (9 and 10), the Peltier element that regulates the temperature (11), and the temperature sensor (12). Scale bar = 10 mm. (b–d) are reproduced from Augustsson *et al.*[54] Copyright 2012 American Chemical Society.

The system was applied to enrichment of prostate cancer cells spiked into blood samples. Red blood cells were lysed and the remaining white blood cell population, including the cancer cells, was subjected to pre-alignment FFA, providing an enriched cancer cell fraction in the central outlet. Again the piezo actuator voltage was tuned to the level of maximal separation. For cells fixed with paraformaldehyde, cancer cell recovery ranged from 93.6% to 97.9% with purity ranging from 97.4% to 98.4%. There was no detectable loss of cell viability or cell proliferation subsequent to the exposure of viable tumor cells to acoustophoresis. For non-fixed, viable cells, tumor cell recovery ranged from 72.5% to 93.9% with purity ranging from 79.6% to 99.7%.

8.6.4 Bi-Directional Fractionation

From eqn (4.23a) it is clear that the radiation force on a particle can be directed either towards a pressure node (channel center) or towards a pressure maximum (channel wall) depending on the relative density and compressibility of the particle with respect to the suspending liquid. This means that in a given media, particles may migrate in opposite directions given that they have different acoustic properties, and that the properties of the media are chosen so that the acoustic contrast factor of the particles (Φ) are of opposite sign. Gupta et al.[55] demonstrated bi-directional acoustophoresis in an early paper, where low density polyethylene particles (size 5 μm to 10 μm) displayed opposite sign acoustic contrast relative to high density polyethylene particles of overlapping size. To accomplish this, the particles were suspended in 30% glycerol thus increasing the density and lowering the compressibility of the liquid.

In the more recent past Petersson et al. reported that erythrocytes and lipid particles can be separated in a microsystem, Figure 8.21(b), extracting erythrocytes free from lipid micro emboli in blood recovered from major surgery.[56,57] The bi-directional separation is possible due to the low density and high compressibility of the lipids relative to that of an aqueous liquid which cause them to have negative contrast. As a means to increase sample throughput, parallel channel configurations can be realized, yet actuated by a single transducer. Along this line, Jönsson et al.[58] demonstrated continuous blood processing for lipid emboli removal at 500 μL min^{-1} in an eight parallel channel configuration, Figure 8.21(c).

Bi-directional separation has also been utilized in applications of food stuff quality control where a system for raw milk sample pretreatment prior to Fourier transform infrared spectroscopy (FTIR) analysis was proposed by Grenvall et al.[59] The acoustophoresis set-up depleted the lipid emulsion from the milk by focusing these in the pressure antinode (white particles, Figure 8.21(a)) of an acoustic standing wave, providing a clarified milk fluid with casein micelles (brown particles). The removal of the lipid emulsion from the milk enabled direct FTIR analysis of the protein and lactose content in the clarified fraction. Also, an improved lipid analysis was obtained in the enriched lipid fraction. The bi-directional separation was performed in a multinode configuration where the milk sample was laminated in the center of the acoustophoresis channel. As the system was actuated in a 3λ/2-resonance mode, lipids were forced to the nearest pressure antinode off the channel center. Thereby lipid aggregation along the channel sidewalls and hence disturbed laminar flow properties were avoided, allowing continuous operation over an extended period of time. In a more recent publication, the same principle was used for sample preconditioning for impedance cytometry of somatic cells in milk. Samples were depleted of fat particles that would otherwise interfere with the measurements.[60]

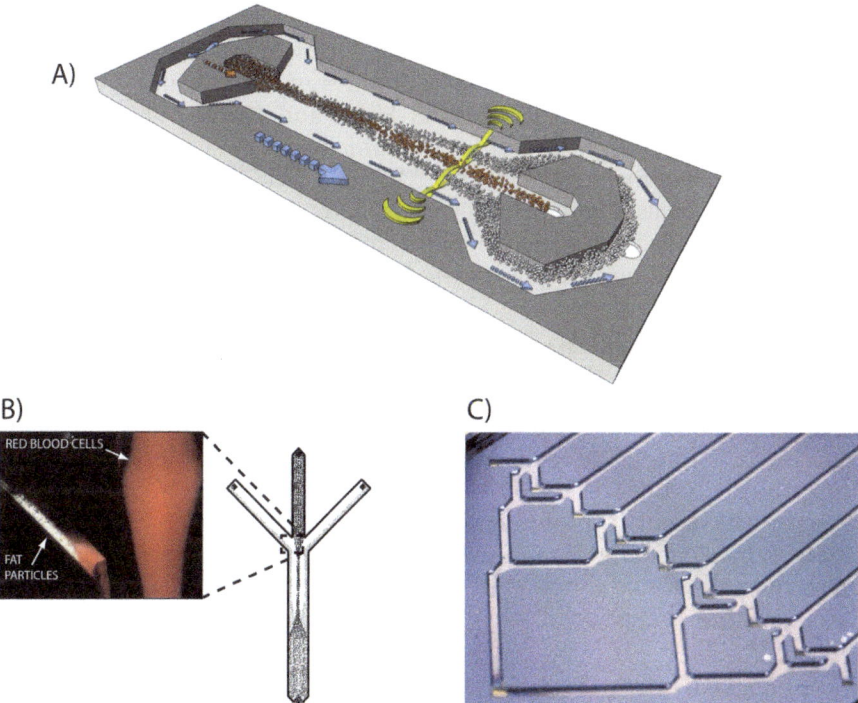

Figure 8.21 (A) Schematic of multinode actuation for the separation of lipid particles in milk, such that lipid particles are directed to the side outlets and the remaining micelles and milk fluid can be accessed for FTIR analysis *via* the central outlet. Image courtesy of Carl Grenvall, Lund University, Sweden. (B) Microscope image at the outlet of an acoustophoresis microfluidic chip for separation of fat particles from red blood cells. Reproduced from Petersson *et al.*[56] with permission from the Royal Society of Chemistry. (C) Increasing throughput in continuous flow acoustophoresis. Close-up of the outlet zone of the silicon microfabricated eight parallel channel acoustophoresis chip. Reproduced from Lenshof *et al.*[45] with permission.

8.6.5 Altering the Acoustic Properties of the Medium

As mentioned in the previous section, not many cells are of negative acoustic contrast when suspended in cell-compatible media such as isotone NaCl, PBS, serum-supplemented cell medium or blood plasma. In contrast to polymer particles, processing of viable cells put a restriction on how the properties of the media can be altered.

Petersson et al.[46] demonstrated that CsCl can be added to the media to alter the separation of polymer beads of equal size to favor a heavier and less compressible polystyrene fraction relative to a lighter and more compressible polyamide fraction. Furthermore, addition of CsCl (20 wt%) to the media was found to improve the separation of blood components. The use of CsCl in high concentrations is, however, problematic when used with cells over a prolonged period of time since it alters the osmolality of the solution, making it hypertonic, which causes cells to shrink and eventually die.

Other means to alter the acoustic properties of the medium are density gradient centrifugation media such as Ficoll and Percoll. These are cell compatible and have low impact on the tonicity of the solution. The main drawback in using these media is that when the concentration is increased, the viscosity increases exponentially. This slows down the acoustophoretic velocity of the cells to levels at which the acoustic streaming dominates the cell motion and no meaningful transport of cells can occur.

8.7 Cells Bound to Beads

Altering the medium may not be sufficient to create a condition where particles of similar acoustic properties and size may be separated. The main factor contributing to acoustic separation is the size, while the acoustic properties are secondary factors. Many cells, for instance, are about the same size and the acoustic properties do not differ much from each other. However, by using affinity-capture beads it is possible to create a condition where the net force on the bead–cell cluster either increases or decreases so that they move at a different rate to that of un-laden cells.

8.7.1 Positive Contrast Particles

By using affinity-capture beads of high acoustic contrast factor, the bead-bound cells can be pulled to the pressure node at a higher rate than the un-laden cells. The size of the affinity beads is important. Larger beads result in a substantial increase in the size of the bead–cell complex and thus a higher acoustic net force. However, as can be seen in Figure 8.22 (left), the number of beads connected to each cell can vary a lot, which will affect the force component and the stability and repeatability of such a system. Smaller beads may be better in terms of being more evenly distributed on the cell surface.

Lenshof et al.[61] used magnetic Dynabeads of 4.5 μm diameter as affinity-capture beads to enable separation of CD4+ T-helper cells from other white blood cells. The magnetic beads where much denser (1.6 g cm^{-3}) than the

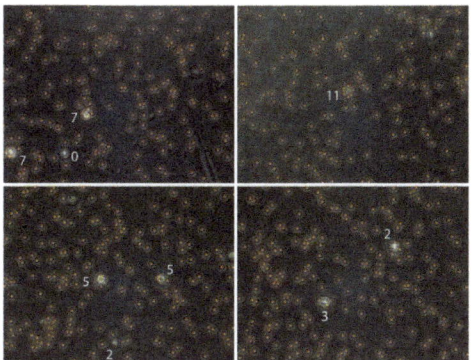
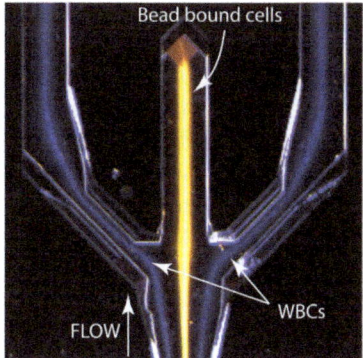

Figure 8.22 (Left) Magnetic beads bound to cells. As seen, the number of beads bound to each cell varies a lot and thereby the acoustic response during separation varies. (Right) Separation of bead-labeled cells exiting the center outlet while unbound white blood cells (WBCs) exit through the side outlets.

cells. However, the increase in size alone was not enough to receive a good separation of the targeted cells. By laminating a solution containing Ficoll histopaque (density 1.077 g cm^{-3}) in the center of the channel (similar set-up[50] as in Figure 8.19), a barrier was set up through which only the bead-bound cells could enter. Thus, this method introduces the possibility of favoring one of the cell types by making the force vanish on the other one by moving them into a medium of higher density and lower compressibility. Figure 8.22 (right) shows the separation of the bead-bound cells exiting through the center outlet while the unbound WBCs exit through the side outlets.

8.7.2 Negative Contrast Particles

Another possibility is to use affinity particles that are lighter than the medium, which causes them to move to the pressure antinodes instead. In a $\lambda/2$-system, such a particle would move the targeted particle towards the channel walls instead. Cushing *et al.*[62] reports such negative contrast particles from a polymer (PDMS). These negative contrast particles were then surface-modified to bind to certain proteins present in blood plasma, effectively separating them as the cellular content migrates to the pressure node while the polymer particles with the bound proteins migrate to the antinodes.

8.8 Frequency Switching

An alternative mode of separation, which like FFA is based on the magnitude of the primary acoustic radiation force, is the use of frequency switching such that the acoustic resonator is switched between different resonance modes at controlled time intervals. If the amplitudes for the employed actuation frequencies and the actuation intervals are tuned accordingly, an alignment of two particle categories into the different pressure nodes of the two frequencies, respectively, can be obtained. This was first explored by Mandralis and Feke in a system that combined an oscillating flow that was synchronized with the frequency switching between a single node and a triple node resonance—so-called synchronized ultrasonic flow field fractionation.[63] If the oscillating flow was defined to have a net flow in one direction, the system was able to drive particles larger than a defined size to one end of the flow system and smaller particles to the other end, enabling a continuous mode of operation.

More recently, Liu and Lim described an approach where controlled switching between $3\lambda/2$ and $\lambda/2$ resonances (1.8 and 5.6 MHz) enabled separation between 10 and 5 micrometre polystyrene beads.[64] Figure 8.23 illustrates the trajectories of the beads as they undergo a repetitive frequency switching.

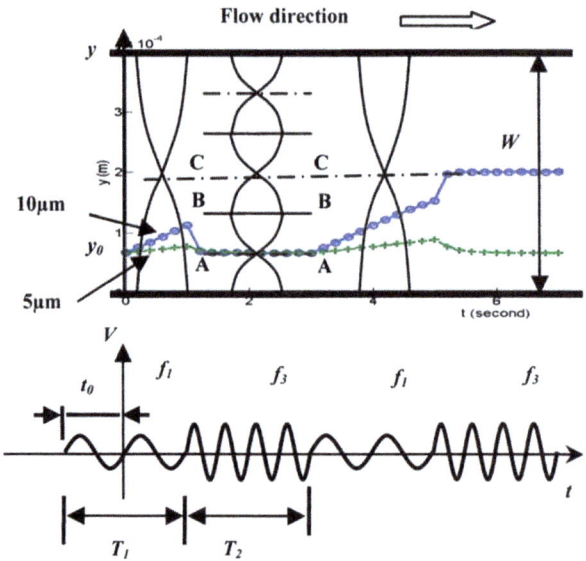

Figure 8.23 Schematic of particle separation by frequency switching as outlined by Liu and Lim.[64] The top figure shows the two operational modes with the fundamental resonance, f_1, initially applied during the period T_1, which is indicated in the amplitude vs. time diagram below. During time period T_2 the frequency is tripled, f_3, generating three nodes that align the cells along the node closest to the side wall indicated by A. As the frequency again is switched to f_1, larger particles (blue) move faster against node C, passing the position indicated by B. As the frequency again is switched to f_3 smaller beads (green) are retrieved into the A node whereas larger particles (blue) progress towards the central line C. Reproduced from ref. 64 with permission.

A general observation when performing separation by frequency switching is that it requires a solid process control if results are to be generated over an extended period of time. Fine tuning of duty cycles and actuation amplitudes is critical. It is also recommended that a temperature control of the chip is implemented. It should be pointed out that temperature control is a general requirement for long term stable operation, as primarily the speed of sound and hence resonance is dependent on temperature. Recently, Augustsson *et al.* demonstrated the importance of a strict temperature control while performing acoustophoresis experiments.[65] The beauty of separation by frequency switching, however, is that it offers a continuous flow based mode of operation, which is less sensitive to flow velocity fluctuations compared to sorting by FFA.

8.9 Acoustic Radiation Forces in Stratified Liquids

When an acoustic sound beam wave is transmitted through an interface between two liquids of different acoustic properties, the interface will deform due to the acoustic radiation pressure,[66] Figure 8.24(a) and (b). The phenomenon is also present in resonant systems, and Johansson *et al.*[67] was first to demonstrate that it could be used to induce mixing and transport of cells at liquid interfaces (see Section 8.4).

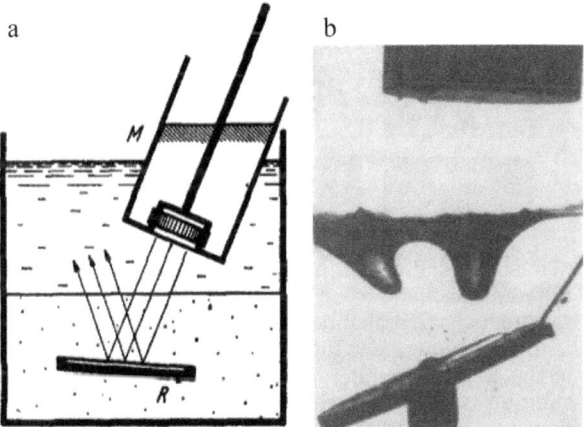

Figure 8.24 (a) Schematic of the experimental set-up of Hertz *et al.*,[66] demonstrating the effect of radiation pressure at the interface between two liquids of different acoustic properties (water above and aniline below). (b) The acoustic force induced on the interface depends on the relative difference in speed of sound in the two media and hence independent of the direction of sound propagation. This is evidenced by the interface being deflected downwards the aniline both for the incoming and reflected sound beam. Reproduced from ref. 66 with permission from Springer Science and Business Media.

Figure 8.25(a)–(c) shows acoustically induced relocation of liquids of different density (ρ) and speed of sound (c) in an acoustophoresis microchannel. The effect is present when the acoustic impedance ($Z = c \cdot \rho$) of the central liquid is lower than that of the side flows.[68] Figure 8.25(d) shows transverse vertical confocal scans of the relocation. When the sound intensity is increased through 4 V_{pp} and 6 V_{pp} the NaCl solutions injected at each side migrate towards the center along the bottom of the channel while the water follows the ceiling towards the side walls. At 10 V_{pp} the liquids have completed the relocation. An initial perturbation is required for the unstable configuration to relocate and in this case the density difference between the liquids causes the heavier liquid to sink.

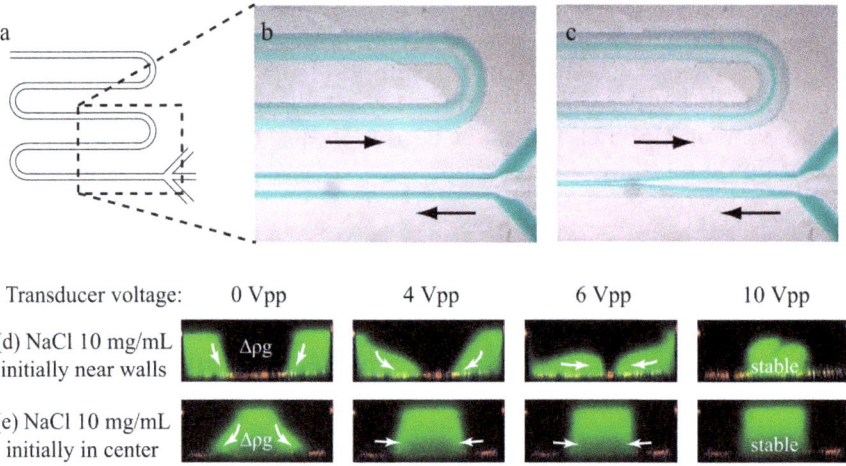

Figure 8.25 Acoustic radiation acting at the interface between two media of different acoustic contrast. (a) Schematic of a trifurcation-inlet meandered microchannel. (b) Saline solution spiked with blue dye is infused along both sides of MilliQ water. Diffusion is pronounced at the second meander. (c) After the onset of ultrasound, the saline solution is focused to the central part of the flow due to its higher density and speed of sound. Reproduced with permission from Carl Grenvall, Lund University, Sweden. (d) Confocal vertical scans near the end of an acoustophoresis channel. Fluorescent liquid (green) of NaCl-concentration (10 mg mL^{-1}) was laminated (d) near the side walls of the channel or (e) in the center of the channel. Pure water occupied the remaining parts of the flow (dark areas). The images show (d) acoustic relocation of liquid for increasing actuator voltages. (e) For negative mismatch the onset of the acoustic resonance does not cause any relocation of the liquids but rather counteracts the initial inclination of the interfaces which is caused by the density difference between the liquids. Reproduced from Deshmukh et al.[68] with permission.

Figure 8.25(e) shows that when the NaCl solution is infused in the central inlet the constellation is stable. Upon increasing the piezo actuator voltage the initial tilt of the liquid interfaces are gradually pushed to a vertical orientation.

Acoustic relocation of liquids has obvious implications for applications wherein liquids are flow laminated. Blood cells can, for example, not be straightforwardly extracted directly from undiluted plasma (*i.e.* whole blood) into PBS due to the difference in acoustic properties between the two liquids. The high protein content of blood plasma makes it less compressible and denser than PBS. For a configuration where blood is laminated along each side of PBS, the onset of an acoustic resonance renders the two liquids to relocate with respect to each other so that the plasma occupies the central part of the flow destroying the intended separation of cells. To circumvent this problem the acoustic properties of the central liquid stream can be altered with *e.g.* gradient density media.[69]

8.10 Considerations

When designing an acoustophoresis system, the specific application and its critical boundaries must be taken into account. It is crucial to maintain a constant acoustic performance of the system over time and hence parameters such as chip temperature and changed buffer compositions must be considered. Although acoustophoresis is a very gentle cell handling method, temperature control is crucial for optimal outcome. When performing acoustophoresis on sample species that are in the size range of 1–2 µm, care must taken in the system design and operation such that acoustic streaming does not override the desired effect of the primary acoustic radiation force.

8.10.1 Variations of the Acoustic Field

In most acoustophoretic devices, the amplitude of the acoustic field is not constant along the whole length of the channel.[65] Moreover, the location of the pressure minima in the width dimension can display a slight variation[3,21] such that the particles wiggle sideways as they flow along the channel. This effect is particularly prominent when the acoustic velocity of the particles is high relative to the flow velocity in the channel. When the flow velocity is substantially higher than the acoustic velocity of the particles, those variations tend to average out. However, in chip regions where the acoustic resonator is not so well defined, *e.g.* in a flow splitter zone, care must be taken that an upstream acoustophoretic process is not disturbed by the arbitrary resonances in this region.

The spatial variation in the acoustic field is highly frequency dependent. By identifying a frequency range for which the channel has multiple resonances, the variations can be averaged out by sweeping the piezo actuator frequency over that range.[2]

8.10.2 Temperature Aspects

Acoustofluidic systems are inherently dependent of temperature fluctuations. As most of the delivered electric power is dissipated in the assembled actuator/microchip device and converted to heat, special attention has to be given when designing bulk acoustic resonators actuated by piezoceramic actuators. Since compressibility and density are temperature dependent parameters, a temperature change will alter the resonance in the cavity. The frequency shift due to a temperature change in an acoustophoresis experiment at a $\lambda/2$-resonance of ~1.95 MHz, at 25 °C, was reported to be approximately 2 kHz per °C.[65] Since the resonance peak width at half maximum (FWHM) was measured to be 8 kHz, a temperature shift of only 1 °C would dramatically affect the acoustic amplitude in the channel. It should be stressed, however, that systems of lower Q-factor, *i.e.* more attenuated and/or bulkier, are likely to be less sensitive to temperature fluctuations. For example, Hawkes *et al.*[12] report a frequency shift of 5 kHz per °C and a corresponding peak width of ~100 kHz. Likewise, Wiklund *et al.*[70] implemented attenuation of the microfluidic manifold to obtain a broader resonance peak which is inherently less sensitive to temperature changes but, on the other hand, requires elevated input power to accomplish sufficient acoustic energy density in the resonator. The broader resonance peak also facilitates the operation of the acoustic resonator in a frequency modulated mode to obtain a more homogeneous time-averaged actuation of an array of micro-cavities. In order to ensure that temperature variations are not affecting the outcome of the acoustofluidic experiment, it is advised to measure the temperature dependency for each platform, or to ensure that the chip temperature is not changing significantly during the experiment. Temperature can be straightforwardly controlled using a Peltier element and a temperature sensor feedback control.[54,65]

8.10.3 Cell Analysis Following Acoustofluidic Handling

When changing from particle- to cell-based acoustic separation, a new set of aspects needs to be considered. Viable cell handling is challenging due to the stringent environmental conditions for cell survival on account of the fragile nature of the cell membrane.[71,72] The maintenance of optimal cell conditions is vital to preserve the expected lifetime of the cells. If viable cells are not needed for subsequent analysis, cell membrane stabilization (fixation) will facilitate the cell handling. Fixation can be performed with cross-linking reagents such as paraformaldehyde (PFA) or with organic solvents *e.g.* methanol. Cell fixation terminates any ongoing biochemical reaction, and stabilizes the cells.[73] For fixed cells the amount of time spent on cell processing is generally of little relevance. However, for viable cells the time needed for subsequent biochemical steps *e.g.* fluorescent staining and washing *etc* should be minimized. Cell mortality generally increases with longer handling times. Depending on cell type, the cells should be handled on ice, in room temperature or at 37 °C. The optimal handling temperature will reduce cell damage and death for sensitive cell populations.

FACS analysis can be used to identify and quantify the separated cell populations.[74] For non-fixed cells it is advisable to analyse the outlet fractions immediately. Fixed cells can be stored in the fridge and analyzed the following day.

For optimal acoustic separation performance, it is best to use fresh cell samples. The cell samples should be prepared directly before the acoustic separation. For example, blood samples stored overnight have a tendency to contain more aggregated cells, which aggravate the acoustic cell separation.

Acoustophoresis is a gentle separation method, and generates no immediate or long-term changes in cell viability or cell proliferation[50,74] when operated under typical acoustophoretic separation conditions with energy densities below 100 J m^{-3}. It should be noted that temperature control is recommended not only to maintain a fixed resonance in the chip but also because chip heating may induce cell damage if the actuator is operated at higher voltages. Acoustophoresis can therefore be considered to be a non-perturbing method for cell separation. A more detailed analysis of cells exposed to acoustic standing wave fields in terms of viability and biocompatibility is discussed in Chapter 21.

8.10.4 Measuring Separation Performance

When developing acoustophoresis-based separation protocols, the quantification of the separations performed by acoustophoresis becomes a challenge in each case as different particle mixtures or cell suspensions compose the sample and on many occasions, the development of protocols to validate the performance of the executed separations has required a significant effort. When working with microparticles in the micrometre size range, Coulter counting is a straightforward method. Coulter counting provides absolute particle counts and yields excellent size distribution data. The lower limit of detection in conventional Coulter counting is in the range of 1 micrometre, and at lower particle size, the noise level rises dramatically. It should be mentioned that the most commonly expressed disadvantage of Coulter counting is the frequent tendency to clog the measuring orifice and hence mainly analytically pure systems without a tendency for particle aggregation should be employed.

When working with cell-based systems, fluorescence activated cell sorting, FACS, is an obvious technique to use when characterizing the content of a sample with mixed cell populations. A drawback, however, is that costly labeling protocols prior to FACS analysis have to be employed and if larger sample volumes are to be processed, this may not be a viable option. The data obtained is composed of measurements of forward- and backscattered light as well as fluorescence intensity and the software included with the instruments commonly allows elaborate data mining. FACS instruments that also measure the sample flow rate during the measurement can also provide data on particle concentrations.

As an alternative to Coulter counting or FACS, counting of cells or particles can be performed manually by traditional counting in a Bürcher chamber. This offers operator control and visual inspection of the sample composition, which sometimes may provide information that is not revealed in FACS or Coulter counting but the drawback is evidently the tedious manual work when extended studies are performed.

More recent instrumentation technology utilizes shadow imaging to precisely estimate particle size down to 1 micrometre, which matches the lower size range for acoustophoresis separations. This method offers a reasonably straightforward means of monitoring size distributions derived from acoustophoresis experiments. A simple but rather crude method to quantify particle concentrations in different outlet fractions of an acoustophoresis experiment is to use standard sedimentation capillaries that are centrifuged and the height of the solid matter in the sample is derived as a measure of the particle concentration. It is a simple and fast method but does not reveal any information about size distribution or different species in the sample.

8.10.5 Size Limitations

When performing acoustophoresis experiments, particle/cell size typically ranges from 1–20 µm. The lowest size limit at which the primary acoustic radiation force is effective is about 1 µm. This is based on the fact that the mobility of a particle driven by the primary acoustic radiation force scales with the square of the radius and at a particle size around 1–2 micrometres (depending on density, and compressibility), hydrodynamic forces generated by acoustic streaming equal the effective primary acoustic radiation force at an actuation frequency of 2 MHz.[75] Means to move beyond this limit can be accomplished by reducing the dimensions of the acoustic resonator, hence increasing the actuation frequency, since the primary acoustic radiation force is proportional to the operating frequency. Reduced influence of acoustic streaming can also be realized by changing the channel aspect ratio such that the Rayleigh streaming rolls are mainly confined to the top and bottom of a narrow channel, Figure 8.26.[75]

Recent reports also show that acoustophoresis systems can be designed to support a net rotational streaming motion in the channel cross section, which does not interfere with the minute primary acoustic radiation force acting on submicron particles. Both bacteria (*E. coli*) and particle sizes down to 600 nm were shown to focus in a microchannel configured to support both a vertical and a horizontal $\lambda/2$-resonance.[76]

By driving the acoustic system at a frequency that sweeps across the optimal resonance, a suppression of streaming artifacts can also be accomplished. Ohlin *et al.*[70] demonstrated that acoustic streaming in microwells can be reduced by scanning the actuation frequency (2.13 MHz) over a 200 kHz bandwidth. The upper size limit in acoustophoresis experiments is commonly set by sedimentation, which becomes a considerable problem for

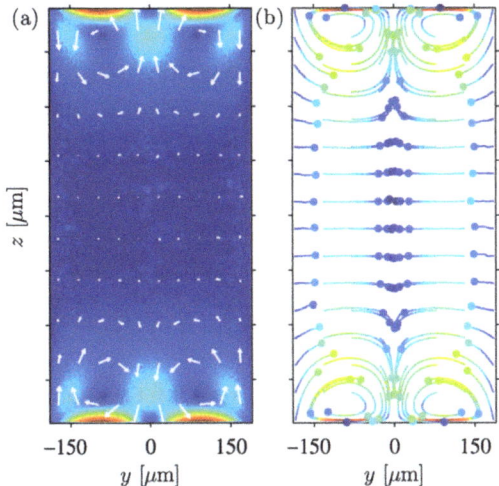

Figure 8.26 Rayleigh streaming in a high aspect ratio microchannel with the standing wave in the y-direction. The influence of the streaming rolls is mainly restricted to regions close to the top and bottom of the channel. Reproduced from Muller et al.[75] with permission.

particles larger than 20 µm. If compensation of the buffer composition in terms of the buoyancy of the particle is performed, an extended working range can be accomplished. Alternatively, the entire fluidic system needs to be aligned vertically such that sedimentation does not influence system performance.

References

1. H. Bruus, *Lab Chip,* 2012, **12**, 20–28.
2. O. Manneberg, B. Vanherberghen, B. Onfelt and M. Wiklund, *Lab Chip,* 2009, **9**, 833–837.
3. S. M. Hagsater, A. Lenshof, P. Skafte-Pedersen, J. P. Kutter, T. Laurell and H. Bruus, *Lab Chip,* 2008, **8**, 1178–1184.
4. K. Yasuda, S. S. Haupt, S. Umemura, T. Yagi, M. Nishida and Y. Shibata, *J. Acoust. Soc. Am.,* 1997, **102**, 642–645.
5. K. Yasuda, S. Umemura and K. Takeda, *Jpn. J. Appl. Phys., Part 1,* 1995, **34**, 2715–2720.
6. G. Goddard and G. Kaduchak, *J. Acoust. Soc. Am.,* 2005, **117**, 3440–3447.
7. G. Goddard, J. C. Martin, S. W. Graves and G. Kaduchak, *Cytometry Part A,* 2006, **69A**, 66–74.
8. N. R. Harris, M. Hill, S. Beeby, Y. Shen, N. M. White, J. J. Hawkes and W. T. Coakley, *Sens. Actuators B,* 2003, **95**, 425–434.
9. N. R. Harris, M. Hill, R. Townsend, N. M. White and S. P. Beeby, *Sens. Actuators B,* 2005, **111**, 481–486.

10. A. Nilsson, F. Petersson, H. Jonsson and T. Laurell, *Lab Chip,* 2004, **4**, 131–135.
11. M. Nordin and T. Laurell, *Lab Chip,* 2012, **12**, 4610–4616.
12. J. J. Hawkes and W. T. Coakley, *Sens. Actuators B,* 2001, **75**, 213–222.
13. S. Kapishnikov, V. Kantsler and V. Steinberg, *J. Stat. Mech.: Theory Exp.,* 2006, P01012.
14. A. Lenshof, A. Ahmad-Tajudin, K. Jaras, A. M. Sward-Nilsson, L. Aberg, G. Marko-Varga, J. Malm, H. Lilja and T. Laurell, *Anal. Chem.,* 2009, **81**, 6030–6037.
15. A. A. Tajudin, K. Petersson, A. Lenshof, A. M. Sward-Nilsson, L. Aberg, G. Marko-Varga, J. Malm, H. Lilja and T. Laurell, *Lab Chip,* 2013, **13**, 1790–1796.
16. P. W. S. Pui, F. Trampler, S. A. Sonderhoff, M. Groeschl, D. G. Kilburn and J. M. Piret, *Biotechnol. Prog.,* 1995, **11**, 146–152.
17. M. Groschl, W. Burger, B. Handl, O. Doblhoff-Dier, T. Gaida and C. Schmatz, *Acustica,* 1998, **84**, 815–822.
18. M. Sundin, A. Nilsson, F. Petersson and T. Laurell, Binary valving of particles using acoustic fields, *The 8th International Conference on Miniaturized Systems for Chemistry and Life Sciences (μTAS)*, The Chemical and Biological Microsystems Society, 2004, vol. 1, pp. 662–664.
19. T. Laurell, F. Petersson and A. Nilsson, *Chem. Soc. Rev.,* 2007, **36**, 492–506.
20. O. Manneberg, S. M. Hagsater, J. Svennebring, H. M. Hertz, J. P. Kutter, H. Bruus and M. Wiklund, *Ultrasonics,* 2009, **49**, 112–119.
21. O. Manneberg, J. Svennebring, H. M. Hertz and M. Wiklund, *J. Micromech. Microeng.*, 2008, 18.
22. L. Johansson, F. Nikolajeff, S. Johansson and S. Thorslund, *Anal. Chem.,* 2009, **81**, 5188–5196.
23. C. Grenvall, M. Carlsson, P. Augustsson, F. Petersson and T. Laurell, Fluorescent activated cell sorter using ultrasound standing waves in micro channels, *The 11th International Conference on Miniaturized Systems for Chemistry and Life Sciences, μTAS,* 2007, The Chemical and Biological Microsystems Society, Paris, France, 2007.
24. O. Jakobsson, C. Grenvall, M. Nordin, M. Evander and T. Laurell, *Lab Chip,* 2014, **14**, 1943–1950.
25. P. P. A. Suthanthiraraj, M. E. Piyasena, T. A. Woods, M. A. Naivar, G. P. Lopez and S. W. Graves, *Methods,* 2012, **57**, 259–271.
26. C. Grenvall, C. Antfolk, C. Zoffmann Bisgaard, S. Kjaer Andersen and T. Laurell, Improved microfluidic impedance cytometer using novel 2D acoustophoresis pre-focusing, *XXVII Congress of the International Society for Advancement of Cytometry – CYTO 2012*, The International Society for Advancement of Cytometry, Leipzig, Germany, 2012.
27. J. J. Hawkes, R. W. Barber, D. R. Emerson and W. T. Coakley, *Lab Chip,* 2004, **4**, 446–452.
28. F. Petersson, A. Nilsson, H. Jönsson and T. Laurell, *Anal. Chem.,* 2005, **77**, 1216–1221.
29. J. Persson, P. Augustsson, T. Laurell and M. Ohlin, *FEBS J.,* 2008, **275**, 5657–5666.

30. A. Lenshof, B. Warner and T. Laurell, Acoustophoretic pretreatment of cell lysate prior to FACS analysis, *The 14th International Conference on Miniaturized Systems for Chemistry and Life Sciences*, The Chemical and Biological Microsystems Society, Groningen, Netherlands, 2010.
31. P. Augustsson, J. Persson, S. Ekstrom, M. Ohlin and T. Laurell, *Lab Chip*, 2009, **9**, 810–818.
32. P. Augustsson, PhD Thesis, Lund University, 2011.
33. P. Augustsson, MSc Thesis, Lund University, 2006.
34. C. Mikkelsen and H. Bruus, *Lab Chip*, 2005, **5**, 1293–1297.
35. P. Augustsson, J. Persson, S. Ekström, M. Ohlin and T. Laurell, *Lab Chip*, 2009, **9**, 810–818.
36. P. Augustsson, L. B. Åberg, A.-M. K. Swärd-Nilsson and T. Laurell, *Microchim. Acta*, 2009, **164**, 269–277.
37. P. Augustsson, T. Laurell and S. Ekström, Flow-through chip for sequential treatment and analyte elution from beads or cells, *The 12th International Conference on Miniaturized Systems for Chemistry and Life Sciences*, The Chemical and Biological Microsystems Society, San Diego, CA, USA, 2008.
38. P. Augustsson, J. Malm and S. Ekström, *Biomicrofluidics*, 2012, **6**, 034115.
39. M. A. H. Weiser and R. E. Apfel, *J. Acoust. Soc. Am.*, 1982, **71**, 1261–1268.
40. D. Hartono, Y. Liu, P. L. Tan, X. Y. S. Then, L.-Y. L. Yung and K.-M. Lim, *Lab Chip*, 2011, **11**, 4072–4080.
41. J. C. Giddings, *Sep. Sci. Technol.*, 1985, **20**, 749–768.
42. D. A. Johnson and D. L. Feke, *Sep. Technol.*, 1995, **5**, 251–258.
43. M. Kumar, D. L. Feke and J. M. Belovich, *Biotechnol. Bioeng.*, 2005, **89**, 129–137.
44. A. Lenshof and T. Laurell, *Chem. Soc. Rev.*, 2010, **39**, 1203–1217.
45. A. Lenshof, C. Magnusson and T. Laurell, *Lab Chip*, 2012, **12**, 1210–1223.
46. F. Petersson, L. Aberg, A. M. Sward-Nilsson and T. Laurell, *Anal. Chem.*, 2007, **79**, 5117–5123.
47. J. D. Adams and H. T. Soh, *Appl. Phys. Lett.*, 2010, 97.
48. P. Thevoz, J. D. Adams, H. Shea, H. Bruus and H. T. Soh, *Anal. Chem.*, 2010, **82**, 3094–3098.
49. A. H. J. Yang and H. T. Soh, *Anal. Chem.*, 2012, **84**, 10756–10762.
50. J. Dykes, A. Lenshof, I. Åstrand-Grundström, T. Laurell and S. Scheding, *PLoS One*, 2011, **6**, e23074.
51. J. Nam, Y. Lee and S. Shin, *Microfluid. Nanofluid.*, 2011, **11**, 317–326.
52. J. D. Adams, P. Thevoz, H. Bruus and H. T. Soh, *Appl. Phys. Lett.*, 2009, 95.
53. J. J. Shi, H. Huang, Z. Stratton, Y. P. Huang and T. J. Huang, *Lab Chip*, 2009, **9**, 3354–3359.
54. P. Augustsson, C. Magnusson, M. Nordin, H. Lilja and T. Laurell, *Anal. Chem.*, 2012, **84**, 7954–7962.
55. S. Gupta, D. L. Feke and I. Manas-Zloczower, *Chem. Eng. Sci.*, 1995, **50**, 3275–3284.
56. F. Petersson, A. Nilsson, C. Holm, H. Jonsson and T. Laurell, *Analyst*, 2004, **129**, 938–943.

57. F. Petersson, A. Nilsson, C. Holm, H. Jonsson and T. Laurell, *Lab Chip,* 2005, **5**, 20–22.
58. H. Jönsson, C. Holm, A. Nilsson, F. Petersson, P. Johnsson and T. Laurell, *Ann. Thoracic Surg.,* 2004, **78**, 1572–1578.
59. C. Grenvall, P. Augustsson, J. R. Folkenberg and T. Laurell, *Anal. Chem.,* 2009, **81**, 6195–6200.
60. C. Grenvall, J. R. Folkenberg, P. Augustsson and T. Laurell, *Cytometry Part A,* 2012, **81A**, 1076–1083.
61. A. Lenshof, A. Jamal, J. Dykes, A. Urbansky, I. Åstrand-Grundström, T. Laurell and S. Scheding, *Cytometry Part A,* 2014, **85**, 933–941.
62. K. W. Cushing, M. E. Piyasena, N. J. Carroll, G. C. Maestas, B. A. Lopez, B. S. Edwards, S. W. Graves and G. P. Lopez, *Anal. Chem.,* 2013, **85**, 2208–2215.
63. Z. I. Mandralis and D. L. Feke, *AIChE J.,* 1993, **39**, 197–206.
64. Y. Liu and K. M. Lim, *Lab Chip,* 2011, **11**, 3167–3173.
65. P. Augustsson, R. Barnkob, S. T. Wereley, H. Bruus and T. Laurell, *Lab Chip,* 2011, **11**, 4152–4164.
66. G. Hertz and H. Mende, *Z. Phys.,* 1939, **114**, 354–367.
67. L. Johansson, S. Johansson, F. Nikolajeff and S. Thorslund, *Lab Chip,* 2009, **9**, 297–304.
68. S. Deshmukh, Z. Brzozka, T. Laurell and P. Augustsson, *Lab Chip,* 2014, **14**, 3394–3400.
69. P. D. Ohlsson, K. Petersson, P. Augustsson and T. Laurell, Acoustophoresis Separation Of Bacteria From Blood Cells For Rapid Sepsis Diagnostics, *The 17th International Conference on Miniaturized Systems for Chemistry and Life Sciences*, The Chemical and Biological Microsystems Society, Freiburg, Germany, 2013.
70. M. Ohlin, A. E. Christakou, T. Frisk, B. Onfelt and M. Wiklund, Controlling acoustic streaming in a multi-well microplate for improving live cell assays, *The 15th International Conference on Miniaturized Systems for Chemistry and Life Sciences*, The Chemical and Biological Microsystems Society, Seattle, Washington, USA, 2011.
71. S. J. Singer and G. L. Nicolson, *Science,* 1972, **175**, 720–731.
72. K. Simons and D. Toomre, *Nat. Rev. Mol. Cell Biol.,* 2000, **1**, 31–39.
73. J. L. Javois, *Immunocytochemical Methods and Protocols, Methods in Molecular Biology,* vol. 115, Humana Press Inc., Totowa, NJ. 1999.
74. P. Augustsson, C. Magnusson, C. Grenvall, H. Lilja and T. Laurell, Extraction of circulating tumor cells from blood using acoustophoresis, *The 14th International Conference on Miniaturized Systems for Chemistry and Life Sciences*, The Chemical and Biological Microsystems Society, Groningen, Netherlands, 2010.
75. P. B. Muller, R. Barnkob, M. J. H. Jensen and H. Bruus, *Lab Chip,* 2012, **12**, 4617–4627.
76. M. Antfolk, P. B. Muller, P. Augustsson, H. Bruus and T. Laurell, *Lab Chip,* 2014, **14**, 2791–2799.

CHAPTER 9

Applications in Acoustic Trapping

MIKAEL EVANDER AND JOHAN NILSSON*

Department of Biomedical Engineering, Lund University, Sweden
*E-mail: johan.nilsson@bme.lth.se

9.1 Introduction

Techniques to immobilise and/or enrich particles and cells in microfluidic systems are often required in lab-on-a-chip systems. Immobilisation can facilitate several different applications in the life science research area, *e.g.* microscopy studies, incubation and washing of cells, enrichment of cells from dilute suspensions and studies on cell–cell interactions. Non-contact immobilisation specifically enables the study of non-adherent cells in a more controlled and *in vivo* like environment, *e.g.* non-contact cell culturing in perfusion chambers.

Immobilisation of particles in microfluidic systems can be realised in several different ways, each with its own advantages and disadvantages.[1] In the well-established patch-clamp technique, particles are normally trapped against an orifice using under-pressure and it is a method that is well suited for electrophysiological studies of single cells. Related systems, combining microfluidics with patch-clamp, have also been presented.[2,3] In optical trapping, particles are trapped at the focal point of a laser beam. While the technique was originally a single-object technique, modern, commercially available systems allow for manipulation of hundreds of objects simultaneously, which combined with the possibility to measure forces, makes optical tweezers a powerful technique.[4] In dielectrophoresis, particles are trapped by dielectric forces in a non-uniform electrical field created by microelectrodes in a fluidic channel. In magnetic trapping, non-uniform magnetic fields are used to trap magnetic particles. Finally, in acoustic trapping, particles are trapped in a non-uniform acoustic field due to the radiation force created by a pressure gradient.

While patch-clamp is a single-object technique, dielectric, magnetic and acoustic trapping can all be used for single particle, array or cluster trapping. Optical trapping can be used for both single particle and array manipulation. Both dielectric and magnetic trapping can be miniaturised and arrayed through the use of MEMS-techniques, while optical trapping requires a microscope set-up. All trapping methods, with patch-clamp as the exception, also have the possibility to be used in non-contact mode where cells can be immobilised without any surface contact. Most trapping techniques have certain demands either on the buffer or on the object that is to be manipulated. In optical trapping, it is desirable to have optically transparent objects to achieve a stable trapping. In magnetic trapping, the object must either be paramagnetic or ferromagnetic (or a paramagnetic buffer can be used).[5] For dielectric trapping, the object's and the media's complex permittivity need to differ while in acoustic trapping it is the density and compressibility that need to differ between the object and the media.

Historically, acoustic trapping was introduced as a tool for containerless processing by NASA and ESA about 40 years ago.[6] The acoustic levitators developed then were typically aimed at levitating liquid droplets in the open air at frequencies at a couple of tens of kilohertz. In the early 1990s, acoustic levitation in liquid was investigated for use in bulk suspension cultures to improve the productivity. Trampler *et al.* showed how a 2.5 MHz resonator could be used with a mammalian cell culture system to increase antibody production.[7] A cell suspension was flowed through an acoustic resonator where cells were trapped and aggregated until they sedimented back into the main culture chamber. Pui and Gaida *et al.* enhanced the system further and added more control and could report retention percentages of viable cells of up to 99%.[8,9] Trapping systems utilising focused transducers were also demonstrated in the early to mid 90s by Wu and Hertz.[10,11] In the early 2000s, more miniaturized trapping systems aimed at studies of small cell clusters were presented by Coakley *et al.*[25] and since then, there have been many interesting publications integrating acoustic trapping with the world of lab on a chip. In this review, we start with a short introduction to acoustic trapping and give an overview of different methods and applications that have been presented in this field.

9.2 Theory

As with acoustophoresis, acoustic trapping relies on the acoustic radiation forces that arise when a standing wave is set up in a microfluidic channel. To create a standing wave, the driving frequency must meet a resonance criterion of the channel; typically with a width or height of: $n \cdot \frac{v}{2f} = n \cdot \frac{\lambda}{2}$ where n is the number of pressure nodes in the standing wave, v is the speed of sound in the fluid, f is the acoustic frequency and λ the wavelength of the sound. Other configurations can also be used, e.g. devices with a $\lambda/4$ resonance where particles are trapped on a surface.[12] For a more detailed description of the different possible resonances, see chapter 7 and 17 in this book.[13]

The acoustic radiation force will move objects either to the point of maximum or minimum acoustic potential in the standing wave. A full derivation of the acoustic radiation force can be found in Chapter 4 and only the conclusion is presented here.[14] The primary acoustic radiation force, F^{rad}, on a small, spherical particle (radius a and $a \ll \lambda$) in an inviscid fluid can be expressed as the gradient of the acoustic potential, U^{rad}:

$$F^{rad} = -\nabla U^{rad}, \tag{9.1}$$

$$U^{rad} = \frac{4\pi}{3}a^3 \left[f_1 \frac{1}{2} \kappa_0 \langle p_{in}^2 \rangle - f_2 \frac{3}{4} \rho_0 \langle v_{in}^2 \rangle \right], \tag{9.2}$$

$$f_1(\tilde{\kappa}) = 1 - \tilde{\kappa} \quad \text{with} \quad \tilde{\kappa} = \frac{\kappa_p}{\kappa_0}, \tag{9.3}$$

$$f_2(\tilde{\rho}) = \frac{2(\tilde{\rho} - 1)}{2\tilde{\rho} + 1} \quad \text{with} \quad \tilde{\rho} = \frac{\rho_p}{\rho_0}. \tag{9.4}$$

where $\langle p_{in}^2 \rangle$ and $\langle v_{in}^2 \rangle$ are the time averages of the incoming pressure and velocity fields squared, κ_0 and ρ_0 are the compressibility and density of the fluid and κ_p and ρ_p are the compressibility and density of the spherical particle. The density and compressibility of the object and the fluid will decide if the object will move to the point of maximum or minimum acoustic potential. Most cells and microparticles with densities higher and/or compressibilities lower than the surrounding medium in standard buffers will move to the point of minimal acoustic potential, which in a perfect standing wave corresponds to the pressure node, see Figure 9.1.

In acoustophoresis, an acoustic potential along the entire length of the channel is desirable in order to be able to efficiently move objects between flow paths at high flow rates.[15] Acoustic trapping, however, requires a local acoustic potential with a high gradient to be able to immobilise objects against a flow, see Figure 9.1. There are also secondary forces that act mainly over short distances and in the pressure node will attract particles to each other and thus help stabilize the cluster.[16]

In some cases, the lateral component of the primary radiation force is mentioned, referring to the part of the gradient that is perpendicular to the primary sound propagation direction. The acoustic radiation force in the

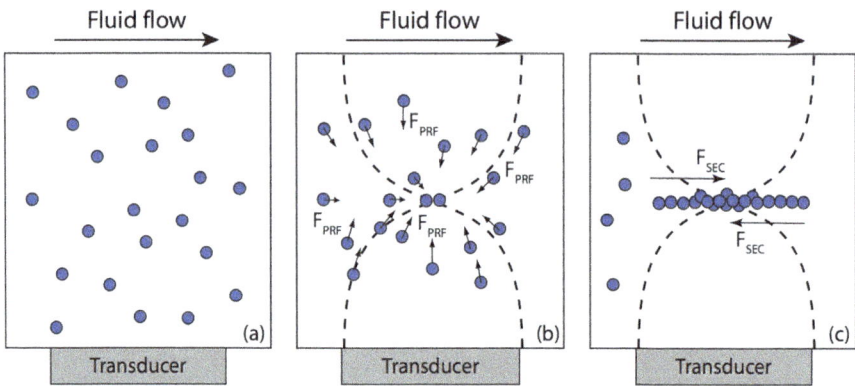

Figure 9.1 A typical trapping configuration where a transducer emits sound in a direction perpendicular to the flow (a). When the transducer is active, a standing wave will form in the fluidic channel and in the case of a $\lambda/2$-resonance, a single pressure node will be created in the centre. The primary radiation forces will cause objects in the fluid to aggregate in the pressure node (b). Once a cluster is formed, secondary forces and the lateral component of the primary force will keep the particles together and counteract the Stokes' drag force on the cluster while new particles can be transported to the pressure node by the fluid flow and trapped (c).

lateral direction can be a couple of orders of magnitude lower than in the sound propagation direction. Trapping systems that are designed to hold objects against a flow typically utilise the lateral component of the radiation force to counteract the Stokes' drag on the object.

Another phenomenon that can affect the performance of a trapping system is acoustic streaming. Due to the pressure/velocity gradient in the viscous boundary layer close to the walls generated by the standing wave, a bulk circulating flow is generated in the fluid and this flow can either help to draw particles into the acoustic trap or prevent them from being trapped.[17] The acoustic streaming affects the particles through Stokes' drag that scales with the radius of the particles. As can be seen in eqn (9.1)–(9.2), the primary radiation force scales with the cube of the radius. This means that the particle size will determine which effect is dominating and there exists a particle size transition region from streaming-dominated to primary force-dominated behaviour of the particles. For boundary driven streaming, this region has been reported to be around 1.3–1.8 μm for typical fluid/particle combinations at low MHz frequencies.[18,19] In planar resonators there is also a four-quadrant streaming pattern in plane with the transducer that can be pronounced in acoustic trapping systems.[20] There are other kinds of acoustic streaming as well and for more details and a thorough review on how the streaming will affect acoustic systems we refer to Chapter 13.[17]

9.3 Methods for Acoustic Trapping

In acoustic trapping, standing waves are used to build up high acoustic potentials in the systems.

A straightforward way of generating a standing wave for acoustic trapping is to use two opposing transducers. Wu *et al.* presented an early example of acoustic tweezers based on this principle in 1991.[10] Two focused transducers opposing each other were operated at 3.5 MHz and used to create a well-defined potential well where latex beads and frog eggs could be trapped. Hertz also presented a similar system where 2.1 μm glass particles could be trapped at a frequency of 11 MHz, see Figure 9.2.[11] The design created lateral forces of the same magnitude as the axial forces to promote 3D-trapping. However, focused transducers tend to lead to more bulky systems, since the focal point is located several wavelengths away from the transducer.

Rather than having two transducers, a single transducer and a reflector may be used to generate the standing wave. Wiklund *et al.* combined the principle with a microfluidic set-up for trapping in quartz capillaries, see Figure 9.3.[21] Several nodes were created along the capillary where particles were trapped. Lee *et al.* have demonstrated trapping of lipid droplets in a system based on a single focused acoustic beam without a reflector. The Gaussian profile of the beam trapped the droplets in the focal spot, similar to optical trapping.[22] It should be noted that the droplets in these systems must be larger than the wavelength of the ultrasound.[23]

Using layered resonators, as described in Chapter 7, smaller trapping systems can be designed where an unfocused transducer is used to generate standing waves in a wavelength-matched structure of coupling, channel and

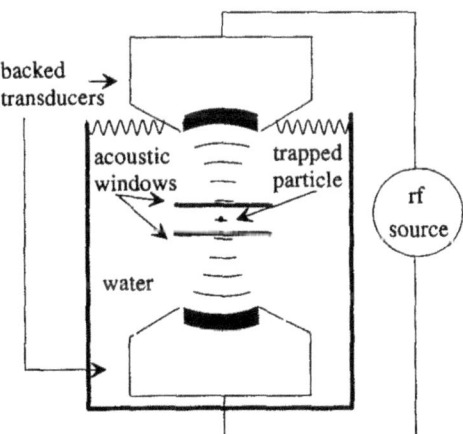

Figure 9.2 Two opposing focused transducers in a confocal arrangement generating a standing wave field where the particles can be trapped. The trapping force is strongest in the focal spot. Reprinted with permission from ref. 11. Copyright 1995, American Institute of Physics.

reflector layers.[13,24] This approach also facilitates the integration of acoustic trapping in microfluidic systems. Spengler and Coakley introduced a system based on this principle in 2001 that has been used in several different cell study applications (see later).[25] Here, the system was designed as a layered resonator to achieve high Q-values. A 3 MHz transducer was used together with a spacer and a matched reflector to create a chamber with a $\lambda/2$-resonance. The lateral component of the primary acoustic radiation force was used for holding objects against a flow in the system. In 2004, an alternative design was introduced, incorporating a circular transducer, see Figure 9.4.[26] Other examples of trapping systems based on a layered resonator are systems

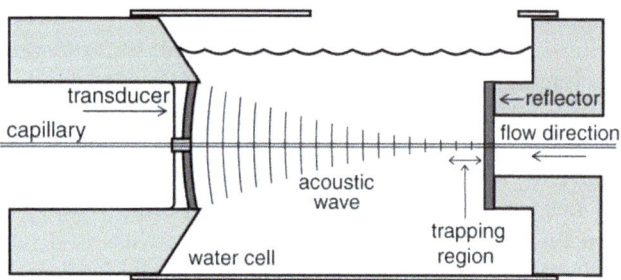

Figure 9.3 Particles are trapped in a standing wave in a capillary using a focused transducer and a reflector. Reprinted with permission from ref. 21. Copyright 2001, American Institute of Physics.

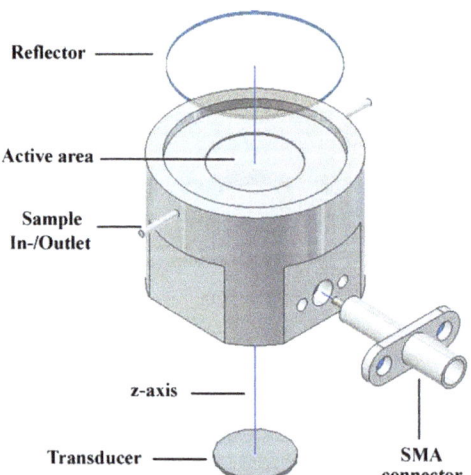

Figure 9.4 An example of a layered resonator, based on a circular transducer, matching layer and a glass reflector. When a flow is applied to the chamber, the lateral component of the primary radiation force will retain the particles in the chamber. Reprinted from ref. 26 with permission from Elsevier.

Applications in Acoustic Trapping

with transducers acting in direct contact with the fluid in a channel[27-29] and the use of rectangular glass capillaries with external transducers.[30]

Rather than using the wave propagation from a transducer directly for generating the standing wave, an alternative is to design a system with a resonant cavity where the energy is built up when the entire system is actuated, see Figure 9.5. Silicon/glass single node devices have been

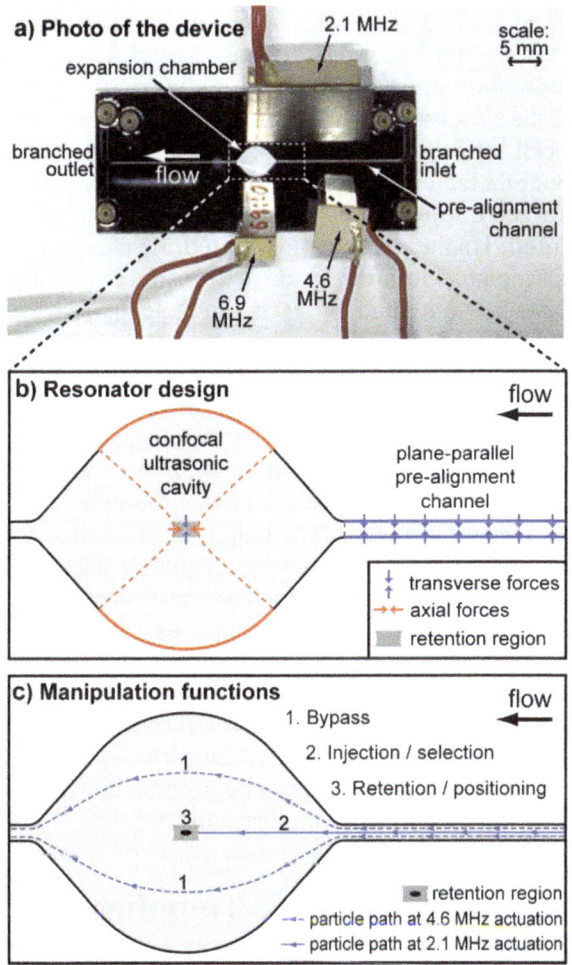

Figure 9.5 An example of a resonant cavity device where a multiple-node standing wave is generated in a confocal chamber. A photo of the device can be seen in (a). The particles entering the chamber are pre-focussed in the supply channel (b). By switching between one and two pressure nodes in the channel, the particles can be addressed to the central trapping position or to follow a trajectory where the trapping force is too weak to retain them in the chamber (c). Reprinted from ref. 33 with permission. Copyright 2009, Wiley Periodicals, Inc.

presented by Evander *et al.*[31] and Manneberg *et al.*[32] The device by Manneberg *et al.* utilised 3D-trapping by using two excitation frequencies and matching the width/height-dimensions to these. The operation of these devices is facilitated by the fabrication in high-Q materials such as glass and silicon. Svennebring *et al.* developed the system further into a confocal cavity where trapping occurs in a multi-wavelength standing wave.[33]

Since resonant cavities are fabricated using MEMS-technologies, they can easily be created as an arrayed structure. An example is the system developed by Vanherberghen *et al.* based on an open array of 100 microwells that all have a resonance around 2.65 MHz.[34] By actuating the entire chip through a frequency modulation approach, cells could be concentrated in the centre of each of the wells simultaneously and facilitate high-resolution microscopy studies of cell–cell interactions.

Layered resonators typically have only one or a few trapping positions and to have an array of trapping positions, a two-dimensional standing wave pattern is required. Haake *et al.* demonstrated 3D trapping using bending waves in a glass plate vibrated by edge transducers.[35] Similar trapping patterns can also be generated by 2D surface acoustic waves (SAW), see Figure 9.6A.[35]

Three-dimensional standing bulk wave patterns have been demonstrated for reconfigurable phononic metamaterial generation using five transducers arranged as two orthogonal pairs and a levitation transducer in the bottom of the device,[36] Figure 9.6B. 90 μm polystyrene spheres were trapped in the nodes of the 3D-acoustic field inside a 5 mm in diameter thin-walled Mylar tube generating a lattice structure. The height of the device was 17 mm and the lattice spacing was set by the driving frequency used, in this example between 2.25–5.25 MHz. Similar set-ups have also been suggested for two-dimensional active patterning of cells on surfaces.[37,38]

Most systems have a fixed position where objects can be trapped. A more flexible system can be designed by utilising phase control of two opposing transducers,[39,40] activation of a variable subset of transducers[41,42] or sweeping of the drive frequency,[43] see Figure 9.7. Using these, techniques, variable trapping positions, moving arrays *etc.* can be generated, which will be needed in future, more complex, trapping applications.

9.4 Applications in Acoustic Trapping

9.4.1 Particle Studies

Spengler and Coakley used a layered resonator ultrasonic trap with a square transducer to facilitate the studies of the morphology and stability of microparticle aggregates.[44] The ultrasound was used to move latex microparticles into the pressure node while optically monitoring the aggregation mechanism for different media. It was shown that latex particles in distilled water aggregated under reaction-limited colloid aggregation (RLCA) meaning that the particles did not form any permanent particle–particle contact and

Applications in Acoustic Trapping 197

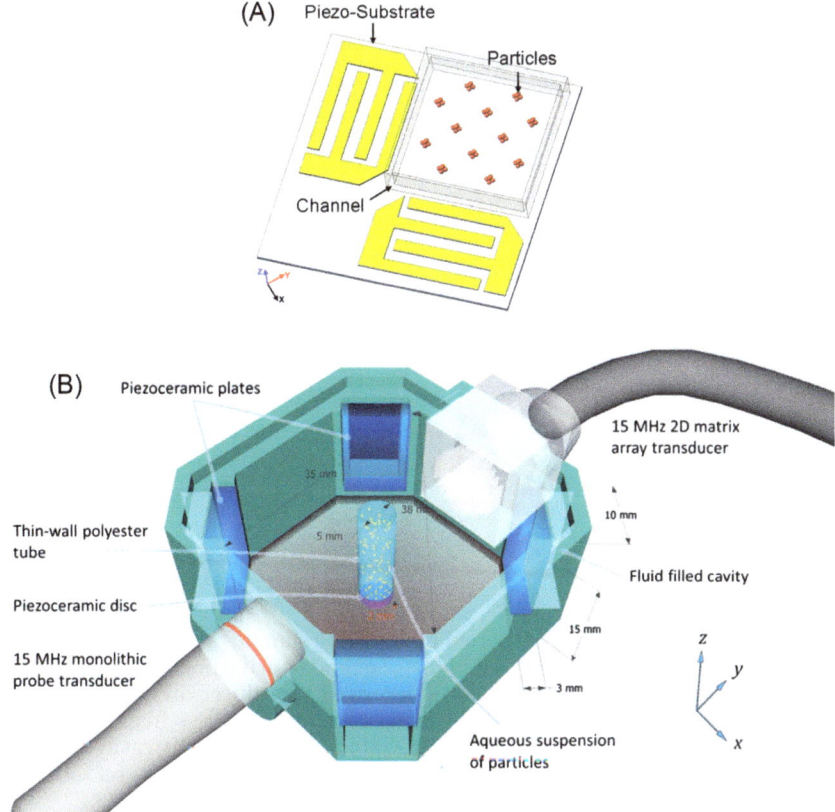

Figure 9.6 (A) A two-dimensional standing surface wave pattern can be generated on a piezoelectric substrate by orthogonal interdigitated electrodes. The two-dimensional wave pattern will define the array trapping positions in a chamber placed on top of the substrate. Reproduced from ref. 38 with permission. (B) A device for the generation of reconfigurable 3D phononic metamaterials. A three-dimensional standing bulk wave field was created by two orthogonal pairs of transducers and a fifth transducer in the bottom of the device. Reproduced from ref. 36 with permission.

that the particles were free to rearrange themselves within the closely-packed cluster. These kinds of clusters also disaggregated when the ultrasound was terminated. When using a suspension of latex microparticles in 10 mM $CaCl_2$, a different particle behaviour was observed. In diffusion-limited colloid aggregation (DLCA), particles formed permanent adhesion to each other upon contact. No rearrangement or merging of subclusters was observed and the cluster formed an "open" dendritic structure that was maintained even after termination of the standing wave, see Figure 9.8. By comparing the magnitude of the acoustic force to the van der Waals

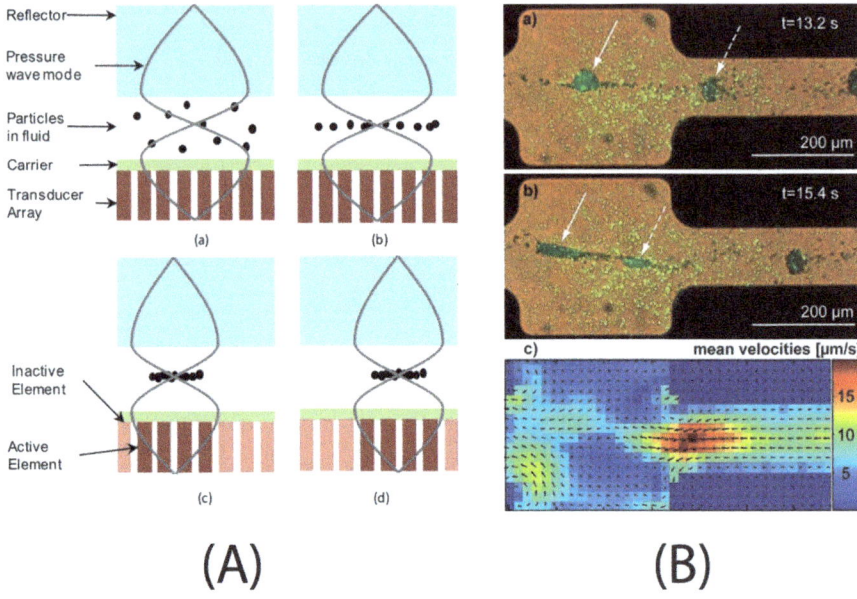

Figure 9.7 Particles can be transported by a moving standing wave field. In (A), the moving field is generated by sequential activation of a sub-set of transducer elements in a miniaturised transducer array.[41] In (B), the moving field is created by slightly sweeping the drive frequency of the drive signal to the transducer vibrating the resonant cavity device.[43] (A) © 2010 IEEE. Reprinted, with permission, from ref. 41. (B) reproduced from ref. 43 with permission.

interaction and the electrostatic repulsion it could be shown that the acoustic forces shouldn't make a significant contribution to the forces that determine the surface–surface interactions. In 2004, the work was expanded by Bazou *et al.* to include a wider range of salt concentrations and tools to analyse the amount of voids in a cluster.[45] Although the results are very interesting and would enable control of the bead adhesion in a trapped cluster through the choice of buffer, no further utilisation of this phenomenon has, to our knowledge, been reported.

By combining ultrasonic trapping with Raman microspectroscopy, Ruedas-Rama *et al.* demonstrated an interesting application with an outlook towards label-free monitoring of bioassays and bead-based chemical synthesis.[46] Particles were trapped in a flow cell for IR spectroscopy using a 2 MHz transducer and Raman spectra were taken from beads of different materials. The combination of acoustic trapping in a microfluidic flow cell enables full control of the chemical microenvironment around the particles. Monitoring of an online synthesis of silver on trapped beads was also demonstrated by the same group through the use of surface-enhanced Raman spectroscopy.

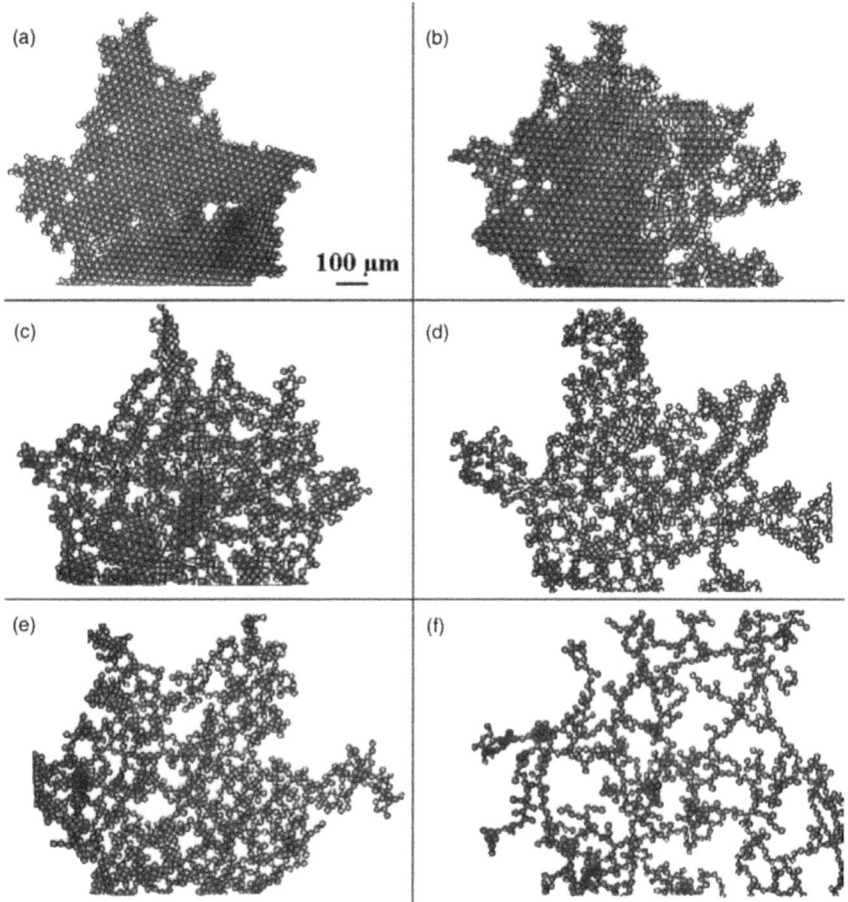

Figure 9.8 The difference in polystyrene particle aggregates depending on the concentration of $CaCl_2$ in the buffer. The salt concentration varied from 0 mM $CaCl_2$ (a), 2 mM $CaCl_2$ (b), 4 mM $CaCl_2$ (c), 8 mM $CaCl_2$ (d), 10 mM $CaCl_2$ (e) to 20 mM $CaCl_2$ (f). Reprinted from ref. 45 with permission from Elsevier.

9.4.2 Bioassays on Trapped Microparticles

Much work has been performed in this field and we will only give a brief overview on some applications that have been presented. However, Chapter 17 will describe this field in much greater detail. Several publications demonstrate how acoustic standing wave fields can increase the sensitivity in traditional latex agglutination tests (LATs). Although LATs have a limited sensitivity compared to more modern techniques (ELISA, MALDI, PCR *etc.*), they provide a quick and easy way of testing the presence of different bacteria. In ultrasonic wave enhanced latex agglutination tests (USELATs), the antibody coated latex beads are brought into close contact through the use of a standing wave, rather than relying on random particle collisions due to Brownian motion or gravity, and the sensitivity can be increased up to 2000-fold.[47] Many of the commercially available LAT-kits (adenovirus, rotavirus, *Neisseria meningitidis*, Legionella *etc.*) have been tested together with an ultrasonic standing wave and an improved sensitivity has been shown.[47–49] Although LATs provide a cheap and fast method for simple diagnostics, these types of assays typically suffer from poor sensitivity and unspecific agglutination due to van der Waals forces, electrostatic forces and hydrophobic interactions.

The early work with USELATs was performed in capillaries subjected to an acoustic field in a water bath and not in lab-on-a-chip systems. Recently, Bavli *et al.* implemented the method in a lab-on-a-chip system to perform real-time monitoring of bacteria in water.[50] Their method includes a separation step to remove unwanted particles and an acoustic trap where bacteria and latex beads are trapped. The amount of trapped bacteria was monitored by measuring the shift in resonance frequency. Apart from LATs, there are other bead-based assay types that have been evaluated together with ultrasonic standing waves in miniaturized systems.[51]

The bead-based doublet assay is essentially focused at the early stages of the agglutination assay and only looks at the binding of two beads to each other. Wiklund *et al.* explored this concept by studying selective trapping/enrichment of doublets in a capillary using a focused ultrasonic transducer and a reflector.[21,52] Although not tested on actual doublets and singlet beads, it was demonstrated that it was possible to achieve size-selective trapping of different sized single particles with high efficiency and that would enable the use of the technique to increase the sensitivity in capillary electrophoresis systems.

The bead-based singlet assay, on the other hand, is used to study the binding of antibodies and molecules to functionalized microparticles. By reducing the amount of beads in the sample, the amount of bound antibodies per bead can be maximized and the signal therefore also maximized.[51] Reducing the amount of beads makes them more difficult to detect though and acoustic trapping can be used to enrich and position the beads for fluorescent microscopy readout. Wiklund *et al.* performed a fluorescent singlet assay in a microtiter plate by building a focused transducer that could be inserted into each of the microwells to create a standing wave using the bottom of the well as a reflector, see Figure 9.9. The approach was demonstrated using a human thyroid stimulating hormone (hTSH) assay where concentrations down to 70 pM could be detected fluorescently.[53]

Lilliehorn *et al.* envisioned a slightly different approach towards fluorescent antibody detection in an acoustic standing wave system, termed dynamic arraying.[28] Microfluidic channels were mounted on top of an array of miniature transducers to create an open platform. The concept would make it possible to move functionalized bead clusters to different transducer locations using microfluidics, trap the particles and perform different steps

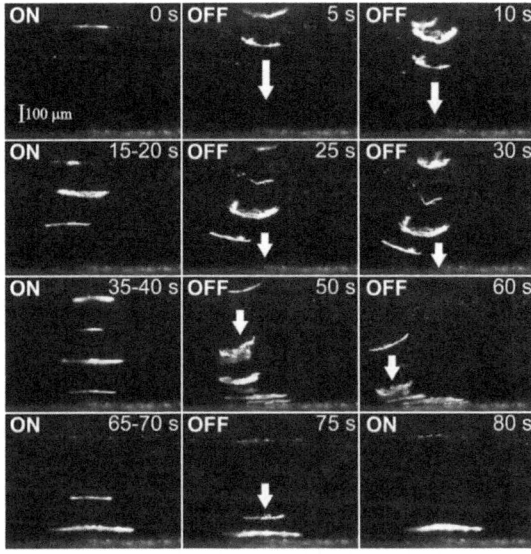

Figure 9.9 Trapping of 4.5 μm fluorescent particles using a focused transducer in a 96-well plate. By sequentially turning the ultrasound on and off, gravitational forces makes it possible to concentrate all particles in one cluster close to the bottom of the well where they can be analysed using microscopy. Reprinted with permission from ref. 53. Copyright 2004, American Institute of Physics.

of an assay at each transducer location. The idea was demonstrated using a single trapping site and an assay showing the binding between fluorescently tagged avidin to biotinylated microparticles was performed.

9.5 Cell Population Studies

Functional cell assays are based on the assumption that cell behaviour during the experiment will be indicative of their behaviour in their natural environment. Non-contact 3D-positioning of small cell populations and retention against a laminar flow make it possible to achieve precise and dynamic control of the cell microenvironment, both in terms of chemical composition and exerted mechanical forces from the flowing liquid. An important question when using acoustic trapping in cell based assays is whether exposure to high frequency ultrasound can be considered a minor disturbance to cell behaviour or not. The topic of biocompatibility and cell viability in acoustic systems is discussed in further detail in Chapter 21.[54]

Bazou et al. examined the physical environment experienced by levitated neural cells in a 1.5 MHz half-wavelength acoustic trap by monitoring the temperature, acoustic streaming, pressure amplitudes, white noise and the interparticular forces acting on the cells.[55] The conclusion was that no adverse effects or changes in the *in vitro* surface receptor interactions could be detected for the cells. Bazou *et al.* also showed that acoustic levitation of embryonic stem cells did not affect the gene expression or the pluripotency of the cells even after 60 min of acoustic radiation at 0.85 MPa.[56] Hultström *et al.* also investigated the viability by culturing COS-7 cells and measuring their doubling time after exposure in an acoustic trap, see Figure 9.10.[57] No direct or delayed damage to the cells could be found after acoustic radiation exposure up to 75 min.

Evander *et al.* used a miniaturized transducer that was integrated in a microfluidic channel to demonstrate an online viability assay on cells.[27] Neural stem cells were trapped and held in the acoustic field for 15 minutes before a viability marker—acridine orange—was introduced into the channel. The viability marker is actively transported into the cell and through a clear increase in fluorescence, the cells can be shown to still be viable.

Vanherbergen *et al.* used non-adherent human B cells that were trapped in wells to study their reaction to a continuous exposure of ultrasound. After 12 hours at a frequency of 2.55–2.65 MHz, no changes in viability could be seen. A continued ultrasound exposure for up to 72 hours showed that the cells were still viable and were able to divide several times.[34]

A very interesting application that also demonstrates the long-term viability of acoustically manipulated cells is the formation of 3D cell aggregate cultures. 3D cell cultures have been shown to have different cell behaviour compared to standard monolayer cultures and more closely resemble the *in vivo* situation. The standard way of forming 3D cell cultures is, however, time-consuming and the use of ultrasonic aggregation can greatly speed up the process.[58] Liu *et al.* and Bazou *et al.* both demonstrated functional 3D HepG2 aggregates formed by an ultrasonic trap,

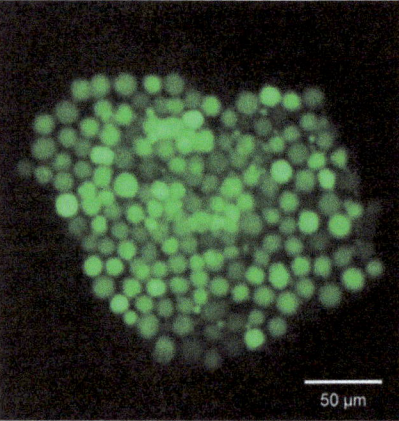

Figure 9.10 COS-7 cells trapped in a standing wave and fluorescently labelled using Calcein AM to indicate viability. The microfluidic chip is constructed from glass–PDMS–glass and uses a 3 MHz transducer to create a single pressure node in the chip. Reprinted from ref. 57 with permission from Elsevier.

see Figure 9.11.[58,59] The ultrasonically formed 3D-clusters showed equivalent or greater albumin kinetics compared to standard gyratory spheroids and remained stable for 20 days.

Acoustic trapping has also been used when studying cell–cell interactions. Coakley *et al.* used a half wavelength acoustic trap to study the aggregation dynamics and cell–cell contacts for chondrocytes, erythrocytes and neural cells.[26] Later, Bazou and Edwards *et al.* used the same system to study cell adhesion in neural cells and HepG2 cell aggregates using fluorescent staining of the trapped cells.[60,61]

Christakou *et al.* demonstrated the possibility of monitoring the interaction between acoustically trapped natural killer cells and cancer cells in an array format.[62] Ultrasound was used to aggregate and position the killer cells and the target cells to enable long-term high-resolution microscopy studies of the immune synapse formed between the cells. The experiments showed that the killer cells maintained their functionality and were able to form immune synapses during 17 hours of continuous ultrasound exposure.

Chemiluminescence can also be used in cell assays and was utilised for the detection of ATP-release from acoustically trapped erythrocytes.[63–65] ATP is released from erythrocytes as part of the vascular regulation system in the human body and the release is triggered by both changes in the microenvironment of the erythrocyte and physical stress. In order to study what may trigger and prevent an ATP-release, a method of studying the erythrocytes without any changes in their environment or physical stimuli is needed. By trapping erythrocytes in a standing wave at 10 MHz in a microfluidic

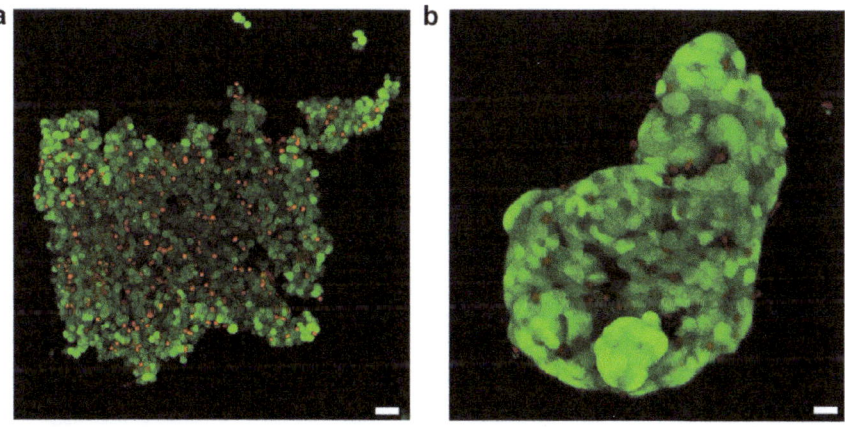

Figure 9.11 Fluorescent images of 3D HepG2 clusters at day 0 (a) and day 3 (b). The cells are stained for viability with Calcein AM (green fluorescence) and for cell death with Ethidium Homodimer (red fluorescence). At day 0, individual cells can easily be distinguished while at day 3, the membrane spreading of the cells was so pronounced that it was hard to distinguish individual cells. Reprinted from ref. 58 with permission from Elsevier.

channel, the cells can be infused with different stimuli and their ATP-response can be monitored by measuring the chemiluminescent response from the reaction of ATP with luciferine and luciferase using a photomultiplier tube (PMT). The system was demonstrated to be able to detect changes in the ATP-release from trapped cells by stimulation using ethanol and epinephrine, see Figure 9.12.

An alternative to microscopy studies of cell clusters was demonstrated by Hammarström *et al.* where drug partitioning in blood cells was addressed by combining an acoustic trap and MALDI-MS analysis.[66] Erythrocytes were incubated with a drug and a peptide, trapped, washed and lysed while elution fractions were collected on a MALDI target. Analysis confirmed cell permeability of the drug into the cells and that the peptide could not penetrate the membrane as expected, providing a way to discriminate between intracellular and extracellular material.

9.5.1 Enrichment/Washing of Cells

Acoustic trapping is very well suited for washing of cells or enrichment of dilute samples or rare event experiments. By utilising a continuous flow and high trapping efficiencies, cells can be trapped and enriched when they enter the acoustic field. By changing the sample flow to a washing buffer, the trapped cluster can be washed and then released in a small volume. This application area is becoming increasingly more interesting as applications targeting bacteria and rare cell enrichment are surfacing.

Hammarström *et al.* demonstrated how small objects such as bacteria or nanometre-sized particles can be trapped and enriched through the use of seeding particles.[67,68] The trapping system consisted of a borosilicate

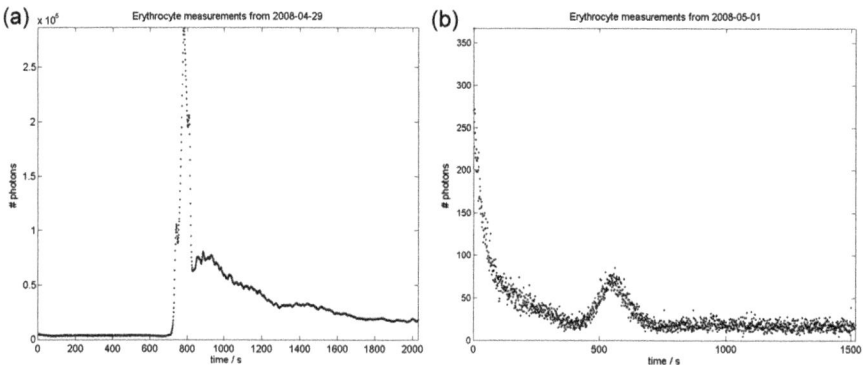

Figure 9.12 Results from the chemiluminescent ATP-assay performed on acoustically trapped erythrocytes. (a) shows the chemiluminescent response of a chemical lysis of the cells. In (b), the cluster is stimulated with a dose of epinephrine and a small response (in comparison to the lysis) can be seen. Reprinted from ref. 65 with permission.

capillary and a 4 MHz lead zirconate titante (PZT) transducer. By initially trapping 10 μm polystyrene particles (seeding particles), objects down to 110 nm could be trapped in the interstitial sites between the seeding particles. Using fluorescent *E. coli* bacteria and continuously monitoring the fluorescent intensity of the cluster, a long-term enrichment of bacteria could be demonstrated, see Figure 9.13. This concept was later used to show how acoustic trapping in combination with MALDI-MS could be used for bacteremia diagnosis.[69] Whole blood from cultured flasks was lysed and aspirated into an acoustic trap pre-filled with seeding particles. The plasma was rinsed away using de-ionized water and the bacteria were released to an ISET sample preparation platform for MALDI-detection. The system was successfully tested with *E. coli* using 20 μl of whole blood.

An example of enrichment/washing is acoustic differential extraction (ADE), presented by Norris *et al.* in 2009.[70] Here, a mixed sample consisting of sperm cells and lysed epithelial cells was separated and enriched as a sample preparation step for sexual assault evidence assays. The ADE system used laminar flow valving in combination with a 10 MHz acoustic trap and by trapping and washing sperm cells, an unidentifiable sample with $26 \pm 2\%$ male DNA could be processed to $92 \pm 7.9\%$ male DNA where a DNA identification could be performed, see Figure 9.14.

Recently, Chen *et al.* demonstrated continuous enrichment of dilute samples using a surface acoustic wave based trap.[71] The system used a one-dimensional SAW-field at 19.6 MHz that was coupled into a disposable, flexible tubing to create a trapping region of 5×0.28 mm^2. Diluted blood cells were trapped and enriched at 7 μl min^{-1} for volumes up to 70 μl.

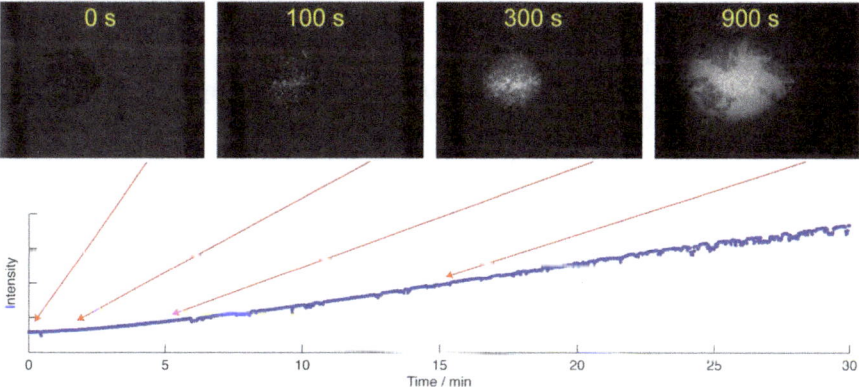

Figure 9.13 Enrichment of *E. coli* over a time period of 30 minutes in an acoustic trap. A seeding cluster of 10 μm polystyrene beads is used to enable the trapping of bacteria (0 s). When bacteria are transported to the seeding cluster by the laminar flow, they are trapped between the seeding particles (100 s and 300 s). As can be seen in the picture at 900 s, the *E. coli* will eventually dominate the trapped cluster.[67] Reproduced from ref. 67 with permission.

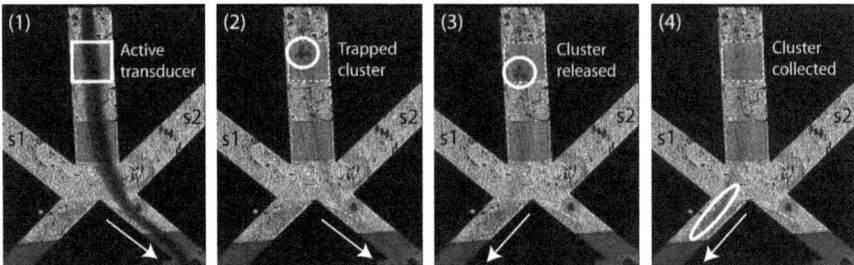

Figure 9.14 A demonstration of the ADE chip using blue dye (female DNA) and polystyrene particles (sperm cells). In (1), the particles are trapped and enriched while the dye exits the chip through the waste outlet. In (2), the sample infusion is stopped and the dye is washed away from the particles using a wash buffer. In (3), the laminar flow valve has been switched and the trapped particles are released into the sample outlet (4). Reproduced with permission from ref. 70. Copyright 2007, American Chemical Society.

Ultrasonic trapping has also been used together with magnetic and gravitational forces for washing and separating paramagnetic particles in an assay for detection of tuberculosis in sputum as shown by Hill et al.[72] The particles are agglomerated in a fluidic chamber using magnetic and acoustic forces and samples can be infused over the beads at high flow rates (ml min^{-1}). After incubating the particles, they are washed in a buffer and released. During the release, ultrasonic forces are used to keep the particles away from the chamber walls to minimize dispersion. The fluidic chamber was fabricated from ceramic and a $3\lambda/2$ standing wave at 1.6 MHz was utilised in this set-up.

Hammarström et al. demonstrated how an acoustic trap can be used for removing unspecific background in immuno-MALDI-MS.[73] The system utilised a borosilicate capillary that was coupled to a transducer to create a standing wave at 4 MHz, see Figure 9.15. The capillary was used to aspirate antibody-coated beads that were trapped, incubated and then washed before being released to an integrated selective enrichment target (ISET). The system was demonstrated with an angiotensin I antibody and compared to a standard magnetic bead assay. The unspecific background from detergents in the magnetic assay could be greatly reduced using the washing in the acoustic trap, facilitating high sensitivity detection.

In a capillary trapping system, the resonance frequency is to a large extent determined by the resonance conditions within the capillary. This enables supervision of the trapping site by monitoring the change in resonance frequency. Hammarström et al. developed an automatic tracking system that was able to match the drive frequency to the system resonance frequency.[74] By recording the resulting drive frequency, changes in the acoustic properties of the trapping site could be detected. The reason for the changes may be

Applications in Acoustic Trapping

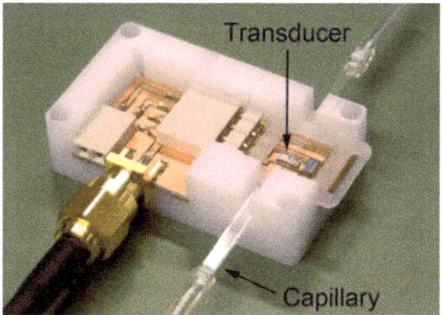

Figure 9.15 The capillary trapping system used by Hammarström *et al.* Flexible tubing is glued to a square, borosilicate capillary that is inserted in a fixture. The fixture holds a circuit board with a transducer that is in close contact with the capillary through a thin layer of glycerol. The capillary ends can either have flexible tubing for infusing clean buffer for washing or introducing reagents to the trapped cluster, or they can be open-ended to allow for sample aspiration.

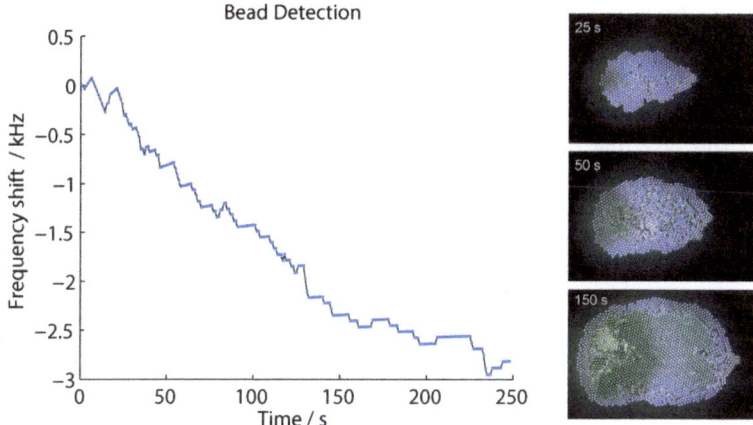

Figure 9.16 The figure shows how the accumulated beads affected the resonance frequency of a capillary trapping system with a kerfed transducer indicating a frequency shift of around 500 Hz per 500 beads. Reproduced from ref. 74 with permission.

varying temperature, liquid media properties, particle content, *etc*. To simplify the frequency matching, a kerfed transducer was developed that decreased the influence from lateral resonance modes in the transducer impedance spectrogram (see Chapter 6 for more details on kerfed transducers). Figure 9.16 shows how accumulation of 12 µm polystyrene beads affected the resonance frequency.

9.5.2 Size Sorting and Separations

Acoustophoresis is known for being able to sort objects based on size and material properties while acoustic trapping, typically, is not used in these kinds of applications. However, Jeong et al. recently demonstrated a particle sorting solution based on acoustic trapping.[75,76] Rather than relying on standing waves, a high intensity, single element transducer was used at 24–30 MHz to draw single objects into the beam axis, similar to how optical tweezers work. The acoustic trap was combined with a PDMS microfluidic channel and 60–70 µm lipid particles were manipulated. After trapping, the transducer was translated across the width of the channel and the particle could then be released into different outlets. Leukaemia cells with a diameter of 15 µm are also reported to have been trapped but no details are given.[77]

Separations are traditionally performed in continuous flow acoustophoresis systems, as presented in Chapter 8 in this book. However, Wang et al. recently demonstrated a separation and detection of low-concentration oil based on a resonance cavity.[78] Oil droplets were trapped using a 1 MHz transducer and after enrichment, the oil droplets could be released and separated into a sample outlet through an additional acoustophoretic separation step prior to the flow-splitter. The system was successfully tested with oil droplets between 14 and 62 µm and flows up to 2 ml per hour.

9.6 Conclusion

During the last few decades, acoustic trapping has developed from larger systems to miniaturised platforms that have been successfully integrated with lab-on-a-chip-based life science research. The technical advancements now enable non-contact single-cell manipulation as well as three-dimensional trapping of clusters that can be moved through phase or frequency control of the driving signal or by using a subset of transducer arrays. Manipulation of nano-scaled objects has also been demonstrated using seeding particles, which opens up applications aimed at bacteria and viruses.

As an outlook, we believe that more three-dimensional systems will be presented, probably with individual acoustic and fluidic control over trapping sites in array formats. An increased use of acoustic trapping as a sample preparation step (washing, enrichment *etc.*) for sensitive detection methods is also expected in the near future. The emergence of SAW-based trapping devices has the potential to simplify interfacing to other microfluidic devices as well as lead to new array concepts.

References

1. J. Nilsson, M. Evander, B. Hammarstrom and T. Laurell, *Anal. Chim. Acta*, 2009, **649**, 141–157.
2. A. Y. Lau, P. J. Hung, A. R. Wu and L. P. Lee, *Lab Chip*, 2006, **6**, 1510–1515.

3. N. Fertig, R. H. Blick and J. C. Behrends, *Biophys. J.*, 2002, **82**, 3056–3062.
4. C. Piggee, *Anal. Chem.*, 2008, **81**, 16–19.
5. N. Pamme, *Lab Chip*, 2006, **6**, 24–38.
6. E. G. Lierke, *Acustica*, 1996, **82**, 220–237.
7. F. Trampler, S. A. Sonderhoff, P. W. S. Pui, D. G. Kilburn and J. M. Piret, *Nat. Biotechnol.*, 1994, **12**, 281–284.
8. P. W. S. Pui, F. Trampler, S. A. Sonderhoff, M. Groeschl, D. G. Kilburn and J. M. Piret, *Biotechnol. Prog.*, 1995, **11**, 146–152.
9. T. Gaida, O. Doblhoff-Dier, K. Strutzenberger, H. Katinger, W. Burger, M. Gröschl, B. Handl and E. Benes, *Biotechnol. Prog.*, 1996, **12**, 73–76.
10. J. Wu, *J. Acoust. Soc. Am.*, 1991, **89**, 2140–2143.
11. H. M. Hertz, *J. Appl. Phys.*, 1995, **78**, 4845–4849.
12. P. Glynne-Jones, R. J. Boltryk, M. Hill, N. R. Harris and P. Baclet, *J. Acoust. Soc. Am.*, 2009, **126**, EL75–79.
13. P. Glynne-Jones, R. J. Boltryk and M. Hill, *Lab Chip*, 2012, **12**, 1417–1426.
14. H. Bruus, *Lab Chip*, 2012, **12**, 1014–1021.
15. A. Lenshof, C. Magnusson and T. Laurell, *Lab Chip*, 2012, **12**, 1210–1223.
16. M. Groschl, *Acustica*, 1998, **84**, 432–447.
17. M. Wiklund, R. Green and M. Ohlin, *Lab Chip*, 2012, **12**, 2438–2451.
18. P. Augustsson, Lund University, PhD thesis, 2006.
19. R. Barnkob, P. Augustsson, T. Laurell and H. Bruus, *The 14th international conference on miniaturized systems for chemistry and life sciences*, Chemical and Biological Microsystems Society, Groningen, Holland, 2010.
20. J. Lei, P. Glynne-Jones and M. Hill, *Lab Chip*, 2013, **13**, 2133–2143.
21. M. Wiklund, S. Nilsson and H. M. Hertz, *J. Appl. Phys.*, 2001, **90**, 421.
22. J. Lee, S. Y. Teh, A. Lee, H. H. Kim, C. Lee and K. K. Shung, *Appl. Phys. Lett.*, 2009, **95**, 73701.
23. J. Lee, C. Lee, H. H. Kim, A. Jakob, R. Lemor, S. Y. Teh, A. Lee and K. K. Shung, *Biotechnol. Bioeng.*, 2011, **108**, 1643–1650.
24. A. Lenshof, M. Evander, T. Laurell and J. Nilsson, *Lab Chip*, 2012, **12**, 684–695.
25. J. F. Spengler, M. Jekel, K. T. Christensen, R. J. Adrian, J. J. Hawkes and W. T. Coakley, *Bioseparation*, 2001, **9**, 329–341.
26. W. T. Coakley, D. Bazou, J. Morgan, G. A. Foster, C. W. Archer, K. Powell, K. A. Borthwick, C. Twomey and J. Bishop, *Colloids Surf., B*, 2004, **34**, 221–230.
27. M. Evander, L. Johansson, T. Lilliehorn, J. Piskur, M. Lindvall, S. Johansson, M. Almqvist, T. Laurell and J. Nilsson, *Anal. Chem.*, 2007, **79**, 2984–2991.
28. T. Lilliehorn, M. Nilsson, U. Simu, S. Johansson, M. Almqvist, J. Nilsson and T. Laurell, *Sens. Actuators, B*, 2005, **106**, 851–858.
29. T. Lilliehorn, U. Simu, M. Nilsson, M. Almqvist, T. Stepinski, T. Laurell, J. Nilsson and S. Johansson, *Ultrasonics*, 2005, **43**, 293–303.
30. B. Hammarstrom, M. Evander, H. Barbeau, M. Bruzelius, J. Larsson, T. Laurell and J. Nilsson, *Lab Chip*, 2010, **10**, 2251–2257.
31. M. Evander, Lund University, PhD thesis, 2008.

32. O. Manneberg, B. Vanherberghen, J. Svennebring, H. M. Hertz, B. Onfelt and M. Wiklund, *Appl. Phys. Lett.,* 2008, **93**, 063901.
33. J. Svennebring, O. Manneberg, P. Skafte-Pedersen, H. Bruus and M. Wiklund, *Biotechnol. Bioeng.,* 2009, **103**, 323–328.
34. B. Vanherberghen, O. Manneberg, A. Christakou, T. Frisk, M. Ohlin, H. M. Hertz, B. Onfelt and M. Wiklund, *Lab Chip,* 2010, **10**, 2727–2732.
35. A. Haake, A. Neild, G. Radziwill and J. Dual, *Biotechnol. Bioeng.,* 2005, **92**, 8–14.
36. M. Caleap and B. W. Drinkwater, *Proc. Natl. Acad. Sci. U. S. A.,* 2014, **111**, 6226–6230.
37. F. Gesellchen, A. L. Bernassau, T. Dejardin, D. R. S. Cumming and M. O. Riehle, *Lab Chip,* 2014, **14**, 2266–2275.
38. J. J. Shi, D. Ahmed, X. Mao, S. C. S. Lin, A. Lawit and T. J. Huang, *Lab Chip,* 2009, **9**, 2890–2895.
39. C. R. Courtney, C. K. Ong, B. W. Drinkwater, P. D. Wilcox, C. Demore, S. Cochran, P. Glynne-Jones and M. Hill, *J. Acoust. Soc. Am.,* 2010, **128**, EL195–199.
40. C. R. P. Courtney, C. K. Ong, B. W. Drinkwater, A. L. Bernassau, P. D. Wilcox and D. R. S. Cumming, *Proc. R. Soc. A,* 2011, **468**, 337–360.
41. C. Demore, Q. Yongqiang, S. Cochran, P. Glynne-Jones, Y. Congwei and M. Hill, *Ultrasonics Symposium (IUS), 2010*, IEEE, 2010.
42. P. Glynne-Jones, C. E. M. Demore, Y. Congwei, Q. Yongqiang, S. Cochran and M. Hill, *IEEE Trans. Sonics Ultrason.,* 2012, **59**, 1258–1266.
43. O. Manneberg, B. Vanherberghen, B. Önfelt and M. Wiklund, *Lab Chip,* 2009, **9**, 833.
44. J. F. Spengler and W. T. Coakley, *Langmuir,* 2003, **19**, 3635–3642.
45. D. Bazou, W. T. Coakley, K. M. Meek, M. Yang and D. T. Pham, *Colloids Surf., A,* 2004, **243**, 97–104.
46. M. J. Ruedas-Rama, A. Dominguez-Vidal, S. Radel and B. Lendl, *Anal. Chem.,* 2007, **79**, 7853–7857.
47. R. W. Ellis and M. A. Sobanski, *J. Med. Microbiol.,* 2000, **49**, 853–859.
48. M. A. Sobanski, J. Stephens, G. A. Biagini and W. T. Coakley, *J. Med. Microbiol.,* 2001, **50**, 203–203.
49. M. A. Sobanski, R. Vince, G. A. Biagini, C. Cousins, M. Guiver, S. J. Gray, E. B. Kaczmarski and W. T. Coakley, *J. Clin. Pathol.,* 2002, **55**, 37–40.
50. D. Bavli, N. Emanuel and Y. Barenholz, *Anal. Methods,* 2014, **6**, 395.
51. M. Wiklund and H. M. Hertz, *Lab Chip,* 2006, **6**, 1279–1292.
52. M. Wiklund, P. Spégel, S. Nilsson and H. M. Hertz, *Ultrasonics,* 2003, **41**, 329–333.
53. M. Wiklund, J. Toivonen, M. Tirri, P. Hänninen and H. M. Hertz, *J. Appl. Phys.,* 2004, **96**, 1242.
54. M. Wiklund, *Lab Chip,* 2012, **12**, 2018–2028.
55. D. Bazou, L. A. Kuznetsova and W. T. Coakley, *Ultrasound Med. Biol.,* 2005, **31**, 423–430.

56. D. Bazou, R. Kearney, F. Mansergh, C. Bourdon, J. Farrar and M. Wride, *Ultrasound Med. Biol.*, 2011, **37**, 321–330.
57. J. Hultström, O. Manneberg, K. Dopf, H. M. Hertz, H. Brismar and M. Wiklund, *Ultrasound Med. Biol.*, 2007, **33**, 145–151.
58. D. Bazou, W. T. Coakley, A. J. Hayes and S. K. Jackson, *Toxicol. In Vitro*, 2008, **22**, 1321–1331.
59. J. Liu, L. A. Kuznetsova, G. O. Edwards, J. Xu, M. Ma, W. M. Purcell, S. K. Jackson and W. T. Coakley, *J. Cell. Biochem.*, 2007, **102**, 1180–1189.
60. D. Bazou, E. J. Blain and W. Terence Coakley, *Mol. Membr. Biol.*, 2008, **25**, 102–114.
61. G. O. Edwards, D. Bazou, L. A. Kuznetsova and W. T. Coakley, *Cell Commun. Adhes.*, 2007, **14**, 9–20.
62. A. E. Christakou, M. Ohlin, B. Vanherberghen, M. A. Khorshidi, N. Kadri, T. Frisk, M. Wiklund and B. Onfelt, *Integr. Biol.*, 2013, **5**, 712–719.
63. M. Evander, K. Mileros, C. Högberg, D. Erlinge, M. Almqvist, T. Laurell and J. Nilsson, *Micro Total Analysis Systems 2007*, Chemical and Biological Microsystems Society, Paris, France, 2007.
64. B. Hammarström, Lund University, MSc thesis, 2008.
65. C.-M. Tran, Lund University, MSc thesis, 2008.
66. B. Hammarström, T. Laurell, M. Evander, J. Nilsson and S. Ekström, *Micro Total Analysis Systems*, Chemical and Biological Microsystems Society, Jeju, Korea, 2009.
67. B. Hammarström, T. Laurell and J. Nilsson, *Micro Total Analysis Systems 2011*, Chemical and Biological Microsystems Society, Seattle, USA, 2011.
68. B. Hammarström, T. Laurell and J. Nilsson, *Lab Chip*, 2012, **12**, 4296–4304.
69. B. Hammarström, B. Nilson, L. Laurell, J. Nilsson and S. Ekström, *Micro Total Analysis Systems*, Chemical and Biological Microsystems Society, Freiburg, Germany, 2013.
70. J. V. Norris, M. Evander, K. M. Horsman-Hall, J. Nilsson, T. Laurell and J. P. Landers, *Anal. Chem.*, 2009, **81**, 6089–6095.
71. Y. Chen, S. Li, Y. Gu, P. Li, X. Ding, L. Wang, J. P. McCoy, S. J. Levine and T. J. Huang, *Lab Chip*, 2014, **14**, 924–930.
72. M. Hill, P. Glynne-Jones, N. R. Harris, R. J. Boltryk, C. Stanley and D. Bond, Optical Trapping and Optical Micromanipulation VII, *Proc. SPIE*, The International Society for Optical Engineering, 2010.
73. B. Hammarström, H. Yan, J. Nilsson and S. Ekström, *Biomicrofluidics*, 2013, 7.
74. B. Hammarström, M. Evander, J. Wahlstrom and J. Nilsson, *Lab Chip*, 2014, **14**, 1005–1013.
75. J. S. Jeong, J. W. Lee, C. Y. Lee, S. Y. Teh, A. Lee and K. K. Shung, *Biomed. Microdevices*, 2011, **13**, 779–788.
76. C. Lee, J. Lee, H. H. Kim, S.-Y. Teh, A. Lee, I.-Y. Chung, J. Y. Park and K. K. Shung, *Lab Chip*, 2012, **12**, 2736–2742.
77. K. K. Shung, *J. Acoust. Soc. Am.*, 2012, **131**, 3533.
78. H. Wang, Z. Liu, S. Kim, C. Koo, Y. Cho, D. Y. Jang, Y. J. Kim and A. Han, *Lab Chip*, 2014, **14**, 947–956.

CHAPTER 10

Ultrasonic Microrobotics in Cavities: Devices and Numerical Simulation

JÜRG DUAL*, PHILIPP HAHN, ANDREAS LAMPRECHT,
IVO LEIBACHER, DIRK MÖLLER, THOMAS SCHWARZ, AND
JINGTAO WANG

Institute of Mechanical Systems, Department of Mechanical and Process Engineering, ETH Zentrum, CH-8092 Zurich, Switzerland
*E-mail: dual@imes.mavt.ethz.ch

10.1 Introduction

Acoustic radiation forces have obtained increased attention recently in the context of handling cells, functionalized beads or other micron sized particles in microfluidic systems. Primary radiation forces are generated when an acoustic wave interacts with a single particle. In addition, secondary acoustic forces arise when several particles are present. Typically a resonance (at lower MHz frequencies) is set up in the system consisting of chip, fluid, particles and piezoelectric transducer. The pattern of the pressure distribution in the fluid then allows the prediction of the location of spherical compressible particles in the bulk of the fluid based on the acoustic field using Gor'kov's potential:[1] the force **F** acting on a spherical particle in an unbounded domain can be calculated from Gor'kov's potential U

$$U = 2\pi r_p^3 \rho_0 \left(\frac{1}{3}\frac{\langle p_1^2 \rangle}{\rho_0^2 c^2} f_1 - \frac{1}{2}\langle v_1^2 \rangle f_2\right)$$

$$\mathbf{F} = -\nabla U \qquad (10.1)$$

$$f_1 = 1 - \frac{\kappa_p}{\kappa_0}, \ f_2 = \frac{2(\rho_p - \rho_0)}{2\rho_p + \rho_0}$$

where r_p, ρ_p and κ_p denote the radius, the density and the compressibility of the particle, respectively, and ρ_0, c and κ_0 the density, the speed of sound and the compressibility of the fluid, respectively. p_1 is the pressure and $\mathbf{v}_1$ the velocity in the fluid due to the acoustic field. The angle brackets denote time averaging over one period, and f_1 and f_2 are dimensionless functions containing the properties (density and compressibility) of fluid and particle. Several fields might be superimposed to produce time independent or time varying patterns of particles in the fluid, resulting in the formation of lines, clumps or even in particle rotation. Excellent qualitative agreement between theory and experiment is generally found, while it might be quite difficult to quantitatively assess the force magnitude due to other effects such as acoustic streaming.[2]

In most situations, acoustic standing waves are used because they can be produced with higher amplitudes and steeper gradients than traveling waves and therefore result in larger forces. The frequencies are typically in the upper kHz to lower MHz range, resulting in water in a wavelength of millimetre size and below. The manipulated particles are normally much smaller in size than the wavelength and must have acoustic properties that differ from the surrounding fluid, such that a scattered wave is produced. In early work, biological cells have been manipulated in macroscopic devices for separation science and in bioreactors.[3,4] For recent reviews on the subject the reader is referred to the current acoustofluidics series and refs. 5–7. Due to the non-contact nature of ultrasonic manipulation, its advantages as compared to other methods soon became clear. Also, it turned out that the viability of the cells is no problem in most cases, as long as cavitation is avoided.[8–10]

For this chapter, the focus is placed on the robotic aspects of ultrasonic particle manipulation, which involve controlled positioning, movement and rotation of particles. The key advantages are: highly parallel operation, non-contact manipulation, isothermal, viability for living particles.

First, the formation of patterns of particles in micromachined chambers is described. Then the in-plane rotation of particles in similar cavities is introduced. For some applications, a more macroscopic application that involves ultrasonic manipulation over larger distances is needed. In the last section, some numerical tools currently being developed are described, which are needed to overcome the limitations of Gor'kov's eqn (10.1), which is limited to the cases of spherical particle shapes, non-viscous fluids and infinite domains. For spherical particle shapes, the effect of viscosity can be incorporated with some effort,[11–13] numerical solutions are the only tool available when it comes to more complicated shapes, used *e.g.* in robotic assembly operations.

10.2 Particle Manipulation in Cavities

In this section, three different particle manipulation tasks will be presented: pattern formation, rotation and transport of microparticles. Both theoretical aspects and their experimental evaluation are covered, whereas differently sized and manufactured devices will be shown.

10.2.1 Pattern Formation in Microchambers

In the following, the agglomeration of particles to two-dimensional particle patterns is outlined. From the perspective of practical applications, particle patterns such as clumps and arrays are a promising technique to align and screen biologically relevant particles (such as cells or antibody-coated beads) in a bioanalytical context. As Lilliehorn *et al.*[14] proposed, targeted perfusion of the separated clustered particles would allow one to perform on-line analysis of biological particles in a controlled parallel manner.

10.2.1.1 Device Description

The experiments in this section build on an ultrasonic particle manipulation microdevice made of silicon. The device has been presented in detail by Oberti *et al.*[15] as illustrated in Figure 10.1. It consists of a fluidic cavity (length and width of 5 mm and a height h of 0.2 mm) where particles will be manipulated, driven by a piezoelectric plate. Since $h < \lambda/2$, where λ means the acoustic wavelength in this work, the third dimension plays a minor role and particles in the cavity will be aligned in two-dimensional patterns between the four side walls of the cavity. As shown in Figure 10.1(c), the bottom electrode of the 0.5 mm thick piezoelectric plate is segmented on the surface by a wafer saw, so that four distinct main electrodes resulted. Electrodes 1 and 2 can be contacted independently in order to generate standing waves in the cavity in the x- and y-directions, respectively. Because of the vibration of the whole structure, it can be noted that an excitation in the x-direction will also excite a standing wave in the y-direction, however, with a much smaller amplitude.

10.2.1.2 Superposition of Two Standing Waves

The cavity introduced above allows for the formation of two-dimensional particle arrays, in contrast to microfluidic channels in which particles are aligned on lines by means of a single one-dimensional standing wave. In a cavity, two one-dimensional standing waves $p_x(x,t)$ and $p_y(y,t)$ in the x- and y-direction, respectively, can be superposed to a two-dimensional first order pressure field:

$$p_1(x,y,t) = p_x(x,t) + p_y(y,t) \qquad (10.2)$$

The superposition principle holds because the wave equation is a *linear* partial differential equation. Assuming harmonic time dependence and hard wall boundary conditions,[16] we can write

Ultrasonic Microrobotics in Cavities: Devices and Numerical Simulation 215

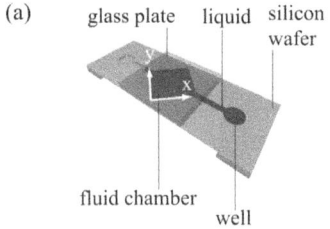

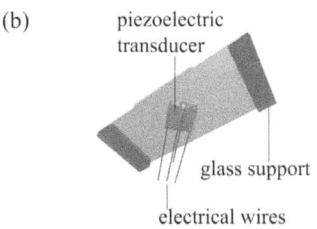

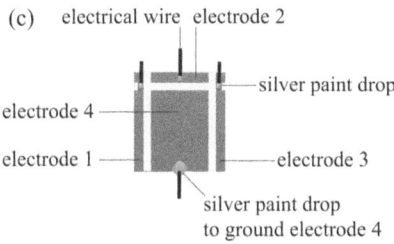

Figure 10.1 Ultrasonic particle manipulation device with a square cavity (5 × 5 mm²) on the top side (a) and a piezoelectric transducer on the backside (b). The piezoelectric transducer has a patterned electrode (c) for the individual excitation in the x- and y-direction with electrodes 1 and 2, respectively. Adapted with permission from ref. 15. © 2007, Acoustical Society of America.

$$p_x(x, t) = A\cos(k_x x)\sin(\omega_x t) \qquad (10.3)$$

$$p_y(y, t) = B\cos(k_y y)\sin(\omega_y t) \qquad (10.4)$$

with the amplitudes A, B, the wave numbers $k_x = 2\pi/\lambda_x$, $k_y = 2\pi/\lambda_y$, and angular frequencies ω_x, ω_y with the frequency difference $\Delta\omega = \omega_y - \omega_x$ between the two orthogonal waves.

For the prediction of the resulting particle forces, the Gor'kov potential[1,17] can be calculated analytically. Hereby the time-averaged pressure squared $\langle p_1^2 \rangle$ is of special interest. It follows:

$$\langle p_1^2 \rangle = \left\langle \left(p_x + p_y\right)^2 \right\rangle = \frac{A^2\cos^2(k_x x)}{2} + \frac{B^2\cos^2(k_y y)}{2} \\ + AB\cos(k_x x)\cos(k_y y)\langle\cos(\Delta\omega t)\rangle \qquad (10.5)$$

The equation above was simplified with $\langle \sin^2(\omega t)\rangle = \frac{1}{2}$ and the following trigonometric relations:

$$\langle \sin(\omega_x t)\sin(\omega_y t)\rangle$$
$$= \langle \sin(\omega_x t)\sin((\omega_x + \Delta\omega)t)\rangle$$
$$= \left\langle -\frac{\cos((2\omega_x + \Delta\omega)t)}{2} + \frac{\cos(\Delta\omega t)}{2}\right\rangle \qquad (10.6)$$
$$= \left\langle \frac{\cos(\Delta\omega t)}{2}\right\rangle$$

Time averaging has not yet been performed on the last remaining term, since it depends on the value of $\Delta\omega$. A case distinction for small and large values of $\Delta\omega$ will follow.

Furthermore, the velocity field $\mathbf{v}_1(x,y,t) = (v_x, v_y, 0)$ can be derived with the expression

$$\rho_0 \frac{\partial \mathbf{v}_1}{\partial t} = -\nabla p_1 \qquad (10.7)$$

which follows from the inviscid first-order Navier stokes equation.[16] Using trigonometric identities, the mean square fluctuation of the velocity $\langle \mathbf{v}_1^2\rangle$ is then given by

$$\langle \mathbf{v}_1^2\rangle = \langle v_x^2 + v_y^2\rangle$$
$$= \frac{1}{2\rho_0^2 c^2} \times \left[-A^2\cos^2(k_x x) - B^2\cos^2(k_y y) + A^2 + B^2\right] \qquad (10.8)$$

It is important to note that $\langle p_1^2\rangle$ depends on the frequency difference $\Delta\omega$, unlike the field $\langle \mathbf{v}_1^2\rangle$.

Based on these derivations, the Gor'kov potential U can now be derived. First, we consider the case of a very small frequency difference of e.g. $\Delta\omega_s = 5$ Hz, whereas $\omega_x \approx 2.6$ MHz. In this case, the period of $\cos(\Delta\omega_s t)$ in eqn (10.5) is so long compared to our time of interest that this term appears almost constant, which means that

$$\left\langle \frac{\cos(\Delta\omega_s t)}{2}\right\rangle \approx \frac{\cos(\Delta\omega_s t)}{2} \qquad (10.9)$$

so that the potential U varies on a timescale which can be recognized by eye. Figure 10.2 shows U at different moments within a period $T_{osc} = \frac{2\pi}{\Delta\omega_s}$ for copolymer particles with $f_1 = 0.7684, f_2 - 0.034$ and a particle radius of $r_p = 9.6$ μm.

As can be seen in the expression

$$\sin(\omega_y t) = \sin(\omega_x t + \Delta\omega_s t) \qquad (10.10)$$

the frequency difference $\Delta\omega_s$ causes a slowly varying phase difference $\varphi(t) = \Delta\omega_s t$ between p_x and p_y. These phase angles have also been noted in Figure 10.2.

Oberti et al.[15] have shown these patterns experimentally with the parameters and particles mentioned. The corresponding continuously changing particle pattern could be observed as shown in Figure 10.3.

In order to align particles in circular, well separated clumps forming an array, a Gor'kov potential as in Figure 10.2(b) is optimal. This potential lets the particles agglomerate at the four points of minimal potential. In order to generate this potential, the third summand in eqn (10.5) has to vanish. There are several ways to let this term vanish. The first way would be to choose a constant phase difference of $\varphi = 90°$ between p_x and p_y. The second way to reach this potential has been shown experimentally by Oberti et al.:[15] $\Delta\omega_l$ can be chosen to be so large that the period T_{osc} is shorter than the timescale of

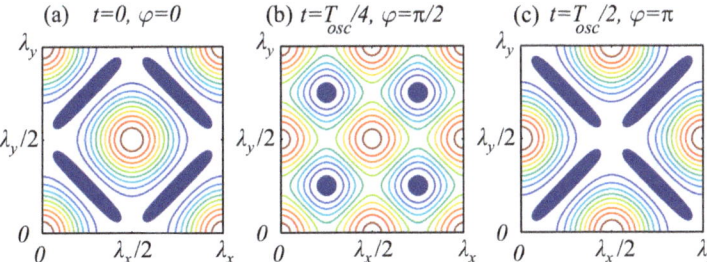

Figure 10.2 Gor'kov potential contours plotted in the x- and y-direction at three different times $t = 0$ (a), $t = T_{osc}/4$ (b), $t = T_{osc}/2$ (c) of the oscillatory period T_{osc} with their corresponding phase differences φ. The minima where the particles will collect are the four ovals/circles in blue, the maxima are in red. Reproduced from ref. 53.

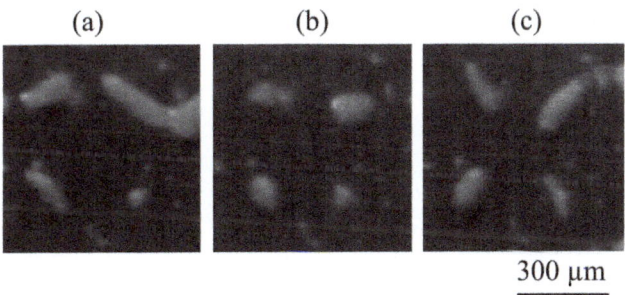

Figure 10.3 Time series of an experiment for an excitation with a very small frequency difference of $\Delta\omega_s = 5$ Hz between the standing wave in the x- and y-direction. Four particle clumps change their shape visibly and continuously according to the three potential plots of Figure 10.2 for the same times $t = 0$, $t = T_{osc}/4$ and $t = T_{osc}/2$ in (a), (b) and (c). Adapted with permission from ref. 15. © 2007, Acoustical Society of America.

interest in the experiment. In this case, the force field varies so fast that the particles' inertia and their viscous drag force does not allow them to follow these changes anymore, so we write

$$\left\langle \frac{\cos(\Delta\omega_l t)}{2} \right\rangle = 0 \qquad (10.11)$$

Then the third summand in eqn (10.5) yields zero as intended.

In Figure 10.4, the resulting $\langle p_1^2 \rangle$ and $\langle \mathbf{v}_1^2 \rangle$ are plotted, whereas the corresponding U is identical to Figure 10.2(b). Finally, for the discussed case of a large frequency difference, Figure 10.5 shows the time series of an experiment with a $\Delta\omega_l = 25$ Hz where the particles agglomerated in a steady array of 16 clumps. With this result, the wide application possibilities of particle arrays mentioned earlier become feasible.

An alternative method for the alignment of particles in clumps by ultrasonic radiation forces has been described by Vanherberghen et al.,[18] where frequency modulation of the excitation signal plays a key role.

Furthermore, the alignment of particles in intrinsically two-dimensional resonance modes offers additional alignment patterns, as described by Leibacher et al.[19] In this work, the particle patterns for the pressure field $p_1(x,y,t) = A\cos(k_x x)\cos(k_y y)\sin(\omega t)$ resulted in particle patterns that are highly dependent on the particle's density and compressibility, unlike the patterns of the formerly described superposed fields. For certain types of hollow particles, the particle pattern of Figure 10.6 resulted, and its usability for e.g. particle characterization, particle separation and the exchange of the particle's suspending medium has been outlined in the cited paper.

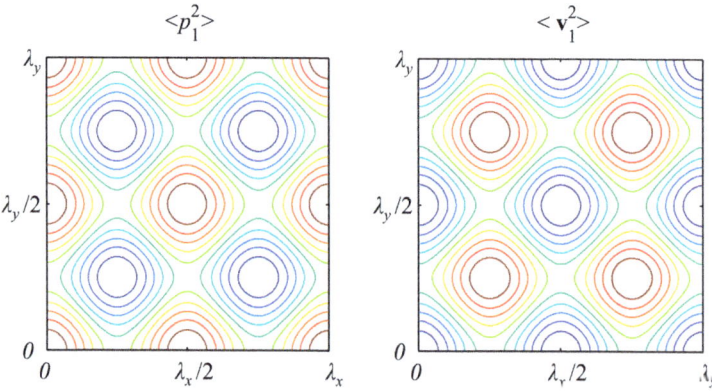

Figure 10.4 The averaged, squared pressure and velocity fields plotted in the x- and y-direction for an excitation with a large frequency difference $\Delta\omega_l$ between p_x and p_y. The same pattern results for a constant phase difference of $\varphi = 90°$ between them. Figure 10.2(b) shows the corresponding Gor'kov potential U. The described particles are expected to be attracted to the minima of $\langle p_1^2 \rangle$ (blue) and to the maxima of $\langle \mathbf{v}_1^2 \rangle$ (red). Reproduced from ref. 53.

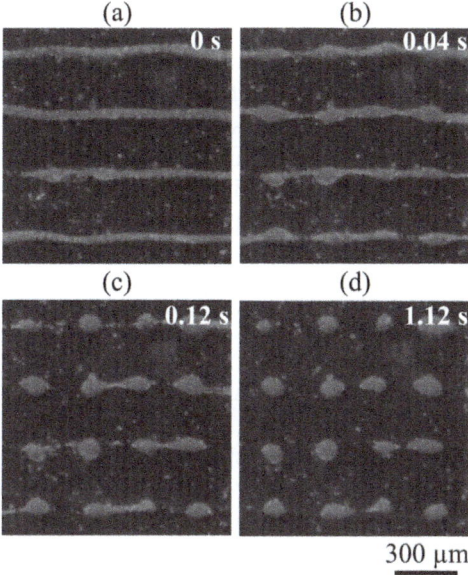

Figure 10.5 A two-dimensional steady array of 16 particle clumps. p_x was excited first at 2.562 MHz, then at the beginning of the image sequence (a), p_y was additionally excited with a frequency of 2.562 MHz + 25 Hz, merging the particles into well-separated, circular clumps (b to d). Reprinted with permission from ref. 15. © 2007, Acoustical Society of America.

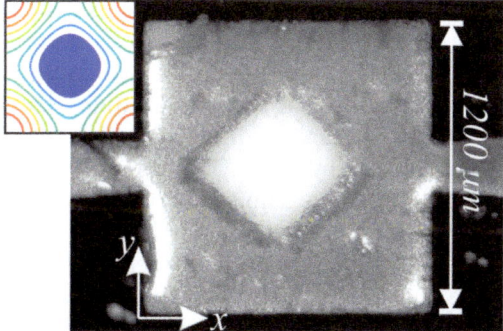

Figure 10.6 In an intrinsically two-dimensional resonance mode at 870 kHz, hollow glass particles of 14 μm diameter are found to be aligned to a particle clump that corresponds to the blue pressure minimum in the inserted plot top left.[19] Adapted from ref. 19 with kind permission from Springer Science and Business Media.

10.2.2 Rotation of Micro-particles

In addition to the movement and positioning of particles, it is also possible to rotate them with acoustic radiation forces. This technique can be used for robotic or lab on a chip applications such as ultrasonically-driven micro machines or orientation and positioning of micro components.

Ultrasonic standing waves can excite two kind of torques—the viscous torque and the acoustic radiation torque.

10.2.2.1 Rotation of Non-spherical Particles with Acoustic Radiation Torque

This section outlines in brief the results of a recently published paper by Schwarz *et al.*[20]

Non spherical particles behave like spheres in an ultrasonic standing wave with an additional torque acting on the particle. Stiff and dense fibers that are shorter than one-fourth of the wavelength are constrained at the pressure node and are oriented along the nodal lines.[21] In addition to alignment it is possible to use this acoustic radiation torque for a continuous and controlled rotation of objects. Therefore a time-varying pressure field with change of orientation of the potential well is needed. Oberti *et al.*[15] studied the different cases of the superposition of two orthogonal standing waves generating a two-dimensional pressure field such as the one described in the previous section.

This method is extended to the amplitude modulation of two orthogonal standing waves. By varying the amplitudes A and B of the standing waves (eqn (10.3) and (10.4)) in the x- and y-direction, the orientation of the force potential minima is rotated. Figure 10.7 shows the Gor'kov force potential for different amplitudes of A and B. In the top row, the amplitude A of the standing wave in the x-direction decreases from $+1$ to -1, while the amplitude in the y-direction is set to a constant value. The bottom row in Figure 10.7 shows the inverse case, with a decrease of the amplitude in the y-direction, while the amplitude in the x-direction is set to a constant value. The location and orientation of a fiber in the potential minima is shown with a black arrow. The combination of these amplitude changes leads to the rotation. When the amplitudes are modified in the manner indicated by the grey arrow, the fiber is rotated by 180°. A full rotation of 360° is possible by running the same sequence twice. The speed of the rotation is governed by the frequency of the amplitude change and the direction depends on the location of the object and the modulation sequence of the amplitudes.

The plots in Figure 10.7 are showing only a "unit cell" for one wavelength in the x- and y-directions. For each unit cell, there exist four potential minima where a fiber could be located and rotate. The fibers at the locations ($\frac{1}{4} \lambda_x$, $\frac{1}{4} \lambda_y$) and ($\frac{3}{4} \lambda_x$, $\frac{3}{4} \lambda_y$) are rotating counter-clockwise, while at ($\frac{1}{4} \lambda_x$, $\frac{3}{4} \lambda_y$) and ($\frac{3}{4} \lambda_x$, $\frac{1}{4} \lambda_y$), the rotations are clockwise. In theory, a device with half a wavelength inside the cavity has only one rotating spot directly in the middle of the chamber.

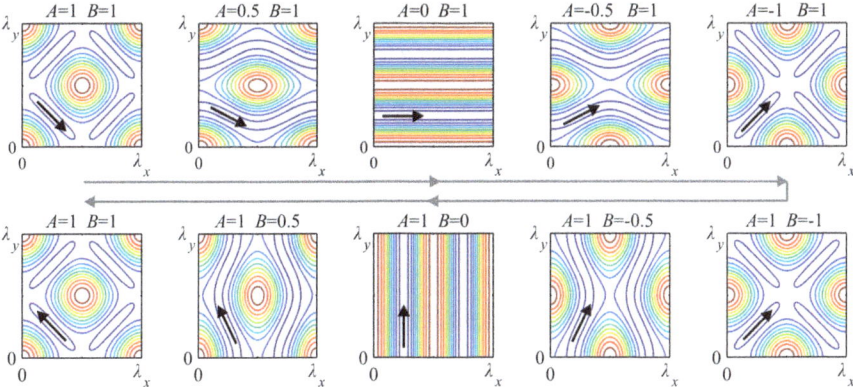

Figure 10.7 Contour plot sequence of the Gor'kov force potential for one wavelength in the *x*- and *y*-directions (unit cell) as a result of the amplitude change of two superimposed cosine functions with identical frequency and phase. In the top row, the amplitude of the standing wave in the *x*-direction decreases from +1 to −1, while the amplitude in the *y*-direction is set to a constant value. The bottom row shows the inverse case, with a decrease of the amplitude in the *y*-direction, while the amplitude in the *x*-direction is set to a constant value. The combination of these amplitude changes leads to the rotation. The location of a fiber in the potential minima is shown with a black arrow. Reproduced from ref. 53.

Experiments on rotating micro-particle clumps and fibers have been performed. Figure 10.8 shows two experiments demonstrating rotation with amplitude modulation. The device used for the experiments consists of a 3 × 3 mm² chamber etched into silicon and covered with a glass plate as shown in Figure 10.1. The actuation is done through a 4 × 4 mm² piezoelectric plate fixed at the back side of the device. By exciting two orthogonally oriented strip electrodes, defined on the surface of the same piezoelectric transducer, a two-dimensional pressure field can be set up, assuming that the chamber depth is kept smaller than half of the acoustic wavelength in the fluid.[15]

The rotation with amplitude modulation is depicted in Figure 10.8(a) with particle clumps formed out of 17 μm copolymer particles at a frequency of 1689 kHz. There are 3.5 wavelengths inside the cavity and so 7 × 7 rotating particle clumps. All clumps are rotating and the white square is showing a unit cell (one wavelength) with the corresponding rotation directions. Figure 10.8(b) shows the rotation of a 210 μm long glass fiber with an actuation frequency of 1085 kHz. The speed of rotation could be increased until about 35 rpm. The fiber can be stopped at any given orientation and it is possible to invert the rotational direction. For all the experiments the amplitudes have been varied piecewise linearly, as is shown in Figure 10.7. The angle of the fiber will not change linearly and the angular velocity of the fiber will not be constant over one complete rotation cycle. For this purpose

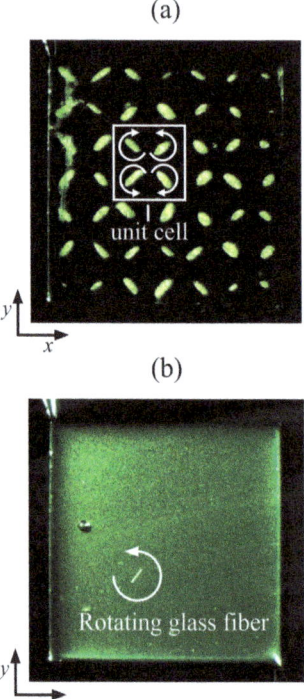

Figure 10.8 (a) Rotation of particle clumps (copolymer particle, 17 μm) at a frequency of 1689 kHz with amplitude modulation inside a 3 × 3 mm² cavity. (b) Rotation of a glass fiber (length 210 μm, diameter 15 μm) at a frequency of 1085 kHz with amplitude modulation. Reproduced from ref. 53.

a sinusoidal variation of the amplitudes is needed. The difference in the maximum amplitude of A and B will also lead to a variation of the angular velocity of the fiber during one rotation cycle.

10.2.2.2 Rotation of Spherical Particles with the Acoustic Viscous Torque

The viscous torque is generated by two orthogonal standing waves at the same frequency with amplitudes A and B and a relative phase shift φ.[22,23] A near boundary streaming spins the spherical or axisymmetric objects around the symmetry axis.

Lee et al.[24] studied the streaming around a fixed sphere due to two orthogonal standing waves by using Nyborg's equations and assumptions.[25] The consideration of a rotating sphere was done by Lamprecht et al.[26] to predict the angular velocity. It was shown that the viscous torque can be a dominating effect for the particle manipulation with ultrasound and high rotational speeds can be obtained.

In the calculations of Lee et al.,[24] the initial torque Γ_0 at an angular velocity $\Omega = 0$ on a rigid sphere within an orthogonal standing wave field is given by

$$\Gamma_0 = \frac{3}{4}\delta S_S AB(\rho_0 c^2)^{-1}\sin\varphi\cos(kx)\cos(ky)$$

where S_S is the sphere surface and δ is the thickness of the viscous boundary layer (Stokes layer).

The necessary condition to realize a particle rotation by the viscous torque is a homogeneous and stable wave field due to two orthogonal standing waves. These two standing waves are shifted in phase with $\varphi = 90°$ or $\varphi = 270°$ and excited with the same frequency f and amplitude ($A = B$).

The phase shift φ and the position of the sphere $\{x,y\}$ in the wave field have an influence on the rotation direction of the particle. In Figure 10.9(a), it can be seen that particles at the pressure nodes will rotate in clockwise and anticlockwise directions depending on the specific pressure node. A phase jump of 180° from $\varphi = 90°$ to $\varphi = 270°$ will lead to a change of the rotation direction as shown in Figure 10.9(b).

The theoretical predictions for the steady state angular velocities and experiments have shown[26] that smaller particles rotate much faster than larger particles because the rotation by the viscous torque is limited by the stokes drag of the particle rotation. In the experiments, particle rotations of over 1000 rpm were observed for particles smaller than 70 µm.

10.2.3 Particle Transport

With ultrasound, particles can not only be aligned to lines or specific patterns, but they can also be moved over a larger distance that covers several wavelengths. One technique to do so is using acoustic streaming to move the fluid and with it the particles by utilizing drag forces. Here the focus is on another technique where the fluid is calm and only the particles are moved utilizing acoustic radiation forces. This could be done e.g. by amplitude or phase modulation of two transducers, by shifting boundaries or by changing the frequency.[27–32] The advantage of the latter is that a large range of motion is achievable and it can be easily used with a closed chamber that can be built independently of the transducer. Apart from more complex ways of changing the frequency, there are two basic modes. One mode is changing the frequency in discrete steps, only using the resonances. The other mode is changing the frequency continuously, also utilizing the frequencies between resonances. Here the particle movement with these modes is presented in a cavity with a hard-wall boundary condition on one side and excitation boundary condition on the other side. Particle paths with discrete steps and four consecutive resonance frequencies are shown in Figure 10.10. Particles can be moved either towards the centre or equally towards the left and right boundary.

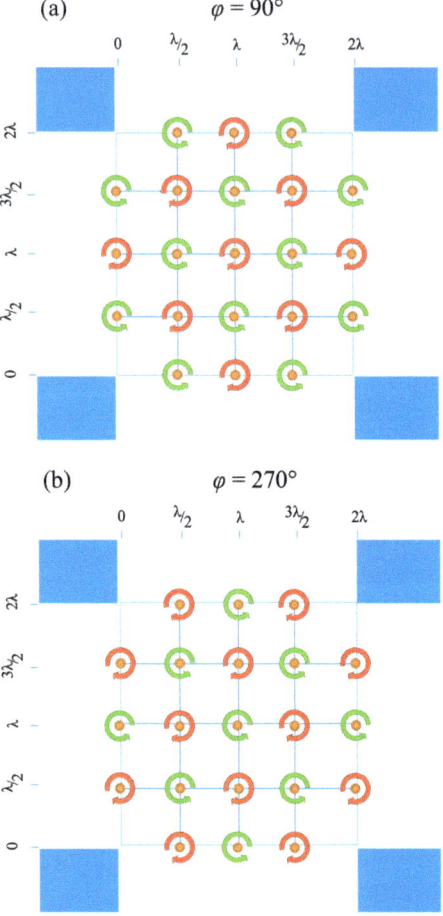

Figure 10.9 Change of rotation direction for different pressure nodes of two orthogonal standing waves for a jump in phase from $\varphi = 90°$ (a) to $\varphi = 270°$ (b). The red arrows are indicating a rotation in a clockwise direction and the green arrows are indicating a counter-clockwise rotation.

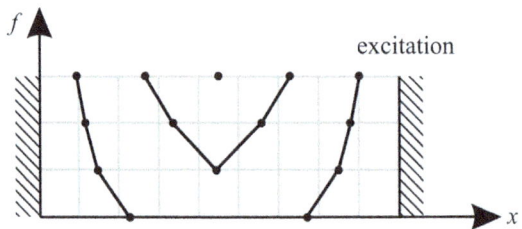

Figure 10.10 Frequency over distance and particle paths with a frequency change in discrete steps. Reproduced from ref. 53.

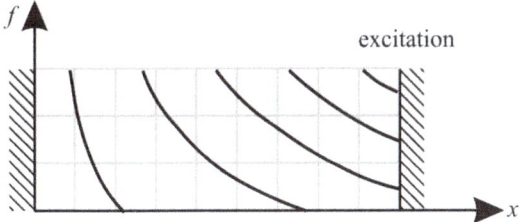

Figure 10.11 Frequency over distance and isolines (particles paths) with a continuous frequency change. Reproduced from ref. 53.

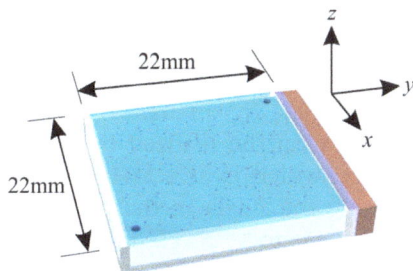

Figure 10.12 Square plastic chamber with particles immersed in water and piezoelectric transducer (brown). Reproduced from ref. 53.

On the other hand, if the frequency is changed continuously, particles can be moved towards or away from the transducer, that is *e.g.* from the right boundary to the left by increasing the frequency or *vice versa* by decreasing the frequency as illustrated in Figure 10.11.

The isolines depict the path particles would travel in an ideal setup without drag forces and the same frequency range as in Figure 10.10. Obviously for the potential field to be strong enough between the resonances, damping needs to be present.

In the continuous case, particles close to the transducer are transported faster than those far away from it, while for a discrete frequency change, the particle transport velocity is the largest in the centre of the cavity. For transport from one boundary to the other the frequency change needs to be repeated, *e.g.* with a saw-tooth profile. There is also a minimum requirement on the frequency range covered to effectively move particles more than one wavelength, or *e.g.* in the case of continuous frequency sweeping, move the particles close to the non-excited boundary.

As an example, continuous frequency sweeping can be used in a macroscopic square plastic chamber, presented here.

10.2.3.1 Device Description

The device is a square chamber as shown in Figure 10.12. The water domain has a side length of 20 mm and a height of 3 mm. The 1 mm thick sidewalls are laser cut from a 3 mm PMMA sheet and the top and bottom are 250 µm foils which are glued to the sidewall frame using GBL (gamma-butyrolactone). In the top cover foil there are two 1 mm holes in opposing corners to fill the chamber.

A 1 mm thick piezoelectric element, Pz26, is either directly glued to one of the chamber walls or first glued to an aluminium waveguide and then clamped to the device.

10.2.3.2 Numerical Simulation

The device has been simulated in 2D using COMSOL Multiphysics. The FEM simulation is fully coupled including the acoustic, linear elastic and piezo-electric domain. For the water and Pz26, the same values as presented in the paper by Dual *et al.*[33] were used but with a frequency dependent Q-factor of 2000*1 MHz/f for the water. The PMMA is modeled with structural damping and an isotropic loss factor of $\mu_{PMMA} = 0.02$.[34]

Large pressure amplitudes can be realized by operating the transducer around its main resonance, which is at 2 MHz. This frequency results in about 30 wavelengths along the x-axis in the chamber. Thus a frequency range of close to 0.5 MHz needs to be covered to move the particles close to the non-excited wall. Here a frequency range of 1.5–2.5 MHz has been used. This is also due to the fact that the pressure field is not as homogeneous as shown in Figure 10.11. Thus particles are not always moving forward at every frequency and position within the chamber. Particularly, the simulation of the pressure field in the xz-plane shows resonances in the z-direction with wavelengths of the same order as the excitation frequency. In the xy-plane there are also resonances perpendicular to the intended excitation direction but typically with a much higher wavelength as shown in Figure 10.13.

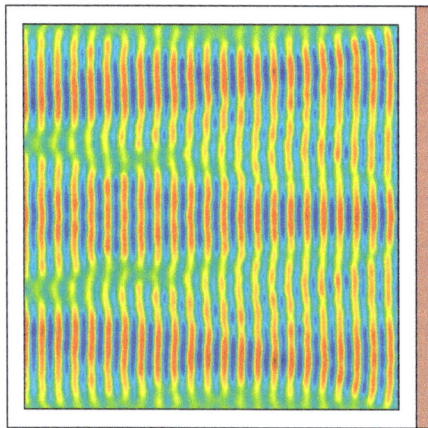

Figure 10.13 Pressure field at 1.59 MHz showing an inhomogeneous field example. Transducer is on the right side (brown). Reproduced from ref. 53.

10.2.3.3 Experimental Results

Repeated frequency sweeping in a range from 1.5 to 2.5 MHz has been used to collect particles along one of the sidewalls as shown in Figure 10.14. Operating with particles with a diameter of 26 µm, about 80% of the particles can be concentrated to the four final nodes in less than 2 min. The number of final nodes the particles are confined to is limited by the amount of particles. With larger particles, the time and yield are significantly increased. The same is true for changing the base material, in particular the bottom cover plate to *e.g.* polypropene or even better aluminium with less damping than PMMA.

As expected from the results of the numerical simulation, particles are only barely transported close to the sidewalls which are perpendicular to the pressure nodal lines. In the experiment, this effect is increased by the presence of the openings in the top cover. The perpendicular resonances shown in the simulation are also observed in the experiment. Particles can locally move along pressure nodes with velocities exceeding the typical velocity in the intended direction.

Problems that might arise are sticking of particles to the surface, convection and traveling wave attenuation streaming.

Finally, the devices described have been experimentally analyzed with the schlieren method in a recent paper by Möller *et al.*[35] The schlieren method provides an optical tool for the visualization of the pressure fields in the water domain.

10.2.4 Particle Positioning Combined with a Microgripper

Many of the acoustofluidic devices presented in the literature aim at the controlled handling of particles for automation purposes, often in combination with up- or downstream procedures. Contributing to such microrobotic applications, here we summarize the combination of an acoustofluidic device with a microgripper for the subsequent single-particle handling.

Figure 10.14 Concentration of 26 µm particles with repeated continuous frequency change (1.5–2.5 MHz). Initial condition is shown on the left, after 40 s in the second image and after 120 s in the third image. Excitation is from the right. Reproduced from ref. 53.

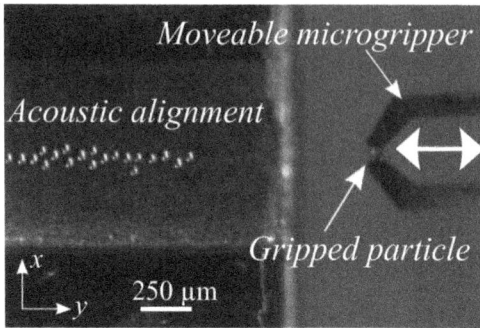

Figure 10.15 The combination of acoustofluidic particle handling in the microchannel (left side) with a moveable microgripper (right side) enables controlled, automated single-particle handling. Adapted from ref. 37 by permission of IOP Publishing.

In Figure 10.15, on the left side, a microfluidic channel of 1 mm width can be seen with an open access on its right side. Within the water-filled channel, 20 particles of 74 μm diameter have been aligned by ultrasonic excitation. First, the particles were prealigned with a $\lambda/2$ ultrasonic standing wave at 780 kHz, then the frequency was changed to 2.08 MHz to tighten the particles in the middle pressure nodal line of this $3\lambda/2$ mode, which exerts higher acoustic radiation forces on the particles according to its Gor'kov potential, eqn (10.1). After this alignment step, a microgripper of 2 mm length and 50 μm finger thickness was inserted into the microchannel. This system allowed the gripper to extract a single particle as visible on the right side of the image. The ultrasonic device and the microgripper have been further described by Neild et al.[36] and Beyeler et al.,[37] respectively.

Whereas ultrasonic manipulation only allowed for particle movement in the x-direction, Oberti et al.[38] further enhanced the device described by fluidic particle/cell movement in the y-direction. As illustrated in Figure 10.16, two additional side channels in the x-direction were fabricated. These side channels were connected to a syringe pump in withdrawal mode, so in the main channel a fluid flow in the y-direction could be generated to move the particles with drag forces (Stokes' law) towards the open channel end, where a microgripper could access them.

Thinking in terms of microrobotic automation, the combined acoustic and fluidic manipulation allowed for a controlled and repeatable particle positioning, which might become a powerful tool when extended with a control loop including e.g. optical feedback. The access of microgrippers or pipettes into open acoustic manipulation channels has a considerable scope for downstream particle/cell processing.

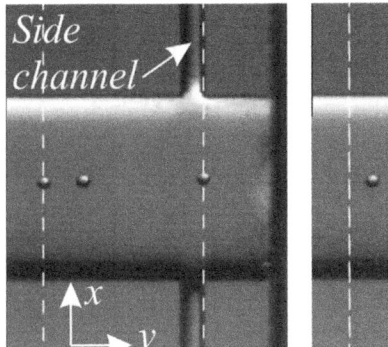

Figure 10.16 Particle alignment in the x-direction combined with fluidic movement in the y-direction by outward flow through two side channels. Particles were moved rightwards in these two time series images with a 0.24 s time difference. Adapted from ref. 38 with permission from Elsevier.

10.3 Numerical Techniques for the Time-averaged Acoustic Effects

When ultrasonic waves propagate in a fluid cavity, not only is a field of acoustic standing waves established but also time-averaged effects, such as acoustic radiation pressure and acoustic streaming. These time-averaged effects are easily observed in experiments, since they vary much slower with time (typically in second scale) than the standing ultrasonic waves in the fluid cavity (in microsecond scale). The phenomena of acoustic radiation pressure and acoustic streaming have been applied widely to engineering fields such as acoustic levitation and acoustic contactless micro particle manipulators.[27,39,40] Numerical simulations aim to better understand and accurately describe the time-averaged effects in order to improve the design of these ultrasonic devices, in particular for nonstandard situations.

Theoretically, the fluid motion in the cavity obeys the equation of continuity and the Navier–Stokes (N–S) equations. The time-averaged phenomena derive from the nonlinear convection terms in the governing equations. If the incident wave amplitude is small enough, the nonlinearity in the governing equations becomes weak, thus the solutions for the N–S equations can be expanded to series based on the dimensionless parameter $\varepsilon = U_0/(\omega a)$, where U_0 is the incident velocity amplitude, ω is the circular frequency and a is the typical size of the particle, by using the perturbation method introduced by Sadhal.[41] Besides, since there exists a huge gap of the characteristic timescales between the ultrasound wave propagation and the time-averaged effects, it is also reasonable to physically separate the governing equations into a periodic (first-order) part and a time-averaged (second-order) part. In the periodic part, the nonlinear convection term is totally eliminated, whereas the time-averaged part is the time-averaging of the N–S equations including the nonlinear terms induced by the periodic variables as illustrated in Chapter 2 by Bruus.[16] Based on these perturbation methods, the time-averaged effects can be determined analytically under specific conditions, *i.e.* the acoustic streaming generated by an oscillating sphere or cylinder in an infinite viscous fluid.

The acoustic radiation pressure is easily obtained in an ideal fluid. In King's formulation,[42] the linear equation of state was adopted, but the relation between the pressure and the density is corrected up to the order of the square of the periodic variables in order to take the convection effects into account. The time-averaged pressure was obtained from this corrected pressure–density relation directly instead of solving the time-averaged equation. Yosioka[43] extended King's theory and presented a general formula for the acoustic radiation force acting on an oscillating boundary in an inviscid fluid. This formula is very convenient for arbitrarily shaped particles in both analytical and numerical modeling. Moreover, Gor'kov[1] proposed an elegant formula for the acoustic force on a small sphere. The Gor'kov formula is so simple and successful that it has been applied widely for the design of ultrasonic devices.[17] For cases in which the viscosity must be taken into account, the situations become more complicated. The Reynolds stresses have to be calculated to build the connections between the periodic variables and the time-averaged ones. Subsequently, the time-averaged equations must be fully solved to gain the acoustic streaming velocity and the time-averaged pressure. Furthermore, the acoustic boundary layers have to be treated very carefully. Riley[44] demonstrated many applications of the perturbation method for the acoustic streaming while Doinikov[45] and Danilov[12] exhibited the detailed calculations of the time-averaged acoustic force on a sphere in a viscous fluid by solving both the periodic and the time-averaged equations.

Numerical simulations are always necessary for systems with complex conditions such as complex particle shapes and chamber boundaries, the requirement for viscous effects, complex incident waves and so on. Generally speaking, we have two approaches to determine the time-averaged effects. The first one is the numerical implementation for the perturbation methods including Riley's and Doinikov's formulations, as well as Gor'kov's and Yosioka's theories neglecting viscosity. Since the periodic equations are equivalent to the traditional sound wave equations, they are able to be solved by commercial software packages such as COMSOL and ANSYS. The time-averaged equation is usually in the form of the Stokes' equation with a source term of Reynolds stress and a thin boundary layer, where the boundary layer mesh should be applied. This equation can certainly be solved by commercial software, but the current authors[46] also tried to obtain the numerical results through BEM (boundary element method). The other numerical approach relies on the full solutions for the governing equations, which contain both the periodic parts and the time-averaged ones, and then taking the average of the solutions over time leads to the time-averaged variables. Although this method is able to gain the time-averaged effects without any restrictions such as small incident magnitude and Newtonian fluid, the computational efforts are relatively large. This idea has been implemented by CFD (computational fluid dynamics) algorithms, i.e. Haydock[47,48] via the LB (lattice Boltzmann) algorithm and Wang et al.[23,49] through the FVM (finite volume method) algorithm.

In the following subsections, the two methods are introduced briefly, and the numerical examples are subsequently presented. The governing equations for a fluid motion without heat conduction are

$$\partial_t \rho + \nabla \cdot \rho \mathbf{v} = 0 \tag{10.12}$$

$$\rho \partial_t \mathbf{v} + \rho \mathbf{v} \cdot \nabla \mathbf{v} = -\nabla p + \mu \nabla^2 \mathbf{v} + \left(\mu_B + \frac{1}{3}\mu\right) \nabla \nabla \cdot \mathbf{v} \tag{10.13}$$

with the equation of state $p = p_0 + c^2(\rho - \rho_0)$. Here, ρ, p and $\mathbf{v}$ are the fluid density, pressure and velocity respectively, ρ_0 and p_0 are the undisturbed fluid density and pressure, and μ and μ_B are the dynamic viscosity and the bulk viscosity.

10.3.1 Perturbation Method

As demonstrated by Bruus[16] and Sadhal,[41] the total field can be expressed in a perturbed series of the small dimensionless parameter ε. The governing eqn (10.12) and (10.13) of order ε^0 are

$$\partial_t \rho_1 + \rho_0 \nabla \cdot \mathbf{v}_1 = 0 \tag{10.14}$$

$$\rho_0 \partial_t \mathbf{v}_1 = -\nabla p_1 + \mu \nabla^2 \mathbf{v}_1 + \left(\mu_B + \frac{1}{3}\mu\right) \nabla \nabla \cdot \mathbf{v}_1 \tag{10.15}$$

with the linear equation of state $p_1 = c^2 \rho_1$. The subscript '1' indicates the periodic variables. The time-averaged governing equations of order ε^1 are

$$\rho_0 \nabla \cdot \mathbf{v}_2 = -\nabla \cdot \langle \rho_1 \mathbf{v}_1 \rangle \tag{10.16}$$

$$\mu \nabla^2 \mathbf{v}_2 + \left(\mu_B + \frac{1}{3}\mu\right) \nabla \nabla \cdot \mathbf{v}_2 - \nabla p_2 + \mathbf{F}_R - \langle \partial_t (\rho_1 \mathbf{v}_1) \rangle = 0 \tag{10.17}$$

where $\mathbf{F}_R = -\rho_0 \langle \mathbf{v}_1 \nabla \cdot \mathbf{v}_1 + \mathbf{v}_1 \cdot \nabla \mathbf{v}_1 \rangle$ is a driving force related to the spacial derivatives of the Reynolds stress. If the viscosity is not significant, the time-average $\langle \rho_1 \mathbf{v}_1 \rangle$ approaches zero approximately since the velocity and the pressure of the first-order are always out of phase in this circumstance. Therefore, eqn (10.16) and (10.17) become

$$\nabla \cdot \mathbf{v}_2 = 0 \qquad (10.18)$$

$$\mu \nabla^2 \mathbf{v}_2 - \nabla p_2 + \mathbf{F}_R = 0. \qquad (10.19)$$

The above governing equations reveal that the time-averaged flow (acoustic streaming) is incompressible without convection and is driven by the Reynolds stress. The time-averaged force and torque exerted on a solid particle are then calculated by[45]

$$\mathbf{F} = \int_{S_0} \langle \boldsymbol{\sigma}_2 - \rho_0 \mathbf{v}_1 \otimes \mathbf{v}_1 \rangle \cdot \mathbf{n} \, dS \qquad (10.20)$$

$$\mathbf{T} = \int_{S_0} \mathbf{r} \times \langle \boldsymbol{\sigma}_2 - \rho_0 \mathbf{v}_1 \otimes \mathbf{v}_1 \rangle \cdot \mathbf{n} \, dS \qquad (10.21)$$

where $\boldsymbol{\sigma}_2 = -p_2 \mathbf{I} + \mu[\nabla \mathbf{v}_2 + (\nabla \mathbf{v}_2)^T]$ is the time-averaged stress tensor and S_0 is referred to as the particle equilibrium surface. In the calculations, it is assumed that the solid particle is movable in the periodic field and must be fixed in the time-averaged field. The time-averaged force computed by eqn (10.20) is the total time-averaged (mean) one which includes the influences of acoustic streaming. If S_0 is taken as an arbitrary surface enclosing the particle, the acoustic radiation force is presented, where only the momentum flux of the sound waves is taken into account.[12] In the case of low viscosity where the acoustic streaming is not significant, these two time-averaged forces differ only slightly. Clearly, the total time-averaged force is more important in practical applications.

While the fluid viscosity is neglected, the time-averaged force and torque can be conveniently calculated by[43]

$$\mathbf{F} = \frac{1}{2} \rho_0 \int_{S_0} \left[\langle \mathbf{v}_1^2 \rangle - \frac{1}{\rho_0^2 c^2} \langle p_1^2 \rangle \right] \mathbf{n} \, dS - \rho_0 \int_{S_0} \langle (\mathbf{n} \cdot \mathbf{v}_1) \mathbf{v}_1 \rangle \, dS \qquad (10.22)$$

$$\mathbf{T} = \frac{1}{2} \rho_0 \int_{S_0} \mathbf{r} \times \left[\langle \mathbf{v}_1^2 \rangle - \frac{1}{\rho_0^2 c^2} \langle p_1^2 \rangle \right] \mathbf{n} \, dS - \rho_0 \int_{S_0} \mathbf{r} \times \langle (\mathbf{n} \cdot \mathbf{v}_1) \mathbf{v}_1 \rangle \, dS \qquad (10.23)$$

Here, due to the absence of acoustic streaming terms, the integrations in eqn (10.22) and (10.23) can take place over an arbitrary S_0 enclosing the particle.[17]

10.3.2 Solving the N–S Equation by FVM

If parameter ε, which indicates the influence of the nonlinear convection term in the N–S equation, is not small enough, *i.e.* a large incident magnitude or a low incident frequency, the perturbation theory becomes invalid. The time-averaged effects can only be obtained by solving the full N–S equation directly. In this subsection, we introduce the FVM algorithm based on Jameson's scheme[50] for the governing equations.

Eqn (10.12) and (10.13) can be expressed in the integral form as

$$\partial_t \int_V \mathbf{U} dV + \oint_{\partial V} \bar{\mathbf{F}} \cdot \mathbf{n} dS = 0 \tag{10.24}$$

where $\mathbf{U}$ is the vector of conservation parameters, $\bar{\mathbf{F}}$ is the vector flux matrix which is a function of $\mathbf{U}$, ∂V is the boundary of a certain computational region and $\mathbf{n}$ is the outward unit normal of ∂V. The conservation law eqn (10.24) at every triangular element shows the semi-discrete equations for element n (Figure 10.17)

$$\frac{d\mathbf{U}_n}{dt} = -\frac{1}{\Delta V_n} \sum_{j=1}^{N_n} \bar{\mathbf{F}}(\mathbf{U})_{n,j} \cdot \mathbf{n}_j S_j \tag{10.25}$$

where ΔV_n is the element volume, S_j is the area of the element side j, $\mathbf{n}_j$ is the outward unit normal vector of the side j and N_j is the number of sides of element n. For example, referring to Figure 10.17, the variables on an element's surface are computed by $\mathbf{U}_{n,j} = (\mathbf{U}_n + \mathbf{U}_p)/2$.

The dissipative terms that are needed to capture shock waves and to suppress numerical oscillations suggested by Jameson[50] are

$$D_n = \sum_{j=1}^{N_n} d_j^{(2)} + \sum_{j=1}^{N_n} d_j^{(4)} \tag{10.26}$$

where $d_j^{(2)}$ and $d_j^{(4)}$ are the second and fourth order dissipative flux on side j, respectively, which are expressed as (Figure 10.17),

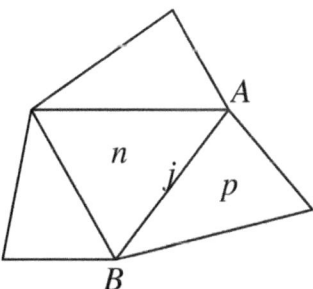

Figure 10.17 Elements and sides. Reproduced from ref. 53.

$$d_j^{(2)} = \varepsilon_j^{(2)}(\mathbf{U}_p - \mathbf{U}_n)_j$$

$$d_j^{(4)} = -\varepsilon_j^{(4)}(\nabla^2\mathbf{U}_p - \nabla^2\mathbf{U}_n)_j$$

Here, the operator ∇^2 means $\nabla^2\mathbf{U}_n = \sum_{p=1}^{N_n}(\mathbf{U}_p - \mathbf{U}_n)$, and $\varepsilon_j^{(2)}$ and $\varepsilon_j^{(4)}$ are the adapted coefficients.

The fourth order Runge–Kutta algorithm can be used to discretize the time derivatives. Rewrite eqn (10.25) as

$$\frac{d\mathbf{U}_n}{dt} = \mathbf{R}_n(\mathbf{U}_n) \tag{10.27}$$

where $\mathbf{R}_n(\mathbf{U}_n)$ denotes the right hand terms of eqn (10.25). The procedure is written as

$$\mathbf{U}^{(0)} = \mathbf{U}^m, \quad \mathbf{U}^{(k)} = \mathbf{U}^{(0)} + \alpha_k \Delta t \mathbf{R}^{(k-1)} \quad \text{and} \quad \mathbf{U}^{m+1} = \mathbf{U}^{(4)}$$

where m denotes the last time step, $m+1$ denotes the current time step, k is the Runge–Kutta step number which varies from 1 to 4, Δt is the time step, and $\alpha_1 = 1/4$, $\alpha_2 = 1/3$, $\alpha_3 = 1/2$ and $\alpha_4 = 1$, respectively.

The time-averaged effects are then calculated by the direct time-average of the computational values at every time step.

10.3.3 Numerical Examples

In this section, a number of two-dimensional simulations have been performed to illustrate the methods mentioned before. In the following simulations, the normalized parameters are used for convenience. The length is normalized by $l_0' = 1 \times 10^{-6}$ m in the first, the third and the fourth examples, and by $l_0' = 1 \times 10^{-7}$ m in the second example. In all numerical cases, the density is normalized by $\rho_0' = 1000$ kg m^{-3} and the velocity $c' = 1400\sqrt{3}$ ms^{-1}. It is calculated that the circular frequency is 8.796 MHz in the first, the third and the fourth examples and 17.59 MHz in the second example.

10.3.3.1 Acoustic Radiation Force on a Fixed Cylinder in Inviscid Fluid

A cylinder is fixed in a rectangular chamber with a standing wave in the horizontal direction (Figure 10.18). The length of the computational region is set as $L_x = \lambda$, where $\lambda = 1000$ is the wavelength and the height L_y varies in different cases. The wave is initialized as $\rho = \rho_0 + \Delta\rho\cos(kx)$ and zero fluid velocity field $\mathbf{u} = 0$, where k is the wave vector, $\rho_0 = 1.0$ and $\Delta\rho = 0.01$. To generate a standing wave, the density on the left boundary is maintained at $\rho = \rho_0 + \Delta\rho\cos(\omega t)$ and the reflecting condition $u_x = 0$ is imposed to the right boundary. The host fluid is inviscid and has a speed of sound $c = 1/\sqrt{3}$.

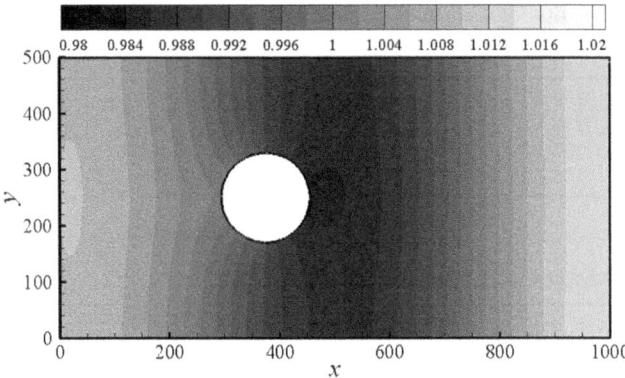

Figure 10.18 Density distribution at time 100 T. Reproduced from ref. 53.

Figure 10.18 plots the density contours computed by the FVM method after 100 periods. The results obtained by different methods are summarized in Table 10.1. In Table 10.1, the cylinder center is located at $(3/8)\lambda = 375$ from the left chamber boundary. The column titled F_{th} shows the results of the analytical formula for an infinite domain provided by Wang et al.,[49] the column with F_{COM} contains the results by COMSOL *via* the perturbation method and F_{FVM} indicates the results from the FVM simulations. It can be seen in Table 10.1 that the results between the FVM program and COMSOL agree with each other well. The analytical results can only provide reference values since the formula does not take the effects of chamber boundaries into account.

10.3.3.2 Time-averaged Force on a Fixed Cylinder in Viscous Fluid

In this example, the parameters are the same as the previous one except for the wavelength (5000) and cylinder radius (500). The PML algorithm[23] is used to suppress the boundary reflections as if the cylinder was located in an infinite region.

Tables 10.2 and 10.3 present the time-averaged acoustic force for the cylinder excited by a traveling and standing wave, respectively, calculated by the FVM method compared with the analytical solutions described in ref. 23. The columns titled F_p, F_s, F_{pFVM} and F_{sFVM} contain the analytical forces due to pressure, the analytical forces due to shear stress, the numerical forces due to

Table 10.1 Results for a fixed cylinder.

R	L_y	$F_{th}\ (10^{-5})$	$F_{COM}\ (10^{-5})$	$F_{FVM}\ (10^{-5})$
5	100	−1.23	−1.25	−1.25
20	200	−19.65	−21.22	−21.17
40	200	−81.02	−106.99	−105.15
80	500	−247.20	−387.74	−375.30

Table 10.2 Results for a traveling wave.

$\nu\ (10^{-3})$	$F_p\ (10^{-5})$	$F_s\ (10^{-5})$	$F_{pFVM}\ (10^{-5})$	$F_{sFVM}\ (10^{-5})$
1.67	1512.7	11.5	1520.8	11.3
4.12	1523.1	18.0	1523.2	17.9
8.35	1535.3	25.7	1527.4	25.9
16.7	1552.2	36.3	1535.4	35.9

Table 10.3 Result for a standing wave.

$\nu\ (10^{-3})$	$F_p\ (10^{-5})$	$F_s\ (10^{-5})$	$F_{pFVM}\ (10^{-5})$	$F_{sFVM}\ (10^{-5})$
1.67	−1680.7	−0.39	−1684.8	−0.62
4.12	−1680.6	−0.64	−1684.3	−0.93
8.35	−1680.4	−0.95	−1683.7	−1.32
16.7	−1680.1	−1.42	−1682.2	−1.91

pressure and the numerical forces due to shear stress, respectively. It is shown in Table 10.2 that the numerical results and the analytical results agree with each other very well. The maximum difference is less than 2.1% and even less than 1% in most of the cases. The acoustic radiation force is given as 1494.5×10^{-5} in an ideal fluid by using the theory in ref. 49, whereas the total acoustic force is found to be 1541.1×10^{-5} for water ($\nu = 0.00412$) and 1588.5×10^{-5} for the liquid with the kinematic viscosity 0.0167. The viscous effects are notable in the traveling wave cases. It is seen from Table 10.3 that the forces due to shear stress are nearly equal to zero with all viscosities, which means that the viscous effects are not remarkable in this case. The forces due to pressure are close to the acoustic radiation force, which is predicted to be 1681.1×10^{-5} by the theory in ref. 49, and decrease slightly with increasing viscosity. The agreement of the numerical simulation and exact solution for the force due to pressure is pretty good, and the relative errors in all cases are less than 0.3%.

10.3.3.3 Acoustic Radiation Force on a Particle Near a Wall

A solid wall will attract particles as predicted theoretically in ref. 51. This subsection exhibits the simulations for a cylinder close to a rigid wall similar to what has been presented in ref. 51. A cylinder located in a closed rectangular chamber near a wall is actuated by a horizontal standing wave as demonstrated in Figure 10.19. The incident wave is the same as the first example, and the cylinder radius $R = 20$, $L_x = 1000$ is the wavelength and the height $L_y = 500$.

The acoustic radiation forces at several values of d (distance of the cylinder center away from the wall) in horizontal and vertical directions calculated by COMSOL based on the perturbation method and the FVM algorithm are listed in Table 10.4.

Ultrasonic Microrobotics in Cavities: Devices and Numerical Simulation 237

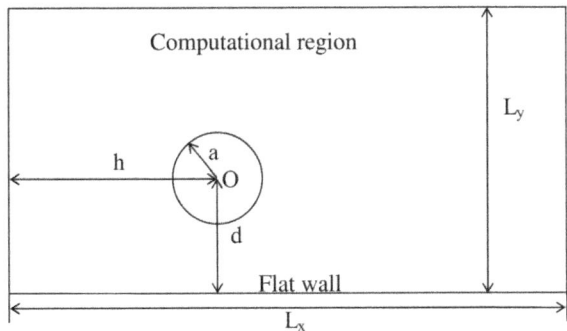

Figure 10.19 Computational geometry. Reproduced from ref. 53.

Table 10.4 Results for a cylinder near a wall.

d	$F_{x\text{COM}}$ (10^{-5})	$F_{y\text{COM}}$ (10^{-5})	$F_{x\text{FVM}}$ (10^{-5})	$F_{y\text{FVM}}$ (10^{-5})
22	−24.71	−68.06	−24.55	−67.50
25	−23.03	−28.27	−22.91	−28.12
30	−21.84	−12.38	−21.57	−12.27
40	−20.90	−4.49	−20.65	−4.45
60	−20.32	−1.37	−20.23	−1.38
100	−20.07	−0.41	−20.00	−0.42

It is seen that the results from the two methods agree with each other very well. The wall influences the forces both in the horizontal and vertical directions. The cylinder is attracted towards the wall and the attraction decays very quickly with increasing distance d. This attraction is also recognized as the secondary acoustic radiation force,[52] which leads to particle clumps in ultrasonic particle manipulators.

10.3.3.4 Acoustic Radiation Torque on Non-spherical Particles

The experiment of a non-spherical particle rotating in orthogonal standing waves has been presented in Section 10.2.2. Here, a series of two-dimensional COMSOL simulations based on the perturbation method as well as the comparable FVM calculations, corresponding to different cylinder orientations, are performed to compute the radiation torques on an elliptic cylinder. The geometric details are illustrated in Figure 10.20(left). Similar to the previous simulations, the cylinder as well as the walls are assumed to be rigid and fixed in space. The boundary conditions and the excitation are implemented in the same way as the first example. Inside the acoustic field, there is an elliptic cylinder with the short semi-axis length 10 and the long semi-axis length 20 located at $h = 250$, and $d = 500$. The height of the domain $L_y = 1000$ and the width $L_x = 1000$.

The acoustic pressure field computed by COMSOL at 100 periods and $\alpha = 45°$ is plotted in Figure 10.20(right). Figure 10.21 shows the acoustic radiation torques computed by COMSOL and FVM acting on the ellipse varying with the orientation angle α from 0 to 360° which is defined in Figure 10.20. In the COMSOL simulations, the orientation angle α has a step of 5°, the radiation torque **T** for each simulation is calculated according to eqn (10.23).

It is seen that the results of COMSOL and FVM coincide with each other very well. The acoustic radiation torque may not vanish even if a non-spherical particle is located at the pressure nodes of the incident standing wave. This implies that a non-spherical particle is able to be rotated by the acoustic radiation torque, as shown in the experiments presented in Section 10.2.2.1.

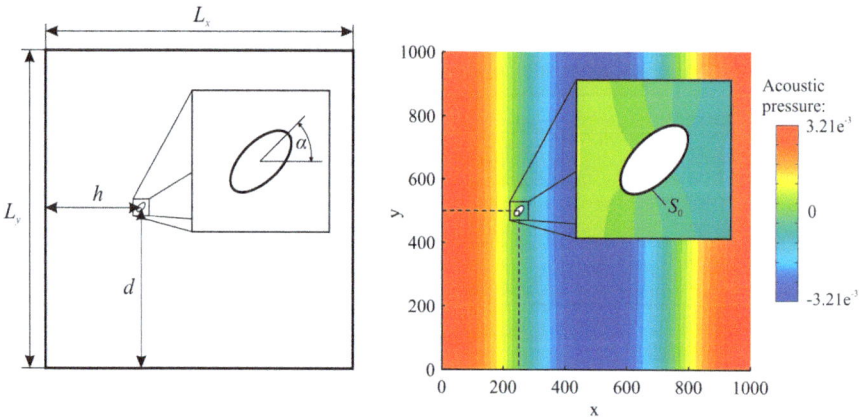

Figure 10.20 Geometry (left) and acoustic pressure distribution (right) at 100 periods and $\alpha = 45°$. The boxes show a magnified view of the particle. Reproduced from ref. 53.

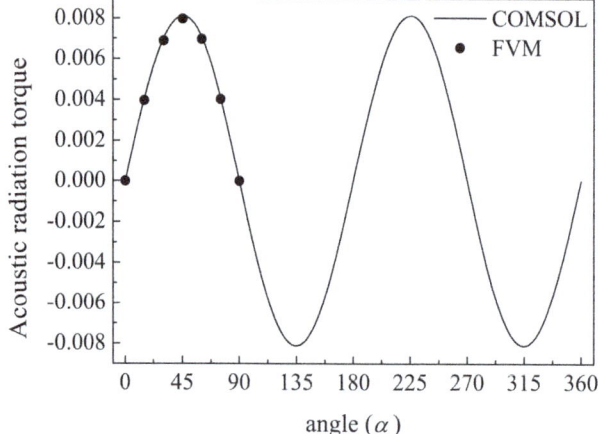

Figure 10.21 Radiation torque on an elliptic cylinder. Reproduced from ref. 53.

10.3.4 Summary of the Numerical Examples

In this section, the numerical approaches for the time-averaged effects occurring in the fluid cavity excited by ultrasound are briefly introduced. The analysis based on the perturbation method is able to obtain accurate results in the cases of weak incident waves and Newtonian fluids with low viscosity. On the other hand, the method of solving the N–S equations directly can be applied in any case, but it requires more computational capability. Both of the methods are able to achieve consistent results in the inviscid cases, which are now widely adopted in engineering. The second example implies that the viscous effects may be ignored in a standing wave and are more important in a traveling wave. A particle close to a rigid wall will be attracted, but the attraction decays quite quickly with increasing distance from the wall. The wall effects can be neglected if the particle is located at a distance from the wall that is more than five times the size of the particle. It is observed that a non-spherical particle might be driven to rotate under the actuation of the standing waves in a fluid cavity. This phenomenon was also observed in experiments (see Section 10.2.2.1).

10.4 Summary and Conclusions

An infinite variety of configurations and combinations of chambers, transducers and frequencies to excite systems containing fluids and particles exist, that can be further combined with other tools like micromechanical tweezers, dielectrophoresis (DEP), optical or magnetic manipulation. This makes ultrasonic manipulation a valuable tool for many applications in basic sciences and towards contributing to human health. The next challenges consist, for example, of quantitative assessment of the relevant forces, their effects on cells or bacteria, as well as possible robotic applications towards automated assembly of microparts. Because then the particles are necessarily of complex shapes, numerical simulations will play a crucial part in this respect.

References

1. L. P. Gor'kov, *Phys. -Dokl.*, 1962, **6**, 773–775.
2. R. Barnkob, P. Augustsson, T. Laurell and H. Bruus, *Lab Chip*, 2010, **10**, 563–570.
3. W. T. Coakley, D. W. Bardsley, M. A. Grundy, F. Zamani and D. J. Clarke, *J. Chem. Technol. Biotechnol.*, 1989, **44**, 43–62.
4. S. Radel, L. Gherardini, A. J. McLoughlin, O. Doblhoff-Dier and E. Benes, *Bioseparation*, 2000, **9**, 369–377.
5. J. Nilsson, M. Evander, B. Hammarström and T. Laurell, *Anal. Chim. Acta*, 2009, **649**, 141–157.
6. S. Oberti and J. Dual, in *Micro and Nano Techniques for the Handling of Biological Samples*, ed. J. Castillo, W. E. Svendsen and M. Dimaki, CRC Press, Taylor Francis Group, 2011.

7. J. Friend and L. Y. Yeo, *Rev. Mod. Phys.*, 2011, **83**, 647–704.
8. J. Hultström, O. Manneberg, K. Dopf, H. M. Hertz, H. Brismar and M. Wiklund, *Ultrasound Med. Biol.*, 2007, **33**, 145–151.
9. D. Bazou, *Cell Biol. Toxicol.*, 2010, **26**, 127–141.
10. M. Wiklund, *Lab Chip*, 2012, **12**, 2018–2028.
11. A. A. Doinikov, *J. Fluid Mech.*, 1994, **267**, 1–22.
12. S. D. Danilov and M. A. Mironov, *J. Acoust. Soc. Am.*, 2000, **107**, 143–153.
13. M. Settnes and H. Bruus, *Phys. Rev. E*, 2012, **85**, No. 016327.
14. T. Lilliehorn, M. Nilsson, U. Simu, S. Johansson, M. Almqvist, J. Nilsson and T. Laurell, *Sens. Actuators B*, 2005, **106**, 851–858.
15. S. Oberti, A. Neild and J. Dual, *J. Acoust. Soc. Am.*, 2007, **121**, 778–785.
16. H. Bruus, *Lab Chip*, 2012, **12**, 20–28.
17. H. Bruus, *Lab Chip*, 2012, **12**, 1014–1024.
18. B. Vanherberghen, O. Manneberg, A. Christakou, T. Frisk, M. Ohlin, H. M. Hertz, B. Onfelt and M. Wiklund, *Lab Chip*, 2010, **10**, 2727–2732.
19. I. Leibacher, W. Dietze, P. Hahn, J. Wang, S. Schmitt and J. Dual, *Microfluid. Nanofluid.*, 2014, **16**, 513–524.
20. T. Schwarz, G. Petit-Pierre and J. Dual, *J. Acoust. Soc. Am.*, 2013, **133**, 1260–1268.
21. S. Yamahira, S.-i. Hatanaka, M. Kuwabara and S. Asai, *Jpn J. Appl. Phys.*, 2000, **39**, 3683–3687.
22. F. H. Busse and T. G. Wang, *J. Acoust. Soc. Am.*, 1981, **69**, 1634–1638.
23. J. Wang and J. Dual, *J. Acoust. Soc. Am.*, 2011, **129**, 3490–3501.
24. C. P. Lee and T. G. Wang, *J. Acoust. Soc. Am.*, 1989, **85**, 1081–1088.
25. W. L. Nyborg, *J. Acoust. Soc. Am.*, 1958, **30**, 329–339.
26. A. Lamprecht, T. Schwarz, J. Wang and J. Dual, *presented in part at the International Congress on Ultrasonics, Singapore, May 2–5*, 2013.
27. A. Haake, A. Neild, G. Radziwill and J. Dual, *Biotechnol. Bioeng.*, 2005, **92**, 8–14.
28. D. Möller and J. Dual, *presented in part at the 7th USWnet meeting, Stockholm, Sweden*, 2009.
29. M. Saito, N. Kitamura and M. Terauchi, *J. Appl. Phys.*, 2002, **92**, 7581–7586.
30. G. Whitworth, M. A. Grundy and W. T. Coakley, *Ultrasonics*, 1991, **29**, 439–444.
31. O. Manneberg, B. Vanherberghen, B. Önfelt and M. Wiklund, *Lab Chip*, 2009, **9**, 833–837.
32. C. R. P. Courtney, C.-K. Ong, B. W. Drinkwater, P. D. Wilcox, C. Demore, S. Cochran, P. Glynne-Jones and M. Hill, *J. Acoust. Soc. Am.*, 2010, **128**, EL195–EL199.
33. J. Dual and D. Möller, *Lab Chip*, 2012, **12**, 506–514.
34. L. Cremer, M. Heckl and E. E. Ungar, *Structure-borne sound: structural vibrations and sound radiation at audio frequencies*, Springer-Verlag, New York, 1988.
35. D. Möller, N. Degen and J. Dual, *J. Nanobiotechnol.*, 2013, **11**, 21.

36. A. Neild, S. Oberti, F. Beyeler, J. Dual and B. J. Nelson, *J. Micromech. Microeng.*, 2006, **16**, 1562.
37. F. Beyeler, A. Neild, S. Oberti, D. J. Bell, S. Yu, J. Dual and B. J. Nelson, *J. Microelectromech. Syst.*, 2007, **16**, 7–15.
38. S. Oberti, D. Möller, A. Neild, J. Dual, F. Beyeler, B. J. Nelson and S. Gutmann, *Ultrasonics*, 2010, **50**, 247–257.
39. T. Laurell, F. Petersson and A. Nilsson, *Chem. Soc. Rev.*, 2007, **36**, 492–506.
40. S. Oberti, Doctoral thesis, ETH Zurich, 2009.
41. S. S. Sadhal, *Lab Chip*, 2012, **12**, 2292–2300.
42. L. V. King, *Proc. R. Soc. London, Ser. A*, 1934, **147**, 212–240.
43. K. Yosioka and Y. Kawasima, *Acustica*, 1955, **5**, 167–173.
44. N. Riley, *Annu. Rev. Fluid Mech.*, 1998, **33**, 43–65.
45. A. A. Doinikov, *J. Acoust. Soc. Am.*, 1997, **101**, 713–721.
46. J. Wang and J. Dual, *Presented in part at the International Congress on Ultrasonics, Gdansk, Poland, September 5–8*, 2011.
47. D. Haydock, *J. Phys. A: Math. Gen.*, 2005, **38**, 3265–3277.
48. D. Haydock and J. M. Yeomans, *J. Phys. A: Math. Gen.*, 2001, **34**, 5201–5213.
49. J. Wang and J. Dual, *J. Phys. A: Math. Theor.*, 2009, **42**, No. 285502.
50. A. Jameson and D. Mavriplis, *AIAA*, 1985, paper 85-0435.
51. J. Wang and J. Dual, *Ultrasonics*, 2012, **52**, 325–332.
52. A. Haake, Doctoral thesis, ETH Zurich, 2004.
53. J. Dual, P. Hahn, I. Leibacher, D. Möller, T. Schwarz and J. Wang, *Lab Chip*, 2012, **12**(20), 4010–4021.

CHAPTER 11

Acoustic Manipulation Combined with Other Force Fields

PETER GLYNNE-JONES AND MARTYN HILL*

Engineering Sciences, University of Southampton, Southampton, SO17 1BJ, UK
*E-mail: m.hill@soton.ac.uk

11.1 Introduction

This book has explored the theory underlying acoustic radiation forces and acoustically-induced streaming, has described experimental techniques for evaluating these effects, and has discussed a variety of applications based on both bulk acoustic waves and surface acoustic waves. It has become apparent that the acoustic phenomena employed have a potentially large scale of action, can trap and manipulate relatively large particles and agglomerates, are appropriate for biological use, and are suitable for integration with many microfluidic fabrication techniques. There are situations in which these properties make the technology attractive for use in combination with other forces, for example when its scale of action complements technologies that generate more localised force fields, or in cases when competition with force fields that scale with other physical properties allows for separation and fractionation.

Integrating acoustophoretic action with other systems within microfluidic devices can create design problems as acoustophoretic systems typically rely on acoustic resonances dependent on the boundaries of the fluid channels, and modifying these boundaries to suit other technologies can interrupt the intended action. In traditional planar resonators (Chapter 7), the transducer is usually placed under the channel, with corresponding primary radiation forces out of the plane of the substrate. The in-plane designs of Petersson et al.,[1,2] along with devices actuated by surface acoustic waves[3] coupling wedges[4] or flexural waves[5] have enabled the acoustic excitation to be placed some distance from the channel, allowing easier integration with other technologies.

For the examples considered here, the way in which the forces scale with particle size and properties is crucial for applications intended to differentiate between particles. For example, in the viscosity dominated regime found in microfluidic applications, the Stokes drag force is proportional to particle radius compared to the acoustic radiation force, which is proportional to the radius cubed. Thus by balancing, or opposing these two forces, it is possible to differentiate between particles of different sizes.[6]

With this in mind, it is instructive to consider the factors that contribute to the acoustic radiation force F_{ac} on a small sphere (that is small in comparison to a wavelength) of radius r, given by:

$$F_{ac}(r) = \frac{4\pi r^3}{3} \nabla \left(\frac{3(\rho_p - \rho_f)}{(2\rho_p + \rho_f)} E_{kin}(r) - \left(1 - \frac{\beta_p}{\beta_f}\right) E_{pot}(r) \right). \qquad (11.1)$$

The force is proportional to the volume of the particle and to the gradients of the time averaged kinetic and potential acoustic energy densities E_{kin} and E_{pot}. The kinetic energy gradient is weighted by a function of the particle and fluid densities ρ_p and ρ_f, while the potential energy gradient is weighted by the particle and fluid compressibilities, β_p and β_f.[7] Thus we have the potential to differentiate between particles based on their size, density, and compressibility (in addition to any other properties to which the second force is sensitive). In practice the effects of varying particle size tend to dominate over the other parameters, which can be problematic if, for example, one wishes to separate two populations of particles whose size distribution overlaps (e.g. separating biological cells).

The following sections describe how different force fields have been used in combination with acoustic forces in a variety of applications. Examples are given of acoustic forces combined with gravitation fields, hydrodynamic forces, electric fields (including dielectrophoresis), magnetic forces and optical forces. At the beginning of each subsection the system properties that influence the complementary force are briefly discussed. For a recent overview of magnetic, dielectrophoretic and optical forces in microfluidic systems, and a comparison with acoustic forces, the reader is referred to Tsutsui and Ho.[8]

11.2 Gravitational Forces

While levitation against the force of gravity is common in acoustic manipulation, we are concerned here with some example microfluidic systems that utilise the gravitational force in competition with acoustic forces. In equilibrium a particle in an ultrasonic force field will settle to a position where the gravitational forces (including the buoyancy) balance with the acoustic forces, *i.e.* when

$$F_{ac} = \frac{4\pi r^3}{3}(\rho_p - \rho_f)g, \quad (11.2)$$

where g is the gravitational constant. The gravitational force is dependent on the densities of the particle and the fluid and its direction of action relative to other forces can be modified by altering the orientation of the acoustofluidic device. Like the acoustic radiation force, gravitational forces are proportional to particle volume (for homogeneous particles), so the equilibrium position is independent of particle size. This size independence can be utilised to provide a method for estimating acoustic energy densities in a manipulation device as described by Spengler *et al.*[9] A test particle is placed in the field; as the acoustic field is decreased the particle equilibrium position sinks, until it reaches the turning point (at a position of one eighth of a wavelength below the pressure node, marked 'O' in Figure 11.1), where further decrease leads to the particle dropping. Since the material properties and hence buoyancy force on the particle are known, the acoustic energy density and hence pressure amplitude can be calculated. However, this experiment is prone to high measurement errors due to the difficulty in distinguishing a particle in equilibrium from one that is slowly dropping, and the uncertainty in material properties for the bead. Much more accurate measurements can be produced in lateral aspect devices using PIV based methods.[10] The general presence of the earth's gravitational field makes it a particularly simple field to use yet its

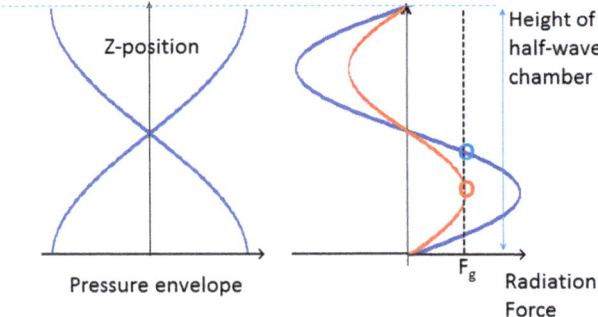

Figure 11.1 Pressure and force distribution in a 1D standing wave for a half-wave device for two different transducer voltages. For each voltage the cross or circle marks the equilibrium position of a particle with the net gravitational force, F_g. When performing a 'voltage drop' experiment, the pressure amplitude is steadily reduced until the test particle reaches the height of the position marked with 'O'. At this stage the gravitational forces are balanced by the maximum available force at any height within the fluid; any further reduction in pressure causes the particle to drop. Since this position is independent of particle size, the pressure at the 'drop voltage' can be calculated. Reproduced from ref. 40 with permission.

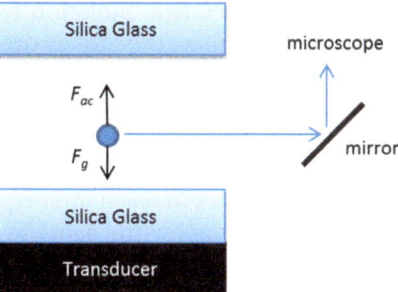

Figure 11.2 Schematic representation of the device used by Masudo and Okada[11] to investigate particle separation in a coupled gravity-acoustic field. F_{ac} and F_g represent the acoustic and net gravitational forces respectively.

relative weakness compared to the acoustic forces in microfluidic systems means that time must be allowed when equilibrium situations have to be reached.

A similar configuration is used by Masudo and Okada,[11] shown here diagrammatically in Figure 11.2, who tested the acoustic field/displacement relation in detail for a range of different particle types, finding good agreement between experimental results and theoretical predictions. In a later paper,[12] they demonstrate the possibility of sorting particles by their acoustic properties in a size independent manner using the gravity/acoustic field balance.

Kanazaki et al.[13] made use of this relation to monitor ion-exchange processes. An acoustically levitated ion-exchange bead was optically monitored in a system similar to that shown in Figure 11.2. In response to a varying chemical environment, counter-ions were replaced in the bead, which led to a change in the bead's physical properties including its density and compressibility. Thus the equilibrium position of the bead was directly related to its chemical environment. In contrast to off-line sensors, the system offers a compact and dynamically sensitive reading.

The combination of sedimentation and acoustic radiation forces underlie a commercial application of acoustic manipulation technology.[14,15] In continuously perfused bioreactors, acoustic forces are used to aggregate cells such as yeast near the outlet of the reactor. As the aggregates reach a certain critical size, they sediment and drop back into the reactor. In this way the perfusion flow can be maintained without loss of the reactor contents. Similar modes of operation can be found in a range of other work.[16,17]

11.3 Hydrodynamic Forces

Particles within a fluid flow experience a number of hydrodynamic forces. The drag force (dependent on particle size and shape, the fluid flow field, and the fluid viscosity) is discussed here as many particle sorting technologies balance the drag force against the acoustic radiation force. For a sphere of radius r in a low Reynold's number flow, the Stokes drag is.

$$F_D = 6\pi\eta rU, \tag{11.3}$$

where η is the dynamic viscosity and U is the relative velocity between the particle and the fluid. In such particle sorting techniques the active section of the devices is too short to allow particles to reach their equilibrium position in the acoustic field (which would be roughly the same for particles of similar density with only a small deviation due to gravitational forces). Typically, particles are initially aligned to a starting position within a flow chamber using a sheath flow,[1,18,19] or an acoustic pre-alignment stage,[20] then deflected under the action of the sorting acoustic field, which can be excited using bulk axial,[18] bulk lateral,[19] or surface acoustic wave excitation.[3] It can be shown that in the low Reynolds number regime typically found in this work, particles reach their terminal velocity within a matter of milli-seconds (see Chapter 1). Since the hydrodynamic drag scales with particle radius (competing with the cubic dependence of radiation force), this terminal velocity (and hence the displacement during the time it is in the active region) scales with the square of particle radius. While the acoustic force also depends upon the acoustic properties of the particle, the size dependence will often dominate. Thus when applying this technique to sorting biological cells, its use is currently restricted to separating cell types whose size distributions do not significantly overlap.[23] A similar principle of competing drag and acoustic forces can be used to separate particles by frequency switching,[24,25] in which the driving frequency of a channel switches between two resonances with a different number of pressure nodes. The time spent at each resonance is chosen so that large particles have time to move to their destination, but the small particles do not move sufficiently far before being pulled back to their original nodal plane as shown in Figure 11.3.

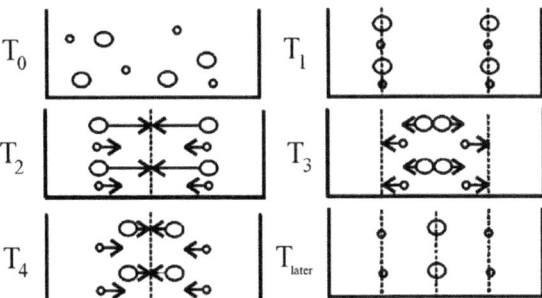

Figure 11.3 Competing drag and acoustic forces used to separate particles by frequency switching. At T_0 particles are dispersed and then (T_1) they align at the nodes of the first harmonic. The frequency is switched between the fundamental (T_2) and the first harmonic (T_3) at a rate such that only the largest particles have time to reach the region where T_3 exerts a force towards the centre, and hence reach the single node of the fundamental. Reproduced from ref. 24 with permission.

Mandralis and Feke[6] reported the incorporation of acoustic radiation forces into a field flow fractionation (FFF) scheme. In field flow fractionation,[26] a force field is used to cause suspended particles to move into differing positions within a laminar flow field. Particles in regions of higher flow are eluted more quickly than those that remain in slower regions near walls. In this case, two acoustic modes were chosen that drove particles to the centre and wall of the channel respectively. By synchronising an oscillating flow field with the switching between the two acoustic modes, they demonstrated effective separation of a range of polystyrene particles in the range 2–30 μm. Bhat and Chakraborty[27] present a theoretical analysis of acoustic flow field fractionation subject to the interaction of near-wall attractive and repulsive forces such as electric double-layer fields, and Van Der Waals forces. They predict enhanced sorting resolution under certain combinations of these fields.

Due to boundary and bulk losses, acoustic streaming flow fields are almost universally present in acoustic radiation force devices (Chapter 13). The hydrodynamic drag force resulting from this streaming often acts in competition with the radiation force. As above, this effect scales with the square of particle radius, and Bruus has shown (Chapter 1) that there is a critical particle radius below which the ultrasonic radiation forces no longer dominate and particle motion is determined primarily by the streaming fields. This phenomenon has been used by Rogers and Neild[28] to selectively trap particles through a combination of secondary radiation forces (Bjerknes forces) and streaming induced drag forces. In their system, bubbles oscillating in an acoustic field generate strong local streaming currents and attract particles through Bjerknes forces. Depending on their size and density, particles are either trapped by the Bjerknes forces or carried away from the bubble by the streaming vortex. Figure 11.4 shows how this was used to selectively trap 5 μm silica particles but not 5 μm polystyrene particles.

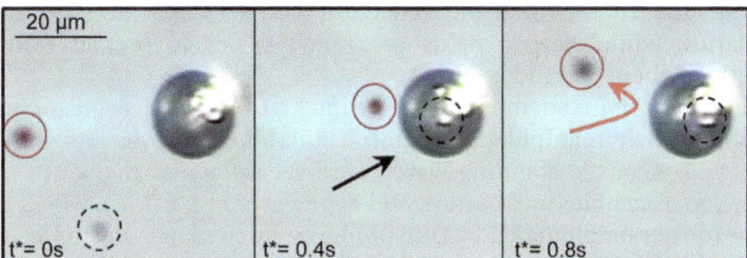

Figure 11.4 Approach used by Rogers and Neild to selectively trap particles. The grey silica particles are trapped in contact with the oscillating bubble, while the polystyrene particle (marked in red) circulates within the streaming field. Reproduced from ref. 28 with permission.

The balancing of viscous drag forces with acoustic forces can also be used to characterise the forces within a resonant acoustic field. Woodside et al.[29] estimated the distribution of acoustic energy density within a resonant field by observing the speed of polystyrene particles. They showed how the lateral variations in energy were related to the variation of amplitude across the face of the transducer boundary.

11.4 Forces Induced by Electrical Fields

An approach to particle manipulation based on an electrical field that has been widely exploited in microfluidic systems uses dielectrophoresis (DEP).[21] The force on a sphere, F_{DEP}, depends on the complex permittivities of the particle, ε_p, and the fluid medium, ε_f, at the angular frequency of operation, ω, such that[22]

$$F_{DEP} = 2\pi r^3 \varepsilon_f \text{Re}\{K(\omega)\} \nabla |E|^2, \qquad (11.4)$$

where E is the r.m.s. electric field and the factor $K(\omega)$ is given by

$$K(\omega) = \frac{\varepsilon_p - \varepsilon_f}{\varepsilon_p + 2\varepsilon_f} \qquad (11.5)$$

The field induces dipoles in neutral particles by moving the mobile charges associated with them. A gradient in the applied field puts an unequal force on the dipole and moves the particles. The process is effective under AC fields which have the advantage of avoiding electrolysis at the electrodes. DEP can be seen as complementary to ultrasonic manipulation in that DEP performs best in terms of fine control of small numbers of particles over relatively short length scales, while ultrasonic forces can easily act on a large number of particles over rather greater length scales. While DEP is typically sensitive to the conductivity of the fluid medium, with potential heating problems due to electrical currents, acoustic manipulation usually creates less heating due to the low acoustic losses found in most host fluids. Acoustic manipulation is often difficult to control precisely due to the multiple reflection of the waves at surrounding boundaries, while electric fields are much easier to predict from the geometry of their electrodes.

Wiklund et al.[30] first demonstrated the integration of both acoustophoretic and DEP particle manipulation within a microfluidic device, as shown in Figure 11.5. Acoustic standing waves were set up across the width of the channel, to assemble linear arrays and aggregates of particles, which could then be further manipulated by DEP produced by co-planar electrodes at the bottom of the channel. Various manipulation operations were demonstrated. In addition to the possibility of sorting cells through balancing forces, the longer range acoustic forces were used to pre-align populations of cells in preparation for more locally specific DEP switching/manipulation operations.

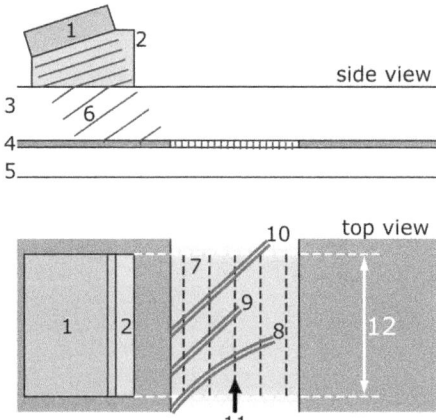

Figure 11.5 Schematic representation of the combined ultrasonic/DEP system described by Wiklund et al. The piezoceramic transducer (1) is coupled into the structure of the chip (3–5) through a PMMA block (2). The refracted ultrasonic field (6) excites the fluid channel adjacent (12) to the transducer and particles brought into the region by bulk fluid flow (11) are driven to the lateral pressure nodes (7). Hence the particles are pre-aligned before crossing the DEP electrodes (8–10) where combined DEP/ultrasonic manipulation can be implemented. Reproduced from ref. 30 with permission.

Ravula et al.[31] describe a layout that utilised the levitation forces generated by ultrasound in a planar resonator coupled to either glass or silicon-substrate DEP chips. Particles are pre-concentrated and aligned by the acoustic radiation forces to increase throughput and reduce the positional variability of particles prior to fine adjustments by the DEP electrodes. The authors note that this pre-alignment also reduces the tendency of particles to clump, or to stick to channel walls.

Wiklund et al.[32] also combined acoustic radiation forces with a capillary electrophoresis (CE) system, to enhance the detection of proteins. The system is designed so that acoustic radiation forces are parallel to the direction of flow, such that viscous drag forces are balanced against the acoustic ones. The electro-osmotic flow (EOF) creates a flow profile that is uniform across the width of the capillary, which ensures that the particle separation is not dependent on lateral particle position. Polystyrene micro-beads functionalised with antibodies indicate the presence of the target protein by forming bead–protein–bead complexes. At the chosen flow rate these complexes tend to be trapped in the acoustic field, while the single beads are carried through by the viscous drag. A CCD detection stage counts the trapped fluorescent beads.

Neale et al.[33] describe a combination of DEP and acoustophoresis in which DEP electrodes within a channel immobilise a population of beads into a line that can be focussed further by the acoustic radiation forces into an aggregate. The strength of this combination is in utilising the DEP forces to hold particles against the flow, which although possible acoustically is not easy to

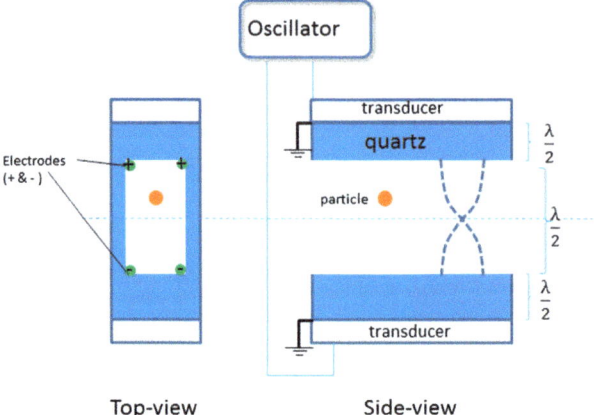

Figure 11.6 Combined electrostatic and acoustic forces (after Yasuda et al.[34]). Electrostatic forces generated by four electrodes are balanced against acoustic radiation forces arising from a standing wave field, permitting accurate measurement of both types of force. Reproduced from ref. 40 with permission.

achieve or control, while simultaneously using the longer range nature of the acoustic forces to bring the particles to a single focussed location. They also remark on the use of the two alternative forces to effect size dependent sorting. Alternatively, they suggest a 2D sorting approach that fractionates particles into spatial locations that indicate the response of particles to forces from both technologies.

Yasuda et al.[34] demonstrated a system that allowed electrophoretic forces on particles in water to compete with acoustophoretic ones as shown in Figure 11.6. Electrodes within a resonant chamber created an electric field of order 10 V mm^{-1}. In the absence of acoustic excitation the effective charge on polystyrene micro-beads was calculated by observing their movement in a low frequency alternating electric field. By then activating the acoustic field, and measuring the position of the beads relative to the pressure node, the acoustic radiation force was determined (in a manner analogous to that described Figure 11.1, but in this case the electric field replaces the gravitational one), permitting verification of analytical expressions for radiation force.

11.5 Magnetic Forces

Magnetic forces, which scale with the difference in susceptibility between the particle, χ_p, and the fluid, χ_f, have been mentioned in the context of enhanced agglutination (Chapter 17) and the force on a sphere in a field of flux density B is[35]

$$F_{mag} = 2\pi r^3 \frac{(\chi_p - \chi_f)}{3\mu_0} \nabla |B|^2 \tag{11.6}$$

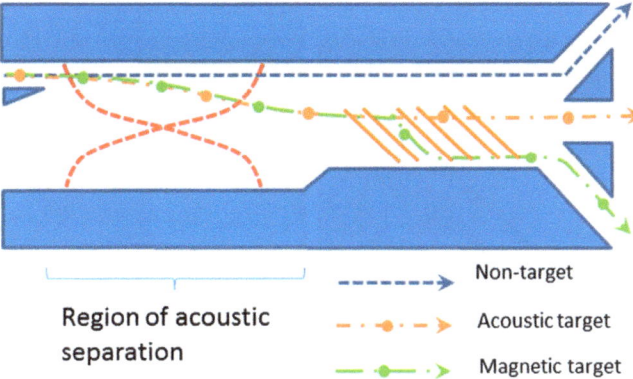

Figure 11.7 Integrated acoustic and magnetic separation, after Adams et al.[36] In two successive stages, acoustic then magnetic forces displace particles from their initial position to create three sorted outlets. The red dotted lines indicate the acoustic standing wave field, and orange diagonal elements represent a magnetic comb. Reproduced from ref. 40 with permission.

where μ_0 is the magnetic permeability of free space. Adams et al.[36] demonstrate a microfluidic sorting system that comprises an acoustic sorting stage followed serially by magnetic sorting that can operate at rates of up to 10^8 particles per hour. As shown in Figure 11.7, the input sample is thus separated into three streams: those having low response to both magnetic and acoustic forces, (b) those that respond to acoustic forces only, and (c) those that respond to both magnetic forces and acoustic forces. This could, for example, correspond to separating a sample of cells into streams that correspond to waste material, un-labelled cells, and cells labelled with magnetic beads.

In a device described by Hill et al.,[37] a combination of magnetic, gravitational, and acoustic forces are used to facilitate stages in the sample preparation and concentration of tuberculosis bacteria. Figure 11.8 shows the device. Initially, the bacteria from a sputum sample are labelled with ferrous beads (TB-Beads, supplied by Microsens Biotechnologies), and then a resonant acoustic field is used to create numerous small aggregates of the labelled bacteria–bead complexes; the combination of both acoustic and magnetic forces helps agglomerates to form more rapidly than would occur with magnetism alone. When the acoustic field is switched off, the aggregates quickly sediment to the bottom, allowing the supernatant to be washed away. When the acoustic field is reactivated, the aggregates are re-suspended, and further washing cycles can take place. Finally, with the acoustic field active, the aggregates can be eluted with a minimum of loss since they are suspended in the centre of the chamber where the flow profile has maximum velocity. The advantage of the acoustic re-suspension over a comparable magnetic stage is that a magnetic process would usually leave the aggregates

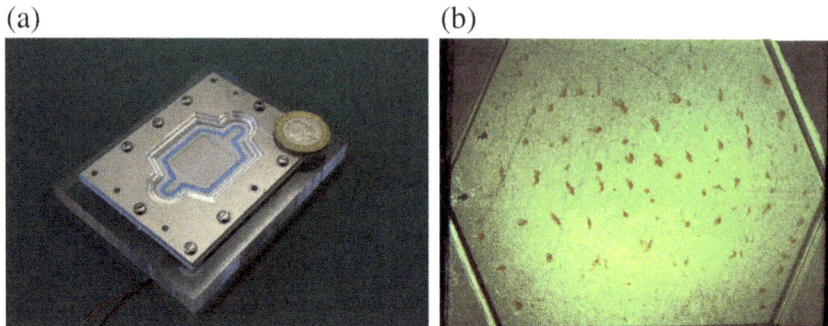

Figure 11.8 (a) Triple force unit prototype, utilising magnetic, acoustic and gravitational forces for sample preparation in a TB detection system. Constructed in Macor, glass and aluminium. (b) Acoustically enhanced aggregation of magnetic beads seen in a disposable chamber created entirely from polymeric materials. Reproduced from ref. 40 with permission.

on a surface, which would cause the aggregate to become disrupted by shearing forces as flow was actuated to remove it.

11.6 Optical Forces

Optical traps based on focused laser beams represent another well-established technique for manipulating microparticles and generate a force dependent on the refractive indices of the particle and the fluid medium. When handling particles that are also amenable to acoustic manipulation, optical traps will be operating in the Mei regime, where the particle cannot be assumed to be much smaller than a wavelength. However, optical gradient forces for small particles in the Rayleigh scattering regime have a similar form to the other gradient forces discussed here. Optical traps, with a wavelength of the order of 1 μm, are able to manipulate single small particles with a very high degree of precision, while acoustic radiation forces are able to handle a large number of particles simultaneously and can levitate relatively large, dense particles efficiently. Thalhammer *et al.*[38] use a combination of acoustical and optical techniques to undertake a variety of manipulation tasks. The ultrasonic field is set up in a square capillary that is held between a glass slide and an optical mirror backed with a PZT transducer, as shown schematically in Figure 11.9. The mirror allows a laser source, placed below this setup, to act like a dual beam trap,[39] and allows optical manipulation from a large working distance. In this trap, radial optical trapping is due to the lateral gradient force, while axially it is the balancing of the scattering forces. Motile micro-organisms within the fluid in the capillary can be confined within the nodal plane of a half wave acoustic resonance. Once the micro-organisms are confined within a plane, an individual organism can be selected, optically trapped, and then moved across the width of the capillary.

Acoustic Manipulation Combined with Other Force Fields 253

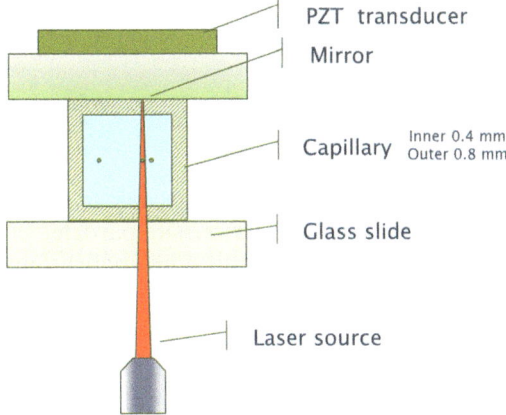

Figure 11.9 Schematic representation of the setup used by Thalhammer *et al.*[38] to demonstrate simultaneous acoustic and optical manipulation. A glass capillary is placed between a glass slide and a mirror, on top of which sits a PZT transducer that excites an ultrasonic standing wave in the layered structure. The clear optical path allows long working distance optical tweezers to further manipulate particles levitated within the ultrasonic field. Reproduced from ref. 40 with permission.

A further advantage of the combined force fields is that relatively large particles, including a 75 μm polystyrene bead and a starch grain, can be manipulated. The force required to levitate the starch grain is of the order of 1 nN, which can be achieved comfortably with the acoustic field for a large, high acoustic contrast factor particle. This then allows a relatively low power optical trap to undertake lateral manipulation of the large particles. Finally, Thalhammer *et al.* demonstrate the use of optical tweezers to measure the trapping forces of the acoustic field at different ultrasonic excitation frequencies.

11.7 Conclusions

As the technologies of acoustofluidics (the combination of microfluidics with ultrasonic particle manipulation) mature, there is increasing scope for integrating other force field technologies with it. As with other approaches, acoustofluidic manipulation has operations that it can perform well and other operations that benefit from complementary force systems. Acoustofluidics is particularly good for:

- The ability to manipulate large numbers of particles simultaneously over a wide range of sizes and types.
- The relatively low complexity of the actuation schemes.
- The ability to continuously manipulate cells and maintain viability.

- The ability to create multi node traps with simple geometric trapping patterns, including points, lines, planes, and cylinders of trapped particles.
- Creating forces that discriminate on volume, density and compressibility.

Combining these strengths with technologies able to apply forces to single particles among a population, or to discriminate particles on other physical bases, will lead to a richer range of microfluidic applications.

References

1. F. Petersson, L. Åberg, A.-M. Swärd-Nilsson and T. Laurell, *Anal. Chem.*, 2007, **79**, 5117–5123.
2. F. Petersson, A. Nilsson, C. Holm, H. Jonsson and T. Laurell, *Lab Chip*, 2005, **5**, 20–22.
3. J. J. Shi, H. Huang, Z. Stratton, Y. P. Huang and T. J. Huang, *Lab Chip*, 2009, **9**, 3354–3359.
4. O. Manneberg, J. Svennebring, H. M. Hertz and M. Wiklund, *J. Micromech. Microeng.*, 2008, 18.
5. S. Oberti, A. Neild and J. Dual, *J. Acoust. Soc. Am.*, 2007, **121**, 778–785.
6. Z. I. Mandralis and D. L. Feke, *AIChE J.*, 1993, **39**, 197–206.
7. L. P. Gor'kov, *Phys. Dokl.*, 1962, **6**, 773–775.
8. H. Tsutsui and C.-M. Ho, *Mech. Res. Commun.*, 2009, **36**, 92–103.
9. J. F. Spengler, M. Jekel, K. T. Christensen, R. J. Adrian, J. J. Hawkes and W. T. Coakley, *Bioseparation*, 2000, **9**, 329–341.
10. P. Augustsson, R. Barnkob, S. T. Wereley, H. Bruus and T. Laurell, *Lab Chip*, 2011, **11**, 4152–4164.
11. T. Masudo and T. Okada, *Anal. Chem.*, 2001, **73**, 3467–3471.
12. T. Masudo and T. Okada, *Anal. Sci.*, 2007, **23**, 385–387.
13. T. Kanazaki, S. Hirawa, M. Harada and T. Okada, *Anal. Chem.*, 2010, **82**, 4472–4478.
14. P. W. S. Pui, F. Trampler, S. A. Sonderhoff, M. Groeschl, D. G. Kilburn and J. M. Piret, *Biotechnol. Prog.*, 1995, **11**, 146–152.
15. *United States Pat.* 1997, **5**, 626–767.
16. J. J. Hawkes, M. S. Limaye and W. T. Coakley, *J. Appl. Microbiol.*, 1997, **82**, 39–47.
17. M. S. Limaye and W. T. Coakley, *J. Appl. Microbiol.*, 1998, **84**, 1035–1042.
18. M. Kumar, D. L. Feke and J. M. Belovich, *Biotechnol. Bioeng.*, 2005, **89**, 129–137.
19. J. D. Adams and H. T. Soh, *Appl. Phys. Lett.*, 2010, 97.
20. P. Augustsson, C. Magnusson, M. Nordin, H. Lilja and T. Laurell, *Anal. Chem.*, 2012, **84**, 7954–7962.
21. A. Castellanos, A. Ramos, A. Gonzalez, N. G. Green and H. Morgan, *J. Phys. D: Appl. Phys.*, 2003, **36**, 2584–2597.

22. T. B. Jones, *Electromechanics of Particles*, Cambridge University Press, Cambridge, 1995.
23. P. Augustsson, R. Barnkob, C. Grenvall, T. Deierborg, P. Brundin, H. Bruus and T. Laurell, Measuring the acoustophorectic contrast factor of living cells in microchannels, in *Proceedings of 14th International Conference on Miniaturized Systems for Chemistry and Life Sciences,* 2010, pp. 1337–1339.
24. T. Laurell, F. Petersson and A. Nilsson, *Chem. Soc. Rev.,* 2007, **36,** 492–506.
25. Y. Liu and K.-M. Lim, *Lab Chip,* 2011, **11,** 3167–3173.
26. J. C. Giddings, K. A. Graff, K. D. Caldwell and M. N. Myers, in *Polymer Characterization*, American Chemical Society, 1983, vol. 203, ch. 14, pp. 257–269.
27. B. Bhat and S. Chakraborty, *Langmuir,* 2010, **26,** 15035–15043.
28. P. Rogers and A. Neild, *Lab Chip,* 2011, **11,** 3710–3715.
29. S. M. Woodside, J. M. Piret, M. Groschl, E. Benes and B. D. Bowen, *AIChE J.,* 1998, **44,** 1976–1984.
30. M. Wiklund, C. Günther, R. Lemor, M. Jäger, G. Fuhr and H. M. Hertz, *Lab Chip,* 2006, **6,** 1537–1544.
31. S. K. Ravula, D. W. Branch, C. D. James, R. J. Townsend, M. Hill, G. Kaduchak, M. Ward and I. Brener, *Sens. Actuators B,* 2008, **130,** 645–652.
32. M. Wiklund, P. Spegel, S. Nilsson and H. M. Hertz, *Ultrasonics,* 2003, **41,** 329–333.
33. S. L. Neale, C. Witte, Y. Bourquin, C. Kremer, A. Menachery, Y. Zhang, R. Wilson, J. Reboud and J. M. Cooper, *presented in part at the SPIE 8251, Microfluidics, BioMEMS, and Medical Microsystems X,* 2012.
34. K. Yasuda, S. Umemura and K. Takeda, *J. Acoust. Soc. Am.,* 1996, **99,** 1965–1970.
35. M. Zborowski, L. P. Sun, L. R. Moore, P. S. Williams and J. J. Chalmers, *J. Magn. Magn. Mater.,* 1999, **194,** 224–230.
36. J. D. Adams, P. Thevoz, H. Bruus and H. T. Soh, *Appl. Phys. Lett.,* 2009, **95,** 254103.
37. M. Hill, P. Glynne-Jones, N. R. Harris, R. J. Boltryk, C. Stanley and D. Bond, *Presented in part at the Optical Trapping and Optical Micromanipulation, San Diego, 1–5 August,* 2010.
38. G. Thalhammer, R. Steiger, M. Meinschad, M. Hill, S. Bernet and M. Ritsch-Marte, *Biomed. Opt. Express,* 2011, **2,** 2859–2870.
39. M. Pitzek, R. Steiger, G. Thalhammer, S. Bernet and M. Ritsch-Marte, *Opt. Express,* 2009, **17,** 19414–19423.
40. P. Glynne-Jones and M. Hill, *Lab Chip,* 2013, **13,** 1003–1010.

CHAPTER 12

Analysis of Acoustic Streaming by Perturbation Methods

SATWINDAR SINGH SADHAL

Department of Aerospace and Mechanical Engineering, University of Southern California, Los Angeles, CA 90089-1453, USA
E-mail: sadhal@usc.edu

12.1 Introduction

Streaming is the phenomenon corresponding to net flow in a fluid medium that is otherwise dominated by oscillatory flow. It is therefore a common occurrence with sound waves, and researchers have classified it into two categories. One happens because of the spatial attenuation of waves in free space, *e.g.*, an attenuating beam of plane traveling wave. This type of streaming is usually associated with a high Reynolds number flow, and is referred to as the 'quartz wind' or Eckart streaming. The physical mechanism for this type of streaming is described very eloquently by Wiklund, Green and Ohlin[1] as well as by Lee and Wang.[2] Briefly, the attenuation causes the pressure and velocity amplitudes to decrease in the direction of the propagation of the wave. On the average (over time), this results in a net force in the same direction, when considering the nonlinear effects, and therefore a net flow. The second mechanism arises from the friction between the fluid medium and a solid wall when the former is vibrating in contact with the latter, *e.g.*, a wave traveling down a wave-guide, a standing wave in a resonant chamber, or a wave scattering off a solid object. It has also been shown that this type of streaming happens due to a sound wave interacting with a fluid–fluid interface, *e.g.*, with liquid drops in a gas, or gas bubbles in a liquid.

Unlike the spatial attenuation mentioned earlier, this effect is largely confined to a thin viscous boundary layer (called the shear-wave layer or the Stokes layer) of thickness $\delta \simeq (\nu/\omega)^{1/2}$ on the surface. Here, ν is the kinematic viscosity and ω is the angular frequency related to the rotational frequency f as $\omega = 2\pi f$. This layer also represents a significant dissipation mechanism, and provides a strong force in driving acoustic streaming. While the medium outside the layer vibrates irrotationally as in a sound field, the one within the layer is forced to vibrate rotationally (*i.e.*, with vorticity) because its motion has to conform to the no-slip condition on the wall. Most of the discussion in this chapter will be on the second mechanism of streaming.

We shall examine various situations in which we can realize streaming, and formulate mathematically the relevant fluid mechanics for understanding and analyzing the associated flow fields. In particular, we take the regular and singular perturbation approach when a small or a large parameter can be identified. Usually, for high-frequency oscillations, we can consider a frequency-based Reynolds number in which the velocity may be considered to be the product of the frequency and a length scale. This would give a large frequency parameter typically denoted by $|M|^2 = \omega a^2/\nu \gg 1$, where a is a length scale. The square root $|M|$ of the frequency parameter is also known as the Womersley number with the notation α. A small parameter for the vibrating systems under consideration can be the ratio of small velocity amplitude in relation to a frequency-based velocity, *i.e.*, $\varepsilon = U_0/\omega a$. We first develop the fundamental equations based on the classical irrotational-flow theory, and follow up with the nonlinear viscous fluid dynamics.

In the absence of solid or liquid boundaries, simple sound waves in a gaseous medium usually have an irrotational character. Interaction with such boundaries generates vorticity whereby nonlinearities set in within a thin layer (called the Stokes layer or the shear-wave layer) at the boundary. As discussed in ref. 3 and 4, nonlinearities can lead to a net steady flow component that we refer to as streaming. This steady streaming persists outside the Stokes layer with vorticity. In order to better understand the streaming phenomenon, let us consider a body of typical dimension a that oscillates with velocity $U_0 \cos(\omega t)$ and angular frequency ω in a viscous fluid. Of course, for the purpose of mathematical analysis, it is often convenient to fix the body to the coordinate system, and let the far-field fluid oscillate. If in this case $\varepsilon = U_0/\omega a \ll 1$, then, although the leading order solution is purely oscillatory, higher order terms include not only higher harmonics but also steady contributions to the velocity. This can be mathematically explained by the existence of the nonlinear terms which may have steady nonzero components. For example, $\cos \omega t \cos \omega t = \cos^2 \omega t = \frac{1}{2}(1 + \cos 2\omega t)$ has 1/2 as a steady component (see also discussions by Bruus[3,5]). Physically, the condition $\varepsilon \ll 1$ implies that the amplitude of the oscillation is small compared with a.

The existence of steady streaming has been pointed out by Rayleigh[6] as early as 1883 and 1896 in his work on Kundt's dust tube and was later studied in a boundary layer context by Schlichting[7] who considered flows with the additional constraint $|M|^2 = \omega a^2/\nu \gg 1$. For such a flow it is now well established that the first order fluctuation vorticity is confined to the Stokes layer region of thickness $O(\nu/\omega)^{1/2}$. We have depicted a schematic of the different flows in Figure 12.1 where on the left we show the unsteady oscillatory flow and on the right we have steady streaming. In reality these flows overlap each other.

The flow field in this thin layer, being of nonlinear character, has a nonzero steady component of $O(\varepsilon)$ which comes about from the Reynolds stresses. This steady flow carries through to the edge of the Stokes layer, and drives the steady streaming of the same order in the bulk of the fluid. Therefore, there is steady flow both in the Stokes layer and the bulk of the fluid. The edge of the Stokes layer can be considered to be the transition point from rotational viscous flow to irrotational flow for the leading order, and also serves to define the inner and outer streaming flows. The outer streaming flow field can be effectively analyzed using Nyborg's method,[8] in which the nonlinear leading-order inertial terms appear in the $O(\varepsilon)$ equation. These can be lumped in the form of a conservative force that drives the outer streaming. In essence, under conditions for $|M| \gg 1$, the streaming velocity distribution at the edge of the thin Stokes layer can be used as a slip velocity on the solid surface for the outer steady streaming flow field. This slip velocity may be calculated without the detailed Stokes layer solutions. The technique has been extended by Lee and Wang[2] to allow for compressibility and employed for the outer streaming calculations for a standing wave between parallel plates, as well as the cases of the cylinder and sphere displaced from the velocity antinode. More recently, Rednikov and Sadhal[9] further extended the idea with allowance for non-adiabatic effects, together

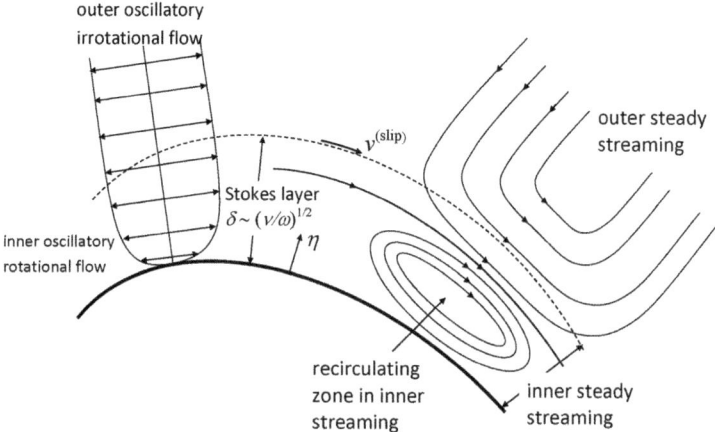

Figure 12.1 A schematic of the various flow characterizations. The region between the heavy line (solid boundary) and the broken line is the Stokes layer, which is the inner region with nonzero vorticity for the leading-order unsteady flow. The inner region contains a steady recirculating zone.

with several examples where corrections to the classical slip-velocity solutions have been provided.

12.1.1 Oscillatory Flows

For sound waves, the basic equations of fluid mechanics are applicable. While our focus is nonlinear acoustics, we shall start with the development of the linear theory. In the general tensor form for a compressible fluid, we have (see, *e.g.* Landau and Lifshitz,[10] Bruus[11]):

CONTINUITY:

$$\frac{\partial \rho}{\partial t} + \frac{\partial}{\partial x_i}(\rho u_i) = 0 \quad \text{or} \quad \frac{\partial \rho}{\partial t} + \nabla \cdot (\rho \boldsymbol{u}) = 0 \tag{12.1}$$

MOMENTUM:

$$\rho\left(\frac{\partial u_i}{\partial t} + u_k \frac{\partial u_i}{\partial x_k}\right) = -\frac{\partial p}{\partial x_i}$$
$$+ \frac{\partial}{\partial x_i}\left[\mu\left(\frac{\partial u_i}{\partial x_k} + \frac{\partial u_k}{\partial x_i} - \frac{2}{3}\delta_{ik}\frac{\partial u_m}{\partial x_m}\right)\right] \tag{12.2}$$
$$+ \frac{\partial}{\partial x_i}\left(\beta' \frac{\partial u_m}{\partial x_m}\right),$$

For most practical situations, the viscosities, μ and β', may be treated as constant, resulting in the following vector form of the momentum equation,

$$\rho\left(\frac{\partial \boldsymbol{u}}{\partial t} + \boldsymbol{u} \cdot \nabla \boldsymbol{u}\right) = -\nabla p + \mu \nabla^2 \boldsymbol{u}$$
$$+ \left(\beta' + \frac{1}{3}\mu\right)\nabla(\nabla \cdot \boldsymbol{u}). \tag{12.3}$$

It should be noted that some authors (*e.g.* ref. 11 and 12) choose to express $\mu\beta = \beta' + \frac{1}{3}\mu$. Sound waves consist of compression and rarefaction of a compressible fluid, which can be characterized by oscillatory motion of small amplitude. Following the standard procedure for acoustics, we begin with the linearized form of the above equations by considering small acoustic disturbances to the pressure and density,[10,13] *i.e.*,

$$p = p_0 + p' \quad \text{and} \quad \rho = \rho_0 + \rho', \tag{12.4}$$
$$\text{with} \quad \rho' \ll \rho_0 \quad \text{and} \quad p' \ll p_0.$$

Similarly, the velocity is taken to be of the form

$$\boldsymbol{u} = \boldsymbol{u}_0 + \boldsymbol{u}' \tag{12.5}$$

but since the undisturbed state here is a quiescent fluid, $\boldsymbol{u}_0 = 0$, the disturbed state velocity $\boldsymbol{u}'$ is the only one that needs to be considered. In addition, $\boldsymbol{u}'$ is considered to be small whereby the inertial effects (the term $\boldsymbol{u} \cdot \nabla \boldsymbol{u}$) can be neglected. This linearization needs to be carefully evaluated in regions of large changes in the velocity (such as a solid boundary) and under many such circumstances, we will need to include this nonlinear term in the analysis. Thus, ignoring any viscous effects, eqn (12.1) and (12.3) take the form,

$$\frac{\partial \rho'}{\partial t} + \rho_0 \nabla \cdot \boldsymbol{u}' = 0, \qquad (12.6)$$

and

$$\rho_0 \frac{\partial \boldsymbol{u}'}{\partial t} = -\nabla p, \qquad (12.7)$$

respectively. Considering the flow to be irrotational, *i.e.*, allowing the velocity to be described by a potential,

$$\boldsymbol{u}' = \nabla \phi, \qquad (12.8)$$

eqn (12.7) becomes

$$\nabla \left(p' + \rho_0 \frac{\partial \phi}{\partial t} \right) = 0, \qquad (12.9)$$

which may be integrated to give

$$p' = -\rho_0 \frac{\partial \phi}{\partial t} \qquad (12.10)$$

We now assume adiabatic compression and rarefactions, and a linear relationship between pressure and density,

$$p' = \left(\frac{\partial p}{\partial \rho} \right)_s \rho' = c^2 \rho' \qquad (12.11)$$

where the subscript *s* corresponds to isentropic changes, and *c* is the speed of sound,

$$c = \sqrt{\left(\frac{\partial p}{\partial \rho} \right)_s}. \qquad (12.12)$$

Next, using eqn (12.8), (12.10) and (12.11) in the continuity eqn (12.6), we obtain the wave equation,

$$\frac{1}{c^2} \frac{\partial^2 \phi}{\partial t^2} = \nabla^2 \phi. \qquad (12.13)$$

This is of course the linearized version which is applicable in many practical circumstances.

One familiar solution to the wave equation is the plane wave (one-dimensional)

$$\phi(x, t) = A e^{i(kx - \omega t)} \qquad (12.14)$$

Analysis of Acoustic Streaming by Perturbation Methods

where $k = \omega/c = 2\pi/\lambda$ is known as the wavenumber, ω is already defined as the wave angular frequency, λ is the wavelength, and $i = \sqrt{-1}$. This expression represents a travelling wave, *i.e.*, a wave travelling with a specific velocity. Admitting other possible forms of solutions and superimposing several solutions with various frequencies, we may write

$$\phi(x,t) = \sum_{n=0}^{\infty} a_n e^{i\omega_n(x/c-t)} + b_n e^{-i\omega_n(x/c-t)}$$
$$+ c_n e^{i\omega_n(x/c+t)} + d_n e^{-i\omega_n(x/c+t)}, \quad (12.15)$$

or in real variables, we have

$$\phi(x,t) = \sum_{n=0}^{\infty} \left[A_n^* \sin(\omega_n x/c) + B_n^* \cos(\omega_n x/c) \right]$$
$$\times \left[C_n^* \sin(\omega_n t) + D_n^* \cos(\omega_n t) \right] \quad (12.16)$$

By introducing parameters α_n and β_n, it is not difficult to see that

$$\phi(x,t) = \sum_{n=0}^{\infty} A_n \cos(\omega_n x/c + \alpha_n) \cos(\omega_n t + \beta_n) \quad (12.17)$$

Each term in the summation represents a standing wave of the given frequency,

$$\phi(x,t) = A\cos(\omega x/c + \alpha)\cos(\omega t + \beta), \quad (12.18)$$

where we have dropped the index n. Here we refer to A as the amplitude of the wave. For a single-frequency wave, the phase-differences α and β can be dropped by just choosing appropriate coordinate reference frames so that

$$\phi(x,t) = A\cos(\omega x/c)\cos(\omega t). \quad (12.19)$$

The velocity now is

$$u(x,t) = \frac{\partial \phi(x,t)}{\partial x} = -\frac{A\omega}{c}\sin(\omega x/c)\cos(\omega t), \quad (12.20)$$

and the pressure is

$$p'(x,t) = -\rho_0 \frac{\partial \phi(x,t)}{\partial t} = \rho_0 A\omega \cos(\omega x/c)\sin(\omega t). \quad (12.21)$$

The velocity takes on zero values at positions $\omega x/c = n\pi$, or $x = n\pi c/\omega$. These points are called the velocity nodes. The velocity has maximum magnitude at $x = (n + \frac{1}{2})\pi c/\omega$. At these points, known as the velocity antinodes, the pressure is zero. These nodes and antinodes are shown in Figure 12.2.

This is an example of a simple standing wave in which the flow field is both inviscid and irrotational. However, as mentioned earlier, there are two types of situations in which nonlinearities can set in. One nonlinearity is due to the

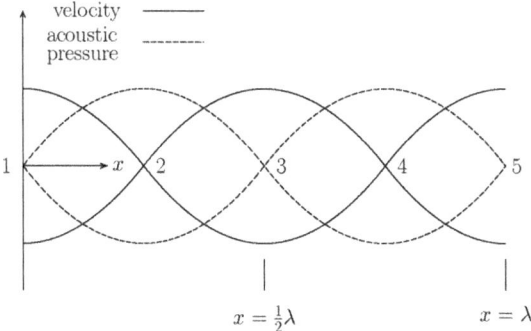

Figure 12.2 Node and antinode identification. The maximum and minimum velocity and acoustic pressure distributions in a standing wave are depicted here. Points 1, 3 and 5 are the velocity antinodes and the pressure nodes; points 2 and 4 are the velocity nodes and pressure antinodes.

presence of solid boundaries which can change this (inviscid and irrotational) characteristic by causing sufficiently large velocity gradients so that viscous forces are significant, and at the same time vorticity $\zeta = \nabla \times u$ is generated at the surface. Mathematically, the term $u \cdot \nabla u$ in eqn (12.3) becomes relevant.

Among the simplest examples to illustrate this concept is Rayleigh streaming which is mentioned later. Another type of nonlinearity comes about without any boundaries when an ultra-high-frequency beam penetrates the half-space $x > 0$. This leads to what is known as the quartz wind which is discussed later in Section 12.3.1.

The solid-boundary effect can be visually observed in Figure 12.3 of the flow around an oscillating cylinder. Here, one can see recirculating zones near the cylinder walls, as well as the outer streaming. One of the earliest examples of this type of streaming is Rayleigh's problem[6] in which standing sound waves between two walls a distance $2a$ apart were considered. Away from the walls, the velocity field is described by the inviscid, irrotational flow wave eqn (12.13). Assuming only an x-component of velocity for the inviscid field, the solution in non-dimensional form is

$$\begin{aligned} \boldsymbol{u} &= [\sin(akx)\cos t]\widehat{\boldsymbol{e}}_x, \\ &= [\sin(2\pi ax/\lambda)\cos t]\widehat{\boldsymbol{e}}_x \end{aligned} \quad (12.22)$$

which, of course, has zero time-average velocity. For this case, there is a streaming velocity at the edge of the Stokes layer given by

$$u^{(\text{slip})} = -\frac{3}{8} ak\sin(2akx), \quad (12.23)$$

As discussed later in Section 12.3.2, detailed calculations yield a higher-order steady streaming velocity field with both x and y components in the form,

Analysis of Acoustic Streaming by Perturbation Methods

Figure 12.3 Secondary streaming induced by an oscillating cylinder in a water-glycerine mixture for $|M|^2 = 70$ and the streamlines are directed towards the cylinder along the axis of oscillation indicated by arrows. Reproduced from ref. 14.

$$\boldsymbol{u}_1^{(s)} = \left[\frac{3}{16}ak(1 - 3y^2)\sin(2akx)\right]\widehat{\boldsymbol{e}}_x$$
$$-\left[\frac{3}{8}a^2k^2(y - y^3)\cos(2akx)\right]\widehat{\boldsymbol{e}}_y \quad (12.24)$$

or equivalently, as a stream function,

$$\psi_1^{(s)}(x,y) = \frac{3}{16}(y - y^3)\sin(2akx), \quad (12.25)$$

where the superscript (s) refers to the steady streaming, *i.e.*, a mean DC component. For proper scaling, these higher-order expressions (12.24) and (12.25) should be multiplied by the small parameter $\varepsilon = U_0/\omega a$. The streamlines are exhibited (schematically) in Figure 12.4 for half the vertical region, below the plane of symmetry. The coordinate y is scaled with half the vertical dimension.

12.1.2 Incompressible Flow Approximation

Several types of flows involving acoustic streaming can be formulated with the incompressible flow approximation whereby $\boldsymbol{\nabla} \cdot \boldsymbol{u} = 0$, and the last term in eqn (12.3) may be dropped. While this is generally applicable in the low Mach number range, there can be circumstances when compressibility needs to considered even otherwise (such as in the case of streaming around a particle at the velocity node[15]).

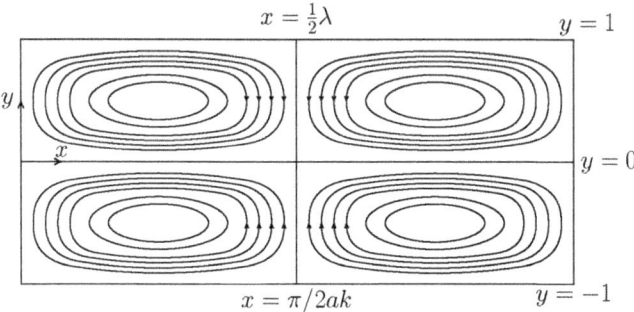

Figure 12.4 A schematic of the streaming flow streamlines in Rayleigh's third problem.[6] The recirculating region in the Stokes layer near $y = \pm 1$ is not shown.

Following the reviews by Riley,[16,17] we write eqn (12.3) for an incompressible fluid in the form.

$$\frac{\partial \boldsymbol{u}'}{\partial t} - \boldsymbol{u}' \times \boldsymbol{\zeta}' = -\frac{1}{\rho}\boldsymbol{\nabla}'\left(p + \frac{1}{2}\rho \boldsymbol{u}' \cdot \boldsymbol{u}'\right) + \boldsymbol{F}' + \nu \nabla^2 \boldsymbol{u}', \qquad (12.26)$$

where $\boldsymbol{\zeta}' = \boldsymbol{\nabla}' \times \boldsymbol{u}'$ is the vorticity. The primes are introduced to denote dimensional quantities, whereas property values ρ, p and ν are understood to be dimensional. In addition, we are using the kinematic viscosity $\nu = \mu/\rho$, and we have introduced $\boldsymbol{F}'$ as a body force per unit mass. With a as a length scale, F_0 as a typical body force per unit mass, and $U_0 = F_0/\omega$ as velocity, we nondimensionalize as follows:

$$\begin{aligned} \boldsymbol{F} = \boldsymbol{F}'/F_0, \quad \boldsymbol{x} = \boldsymbol{x}'/a, \quad t = \omega t', \\ \boldsymbol{u} = \boldsymbol{u}'/U_0, \quad \boldsymbol{\zeta} = a\boldsymbol{\zeta}'/U_0, \quad \boldsymbol{\nabla} = a\boldsymbol{\nabla}', \end{aligned} \qquad (12.27)$$

along with

$$\varepsilon = \frac{U_0}{\omega a}, \quad R = \frac{U_0 a}{\nu}, \quad \text{and} \quad M^2 = \frac{i\omega a^2}{\nu} = \frac{iR}{\varepsilon} \qquad (12.28)$$

where $\varepsilon \ll 1$ is a small parameter that will be used in perturbation expansions, R is the Reynolds number, and M^2 is the frequency parameter. Using these parameters, with the density ρ being treated as constant, and taking the curl of eqn (12.26) to eliminate the terms within the $\boldsymbol{\nabla}$ operator, we obtain

$$\frac{\partial \boldsymbol{\zeta}}{\partial t} - \varepsilon \boldsymbol{\nabla} \times (\boldsymbol{u} \times \boldsymbol{\zeta}) = \frac{\varepsilon}{R}\nabla^2 \boldsymbol{\zeta}, \qquad (12.29)$$

where $\boldsymbol{\nabla} \times \boldsymbol{F}$ is eliminated on account of $\boldsymbol{F}$ being considered conservative.

12.1.2.1 Two-Dimensional Flows

For the purpose of illustration, we consider two-dimensional flows (see Riley[16,17]) so that

$$\boldsymbol{\zeta} = \boldsymbol{\nabla} \times \boldsymbol{u} = \zeta \widehat{\boldsymbol{e}}_z \quad \text{with} \quad \zeta = -\nabla^2 \psi, \tag{12.30}$$

where $\psi(x,y,t)$ is the stream function that takes the role of a vector potential defined by

$$\boldsymbol{u}(x,y,t) = \boldsymbol{\nabla} \times [\psi(x,y,t)\widehat{\boldsymbol{e}}_z]. \tag{12.31}$$

As a result, eqn (12.29) becomes

$$\frac{\partial \zeta}{\partial t} + \varepsilon(\boldsymbol{u}\cdot\boldsymbol{\nabla})\zeta = \frac{\varepsilon^2}{R_s}\nabla^2 \zeta, \tag{12.32}$$

where $R_s = \varepsilon R$ is the streaming Reynolds number. This can be written fully in terms of the stream function using eqn (12.31) as

$$\frac{\partial \nabla^2 \psi}{\partial t} - \varepsilon \frac{\partial(\psi, \nabla^2 \psi)}{\partial(x,y)} = \frac{\varepsilon^2}{R_s} \nabla^4 \psi,$$
$$= \frac{1}{|M|^2}\nabla^4 \psi, \tag{12.33}$$

where

$$\frac{\partial(P,Q)}{\partial(x,y)} = \frac{\partial P}{\partial x}\frac{\partial Q}{\partial y} - \frac{\partial P}{\partial y}\frac{\partial Q}{\partial x} \tag{12.34}$$

is the Jacobian operator. With ε as a small parameter, the system is set for solution by perturbation methods.

12.1.2.2 Three-Dimensional Axisymmetric Flows

Many of the examples we consider are axially symmetric, and when the incompressible flow approximation is applicable, we can formulate such cases as well in terms of the stream function. In terms of the spherical coordinate system (r, θ), we may write the velocity $\boldsymbol{v}(r,\theta,t) = v_r\widehat{\boldsymbol{e}}_r + v_\theta\widehat{\boldsymbol{e}}_\theta$ as the curl of a vector potential,

$$\boldsymbol{v}(r,\theta,t) = \boldsymbol{\nabla} \times \left[\frac{\Upsilon(r,\theta,t)}{r\sin\theta}\widehat{\boldsymbol{e}}_\phi\right], \tag{12.35}$$

where $\widehat{\boldsymbol{e}}_\phi$ is a unit vector in the azimuthal direction. In component form, we have

$$v_r = \frac{1}{r^2\sin\theta}\left(\frac{\partial \Upsilon}{\partial \theta}\right) \quad \text{and} \quad v_\theta = -\frac{1}{r\sin\theta}\left(\frac{\partial \Upsilon}{\partial r}\right). \tag{12.36}$$

To use this formulation in the incompressible form of the momentum eqn (12.3), we drop the $\nabla \cdot \mathbf{u}$ term, and take the curl of the equation to eliminate the pressure gradient. The result is a fourth-order nonlinear scalar partial differential equation (dimensionless),

$$\frac{\partial}{\partial t}(D^2 \Upsilon) + \varepsilon \left[\frac{1}{r^2} \frac{\partial(\Upsilon, D^2 \psi)}{\partial(r, \bar{\mu})} + \frac{2}{r^2} D^2 \Upsilon L \Upsilon \right] = \frac{1}{|M|^2} D^4 \Upsilon, \tag{12.37}$$

where $\bar{\mu} = \cos\theta$, D^2 is the Stokes operator,

$$D^2 = \frac{\partial^2}{\partial r^2} + \frac{(1-\bar{\mu}^2)}{r^2} \frac{\partial^2}{\partial \bar{\mu}^2}, \tag{12.38}$$

and the operator L is given by

$$L = \frac{\bar{\mu}}{(1-\bar{\mu}^2)} \frac{\partial}{\partial r} + \frac{1}{r} \frac{\partial}{\partial \bar{\mu}}. \tag{12.39}$$

Here, for the sake of compactness, we are using the notation Υ, v_r and v_θ as global variables for the flow. For the various situations, and for different perturbation expansion orders, we will replace these with proper applicable notation, for example ψ or ψ_1 for Υ. It should be noted that the stream function as defined in eqn (12.35) is already dimensionless. It is related to the dimensional stream function as

$$\Upsilon = \frac{\Upsilon'}{U_0 a^2}. \tag{12.40}$$

12.2 One- and Two-Dimensional Cases

We shall examine several examples in relatively simple geometries. The first one is a one-dimensional case dealing with the quartz wind that was treated in detail by Nyborg.[18] The second has to do with Rayleigh's third problem,[6] which concerns streaming from a standing wave between two parallel plates. The third example involves vertical and horizontal oscillations of a plate parallel to another plate below. For the first two problems, we closely follow Riley's[17] development. We will subsequently present three-dimensional axially symmetric cases with spherical drops and bubbles, as well as the limiting cases of solid particles.

12.2.1 The Quartz Wind

In the half-space $x > 0$, consider a one-dimensional beam of frequency ω. Taking $1/\omega$ as a time scale, c/ω as a length scale, the velocity amplitude U_0 as a velocity scale, and $\rho_0 c^2$ for pressure, we obtain from eqn (12.3) and (12.1)

$$\frac{\partial u}{\partial t} + \varepsilon u \frac{\partial u}{\partial x} = -\frac{1}{\varepsilon}\frac{\partial p}{\partial x} + \frac{4}{3}\alpha \frac{\partial u}{\partial x},$$
$$\frac{\partial \rho}{\partial t} + \varepsilon \frac{\partial u}{\partial x} = 0, \tag{12.41}$$

where $\varepsilon = U_0/c \ll 1$, $\frac{4}{3}\alpha = \omega(\beta' + \frac{4}{3}\mu)/\rho_0 c^2 \ll 1$. Using the adiabatic linear relationship between pressure and density (12.11), and combining the set of eqn (12.41), we obtain

$$\frac{\partial^2 u}{\partial t^2} - \frac{\partial^2 u}{\partial x^2} - \frac{4}{3}\alpha \frac{\partial^3 u}{\partial x^2 \partial t} = -\varepsilon \frac{\partial}{\partial t}\left(u \frac{\partial u}{\partial x}\right). \tag{12.42}$$

Now, if we expand $u(x,t)$ in powers of ε,

$$u(x,t) = u_0(x,t) + \varepsilon u_1(x,t) + \varepsilon^2 u_2(x,t) + \cdots,$$

then the leading-order form of eqn (12.42) is

$$\frac{\partial^2 u_0}{\partial t^2} - \frac{\partial^2 u_0}{\partial x^2} - \frac{4}{3}\alpha \frac{\partial^3 u_0}{\partial x^2 \partial t} = 0. \tag{12.43}$$

This has the solution

$$u_0(x,t) = e^{-2\alpha x/3} \cos(x - t), \tag{12.44}$$

for a beam originating at $x = 0$. Now regarding the nonlinear term in eqn (12.42) as a Reynolds stress, we may consider its time-average as the net force,

$$-\left\langle \varepsilon u_0 \frac{\partial u_0}{\partial x}\right\rangle = \frac{1}{3}\varepsilon \alpha e^{-4\alpha x/3}, \tag{12.45}$$

where $\langle \cos^2(x-t)\rangle = \frac{1}{2}$ has played in. Therefore, to $O(\varepsilon)$, this term will act like body force and result in a nonzero time-averaged velocity field, i.e.,

$$u_1(x,t) = u_1^{(u)}(x,t) + u_1^{(s)}(x), \tag{12.46}$$

where the superscripts (u) and (s) correspond to unsteady and steady components of the $O(\varepsilon)$ velocity field. The detailed derivation has been given by Nyborg.[18]

In microfluidic systems, the existence of Eckart streaming has been reported (see Wiklund, Green and Ohlin[1]) when the dimension of the fluid channel parallel to the direction of propagation of the acoustic wave is greater than 1 mm. Streaming velocities up to 1 cm s^{-1} can be generated for such channels.

12.2.2 Rayleigh Streaming

For $\varepsilon \ll 1$, we apply a perturbation expansion of the type

$$\boldsymbol{u}(x,y,t) = \boldsymbol{u}_0(x,y,t) + \varepsilon \boldsymbol{u}_1(x,y,t) + \varepsilon^2 \boldsymbol{u}_2(x,y,t) + \cdots$$
$$\zeta(x,y,t) = \zeta_0(x,y,t) + \varepsilon \zeta_1(x,y,t) + \varepsilon^2 \zeta_2(x,y,t) + \cdots \quad (12.47)$$
$$\psi(x,y,t) = \psi_0(x,y,t) + \varepsilon \psi_1(x,y,t) + \varepsilon^2 \psi_2(x,y,t) + \cdots$$

Substitution of the expansion for $\psi(x,y,t)$ into eqn (12.33) gives to the leading order,

$$\frac{\partial \nabla^2 \psi_0}{\partial t} = 0. \quad (12.48)$$

Since the applied force is considered conservative, the leading-order flow can in fact be regarded as irrotational ($-\nabla^2 \psi_0 = \zeta_0 = 0$). Further, when considering the applied force to have an oscillatory character, $F(x,t) = f(x)e^{it}$, the leading-order solution may be written as $\psi_0(x,y,t) = \psi_{0f}(x,y)e^{it}$, where $\psi_{0f}(x,y)$ is the stream function corresponding to the force f. With irrotational flow, we cannot satisfy the no-slip condition, and therefore, we tolerate a slip velocity [see later discussion following eqn (12.51)],

$$u(x,y,t) = \frac{\partial \psi(x,y,t)}{\partial y} = U(x)e^{it} \quad \text{at } y = 0. \quad (12.49)$$

To properly deal with the flow field at the boundary (and satisfy the no-slip condition), we define inner variables in the Stokes layer,

$$\psi = \left(\frac{2}{R_s}\right)^{\frac{1}{2}} \varepsilon \Psi, \quad y = \left(\frac{2}{R_s}\right)^{\frac{1}{2}} \varepsilon \eta, \quad (12.50)$$

or equivalently,

$$\psi = \varepsilon' \Psi, \quad y = \varepsilon' \eta, \quad (12.51)$$

where we define $\varepsilon' = \sqrt{2}/|M|$. Mathematically, the transformation of y to the inner variable η stretches the coordinate normal to the solid surface. This has the effect of bringing in the important role of the highest derivative by shifting its order lower. As we see in eqn (12.33), the small parameter $1/|M|^2$ multiplies the highest derivative term $\nabla^4 \psi$. Thus, as it is with typical boundary-layer analysis, eliminating the highest derivative to the leading order leaves too many boundary conditions, and the no-slip condition needs to be given up. Physically, we recognize that the normal derivatives in the Stokes layer region are large and this effect needs to be mathematically implemented. The transformation (12.51) leads to the following form of the momentum eqn (12.33),

$$\frac{\partial}{\partial t}\left(\frac{\partial^2 \Psi}{\partial \eta^2}\right) - \varepsilon \frac{\partial(\Psi, \partial^2 \Psi/\partial \eta^2)}{\partial(x, \eta)} = \frac{1}{2}\frac{\partial^4 \Psi}{\partial \eta^4} + O(\varepsilon^2). \quad (12.52)$$

Analysis of Acoustic Streaming by Perturbation Methods

Upon expanding this inner variable in the same form as (12.47),

$$\Psi(x,\eta,t) = \Psi_0(x,\eta,t) + \varepsilon\Psi_1(x,\eta,t) + \varepsilon^2\Psi_2(x,\eta,t) + \cdots, \qquad (12.53)$$

we have for the leading order,

$$\frac{\partial}{\partial t}\left(\frac{\partial^2\Psi_0}{\partial\eta^2}\right) = \frac{1}{2}\frac{\partial^4\Psi_0}{\partial\eta^4}, \qquad (12.54)$$

where we see that the fourth-derivative term is at the leading order. This equation has the classical Stokes layer solution,

$$\Psi_0(x,y,t) = U(x)\left[\eta - \frac{1}{2}(1-i)\left\{1 - e^{-(1+i)\eta}\right\}\right]e^{it}, \qquad (12.55)$$

where, as $\eta \to \infty$, the solution approaches the slip-velocity condition given in eqn (12.49).

Next, to $O(\varepsilon)$, from eqn (12.52) and (12.53), we obtain

$$\frac{1}{2}\frac{\partial^4\Psi_1}{\partial\eta^4} - \frac{\partial}{\partial t}\left(\frac{\partial^2\Psi_1}{\partial\eta^2}\right) = \frac{\partial(\partial^2\Psi_0/\partial\eta^2,\Psi_0)}{\partial(x,\eta)}. \qquad (12.56)$$

In spite of the oscillatory character of the flow field, we can expect a nonzero time-average contribution from the term on the right-hand side when real parts of the solutions are used. This is because it involves the products leading to $\sin^2 t$ and $\cos^2 t$ types of terms which have a nonzero mean. If we are concerned only with the time-averaged part, we can decompose $\Psi_1(x,\eta,t)$ into steady and unsteady parts, i.e.,

$$\Psi_1(x,\eta,t) = \Psi_1^{(u)}(x,\eta,t) + \Psi_1^{(s)}(x,\eta), \qquad (12.57)$$

where the superscripts (u) and (s) refer to unsteady and steady parts, respectively. Taking the time-average of eqn (12.56), we obtain

$$\frac{1}{2}\frac{\partial^4\Psi_1^{(s)}}{\partial\eta^4} = \left\langle\frac{\partial(\partial^2\Psi_0/\partial\eta^2,\Psi_0)}{\partial(x,\eta)}\right\rangle. \qquad (12.58)$$

On the right-hand side, the x-dependence will emanate from $U(x)$ in the expression for Ψ_0 in eqn (12.55). This will contain terms involving $U(x)$ and $U'(x)$, as well as their complex conjugates, $U^*(x)$ and $U^{*'}(x)$. The solution to eqn (12.58) can be written as

$$\Psi_1^{(s)}(x,\eta) = \frac{d}{dx}(UU^*)f(\eta) + U^*\frac{dU}{dx}g(\eta). \qquad (12.59)$$

The successive integration of eqn (12.58), requiring no-slip at the surface ($\eta = 0$) and boundedness as $\eta \to \infty$ yields:[17]

$$\frac{\partial \Psi_1^{(s)}(x,\eta)}{\partial \eta} = \frac{1}{2}\frac{d}{dx}(UU^*)\left(e^{-\eta}\sin\eta + \frac{1}{4}e^{-2\eta} - \frac{1}{4}\right)$$
$$- U^*\frac{dU}{dx}\left[\left\{\frac{1}{2}(1+i)\eta + i - \frac{1}{2}\right\}e^{-(1-i)\eta}\right. \qquad (12.60)$$
$$\left. -\frac{1}{4}ie^{-2\eta} - \frac{3}{4}i + \frac{1}{2}\right].$$

This represents the tangential component (x-direction) of the streaming velocity within the Stokes layer. The interesting aspect here is that velocity continues beyond the Stokes layer, i.e.,

$$\lim_{\eta \to \infty}\left[\text{Re}\left(\frac{\partial \Psi_1^{(s)}(x,\eta)}{\partial \eta}\right)\right]$$
$$= -\frac{3}{8}\left[(1-i)U^*\frac{dU}{dx} + (1+i)U\frac{dU^*}{dx}\right] \qquad (12.61)$$
$$= u^{(\text{slip})}.$$

This slip velocity is considered to be the driving mechanism for steady streaming in the bulk. The structure of this outer streaming is considered next.

12.2.3 Stokes Drift

Again, following Riley,[16,17] and taking the perturbation expansions (12.47) together with eqn (12.32), to $O(\varepsilon^2)$, we obtain

$$\frac{\partial \zeta_2}{\partial t} = -(\boldsymbol{u}_0 \cdot \boldsymbol{\nabla})\zeta_1^{(s)}, \qquad (12.62)$$

which integrates to

$$\zeta_2 = -\left[\left(\int^t \boldsymbol{u}_0 dt\right) \cdot \boldsymbol{\nabla}\right]\zeta_1^{(s)}. \qquad (12.63)$$

Next, to $O(\varepsilon^3)$, we have

$$\frac{1}{R_s}\nabla^2\zeta_1^{(s)} - (\boldsymbol{u}_1 \cdot \boldsymbol{\nabla})\zeta_1^{(s)} = \frac{\partial \zeta_3}{\partial t} + (\boldsymbol{u}_0 \cdot \boldsymbol{\nabla})\zeta_2, \qquad (12.64)$$

which, upon taking the time average yields

$$\frac{1}{R_s}\nabla^2\zeta_1^{(s)} - (\boldsymbol{u}_1^{(s)} \cdot \boldsymbol{\nabla})\zeta_1^{(s)} = \langle(\boldsymbol{u}_0 \cdot \boldsymbol{\nabla})\zeta_2\rangle. \qquad (12.65)$$

With the use of eqn (12.63),

$$\langle(u_0 \cdot \nabla)\zeta_2\rangle$$
$$= \left\langle\left(\int^t u_0 dt\right) \cdot \nabla u_0\right\rangle \cdot \nabla \zeta_1^{(s)} \qquad (12.66)$$
$$= (u_d \cdot \nabla)\zeta_1^{(s)},$$

where

$$u_d = \left\langle\left(\int^t u_0 dt\right) \cdot \nabla u_0\right\rangle \qquad (12.67)$$

is the Stokes drift velocity. Now, by defining $u_L^{(s)} = u_1^{(s)} + u_d$, we may write eqn (12.65) as

$$\frac{1}{R_s}\nabla^2 \zeta_1^{(s)} - (u_L^{(s)} \cdot \nabla)\zeta_1^{(s)} = 0, \qquad (12.68)$$

where, as Riley[17] has pointed out, the vorticity in the outer region is convected with the mean Lagrangian velocity, and R_s serves as a streaming Reynolds number.

Using the development in this section, it is possible to formally obtain the results for the Rayleigh problem discussed earlier. We therefore reconsider the velocity field given earlier by eqn (12.22)

$$u = (\sin(akx)\cos t, 0). \qquad (12.69)$$

By comparing with eqn (12.49), we may identify $U(x)$ as

$$U(x) = \sin(akx). \qquad (12.70)$$

Using this in eqn (12.61), we obtain

$$u^{(\text{slip})} = -\frac{3}{8}\text{Re}\left[(1-i)ak\sin(akx)\cos(akx) + (1+i)\sin(akx)\cos(akx)\right]$$
$$= -\frac{3}{8}\sin(2akx). \qquad (12.71)$$

With this velocity applied at the wall ($y = -1$, in scaled variables), and symmetry about $y = 0$, together with $R_s \ll 1$ in (12.68), the result (12.25) can be derived.

The number of simple examples is limited, and for the purpose of illustration of streaming in confined spaces with potential microfluidic application, the case of the vibration of an infinite plate parallel to another plate is considered. Here, the development of Wang and Drachman[19] is followed.

12.2.4 Streaming between Two Plates

In this example, there is a viscous fluid between two parallel plates, a distance a apart. The lower plate is held fixed while the upper plate is oscillated with motion both normal and parallel to the plate with a phase difference. It is assumed that the problem is two-dimensional while realizing that for large plates, the motion could be three-dimensional axisymmetric (see, for example Ishizawa[20]). The two-dimensional flow is physically realizable if the fluid motion is restricted by fixed sidewalls on two opposite gaps between the two plates. This problem has been dealt with by Wang and Drachman[19] using the regular perturbation method without restriction on the frequency parameter $|M|^2 = \omega a^2/\nu$. In the limit of $|M| \gg 1$, the singular perturbation approach gives an elegant leading-order result, which is the approach we take here. The results of Wang and Drachman[19] will be discussed subsequently.

12.2.4.1 Problem Statement

The two-dimensional problem is depicted in Figure 12.5. The top plate located at $y = a$ is oscillated vertically with velocity,

$$v' = -U_0 \sin(\omega t'), \tag{12.72}$$

and also horizontally as

$$u' = -\alpha U_0 \sin\left[\omega(t' + \beta')\right], \tag{12.73}$$

so that there is a phase lag β'.

12.2.4.2 Outer Solution

With the usual scaling in eqn (12.27) and (12.28), together with the flow description given by eqn (12.33), and expansions of the type (12.47), the leading order outer flow is described by

$$\frac{\partial \nabla^2 \psi_0}{\partial t} = 0, \tag{12.74}$$

with the normal-velocity boundary conditions,

$$\left.\frac{\partial \psi_0}{\partial x}\right|_{y=0} = 0, \quad \text{and} \quad \left.\frac{\partial \psi_0}{\partial x}\right|_{y=1} = -ie^{it}. \tag{12.75}$$

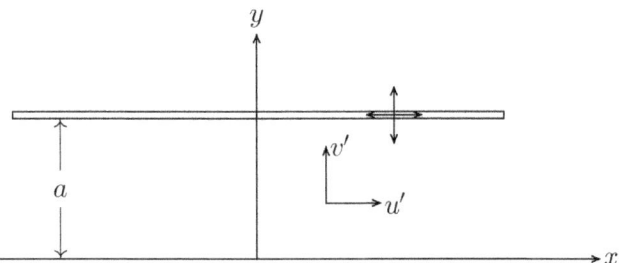

Figure 12.5 A schematic of an oscillating plate parallel to a fixed plate. Adapted from ref. 19.

Here and henceforth, the real part is implied. With the irrotational approximation at this stage, the tangential velocity conditions cannot be satisfied at either of the plates. The solution to this set of equations is

$$\psi_0(x,y) = -ixye^{it}. \tag{12.76}$$

12.2.4.3 Inner Solutions

The boundary conditions for the horizontal velocity cannot be satisfied by the irrotational flow at either of the plates and boundary-layer analysis is needed at both $y = 0$ and $y = 1$. In the inner region near $y = 0$, the variables scale as per eqn (12.51) with a slight change,

$$(1-y) = \left(\frac{2}{R_s}\right)^{\frac{1}{2}} \varepsilon\eta = \varepsilon'\eta, \tag{12.77}$$

where the stretched coordinate η runs downwards from the upper plate. The differential eqn (12.52) applies in the Stokes layer region. Once again, we have an inner expansion of the type (12.53), and the ensuing leading-order eqn (12.54). We write the solution to this in a somewhat general form as

$$\Psi_0^{(1)}(x,\eta,t) = \left[A(x)\eta + B(x) + C(x)e^{-(1+i)\eta}\right]e^{it}, \tag{12.78}$$

where the superscript (1) refers to the upper boundary layer. The boundary conditions at $\eta = 0$ $(y = 1)$ require that the conditions (12.72) and (12.73) be satisfied, i.e.,

$$\varepsilon'\frac{\partial \Psi_0}{\partial x}\bigg|_{\eta=0} = -ie^{it}, \tag{12.79}$$

and

$$\frac{\partial \Psi_0^{(1)}}{\partial \eta}\bigg|_{\eta=0} = -i\alpha e^{i(t+\beta)}, \tag{12.80}$$

where $\beta = \omega\beta'$. These conditions, together with outer-field matching [eqn (12.76)], lead to

$$A(x) = ix, \tag{12.81}$$

$$B(x) = -\frac{|M|ix}{\sqrt{2}} + \frac{1}{2}(1+i)(\alpha e^{i\beta} + x) \tag{12.82}$$

$$C(x) = -\frac{1}{2}(1+i)(\alpha e^{i\beta} + x) \tag{12.83}$$

resulting in

$$\Psi_0^{(1)}(x,\eta,t) = \left\{ix\eta - \frac{ix}{\varepsilon'} - \frac{1}{2}(1+i)(x+\alpha e^{i\beta})\left(1 - e^{-(1+i)\eta}\right)\right\}e^{it} \tag{12.84}$$

for the inner region near $y = 1$. This solution satisfies both the boundary conditions at $y = 1$ ($\eta = 0$) exactly, and the far-field matches $\psi_0(x,y)$ to $O(1)$. It should be noted that the horizontal motion (the term containing α) contributes only to the inner part.

For the inner region near $y = 0$, we define the inner variable η' according to (12.51) so that

$$\psi = \varepsilon' \Psi^{(0)}, \quad y = \varepsilon' \eta', \tag{12.85}$$

with the superscript (0) corresponding to $y=0$. As in the case of the upper inner region, the lower boundary layer also satisfies the differential eqn (12.52) that, after an inner expansion of the type (12.53), leads to leading-order eqn (12.54). The solution has the same form as (12.78), and upon matching with the outer solution (12.76), we obtain

$$\Psi_0^{(0)}(x, \eta, t) = x\left[i\eta - \frac{1}{2}(1+i)\left(1 - e^{-(1+i)\eta}\right)\right] e^{it} \tag{12.86}$$

which satisfies the no-slip condition at $y=0$.

12.2.4.4 Steady Streaming

To the next order, eqn (12.56) applies, and since we are interested only in the steady streaming, the time-averaged eqn (12.58) can be used to calculate the steady streaming flow field.

It is worth mentioning that this problem can be solved fully for $\varepsilon \ll 1$ without requiring that $|M|^2 = \omega a^2/\nu \gg 1$. As stated earlier, this has, in fact, been carried out by Wang and Drachman.[19] They solved for the velocity field directly without using the stream function formulation. Since the calculations are available, we shall not repeat these but focus on the results. The problem while being a simple two-dimensional case in cartesian geometry represents several interesting characteristics relevant to streaming. The phase difference (β) between the vertical and the horizontal oscillations creates typically an elliptical pattern on the upper plate motion, and various types of characterizations of the streaming flow are seen with variations of this parameter.

The streaming flow pattern for vertical-only oscillations with the frequency parameter $|M|^2 = 100$ is shown in Figure 12.6. Here, we see a stagnation point between the plates, somewhat closer to the upper plate. It is important to note that the horizontal oscillations on their own would not generate streaming but the nonlinear interaction of the oscillatory fields indeed brings about steady flow fields. With a phase angle of $\beta = \frac{1}{2}\pi$, and horizontal amplitude of $\alpha=0.25$, the symmetry breaks, and the stagnation point shifts right as seen in Figure 12.7.

In Figure 12.8, we see that an increase in the horizontal amplitude causes the mean flow motion from left to right, representing a pumping action. This pumping rate intensifies with increasing α. The net flux calculations given by Wang and Drachman[19] show that for values of $|M|^2 > 10$, the flux increases with increasing phase angle β up to a peak value and then drops off, as shown in Figure 12.9. For lower values of $|M|^2$ in the approximate range $5 < |M|^2 < 10$, the flux decreases with increasing β. Interestingly, for $|M|^2 = 1$, the flux decreases with β to a minimum and then increases.

Analysis of Acoustic Streaming by Perturbation Methods

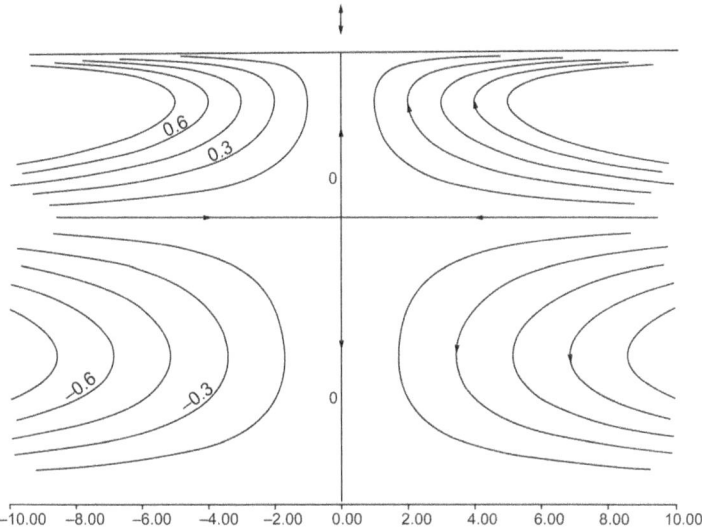

Figure 12.6 Steady streamlines with normal oscillations only ($\alpha = 0$ and $|M|^2 = 100$). Reproduced from ref. 19.

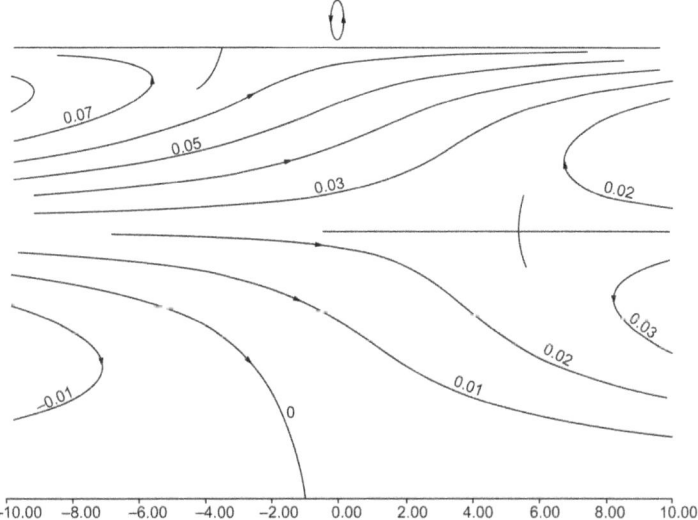

Figure 12.7 Steady streamlines with normal and lateral oscillations ($\alpha = 0.25$, $\beta = \frac{1}{2}\pi$ and $|M|^2 = 1$). Reproduced from ref. 19.

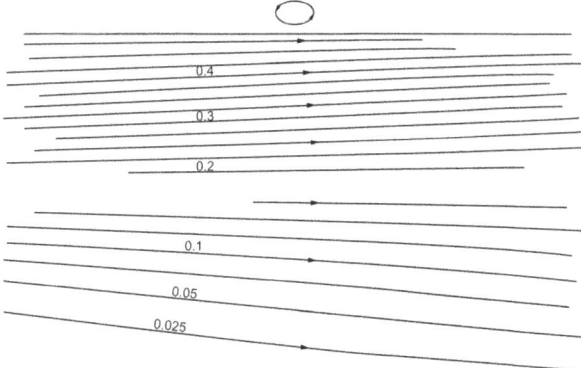

Figure 12.8 Steady streamlines with normal and lateral oscillations ($\alpha = 2$, $\beta = \frac{1}{2}\pi$ and $|M|^2 = 1$). Reproduced from ref. 19.

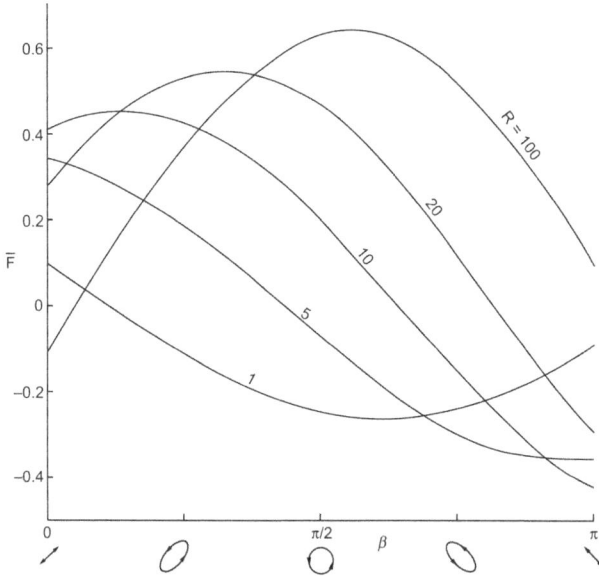

Figure 12.9 Time-averaged flux variation with phase angle β and $|M|^2 = \omega a^2/\nu$. Reproduced from ref. 19.

12.3 Interaction of Solid Particles and Drops with Ultrasound

From a fundamental standpoint, as well as for application to levitated drops and particles, there is interest in studying the streaming phenomenon when a standing wave interacts with a solid or a fluid interface. As mentioned in the Introduction, when particles interact with high-frequency sound waves, streaming occurs. It has also been established that liquid drops interacting with ultrasound exhibit streaming characteristics. The results are very interesting and show that there can be a marked difference in the characterization of the streaming by simply allowing some degree of interfacial mobility. The interaction of ultrasound with drops and solid particles is very common with acoustic levitation devices. Such systems are used for containerless processing, and applications include noncontact trapping of cells and particle-based assays in continuous flow microsystems. For example, an acoustic standing wave is generated in etched glass microchannels by miniature ultrasonic transducers, and particles or cells passing the transducer can be retained and levitated in the center of the channel without any contact with the channel walls.[21] The potential of ultrasonic standing wave fields to facilitate viral transduction rates has been demonstrated by Lee and Peng.[22] Under acoustic exposure, suspended cells move to the pressure nodal planes first and form cell clusters. Then, viruses circulated between nodal planes use the pre-formed cell clusters as nucleating sites to attach to. In the past, several macroscale applications of acoustic levitation have been made, including the non-contact thermophysical property measurement of liquids.[23,24]

The descriptive characterization of streaming has been discussed in the Introduction where we consider a body of typical dimension a that oscillates with angular frequency ω and velocity $U_0 \cos(\omega t)$ in a viscous fluid. We recall that if the parameter $\varepsilon = U_0/\omega a \ll 1$, then, although the leading order solution in powers of ε is oscillatory, higher order terms include not only higher harmonics but steady contributions to the flow field as well.

The analytical procedure that we follow consists of perturbation expansions in small ε and large $|M|$. As discussed in Section 12.2.2, a small parameter appears in front of the highest derivative in the momentum equation, and the expansion procedure needs to be singular in character. This requires inner and outer expansions with stretching of the inner variable. However, it should be noted that if we were interested only in the outer-region streaming, the procedure developed by Nyborg[8] could be employed. With this procedure, the leading-order nonlinear terms when time-averaged appear effectively as a conservative force in the next order. This method has been applied by Lee and Wang[2] for outer streaming associated with flow between parallel plates, as well as the sphere and the cylinder placed between velocity node and antinode. A further extension of Nyborg's procedure,[8] has been recently carried out by Rednikov and Sadhal,[9] with the inclusion of non-adiabatic effects. Nevertheless, in order to fully understand streaming within the Stokes layer, inner and outer perturbation analysis is necessary.

Microfluidic applications and the relevance of streaming in liquids have been discussed by Bruus,[3] as well as in ref. 4 and 1. For micron-sized particles, the flow-visualization work of Hagsäter et al.[25] is of importance in contrasting the streaming-based Stokes drag with the radiation force. They observed that at 2.17 MHz on 1 μm polystyrene beads, the Stokes drag is higher than the radiation force, while on 5 μm beads, the latter is dominant. It should be noted here that the Stokes layer thickness for this frequency is close to 1 μm, *i.e.*, $(\nu/\omega)^{1/2} \sim 1$ μm, corresponding to $|M| \sim 1$ in the case of a 1 μm particle.

Streaming is also prevalent in acoustic levitation apparatus. These are devices that utilize the radiation pressure of a standing wave to levitate small particles and drops in air or gas bubbles in a liquid. This topic is discussed next.

12.3.1 Acoustic Levitators

A typical desktop levitator is shown in Figure 12.10. The main physical principle involved here is that the acoustic field provides the radiation pressure necessary to levitate a liquid drop in a gravitational field. The studies on the effects of radiation pressure on spheres and disks goes as far back as the 1930s.[26,27] With the application of this principle, ultrasound levitators have been in use for many years in ground-based experiments (as opposed to space-based).

Some of the earlier work to characterize this flow includes the developments of Trinh and Robey.[28] One of their works on streaming flow visualization around a levitated drop is given in Figure 12.11. With widespread application of levitation systems in the 1980s and the 90s, there has been an interest in understanding the fluid-flow fundamentals associated with these systems.

Among the items of interest is the information about the characteristics of the levitation process. For example, with acoustically-levitated particles, there is a residual flow field including solid-body rotation for drops. For levitation under microgravity conditions, the drop assumes an equilibrium position at the velocity antinode when the external medium is a gas (see Figure 12.2 for node and antinode definitions). If the particle phase is a fluid with higher compressibility than the external phase (*e.g.*, a gas bubble in

Figure 12.10 Ultrasound levitation apparatus. The picture shows a levitated water–alcohol drop, approximately 3 mm across and 1 mm high. The bottom plate is an ultrasound transducer operating at 20 kHz, and the top is a slightly cupped reflector so that the system produces a standing wave.

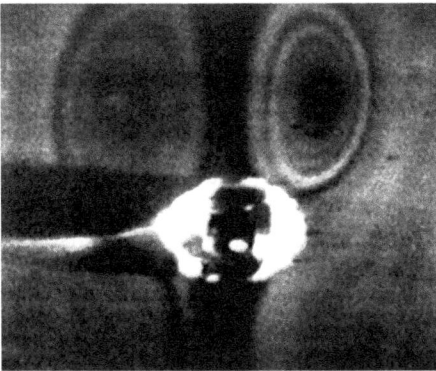

Figure 12.11 Visualization of streaming around a levitated particle. The tested particle is a drop of water with diameter 1.8–1.85 mm. The acoustic frequency is 37 kHz, corresponding to $|M| \simeq 110$. Reproduced from ref. 28.

a liquid), the equilibrium position can occur at the velocity node. While the antinode solution has been available from Riley's[29] classical work as well as a study by Wang,[30] the node solution is relatively more recent.

For a levitated spherical particle positioned at the velocity antinode, Riley's solution[29] for a vibrating sphere in an otherwise quiescent fluid can be accommodated for by $a/\lambda = a\omega/c \ll 1$, *i.e.*, when the particle size is small compared with the wavelength of the standing wave. Generalizations of this work have been carried out by Rednikov *et al.*[31] to include the particle liquidity and its placement away from the velocity antinode. In the limit of infinite drop viscosity and zero displacement from the velocity antinode, the solution collapses to Riley's[29] for the steady streaming part to $O(\varepsilon)$. Therefore, it is meaningful to begin with the general case and consider the various limiting special cases.

12.4 Solid Particles and Drops Placed Between Nodes

In this section, we discuss the flow field dealing with a spherical liquid drop in a gas medium displaced between the velocity node and the antinode in an acoustic levitator. The analysis is carried out for a high-frequency standing acoustic wave, which is useful to levitate particles in Earth gravity or to stabilize particles in low-gravity situations. The drop is considered to have sufficient mass so that it occupies a stable position in the acoustic field and it does not experience significant body oscillations. We depend a great deal on the perturbation procedure based on small-amplitude and high-frequency assumptions for the acoustic fields. The coordinate system is defined so that the z-axis passes through the center of the sphere and points along the direction of vibration, and $z = 0$ represents the velocity antinode closest to the sphere, while $z = z_0$ is the center of the sphere. We use axially symmetric spherical polar coordinates (r, θ) with the origin at the center of the sphere.

We scale the coordinates, time, velocity, stream function, and the gradient operator according to eqn (12.27) and (12.28), where again $\varepsilon \ll 1$ is a small parameter that will be used in perturbation expansions, ν is the kinematic viscosity, R is the Reynolds number, and M^2 is the frequency parameter related to the Womersley number $\alpha = |M|$. We consider the situation $|M|^2 \gg 1$ besides $\varepsilon \ll 1$. Here, the square root of M^2 is $M = (1+i)a\sqrt{\omega/2\nu}$ with $i = \sqrt{-1}$. Once again, the primed variables are the dimensioned quantities while the system parameters and the constant properties (such as U_0, a, ω, μ, ρ_0 and ν) do not have primes. Another dimensionless quantity is the streaming Reynolds number $R_s = \varepsilon R = U_0^2/\nu\omega$ which is also a small parameter. In addition, we use

$$\bar{k} = ka, \quad \phi = \frac{\phi'}{U_0 a}, \tag{12.87}$$

$$p = \frac{p'}{\rho_0 U_\infty \omega a}, \quad \text{and} \quad \rho = \frac{\rho' c^2}{\rho_0 U_0 \omega a}, \tag{12.88}$$

where $k = 2\pi/\lambda$ is the wavenumber, ρ_0 is the unperturbed medium density, c is the speed of sound, ρ' is the density perturbation due to the acoustic wave, and p' is the acoustic pressure. The flow field will be described by the stream function ψ' and the velocity potential ϕ'. Besides the conditions on $|M|$, ε and R_s, we assume that the particle size is much smaller than the acoustic wavelength, i.e., $a/\lambda \ll 1$, or for that matter, $\bar{k} \ll 1$. We shall dispense with much of the details which are available in ref. 31.

A particle levitated in a gravity field would position itself between the velocity node and the antinode. To consider such a problem, we need to expand the dimensionless standing wave velocity u_z such that

$$u_z = \cos(\bar{k}z)e^{it} = \left[\cos(\bar{k}z_0) - \bar{k}(z - z_0)\sin(\bar{k}z_0) + O\left(\bar{k}^2(z - z_0)^2\right)\right]e^{it}, \tag{12.89}$$

represents the local velocity in the neighborhood of the sphere centered at $z = z_0$. The first term in the expanded version of eqn (12.89) is just the far-field velocity for the situation when the sphere is at the velocity antinode of a standing wave, and the second term is the far-field velocity for the case when the sphere is positioned at the node. The cases $\bar{k}z_0 = 0$ and $\bar{k}z_0 = \frac{1}{2}\pi$ would correspond to cases of a sphere placed at the velocity antinode and node, respectively. The displaced sphere problem is a combination of the solutions about the node and the antinode, together with additional nonlinear terms.

It has been established that the flow about a particle at the velocity antinode may be approximated as incompressible.[32] However, near the velocity node, compressibility effects come into play,[15] and stream function formulation given by eqn (12.35)–(12.40) is not suitable. Instead, we work with continuity and momentum eqn (12.1) and (12.3), and with the dimensionless scaling discussed earlier (eqn (12.27) and (21.28)), these may be written as

Analysis of Acoustic Streaming by Perturbation Methods

CONTINUITY:

$$\bar{k}^2 \frac{\partial \rho}{\partial t} + \nabla \cdot \boldsymbol{u} + \varepsilon \bar{k}^2 \nabla \cdot (\rho \boldsymbol{u}) = 0, \qquad (12.90)$$

MOMENTUM:

$$\left(1 + \rho \varepsilon \bar{k}^2\right) \frac{\partial \boldsymbol{u}}{\partial t} + \varepsilon \left(1 + \rho \varepsilon \bar{k}^2\right) \boldsymbol{u} \cdot \nabla \boldsymbol{u} = -\nabla p + \frac{1}{|M|^2} \nabla^2 \boldsymbol{u}, \qquad (12.91)$$

respectively. By using adiabatic relation $\rho' = p'/c^2$, the dimensionless acoustic pressure and density can be shown to be equal for $\gamma = c_p/c_v \simeq 1$, i.e.,

$$p = \rho \qquad (12.92)$$

to the leading order.

While the fluid within the Stokes layer near the surface of the sphere has vorticity to meet the continuity conditions on the interface, the flow outside the layer, to the leading order, behaves irrotationally as in a sound field. This outer flow field can therefore be expressed as a velocity potential,

$$\boldsymbol{u} = \nabla \phi. \qquad (12.93)$$

The dimensionless far-field potential function, corresponding to eqn (12.89), takes the form.

$$\begin{aligned}
\phi_\infty &= \left[\frac{1}{\bar{k}} \sin(\bar{k} z_0) + (z - z_0)\cos(\bar{k} z_0) \right. \\
&\quad \left. - \frac{1}{2}\bar{k}(z - z_0)^2 \sin(\bar{k} z_0) + O(\bar{k}^2) - \right] e^{it} \\
&= \left[\frac{1}{\bar{k}} \sin(\bar{k} z_0) + r\cos(\bar{k} z_0) P_1(\bar{\mu}) - \frac{1}{6}\bar{k} r^2 \sin(\bar{k} z_0) \right. \\
&\quad \left. - \frac{1}{3}\bar{k} r^2 \sin(\bar{k} z_0) P_2(\bar{\mu}) + O(\bar{k}^2)\right] e^{it}, \qquad (12.94)
\end{aligned}$$

where $P_n(\bar{\mu})$ denotes Legendre polynomials, and $\bar{\mu} = \cos\theta$. With the spherical coordinate system centered at $z = z_0$, it should be noted that $z - z_0 = r\bar{\mu}$. From momentum eqn (12.91), we have as before

$$\frac{\partial \boldsymbol{u}_\infty}{\partial t} = -\nabla p_\infty, \qquad (12.95)$$

and using eqn (12.93), we obtain

$$p_\infty = \rho_\infty = -i\left[\frac{1}{\bar{k}} \sin(\bar{k} z_0) + r\cos(\bar{k} z_0) P_1(\bar{\mu}) - \frac{1}{6}\bar{k} r^2 \sin(\bar{k} z_0) \right. \\
\left. - \frac{1}{3}\bar{k} r^2 \sin(\bar{k} z_0) P_2(\bar{\mu})\right] e^{it}, \qquad (12.96)$$

where the real part applies.

12.4.1 Solution

As with the previous section, we solve for the flow field by singular perturbation using $\varepsilon = U_0/(\omega a) \ll 1$. In the Stokes layer, we refer to the flow variables with a superscript b instead of capitalizing them. We refer to the liquid drop region as the dispersed phase and the surrounding region as the continuous phase. Within the latter, the boundary-layer is defined as the inner region and the outer region corresponds to the remainder of the continuous phase.

By applying the perturbation method, we expand the velocity, acoustic pressure and density outside the boundary layer in powers of ε as follows:

$$\boldsymbol{u} = \boldsymbol{u}_0 + \varepsilon \boldsymbol{u}_1 + O(\varepsilon^2), \tag{12.97}$$

$$p = \bar{p}_0 + \varepsilon p_1 + O(\varepsilon^2), \tag{12.98}$$

and

$$\rho = \bar{\rho}_0 + \varepsilon \rho_1 + O(\varepsilon^2), \tag{12.99}$$

where the overbars on $\bar{p}_0$ and $\bar{\rho}_0$ are used to avoid confusion with the background pressure and density. These expansions are substituted into eqn (12.90) and (12.91) to form a hierarchy of equations in orders of ε. We treat here the leading and first orders, referred to as $O(1)$ and $O(\varepsilon)$, respectively.

12.4.1.1 Leading-Order Solutions

12.4.1.1.1 Outer Region

The leading-order velocity potential ϕ_0 takes the form

$$\phi_0 = \left\{ \frac{1}{\bar{k}} \sin(\bar{k}z_0) + \cos(\bar{k}z_0) \left(r + \frac{1}{2}r^{-2} \right) P_1(\bar{\mu}) - \frac{1}{3}\bar{k}\sin(\bar{k}z_0) \left[\left(r^{-1} + \frac{1}{2}r^2 \right) \right. \right. $$
$$\left. \left. + \left(r^2 + \frac{2}{3}r^{-3} \right) P_2(\bar{\mu}) \right] + O(\bar{k}^2) \right\} e^{it}. \tag{12.100}$$

The leading-order acoustic pressure $\bar{p}_0$ and density $\bar{\rho}_0$ are then given by

$$\bar{p}_0 = \bar{\rho}_0 = -\frac{\partial \phi_0}{\partial t} = -i\left\{ \frac{1}{\bar{k}} \sin(\bar{k}z_0) + \cos(\bar{k}z_0) \left(r + \frac{1}{2}r^{-2} \right) P_1(\bar{\mu}) \right.$$
$$\left. - \frac{1}{3}\bar{k}\sin(\bar{k}z_0) \left[\left(r^{-1} + \frac{1}{2}r^2 \right) \right. \right. \tag{12.101}$$
$$\left. \left. + \left(r^2 + \frac{2}{3}r^{-3} \right) P_2(\bar{\mu}) \right] + O(\bar{k}^2) \right\} e^{it}.$$

In the dispersed phase, the leading-order solution is

$$\widehat{\psi}_0 = 0, \tag{12.102}$$

Analysis of Acoustic Streaming by Perturbation Methods

implying no flow at all to the leading order. This also means that to this order, the solution should satisfy the no-slip conditions as for a solid sphere. However, being potential flow, the no-slip condition cannot be satisfied, and detailed development in the Stokes layer is needed. This is done next.

12.4.1.1.2 Continuous-Phase Stokes Layer

In this boundary layer, we write the velocity field in terms of normal (radial) and tangential components,

$$\boldsymbol{u}^b = u_r^b \widehat{\boldsymbol{e}}_r + u_\theta^b \widehat{\boldsymbol{e}}_\theta. \tag{12.103}$$

As usual, with $|M|^2 \gg 1$, the vorticity generated at the surface of the sphere is confined to a thin Stokes layer of dimensional thickness $O(a|M|^{-1})$. We scale the inner variables within the Stokes layer as

$$\varepsilon' \eta = (r-1), \quad \text{and} \quad \varepsilon' u_\eta^b = u_r^b, \tag{12.104}$$

where $\varepsilon' = \sqrt{2}/|M|$ as defined earlier. Again, perturbing in powers in ε,

$$\boldsymbol{u}^b = \boldsymbol{u}_0^b + \varepsilon \boldsymbol{u}_1^b + O(\varepsilon^2), \tag{12.105}$$

$$p^b = p_0^b + \varepsilon p_1^b + O(\varepsilon^2), \tag{12.106}$$

$$\rho^b = \rho_0^b + \varepsilon \rho_1^b + O(\varepsilon^2), \tag{12.107}$$

and using these expansions in the momentum eqn (12.91), we have for the leading-order normal and tangential velocities,

$$\frac{\partial u_{r0}^b}{\partial t} = -\frac{\partial p_0^b}{\partial r} = -\frac{1}{\varepsilon'} \frac{\partial p_0^b}{\partial \eta} \tag{12.108}$$

and

$$\frac{\partial u_{\theta 0}^b}{\partial t} = -\frac{\partial p_0^b}{\partial \theta} + \frac{1}{2} \frac{\partial^2 u_{\theta 0}^b}{\partial \eta^2}, \tag{12.109}$$

respectively. With $\varepsilon' \ll 1$, and using eqn (12.108), we may deduce that the leading-order acoustic pressure p_0^b in the boundary layer is a function of θ and t only. Therefore, $\partial p_0^b/\partial \eta = 0$, and using this information in eqn (12.101), we find

$$p_0^b = -i \left\{ \sin(\bar{k}z_0) \left(\frac{1}{\bar{k}} - \bar{k} \left[\frac{1}{2} + \frac{5}{9} P_2(\bar{\mu}) \right] \right) + \cos(\bar{k}z_0) \frac{3}{2} P_1(\bar{\mu}) \right\} e^{it} + O(\bar{k}^2). \tag{12.110}$$

With the pressure distribution known at the leading order, its usage in eqn (12.108) and (12.109) yields

$$u_{\theta 0}^b = \left\{ -\frac{3}{2} \cos(\bar{k}z_0) \sin\theta + \frac{5}{3}\bar{k} \sin(\bar{k}z_0) \sin\theta \cos\theta \right\} \left(1 - e^{-(1+i)\eta}\right) e^{it}. \tag{12.111}$$

and

$$u_{\eta 0}^b = \left\{ 3\cos(\bar{k}z_0)\left[\eta - \frac{1}{2}(1-i)\left(1 - e^{-(1+i)\eta}\right)\right]P_1(\bar{\mu})\right.$$
$$- \bar{k}\sin(\bar{k}z_0)\left(\eta + \frac{10}{3}\left[\eta - \frac{1}{2}(1-i)\right.\right.$$ (12.112)
$$\left.\left.\left. \times \left(1 - e^{-(1+i)\eta}\right)\right]P_2(\bar{\mu})\right)\right\}e^{it}.$$

Here, we notice that all the leading-order solutions, including velocity, pressure and density, are just the linear combination of two groups of results. One is for the drop placed at the velocity antinode, and the other one is for the node.

12.4.1.2 $O(\varepsilon)$ Solutions

As with most problems in this class, the first-order solution is much more complex than the leading order. Our interest is mainly in understanding the steady streaming outside the sphere. We therefore consider here only the steady-state components of the solutions. In this section, therefore, all the first-order variables are time-independent, and we shall refer to them with the superscript (s) for the steady part. Also, noting the fact that first order is indeed $O(\varepsilon)$, the leading-order of the steady part is $O(\varepsilon)$.

By taking the time-average of the continuity eqn (12.90), it is not difficult to show that the first-order velocity field is incompressible,[31] i.e., $\nabla \cdot u_1^b = 0$. Making use of eqn (12.105)–(12.107) in the momentum eqn (12.91), equating both sides in the order of ε, and taking the time average, we have

$$\left\langle \rho_0^b \bar{k}^2 \frac{\partial u_{\eta 0}^b}{\partial t}\right\rangle + \left\langle u_{\eta 0}^b \frac{\partial u_{\eta 0}^b}{\partial \eta}\right\rangle + \left\langle u_{\theta 0}^b \frac{\partial u_{\eta 0}^b}{\partial \theta}\right\rangle = -\frac{1}{\varepsilon'^2}\frac{\partial p_1^{b(s)}}{\partial \eta} + \frac{\partial^2 u_{\eta 1}^{b(s)}}{\partial \eta^2},\quad (12.113)$$

and

$$\left\langle \rho_0^b \bar{k}^2 \frac{\partial u_{\theta 0}^b}{\partial t}\right\rangle + \left\langle u_{\eta 0}^b \frac{\partial u_{\theta 0}^b}{\partial \eta}\right\rangle + \left\langle u_{\theta 0}^b \frac{\partial u_{\theta 0}^b}{\partial \theta}\right\rangle = -\frac{\partial p_1^{b(s)}}{\partial \theta} + \frac{1}{2}\frac{\partial^2 u_{\theta 1}^{b(s)}}{\partial \eta^2},\quad (12.114)$$

for the first-order normal and tangential velocities, respectively, in the boundary layer. Recognizing once again that $\varepsilon'^2 \ll 1$, whereby in eqn (12.113) the pressure derivative term is dominant, we end up with

$$\frac{\partial p_1^{b(s)}}{\partial \eta} = 0, \quad (12.115)$$

which means the $O(\varepsilon)$ time-independent pressure in the boundary layer is a function of θ only. Using this information to eliminate the pressure term in (12.114), and applying the leading order parameters (eqn (12.110)–(12.112)) in the time-averaged terms, we obtain after successive integration

$$u_{\theta 1}^{b(s)} = \bar{k}\sin(\bar{k}z_0)\cos(\bar{k}z_0)\left\{\sin\theta\left[\frac{5}{8}e^{-2\eta} + \frac{5}{4}e^{-\eta}\cos\eta\right.\right.$$

$$\left.+\frac{17}{4}e^{-\eta}\sin\eta + \frac{1}{2}\eta e^{-\eta}(\sin\eta - \cos\eta) + Q_1\right]$$

$$+\sin\theta\cos^2\theta\left[-\frac{15}{8}e^{-2\eta} - 20e^{-\eta}\sin\eta\right.$$

$$\left.\left.+\frac{25}{4}e^{-\eta}(\eta\cos\eta - \eta\sin\eta - \cos\eta) + Q_3\right]\right\}$$

$$+\cos^2(\bar{k}z_0)\sin\theta\cos\theta\left\{\frac{9}{16}e^{-2\eta} + \frac{27}{4}e^{-\eta}\sin\eta\right.$$

$$\left.+\frac{9}{4}e^{-\eta}(\eta\sin\eta - \eta\cos\eta + \cos\eta) + Q_2\right\} + O(\bar{k}^2), \quad (12.116)$$

where Q_1, Q_2, and Q_3 are constants found to be

$$Q_1 = \frac{23}{168}\sqrt{2}M_\mu - \frac{15}{8}, \quad (12.117)$$

$$Q_2 = \frac{9}{80}\sqrt{2}M_\mu - \frac{45}{16}, \quad (12.118)$$

$$Q_3 = -\frac{15}{56}\sqrt{2}M_\mu + \frac{65}{8}, \quad (12.119)$$

with $M_\mu = |M|\mu/\hat{\mu}$. The component $u_{1\eta}^{b(s)}$ can also be found by using $\nabla\cdot\mathbf{u}_1^b = 0$. However, at this stage, it is better to construct the stream function since the flow is considered incompressible. Within the boundary layer, eqn (12.36) with $\Upsilon = \psi_1^{b(s)}$ takes the form

$$u_{\eta 1}^{b(s)} = \frac{1}{\sin\theta}\left(\frac{\partial\psi_1^{b(s)}}{\partial\theta}\right) \quad \text{and} \quad u_{\theta 1}^{b(s)} = -\frac{1}{\sin\theta}\left(\frac{\partial\psi_1^{b(s)}}{\partial\eta}\right). \quad (12.120)$$

The resulting expression for the stream function is

$$\psi_1^{b(s)} = -\left\{\bar{k}\sin(\bar{k}z_0)\cos(\bar{k}z_0)\left[\left(-\frac{5}{16}e^{-2\eta} - 3e^{-\eta}\cos\eta\right.\right.\right.$$

$$\left.-\frac{7}{4}e^{-\eta}\sin\eta - \frac{1}{2}\eta e^{-\eta}\sin\eta + Q_1\eta + \frac{53}{16}\right)(1-\bar{\mu}^2)$$

$$+\left(\frac{15}{16}e^{-2\eta} + \frac{65}{4}e^{-\eta}\cos\eta + 10e^{-\eta}\sin\eta + \frac{25}{4}\eta e^{-\eta}\sin\eta\right.$$

$$\left.\left.+Q_3\eta - \frac{275}{16}\right)\bar{\mu}^2(1-\bar{\mu}^2)\right] + \cos^2(\bar{k}z_0)\left(-\frac{9}{32}e^{-2\eta} - \frac{45}{8}e^{-\eta}\cos\eta\right.$$

$$\left.\left.-\frac{27}{8}e^{-\eta}\sin\eta - \frac{9}{4}\eta e^{-\eta}\sin\eta + Q_2\eta + \frac{189}{32}\right)\bar{\mu}(1-\bar{\mu}^2)\right\} + O(\bar{k}^2). \quad (12.121)$$

This inner solution is valid only in the Stokes layer, and to complete the analysis, we must seek the steady streaming in the outer region as well. In the limit of small streaming Reynolds numbers ($R_s \ll 1$), the outer streaming satisfies the Stokes equation

$$D^4 \psi_1^{(s)} = 0, \tag{12.122}$$

where the relationships in eqn (12.36) with $\Upsilon = \psi_1^{(s)}$, $v_r = u_{r1}^{(s)}$ and $v_\theta = u_{\theta 1}^{(s)}$ apply. To obtain the solution to eqn (12.122), the angular eigenfunctions are chosen to be the same as the inner solution, ψ_1^b, given by eqn (12.121). The far-field behaviour of the solution requires the flow velocity to fade away. At the surface we require matching with the Stokes layer solution (12.121). Thus, we obtain

$$\psi_1^{(s)} = \bar{k}\sin(\bar{k}z_0)\cos(\bar{k}z_0)\left[\left(\frac{1}{8} - \frac{1}{24}\sqrt{2}M_\mu\right)\left(r - \frac{1}{r}\right)(1 - \bar{\mu}^2)\right.$$
$$\left. -\left(\frac{13}{16} - \frac{3}{112}\sqrt{2}M_\mu\right)\left(\frac{1}{r} - \frac{1}{r^3}\right)(5\bar{\mu}^2 - 1)(1 - \bar{\mu}^2)\right]$$
$$+\cos^2(\bar{k}z_0)\left(\frac{45}{32} - \frac{9}{160}\sqrt{2}M_\mu\right)\left(1 - \frac{1}{r^2}\right)\bar{\mu}(1 - \bar{\mu}^2) + O(\bar{k}^2). \tag{12.123}$$

Within the drop, assuming that the Reynolds number is small, the steady streaming satisfies the Stokes equation, i.e.,

$$D^4 \widehat{\psi}_1^{(s)} = 0, \tag{12.124}$$

where $\widehat{\psi}_1^{(s)}$ takes the role of Υ in eqn (12.36) together with $v_r = \widehat{u}_{r1}$ and $v_\theta = \widehat{u}_{\theta 1}$. Eqn (12.124) has the solution

$$\widehat{\psi}_1^{(s)} = \sqrt{2}M_\mu\left\{\bar{k}\sin(\bar{k}z_0)\cos(\bar{k}z_0)\left[\frac{1}{24}\sqrt{2}M_\mu(r^2 - r^4)(1 - \bar{\mu}^2)\right.\right.$$
$$\left.\left. -\frac{3}{112}(r^4 - r^6)(5\bar{\mu}^2 - 1)(1 - \bar{\mu}^2)\right] + \cos^2(\bar{k}z_0)\frac{9}{160}(r^3 - r^5)\bar{\mu}(1 - \bar{\mu}^2)\right\} + O(\bar{k}^2). \tag{12.125}$$

It should be noted that this solution in the drop region is uniformly valid as there is no internal Stokes layer. To sum it up, the internal circulation within the drop is defined by eqn (12.125), while the steady streaming in the continuous phase is given by (12.121) for the Stokes layer, and by (12.123) for the outer region. We note that the case of the solid sphere is recovered in all these expressions by letting the liquid-phase viscosity go to infinity, i.e., setting $M_\mu = 0$.

12.4.2 Discussion

In the above analysis, we find that the leading-order solution is a linear combination of the two groups of fundamental solutions corresponding to the sphere being placed at the node and antinode, respectively, of a standing wave. At higher orders, nonlinear effects become important and additional terms besides the two fundamental solutions are needed for the proper description of the flow.

The leading-order oscillatory flow for a liquid sphere is essentially the same as for a solid one, in view of the high viscosity of the liquid as compared to the surrounding gas medium. However, this is not the case for the steady streaming, when the difference between the liquid and the solid spheres may be appreciable. The effect of the 'liquidity' on the streaming is measured by the parameter M_μ. Typical flow patterns associated with the steady streaming of liquid spheres displaced between velocity node and antinode are displayed in Figures 12.12(a)–(d). The axially symmetric flow streamlines are provided for an arbitrary plane through the polar axis with $\bar{k} = 0.3$, and $\bar{k}z_0 = \pi/4$. The coordinate values represent distances scaled with the drop radius. Eqn (12.121) contains three terms that are linear in η, with coefficients Q_1, Q_2 and Q_3 given by eqn (12.117)–(12.119). These three factors divide the values of M_μ into four ranges. Each of these is discussed next.

For

$$Q_1 < 0 \quad \text{or} \quad M_\mu < \frac{315}{46}\sqrt{2}, \tag{12.126}$$

there are vortices near the surface of the drop on the side of the velocity node, i.e., the upper side, as shown in Figure 12.12(a). Here, there is a large recirculatory region on the 'front side' of the drop with respect to the outer streaming, which is downward. This may appear to be unusual from the standpoint of flows past obstacles that have a rear-side wake. However, with levitation, there is a low-pressure region on the top, and therefore it is possible for recirculation in that region. This is qualitatively consistent with the experimental observation (see Figure 12.11). While the experiment corresponds to $|M| = 113$, theoretical calculations at such a low value do not show a 'front-side' recirculatory region. However, in the experiment, there were some effects such as those from the chamber walls that have not been accounted for (see discussion by Rednikov and Riley[33]). There are some other interesting features in this flow field. There exist very thin recirculatory regions in the gas phase on the lower side of the drop. These are difficult to resolve graphically, except on a stretched scale (see Figure 12.12(c), for example). With an increase in M_μ, when

$$Q_1 > 0 \text{ and } Q_2 < 0, \quad \text{or} \quad \frac{315}{46}\sqrt{2} < M_\mu < \frac{25}{2}\sqrt{2}, \tag{12.127}$$

the vortices disappear. While the streamlines inside the shear-wave layer join the outside ones smoothly, as shown in Figure 12.12(b), the small recirculatory region still persists. The detailed depiction of this zone is given in Figure 12.12(c). With a further increase in M_μ, when

$$Q_2 > 0 \text{ and } Q_3 > 0, \quad \text{or} \quad \frac{25}{2}\sqrt{2} < M_\mu < \frac{91}{6}\sqrt{2}, \tag{12.128}$$

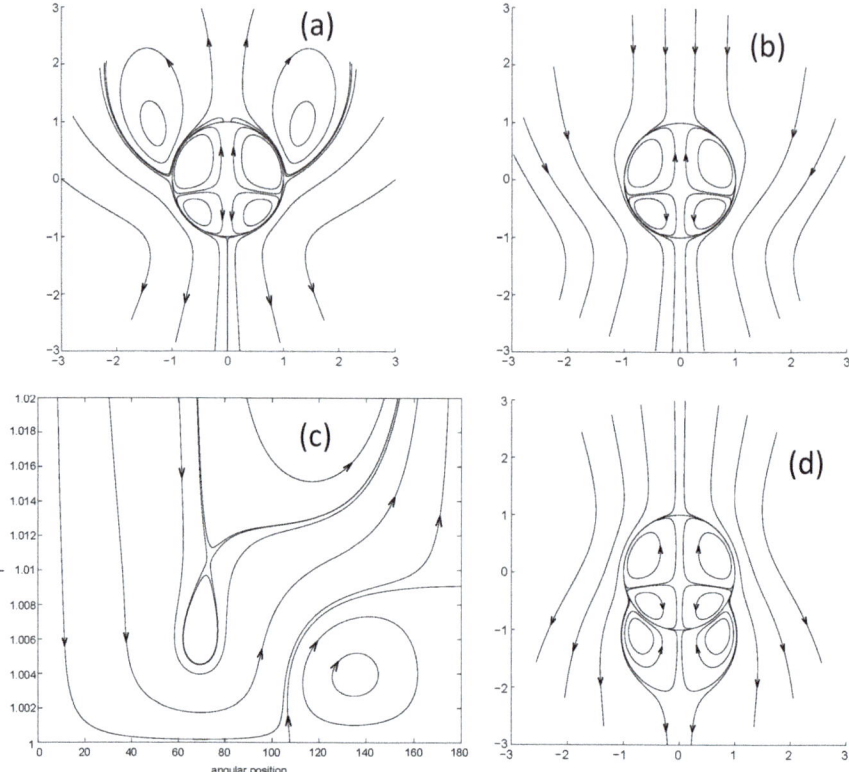

Figure 12.12 Streaming about a drop displaced between velocity node and antinode for $\bar{k}z_0 = \pi/4$, $\bar{k} = 0.3$ at various values of M_μ. The node is on the upper side and the antinode on the lower. (a) $M_\mu = 9.36$ and (b) $M_\mu = 12.48$. The outer streaming along the axis is away from the south pole. Within the drop, the flow on the lower side is in the opposite sense, i.e., away from the middle towards the poles, and towards the middle from the equatorial region. This is owing to the complex recirculating region in the Stokes layer, which is detailed in (c) for the case (b). The case (d) corresponds to $M_\mu = 28.08$ where we see the lower-side vortex enlarging considerably into the outer streaming. Reproduced from ref. 31.

the thin layer of recirculation becomes apparent near the lower surface of the drop (for details, see Figure 12.12 in ref. 31). When M_μ is very large, corresponding to

$$Q_3 < 0 \quad \text{or} \quad M_\mu > \frac{91}{6}\sqrt{2}, \tag{12.129}$$

the thin layer of recirculation becomes enlarged and vortices are created on the side of the velocity antinode as shown in Figure 12.12(d).

We next consider the limiting case of a solid sphere placed between nodes.

12.4.3 Streaming around a Solid Sphere Placed between Nodes

The results given by eqn (12.116)–(12.123) collapse to the solid-sphere case when $\mu/\hat{\mu} = 0$, or equivalently, $M_\mu = 0$. This limit is consistent with the outer solution of Lee and Wang.[2] In Figures 12.13 through 12.15, we can see the streamlines for a solid sphere with $\bar{k} = 0.3$. It is apparent that the asymmetry about the equator in the streaming pattern when the sphere is away from the velocity antinode is because of the asymmetric distribution of the undisturbed flow. There is stronger streaming on the velocity antinode side where the fluid velocity tends to be higher. Away from the surface of the sphere, the flow pattern does not depend on $|M|$ of course, but on the displacement $\bar{k}z_0$. It is noted that there is a transition value $\bar{k}z_0 = K_0$ (with $5\pi/16 < K_0 < 3\pi/8$) in the flow pattern. When $\bar{k}z_0 < K_0$, there exists a thin recirculating region, limited to the Stokes layer adjacent to the surface, quite similar to that for a solid particle at the velocity antinode. Since this region is quite thin, it is not clearly visible in Figures 12.13 through 12.14. However, when $\bar{k}z_0 > K_0$, larger vortices appear around the north pole region as shown in Figure 12.15.

One of the important findings of these studies[31,32] is the significant difference in the streaming flow behaviour about a liquid drop from that about a solid sphere. This is the case even when the liquid viscosity is quite high. It is apparent that the flow characterization is sensitive to surface

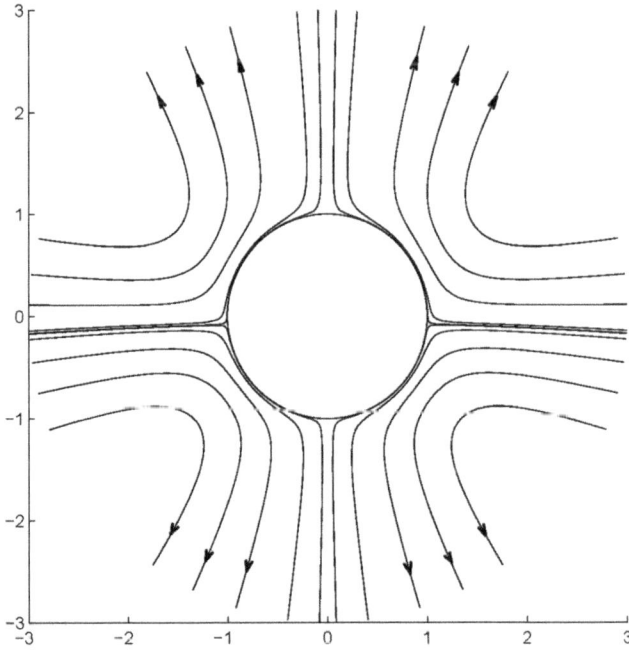

Figure 12.13 Streaming about a solid sphere displaced between velocity node and antinode for $\bar{k}z_0 = \pi/8$, $\bar{k} = 0.3$, and $|M| = 800$. With small displacement from the antinode, the flow is nearly symmetric about the equator. Reproduced from ref. 31.

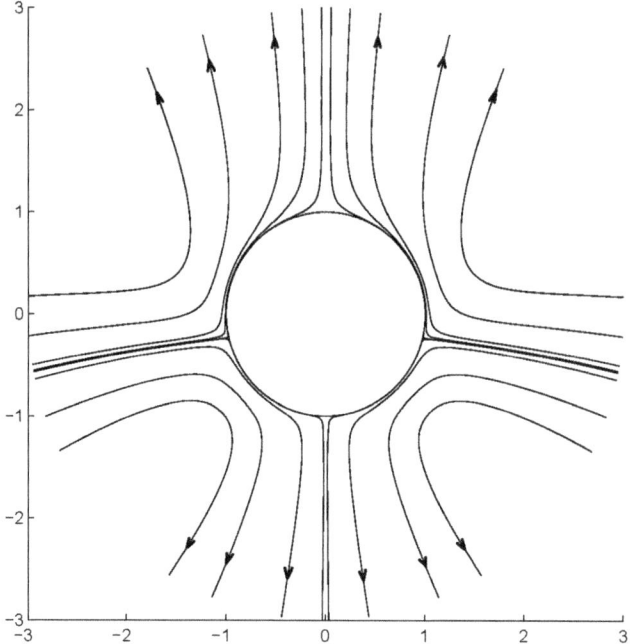

Figure 12.14 Streaming about a solid sphere displaced between velocity node and antinode for $\bar{k}z_0 = 5\pi/16$, $\bar{k} = 0.3$, and $|M| = 800$. With increasing $\bar{k}z_0$, the equatorial symmetry is broken. Reproduced from ref. 31.

mobility, which affects the interaction of the acoustic wave with that surface. As shown by Schlichting[7] and by Riley,[29] the interaction with a solid surface produces recirculating regions adjacent to the surface. This aspect can be viewed from a better perspective in the case of a liquid drop at the velocity antinode. In the next section, more discussion is given with parametric comparison of the shear forces in the liquid and the gas regions.

12.4.4 Solid Particle/Liquid Drop at the Velocity Antinode

In the limit of $z_0 \to 0$, the sphere is considered to be placed at the velocity antinode [see discussion following eqn (12.89)]. The far-field dimensionless velocity can be taken to be effectively $u_z = e^{it}$, and the flow field may be approximated as incompressible. We focus only on the steady streaming results which correspond to $O(\varepsilon)$, and in this case the solutions are given as follows. The stream function in the boundary layer (eqn (12.121)) becomes

$$\psi_1^{b(s)} = -\left(-\frac{9}{32}e^{-2\eta} - \frac{45}{8}e^{-\eta}\cos\eta - \frac{27}{8}e^{-\eta}\sin\eta - \frac{9}{4}\eta e^{-\eta}\sin\eta + 2B\eta \right.$$
$$\left. + \frac{189}{32}\right)\mu(1-\mu^2) + O(\bar{k}^2),$$

(12.130)

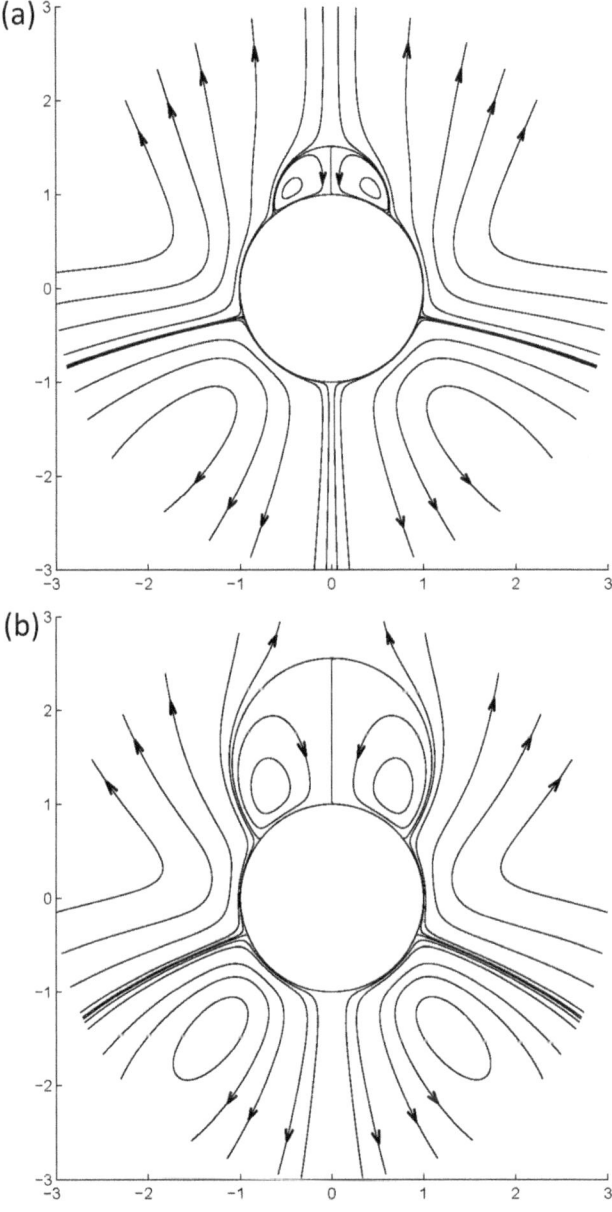

Figure 12.15 Streaming about a solid sphere displaced between velocity node and antinode for $\bar{k} = 0.3$, and $|M| = 800$. Close to the node ($\bar{k}z_0 = \pi/2$), vortices develop on the north pole side. (a) $\bar{k}z_0 = 3\pi/8$; (b) $\bar{k}z_0 = 7\pi/16$. Reproduced from ref. 31.

where

$$B = -\frac{45}{32}\left(1 - \frac{1}{25}\sqrt{2}M_\mu\right). \tag{12.131}$$

The steady streaming in the outer region given by eqn (12.123) takes the form

$$\psi_1^{(s)} = -B\left(1 - \frac{1}{r^2}\right)\bar{\mu}(1 - \bar{\mu}^2) + O(\bar{k}^2). \tag{12.132}$$

Inside the drop, eqn (12.125) reduces to

$$\hat{\psi}_1^{(s)} = \frac{9}{160}\sqrt{2}M_\mu(r^3 - r^5)\bar{\mu}(1 - \bar{\mu}^2) + O(\bar{k}^2). \tag{12.133}$$

In the next section, we discuss the solution developed by Riley[29] for a solid particle at the velocity antinode together with that of Zhao et al.,[32] both of which are recoverable in part from the general case above.

The flow field represented by eqn (12.130)–(12.133) corresponds to the results given by Zhao et al.[32] Since here we have considered the fact that with $\kappa \ll 1$, the form is slightly different from ref. 32 notably the expression for B. These flow descriptions show that the steady inner and outer fields depend on the frequency parameter $|M|$ quite directly, besides being in the stretched coordinate η. At the macroscale, with many of the levitation experiments, the wave frequency is 20–40 kHz, and the diameter of the sphere is 3–8 mm. For a gaseous medium outside the liquid drop (water), $k \simeq 0.016$. In this case, M_μ ranges from approximately 2 to 10, and we find that

$$B < 0 \quad \text{or} \quad M_\mu < \frac{25}{2}\sqrt{2}. \tag{12.134}$$

In this range, there is a recirculating layer around the drop similar to the case of a solid sphere as shown in Figure 12.16(a). With increasing M_μ, the parameter B in eqn (12.132) vanishes and then reverses sign. At that point, the recirculation in the shear layer ceases, and a further increase in M_μ leads to streaming in the opposite direction without a recirculating region as shown in Figure 12.16(b).

While this theory that predicts the cessation of recirculation in the shear-wave layer has been reported for over ten years now, there is still a need for conclusive experimental verification. In recent experiments,[34] the strong dependence of the internal and external flow characteristics on the liquid viscosity have been reported. For a bubble in an acoustic field, experiments show that the streaming in the continuous phase is in the opposite direction to that for a solid sphere.[35] While we are still seeking detailed theoretical explanations for this phenomenon, we should mention that the vorticity generated by the acoustic field interacting with an interface is manifested in the form of recirculation (see Schlichting[7]). This takes place when an acoustic

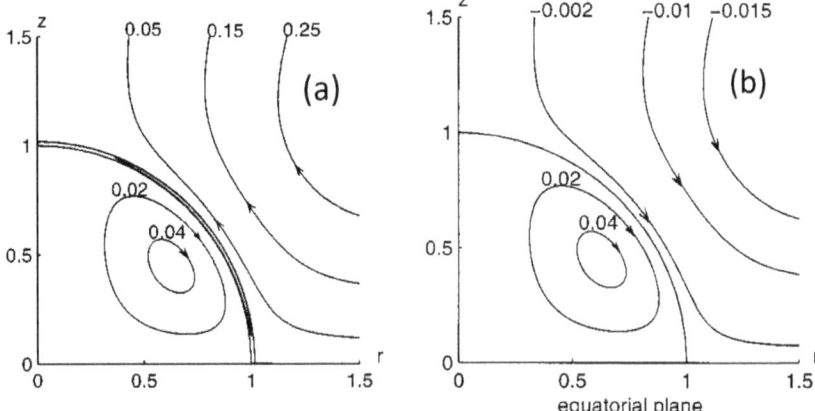

Figure 12.16 Streaming about a drop at the velocity antinode for $\kappa=0.0156$ with (a) $B < 0$, for $|M| = 200$ and (b) $B > 0$, with $|M| = 1200$ and $\kappa=0.0156$. The parameter B is the constant factor in $\psi_1^{(s)}$ (eqn (12.132)). Reproduced from ref. 32.

field interacts with a solid surface. However, in the case of a fluid surface, interfacial mobility is likely to reduce this effect. With decreasing drop-phase viscosity, the strength of the shear-wave layer recirculation diminishes and could vanish when the parameter B in eqn (12.131) is zero. The effect is particularly pronounced because the thinness of the recirculating layer affords a great deal of shear, and fluidity at the interface appears to mitigate that effect. It should be noted that at the cessation point, ψ_1 also vanishes, and higher-order solutions are needed for a valid description.

An important comparison is that with Riley's[29] result corresponding to a solid sphere. The steady-state flow pattern outside the liquid sphere for $B > 0$ is quite similar to that of the solid sphere obtained by Riley.[29] The Stokes layer in this case has a region of recirculation (see Figure 12.16(a)). In comparison with a liquid drop, it is interesting that just a small degree of liquidity dramatically changes the flow characterization. This is owing to the interplay between the shear in the Stokes layer competing with the viscous resistance due to the high-viscosity drop. Thus the competing shear forces are

$$\mu \frac{U_0}{a/|M|} \quad \text{and} \quad \hat{\mu}\frac{U_0}{a}, \tag{12.135}$$

respectively, where $a/|M|$ represents the Stokes layer thickness. Thus, the ratio of these two parameters,

$$M_\mu = |M|\frac{\mu}{\hat{\mu}} \tag{12.136}$$

is meaningful here, and an important transition takes place at a rather moderate value $M_\mu = \frac{25}{2}\sqrt{2}$.

While Riley's[29] results for the steady part of the streaming have been derived from the general case given by eqn (12.121)–(12.125), it should be mentioned that Riley[29] rigorously went quite a bit further with the solution. In particular, he kept the unsteady terms to $O(\varepsilon)$, and formally established that the steady outer streaming to this order obeys Stokes flow by examining higher order equations.

In principle, the solution given by eqn (12.121)–(12.125) should also go to the case of drop or a solid sphere at the velocity node with $\bar{k}z_0 = \tfrac{1}{2}\pi$. However, the solution vanishes to $O(\bar{k})$, and to have a meaningful solution in that region, we need to go to $O(\bar{k}^2)$.[15] The case of a solid sphere is discussed next.

12.4.5 Solid Sphere at the Velocity Node

For a solid sphere positioned at the velocity node, with the parametric restrictions $|M| \gg 1$ and $\varepsilon \ll 1$, the leading order solution can be taken as a special case of the general-case result given by eqn (12.110)–(12.112) where we let $\bar{k}z_0 = \tfrac{1}{2}\pi$. At this stage, the infinite viscosity limit $\mu/\hat{\mu} \to 0$ is not needed since at $O(\varepsilon^0)$, the liquid drop effectively behaves like a solid sphere.

12.4.5.1 The Leading-Order Solution

The leading-order pressure distribution, as well as the normal and tangential velocities in the Stokes layer are

$$p_0^b = -\frac{i}{\bar{k}}\left[1 - \frac{1}{2}\bar{k}^2 - \frac{5}{9}\bar{k}^2 P_2(\bar{\mu})\right]e^{it}, \tag{12.137}$$

$$u_{\theta 0}^b = \frac{5}{3}\bar{k}\sin\theta\cos\theta(1 - e^{-(1+i)\eta})e^{it}, \tag{12.138}$$

and

$$u_{\eta 0}^b = \left\{-\bar{k}\eta + \frac{10}{3}\bar{k}\left[-\eta + \frac{1}{2}(1-i)(1 - e^{-(1+i)\eta})\right]P_2(\bar{\mu})\right\}e^{it}. \tag{12.139}$$

Here, it should be noted that the first term $-\bar{k}\eta e^{it}$ represents the compressibility in the boundary layer.

12.4.5.2 The First-Order Solution $[O(\varepsilon)]$

At $O(\varepsilon)$, the solution for the general case given by eqn (12.121) vanishes to $O(\bar{k})$ when we set $\bar{k}z_0 = 0$. For a particle placed at the velocity node, the steady streaming requires analysis to $O(\bar{k}^2)$. Following Zhao et al.,[15] and considering that the steady flow in the boundary layer may be regarded as incompressible, the velocity field can be written in terms of the stream function $\psi_1^{b(s)}$ with the same set of relationships as in eqn (12.120). Using the expression for $u_{\theta 1}^{b(s)}$ in eqn (12.114), and differentiating with respect to η once to eliminate the pressure derivative, results in a fourth-order differential equation for ψ_1^b. After successive integration, and applying the limit $\psi_1^b = o(\eta^2)$ as $\eta \to \infty$, together with the boundary conditions,

$$\psi_1^b = 0 \quad \text{and} \quad \frac{\partial \psi_1^b}{\partial \eta} = 0 \quad \text{at } \eta = 0,$$

we obtain the solution for ψ_1^b as

$$\psi_1^b = -\varepsilon' \bar{k}^2 \left\{ \left(\frac{25}{72} e^{-2\eta} + \frac{10}{3} e^{-\eta} \cos \eta + \frac{35}{18} e^{-\eta} \sin \eta + \frac{5}{9} \eta e^{-\eta} \sin \eta + \frac{25}{12} \eta \right. \right.$$
$$\left. - \frac{265}{72} \right) \bar{\mu}(1 - \bar{\mu}^2) + \left(-\frac{25}{36} e^{-2\eta} - \frac{100}{9} e^{-\eta} \cos \eta - \frac{125}{18} e^{-\eta} \sin \eta \right.$$
$$\left. \left. - \frac{25}{6} \eta e^{-\eta} \sin \eta - \frac{50}{9} \eta + \frac{425}{36} \right) \bar{\mu}^3 (1 - \bar{\mu}^2) \right\}.$$

(12.140)

The perturbation solution (12.140) represents an inner solution, corresponding to the Stokes layer. For the outer region where $(r-1) = O(1)$, we need to construct another asymptotic solution. Again, it is not difficult to demonstrate incompressibility[15] so that $\nabla \cdot \boldsymbol{u}_1 = 0$. Therefore, once more, we introduce the stream function ψ_1, for the outer region this time, using eqn (12.36) with $\Upsilon = \psi_1$, $v_r = u_{r1}$ and $v_\theta = u_{\theta 1}$. Equating coefficients of powers of ε in the momentum eqn (12.91), and using the above stream function relationship, followed by a time average, we obtain the Stokes flow equation.

$$D^4 \psi_1 = 0, \quad (12.141)$$

where D^2 is the Stokes operator given by eqn (12.38). After asymptotic matching, we obtain the following expression for ψ_1:

$$\psi_1 = \bar{k}^2 \left[\frac{25}{168}(-r^{-2} + 1)\bar{\mu}(1 - \bar{\mu}^2) + \frac{25}{63}(-r^{-4} + r^{-2})(7\bar{\mu}^3 - 3\bar{\mu})(1 - \bar{\mu}^2) \right],$$

(12.142)

again demonstrating the persistence of streaming outside the Stokes layer. As in Section 12.2.3, we can obtain the slip velocity

$$u_\theta^{(\text{slip})} = -\varepsilon \frac{(1 - \bar{\mu}^2)^{-\frac{1}{2}}}{r} \frac{\partial \psi_1}{\partial r} \bigg|_{r=1}$$

$$= -\varepsilon \bar{k}^2 \left[\frac{25}{84} \bar{\mu}(1 - \bar{\mu}^2)^{1/2} + \frac{50}{63}(7\bar{\mu}^3 - 3\bar{\mu})(1 - \bar{\mu}^2)^{1/2} \right]. \quad (12.143)$$

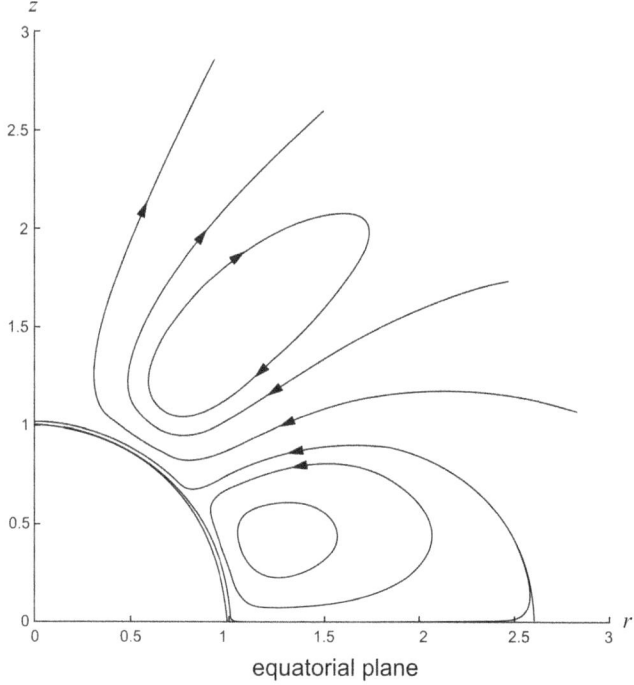

Figure 12.17 Streaming in the outer region for a sphere placed at the velocity node. The detail in the Stokes layer is shown in Figure 12.18. Reproduced from ref. 15.

12.4.6 Discussion

The streaming flow field in the outer region is depicted in Figure 12.17. Here, unlike the sphere at the velocity antinode, the outer region has a pair of toroidal vortices (only one is shown) symmetrical about the equatorial plane. The recirculating part of the Stokes layer does not cover the entire sphere but just the equatorial belt. Over the remaining region in the Stokes layer, the outer flow continues into the Stokes layer. The detail in the Stokes layer region is not clear in this figure, and is shown on a stretched radial scale in Figure 12.18.

12.5 Bubbles in Acoustic Fields

In this section, we shall examine the phenomenon of microstreaming with bubbles in two types of situations. One arises due to a spherical bubble when its surrounding liquid is undergoing steady vibrations, and another recent development deals with a two-dimensional semi-cylindrical microbubble. For the spherical bubble, we consider first the case of a fixed bubble volume

Analysis of Acoustic Streaming by Perturbation Methods

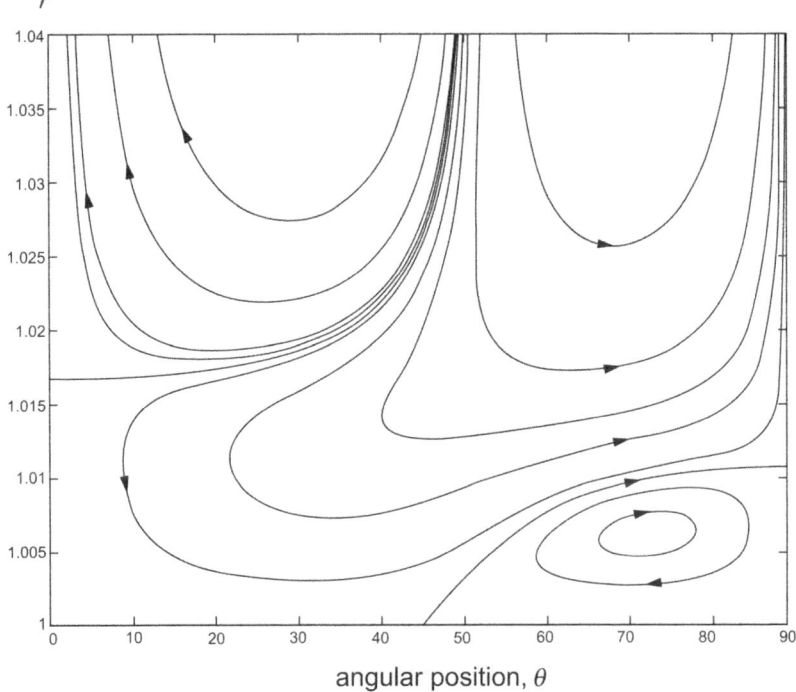

Figure 12.18 Detailed flow field in the Stokes layer on the surface with a stretched radial scale. Reproduced from ref. 15.

followed by the analysis for a radially pulsating bubble. Next, we discuss the experimental and theoretical developments for the cylindrical microbubble. We shall not discuss other phenomena such as sonoluminescence.

12.5.1 Transverse Oscillations with Fixed Volume

As mentioned in Section 12.4.4, early experiments by Elder[35] demonstrated that the streaming observed around the bubble is opposite to that for a solid particle. This is indeed consistent with the absence of the recirculating layer for the bubble, as well as the theoretical prediction for a droplet when its viscosity is low enough for the recirculation in the Stokes layer to be absent. In fact, for a surface-contaminated bubble that behaves like a rigid particle, the recirculating layer was observed. Interestingly, when the driving amplitude was increased, the surface skin broke and the outer streaming changed sign. The analysis of this type of flow was conducted by Davidson and Riley.[36] Among the solutions obtained were cases for $|M| = \sqrt{(\omega/\nu)}a \gg 1$, $|M| = O(1)$ and $|M| \ll 1$. Since the groundwork for the basic perturbation procedure is similar to Riley's[29] classical solution, we shall give only the results. The analysis was carried out for a bubble at the velocity antinode. Keeping in mind that the flow field is axisymmetric and that the incompressible flow approximation applies, the stream function formulation in eqn (12.35)–(12.39) applies with $\Upsilon = \psi$ and $\mathbf{v} = \mathbf{u}$. The boundary conditions are as follows:

Far-Field

As in the solid-sphere and liquid-drop cases,[29,32] with the long-wavelength assumption ($\lambda \gg a$), the far-field around the drop is a simple oscillating field around the velocity antinode,

$$\psi(r, \bar{\mu}, t)|_{r \to \infty} = \frac{1}{2} r^2 (1 - \bar{\mu}^2) e^{it}. \qquad (12.144)$$

Interface $r = 1$

At the interface, we have zero normal velocity,

$$\psi(1, \bar{\mu}, t) = 0, \qquad (12.145)$$

and the shear-free condition, which may be expressed by

$$\frac{\partial^2 \psi}{\partial r^2} - 2 \frac{\partial \psi}{\partial r} = 0 \quad \text{at} \quad r = 1. \qquad (12.146)$$

The perturbation expansion used in this analysis employed both $\varepsilon = U_0/\omega a \ll 1$ and $\varepsilon' = \sqrt{2}/|M| \ll 1$. First, with the expansion in ε,

$$\psi = \psi_0 + \varepsilon \psi_1 + \cdots, \qquad (12.147)$$

the momentum eqn (12.37) to the leading order becomes

$$\frac{\partial}{\partial t}(D^2 \psi_0) = \frac{1}{2} \varepsilon'^2 D^4 \psi_0. \qquad (12.148)$$

Further expansion in ε',

$$\psi_0 = \psi_{00} + \varepsilon' \psi_{01} + \varepsilon'^2 \psi_{02} + O(\varepsilon'^3)$$

$$\psi_1 = \psi_{10} + \varepsilon' \psi_{11} + \varepsilon'^2 \psi_{12} + O(\varepsilon'^3) \qquad (12.149)$$

for the outer region leads to

$$\frac{\partial}{\partial t}(D^2 \psi_{0n}) = 0, \quad n = 0, 1, 2, \qquad (12.150)$$

with solutions,[36]

$$\psi_{00} = \frac{1}{2} r^2 (1 - 1/r^3)(1 - \bar{\mu}^2) e^{it}, \qquad (12.151)$$

$$\psi_{01} = 0, \qquad (12.152)$$

$$\psi_{02} = \frac{3i}{2r}(1 - \bar{\mu}^2) e^{it}, \qquad (12.153)$$

Analysis of Acoustic Streaming by Perturbation Methods

where the boundary conditions (12.145) and (12.146) cannot be completely satisfied since the highest derivatives in (12.148) are lost in the outer expansion with $\varepsilon' \ll 1$. Using the inner radial coordinate η defined by eqn (12.104), and the stream function according to eqn (12.51), the expansion procedure

$$\Psi = \Psi_{00} + \varepsilon'\Psi_{01} + \varepsilon'^2\Psi_{02} + O(\varepsilon'^3) + \varepsilon\left[\Psi_{10} + \varepsilon'\Psi_{11} + \varepsilon'^2\Psi_{12} + O(\varepsilon'^3)\right] + \ldots \tag{12.154}$$

is used. The result is given as[36]

$$\Psi_{00} = \frac{3}{2}\eta(1-\bar{\mu}^2)e^{it} \tag{12.155}$$

$$\Psi_{01} = \frac{3}{2}\left(1 - e^{-(1+i)\eta}\right)(1-\bar{\mu}^2)e^{i(t+\pi/2)} \tag{12.156}$$

$$\Psi_{02} = \frac{1}{2}\left[\eta^3 - 3i\eta - 3(1+i)\left(1 - e^{-(1+i)\eta}\right)\right](1-\bar{\mu}^2)e^{it}. \tag{12.157}$$

To $O(\varepsilon)$, the steady components are given as

$$\psi_{11}^{(s)} = \frac{27}{40}\left(1 - \frac{1}{r^2}\right)\bar{\mu}(1-\bar{\mu}^2)$$

$$\psi_{11}^{(s)} = -9\left\{1 + \frac{3}{20}\eta - \frac{1}{4}\left[4\cos\eta e^{-\eta} + \eta e^{-\eta}(\cos\eta + \sin\eta)\right]\right\}\bar{\mu}(1-\bar{\mu}^2) \tag{12.158}$$

with $\psi_{10}^{(s)} = \psi_{12}^{(s)} = \psi_{10}^{(s)} = \psi_{12}^{(s)} = 0$. As with the case of the solid sphere, the nonzero expression for $\psi_{11}^{(s)}$ indicates the persistence of the streaming outside the Stokes layer. The order of the streaming velocities in this case are $O(\varepsilon\varepsilon')$ as compared with $O(\varepsilon)$ for the solid sphere. The outer streaming pattern is shown in Figure 12.19(a). Of course, with the absence of closed streamlines in the Stokes layer (see Figure 12.16), the outer streaming pattern is the reverse of the high-viscosity droplet case (or a solid sphere case). Not surprisingly, the direction is consistent with the low-viscosity droplet case.

12.5.2 Radial and Transverse Oscillations of Bubbles

If we include radial oscillations besides the transverse ones, interesting streaming patterns emerge.[37,38] This happens, for example, in the case of a bubble undergoing sonoluminescence.[39] Longuet-Higgins[37] tackled this

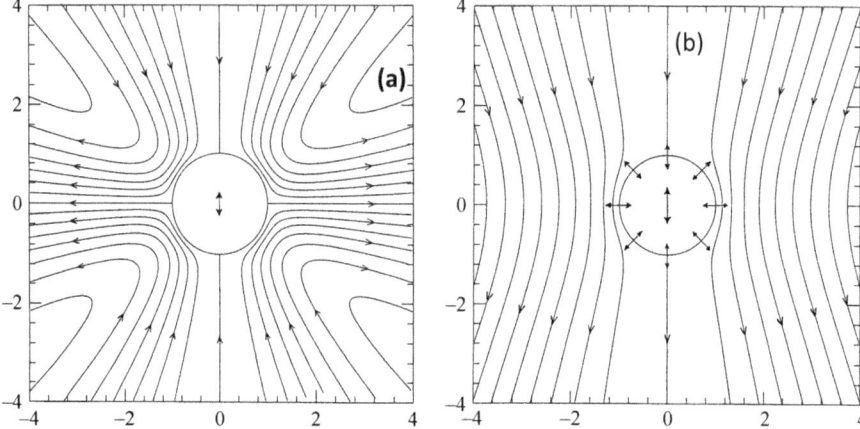

Figure 12.19 The outer steady streaming pattern for a bubble with (a) only transverse oscillations, and (b) radial and transverse oscillations. Reproduced from ref. 37.

problem by allowing a pulsating normal velocity at the bubble surface at the same frequency as the transverse oscillations but out of phase. He assumed the bubble surface to be defined in dimensional form by

$$r' = \tilde{R}(t') = a\left[1 + i\tilde{\varepsilon}e^{i(\omega t' + \beta)}\right], \quad (12.159)$$

where $\tilde{\varepsilon}a\omega$ represents the velocity amplitude of the radial oscillations, and β is the dimensionless phase difference with the transverse oscillations. To accommodate the radial oscillations, we replace the condition (12.145) with

$$\left.\frac{\partial \psi}{\partial \bar{\mu}}\right|_{r=1} = -\tilde{\varepsilon}e^{i(t+\beta)}, \quad (12.160)$$

which is dimensionless. The zero-shear stress condition (12.146) remains unchanged. We now apply the perturbation expansion,

$$\psi = \psi_0 + \frac{\tilde{\varepsilon}}{\varepsilon}\tilde{\psi}_0 + \left[\varepsilon\psi_1 + \tilde{\varepsilon}\tilde{\psi}_1\right] + O(\varepsilon^2) + O(\tilde{\varepsilon}^2), \quad (12.161)$$

and substitute into the scalar momentum eqn (12.37), to yield the same linearized differential eqn (12.148) to the leading order in ε and ε'. Longuet-Higgins[37] chose not to perturb this in powers of $\varepsilon' = \sqrt{2}/|M|$, and used the full solution in terms of spherical Bessel functions. Either way, the effect of radial oscillations is simply the addition of a solution corresponding to just radial oscillations,

$$\frac{\tilde{\varepsilon}}{\varepsilon}\tilde{\psi} = \frac{\tilde{\varepsilon}}{\varepsilon}\bar{\mu}c^{i(t+\beta)}, \quad (12.162)$$

Analysis of Acoustic Streaming by Perturbation Methods

to the leading-order solution which Davidson and Riley[36] have given by equations (12.151)–(12.153) and (12.155)–(12.157). The far-field then should satisfy

$$\psi_0 + \frac{\tilde{\varepsilon}}{\varepsilon}\tilde{\psi}_0 \sim \left[\frac{1}{2}r^2(1-\bar{\mu}^2) - \frac{\tilde{\varepsilon}}{\varepsilon}e^{i\beta}\bar{\mu}\right]e^{it}. \qquad (12.163)$$

To the next order, the steady streaming manifests itself through the nonlinear interaction described by the ε-terms in eqn (12.37) after a time-average.

12.5.2.1 Streaming Results

Longuet-Higgins[37] has given the steady parts of the $O(\varepsilon)$ and $O(\tilde{\varepsilon})$ solutions for the Stokes layer region in terms of the Stokes layer thickness, which can be represented by the parameter ε'. The result is

$$\tilde{\psi}_1^{(s)} = \frac{3}{4}e^{-i\beta}\left\{\left(i + \varepsilon'^2 - 2i\varepsilon'\eta + 4i\varepsilon^2\eta'^2\right) - 6\varepsilon'^2 e^{-(1+i)\eta} - i + 5\varepsilon'^2\right.$$
$$\left. -\left[\frac{1}{3}i + 4(1+i)\varepsilon'\right]\varepsilon'\eta + \left[-\frac{1}{3}i + 2(1+i)\varepsilon'\right]\varepsilon'^2\eta^2\right\}(1-\bar{\mu}^2). \qquad (12.164)$$

Outside the Stokes layer, he gave in real variables, to the leading order,

$$\tilde{\psi}_1^{(s)} = -\frac{1}{2}\sin\beta\left(r - \frac{1}{2}r^{-1} - \frac{1}{2}r^{-4}\right)(1-\mu^2). \qquad (12.165)$$

It is important to note here that the leading order streaming flow is $O(\tilde{\varepsilon})$ as compared with $O(\varepsilon\varepsilon')$ for the case with only the transverse oscillations. The presence of radial oscillations therefore produces a flow field behavior quite different than without. This is evident through the streamline plot given in Figure 12.19(b).

Longuet-Higgins[37] attributed the dominance of the radial effect to stronger stretching of the vortex lines in the neighborhood of the spherical surface, as compared with ones caused by purely transverse oscillations. While simple transverse oscillations, and purely radial oscillations, do not produce any drift, the combination of these [in eqn (12.165)] produces terms proportional to r (Stokslet) and $1/r$. The latter is a dipole potential (dimensional),

$$\Phi = \frac{S\cos\theta}{2r'^2}, \qquad (12.166)$$

where the dipole strength S is related to the oscillatory parameters by[37]

$$S = \frac{1}{4}\varepsilon\tilde{\varepsilon}a^4\omega\sin\beta. \qquad (12.167)$$

12.5.3 Semi-cylindrical Bubble

This research has been motivated by the possibility of using ultrasound-driven oscillating microbubbles to serve as actuators to induce steady microstreaming.[40] From a practical standpoint, acoustic excitation can be easily provided with a commercially available transducer glued on a substrate. With the gas in the bubble being compressible, interfacial oscillations occur as the liquid undergoes vibrations. The analysis is based on the two-dimensional theory given in Section 12.1.2.1 with the momentum conservation described by eqn (12.33). While the work is quite extensive, we give a basic overview of the analysis and experiments. The problem formulation and some results are given next.

12.5.3.1 Analysis

The geometry for the semi-cylindrical bubble is described in Figure 12.20. For the two-dimensional momentum eqn (12.33) where, in polar coordinates,

$$\nabla^2 = \frac{\partial^2}{\partial r^2} + \frac{1}{r}\frac{\partial}{\partial r} + \frac{1}{r^2}\frac{\partial^2}{\partial \theta^2},$$

the boundary conditions on the bubble surface are as follows:[40]

Radially Pulsating Surface

$$\varepsilon\frac{1}{r}\frac{\partial \psi}{\partial \theta}\bigg|_{r=1} = \dot{R} = \varepsilon\chi(\theta)e^{it}, \qquad (12.168)$$

Shear-Free Surface:

$$\left[\frac{\partial^2 \psi}{\partial r^2} - \frac{1}{r}\frac{\partial \psi}{\partial r} - \frac{1}{r^2}\frac{\partial^2 \psi}{\partial \theta^2}\right]_{r=1} = 0. \qquad (12.169)$$

No-Slip Wall:
On the solid wall, we have the no-slip conditions

$$\frac{1}{r}\frac{\partial \psi}{\partial \theta} = \frac{\partial \psi}{\partial r} = 0 \quad \text{at} \quad \theta = 0, \pi \qquad (12.170)$$

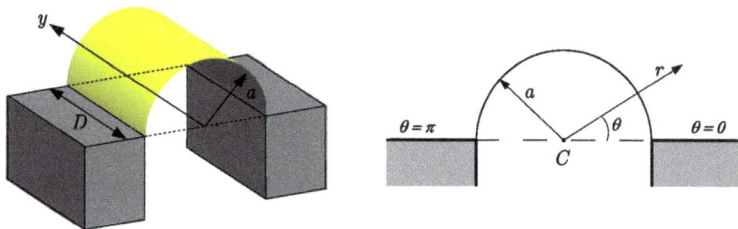

Figure 12.20 Schematic of a semi-cylindrical bubble. Reproduced from ref. 40.

The leading-order outer solution is found to be[40]

$$\psi_0(r,\theta) = \sum_{n=0}^{\infty} A_n \left(\frac{C_{2n}}{r^{2n}} + d_{2n}K_{2n}(\alpha r) \right) \sin(2n\theta)e^{it}, \qquad (12.171)$$

where

$$c_j = \frac{1}{j} - d_j K_j(\alpha) \qquad (12.172)$$

and

$$d_j = -\frac{2(j+1)}{\alpha^2 K_{j-2}(\alpha) + 2jMK_{j-1}(\alpha)}, \qquad (12.173)$$

with $K_j(z)$ representing the modified Bessel function of the second kind, and $\alpha = |M|$. The bubble shape profile is given by

$$\chi_0(\theta) = \sum_{n=0}^{\infty} A_n \cos(2n\theta). \qquad (12.174)$$

This solution does not satisfy the no-slip condition (12.170). For the boundary-layer region at $\theta = 0$, Wang, Rallabandi and Hilgenfeldt[40] constructed an inner solution to the leading order as

$$\Psi_0(r,\xi) = U(r)r\left[\xi - \frac{1-i}{2r}\left(1 - e^{-(1+i)\xi r}\right)\right]e^{it}, \qquad (12.175)$$

where the inner and the outer stream functions are related by (12.51), $\xi = \varepsilon'\theta$ and the velocity component $U(r)$ is slip velocity given by

$$U(r) = \sum_{n=0}^{\infty} 2nA_n \frac{C_{2n}}{r^{2n+1}} = \sum_{n=0}^{\infty} \frac{A_n}{r^{2n+1}}. \qquad (12.176)$$

With further calculations, a uniformly valid (composite of inner solutions at $\theta = 0$ and $\theta = \pi$ and the outer solution) radial velocity can be derived from here to yield

$$u_{r0}^c(r,\theta,t) = \sum_{n=0}^{\infty} \frac{A_n}{r^{2n+1}} \left[\cos(2n\theta) - \left(e^{-\alpha r\theta} + e^{-\alpha r(\pi-\theta)}\right)\right]e^{it}. \qquad (12.177)$$

The shape profile is found to be

$$\chi_0^c(\theta) = \sum_{n=0}^{\infty} A_n \left[\cos(2n\theta) - \left(e^{-\alpha r\theta} + e^{-\alpha r(\pi-\theta)}\right)\right] \qquad (12.178)$$

From this shape profile, Wang, Rallabandi and Hilgenfeldt[40] have further calculated the curvature and applied it for the normal stress balance across the liquid–gas interface and found expressions for the corresponding contributions for pressure, viscous damping and curvature. The details are not discussed here. These researchers have conducted experiments on flow visualization and obtained very interesting results.

12.5.3.2 Experiments

The oscillations of a sessile semi-cylindrical bubble in 23% glycerol sandwiched between two plane walls, and attached to a side wall have been studied under different driving frequencies. The streaming flow, which has a nonzero time-average, has been visualized with density-matched polystyrene microparticles (0.5–1 µm radius). The driving frequency ranged from 1.6 to 103.6 kHz in increments of 0.5 kHz, keeping the input voltage to the piezoelectric transducer constant. The streaming flow pattern has been presented for several frequencies between 9.5 and 111 kHz. These results are exhibited in Figure 12.21.

At low frequencies (10–20 kHz), a single pair of vortices is observed. With increasing frequency, a second pair of vortices emerges (50 kHz) with opposite orientation. The lower set of images shows the oscillatory modes of the bubble at the different frequencies.

The use of microstreaming around a bubble has been proposed as a mixing device by combining such streaming with Poiseuille flow after merging from a T-junction. The investigators[40] found that a very effective mixing mechanism is attained by switching the bubble frequency at intervals approximating the clearance time for the channel flow to pass over the full bubble diameter. Thus for an average channel velocity $\bar{u}_p$ and bubble diameter d, time intervals $t_p = d/\bar{u}_p$ were the best values. For the specific case under consideration, $\bar{u}_p \simeq 1.3 \text{ ms}^{-1}$ and $d = 80$ µm, $t_p = 60$ ms, frequency switching between 27.1 kHz and 91.3 kHz significantly enhanced the mixing quality. They have noted that this technique only works when the switch frequencies belong to different oscillation modes and correspond to substantially different flow fields.

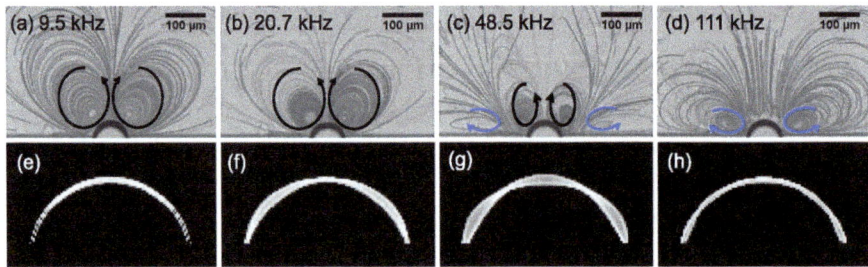

Figure 12.21 Streaming flow pattern around the bubble at different driving frequencies. The lower images are the outlines showing the bubble oscillation modes: (e) 9.6 kHz, (f) 20.6 kHz, (g) 48.6 kHz, (h) 100.3 kHz. Reproduced from ref. 40.

In earlier work, Tho, Manasseh and Ooi[41] experimentally observed microstreaming around single and multiple hemispherical bubbles on solid surfaces. They considered various modes of oscillation as special cases of generalized elliptical oscillations, including translational and circular. Also, they considered cases with purely volumetric oscillations and combinations with translational as well as shape mode oscillations. Translational oscillations along the plane of the solid surface produced a quadrupole streaming pattern. Upon introducing ellipticity in the oscillations, and gradually shifting towards circular oscillations, the quadrupole pattern became a skewed quadrupole, and then elliptical and finally circular. Volumetric oscillations indicated a dipole pattern.

12.6 Concluding Remarks

We have attempted here to present the acoustic streaming phenomenon with some generalized sets of principles, and approached the analytical aspects from a perturbation standpoint. Recognizing that the number of analytically solvable problems is limited, and at the same time setting the goals of providing a depth of understanding of the subject matter, we have resorted to a simple class of problems that can provide an illustration of the streaming phenomenon. Clearly, it is the rotational aspects of the flow fields that tend to become significant near solid boundaries, and ensure the nonlinear interactions that bring about the streaming phenomenon. Perturbation theory provides an excellent avenue for bringing out the fundamental core of this subject while retaining a relative simplicity in the mathematical formulation. We have relied heavily on Riley's[17] review where he has provided a generalized approach to several types of streaming problems and applied it to, for example, the Rayleigh problem. The generalization has been taken further by Rednikov and Sadhal.[9] The example of an oscillating flat plate parallel to another flat plate is considered particularly instructive, especially since it exhibits a wide variety of streaming flow characteristics. As noted earlier, oscillations purely in the plane of the plates produces no streaming. There are several examples where two types of oscillatory motions do not independently generate streaming but together, they interact nonlinearly and cause streaming to take place. This point has been noted by Kelly[42] with regard to oscillations involving a cylinder. Neither the rotational oscillations nor the longitudinal ones generate streaming but the combination of these two indeed produce steady flow.

In terms of microfluidic applications, the importance of streaming in liquids is elucidated by Bruus[3] (see Figure 1 in Bruus's paper). In an example taken from earlier work,[25] the relevance of acoustic streaming providing Stokes drag *versus* the radiation force is exhibited in a series of flow visualization pictures. For water in a 2 mm × 2 mm × 0.2 mm chamber at 2.17 MHz, the Stokes drag on 1 μm polystyrene beads is seen to be higher than the radiation force, while on 5 μm beads, the latter dominates. It should be noted here that the Stokes layer thickness for this frequency is close to 1 μm.

Similarly, Wiklund, Green and Ohlin[1] have given the example of PIV analysis by Spengler et al.[43,44] on the motion of 1 to 25 µm beads at 3 MHz where the larger particles agglomerate while the 1 µm ones were not noticeably affected by the radiation force. Other recent works on size-dependent transition from radiation force dominance to streaming include ref. 45. Similar results have been given by Kuznetsova and Coakley[46] who also measured streaming velocities. For example, at 1.38 MHz with half wavelength standing wave in a cylindrical resonator with pressure amplitudes 310–600 kPa, average streaming velocities ranged from below 100 µm s^{-1} to approximately 350 µm s^{-1} (see Figure 3 in ref. 46). The average velocity was seen to vary fairly linearly with the square of the pressure amplitude.

While the presence of streaming has for a long time been seen as an interfering nuisance, Wiklund, Green and Ohlin[1] have discussed ways of designing systems whereby streaming can be used to enhance particle manipulation through choices of microchannel widths and frequency.

The theoretical work on streaming around solid particles and liquid drops has also been discussed. We have provided techniques for tackling streaming-flow problems that arise when solid particles interact with sound waves, and the detailed inner and outer streaming calculation methods around solid particles have been presented. The cases of a sphere at the velocity antinode (pressure node) and the velocity node (pressure antinode) of a standing wave have been discussed, and dealt with analytically by the singular perturbation method. With all cases considered, a typical small dimensionless parameter $\varepsilon = (U_0/\omega a) \ll 1$, coupled with the frequency parameter $|M|^2 = \omega a^2/\nu \gg 1$, facilitate the analysis. A considerable depth of understanding is achieved with limiting cases of the streaming Reynolds number $R_s = U_0^2/\omega \nu = \varepsilon^2 |M|^2$ being small. This is the Reynolds number associated with the mean time-independent flow that arises as a result of the nonlinear interaction of the sound wave with a solid surface. It has been established that for the flow parameters under consideration, the leading-order streaming flow is typically an order higher than the primary oscillatory flow.

We have presented a set of results concerned with acoustic streaming that takes place when ultrasound waves interact with solid particles and liquid drops. In particular, for $\varepsilon \ll 1$ and $|M| \gg 1$, there is intense streaming in the thin Stokes layer near the boundary. The streaming propagates into the bulk of the fluid with intensities an order higher in ε than the leading-order flow. It should be noted that the bulk streaming (i.e., outside the Stokes layer) can be analyzed by considering an effective slip velocity on the boundary.[8,9,47] The fundamental case of a particle at the velocity antinode is mathematically equivalent to a sphere vibrating in an otherwise quiescent fluid, provided the wavelength is large compared with the particle size. The analysis can be carried out with the incompressible flow approximation. The interesting aspect of the streaming flow field is the existence of a thin recirculating zone at the surface of the sphere. Such a zone is also present in the case of a sphere placed at the velocity node. However, the recirculating region does not envelope the entire sphere but rather just an equatorial belt with coverage depending on the flow parameters. The calculations for this flow field demand the compressible flow analysis which does not lend itself so easily to the usual stream function formulation that works for axisymmetric flow. As far as the streaming phenomenon is concerned, there are several other interesting results such as the effect of two orthogonal ultrasound beams interacting with a particle.[47,48] With the beams being coherent but out of phase by $\pi/2$, there is a net torque on the particle which, if free to move, rotates. This mechanism has been used with levitation work when it is desired to spin a levitated particle.[49] Streaming is also seen to occur when a particle experiences rotational oscillations in an otherwise quiescent fluid.[50] However, it is not immediately obvious if an ultrasound wave would produce an equivalent flow field.

The flow field for the case of the sphere between nodes is a nonlinear combination of the node and antinode solutions. The results are more interesting with the presence of vortices corresponding to the streaming in the Stokes layer as well as the bulk region. In particular, there are full toroidal vortices present near the polar region of the oncoming streaming flow, qualitatively resembling the visualization exhibited in Figure 12.11. However, it is possible to attribute such vortices to the presence of the chamber walls of a levitation system.[33]

In the analytical treatment of liquid droplets encountering an ultrasound standing wave, it has been shown that the interfacial mobility significantly alters the streaming flow pattern. This happens even if the droplet medium viscosity is much higher than the surrounding medium (such as a gas). With a solid particle, the vorticity generation takes place in the thin Stokes layer region adjacent to the particle. The layer being very thin affords large shear stresses but even a small amount of mobility mitigates this effect. Therefore, for the simple case of a drop at the velocity antinode of the wave, there is a thin recirculating region in the Stokes layer, just as for a solid particle. However, with a reduction of the drop viscosity, a point is reached where there is a cessation of the recirculation, which leads to a reversal of the outer streaming flow pattern. The outer streaming being the reverse of the solid-particle case has indeed been observed in the case of bubbles in liquids.[35] With a drop between nodes, the drop viscosity plays an important role in determining the outer streaming flow characteristics, and this has been classified into various parametric ranges.

The analytical result for non-pulsating bubbles at the velocity antinode exhibits vorticity near the interface but substantially lower than that for the solid particle. The resulting outer streaming is found to be an order weaker than that for a solid sphere.[36] When the bubble is allowed to pulsate volumetrically at the same frequency as the ultrasound wave, leading-order streaming is significantly different. The radial oscillations contribute without any angular dependence towards the Reynolds stresses, and create a dipole flow structure as compared with a quadrupole for the non-pulsating case. Additionally, along with the dipole, there is the presence of a Stokeslet, which corresponds to the existence of a net force, or for a free bubble, a drift velocity. The magnitude of the streaming is an order higher than for the non-pulsating bubble.

As for experimental work on microstreaming with bubbles, Wiklund, Green and Ohlin[1] have discussed several works in this area. They cite the recent experiments by Tho, Manasseh and Ooi,[41] who carried out PIV measurements of flow velocities around air bubbles of approximately 250 μm radius. Their experiments included a number of different volumetric and transverse modes of oscillations (approximately 0.5–20 kHz) while the bubble rested on a solid boundary. Maximum streaming velocities of 100–400 μm s^{-1} were measured. While the dipole structure predicted by Longuet-Higgins[37] has been observed, oscillations with higher modes than these simple ones produce streaming velocities several times higher. Microstreaming with two-dimensional microbubbles on solid surfaces has been studied experimentally by Wang, Rallabandi and Hilgenfeldt.[40] For 80 μm radius bubbles, they measured streaming velocities of 10 mm s^{-1} near the bubble surface.

Acknowledgements

The author expresses his deep gratitude to Prof. Norman Riley (University of East Anglia) and Dr Alexey Rednikov (Université Libre de Bruxelles) both of whom have shared their immense expertise with him.

Nomenclature

Roman Symbols

a	length scale, particle radius
c	speed of sound
c_p, c_v	specific heats
D^2	Stokes operator, eqn (12.38)
$\hat{e}$	unit vector
F	body force per unit volume
i	$\sqrt{-1}$
k	wavenumber, $2\pi/\lambda$
L	differential operator (eqn (12.39))
M^2	frequency parameter, $i\omega a^2/\nu$
$|M|$	$a\sqrt{\omega/\nu}$, Womersley number
M_μ	$|M|\mu/\hat{\mu}$
p	pressure
$P_n(\bar{\mu})$	Legendre polynomials
r	radial coordinate (spherical and cylindrical)
R	Reynolds number
R_s	streaming Reynolds number

$\tilde{R}$	radial position of bubble surface
t	time
u, U	velocity
v	global velocity
x, y, z	cartesian coordinates
z_0	particle position

Greek Letters

α	Womersley number, $a\sqrt{\omega/\nu} = |M|$, plate velocity scale
α'	$\omega\left(\frac{3}{4}\beta' + \mu\right) \Big/ (\rho_0 c^2)$
β'	second viscosity, phase difference
β	dimensionless phase difference, $\omega\beta'$
γ	c_p/c_v
δ	$\sim(\nu/\omega)^{1/2}$, Stokes layer thickness
ε	$U_0/(\omega a)$, scaled velocity
ε'	$\sqrt{2}/|M|$
$\tilde{\varepsilon}$	radial displacement parameter, eqn (12.159)
ζ	vorticity
η, η'	scaled inner variables
θ	angular coordinate
κ	viscosity ratio $\mu/\hat{\mu}$
λ	wavelength
μ	continuous-phase viscosity
$\hat{\mu}$	dispersed-phase viscosity (liquid drop)
$\bar{\mu}$	$\cos\theta$
ν	kinematic viscosity, μ/ρ
ξ	inner variable, $\varepsilon' r\theta$
ρ	density
Υ	global stream function variable
ϕ	velocity potential
χ	bubble oscillation profile
ψ	stream function
Ψ	inner stream function

Subscripts, Superscripts and Accents

0	leading order
1	order ε
∞	far-field
b	boundary layer
r	radial direction

(slip)	slip velocity
(s)	steady
s	isentropic changes, streaming
θ	angular direction
(^)	dispersed phase (liquid drop)

References

1. M. Wiklund, R. Green and M. Ohlin, *Lab Chip,* 2012, **12**, 2438–2451.
2. C. P. Lee and T. G. Wang, *J. Acoust. Soc. Am.,* 1990, **88**, 2367–2375.
3. H. Bruus, *Lab Chip,* 2012, **12**, 20–28.
4. S. S. Sadhal, *Lab Chip,* 2012, **12**, 2292–2300.
5. H. Bruus, *Lab Chip,* 2012, **12**, 1014–1021.
6. L. Rayleigh, *Philos. Trans. R. Soc. London A,* 1884, **175**, 1–21.
7. H. Schlichting, *Phys. Z.,* 1932, **23**, 327–335.
8. W. L. Nyborg, *J. Acoust. Soc. Am.,* 1958, **30**, 329–339.
9. A. Rednikov and S. S. Sadhal, *J. Fluid Mech.,* 2011, **667**, 426–462.
10. L. D. Landau and E. M. Lifshitz, *Fluid Mechanics* (English Edition), Pergamon Press, 1959.
11. H. Bruus, *Lab Chip,* 2011, **11**, 3742–3751.
12. H. Bruus, *Theoretical Microfluidics,* Oxford University Press, 2008.
13. A. D. Pierce, Acoustics: An Introduction to its Physical Principles and Applications, *Acoustical Society of America.,* 1991.
14. M. Tatsuno, Secondary streaming induced by an oscillating cylinder, in *An Album of Fluid Motion,* ed. M. Van Dyke, Parabolic Press, 1982, p. 31.
15. H. Zhao, S. S. Sadhal and E. H. Trinh, *J. Acoust. Soc. Am.,* 1999, **106**, 589–595.
16. N. Riley, *Theor. Comput. Fluid Dyn.,* 1998, **10**, 349–356.
17. N. Riley, *Annu. Rev. Fluid Mech.,* 2001, **33**, 43–65.
18. W. L. Nyborg, *J. Acoust. Soc. Am.,* 1953, **49**, 68–75.
19. C. Y. Wang and B. Drachman, *Appl. Sci. Res.,* 1982, **39**, 55–68.
20. S. Ishizawa, *Bull. Jpn. Soc. Mech. Eng.,* 1966, **9**(35), 533–550.
21. M. Evander, L. Johansson, T. Lilliehorn, J. Piskur, M. Lindvall, S. Johansson, M. Almqvist, T. Laurell and J. Nilsson, *Anal. Chem.,* 2007, **79**, 2984–2991.
22. Y.-H. Lee and C.-A. Peng, *Gene Ther.,* 2005, **12**, 625–633.
23. K. Ohsaka, A. Rednikov, S. S. Sadhal and E. H. Trinh, *Rev. Sci. Instrum.,* 2002, **75**, 2091–2095.
24. K. Ohsaka, S. S. Sadhal and A. Rednikov, *J. Heat Transfer,* 2002, **124**, 599.
25. S. M. Hagsäter, T. G. Jensen, H. Bruus and J. P. Kutter, *Lab Chip,* 2007, 7, 1336–1344.
26. L. V. King, *Proc. R. Soc. London A,* 1934, **147**, 212–240.
27. L. V. King, *Proc. R. Soc. London A,* 1935, **153**, 1–16.
28. E. H. Trinh and J. L. Robey, *Phys. Fluids,* 1994, **6**, 3567–3579.

29. N. Riley, *Q. J. Mech. Appl. Math.*, 1966, **19**, 461–472.
30. C. Y. Wang, *J. Sound Vib.*, 1965, **2**, 257–269.
31. A. Rednikov, H. Zhao, S. S. Sadhal and E. H. Trinh, *Q. J. Mech. Appl. Math.*, 2006, **59**, 377–397.
32. H. Zhao, S. S. Sadhal and E. H. Trinh, *J. Acoust. Soc. Am.*, 1999, **106**, 3289–3295.
33. A. Rednikov and N. Riley, *Phys. Fluids*, 2002, **14**, 1502–1510.
34. K. Hasegawa, Y. Abe, A. Kaneko and K. Aoki, PIV measurement of internal and external flow of an acoustically levitated droplet, in *The Sixth Interdisciplinary Transport Phenomena Conference*, Volterra, Italy, October 4–9, 2009, paper 24.
35. S. A. Elder, *J. Acoust. Soc. Am.*, 1959, **31**, 54–64.
36. B. J. Davidson and N. Riley, *J. Sound Vib.*, 1971, **15**, 217–233.
37. M. S. Longuet-Higgins, *Proc. R. Soc. London A*, 1998, **454**, 725–742.
38. M. S. Longuet-Higgins, *Proc. R. Soc. London A*, 1997, **453**, 1551–1568.
39. D. F. Gaitan, L. A. Crum, R. A. Roy and C. C. Church, *J. Acoust. Soc. Am.*, 1992, **91**, 3166–3183.
40. C. Wang, B. Rallabandi and S. Hilgenfeldt, *Phys. Fluids*, 2013, **25**, 022002-22011-16.
41. P. Tho, R. Manasseh and A. Ooi, *J. Fluid Mech.*, 2007, **576**, 191–233.
42. R. E. Kelly, *Q. J. Mech. Appl. Math.*, 1966, **19**, 473–484.
43. J. F. Spengler and W. T. Coakley, *Langmuir*, 2003, **19**, 3635–3642.
44. J. F. Spengler, W. T. Coakley and K. T. Christensen, *AIChE J.*, 2003, **116**, 2773 2782.
45. R. Barnkob, P. Augustsson, T. Laurell and H. Bruus, An automated full-chip micro-PIV setup for measuring microchannel acoustophoresis: Simultaneous determination of forces from acoustic radiation and acoustic streaming, in *Proc. 14th Int. Conference on Miniaturized Systems for Chemistry and Life Sciences (µTAS 2010), 3–7 October 2010*, pp. 1247–1249.
46. L. A. Kuznetsova and W. T. Coakley, *J. Acoust. Soc. Am.*, 2004, **116**, 1956–1966.
47. C. P. Lee and T. G. Wang, *J. Acoust. Soc. Am.*, 1989, **85**, 1081–1088.
48. A. Rednikov, N. Riley and S. S. Sadhal, *J. Fluid Mech.*, 2003, **486**, 1–20.
49. W. K. Rhim, S. Chung and D. Elleman, Experiments on rotating charged liquid drops, in *AIP Conf. Proc. 197, Drops and Bubbles, Third International Colloquim*, American Institute of Physics, Monterey, California, 1988, pp. 91–105.
50. A. Gopinath and A. F. Mills, *J. Heat Transfer*, 1993, **115**, 332–341.

CHAPTER 13

Applications of Acoustic Streaming

ROY GREEN[a], MATHIAS OHLIN[b], AND MARTIN WIKLUND*[b]

[a]Faculty of Engineering and the Environment, University of Southampton, Southampton SO17 1BJ, UK; [b]Department of Applied Physics, Royal Institute of Technology, KTH-Albanova, SE-106 91 Stockholm, Sweden
*E-mail: martin.wiklund@biox.kth.se

13.1 Introduction

Acoustic streaming is widely recognised among researchers as a phenomenon responsible for inadvertent and unwanted generation of fluid flow inside acoustofluidic devices. Nuisance flows can be disruptive, particularly during acoustophoresis. However, acoustic streaming can also be extremely useful as a method for overcoming many of the challenges posed by low Reynolds number flows in microfluidics. Of particular importance is the use of acoustic streaming for mixing and pumping fluid in microfluidic devices.

Due to the many forms in which acoustic streaming arises, it is often misunderstood. Broadly speaking, it may be regarded as a flow formed as a result of the steady momentum flux that arises due to a gradient in an oscillatory momentum flux,[1] where the oscillatory momentum flux is the acoustic wave. Less explicitly, acoustic streaming may be defined as a fluid flow generated by the attenuation of an acoustic wave. The first theoretical model to thoroughly describe acoustic streaming was derived by Rayleigh in 1884.[2] Since Rayleigh's landmark paper, a wealth of literature has been published on acoustic streaming. The flows described in the literature include those arising from acoustic attenuation in the bulk of a fluid, at solid–liquid interfaces and due to the oscillation of microbubbles. For those less familiar with acoustic streaming it is often difficult to grasp that the flows arise from the same phenomenon. In fact, not only do they arise from the same phenomenon, they can also be described by the same governing equation, as derived by Nyborg.[3]

The chapter provides a qualitative description of acoustic streaming and an appreciation of the ways in which it can be harnessed for useful microfluidics applications. A more theoretical approach to acoustic streaming is provided in Chapter 12. The qualitative description includes illustrations of the different forms of acoustic streaming that arise. Furthermore, different applications of each form of acoustic streaming are reviewed and discussed.

13.2 A Qualitative Description of Acoustic Streaming Phenomena

The dynamics of a small volume element within a fluid, through which an acoustic wave propagates, can be defined by pressure and velocity oscillations. In an ideal fluid the time-averaged displacement of the volume element, *i.e.* the net displacement over multiple cycles, is zero. However, in a real fluid such as water, viscous attenuation results in a net displacement of the volume element during each cycle of oscillation as a direct result of the diminishing amplitude of its oscillation. The net displacement of many such volume elements in the fluid manifests itself as a global streaming flow.[4] Thus, acoustic streaming can be described as a steady fluid flow formed by the viscous attenuation of an acoustic wave.[5] Using this concept, Lighthill[1] derived a Navier–Stokes based equation in which acoustic streaming is described as a Reynolds stress (*i.e.* a time-averaged momentum flux in a fluid) generated by the presence of a spatial gradient in the oscillatory momentum flux. The oscillatory momentum flux is the acoustic wave and the presence of a gradient due to viscous attenuation results in a net transfer of momentum into the fluid.

Streaming flows vary greatly depending on the mechanism behind the viscous attenuation of the acoustic wave. The variations include the velocity, length scale and geometry of the flow. The velocity varies from the order of $\mu m\ s^{-1}$ up to the order of $cm\ s^{-1}$ or more.[6] The length scale varies from the order of μm in the case of microstreaming up to the order of cm in bulk-streaming.[6] The flow geometry may take the form of a jet or of vortices.[6] The above scales are not definitive but rather the scales at which acoustic streaming flows are most commonly observed. They are therefore a reflection on the frequencies and acoustic intensities adopted by researchers more than they are a reflection on what is possible. Inner and outer boundary layer streaming, Eckart streaming and cavitation microstreaming are now described to provide the reader with a grasp of the most common forms of acoustic streaming flows generated by the viscous attenuation of an acoustic wave.

13.2.1 Inner and Outer Boundary Layer Acoustic Streaming

Boundary layer driven acoustic streaming is the flow formed by the viscous attenuation of an acoustic wave in a solid–liquid boundary layer running parallel to the direction of acoustic propagation. The dissipation of acoustic energy into the boundary layer is large because there exists a non-slip condition at the solid–liquid interface, resulting in a large velocity gradient perpendicular to the boundary (Figure 13.1). The peak oscillatory particle velocity varies from zero at the solid surface, to its free stream value over a distance of the order of 1 μm (for low MHz ultrasound in water). The large viscous forces that develop between adjacent layers of fluid result in attenuation that drives boundary layer induced acoustic streaming.

Boundary layer acoustic streaming is most commonly encountered in standing waves, where the viscous attenuation of the acoustic wave results in a steady momentum flux along the solid boundary, oriented from the pressure antinode to the pressure node. Due to the spatially fixed pressure nodes and antinodes, this generates boundary layer vorticity termed inner boundary layer streaming or 'Schlichting streaming',[7] named after Hermann Schlichting who first modelled it mathematically. Once established, the powerful inner boundary layer streaming flow drives counter rotating streaming vortices within the bulk of the fluid called outer boundary layer streaming or 'Rayleigh streaming'[2] named after Rayleigh who first modelled them mathematically. In Figure 13.2, inner (Schlichting) and outer (Rayleigh) boundary layer streaming in a standing wave is illustrated schematically. It can be seen that there is a vortex–antivortex pair per half wavelength along the direction of acoustic propagation. Accordingly, the length of the solid boundary must be at least a quarter of the acoustic wavelength for such flows to develop.[8]

Boundary layer driven acoustic streaming is particularly pronounced under the condition that

$$\lambda \gg h \gg \delta_v, \tag{13.1}$$

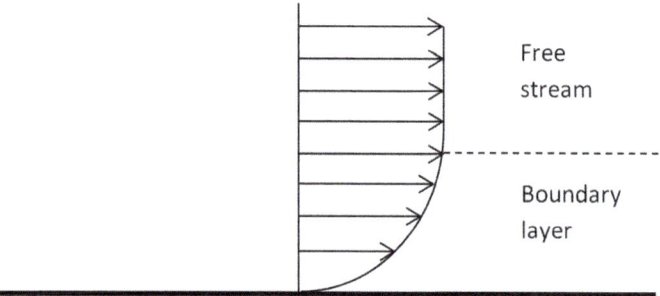

Figure 13.1 Diagram showing the peak magnitude of the oscillatory first order acoustic velocity in the boundary layer. The velocity gradient is normal to the boundary and the velocity falls from its free stream value to 0. The thickness of the boundary layer is typically 1 μm.

Applications of Acoustic Streaming 315

where λ is the acoustic wavelength, h is the characteristic length scale of the fluid chamber (*cf.* Figure 13.2) and δ_v is the viscous penetration depth, which in an oscillating flow is given by[9]

$$\delta_v = \sqrt{2\nu/\omega} \qquad (13.2)$$

where ν is the kinematic viscosity and ω is the angular frequency of the acoustic wave. Thus, for an ultrasound wave in water at the low MHz range, λ is of the order of 1 mm and δ_v is of the order of 1 µm. Owing to this, outer boundary layer streaming is more pronounced for chambers with length scales that are fractions of a wavelength, such as the $\lambda/2$-channel typically employed in acoustophoresis. Importantly, since the source of outer boundary layer streaming is within the fluid boundary layer parallel to the standing wave (along x in Figure 13.2), the perpendicular dimension (h, along y in Figure 13.2) must be of the order of several λ or smaller to obtain significant outer boundary layer streaming.

When comparing the orientation of the outer boundary layer streaming vortices with the position of the pressure nodes of the standing wave parallel to the boundary, the flow is in most cases divergent within the pressure nodal plane relative to the central axis of the channel. This phenomenon is discussed in more detail in Section 13.3.1. In Figure 13.2, this corresponds to a pressure node located at $x = 0$, with its centre at $y = 0$.

It has been demonstrated in a theoretical study[10] assessing the effect of the fluid chamber size on acoustic streaming velocity, that as the size of the fluid chamber is reduced, the inner boundary layer streaming becomes a larger

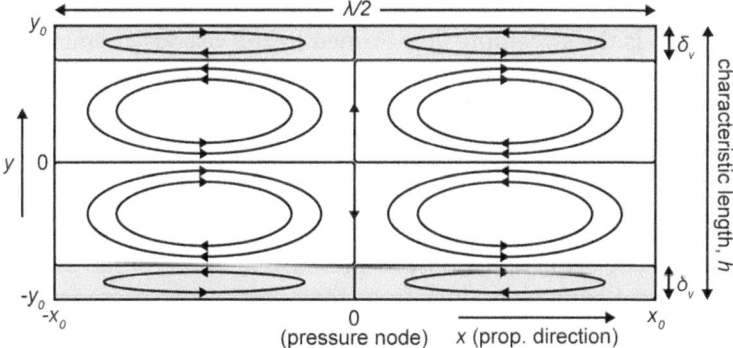

Figure 13.2 A system of inner (Schlichting) boundary layer streaming vortices within the viscous boundary layer of thickness δ_v and outer (Rayleigh) boundary layer streaming vortices in a channel with a standing wave propagating along x. The pressure node is located at $x = 0$, which is the plane in which small particles and cells would be trapped by acoustic radiation forces. Note that the streaming is divergent within this plane and is therefore counteracting the lateral radiation force that attempts to aggregate suspended particles in the pressure nodal plane. Figure 13.2 is reconstructed from ref. 10.

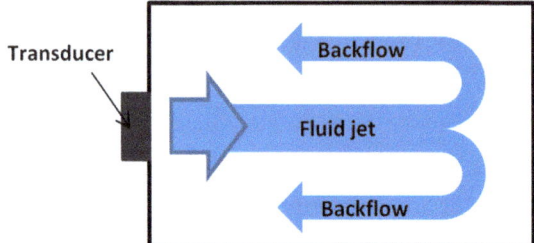

Figure 13.3 A typical Eckart streaming flow including backflow that arises due to the occurrence of a low static pressure at the base of the fluid jet. The fluid jet is more pronounced if the opposite wall of the chamber is acoustically absorbent and if the dimension of the fluid chamber parallel to the fluid jet is comparable or greater than the acoustic attenuation length.

proportion of the overall streaming flow. However, the combined streaming velocity declines. Additionally, it was found that as the acoustic frequency increased so too did the inner boundary layer streaming velocity, though the boundary layer became significantly thinner. A streaming velocity of over 100 μm s^{-1} was predicted[10] in a 1 mm thick fluid channel sonicated at an acoustic pressure amplitude of 40 kPa and frequencies in the low MHz range. This result is in agreement with similar analytical solutions obtained by other researchers.[11]

13.2.2 Eckart Streaming

Eckart streaming, also known as bulk streaming and formerly referred to as 'quartz wind', is the streaming flow formed by the viscous attenuation of an acoustic wave in the bulk of a fluid.[12] As the acoustic wave propagates through the fluid, it is attenuated at a rate proportional to the square of its frequency (as specified by Stokes' law of sound attenuation for an ideal fluid). The attenuation of the acoustic wave causes a decrease in its amplitude with distance downstream. The loss of acoustic energy manifests itself as a steady momentum flux, resulting in a jet of fluid in the direction of acoustic propagation. The reduction in static pressure at the base of the fluid jet causes a vortex-like flow as fluid from the sides of the chamber is drawn towards the transducer to replace the fluid forced away by the streaming jet (Figure 13.3).

A practical example of Eckart streaming is the jet generated in water by Matsuda *et al.*[13] at an acoustic pressure amplitude of 40 kPa and a frequency of 3.45 MHz and characterised by Doppler velocimetry. The fluid jet was approximately 1 cm in diameter, with a maximum velocity in excess of 6 cm s^{-1}. Along the radial axis, velocity was greatest at the centre of the jet whilst axially velocity peaked at approximately 6 cm from the transducer surface. The radial and axial distribution of streaming velocity is in good agreement with a similar characterisation performed by Cosgrove *et al.*[14]

In order to generate significant Eckart streaming, the chamber dimension forming the length through which the acoustic wave propagates must be comparable to or greater in length than the acoustic attenuation length.[15] As an example, the acoustic attenuation length is 8.3 mm in water at 50 MHz and is inversely proportionate to the square of the frequency. Thus, Eckart streaming occurs predominantly in microfluidics devices in which high frequency ultrasound is propagated along a dimension on the order of millimetres or greater. Eckart streaming can be generated in both standing and travelling waves though occurs at significantly lower velocities in the former due to the steady Reynolds stress being generated in both directions, partially cancelling itself out.

13.2.3 Cavitation Microstreaming

Cavitation microstreaming is the streaming flow formed by the viscous attenuation of an acoustic wave in the fluid boundary layer surrounding a stably oscillating microbubble.[16] This should not be confused with the fluid jet formed by the destruction of a microbubble, which strictly speaking is not a form of acoustic streaming. Cavitation microstreaming, being a form of boundary layer induced acoustic streaming, is generated by the attenuation mechanism described in Section 13.2.1. The key difference between cavitation microstreaming and other forms of boundary layer induced streaming is that the oscillation of a microbubble, sonicated at or near its resonance frequency, results in the local amplification of the first order velocity.[8] It is this high amplitude bubble scattered acoustic field, interacting with the fluid–bubble interface that is primarily responsible for cavitation microstreaming.[3] The local amplification of the acoustic field (due to scattering) results in streaming flows up to several orders of magnitude faster than those that would arise around incompressible particles of comparable size.

The geometry of the flow field produced by cavitation microstreaming is dependent on the mode of bubble oscillation. Since there are numerous modes of oscillation there is no single flow geometry associated with cavitation microstreaming. The flow geometry most often associated with an unconfined gas bubble undergoing spherical volume oscillations is that of two toroidal vortices, which when viewing a 2-D plane through the centre of the bubble appear as four individual vortices. Half of such a pattern is shown in Figure 13.4, taken from the well-known experiments performed by Elder.[17] Here, the bubble is resting on a solid boundary and therefore takes on a hemispherical shape, as a result of which only one of the two toroidal streaming vortices are formed.

Cavitation microstreaming flow can be characterised by particle image velocimetry (PIV), a technique in which two images of the flow, seeded with tracer particles, are captured in quick succession and a statistical method is used to compute a vector field of the flow. Such characterisations of microstreaming have been used to demonstrate the generation of various flow patterns at different excitation frequencies.[18] They have also been used to

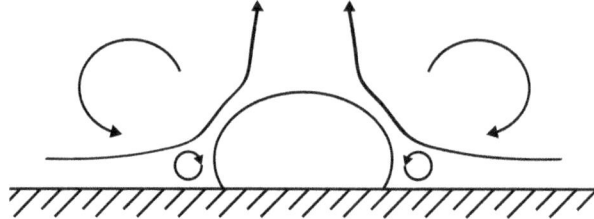

Figure 13.4 Cavitation microstreaming flow around an air bubble with a surface skin. Reproduced from ref. 17.

demonstrate control over the average microstreaming speed up to a maximum speed of 30 mm s^{-1} in the bulk of the fluid.[19]

In practice, due to issues pertaining to the low stability of air bubbles in water, encapsulated microbubbles are usually preferred over free air bubbles for generating microstreaming. Encapsulated microbubbles consist of a gas core, typically a fluorocarbon because of their low coefficient of diffusivity, encapsulated by a thin polymer or lipid shell, thereby increasing the bubbles' stability against dissolution. Using streak photography[20] the microstreaming velocity in the vicinity of Albunex encapsulated microbubbles (1–10 μm radius) was estimated to be in the range of 50–100 μm s^{-1}. Sonication was carried out at an acoustic pressure amplitude of 500 kPa and a frequency of 160 kHz, well below the estimated resonance frequency of 800 kHz for Albunex. Numerical calculations have predicted[21] that microbubbles encapsulated by thin and low stiffness shells are able to produce significantly faster microstreaming than unencapsulated microbubbles because of the difference in boundary conditions (liquid–solid and liquid–air respectively).

13.3 Microfluidic Applications of Acoustic Streaming

In this section we review different microfluidic applications and the observed consequences of acoustic streaming. The applications are classified into Rayleigh streaming, Eckart streaming, cavitation microstreaming and surface acoustic wave induced streaming.

13.3.1 Applications of Rayleigh Streaming

Rayleigh streaming has been discussed extensively within the field of acoustic particle trapping and manipulation because of the practical lower limit that it places on the size of particles that can be manipulated by primary acoustic radiation force in a standing wave. In addition to particle manipulation, there are several examples of the use of Rayleigh streaming in mixing applications. Bengtsson and Laurell[22] used Rayleigh streaming for efficient fluid mixing and stirring in resonant microchannels. They evaluated two different systems: a device for the mixing of two parallel flows in a single

Applications of Acoustic Streaming

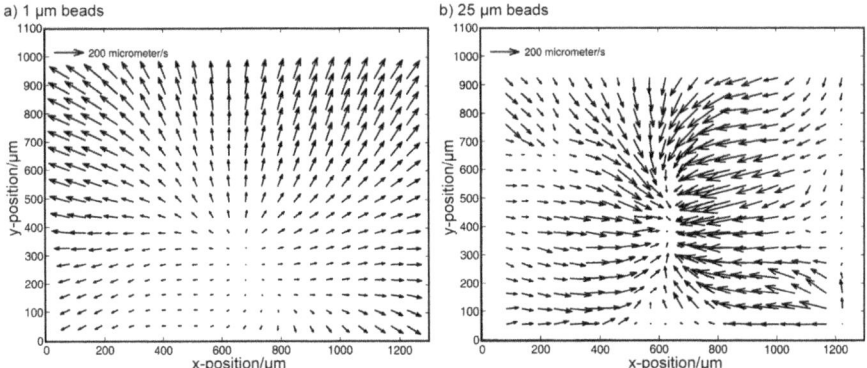

Figure 13.5 Particle image velocimetry (PIV) analysis of the lateral motion of (a) 1 µm and (b) 25 µm particles within the pressure nodal plane (500 kPa pressure amplitude). The diagrams are taken from ref. 23.

channel and a device for mixing in order to increase enzymatic reactions taking place on the chamber walls in a 32-channel microstructure. Both devices had channel heights corresponding to a full wavelength (300 µm at the driving frequency of ~5 MHz). The channel width corresponded to a quarter of a wavelength in the parallel flow device (75 µm) and less (25 µm) in the 32-channel enzyme reactor.

The present section is dedicated to the control of Rayleigh streaming for particle manipulation. The first in-depth studies were carried out by Spengler et al.[23,24] who investigated the effect of Rayleigh streaming on the agglomeration of yeast cells and polymer particles of 1 to 25 µm. They studied the lateral motion of particles within the pressure nodal plane in a half wavelength standing wave device with an acoustic pressure amplitude of 500 kPa and a frequency of 3 MHz. They observed that whilst larger (>10 µm) particles agglomerated in the centre of the pressure nodal plane, 1 µm particles did not because the drag force from the streaming flow overcame the lateral radiation force (Figure 13.5).

It can be seen (Figure 13.5(a)) that streaming was the dominant force on 1 µm particles. The orientation of Rayleigh streaming within the pressure nodal plane is divergent, in agreement with the streaming orientation in the corresponding plane ($x = 0$) in the schematic illustration in Figure 13.2. On the other hand, acoustic radiation force was the dominant force on 25 µm particles (Figure 13.5(b)), resulting in their agglomeration in the pressure nodal plane. Spengler, Coakley and Christensen[23] also observed that intermediate sized particles (10 µm) were only able to overcome the Rayleigh streaming induced drag force after forming mini- aggregates off-axis at the edges of the acoustic field (Figure 13.6). Thus, although the final result was the formation of a large aggregate in the centre of the pressure nodal plane, the process began with the depletion of particles in the nodal plane.

Spengler, Coakley and Christensen[23] explained that this was caused by the increase in the effective volume of the mini-aggregates relative to single particles. Thus, the increase in effective volume has a stronger effect on the volume dependent radiation force than on the cross-sectional area dependent viscous drag force from Rayleigh streaming.

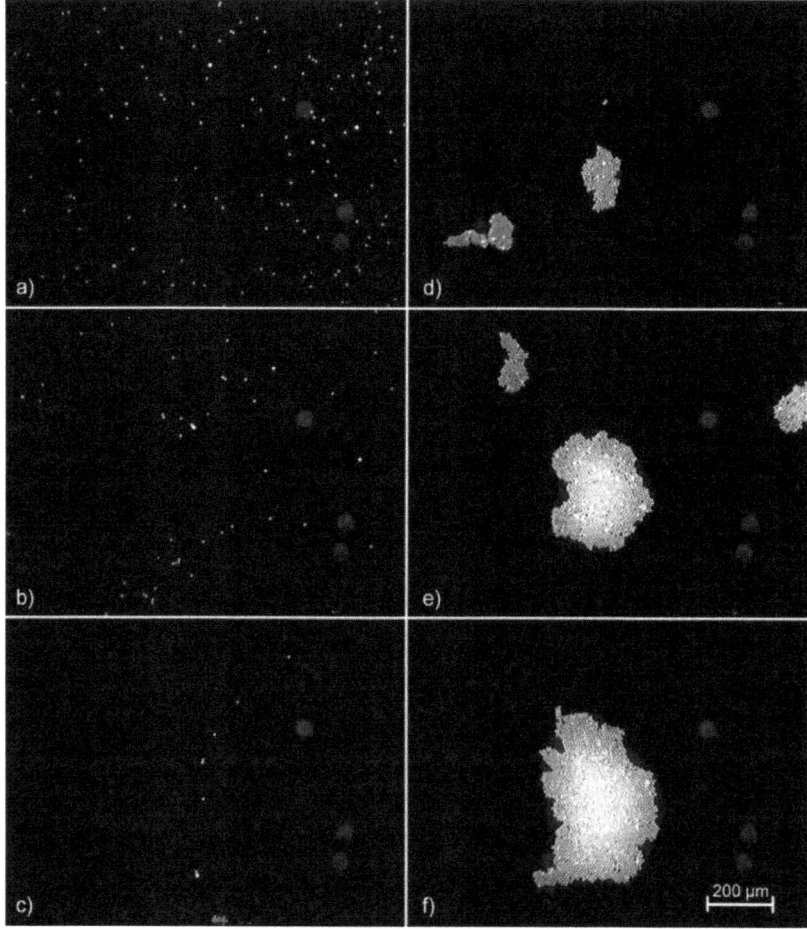

Figure 13.6 In-plane development of an aggregate of 10 μm polymer particles at times (a) 0.2 s, (b) 5 s, (c) 15 s, (d) 60 s, (e) 130 s, and (f) 190 s. Once driven to the pressure nodal plane, the particles initially move away from the center of the field of view due to Rayleigh streaming (a–c). They interact off camera and return as compact mini-aggregates (d–e). The packing of the growing central aggregate adjusts to incorporate the merging mini-aggregates (f). The figure is taken from ref. 24.

Kuznetsova and Coakley[25] generated and measured Rayleigh streaming in circular and rectangular water filled acoustic chambers of varying thicknesses. The chambers were sonicated at acoustic frequencies of 1.5–3 MHz in order to set up either quarter or half wavelength standing wave resonances with pressure amplitudes in the range of 310–600 kPa. They measured maximum streaming velocities ranging from below 100 μm s^{-1} to approximately 500 μm s^{-1}. They observed, in agreement with Frampton, Martin and Minor,[11] that increasing either the frequency or pressure amplitude resulted in a considerable increase in the streaming velocity. However, halving the chamber thickness so as to obtain a quarter wavelength resonance had the effect of reducing the streaming velocity. Through careful device design, Kuznetsova, Martin and Coakley[26] demonstrated that radiation force may dominate over streaming for 1 μm particles. Comparing the results of Spengler et al.[23,24] and Kuznetsova et al.,[25,26] it is apparent that the ability to trap small particles is dependent on the streaming direction relative to the direction of the radiation force. The fluid chambers used both by Spengler et al.[23,24] and Kuznetsova et al.[25,26] had lateral dimensions (h) that did not fulfil the relationship in eqn 13.1. Instead, h was typically much larger than λ in their devices. This indicates that the source of Rayleigh streaming was most likely not solely located in the boundary layer parallel to the standing wave (i.e. parallel to x in Figure 13.2). A possible explanation is that flexural vibrations in the carrier and reflector layers of the $\lambda/2$-chambers generated streaming in the boundary layers perpendicular to the standing wave (i.e. parallel to y in Figure 13.2).

A more detailed analysis of the size-dependence of radiation force and Rayleigh streaming was carried out by Barnkob et al.[27] who inputted experimental data of particle motion into a theoretical model that estimated the relative impact of radiation force and Rayleigh streaming on particle motion. Their theoretical model is based on the decomposition of the particle velocity field into a gradient component and a rotational component in order to derive the contributions of radiation force (gradient-type) and acoustic streaming (rotation-type) to the overall particle motion. They concluded that the theoretical threshold, defined as the particle size for which the two forces were equal in magnitude, corresponded to a particle with a diameter of 2.8 μm at an acoustic frequency of 2 MHz and that the threshold particle size is proportional to $1/\sqrt{f}$, where f is the acoustic frequency. The result is in agreement with general experimental observations, where a 2 MHz acoustophoretic device is suitable for ultrasonic manipulation of cells (which are typically larger than 3 μm)[28] and a >10 MHz-device is suitable for ultrasonic manipulation of objects with sizes down to approximately 1 μm and potentially smaller.[29] However, it should be noted that in fluid exchange applications with larger particles (typically cells with sizes ~10 μm), acoustic streaming may reduce wash efficiency even if particle manipulation is not directly affected by streaming.[28] This is because such applications are dependent not only on accurate displacement of particles or cells but also on achieving a laminar flow.

Several strategies have been suggested for manipulating particles smaller than the theoretical threshold defined by Barnkob et al.[27] It was seen (Figure 13.6) that at sufficiently high particle concentrations the acoustic particle–particle interaction force resulted in the formation of mini-aggregates that could be manipulated due to their larger combined size. This effect is enhanced when other particle–particle interaction forces such as van der Waals interactions, electrostatic interactions and hydrophobic/hydrophilic effects are present. Furthermore, streaming may cause a local depletion of particles from vortex regions. This is shown in Figure 13.7, where the interaction between the radiation force and acoustic streaming causes a redistribution of 400 nm polystyrene particles suspended in water in a 750 µm wide channel. At a frequency of 2.11 MHz (Figure 13.7(a), a channel width corresponding to λ), the characteristic $\lambda/4$-scale Rayleigh streaming vortices cause a local depletion of particles within each vortex. Here, the global effect is a decrease in particle concentration along the centre of the channel. The effect becomes more pronounced at a frequency of 6.62 MHz (Figure 13.7(b), a channel width corresponding to 3 λ) where the radiation

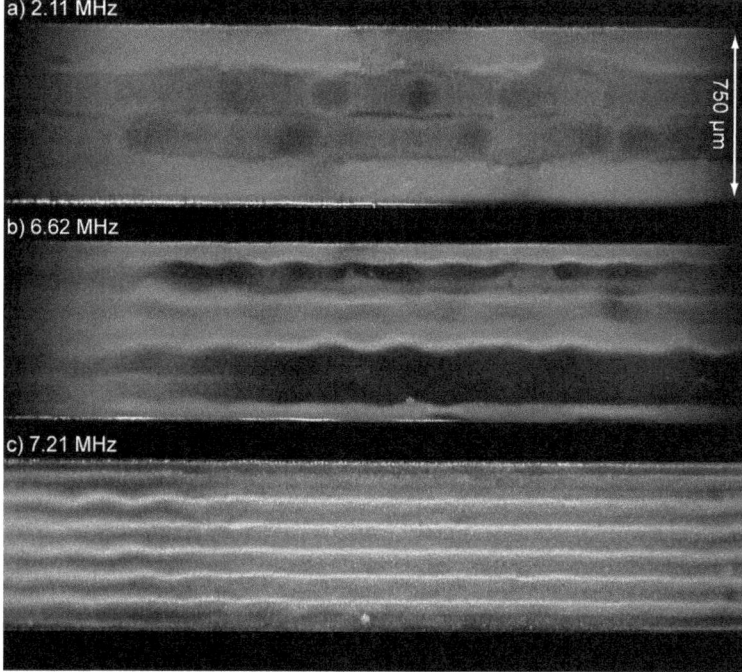

Figure 13.7 Redistribution of highly concentrated 400 nm polystyrene particles suspended in water in a 750 µm wide channel due to the interaction between the radiation force and acoustic streaming. The driving voltage is 7 V_{rms} and the frequency is (a) 2.11 MHz, (b) 6.62 MHz and (c) 7.21 MHz. Experiments by Martin Wiklund.

Applications of Acoustic Streaming 323

force appears to be more significant than at 2.11 MHz. At 7.21 MHz (Figure 13.7(c), a channel width corresponding to 3.5 λ) the pattern of particle alignment is in agreement with standard acoustophoretic operation (*i.e.* $\lambda/2$-distance between nodes) and there is no evidence of Rayleigh streaming. The experiment demonstrates that it should be possible to design systems where acoustic streaming enhances rather than inhibits particle manipulation in a standing wave manipulation device. A further strategy for the manipulation of sub-micron particles is the use of larger (>1 µm) seeding particles to trigger the aggregation of sub-micron particles. The orientation and size of the streaming vortices has been a matter of debate because the theoretically predicted flow (*cf.* Figure 13.2 and 13.7(a)) is not always confirmed experimentally.

A further example of the use of acoustic streaming to enhance particle manipulation was demonstrated by Möller *et al.*[30] They used boundary layer driven streaming to move particles in a circular flow in a PMMA device driven at an acoustic frequency of 2 MHz, achieving a maximum flow velocity of 0.7 mm s^{-1} (Figure 13.8). A potential application for such a device could be to move biological material (*e.g.* bacteria) past a sensor surface.

In some cases the Rayleigh streaming vortices may rotate in the opposite direction or may be of $\lambda/2$-size rather than $\lambda/4$-size. Examples include the 6 × 6-array of Rayleigh vortices observed in a $3\lambda \times 3\lambda$-chamber[31] and the

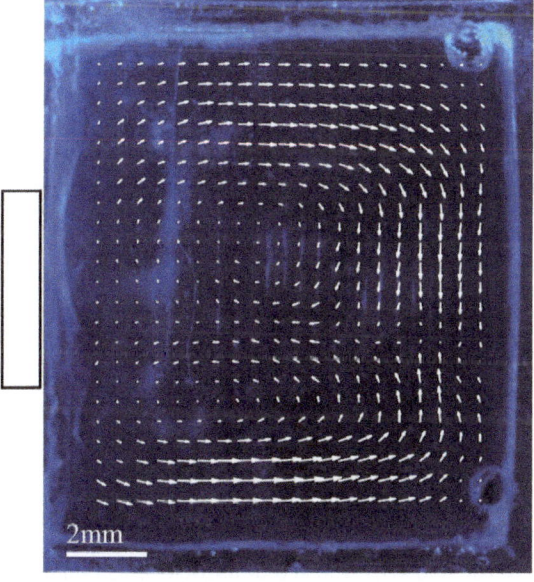

Figure 13.8 PIV vector field of the motion of 20 µm particles due to acoustic streaming, superimposed over an image of the device. The device was driven at an acoustic frequency of 2 MHz. The figure is taken from ref. 30.

1 × n-array of Rayleigh vortices observed in a λ/2 wide microchannel[32] (Figure 13.9). A possible explanation for the mismatch between theory and experiment is based on the hypothesis that it not only losses in the viscous boundary layer, but also transmission losses in the solid structure of the acoustophoretic chip that contribute to acoustic streaming.[33] Thus, while the acoustic resonance modes in an acoustophoresis chip can be modelled in good agreement with experimental observations by only considering the channel geometry,[34] accurate modelling of acoustic streaming may require that the entire chip, including the solid structures surrounding the channel, are incorporated into the model. Another possible explanation for the mismatch is the contribution of flexural vibrations in the solid structure facing the fluid chamber to the losses in the viscous boundary layer.

In addition to the boundary conditions and loss mechanisms involved, the geometry of the microchannel has a large impact on the magnitude and geometry of acoustic streaming. Experimental examples of acoustic streaming observed in micro-scale resonance cavities of different geometries are shown in Figure 13.10. The characteristic ∼λ/4-scale Rayleigh streaming vortices are seen in circular (Figure 13.10(a)) and square (Figure 13.10(b)) chambers inside a flow-through chip. A single λ/2-scale vortex is seen in a single well of a 100-well micro-plate (Figure 13.10(c)). Finally, a well with sharp edges can be used for generating localized streaming vortices originating from the tip of sharp structures (Figure 13.10(d)). The effect is well-known in acoustophoresis chips possessing branched channel outlets. Thus, sharp edges should be avoided in applications where it is desired to minimise streaming. However, the device in Figure 13.10(d) can be used for fluid mixing in the vicinity of trapped particles or cells.[35] This idea has also been implemented in a microchannel for continuous-flow micromixing.[36] In summary, Figure 13.10 demonstrates the complexity of acoustic streaming in microfluidic devices and gives an indication of the difficulty involved in accurately modelling and predicting streaming.

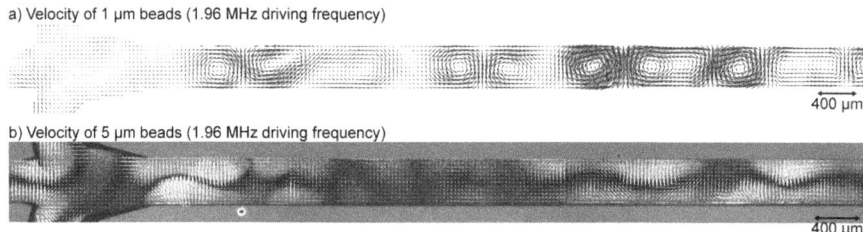

Figure 13.9 (a) λ/2-scaled Rayleigh streaming vortices seen in a half wave channel driven at 1.96 MHz, tracked by particle image velocimetry (PIV) of 1 μm particles. (b) The corresponding motion of 5 μm particles at the same actuation parameters, tracked by PIV, plotted on a background image showing the final distribution of particles. The streaming velocity is of the order of 10 μm s^{-1} in both experiments. The figures are combined parts from two separate figures in ref. 32.

Applications of Acoustic Streaming 325

An interesting method for controlling acoustic streaming inside multi-well plates designed for cell trapping is through the use of frequency modulation.[37] This method is based on the fact that each operating frequency of the device in Figure 13.10(c) causes a $\lambda/2$-scaled vortex with either clockwise or counter-clockwise direction. By sweeping the frequency within a suitable bandwidth, the net effect causes significant suppression of streaming without interfering with the trapping function of the device (Figure 13.11). Ohlin *et al.*[37] used this strategy for switching on and off the acoustic streaming around a continuously trapped cell aggregate in a multi-well microplate.

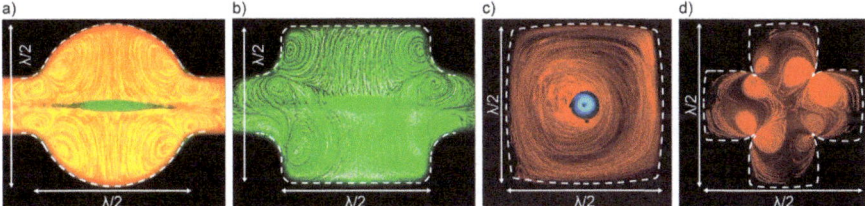

Figure 13.10 Acoustic streaming patterns for different channel geometries. The width of the channels is between 300 and 350 µm and the driving frequency is between 2.1 and 2.6 MHz. The streaming is tracked by 1 µm fluorescent particles and the trapped cluster in the centre of each cavity contains 5 µm particles in (a) and (b), and 10 µm cells in (c). There are only 1 µm particles present in (d). Experiments by Otto Manneberg (a–b) and Mathias Ohlin (c–d).

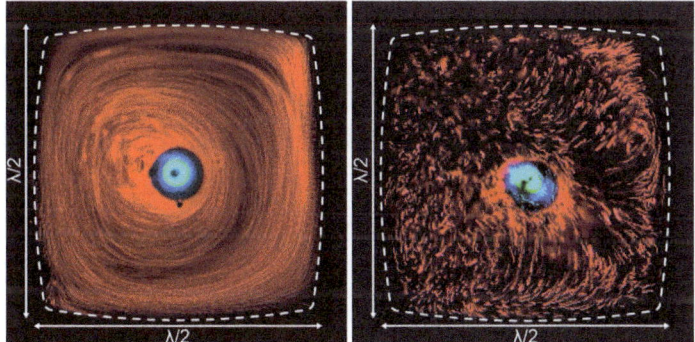

Figure 13.11 Demonstration of the acoustic streaming obtained at single frequency (left) and with frequency modulation (right) inside one 350 µm wide well in a multi-well microplate. The left image is the same as in Figure 13.10(c). The frequency modulation method decreases acoustic streaming by a factor of ∼30 without affecting the positioning of the cluster of 10 µm particles in the centre of the well (blue cluster) by acoustic radiation forces. Further experimental details are found in ref. 36.

13.3.2 Applications of Eckart Streaming

Eckart streaming of significant velocity is rarely generated in microfluidic devices because acoustic attenuation must occur over a relatively large distance. However, if the dimension of the fluid channel in the direction of acoustic propagation is > 1 mm, Eckart streaming may still occur at sufficiently high acoustic frequencies. This is particularly true when more than one dimension is of a scale of >1 mm and if the acoustic chamber is lossy (*e.g.* due to the use of acoustically absorbing materials such as plastics).

Hertz[38] observed Eckart streaming in an acoustic trap consisting of two counter-propagating and strongly focused 11 MHz acoustic waves in a ~10 × 10 × 44 mm^3 fluid chamber. He explained the effect as the result of transducer misalignment and mismatch in the emitted beam profiles causing a small residual propagating wave around the beam axis. In practice, Hertz eliminated the streaming using two acoustically transparent 2 μm thick plastic films positioned in the focal region 1.5 mm apart. A corresponding streaming pattern was observed by Wiklund, Hänninen and Hertz[39] in a device of similar dimension to Hertz's based on a single focusing transducer and a plane acoustic reflector (Figure 13.12). The device was developed on a 96-well plate with the aim of concentrating and positioning beads in a bead-based immunoassay.[40] As can clearly be seen in Figure 13.12, Eckart streaming can be generated over a distance of about 1 mm, resulting in fluid velocities of up to 1 mm s^{-1}. The figure shows two different orientations of

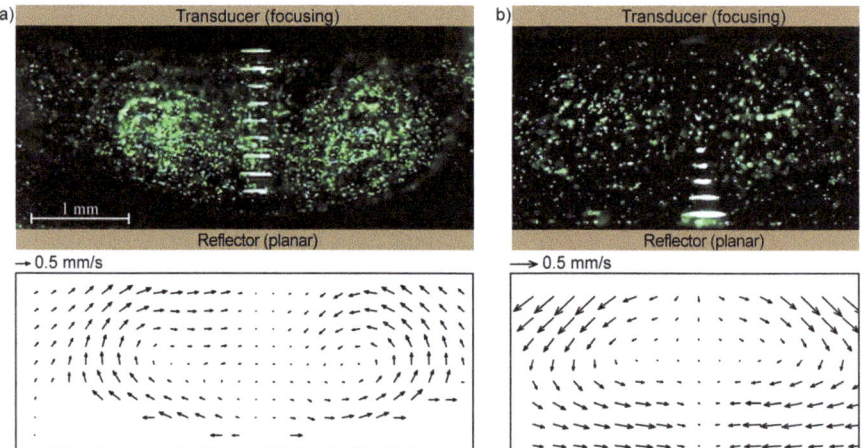

Figure 13.12 Eckart streaming generated between a 4 MHz focused ultrasound transducer and a plane reflector made of polystyrene (a) and molybdenum (b). The upper panel shows experimental side-view snapshots of 4 μm fluorescent particles suspended in water and the lower panel shows the corresponding PIV velocity field. Along the axis of symmetry, 4 μm particles are trapped in several pressure nodes. Further experimental details are found in ref. 38.

Eckart streaming, either away from the transducer along the axis of symmetry when using a soft-wall reflector made of polystyrene (Figure 13.12(a)), or towards the transducer when using a hard-wall reflector made of molybdenum (Figure 13.12(b)). The former orientation (Figure 13.12(a)) is in agreement with the schematic illustration in Figure 13.3. The reason for the different direction in Figure 13.12(b) is due to the focusing geometry; the focal length of the transducer is longer than the fluid chamber length in both experiments. The device in Figure 13.12 could be used for sweeping a fluid sample across trapped aggregates of particles confined close to the axis of symmetry of the chamber. The method has been proposed for enhancing bead-based immunoassays.[41]

Möller et al.[30] investigated the orientation of Eckart streaming in a 1 mL (11.5 mm long) PMMA chamber at two different acoustic frequencies. They noticed that a 6.5 MHz acoustic frequency resulted in an axial flow away from the transducer and a peripheral backflow (similar to Figures 13.3 and 13.12(a)) while a 2 MHz acoustic frequency resulted in the opposite flow pattern (similar to Figure 13.12(b)). They demonstrated that their experimental observations were in good agreement with a theoretical model that included the piezoelectric behaviour of the transducer and the structural mechanics of the PMMA chamber as well as the fluid domain. They explained the effect of the reversed direction at the lower frequency as a result of radiation taking place from the entire PMMA structure and not only from the transducer, resulting in a complicated acoustic field inside the chamber. The suggested application of the device was in purification processes if combined with a particle trap.

Bernassau et al.[42] demonstrated a method for reducing the prevalence of Eckart streaming in a 13 mm × 13 mm heptagonal cell manipulation device driven at a frequency of 4 MHz. This was achieved by placing an agar layer in the fluid channel, reducing its effective thickness. The agar layer did not greatly affect the acoustic propagation due to it having similar acoustic properties to water. Particle velocities of 1–10 µm particles due to Eckart streaming were all reduced by at least one order of magnitude.

13.3.3 Applications of Cavitation Microstreaming

Cavitation microstreaming flows are up to several orders of magnitude faster than streaming flows formed in the vicinity of incompressible spheres of a similar size to the microbubbles. It is possible to achieve flow circulation over a large range of scales from nanolitre up to millilitre volumes, depending on the acoustic pressure and the size of the microbubble. As a consequence of the wide range of scales, cavitation microstreaming is applicable to the generation of both large scale flows (by microfluidic standards) and highly targeted flows. The two applications of cavitation microstreaming that have received the greatest attention are micro-mixing and cell membrane poration; these are discussed alongside several additional novel applications.

The mixing of minute volumes of fluid (micro-mixing) is impeded by the low Reynolds numbers encountered in microfluidics. Micro-mixing poses a technical challenge and yet is vital in order to ensure fluid homogeneity, increase rates of chemical reactions and for the purpose of heat transfer. There are numerous methods of mixing reported in the literature, a review of which will not be carried out here. Cavitation microstreaming has been demonstrated as a promising method for the rapid mixing of fluids and does not require highly complex fluid channel geometries for its implementation.

Microbubbles used for micro-mixing are typically formed by trapping air in hydrophobic grooves cut into the sides of fluid channels. The cavitation microstreaming flows that form upon acoustic excitation have been shown to lead to the complete mixing of fluids within several microseconds in flow-through devices operated at flow rates in excess of 10 μL min^{-1}.[43] Within a static fluid volume it has been demonstrated that a 50 μL volume can be mixed in 6 seconds.[44] Taking an example from the literature, a horseshoe shaped feature (Figure 13.13) was used to trap air inside the microfluidic device so as to mix two laminar flows as they combine inside a larger channel with a cross-section of 240 μm × 155 μm. It was reported that at a flow rate of 8 μL min^{-1} the velocity of the laminar flow past the horseshoe feature was 7.2 mm s^{-1} and that at this velocity the two laminar flows became completely mixed within a 50 μm distance from the start of the mixing region, a distance that takes the fluid 7 ms to traverse at the reported velocity. An alternative design based on the same concept was presented by Tovar and Lee[45] who named their device a lateral cavity acoustic transducer. The device is capable of pumping or mixing fluids in a microchannel at a flow rate of approximately 250 nL min^{-1}. Recently, the same concept has been used for switching cells or particles into bifurcating microchannels.[46]

A further demonstration of cavitation microstreaming enhanced micro-mixing is the five-fold increase in the rate of DNA hybridisation that would otherwise be severely limited by the time taken for diffusion to occur.[44]

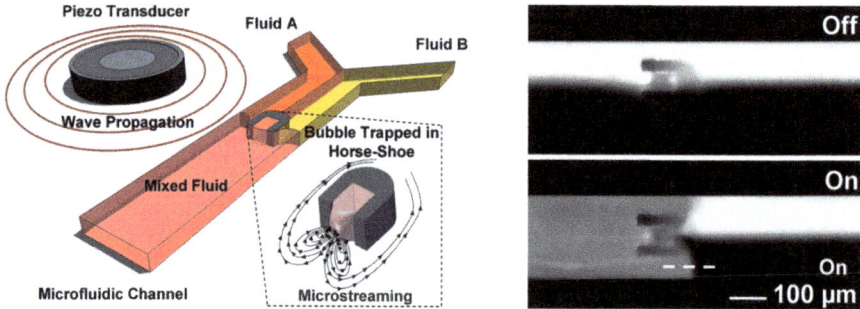

Figure 13.13 A schematic of the experimental setup of Ahmed *et al.*[43] (left) and a demonstration of the device for mixing of water and a fluorescent dye solution (right). Figure 13.13 is based on two separate figures from ref. 43.

Applications of Acoustic Streaming 329

Despite the high success of many of the devices reported in the literature, there do not appear to be any commercial cavitation microstreaming based micro-mixers on the market to date.

In addition to micro-mixing, pumping of fluid is also vital for many microfluidics applications. Due to the flow vorticity of cavitation microstreaming, it has received little attention as a method for micro-pumping. However, cavitation microstreaming has many advantages such as not relying on the chemical/ionic composition of the fluid or on complex channel geometries. In one example of pumping, Ryu, Chung and Cho[47] generated a maximum flow velocity of 5.3 mm s^{-1} and a corresponding flow rate of 0.19 mL s^{-1} with a pressure load of 253 Pa in a 300 µm diameter capillary. This was achieved by fixing a capillary tube directly above a 400 µm diameter oscillating air bubble (Figure 13.14) so as to direct a proportion of the flow into the capillary. As well as the net movement of fluid, cavitation microstreaming can be used to transport particles within a fluid. Marmottant and Hilgenfeldt[48] have demonstrated that by using a bubble–particle doublet, microstreaming flows can be generated with a directional component that can be used to transport cells. Won *et al.*[49] demonstrated the use of cavitation microstreaming for propelling floating particles. Furthermore, competition between microstreaming and acoustic radiation forces has been used to trap particles according to their size.[50]

A further novel application of cavitation microstreaming was its use by Wang, Attinger and Moraga[51] for driving a self-aligning micro-rotor. They were able to drive the rotor at different speeds by varying the acoustic frequency, achieving a maximum angular speed of over 700 rpm. Although

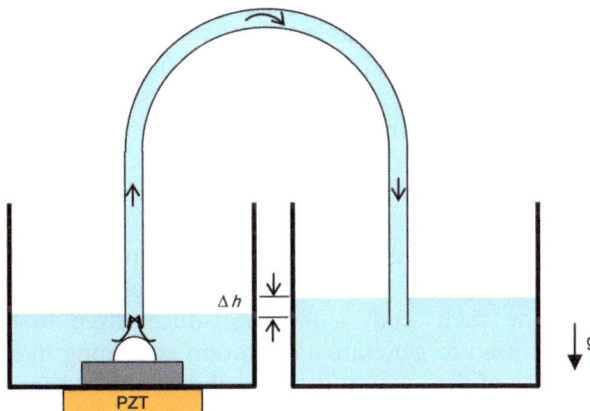

Figure 13.14 A schematic of the experimental setup for a single bubble micropump by Ryu *et al.*[47] A 400 µm diameter air bubble is excited acoustically in order to generate microstreaming flow that drives a flow of 0.19 mL s^{-1} through a 300 µm diameter capillary. The figure is taken from ref. 47.

the rotor's power output was estimated to lie in the femtoWatt range, its power density was 100 MW m^{-3}. Further work demonstrated a linear relationship between the driving voltage of the transducer and the rotational speed of the rotor.[52] The authors did not discuss the possible applications of the micro-rotor though presumably they could include pumping, centrifugation and micro-propulsion. In addition to the high energy density achieved, the advantage of a bubble driven micro-rotor is the absence of the need for a physical connection to a power supply.

The final application of cavitation microstreaming to be discussed is its use in biomedical research, where it may be used to exert a shear force on the cellular membrane of biological cells. Shear stress results in significant physiological effects which, described in more detail by Wiklund,[53] include but are not limited to an increase in cell metabolism,[54] cellular differentiation,[55] endothelial cell elongation[56] and cellular membrane poration.[57] In the majority of studies into the generation of shear stress on cell membranes, encapsulated microbubbles (particularly medical acoustic contrast agents) have been preferred over free air bubbles. The rationale being that the former are more stable and can be more easily targeted to cells.

Of the aforementioned physiological effects of shear stress on cells, the permeabilisation of the cell membrane has received the greatest attention. Much of the research has involved the use of inertial cavitation (a violent event that leads to high levels of cell death[58] and strictly speaking is not a form of streaming). Although cavitation microstreaming has also been used,[19] most research carried out to date has not made a distinction between the two forms of cavitation, such that both forms are likely to play a role in membrane poration. An example of a typical study is that carried out by Bao, Thrall and Miller[59] in which it was reported that almost 40% of Chinese hamster ovary cells took up FITC-dextran when sonicated at 0.8 MPa in the presence of Albunex contrast agent, compared to 0% in their negative controls. The majority of researchers have used arbitrary environments such as petri dishes or well plates for their experiments, though there are examples of purpose built acoustofluidic chambers being used for the generation of cavitation microstreaming.[18,19]

Several studies have demonstrated that the magnitude of shear stress generated during cavitation microstreaming is applicable to biomedical applications. In one such study a PZT transducer fixed to the back of a Mason horn was used to generate a cavitation streaming like flow.[60] The rationale for the use of the Mason horn was that it allowed for the spacing between the horn tip and the cells to be easily determined and that there would be little chance of inadvertently generating inertial cavitation. By utilising theoretical models in combination with experimental data, it was estimated that a threshold shear stress of 12 Pa is required to porate the cellular membrane, a figure that has been advocated by additional researchers.[61-63]

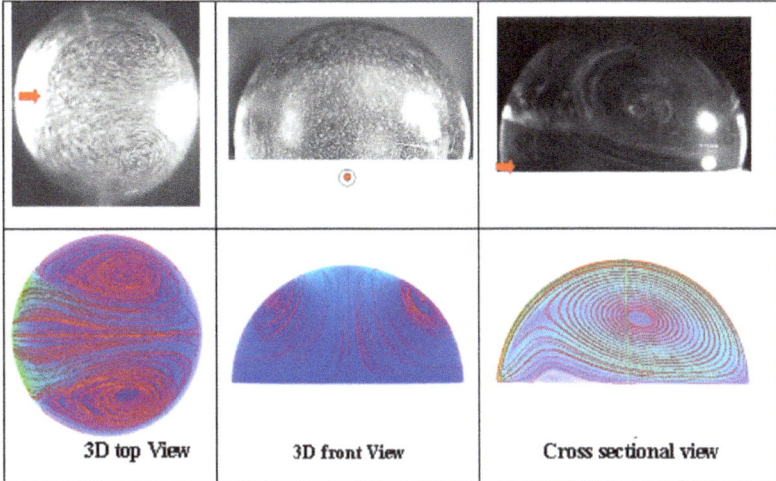

Figure 13.15 Comparison of experimental and numerical modelling for a 30 µl droplet positioned at the centre of a SAW device. The upper row shows pictures of particle trajectories; the bottom row shows the corresponding streaming patterns from numerical simulations. The red arrow indicates the SAW propagation direction. The figure is taken from ref. 65.

13.3.4 Surface Acoustic Wave Induced Streaming Applications

Surface acoustic wave (SAW) streaming is not classified as either boundary layer driven or bulk driven streaming and is therefore discussed separately. Acoustic streaming generated by SAWs has numerous important applications in microfluidics. The SAW is initiated by an interdigitated transducer (IDT) which forms a pressure oscillation inside a fluid compartment such as a droplet. Viscous attenuation of the SAW leads to the generation of acoustic streaming.[64] At low acoustic amplitudes the phenomenon has been utilised for rapid mixing within a droplet (Figure 13.15).

Wixforth[66] has demonstrated that a fluorescent dye can be mixed with a 50 nL droplet of water in under a second. At high acoustic amplitudes the steady momentum flux is capable of displacing the droplet, a phenomenon that has been utilised for the formation and actuation of micro-droplets. The technique is also applicable to mixing in microtiter plates (Figure 13.16). This device has been commercialized under the name PlateBooster™ by Beckman Coulter (formerly by Advalytix).

Using a device similar to Wixforth's,[66] Luong, Phan and Nguyen[67] reported that operating an acoustic micro-mixer at a high voltage could lead to significant heating. For example, using voltages in the interval of 35–75 V

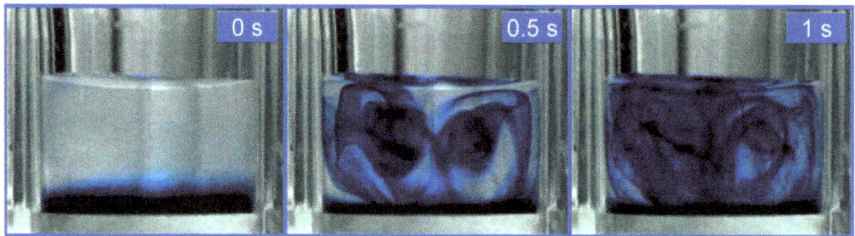

Figure 13.16 Surface acoustic wave (SAW) induced mixing in a 96-well plate. The figure is taken from ref. 66.

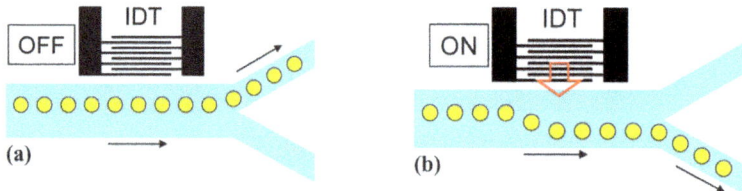

Figure 13.17 A surface acoustic wave used to deflect water droplets. The figure shows the effect when the IDT is turned off (a) and on (b). The figure is taken from ref. 68.

resulted in a maximum temperature in the range ~30–70 °C. Similar results were previously reported by Tseng *et al.*[68]

In a microfluidic device created by Franke *et al.*[69] (Figure 13.17), water droplets of 20 μm in diameter were displaced away from an interdigitated transducer (IDT), providing control over which outlet channel the droplets entered. In a follow-up on this work, cell sorting was demonstrated at a rate of several kHz.[70] In another experiment it was shown that a water droplet linear actuation velocity of 12 mm s^{-1} could be achieved as long as the acoustic wavelength was smaller than the diameter of the droplet.[71]

Strobl, von Guttenberg and Wixforth[71] demonstrated that the SAW-induced displacement of a large fluid droplet over hydrophilic sites on an otherwise hydrophobic substrate can be used to form droplets of a precise volume. The large droplet contained within a hydrophilic reservoir is forced over a smaller hydrophilic anchor site that defines the size of the microdroplet to be formed (Figure 13.18). Upon cessation of the SAW excitation, the large droplet retracts into the reservoir, leaving a precise volume of fluid in the anchor site. The droplet can then be removed from the anchor. In the work being discussed, this is carried out by a second transducer that displaces the droplet into a third site. An advantage of using a SAW device is that the droplet can be positioned anywhere on a 2-D substrate using only four transducers.

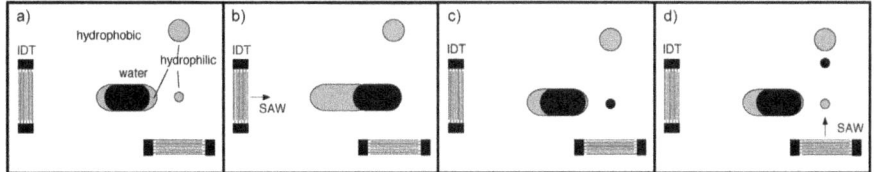

Figure 13.18 Device for forming water droplets of precise volume based on two crossing SAWs. (a) Initial position with a large, water filled hydrophilic anchor site and an empty small one. (b) Left–right propagating SAW pushes large water droplet onto the small anchor site. (c) After turning off the SAW, the large water droplet retreats back to its initial position, leaving the small anchor site filled with a precise minute volume of water. (d) Bottom–top propagating SAW pushes the small water droplet towards the output position. The figure is taken from ref. 71.

13.4 Conclusion

Applications of acoustic streaming can be divided into methods for (1) suppressing or overcoming the streaming, and (2) methods where the streaming is actively used. The first category contains primarily methods based on cell or particle manipulation in standing wave fields. Here, the streaming typically places a practical lower limit on the particle size that can be manipulated by the acoustic radiation force. Depending on the device design, this limit is typically around 1–2 µm. An important conclusion concerning the size and orientation of acoustic streaming is that for Rayleigh streaming (the most common streaming type in microfluidic devices), the vortices are not exclusively $\lambda/4$-sized as often described in the literature, but are sometimes $\lambda/2$-sized or even larger. In addition, the orientations of the vortices are often difficult to predict, although the most commonly observed orientation is according to Figure 13.2. Therefore, reported experimental works need to be complemented with more accurate theoretical modelling in the future.

In the other category (2), the number of lab on a chip devices taking advantage of acoustic streaming is steadily increasing. Here, applications include fluid mixing, fluid pumping, particle/cell sorting, droplet displacement, cellular membrane poration and cell lysis. Currently, this research field is very active and several products based on acoustic streaming have already been commercialised.

References

1. J. Lighthill, *J. Sound Vib.*, 1978, **61**, 391–418.
2. Lord Rayleigh, *Phil. Trans. R. Soc. London*, 1884, **175**, 1–21.
3. W. L. Nyborg, Acoustic streaming, in *Physical Acoustics IIB*, ed. W. P. Mason, Academic Press, New York, 1965, pp. 265–331.

4. H. Feng, *Ultrasound Technologies for Food and Bioprocessing*, Springer, New York, 2011.
5. N. Riley, *Annu. Rev. Fluid Mech.*, 2001, **33**, 43–65.
6. S. Boluriaan and P. J. Morris, *Aeroacoustics*, 2003, **2**, 255–292.
7. H. Schlichting, *Phys. Z.*, 1932, **33**, 327–335.
8. W. L. Nyborg, *J. Acoust. Soc. Am.*, 1958, **30**, 329–339.
9. L. D. Landau and E. M. Lifshitz, *Fluid Mechanics, Course of Theoretical Physics*, Butterworth-Heinemann, 2nd edn, 2006, vol. 6.
10. M. F. Hamilton, Y. A. Ilinskii and E. A. Zabolotskaya, *J. Acoust. Soc. Am.*, 2003, **113**, 153–160.
11. K. D. Frampton, S. E. Martin and K. Minor, *Appl. Acoust.*, 2003, **64**, 681–692.
12. C. Eckart, *Phys. Rev.*, 1948, **73**, 68–76.
13. K. Matsuda, T. Kamakura and M. Maezawa, *Jpn. J. Appl. Phys.*, 2006, **45**, 4448–4452.
14. J. A. Cosgrove, J. M. Buick, S. D. Pye and C. A. Greated, *Ultrasonics*, 2001, **39**, 461–464.
15. T. M. Squires and S. R. Quake, *Rev. Mod. Phys.*, 2005, **77**, 977–1026.
16. X. Liu and J. Wu, *J. Acoust. Soc. Am.*, 2009, **125**, 1319–1330.
17. S. Elder, *J. Acoust. Soc. Am.*, 1959, **31**, 54–64.
18. P. Tho, R. Manasseh and A. Ooi, *J. Fluid Mech.*, 2007, **576**, 191–233.
19. R. Green, R. J. Boltryk, D. Ankrett, P. Glynne-Jones, P. A. Townsend and M. Hill, *Proc. of 8th USWNet Conference, Gdansk, Poland, 4–7 September 2011*.
20. G. Gormley and J. Wu, *J. Acoust. Soc. Am.*, 1998, **104**, 3115–3118.
21. A. A. Doinikov and A. Bouakaz, *J. Acoust. Soc. Am.*, 2010, **127**, 1218–1227.
22. M. Bengtsson and T. Laurell, *Anal. Bioanal. Chem.*, 2004, **378**, 1716–1721.
23. J. F. Spengler, W. T. Coakley and K. T. Christensen, *AIChE J.*, 2003, **49**, 2773–2782.
24. J. F. Spengler and W. T. Coakley, *Langmuir*, 2003, **19**, 3635–3642.
25. L. A. Kuznetsova and W. T. Coakley, *J. Acoust. Soc. Am.*, 2004, **116**, 1956–1966.
26. L. A. Kuznetsova, S. P. Martin and W. T. Coakley, *Biosens. Bioelectron.*, 2005, **21**, 940–948.
27. R. Barnkob, P. Augustsson, T. Laurell and H. Bruus, *Phys. Rev. E*, 2012, **86**, 056307.
28. P. Augustsson, *On microchannel acoustophoresis*, Doctoral Thesis, Lund University, 2011.
29. L. Johansson, *Acoustic manipulation of particles and fluids in microfluidic systems*, Doctoral Thesis, Uppsala University, 2009.
30. D. Möller, T. Hilsdorf, J. Wang and J. Dual, *Proc. of 8th USWNet Conference, Gdansk, Poland, 4–7 September 2011*.
31. S. M. Hagsäter, T. Glasdam Jensen, H. Bruus and J. P. Kutter, *Lab Chip*, 2007, **7**, 1336–1344.
32. S. M. Hagsäter, A. Lenshof, P. Skafte-Pedersen, J. P. Kutter, T. Laurell and H. Bruus, *Lab Chip*, 2008, **8**, 1178–1184.

33. P. Skafte-Pedersen and H. Bruus, *Proc. of 6th USWNet Conference, Zurich, Switzerland, 13–14 November 2008*.
34. H. Bruus, *Lab Chip*, 2012, **12**, 20–28.
35. M. Ohlin, A. E. Christakou, T. Frisk, B. Önfelt and M. Wiklund, *Proc. of 8th USWNet Conference, Gdansk, Poland, 4–7 September 2011*.
36. P.-H. Huang, Y. Xie, D. Ahmed, J. Rufo, N. Nama, Y. Chen, C. Yu Chana and T. Jun Huang, *Lab Chip*, 2013, **13**, 3847–3852.
37. M. Ohlin, A. E. Christakou, T. Frisk, B. Önfelt and M. Wiklund, *Proc. of 15th Int. Conference on Miniaturized Systems for Chemistry and Life Sciences (µTAS 2011), 2–6 October 2011, Seattle, USA*, pp. 1612–1614.
38. H. M. Hertz, *J. Appl. Phys.*, 1995, **78**, 4845–4849.
39. M. Wiklund, P. Hänninen and H. M. Hertz, *Proc. of 2003 IEEE International Ultrasonics Symposium, 5–8 October, 2003, Honolulu, Hawaii, USA*.
40. M. Wiklund, J. Toivonen, M. Tirri, P. Hänninen and H. M. Hertz, *J. Appl. Phys.*, 2004, **96**, 1242–1248.
41. M. Wiklund and H. M. Hertz, *Lab Chip*, 2006, **6**, 1279–1292.
42. A. L. Bernassau, P. Glynne-Jones, F. Gesellchenc, M. Riehlec, M. Hill and D. R. S. Cumming, *Ultrasonics*, 2014, **54**, 268–274.
43. D. Ahmed, X. L. Mao, J. J. Shi, B. K. Juluri and T. J. Huang, *Lab Chip*, 2009, **9**, 2738–2741.
44. R. H. Liu, R. Lenigk and P. Grodzinski, *J. Microlithogr. Microfabr. Microsyst.*, 2003, **2**, 178–184.
45. A. R. Tovar and A. P. Lee, *Lab Chip*, 2009, **9**, 41–43.
46. M. V. Patel, A. R. Tovar and A. P. Lee, *Lab Chip*, 2012, **12**, 139–145.
47. K. Ryu, S. K. Chung and S. K. Cho, *JALA*, 2010, **15**, 163–171.
48. Marmottant and S. Hilgenfeldt, *Proc. Natl. Acad. Sci. U. S. A.*, 2006, **101**, 9523–9527.
49. J. M. Won, J. H. Lee, K. H. Lee, K. Rhee and S. K. Chung, *Int. J. Precis. Eng. Man.*, 2011, **12**, 577–580.
50. P. Rogers and A. Neild, *Lab Chip*, 2011, **21**, 3710–3715.
51. X. L. Wang, D. Attinger and F. Moraga, *Nanoscale Microscale Thermophys. Eng.*, 2006, **10**, 379–385.
52. J. Kao, X. L. Wang, J. Warren, J. Xu and D. Attinger, *J. Micromech. Microeng.*, 2007, **17**, 2454–2460.
53. M. Wiklund, *Lab Chip*, 2012, **12**, 2018–2028.
54. J. A. Frangos, L. V. McIntire and S. G. Eskin, *Biotechnol. Bioeng.*, 1988, **32**, 1053–1060.
55. K. Yamamoto, T. Sokabe, T. Watabe, K. Miyazono, J. K. Yamashita, S. Obi, N. Ohura, A. Matsushita, A. Kamiya and J. Ando, *Am. J. Physiol. Heart Circ. Physiol.*, 2005, **288**, H1915–H1924.
56. M. J. Levesque and R. M. Nerem, *J. Biomech. Eng.*, 1985, **107**, 341–347.
57. D. L. Miller and J. Quddus, *Ultrasound Med. Biol.*, 2000, **26**, 661–667.
58. C. D. Ohl, M. Arora, R. Ikink, N. De Jong, M. Versluis, M. Delius and D. Lohse, *Biophys. J.*, 2006, **91**, 4285–4295.

59. S. Bao, B. D. Thrall and D. L. Miller, *Ultrasound Med. Biol.,* 1997, **23**, 953–959.
60. J. Wu, *Prog. Biophys. Mol. Biol.,* 2007, **93**, 363–373.
61. W. J. Greenleaf, M. E. Bolander, G. Sarkar, M. B. Goldring and J. F. Greenleaf, *Ultrasound Med. Biol.,* 1998, **24**, 587–595.
62. D. L. Miller, S. Bao and J. E. Morris, *Ultrasound Med. Biol.,* 1999, **25**, 143–149.
63. D. L. Miller and J. Quddus, *Ultrasound Med. Biol.,* 2001, **27**, 1107–1113.
64. K. Sritharan, C. J. Strobl, M. F. Schneider, A. Wixforth and Z. Guttenberg, *Appl. Phys. Lett.,* 2006, **88**, 054102.
65. M. Alghane, B. X. Chen, Y. Q. Fu, Y. Li, J. K. Luo and A. J. Walton, *J. Micromech. Microeng.,* 2011, **21**, 015005.
66. A. Wixforth, *JALA,* 2006, **11**, 399–405.
67. T. D. Luong, V. N. Phan and N. T. Nguyen, *Microfluid. Nanofluid.,* 2011, **10**, 619–625.
68. W. K. Tseng, J. L. Lin, W. C. Sung, S. H. Chen and G. B. Lee, *J. Micromech. Microeng.,* 2006, **16**, 539–548.
69. T. Franke, A. R. Abate, D. A. Weitz and A. Wixforth, *Lab Chip,* 2009, **9**, 2625–2627.
70. T. Franke, S. Braunmüller, L. Schmid, A. Wixforth and D. A. Weitz, *Lab Chip,* 2010, **10**, 789–794.
71. C. J. Strobl, Z. von Guttenberg and A. Wixforth, *IEEE Trans. Sonics Ultrason.,* 2004, **51**, 1432–1436.

CHAPTER 14

Theory of Surface Acoustic Wave Devices for Particle Manipulation

MICHAEL GEDGE AND MARTYN HILL*

Engineering Sciences, University of Southampton, Southampton SO17 1BJ, UK
*E-mail: m.hill@soton.ac.uk

14.1 Introduction

The theory of surface waves was first discussed by Lord Rayleigh in the late 19th century.[1] It was shown that their effect decreased rapidly with depth and that their velocity of propagation is of a lower magnitude than that of body waves.[2] As these waves spread only in two dimensions their amplitude decays more slowly with distance than bulk elastic waves in similar media. At the time they were recognised as important in seismology but they have since found use at much smaller scales, such as ultrasonic surface testing and in microfluidics. They also have significant use in electronic filters and sensors.[3] It will be shown that Rayleigh waves are a combination of longitudinal motion and transverse motion confined to the surface of an elastic medium, penetrating only to about one wavelength in depth. Surface waves on a solid are similar to surface waves on a liquid, in that particle motion is elliptical; however, there are differences in direction and restoring forces. In solids it is elastic forces while in liquids it is gravity and surface tension that supply that restoring force.[2]

Lamb waves will not be considered here, but it can be shown that a Lamb wave transforms into a Rayleigh wave as the thickness of the substrate increases.[4] Key equations describing leaky Rayleigh waves and associated attenuation will be presented and discussed. They will show that shear based attenuation and *solid–gas* interface losses can largely be ignored.

Other Rayleigh-like waves, such as Scholte waves, interface waves and Stoneley waves, will be presented and the chapter then discusses the effects of piezoelectric substrates and anisotropy before detailing SAW generation methods.

The choice of using surface acoustic wave (SAW) devices as opposed to the alternative bulk acoustic wave (BAW) devices for ultrasonic particle manipulation depends on a number of factors. SAW devices have the ability to work at very high frequencies, giving the potential for tighter control of particles. Many SAW devices use standing surface acoustic waves to create the potential landscapes required for manipulation so they do not rely on highly resonant chambers to provide the energy gradients. This allows the use of common but acoustically damped microfluidic materials such as PDMS. As the energy within SAWs is inherently confined to the substrate surface prior to launching into the chamber, it is likely that they dissipate less energy in the microfluidic structure as a whole than is the case with BAW devices. While SAW fabrication lends itself to microfluidic integration there is an additional cost and complexity involved than in the construction of BAW devices. Further, the resonant nature and lower frequencies of most BAW devices make them more suitable for handling larger volumes of fluid.

14.2 Rayleigh Waves

In an unbounded isotropic solid only two types of elastic wave can propagate, as shown in Chapter 3. However, surface and interface waves exist at free boundaries and interfaces respectively, allowing Rayleigh waves to be used to manipulate particles and fluids when used in a microfluidic device.

The equations of motion used to describe longitudinal and transverse waves in an isotropic unbounded medium will be manipulated to generate equations detailing the displacements of a surface wave propagating along a half space. Rotation phenomena and dilation phenomena will be decoupled and the Rayleigh wave number defined in terms of longitudinal and transverse wave numbers.

The equation of motion for an unbounded isotropic elastic solid in which body forces are absent can be written as:[2]

$$\rho \frac{\partial^2 u}{\partial t_2} = (\lambda + \mu)\frac{\partial \Delta}{\partial x} + \mu \nabla^2 u, \qquad (14.1)$$

where u defines displacement in the x direction. Similarly v and w are displacements in the y and z directions respectively. ∇^2 is the Laplace operator whilst ρ, λ and μ define the density, Lamé's constant and shear modulus of the medium. Δ, see eqn (14.3), defines the volume strain of the medium.

Theory of Surface Acoustic Wave Devices for Particle Manipulation

As an isotropic, unbounded solid is being considered, similar expressions can be written for v and w. The solutions to these equations define the longitudinal wave speed (C_L) and transverse, or shear, wave speed (C_T) in the x direction:

$$C_L^2 = \frac{\lambda + 2\mu}{\rho}, \quad C_T^2 = \frac{\mu}{\rho}. \tag{14.2}$$

Following the approach of Kolsky,[2] the same equations will be used to generate the equations of motion for a surface wave propagating in the x direction on a half space. Take z to be positive towards the interior of the half space and make the xy plane the free boundary of that half space. The wave will be polarized in the xz plane, so there are no displacements in the y direction, see Figure 14.1. This leads to the following equations describing the volume strain, or dilation Δ of the medium,

$$\Delta = \frac{\partial u}{\partial x} + \frac{\partial w}{\partial z}, \tag{14.3}$$

and the rotation Ω_y in the xz plane

$$\Omega_y = \frac{\partial u}{\partial z} - \frac{\partial w}{\partial x}. \tag{14.4}$$

From here it is possible to define two scalar functions φ and Ψ so that dilation and rotation effects can be decoupled

$$u = \frac{\partial \varphi}{\partial x} + \frac{\partial \Psi}{\partial z}, \tag{14.5a}$$

$$w = \frac{\partial \varphi}{\partial z} - \frac{\partial \Psi}{\partial x}. \tag{14.5b}$$

This allows for dilation and rotation to be defined as $\nabla^2 \varphi$ and $\nabla^2 \Psi$ respectively. By substituting eqn (14.5a) and (14.5b) into eqn (14.1), it is seen that the equations of motion for u and w are satisfied given that.

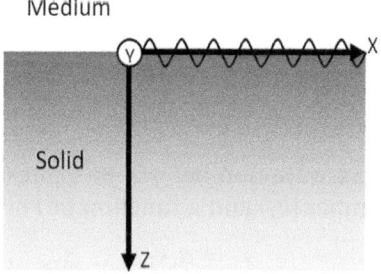

Figure 14.1 A schematic showing the propagation of a surface wave along the xy plane of a half space. Reproduced from ref. 22 with permission.

$$\frac{\partial^2 \varphi}{\partial t^2} = \frac{\lambda + 2\mu}{\rho}\nabla^2\varphi = C_L^2 \nabla^2 \varphi, \tag{14.6a}$$

$$\frac{\partial^2 \Psi}{\partial t^2} = \frac{\mu}{\rho}\nabla^2\Psi = C_T^2 \nabla^2 \Psi. \tag{14.6b}$$

Now consider a sinusoidal wave of angular frequency ω and wave number k propagating in the x direction and look for solutions of φ and Ψ of the form.

$$\varphi = F(z)e^{i(\omega t - kx)}, \tag{14.7a}$$

$$\Psi = G(z)e^{i(\omega t - kx)}. \tag{14.7b}$$

These trial solutions can be substituted into eqn (14.6a) and (14.6b), which after re-arranging give the following.

$$\frac{\partial^2}{\partial z^2} F(z) - \left(k^2 - k_L^2\right) F(z) = 0, \tag{14.8a}$$

$$\frac{\partial^2}{\partial z^2} G(z) - \left(k^2 - k_T^2\right) G(z) = 0, \tag{14.8b}$$

where k_L and k_T are the longitudinal and transverse wave numbers respectively.

The longitudinal and transverse wave numbers can be expressed in the following format:

$$k_L = \frac{\omega}{C_L}, \tag{14.9a}$$

$$k_T = \frac{\omega}{C_T}. \tag{14.9b}$$

Hence:

$$\frac{k_L^2}{k_T^2} = \frac{\mu}{\lambda + 2\mu} = \frac{1 - 2\nu}{2 - 2\nu} = \alpha_1^2, \tag{14.10}$$

where in the third expression we have introduced the Poisson's ratio given by

$$\nu = \frac{\lambda}{2(\lambda + \mu)}. \tag{14.11}$$

Hence the longitudinal wave number can be expressed as the product of the transverse wave number (k_T) and a function of Poisson's ratio (α_1),

$$k_L = \alpha_1 k_T. \tag{14.12}$$

It is known that $k_L^2 < k_T^2$ and it is assumed that $k_T^2 < k^2$ (this will be confirmed later).

To determine the dispersion relation of ω in terms of k, we seek solutions to eqn (14.8a) and (14.8b) of the form.

$$F(z) = Ae^{(-\gamma_L z)} + A'e^{(\gamma_L z)}, \quad (14.13a)$$

$$G(z) = Be^{(-\gamma_T z)} + B'e^{(\gamma_T z)}, \quad (14.13b)$$

where $\gamma_L^2 = k^2 - k_L^2$ for the *longitudinal-Rayleigh* combination and $\gamma_T^2 = k^2 - k_T^2$ for the *transverse-Rayleigh* combination, and where A, B, A' and B' are constants. The A' and B' components of these equations are physically unrealistic as they would lead to an increase in displacement amplitude with increasing depth, z. This leads to the following equations describing φ and Ψ

$$\varphi = Ae^{(-\gamma_L z)}e^{i(\omega t - kx)}, \quad (14.14a)$$

$$\Psi = Be^{(-\gamma_T z)}e^{i(\omega t - kx)}. \quad (14.14b)$$

Boundary conditions can now be applied that allow the elimination of A and B, which ultimately leads to the determination of the displacements in the x and z directions. Plane stress and shear stress will be equal to zero at the free surface, *i.e.* when z is equal to zero. For plane stress:

$$\sigma_{zz} = \lambda \Delta + 2\mu \frac{\partial w}{\partial z}, \quad (14.15)$$

which can be expressed in terms of φ and Ψ:

$$\sigma_{zz} = (\lambda + 2\mu)\frac{\partial^2 \varphi}{\partial z^2} + \lambda \frac{\partial^2 \varphi}{\partial x^2} - 2\mu \frac{\partial^2 \Psi}{\partial x \partial z}. \quad (14.16)$$

Inserting the decaying components of eqn (14.14a) and (14.14b) yields for z equals zero:

$$A[(\lambda + 2\mu)\gamma_L + \lambda k^2] - 2\mu Bi\gamma_T k = 0. \quad (14.17)$$

A similar approach is then taken for shear stress in the half space.

$$\sigma_{zx} = \mu \left(\frac{\partial u}{\partial z} + \frac{\partial w}{\partial x}\right), \quad (14.18)$$

which again can be expressed in terms of φ and Ψ

$$\sigma_{zx} = \mu \left(2\frac{\partial^2 \varphi}{\partial x \partial z} - \frac{\partial^2 \Psi}{\partial x^2} + \frac{\partial^2 \Psi}{\partial z^2}\right). \quad (14.19)$$

Inserting the decaying components of eqn (14.14a) and (14.14b) yields for z equals zero.

$$2i\gamma_L kA + (\gamma_T^2 + k^2)B = 0. \quad (14.20)$$

Equation (14.17) and (14.20) can be combined to eliminate A and B, yielding.

$$4\mu\gamma_L\gamma_T k^2 = \left[(\lambda + 2\mu)\gamma_L^2 - \lambda k^2\right](\gamma_T^2 + k^2) \quad (14.21)$$

Squaring both sides of the equation and inserting expressions for γ_L and γ_T yields:

$$16\left(1 - \frac{k_L^2}{k^2}\right)\left(1 - \frac{k_T^2}{k^2}\right) = \left[2 - \frac{(\lambda + 2\mu)}{\mu}\frac{k_L^2}{k^2}\right]^2\left(2 - \frac{k_T^2}{k^2}\right)^2. \quad (14.22)$$

Using eqn (14.10) and (14.12), this expression can be written as.

$$16\left(1 - \frac{\alpha_1^2 k_T^2}{k^2}\right)\left(1 - \frac{k_T^2}{k^2}\right) = \left(2 - \frac{k_T^2}{k^2}\right)^4 \quad (14.23)$$

Expanding and replacing k_T/k with s yields.

$$s^6 - 8s^4 + (24 - 16\alpha_1^2)s^2 + 16\alpha_1^2 - 16 = 0. \quad (14.24)$$

If the Poisson's ratio of the half space is known, this can be solved numerically and used to find the Rayleigh wave speed with the following.

$$s = \frac{k_T}{k} = \frac{\omega}{kC_T}. \quad (14.25)$$

The Rayleigh wave speed has been defined as $C_R = \omega/k$, meaning that $s = C_R/C_T$. The velocity of propagation is thus independent of the frequency and the waves are therefore non-dispersive. The velocity of the wave depends solely on the elastic constants of the material. The rate at which the wave amplitude reduces with the depth z depends on the values of the factors γ_L and γ_T which are given by:

$$\frac{\gamma_L^2}{k^2} = 1 - \alpha_1^2 s^2, \quad (14.26a)$$

$$\frac{\gamma_T^2}{k^2} = 1 - s^2. \quad (14.26b)$$

If $\nu > 0.263$ there are two complex conjugate roots and one real root. If $\nu \leq 0.263$ then there will be three real roots.[5] However, only one of these roots will be realistic. The change of amplitude with depth for a Rayleigh wave can be calculated from:

$$u_R = \frac{\partial \varphi}{\partial x} + \frac{\partial \Psi}{\partial z} = -(Aike^{-\gamma_L z} + B\gamma_T e^{-\gamma_T z})e^{[i(\omega t - kx)]}, \quad (14.27a)$$

$$w_R = \frac{\partial \varphi}{\partial z} - \frac{\partial \Psi}{\partial x} = -(A\gamma_L e^{-\gamma_L z} - Bike^{-\gamma_T z})e^{[i(\omega t - kx)]}. \quad (14.27b)$$

Theory of Surface Acoustic Wave Devices for Particle Manipulation

These reduce to the following if the real parts are taken and B is substituted for A:

$$u_R = Akf_{u(z)}\sin(\omega t - kx), \quad (14.28a)$$

$$w_R = A\gamma_L f_{w(z)}\cos(\omega t - kx). \quad (14.28b)$$

where we have introduced the decay functions $f_{u(z)}$ and $f_{w(z)}$ given by

$$f_{u(z)} = e^{-\gamma_L z} - \frac{2\gamma_L \gamma_T}{(\gamma_T^2 + k^2)} e^{-\gamma_T z}, \quad (14.29a)$$

$$f_{w(z)} = e^{-\gamma_L z} - \frac{2k^2}{(\gamma_T^2 + k^2)} e^{-\gamma_T z}, \quad (14.29b)$$

which describe the rate at which the displacement amplitudes along the direction of propagation change with depth. Displacement in the x (horizontal) direction decreases rapidly as z increases; this can be seen in Figure 14.2(a) which shows an example of displacements as a function of depth. The functions are normalised against the amplitude of the motion perpendicular to the propagation direction at the surface. At a depth of about $z/\lambda = 0.2$ the displacement in the x direction passes through zero, and changes polarity. Movement perpendicular to the surface, *i.e.* in the z direction, increases slightly before reaching a maximum at a depth of about 0.076 wavelengths and then falls away but does not change polarity.

From eqn (14.28a) and (14.28b), it can be seen that the displacements u and w are functions of *sine* and *cosine* respectively. Hence the motion of particles is rotational about the y axis, rather than traversing back and forth along a curved path. This can be seen in Figure 14.2(b) which shows the paths of particles (arbitrary amplitude) at depths corresponding to Figure 14.2(a).

Particle motion can be thought of as planes of elliptical motion, whose shape and phase at any given time depends upon the depth. The motion is anticlockwise near the surface but reverses to a clockwise motion below the zero crossing of the u velocity component. An alternative, simpler approximation for the speed of Rayleigh waves (*i.e.* the root of eqn (14.24)), which expresses s solely in terms of Poisson's ratio, is quoted by Viktorov:[6]

$$s = \frac{0.87 + 1.12\nu}{1 + \nu}. \quad (14.30)$$

Thus as ν varies from 0 to 0.5, C_R varies from $0.87C_T$ to $0.96C_T$. This is shown in Figure 14.3.

It is worth noting that actual displacements are very small. As an example,[7] in a device operating at 100 MHz with 10 mW average power in a beam 1 cm wide on a substrate with SAW velocity 3 km s^{-1}, the wavelength is 30 μm and the peak vertical displacement of the order of 10^{-10} m.

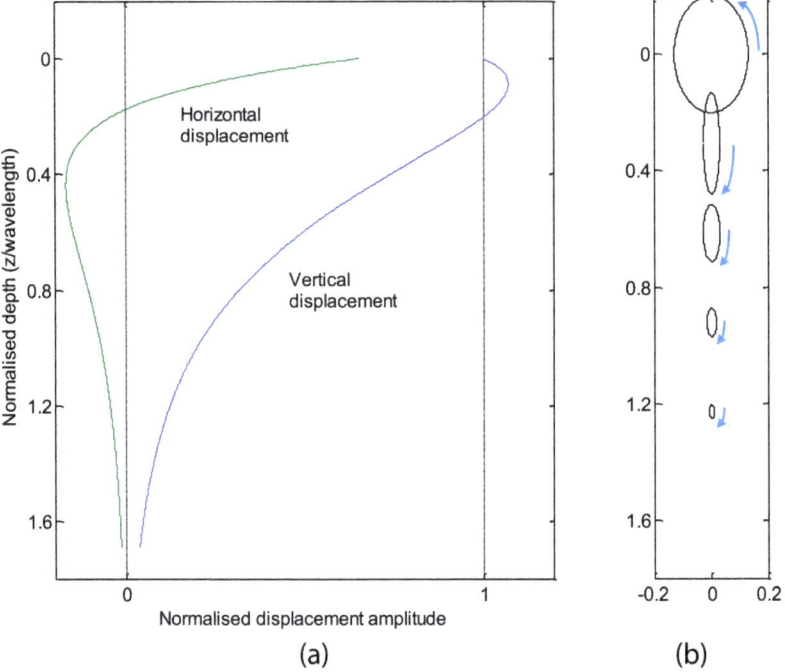

Figure 14.2 (a) Displacement amplitudes for a Rayleigh wave as a function of depth for an example isotropic material based on eqn (14.28). (b) Elliptical particle motions for different depths. Each ellipse corresponds to the motion of a particle with its equilibrium position at its centre. The top ellipse will rotate in the opposite direction to the lower ellipses as indicated by the blue arrows. Reproduced from ref. 22 with permission.

This section has discussed the mechanism of surface wave propagation and the following section describes more general conditions that are of importance to ultrasonic particle manipulation.

14.3 Leaky Rayleigh Waves

A pure Rayleigh wave (with a completely free surface) is of less interest in microfluidics than a surface bounded by a fluid or solid into which the surface wave will deliver energy. This will lead to an exponentially decaying wave propagating along the *fluid–solid* interface.[8] In practice this attenuation is relatively small for *solid–gas* interfaces. In the following, a fluid loaded structure is considered. The fluid is assumed to be unbound. As the impedance of the fluid rises from zero, the wave will behave less like a pure Rayleigh wave and will transform into a "leaky" Rayleigh wave.

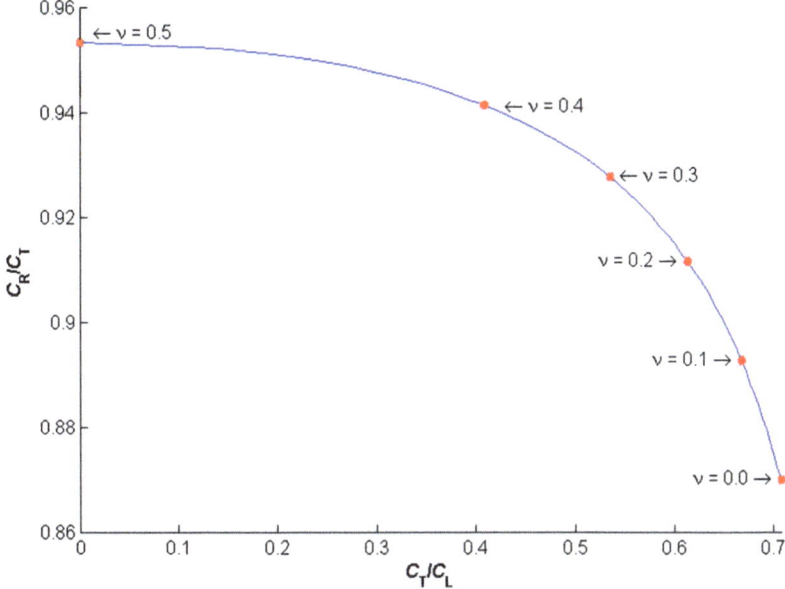

Figure 14.3 Variation in the ratios C_R/C_T and C_T/C_L for different Poisson's ratios (ν).[10] Reproduced from ref. 22 with permission.

The components of a Rayleigh wave can be broken down into normal and tangential displacements and arranged into the following form:[9]

$$4k^2 \gamma_L \gamma_T - \left(k^2 + \gamma_T^2\right)^2 = 0. \tag{14.31}$$

When the surface is loaded with a fluid, the boundary conditions change. By including the continuity of normal stress into the above equation, the following equation is generated:

$$4k^2 \gamma_L \gamma_T - \left(k^2 + \gamma_T^2\right)^2 = i\frac{\rho_F}{\rho_R} \frac{\gamma_L k_T^4}{\sqrt{k_L^2 - k^2}}, \tag{14.32}$$

where ρ_F and ρ_R are the densities of the fluid and the solid half space respectively. Likewise, C_F and C_R are the wave speeds of the two media. Tangential displacement is assumed not to be transferred to the fluid, since it does not support shear modes but more information on this matter can be found in a paper by Vanneste and Bühler.[10] Upon inspection it can be seen that as ρ_F tends to zero, the equation reverts back to a pure Rayleigh wave. This equation has one real root and one complex root. The real root corresponds to a Scholte wave and will be discussed later. The complex root corresponds to a modified Rayleigh wave. It can be shown that given the condition $C_F < C_R$, which is true for most media, then eqn (14.32) has a complex root that corresponds to a system of three waves.[11] One wave is found in the fluid and two in the solid. A simple physical interpretation of

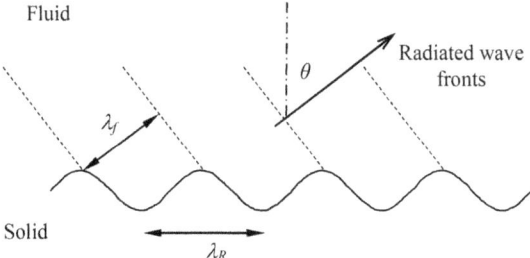

Figure 14.4 A schematic showing a SAW wave radiating into a fluid. Reproduced from ref. 22 with permission.

this is that the Rayleigh wave radiates energy at an angle into the fluid. This must be the case since we are dealing with a complex root, which must be losing energy, and this is the radiation away from the boundary into the fluid.[9] It can be shown that the velocity of the loaded Rayleigh wave is higher than that of the unloaded wave.[6] However, upon further analysis the difference can be shown to be very small. As the fluid has little effect on the velocity of the Rayleigh wave, the following approximation (essentially a statement of Snell's law) can be made about the angle at which the wave propagates into the material, see Figure 14.4:

$$\sin\theta = \frac{C_F}{C_R}. \quad (14.33)$$

This is found by phase matching the wave in the fluid with the Rayleigh wave. As the wave travels along the interface, it can be seen to "leak" energy into the fluid at this angle. For common gases such as air this is near normal to the surface. It should be noted that a Rayleigh wave can be generated on the surface of a substrate if a wave travels through the fluid at the correct angle towards the interface.

In contrast to velocity, the attenuation of the wave is greatly affected by the presence of the fluid. It is this characteristic that makes Rayleigh waves excellent choices for the efficient transfer of energy into cavities or droplets. The attenuation coefficient for a leaky Rayleigh wave travelling along a *solid–fluid* interface is given by[9]

$$\alpha_L = \frac{\rho_F C_F}{\rho_R C_R \lambda_R} \, \text{m}^{-1}, \quad (14.34)$$

where λ_R is the Rayleigh wavelength. The coefficient defines the energy loss resulting from the transmission of a bulk wave into the fluid over a distance x as being $e^{-\alpha_L x}$. The attenuation of the surface wave per wavelength travelled is proportional to the ratio of the fluid and solid impedances. The attenuation coefficient is relatively small for atmospheric air, so for most practical purposes can be ignored.[12]

Theory of Surface Acoustic Wave Devices for Particle Manipulation

Table 14.1 Calculated values for the attenuation of Rayleigh waves in lithium niobate by ambient media.

		Attenuation coefficient (m^{-1}). Bracketed values in dB cm^{-1}.		
Constant	Medium	10 MHz	100 MHz	300 MHz
$\alpha_L{}^a$	Atmospheric air	7.4×10^{-2} (0.0032)	0.74 (0.032)	2.2 (0.096)
$\alpha_L{}^a$	Water	260 (11)	2600 (110)	7800 (330)
$\alpha_s{}^b$	Water	0.16 (0.0067)	5.0 (0.21)	26 (1.1)

[a] Attenuation due to radiation of longitudinal waves.
[b] Attenuation due to frictional loss.

So far it has been assumed that the attenuation of the leaky Rayleigh wave is due to the radiation of a compression wave into the fluid; however, there will also be frictional losses from the transverse motion of the surface. This can be given by the following equation:[9]

$$\alpha_s = \frac{\left(\rho_F \eta \frac{\omega^3}{2}\right)^{\frac{1}{2}}}{4\pi^2 \rho_R C_R^2} \, m^{-1} \quad (14.35)$$

where η is the viscosity of the medium and ω is the angular frequency of the leaky Rayleigh wave. These forces are typically small compared to the longitudinal contribution to attenuation. Typical calculated attenuation factors for leaky Rayleigh waves in LiNbO$_3$ are shown in Table 14.1.

14.4 Scholte Waves

As mentioned earlier, there is a real solution to eqn (14.32). Further analysis shows that it is quite different from a Rayleigh wave and that such solutions exist for any combination of fluid and solid.[13] It can be shown that the wave corresponds to a surface wave travelling along the boundary with a velocity less than the wave velocity of the fluid and the longitudinal and transverse wave speeds of the solid.[14] For lossless media they do not attenuate and simply propagate along the interface: they are therefore pure interface waves. For fluids where the acoustic impedance is less than that of the solid, the decrease in displacement amplitude with depth is significantly higher in the solid than the fluid and most of the energy is contained within the fluid, in contrast to the leaky Rayleigh wave.[6] Scholte wave generation efficiency increases as the acoustic impedance of the fluid increases so little Scholte wave generation is seen when air is the fluid. As the impedance of the fluid increases compared to the solid, the Scholte wave penetrates deeper into the solid.[15]

14.5 Interface Waves

All waves described so far have propagated along a *solid–vacuum* or *solid–fluid* interface. A solid–solid interface will now be considered. For the purpose of this discussion only Rayleigh-like modes will be considered. Love modes, with motion perpendicular to the direction of propagation, are currently of little interest for ultrasonic particle manipulation.

The addition of a thin layer (thickness h) on a substrate carrying a Rayleigh wave provides geometric constraints that will alter the characteristics of propagation, making the wave dispersive. This property has been used to great effect for SAW devices, as particular dispersion characteristics can be designed for.[4] As the layer thickness tends to zero, the wave transforms into a standard Rayleigh wave. The added thin layer can have several different effects on the wave velocity. The following inequality can be used to predict the behaviour of the dispersion curve as kh tends to zero:[16]

$$\frac{C_{T2}}{C_{T1}} > \sqrt{\frac{1 - \left(\frac{C_{T1}}{C_{L1}}\right)^2}{1 - \left(\frac{C_{T2}}{C_{L2}}\right)^2}}, \qquad (14.36)$$

where subscripts *1* and *2* represent the substrate and added layer respectively, whilst ₜ and ₗ represent transverse and longitudinal as before. The limits of the right hand side are $\sqrt{2}$ and $1/\sqrt{2}$. If the above inequality is true, the dispersion curve will be positive as the layer thickness approaches zero. If the above inequality is false, the dispersion curve will have a negative gradient as the layer thickness approaches zero. The case in which the solution lies between these boundaries is more difficult and shall be covered later. This inequality can be presented using shear moduli and density ratios as the ordinate and abscissa axis respectively, see Figure 14.5.[4]

Should the inequality be true, then the layer is said to stiffen the substrate because the presence of the substrate leads to an increase in the surface wave velocity above the unloaded Rayleigh wave velocity. As the layer thickness increases the wave speed increases monotonically up to the shear velocity of the substrate.

As kh increases further, the wave leaks into the substrate so it is no longer a true interface wave. Therefore Rayleigh type propagation is only possible for small values of kh. The variation in wave speed is minimal for these material parameters, as it is dominated by the substrate material parameters. It should be noted that this set of material conditions is an unlikely scenario for ultrasonic particle manipulation devices.

Should $C_{T2} < C_{T1}/\sqrt{2}$ then the layer loads the substrate and the phase velocity of the Rayleigh-like mode decreases as kh increases, until it approaches the Rayleigh velocity of the layer. This is for a layer thickness that is large in comparison to the wavelength. At higher values of kh, the group velocity and phase velocity converge, as characteristics imposed by the layer are less important with regards to the wavelength. The interface waves described so far radiate energy away from the interface.

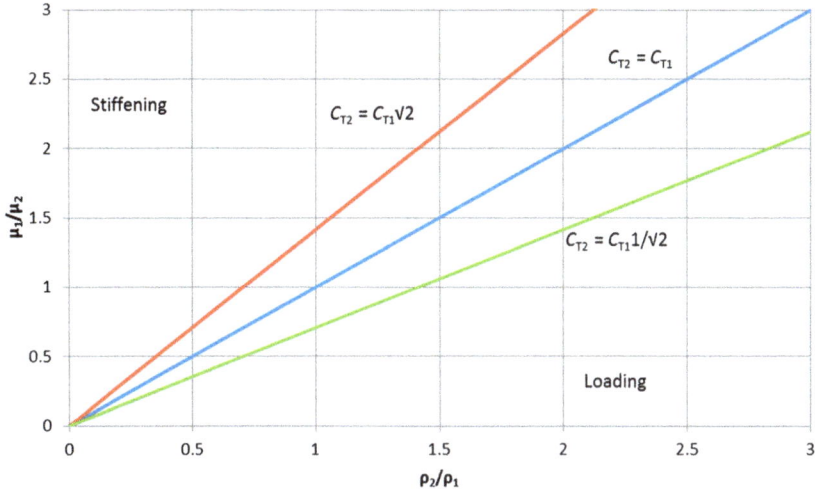

Figure 14.5 The limits of the inequality that determines stiffening or loading of a Rayleigh wave.[4] A stiffening substrate will cause an increase in the Rayleigh wave speed. A loading substrate will cause a drop in the Rayleigh wave speed. It is not possible to generalise on the influence of the layer on the Rayleigh wave speed if the properties lie between the two solid lines. Reproduced from ref. 22 with permission.

14.6 Stoneley Waves

There exists a particular set of conditions such that a Rayleigh-like wave can propagate at the interface between two solids; this type of wave is called a Stoneley wave which is evanescent in both media.[4] For a Stoneley wave to exist both the longitudinal and transverse displacements must be transferred in such a way that energy is confined to the interface. The derivation of a Stoneley wave is similar to that of a Rayleigh wave, except for the transmission of displacement and stress across the boundary. The velocity of the Stoneley wave must be less than the transverse wave speeds of both media but more than the Rayleigh wave speed of the denser medium. The transverse wave speeds of both media need to be similar for Stoneley waves to exist. Stoneley wave velocity is not dependent on frequency, so the wave is therefore non-dispersive. Stoneley waves do not attenuate along the path of propagation so are ideal for the efficient transfer of energy into a sealed fluid chamber.

14.7 Anisotropic Media and Piezoelectric Considerations

The description so far has been based on isotropic non-piezoelectrically active materials. Most materials used for surface acoustic wave devices are anisotropic and in many cases single crystal. The Rayleigh wave and associated Rayleigh-like waves still exist for any given direction of propagation but the characteristics will differ based on orientation. LiNbO$_3$ is often used and it has a typical Rayleigh wave speed between 3500 m s^{-1} and 4000 m s^{-1} depending upon orientation. Depending on the materials used, the ellipse of motion need not be normal to the surface and displacement in the x and z direction may oscillate with depth.[4,17] Wave propagation in anisotropic materials is less straightforward than isotropic materials, however, gains can be made in piezoelectric coupling and the correct orientation will suppress other undesirable wave modes.[4] In an anisotropic solid there are two different transverse modes as well as the longitudinal mode, and as a result a more general surface wave is produced with displacements in three dimensions. This type of wave can be regarded as a combination of Rayleigh and Love waves. It is possible to find orientations that decouple these waves into their respective pure Rayleigh and Love components.[18]

14.8 Generation of Surface Acoustic Waves

There are several methods of generating surface waves. A transducer coupled through a wedge can be designed using the inverse of eqn (14.33), see Figure 14.6.

In this case the Rayleigh wave generated will propagate in one direction only and the coupling angle is independent of the required frequency.[5] The wedge material must have a low longitudinal wave speed for this to work, requiring that $C_W < C_R$. Alternatively, a periodic transducer array, with contacts coupling to the surface at a spacing of λ_R can be used; this will radiate both to the left and the right, see Figure 14.7.

These transduction methods rely on coupling the energy from an active transducer into a passive substrate, but the most common means of generating SAWs is to use a piezoelectrically active substrate or a thin piezoelectric layer on a substrate. In such a case the excitation (and reception) of the SAW is provided by sets of interdigital transducers (IDTs) which comprise metallic

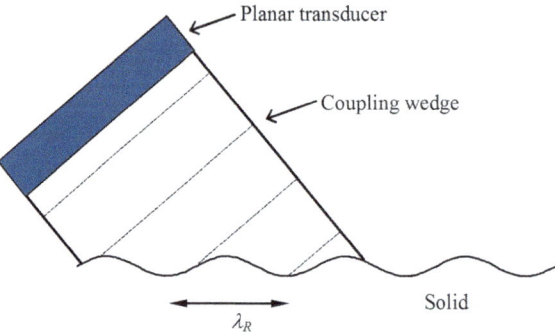

Figure 14.6 Wedge coupling to generate Rayleigh waves in a solid; the waves produced will be unidirectional. Reproduced from ref. 22 with permission.

strips deposited directly onto the piezoelectric surface. Figure 14.8 shows a simple IDT electrode pattern with two sets of interleaved combs extending from the two electrode rails. When a signal is applied to these rails an electric field is established with the spatial period of the comb. This then excites a strain field *via* piezoelectric coupling. The response will be strongest when the spatial period of the electrodes equals the SAW wavelength in the substrate. For a single electrode pair, the Q factor is small and the response is broadband, but as more identically spaced electrode pairs are added, the Q factor increases and the IDT develops a sharp resonance peak. Such an IDT will generate a SAW that will propagate in both directions. If designing for a wave that travels in one direction only attention must be given to absorbing (or reflecting) the opposite-going wave.

A number of additional practical considerations need to be taken into account when designing SAW systems, including the facts that: temperature effects may be significant and may also vary with orientation, and models for SAW propagation in passive substrates do not fully describe SAW behaviour in a piezoelectric substrate as the electromechanical coupling complicates the material's mechanical characteristics. When generating waves it is important to note that electrodes themselves modify the impedance presented to the SAW, through mechanical loading and modification of the

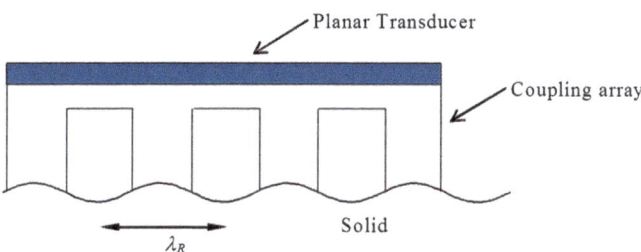

Figure 14.7 A periodic array for the generation of SAWs. The waves produced will be bidirectional and the frequency efficiency. The coupling array is usually made from metal.[7] Reproduced from ref. 22 with permission.

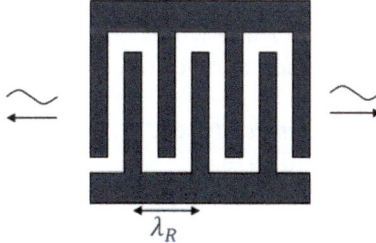

Figure 14.8 Uniform, metallic, periodic IDT array deposited on a piezoelectric substrate. Reproduced from ref. 22 with permission.

electromechanical coupling in the substrate adjacent to the electrode. For this reason each electrode finger can be broken down into two, maintaining the periodicity in terms of the electric field but breaking the periodicity in terms of reflection characteristics. It is normal for the spacing between fingers to be of equal width to the electrodes. This means they are $\lambda/4$ for single electrode configurations or $\lambda/8$ for double electrode configurations. Devices for ultrasonic particle manipulation tend to work at frequencies of 20–40 MHz, which corresponds to electrode widths of approximately 50–25 μm for a typical $LiNbO_3$ device. Most devices are no bigger than a standard microscope glass slide. Electrode fingers need not be parallel to each other[19,20] and can even be concentrically curved to allow for tight focussing into a point on the surface.[21]

14.9 Conclusions

This chapter has derived the equations governing surface acoustic waves in their most basic *solid–vacuum* state before describing the general behaviour of pure Rayleigh waves. The work has also shown how Rayleigh waves differ when propagating along a solid–fluid interface. The leakey SAWs generated from such an interface are described and the associated losses are shown with examples. Scholte waves have been discussed before moving onto *solid–solid* interfaces and the effects these have upon wave propagation velocities. Stonely waves and their potential for efficient transfer of energy along an interface are then discussed.

References

1. Rayleigh, *Proc. London Math. Soc.,* 1885, **17**, 4–11.
2. H. Kolsky, *Stress Waves in Solids*, Clarendon Press, Oxford, 1953.
3. A. A. Oliner, G. W. Farnell, H. M. Gerard, E. A. Ash, J. A. J. Slobodnik and H. I. Smith, *Acoustic Surface Waves*, Springer-Verlag, Berlin, Heidelberg, 1978.
4. G. W. Farnell and E. L. Adler, in *Physical Acoustics 9*, ed. W. P. Mason and R. N. Thurston, Academic Press, New York, 1972.
5. J. L. Rose, *Ultrasonic Waves in Solid Media*, Cambridge University Press, Cambridge, 1999.
6. I. A. Viktorov, *Rayleigh and Lamb Waves*, Plenum Press, New York, 1967.
7. V. M. Ristic, *Principles of Acoustic Devices*, John Wiley & Sons, New York, 1983.
8. H. Überall, in *Physical Acoustics 10*, ed. W. P. Mason and R. N. Thurston, Academic Press, New York, 1973.
9. J. D. N. Cheeke, *Fundamentals and Applications of Ultrasonic Waves*, CRC Press, 2002.
10. J. Vanneste and O. Bühler, *Proc. R. Soc. A*, 2011, **467**, 1779–1800.
11. L. M. Brekhovskikh, *Waves in Layered Media*, Izdatelstvo "Nauka", Moscow, 1957.

12. K. Dransfield and E. Salzmann, in *Physical Acoustics 7*, ed. W. P. Mason and R. N. Thurston, Academic Press, New York, 1970.
13. W. M. Ewing, *Elastic Waves in Layered Media*, McGraw-Hill, New York, 1957.
14. V. G. Gogoladze, Trudy Seismolo, *Inst. Acad. Nauk. USSR,* 1948, **127**, 27–32.
15. C. Glorieux, K. Van de Rostyne, K. Nelson, W. Gao, W. Lauriks and J. Thoen, *J. Acoust. Soc. Am.,* 2001, **110**, 1299–1306.
16. H. F. Tiersten, *J. Appl. Phys.,* 1969, **40**, 770–789.
17. G. W. Farnell, in *Physical Acoustics 6*, ed. W. P. Mason and R. N. Thurston, Academic Press, New York, 1970.
18. H. F. Pollard, *Sound Waves in Solids*, Pion Limited, London, 1977.
19. T. Franke, S. Braunmuller, L. Schmid, A. Wixforth and D. A. Weitz, *Lab Chip,* 2010, **10**, 789–794.
20. T. Frommelt, M. Kostur, M. Wenzel-Schäfer, P. Talkner, P. Hänggi and A. Wixforth, *Phys. Rev. Lett.,* 2008, **100**, 034502.
21. T.-D. Luong, V.-N. Phan and N.-T. Nguyen, *Microfluid. Nanofluid.,* 2010, **10**, 619–625.
22. M. Gedge and M. Hill, *Lab Chip,* 2012, **12**, 2998–3007.

CHAPTER 15

Lab-on-a-chip Technologies Enabled by Surface Acoustic Waves

XIAOYUN DING[a], PENG LI[a], SZ-CHIN STEVEN LIN[a], ZACKARY S. STRATTON[a], NITESH NAMA[a], FENG GUO[a], DANIEL SLOTCAVAGE[a], XIAOLE MAO[b], JINJIE SHI[a], FRANCESCO COSTANZO[a], THOMAS FRANKE[c], ACHIM WIXFORTH[d], AND TONY JUN HUANG*[a,b]

[a]Department of Engineering Science and Mechanics, The Pennsylvania State University, University Park, PA 16802, USA; [b]Department of Bioengineering, The Pennsylvania State University, University Park, PA 16802, USA; [c]School of Engineering, University of Glasgow, Glasgow G12 8LT, UK; [d]University of Augsburg, Augsburg, Germany
*E-mail: junhuang@psu.edu

15.1 Introduction

During the past two decades, microfluidics has emerged as an important platform in chemistry, biology, and medicine.[1-9] The introduction of "lab-on-a-chip" (*i.e.*, small-sized analytical devices) has spurred steadily increasing interest and research efforts in the microfluidics discipline (upon which lab-on-a-chip devices are based). The promise of such lab-on-a-chip devices is grounded in the advantages of microfluidics relative to traditional laboratory techniques used in chemistry and biomedicine: system miniaturization, automation, low cost, reduced reagent and sample consumption, rapid turnaround time, and precise microenvironment control. With an aim to improve the functionality, versatility, and performance of lab-on-a-chip devices, researchers have continued to integrate new physics into the microfluidic platform. For example, the incorporation of electrowetting in microfluidics has enabled effective on-chip manipulation of droplets, *i.e.*, digital microfluidics.[10,11] Incorporation of magnetics in microfluidics has led to the demonstration of highly efficient cell manipulation and separation[12–14] and incorporating optics in microfluidics has spurred the development of fluid-based optical components such as optofluidic lasers, lenses, and modulators with unprecedented tunability.[15–19]

In recent years, surface acoustic wave (SAW) technologies have begun to receive significant attention in the microfluidics community. SAWs are acoustic waves that propagate along the surface of an elastic material. To date, SAW technologies have been used extensively in the telecommunication industry (*e.g.*, cell phones) for signal processing and filtering.[20] Other established applications of SAW technologies include touch-sensitive screens and biological/chemical sensing.[21–25] Recent research demonstrates that SAWs provide an effective means to control fluids and particles in lab-on-a-chip devices.[26] These SAW-based microfluidic devices offer the following useful and inimitable combination of features:

(a) Simple, compact, inexpensive devices and accessories: SAW devices have been used extensively in various compact commercial electronic systems, such as cell phones (each cell phone manufactured today contains multiple SAW devices which, along with their accessories, occupy only a small portion of the phone volume). This widespread commercial deployment demonstrates that SAW devices—and the accessories needed to drive them (such as driving circuits and power supplies)—are compact, inexpensive, and highly reliable.[20] SAW-based microfluidic devices would be simple and inexpensive to fabricate and integrate with other on-chip components in mass production.

(b) High biocompatibility: the acoustic power intensity and frequency used in many SAW-based microfluidic devices are both in a range similar to those used in ultrasonic imaging, which has been used extensively for health monitoring during various stages of pregnancy and proven to be extremely safe.[27] Therefore, we expect that with proper design, SAW-based microfluidic devices will also be safe and biocompatible with cells, molecules, and other biological samples. This expectation has been partially confirmed by cell viability and proliferation tests using existing acoustofluidic devices.[28–30]

(c) Fast fluidic actuation and large forces: current microfluidic technologies have difficulty generating fast fluidic actuation or large forces on particles. These drawbacks have limited their applications in medical diagnostics and biochemical studies. As indicated in a review article on microfluidics,[5] "small feature sizes typically prevent flow velocities from being high enough to yield high (Reynolds) numbers. High-frequency acoustic waves, however, can circumvent such difficulties." SAW-based devices can be used to selectively introduce chaotic advection[31] to a microfluidic system (in which laminar flow generally dominates), thereby enabling fast, effective manipulation of fluids and particles. Thus far, SAW-based microfluidic devices have proven to be able to pump fluids at $1-10$ cm s^{-1} (ref. 31) and manipulate mm-scale objects (*e.g.*, *C. elegans*).[29] Neither of these two features can be readily achieved by other microfluidic techniques.

(d) Versatility: SAW technologies enable biological/chemical detection, fluidic control (*e.g.*, fluid mixing, translation, jetting, and atomization), and particle manipulation (*e.g.*, focusing, patterning, separation, sorting, concentration, and re-orientation). SAW technologies are capable of manipulating most microparticles, regardless of their shape, electrical, magnetic, or optical properties; they are capable of manipulating objects with a variety of length scales, from nm to mm; and they are capable of manipulating a single particle or groups of particles (*e.g.*, tens of thousands of particles).

(e) Contact-free manipulation: SAWs manipulate particles and cells by means of the primary acoustic radiation force applied by the surrounding fluid; and SAWs manipulate fluid by means of acoustic waves leaked into the fluid. This represents contact-free manipulation, eliminating the potential for sample contamination.

(f) Convenient on-chip integration with SAW-based sensors: SAW microfluidic devices can perform not only precise manipulation and control of both fluids and particles, but can also perform sensitive detection and sensing. SAW-based microsensors have matured over the years and can be integrated with other SAW-based microfluidic components.[21-23] These characteristics make it feasible to realize SAW-based, fully integrated, true lab-on-a-chip systems that can be launched into practical settings, rather than the chip-in-a-lab systems that we often encounter in the microfluidics community.

When compared with bulk acoustic wave (BAW) microfluidics,[33] another subset of acoustofluidics (*i.e.*, the fusion of acoustics and microfluidics), SAW microfluidics has its advantages and limitations. While BAW microfluidics is more mature, better understood, and has demonstrated higher throughput, SAW microfluidics has the following advantages:

(a) It allows one to better control excitation frequencies in a wider range and utilize higher excitation frequencies whenever needed. As a result, SAW microfluidics is more versatile and flexible and can achieve more precise and controllable manipulation of fluids and particles.

(b) It does not require that fluid channels be made of materials with high acoustic reflection, making it possible to employ microfluidic devices made of polymers, which generally have low acoustic reflection.

(c) Because SAWs confine most of their energy to the surface of a microfluidic device substrate (whereas BAWs expend energy travelling through the bulk of the device substrate), they require less power than BAWs to achieve the same acoustic effects.

(d) SAW devices can be conveniently fabricated on the same chip as many other microfluidic components through standard micro/nano fabrication processes in a mass-producible fashion. In this regard, SAW microfluidics appear to be more amenable to system integration and mass production than BAW microfluidics.

These recognized features and advantages of SAW microfluidics (relative to BAW microfluidics and other microfluidic approaches) have rendered it an attractive platform for many lab-on-a-chip applications. Although some aspects of SAW microfluidics have been reviewed elsewhere[34–39] (such as travelling SAW based droplet microfluidics, SAW based particle manipulation, and SAW based biosensing in microfluidics), we provide in this book chapter a comprehensive review of SAW microfluidics. We propose the decomposition of SAW microfluidics into two types: travelling SAW (TSAW) and standing SAW (SSAW). For both SAW types, we first introduce the theory before summarizing microfluidic applications to date. Finally, we look forward to identify areas with space for research innovations.

15.2 Travelling Surface Acoustic Wave (TSAW) Microfluidics

15.2.1 Theory Involved with TSAW

The best known form of SAW, Rayleigh SAW, is composed of a longitudinal and a vertically polarized shear component.[38] Rayleigh SAW strongly couples with media in contact with the wave propagation surface, enabling the sensing of mass perturbation and elastic properties of a medium introduced on the wave's propagation path. Other SAW modes exist in elastic materials of different compositions. For example, Love SAW is a guided shear-horizontal wave that propagates in a thin layer on top of a substrate. However, as most SAW-based lab-on-a-chip technologies utilize Rayleigh SAWs, we use "SAW" to specifically indicate Rayleigh SAW in this review.

SAWs are generally produced by applying an appropriate electric field to a piczoelectric material. The piezoelectric material, in turn, generates propagating mechanical stress. A typical SAW device uses at least one set of metallic interdigital transducers (IDTs) fabricated on the surface of a piezoelectric substrate. The IDT then introduces the electric field, generating a SAW displacement amplitude on the order of 10 Å. An IDT consists of a set of connected metallic fingers interspaced with an opposite set of connected metallic fingers; an alternating current electrical signal in the radio frequency (RF) range is applied across the two sets of connected fingers (Figure 15.1). The structure of the IDT determines the bandwidth and directivity of the generated SAW. By changing the number, spacing, and aperture (overlapping length) of the metallic fingers, one can change the characteristics of the resulting SAW. For example, a focused IDT consists of pairs of annular electrodes that can focus SAW energy to a spatially small

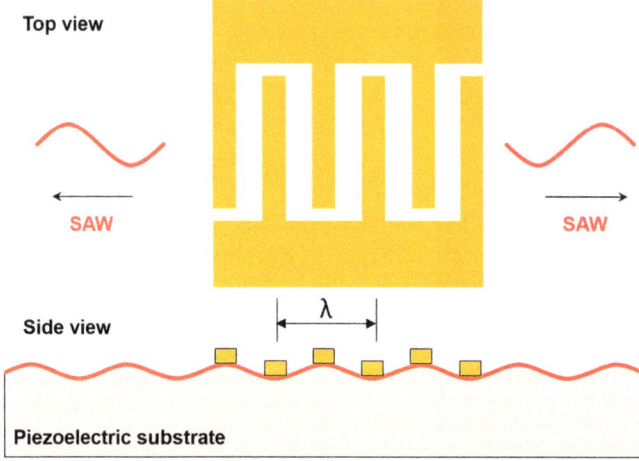

Figure 15.1 A metallic IDT deposited on the piezoelectric substrate generates SAWs that propagate along the substrate surface in both directions.

focal point.[40] A chirped IDT has a gradient of the electrode finger width directed along the SAW propagation direction, allowing it to generate SAWs over a wide frequency range.[29] A slanted finger IDT has a gradient of electrode finger width directed perpendicular to the SAW propagation direction, allowing it to generate narrow SAW beams of varying frequency along its finger length.[41] Each IDT variant has its own set of advantages and disadvantages and the choice of IDT type for use in a SAW-based microfluidic device depends on the device requirements. Additional IDT design types will be discussed in the later portion of this review article.

Generation of SAWs on a piezoelectric material can be mathematically modeled, but is complicated by (1) the anisotropy exhibited in most piezoelectric materials and (2) the intrinsic electromechanical coupling in piezoelectric media. The anisotropy of a piezoelectric material dictates whether the piezoelectric material generates shear-horizontal SAW,[42] leaky SAW,[43] or psuedo-SAW.[44] As such, consideration must be given to anisotropy to ensure the generation of the desired type of wave. The full mathematical modeling of SAW propagation requires the analysis of the link between electrical signal and deformation in piezoelectric media, referred to as electromechanical coupling.[45–72] For more details on theory, see Chapter 14.

15.2.2 Microfluidic Technologies Enabled by TSAWs

In this section, we review the application of TSAWs in both open and confined microfluidic geometries to accomplish (1) fluid mixing, (2) fluid translation, (3) jetting and atomization, (4) particle/cell concentration, (5) droplet and cell sorting, and (6) re-orientation of nano objects. These examples demonstrate the growth of TSAW into a key component for many emerging on-chip applications.

Lab-on-a-chip Technologies Enabled by Surface Acoustic Waves 359

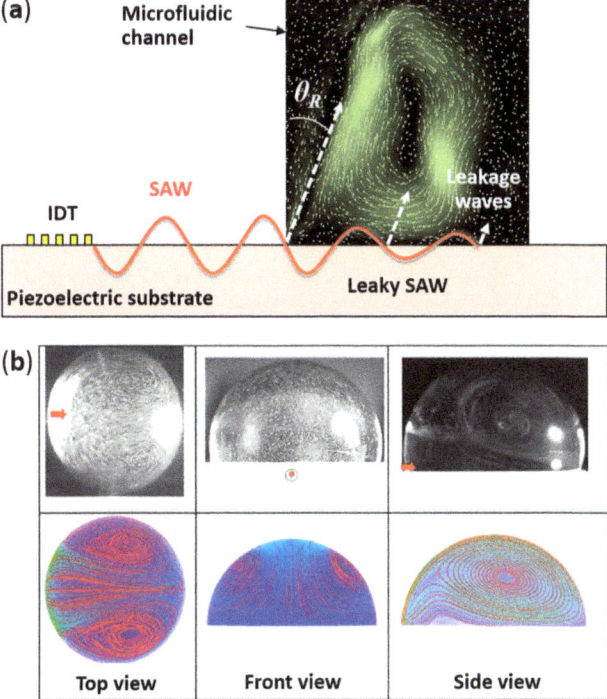

Figure 15.2 SAW-induced acoustic streaming. (a) Rayleigh SAW is excited by means of an IDT. Underneath a liquid, the SAW turns into a leaky SAW, radiating pressure waves at the Rayleigh angle θ_R into the fluid. The streaming map is from Koster. Reprinted with permission from ref. 46. (b) Comparison of experimental results and numerical modeling for a hemispherical droplet positioned at the center of the SAW propagation direction. Reprinted with permission from ref. 52.

15.2.2.1 Fluid Mixing

Many lab-on-a-chip applications require the mixing of two or more fluids. However, the laminar flows that predominate at the micro-scale result in mixing that occurs *via* diffusion. Such diffusion-based mixing is too slow for most lab-on-a-chip applications, so researchers have looked to SAW-induced streaming for its ability to mix fluids quickly by generating chaotic advection (Figure 15.2).

A few groups have demonstrated methods using TSAWs to mix fluids in an unconstrained droplet. Shilton *et al.*[56] generated a TSAW using a single-phase unidirectional transducer (SPUDT) to induce liquid recirculation inside a droplet. Their results, shown in Figure 15.3(a), demonstrate the fast mixing of dyed water and dyed glycerin solution. Frommelt *et al.*[73] demonstrated a more refined TSAW-based droplet mixing, using a pair of tapered IDTs (TIDTs) to generate a narrow SAW beam with a tunable launching point.

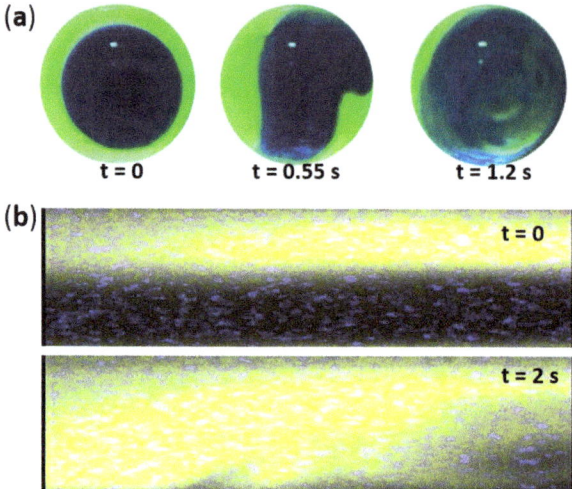

Figure 15.3 Fluid mixing by SAW-induced acoustic streaming. (a) Rapid mixing of glycerin (light) and water (dark) in a droplet. (b) Mixing of fluorescence dyes (light) with water (dark) in a rectangular microfluidic channel. Reprinted with permission from refs. 56 and 54.

By individually modulating the input signals of the two TIDTs, they demonstrated the ability to temporally modulate the flow patterns generated by each IDT to achieve efficient mixing. They also demonstrated that they could control mixing speed (within the droplet) by adjusting SAW amplitude and frequency.

Tseng et al.[54] also used IDTs to demonstrate TSAW-based mixing. Instead of mixing fluids in an unconstrained droplet, they mixed fluids inside a microchannel (Figure 15.3(b)). They performed a comprehensive experimental study on the effects of various operational parameters, showing that mixing performance could be significantly improved by applying higher voltage signals to the IDTs. Luong et al.[74] used a curved IDT design to focus the generated acoustic energy, resulting in considerably improved mixing performance relative to the parallel IDT design employed by previous groups.

Recently, Rezk et al.[75] incorporated SAW-induced mixing in a paper-based microfluidic device. They utilized a hue-based colorimetric technique to compare the mixing efficiency of their device with that of capillary-based mixing: the SAW-based mixing showed greatly enhanced consistency and speed.

By adjusting IDT design and input signal parameters, researchers have proven that TSAWs can effectively and precisely mix fluids in both open and confined fluid geometries. This versatility makes TSAW-based mixing techniques extremely attractive in microfluidics.

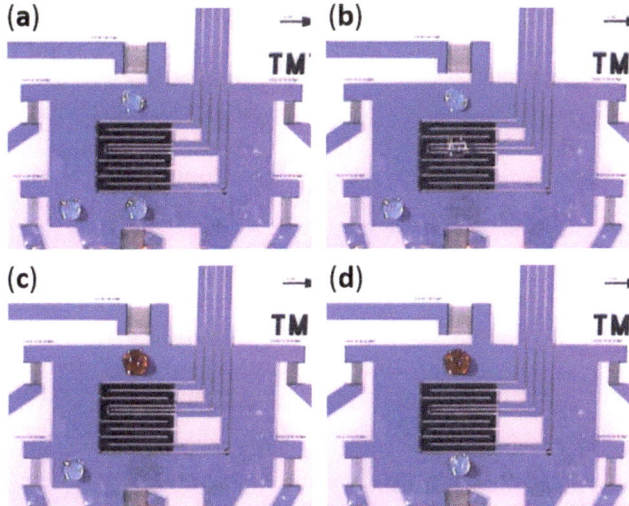

Figure 15.4 TSAW-driven programmable bioprocessors. (a–d) Three droplets are moved individually and with precise control. Reprinted with permission from ref. 78.

15.2.2.2 Fluid Translation

15.2.2.2.1 Fluid Translation in Open Space

When a liquid droplet is placed within the propagation path of TSAWs on a piezoelectric substrate, leaky SAW will be diffracted into the droplet at the Rayleigh angle. If the SAW has low amplitude, it will induce acoustic streaming within the droplet. If the SAW has sufficiently intermediate amplitude, the leaky acoustic energy generates an acoustic force on the droplet along the SAW propagation path, causing the droplet to deform into an axisymmetrical conical shape and translate across the substrate.[49,76,77] Liquid droplet speeds of 1–10 cm s^{-1} can be achieved using this TSAW-based actuation; this is more than an order of magnitude faster than other current micro-pumping actuation schemes.[32]

Moreover, as electrical actuation of IDTs can be programmed, droplet movements can be automated. Through the automated control of multiple droplets, merging, mixing, splitting, and chemical or biochemical reactions can be performed in so-called "programmable bioprocessors". A simple example of a programmable bioprocessor is shown in Figure 15.4, in which three droplets with different fluid content are moved independently in any desired direction and can be made to join together.[78]

Many chemical and biological applications have been demonstrated with TSAW-driven droplets. Guttenberg *et al.* precisely actuated oil-covered aqueous droplets between sinkers and heaters to perform a highly sensitive, fast, and specific DNA amplification reaction with droplet volumes as low as 200 nL.[79] Similarly, Tan *et al.* rapidly and effectively collected and removed micro-particles using this TSAW-driven droplet translation technique.[80] Finally, Li et al. used a similar mechanism to transport cells into tissue scaffolds to enhance cell seeding for tissue engineering studies.[81]

Rezk et al. subjected a droplet of silicone oil to TSAW excitation, observed behavior substantially different from that of water droplets, and developed a theory for the behavior (Figure 15.5(a)).[82] Silicone oil has a much smaller contact angle with a $LiNbO_3$ piezoelectric substrate than does water. When a silicone oil droplet on such a substrate was exposed to TSAW, this small contact angle led to the majority of the droplet being propelled in the SAW propagation direction (Figure 15.5(b)), while a thin film spread in the opposite direction (Figure 15.5(c)). As the thin film advanced, it formed finger-shaped patterns (Figure 15.5(d)); eventually, soliton-like wave pulses appeared above the fingers and propagated along the TSAW direction

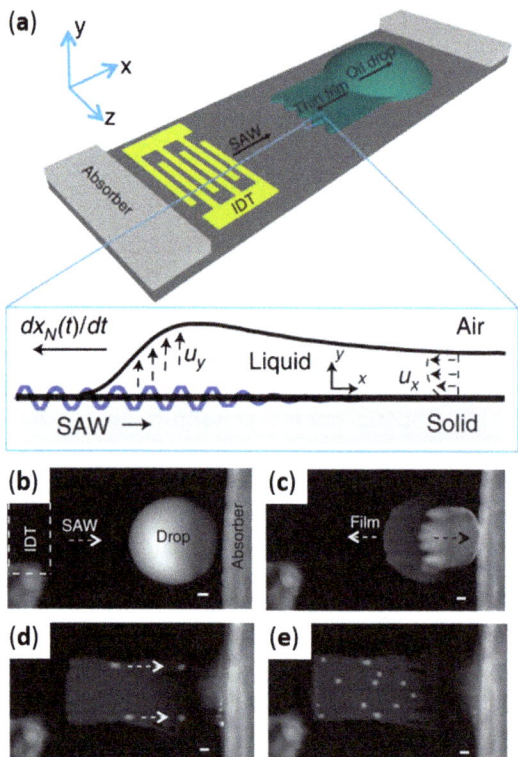

Figure 15.5 (a) Schematic of experimental setup depicting the emergence of a thin film from a standing oil drop due to TSAW exposure. (b) Initial state of the oil droplet. (c) After excitation with TSAW, the bulk of the oil droplet is displaced along the SAW propagation direction while a thin oil film advances in the opposite direction. (d) Finger patterns form in the thin film. (e) Soliton-like wave pulses subsequently appear above the fingers and translate in the SAW propagation direction. Reprinted with permission from ref. 82.

(Figure 15.5(e)). The authors suggest that this TSAW-induced thin film spreading may have applications in film coating and microfluidic actuation.

Historically, most SAW microdevices have been fabricated on bulk LiNbO$_3$ or quartz substrates; these bulk piezoelectric substrates may require some extra processing in order to integrate control electronics (*e.g.*, for the IDTs). To facilitate further advancements in SAW microfluidics, Du *et al.* presented a thin film piezoelectric material for translation of liquid droplets with volumes up to 10 μl.[53] They first deposited a ZnO thin film on a plain silicon substrate. This hydrophilic ZnO thin film layer prevents effective droplet translation. To circumvent this, they treated the ZnO thin film with a self-assembled monolayer of octadecyltrichlorosilane (OTS), which made the substrate surface hydrophobic while producing no measurable acoustic damping. The new substrate is cheaper than bulk piezoelectric substrates, and the silicon base can easily be integrated with control electronics for the IDTs, enabling the potential for a fully automated microsystem.

15.2.2.2.2 Microfluidic Pumping in Enclosed Channels

In addition to translating liquid droplets in open space, TSAWs can be employed to pump fluid through enclosed channels.[83–89] Cecchini *et al.*[86] bonded a straight PDMS microchannel between two IDTs on a LiNbO$_3$

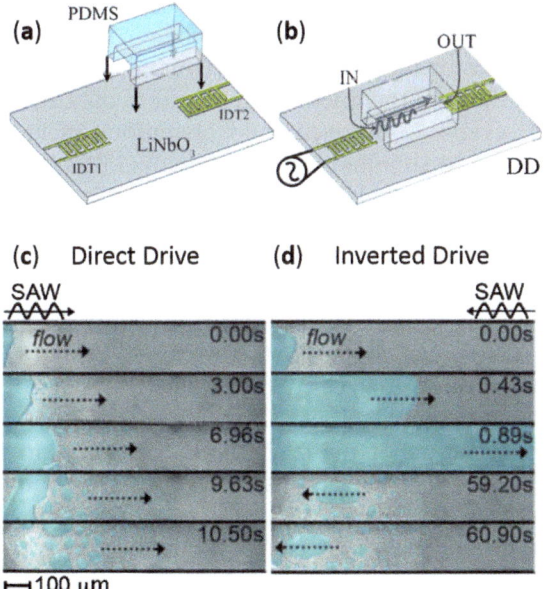

Figure 15.6 (a) Schematic of the assembly of the microfluidic device and (b) activation of IDT1, called direct drive (DD) mode; activation of ID2, called inverted drive (ID) mode, results in TSAW propagation opposite that shown in this image. (c) Photographs of the (ineffective) water filling process under DD mode at different times. (d) Photographs of the water filling process under the ID mode at different times. Reprinted with permission from ref. 86.

substrate (Figure 15.6(a)). After placing a water droplet between one IDT and the channel inlet (this droplet was termed the water "reservoir"), two different operation modes were tested: direct drive mode and inverted drive mode. In direct drive mode, SAWs were excited by the IDT near the channel inlet and propagated from inlet to outlet (Figure 15.6(b)), whereas inverted drive mode had SAWs excited by the IDT near the channel outlet and propagated from outlet to inlet. In direct drive mode, SAW excitation caused significant atomization in the water reservoir, resulting in rapid evaporation of the water and preventing the channel from being filled (Figure 15.6(c)). In inverted drive mode, the water reservoir quickly translated into the microchannel, filling it at a flow speed as high as 1.24 mm s^{-1}. The authors propose that this flow rate, occurring opposite the TSAW propagation direction, is attributable to TSAW-induced atomization at the leading edge of the water. These droplets continuously form, coalesce, and rejoin the water–air meniscus, thereby dragging the water through the channel and resulting in net motion opposite the direction of TSAW propagation (the water reservoir is not exposed to significant atomization under the inverted drive configuration because the TSAW power is sufficiently diminished by the time it reaches the reservoir location at the channel inlet).

Based on this TSAW inverted drive mechanism, Girardo *et al.*[87] developed a fully controlled low-voltage micro pump in a two-dimensional microchannel array. By combining a 5 × 5 orthogonal array of PDMS microchannels on a LiNbO$_3$ substrate with 20 IDTs, the researchers could selectively generate either a single TSAW or multiple TSAWs. Thus, droplets at channel inlets could be directed through the microchannel grid to the desired outlets. This device achieved several SAW-driven fluid operations including extraction, deviation, splitting, and simultaneous multichannel filling.

TSAW-driven pumping has also been demonstrated in a closed microfluidic channel (with no liquid–air interface) by Fallah *et al.*[88] This device featured a microfluidic channel in a square-shaped loop, with an IDT to generate SAWs travelling along one of the four sides. When a TSAW reaches the fluid–substrate interface, the acoustic energy leakage profile decays exponentially with continued wave travel along the fluid channel. With a sufficiently long channel, the location of initial TSAW contact with the fluid essentially acts like a point pressure source, driving the fluid through the channel loop according to conservation of mass. With this setup, fluid flow could be continuously actuated with a very fast response. Fluid pumping in such a closed-loop chamber can be used to mimic the action of small blood vessels, making it useful for clinical applications and biomedical studies.[83,84,89]

15.2.2.2.3 Microfluidic Rotational Motor

Shilton *et al.* utilized SAW-induced acoustic streaming to drive a rotary micromotor in a microfluidic chip (Figure 15.7(a)).[90,91] A pair of IDTs were patterned on a LiNbO$_3$ substrate, and a layer of hydrophobic Teflon film was coated in the space between the two IDTs in order to hold a fluid droplet in place. Silicone gel was placed on the SAW's path to absorb half of the energy from each IDT, creating a centrosymmetric TSAW exposure in the droplet, which resulted in centrosymmetric acoustic streaming inside the droplet. When a Mylar disc with a 5 mm diameter and 100 μm thick was placed on top of the droplet, the droplet's rotational inertia actuated rotational motion in the disc (shown in Figure 15.7(b)). The disk's angular velocity under a range of SAW amplitudes was investigated. In the case in which a 50 μl water drop was used as the fluid-coupling layer, a maximum disk rotation speed of 2250 rpm was achieved with a maximum torque of around 69 nNm. The authors found that increasing SAW amplitude to surpass ∼3 nm would reduce angular speed due to the unstable disc rotation caused by asymmetric flow in the drop.

By replacing the original Mylar micromotor with a patterned disk, Glass et al. successfully used this setup as a miniaturized lab-on-a-disc.[92] These discs were patterned with microfluidic channels, enabling SAW-induced microcentrifugation. With this setup, the authors demonstrated functions such as capillary valving, mixing, and particle concentration and separation.

As described in the above examples, SAW-induced acoustic streaming enables a simple and miniaturized on-chip rotational motor without the need for moving mechanical parts. In future studies, researchers will likely examine fluid-coupling layers other than water in order to circumvent the potential for evaporation.

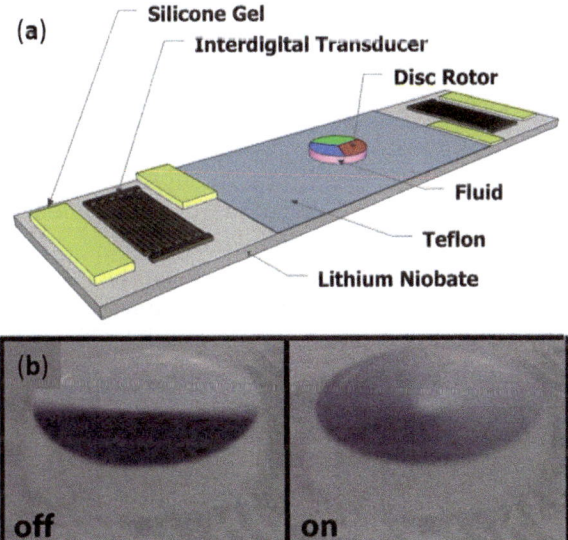

Figure 15.7 (a) Schematic of the SAW-induced rotary micromotor. Silicone gel was used to break the axisymmetry of the planar SAW and generate a centrosymmetric SAW, resulting in centrosymmetric acoustic streaming inside the drop. (b) The static and spinning states of the Mylar disc placed on top of the drop. Reprinted with permission from ref. 90.

15.2.2.3 Jetting and Atomization

A fluid jet is commonly defined as a coherent stream of fluid projected into a surrounding medium. For a fluid to undergo jetting phenomena, it must possess sufficient inertia to overcome the restoring capillary forces acting on the interface of the fluid and surrounding media.[31,93] Jetting at micro-scales finds numerous applications in hand-held ink jet printers, ink jet highlighters, and ink jet brushes.[94,95] Such micro-scale jetting can be accomplished by exposing small fluid volumes to TSAW (with the SAW power higher than that used for droplet translation), Figure 15.8. Recent research has demonstrated that SAW-based fluid jet production offers benefits over existing jetting techniques as it concentrates mechanical energy into a small drop, generating jets without the narrow confinement typically necessary to accelerate fluid.

Recently, Tan *et al.*[96] studied the nature of jet formation using TSAW devices, as shown in Figure 15.8. They characterized jetting length as a function of the driving force and the jet Weber number. They also predicted the jet's velocity as a function of acoustic Reynolds number using the jet momentum equation of Eggers,[93] and verified their results with the experimental data. These rigorous characterization efforts have helped to facilitate future TSAW-based jetting technologies.

Atomization is the making of an aerosol of small solid particles or liquid droplets. It has long been applied in numerous areas, such as internal combustion engines, medicine, agriculture, and cosmetics.[97] Kurosawa *et al.* first proposed and constructed a novel ultrasonic atomizer using a TSAW device.[98] Since then, SAW-based atomization has been used for numerous applications such as protein extraction and characterization for paper-based diagnostics,[99] portable pulmonary delivery of asthmatic steroids,[100] and mass spectrometry interfacing with microfluidics.[101,102] Qi *et al.*[103] proposed a miniature inhalation device based on SAW atomization, and Ho *et al.*[104] merged SAW atomization with a paper-based sample delivery system to detect the presence of caffeine in human whole blood.

TSAW atomization has also been utilized for producing micro and nano particles. Alvarez *et al.*[105] demonstrated the generation of protein aerosols and nanoparticles using SAW atomization, while Friend *et al.*[106] utilized a similar technique to synthesize polymeric nanoparticles. The small footprint of its

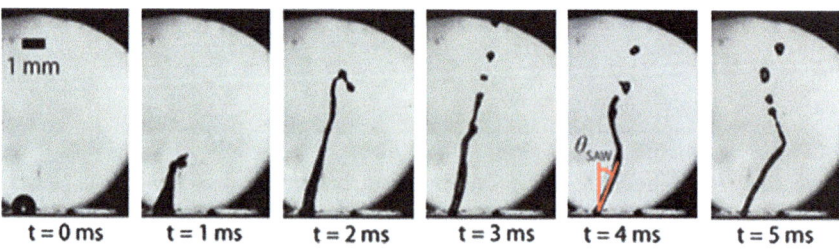

Figure 15.8 Jetting generation resulting from exposure to travelling SAW. Reprinted with permission from ref. 96.

generating mechanism, low power consumption, biocompatibility, and control over aerosol size make SAW an attractive method for atomization and a potential enabler for next-generation drug delivery devices.

TSAW-induced atomization is a result of capillary waves on the air–liquid interface.[107] The mathematical analysis of this phenomenon is challenging due to the presence of a free interface and a non-linear, two-dimensional wave interaction coupled with vastly varying time scales. Despite the associated difficulty, numerous efforts have been made to understand this phenomenon from a theoretical perspective. Köster[57] used a perturbation approach to investigate flow field inside a droplet, while Dong et al.[108] approached the problem by solving three-dimensional Navier–Stokes equations using a volume-of-fluid method to determine the free interface. Tan et al.[109] used a coordinate transformation approach to model the deformation of the interface, deriving the bulk deformation and unsteady capillary wave from the most basic principles, thus allowing their approach to correctly calculate acoustic wave reflections. Most recently, Collins et al.[110] investigated the hydrodynamics associated with SAW atomization, and developed a model for thin film spreading behavior under SAW excitation. Despite recent progress in describing the physics governing SAW-induced atomization, much remains to be understood.

15.2.2.4 Particle/Cell Concentration

Concentrating particle/cell suspensions is a basic but critical operation in many applications in chemistry and biomedicine. On the macro-scale, particles can be easily concentrated using centrifugation. On the micro-scale, however, concentrating cells or particles can be difficult. As volume decreases, surface forces acting on the particles/cells begin to dominate over body forces. Lack of significant body forces makes standard centrifugation impractical, so researchers have explored SAW-induced acoustic streaming to concentrate cells and particles for low-volume systems.[56,111–113]

To initiate rotational fluid motion, Shilton et al.[56] placed a droplet within a portion of a SAW propagation pathway in such a way that the droplet experienced TSAW exposure across a fraction of its width. This resulted in a circular pattern of SAW-induced acoustic streaming within the droplet (Figure 15.9(a)). When microparticles in aqueous suspension were subjected to this rotational pattern of SAW-induced streaming, they were concentrated and deposited in the center of the droplet. The researchers attributed the concentration effects to a shear gradient between regions of high shear at the droplet's periphery and regions of low shear at the droplet's center. Particles tended to move towards the region of low shear, where the linear velocities of the fluid approached zero.

Figure 15.9(b) shows sequential images of the SAW-induced streaming for a water sample containing 0.5 μm white fluorescent beads. The progression from distributed individual dots at $t = 0$ to the central bright spot at $t = 1$ s demonstrates how quickly the rotational SAW-induced streaming moves the

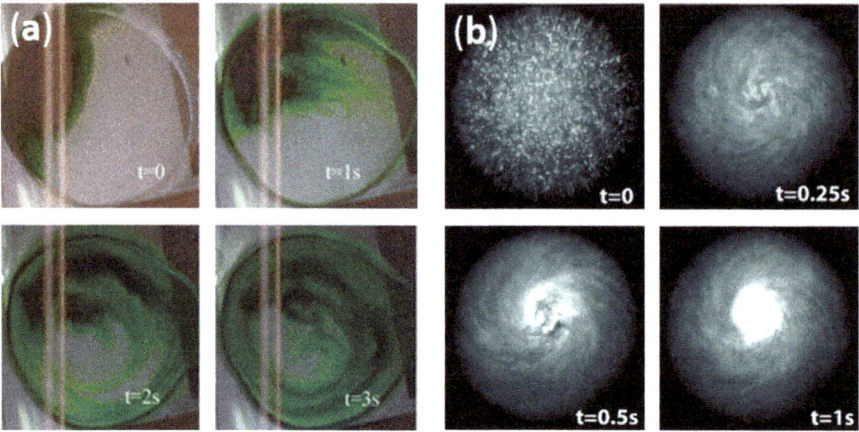

Figure 15.9 (a) Sequential images of rotational acoustic streaming in a drop resulting from asymmetric exposure to TSAWs, with flow streamlines visualized using dye. (b) Concentration of particles in a 0.5 μl droplet *via* such rotational acoustic streaming. Reprinted with permission from ref. 56.

fluid and concentrates the particles. In this way TSAWs allow for quick and efficient particle concentration at the micro-scale with a simple droplet-based device setup.

15.2.2.5 Droplet/Cell Sorting

The ability to precisely sort individual droplets of interest from other droplets is extremely important for various chemical and biological screenings. Similarly, cell sorting is an important task for many disciplines, ranging from basic cell biology to clinical diagnosis. Recently, several groups have shown that SAW-induced acoustic streaming can selectively sort droplets or cells.[114-116] More details on this subject are to be found in Chapter 16.

15.2.2.6 Reorientation of Nano-objects

15.2.2.6.1 Carbon Nanotube Alignment

Carbon nanotubes (CNTs) have been a focus of nanotechnology research for over two decades.[117] Their mechanical strength and unique electrical properties make CNTs attractive for a wide variety of applications ranging from sports equipment to electronics. However, aligning large quantities of CNTs for use in these applications often proves challenging. TSAWs present a practical solution, allowing for the simultaneous alignment of many CNTs.

Strobl *et al.* demonstrated the alignment of multi-walled carbon nanotubes (MWNTs) on a LiNbO$_3$ substrate using TSAWs.[118] Figure 15.10(a) shows a typical setup for the SAW-induced nanotube alignment device. After activating the SAWs, the device aligned carbon nanotubes 25–45° relative to the SAW field (shown in Figure 15.10(b)). To discern between the effects of fluid

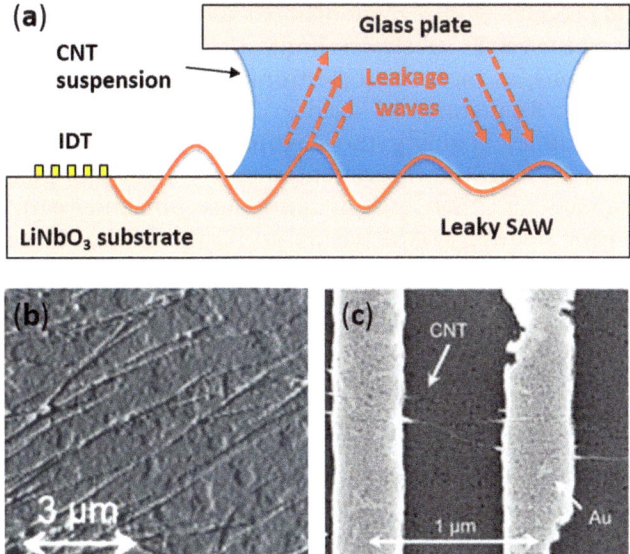

Figure 15.10 (a) Schematic of device setup for SAW-based carbon nanotube alignment. (b) AFM image of aligned multi-walled carbon nanotubes. Reprinted with permission from ref. 118. (c) SEM image of aligned single-walled carbon nanotubes between pre-patterned gold electrodes on LiNbO$_3$. Reprinted with permission from ref. 119.

movement and piezoelectric field in the alignment of these MWNTs, the authors covered the LiNbO$_3$ substrate with a 20 nm layer of NiCr. This thin metallic film then screened the piezoelectric field of the TSAWs. Experiments conducted on this metal-coated substrate device yielded no observable alignment of the MWNTs. To confirm this observed alignment is dependent on the piezoelectric field, the authors then used a LiTaO$_3$ substrate to propagate a shear-wave SAW (as opposed to the Rayleigh SAW propagated on a LiNbO$_3$ substrate). The mechanical component of a shear-wave SAW exists in the plane of the crystal, yielding minimal fluidic coupling relative to that of Rayleigh SAWs. When using LiTaO$_3$ substrate, the MWNTs aligned parallel to the SAW field, showing improved alignment relative to the LiNbO$_3$ device. Thus the authors concluded that the MWNT alignment is attributable to the piezoelectric field associated with SAW propagation, and the acoustic streaming generated by Rayleigh SAWs actually shifted the MWNTs from being directly aligned with the piezoelectric field.

Smorodin *et al.* improved upon the alignment performance of the SAW-only approach by first thiolating single-walled carbon nanotubes (SWNTs) and then placing them on a substrate with patterned gold electrodes (Figure 15.10(c)).[119] By applying SAWs across this setup, the authors found that a majority of SWNTs aligned at angles between 0 and 20° relative to the SAW field. They attributed this improvement to the rapid attachment of thiolated carbon nanotubes to the gold electrodes, reducing the impact of acoustic streaming on the nanotube alignment.

Though the previous studies demonstrated successful alignment of CNTs, their use of piezoelectric substrates limits their usefulness in microelectronics applications, as most microelectronic devices use non-piezoelectric silicon substrates. Seemann *et al.* used a "flip-chip" configuration to overcome this limitation, enabling the alignment of carbon nanotubes on a silicon substrate.[120] In the experiment, a silicon substrate was patterned with gold contacts and then brought into close proximity with a $LiNbO_3$ substrate. A CNT solution was deposited to fill the space between the two substrates. Application of TSAWs to the CNT solution (by means of the $LiNbO_3$ substrate) resulted in a combination of piezoelectric field and acoustic streaming effects that successfully aligned CNTs between the pre-structured gold electrodes on the silicon substrate.

15.2.2.6.2 Liquid Crystal Reorientation

Polymer dispersed liquid crystals (PDLCs) consist of liquid crystal droplets randomly dispersed in a transparent polymer matrix. They are widely used in displays and optical elements, where the application of an electric field to the PDLCs readily adjusts the light transmittance through the material. In addition to PDLC realignment *via* an electric field, several reports have shown that liquid crystals can also be realigned using acoustics.[121] Recently, Liu *et al.* reported a light shutter effect induced by TSAWs applied to PDLCs.[122] As shown in Figure 15.11(a), PDLCs were placed between two IDTs on a $LiNbO_3$ substrate. In this setup, one IDT generated SAWs and the other detected them. Results demonstrated that SAW-induced acoustic streaming successfully aligned PDLCs parallel to the SAW propagation direction, changing the PDLCs from opaque to transparent (Figure 15.11(b)).

During experimentation, SAW propagation on the substrate heated the PDLCs to 40 °C. To verify that the PDLC realignment was attributable to SAW-induced acoustic streaming and not to thermal effects, the researchers set up an experiment using standing SAWs rather than TSAWs. This standing SAW field minimized the acoustic streaming effect while still heating the substrate and PDLCs; no PDLC realignment was observed. Therefore, the authors concluded that SAW-induced acoustic streaming is a viable actuation method for PDLC realignment.

Most recently, Liu *et al.* demonstrated the robustness of this effect by showing that application of a SAW field can also be used to control light transmission of holographic polymer-dispersed liquid crystals.[123]

15.2.3 Microfluidic Technologies Enabled by Phononic Crystal-assisted TSAWs

15.2.3.1 Introduction to Phononic Crystals

In the past decade, there has been a tremendous growth of interest in two- and three-dimensional periodic structures due to their ability to manipulate the propagation of acoustic waves on the wavelength scale. These artificial periodic structures, known as phononic crystals (PCs), are composed of

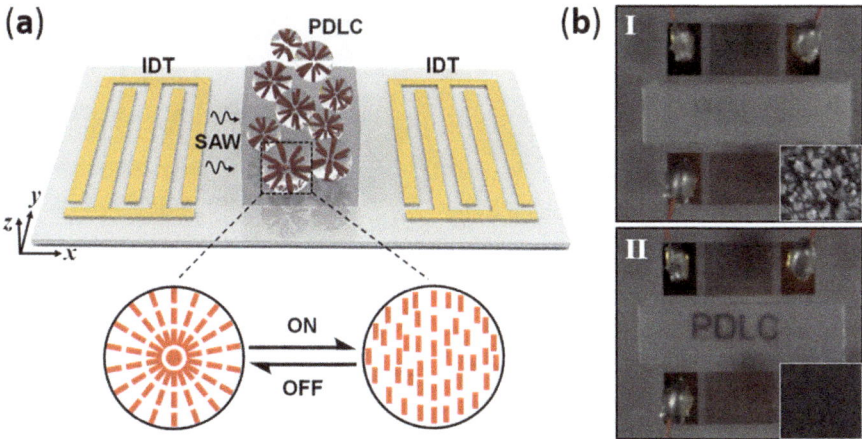

Figure 15.11 (a) The device structure and working principle for the SAW-driven PDLC light shutter. The magnified part shows a reversible switching process between two different liquid crystal droplet configurations. (b) Transmission response of the PDLCs under different acoustic powers. The insets I and II show the imaging quality at the off and on states of the IDT, respectively. Reprinted with permission from ref. 122.

arrays of elastic scatterers embedded in elastic host materials with properties different from those of the scatterers. PCs exhibit several unique behaviors as a result of their structures. Of greatest interest are so-called "phononic band gaps", which occur when phonon wavelengths correspond to the scale of the PC's periodicity. In phononic band gaps, acoustic waves in any vibration mode cannot penetrate the PC's structure in any direction due to destructive interference caused by phonon reflections from the periodic scatterers.[124,125]

The properties that result from PCs' periodic structure allow them to function in a wide variety of applications. The predictable responses produced by phononic band gaps make PCs promising candidates for perfect acoustic mirrors and for acoustic filters at designated frequencies. Likewise, PCs can serve as resonators and acoustic waveguides because any defect in a PC will contain acoustic wave vibrations when operating within the phononic band gap. Most importantly, PCs can be used to achieve super-resolution acoustic imaging when they exhibit negative refraction (the phenomenon occurring at the interface of materials in which acoustic waves are refracted opposite to the typically expected refraction).[126] Super-resolution acoustic imaging has been obtained because of the PC-based acoustic lens' ability to overcome the diffraction limit. Recently, with the introduction of a gradient-index (GRIN) concept to PC, a new class of PC was born. GRIN PC adds the ability to bend, focus, and modify the aperture of an acoustic beam to the already powerful capabilities of PCs.[127–131]

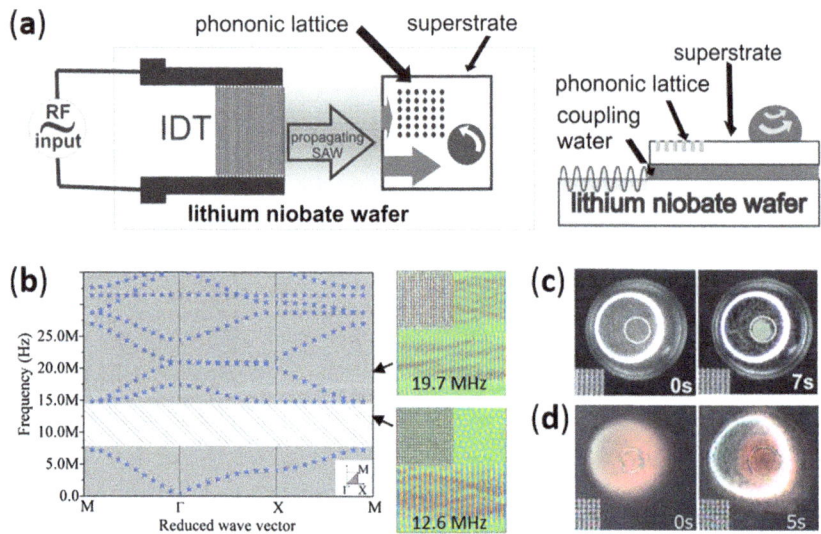

Figure 15.12 (a) Schematic of SAW device using a superstrate with embedded PC. TSAWs generated by IDT on the LiNbO$_3$ substrate leak into a water-coupling layer, inducing Lamb waves in the silicon superstrate that are then spatially filtered by the phononic lattice to provide asymmetric exposure to the droplet. (b) Left panel shows the band structure of the designed phononic lattice; the shaded area depicts the absolute phononic band gap. Right panel shows simulations at two different frequencies. (c) Concentration of 10 μm polystyrene beads at the center of the droplet due to circular acoustic streaming. (d) Concentration of human blood cells at the center of the droplet. Reprinted with permission from ref. 132.

PCs and GRIN PCs are considered the most promising candidates for solving low-frequency noise reduction issues and for providing ideal acoustic isolation for communication-band SAW devices. Moreover, the combination of PCs with SAW-based microfluidic devices can improve the performance of existing SAW microfluidic technologies and lead to novel applications. Though the field of PCs is still relatively new, it offers tremendous potential. In the following sections, we introduce a few examples of the successful technologies that arise from merging PCs with TSAWs.

15.2.3.2 Particle/Cell Concentration

To accomplish particle concentration in a droplet on the surface of a superstrate, Wilson *et al.* exploited the absolute phononic band gap of a PC.[132] As shown in Figure 15.12(a), an IDT was fabricated on a LiNbO$_3$ wafer to generate TSAWs. A PC consisting of circular holes in a square array was designed to exhibit a wide phononic band gap and fabricated on a silicon wafer (Figure 15.12(b)). When a SAW is generated at the proper frequency by the IDT, it travels along the surface of the LiNbO$_3$ substrate, leaks into a water-coupling layer, and excites Lamb waves across the thickness of the silicon superstrate that propagate in the same direction as the SAWs. Half of the excited Lamb waves encounter and are reflected by the PC, and the other half travel through the superstrate undisturbed. Thus the water droplet experiences an asymmetric Lamb wave exposure, and an in-plane counter-clockwise acoustic streaming pattern is induced. Figure 15.12(c) shows the induced circular acoustic streaming being used to focus small particles inside a droplet. The Cooper group also used this method to concentrate human blood cells from a diluted blood sample, demonstrating its applicability in biology and medicine (Figure 15.12(d)).

Despite the added complexity to the manufacturing processes, the use of PCs in acoustic streaming applications enables some notable advantages. For example, PCs with different design characteristics can be used to tune the frequency and amplitude of a single SAW, allowing users to program fluid manipulation without altering the input electrical signal. In addition, PCs and IDTs can be fabricated independently on separate units (*e.g.*, substrates and superstrates, coupled by an intermediate fluid layer), enabling device configurability and disposability. Several studies have applied these advantages to create useful technologies. For example, Bourquin *et al.* combined droplet manipulation *via* acoustic streaming with a lens-free detection system to create a disposable device for immuno-assay.[133] Applying their immunoassay to tuberculosis diagnosis, the researchers detected Interferon γ at pM concentrations within minutes. Reboud *et al.* developed a sophisticated SAW-based fluid manipulation tool for the detection of rodent malaria parasite *Plasmodium berghei* in blood.[134] Red blood cells and the parasitic *Plasmodium berghei* in a drop of blood were mechanically lysed *via* the strong acoustic rotation vortices generated by PC-aided acoustic streaming. The researchers then amplified the parasitic genomic sequence by heating the sample using different acoustic field and frequencies. Their results demonstrated that this PC-assisted device has the ability to detect around 0.07% parasite DNA in a microlitre-size blood sample and, more broadly, that it has potential for disease diagnosis in the developing world.

15.2.3.3 Jetting and Nebulization

Recently Bourquin *et al.* reported an interfacial jetting phenomenon by fabricating a conic PC on a superstrate.[135] Though jetting of a sessile droplet had been previously demonstrated using focused IDTs on a piezoelectric surface,[100] it had not been shown on a superstrate. In this work, the authors forced a SAW in a piezoelectric substrate to couple with a superstrate and excite Lamb waves (Figure 15.13(a)). The cone-shaped design of the PC they used provided a means to effectively focus acoustic energy to different hot spots, depending on the frequency of the SAW, as shown in Figure 15.13(b). They used this spatial control of acoustic energy to dictate the location and direction of the interfacial jetting behavior; Figure 15.13(c) shows that this focused acoustic energy can efficiently produce an interfacial jetting phenomenon.

The researchers used a similar conical PC superstrate setup to selectively nebulize drops, as shown in Figure 15.13(d) and (e).[136] In Figure 15.13(d), a droplet was nebulized at position 3 when a SAW at 12.6 MHz excited the phononic structure. Meanwhile in Figure 15.13(e), a SAW frequency of 9.4 MHz caused the droplet nebulization to occur at position 1. This PC-assisted selective jetting and nebulization technology adds versatility to the applications of TSAWs and may be useful for the nebulization of drugs.

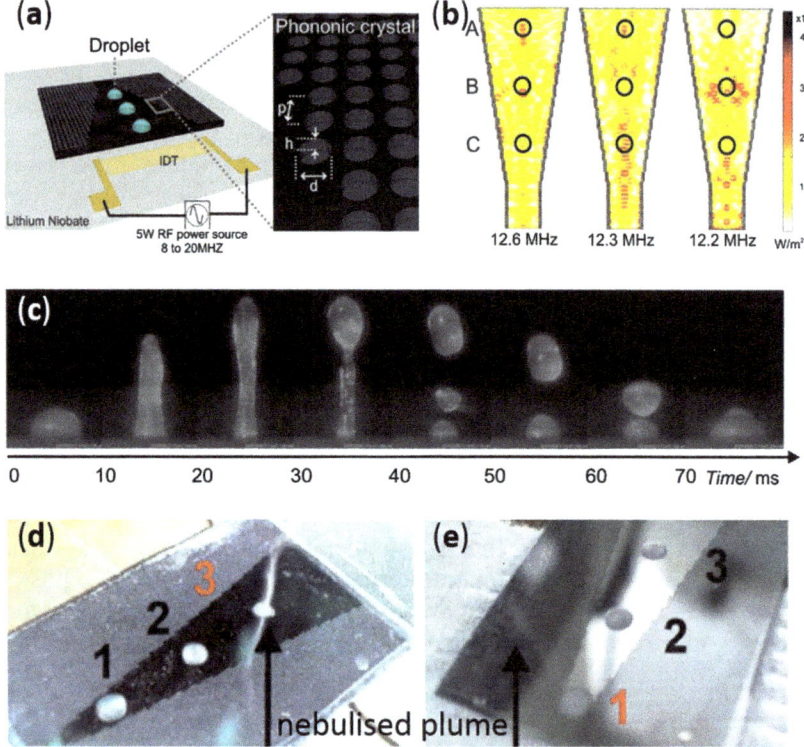

Figure 15.13 (a) Schematic of the SAW-induced interfacial jetting device. (b) Simulations of the conic structure at three different input frequencies. The waves are focused at different positions depending on the frequencies. (c) Continuous microscopic images of the jetting phenomenon on a PC superstrate for a sessile droplet of 10 μL. The drop elongates to form a column of water and breaks up into droplets. Images in (d) and (e) show, alternately, drops at position 3 (excited at 12.6 MHz) and position 1 (excited at 9.4 MHz) being selectively nebulized on the conic phononic superstrate (at a power of 3 W) while the other two drops remain. Reprinted with permission from refs. 135 and 136.

15.3 Standing Surface Acoustic Wave (SSAW) Microfluidics

15.3.1 Theory Involved with SSAW

The basic principles that govern the interference of all waves give rise to the formation of SSAWs.[137–144] If a pair of identical IDTs fabricated on a piezoelectric material generates two travelling SAWs propagating toward each other, the SAWs' interference will result in a one-dimensional SSAW field as shown in Figure 15.14(a). Computational analysis of a one-dimensional SSAW on the surface of a substrate has led to the simulated interference pattern shown in Figure 15.14(b). The light and dark regions represent weak and strong amplitudes of the sound pressure field, respectively. The lightest and darkest lines show that SSAWs have a series of nodes (minimum field) and anti-nodes (maximum field) at fixed locations on the substrate surface. The distance between neighboring nodes—and also the distance between neighboring anti-nodes—is always half of the SAW wavelength on the

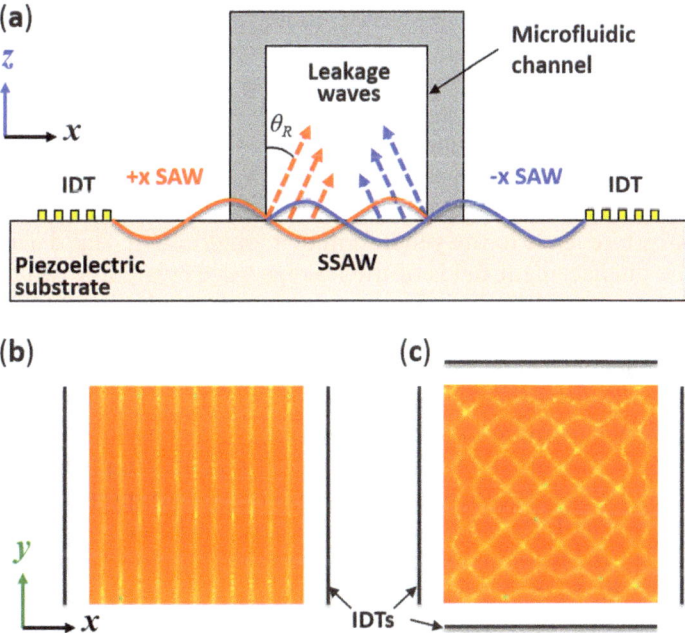

Figure 15.14 (a) A cross-section schematic of a microfluidic channel and the two IDTs used to generate travelling SAWs that propagate in opposite directions, establishing a SSAW across the channel width. This SSAW radiates acoustic leakage waves into the liquid at the Rayleigh angle. (b) and (c) show the simulated interference pattern of a one-dimensional and two-dimensional SSAW field on the substrate surface, respectively.

substrate. BAW-based approaches typically use acoustically reflective materials in channel walls to form standing waves within the channel (or cavity), which inherently gives a precise spatial definition of the standing wave pattern. In contrast, the SSAW-based approaches create standing waves through direct interference of two or more identical SAWs and do not require channel walls to be made of acoustically reflective materials. It should be noted that the phase of the two counter propagating waves has to be accurately controlled to ensure a fixed interference pattern over time.

Similar to the one-dimensional case, a two-dimensional SSAW field can be formed on the surface of a substrate by generating SAWs from two pairs of IDTs arranged orthogonally, as shown in Figure 15.14(c). Note that the two-dimensional interference pattern is tilted 45° with respect to the propagation direction of each SAW, and the shortest distance between nodes—and also the shortest distance between anti-nodes—in this two-dimensional patterned array is $\sqrt{2}/2$ times the SAW wavelength on the substrate.

SSAW-based microfluidic devices establish a standing acoustic field in fluid by means of IDT pairs forming a SSAW on a piezoelectric substrate in contact with the fluid (such as the setup in Figure 15.14(a)). As discussed in Section 15.3.1, we have good understanding of the SSAW field present on the piezoelectric substrate. However, if we are to explain the working mechanism of SSAW-based microfluidic devices, we must understand the standing acoustic field present in the fluid (as induced by acoustic energy leakage from the contacting substrate).

A recent study by Shi and colleagues[138] details a computational method used to model the SSAW-induced pressure field in the microfluidic channel depicted in Figure 15.14(a). The pressure gradient above the surface of the piezoelectric substrate leads to the generation of a longitudinal sound wave in the liquid. The displacement fields of the leakage wave created by two opposing travelling SAWs are plotted in Figure 15.15(a) and (b). Figure 15.15(c) and (d) show the pressure fields corresponding to the combined displacement fields at two time snapshots with half a period time difference. During the time between the two snapshots, the amplitude of the pressure in the vertical center of the channel remains zero (indicated by the dashed line in Figure 15.15(c) and (d)), while pressure in the other channel regions simply changes sign. These results indicate the existence of a standing acoustic pressure field in the fluid of the channel. Furthermore, the pressure nodal plane of this standing acoustic field sits immediately above the pressure node of the substrate SSAW that induced it. Several groups have experimentally confirmed the correspondence between pressure nodes of SSAW on the substrate and the acoustic field in the fluid.[139–143] The authors' computational model of the SSAW also demonstrated that the resulting primary acoustic radiation forces consist of axial mode (acting along the channel width) and transverse mode (acting along the channel height). The axial primary force is stronger than the transverse primary force. The partially reflected waves from the PDMS channel walls are included in the model to calculate the transverse primary force. For more details on SSAW theory, see Chapter 14.

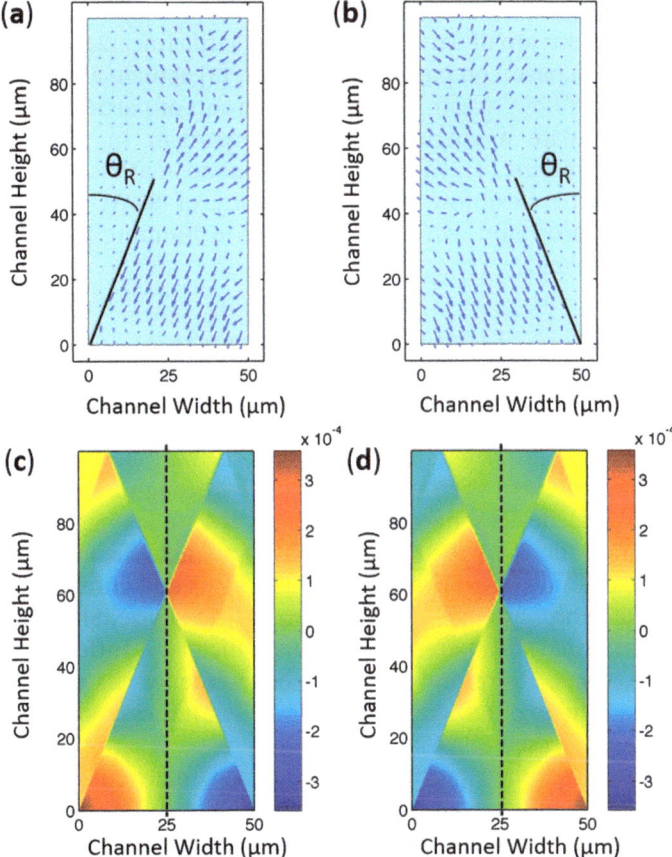

Figure 15.15 Time snapshots of the displacement field of the longitudinal-mode leakage wave from a SAW propagating in (a) the positive x-direction and (b) the negative x-direction. The width of the channel is half the SAW wavelength and the height is one SAW wavelength. The pressure field resulting from the combined displacement fields (*i.e.* resulting from a SSAW on the substrate) is shown in the cross-section of the channel at (c) $t = t_0$ and (d) $t = t_0 + \frac{\tau}{2}$. The dotted line denotes the pressure nodal plane in the middle of the channel. Reprinted with permission from ref. 138.

15.3.2 Microfluidic Technologies Enabled by SSAWs

Section 15.2.2 detailed the notable on-chip technologies and applications of TSAWs. These applications harnessed the acoustic streaming induced by the TSAWs to perform useful functions such as fluid mixing and cell sorting. In this section, we review the latest on-chip innovations enabled by SSAWs. Instead of harnessing the acoustic streaming, SSAW-based devices use the primary acoustic radiation forces acting on particles *via* the surrounding fluid. We will detail devices that use these primary acoustic radiation forces to (1) focus a flow stream of particles into a single-file line, (2) separate a flow stream of particles based on particle properties, (3) actuate a single particle/cell moving with a flow stream, (4) pattern a group of particles in stagnant fluid, (5) manipulate a single particle/cell/organism in stagnant fluid, (6) manipulate proteins, and (7) align micro/nano materials. These technologies enrich the flexibility and functionality of SSAW-based, on-chip particle manipulation and will prove essential in building next-generation lab-on-a-chip systems.

15.3.2.1 Focusing of Particles in a Flow Stream

Microfluidic focusing of particles in a liquid flow stream has attracted attention mainly due to its direct applicability to on-chip flow cytometry.[145–149] To detect particles *via* flow cytometry, the particles need to be lined up single-file so they can each pass individually through a detection point. Many microfluidic devices have achieved particle focusing hydrodynamically, using sheath flows to align a particle stream single-file in the middle of a microfluidic channel. However, these sheath flows add substantial bulk to the microfluidic chip, increasing chip complexity and size. Sheath flow also introduces additional shear stress to the cells, which could affect cell viability and other functions. Therefore, a number of researchers have been investigating particle focusing methods that eschew sheath flows. Some of the best methods have harnessed SSAWs.

Shi *et al.* accomplished particle focusing in a microfluidic channel by situating the channel between two IDTs such that a SSAW was established across the channel width with a single pressure node that was located at the channel center.[139] When particles passed through the region of SSAW exposure, the acoustic radiation force pushed them to the center of the microfluidic channel width (*i.e.*, two-dimensional focusing), as shown in Figure 15.16. Following this original work on SSAW-based particle focusing, Zeng *et al.* reported that adding Bragg reflectors inside or outside of the IDTs enhanced particle focusing effects.[150] These SSAW-based, sheathless particle-focusing devices can be conveniently fabricated and operated.

Though focusing of particles within the channel width satisfies the needs of some applications, others require that particles be focused in both the channel width and height (*i.e.*, three-dimensional focusing). 3D focusing is especially important for flow cytometry applications, as fixing particle position along the channel height minimizes fluorescence variations caused by varying focal depths. Shi *et al.* showed that SSAW can effectively achieve 3D focusing.[138] Applying a SSAW to fluid in a microchannel, the researchers realized that a non-uniform acoustic field generates a primary acoustic radiation force transverse to the particles in the channel (*i.e.*, in the z-direction, relative to the device plane). This force will direct all of the particles towards the point of maximum acoustic kinetic energy. However, the acoustic radiation force acting in the z-direction is weaker than that acting in

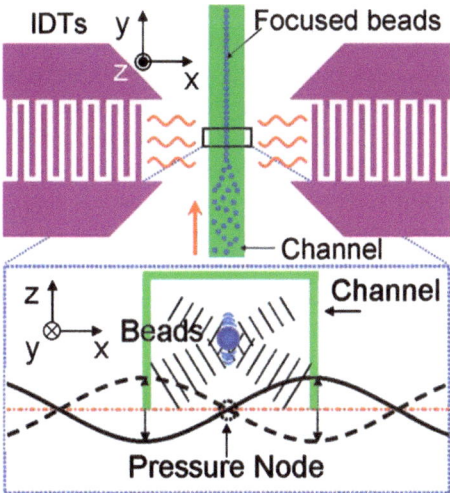

Figure 15.16 Schematic and working mechanism of SSAW-based focusing of particles in a liquid flow stream. Reprinted with permission from ref. 139.

the device plane. Therefore, particles focus first along the channel width and then migrate to a focal point along the channel height. The researchers used a prism placed adjacent to the microchannel to verify experimentally that particles did, in fact, focus in the z-direction; they then validated this with theoretical calculations.

15.3.2.2 Continuous Separation of Particles in a Flow Stream

A standing acoustic wave field exerts a primary acoustic radiation force whose magnitude and direction depend on particle size, density, and compressibility. Therefore, SSAW fields can differentiate particles or cells based on their physical properties. Shi *et al.* first reported SSAW-based continuous separation of particles in a flow stream in a PDMS microfluidic device.[151] The researchers situated a microfluidic channel between two IDTs such that a SSAW was established across the channel width with half a wavelength spanning the channel and a single pressure node at the channel center. The researchers introduced a particle mixture into the channel from side inlets while simultaneously injecting a sheath flow from the center inlet (Figure 15.17(a)). Because larger particles tend to experience larger primary acoustic radiation force, they migrate to the pressure node faster than smaller particles. Therefore, with appropriate length of the SSAW exposure region, SSAW enabled size-based particle separation. Larger particles move to the center outlet, while smaller particles remain in the side streams. This device demonstrated a simple, efficient, and cost-effective method for size-based particle separation.

Following the work of Shi *et al.*, Nam *et al.* developed a similar device setup and applied it to sorting blood cells from platelets.[152] First, to demonstrate the device, the researchers pumped a mixture of large and small particles through a separation channel from the device's center inlet while introducing sheath flows from side inlets. By positioning SSAW pressure nodes near the channel walls, the researchers managed to move larger particles from the center stream towards the channel sides. Smaller particles could then be collected from a middle outlet, while larger particles were collected from side outlets.

Nam *et al.* then applied this device setup to blood samples in order to separate platelets from red and white blood cells.[153] Because platelets are smaller than red and white blood cells, SSAW exposure in the separation channel resulted in large primary acoustic radiation forces pushing the red and white blood cells to the channel sides while platelets remained in the channel center. The SSAW-based sorting process effectively removed 99% of blood cells and achieved 74% separation efficiency for platelets. The work of Nam *et al.* demonstrated that SSAW not only effectively sorts particles, but also efficiently differentiates components of biological samples.

In addition to separating a particle flow stream based on particle size, SSAW devices can separate based on particle density. Cell-encompassing polymer beads provide one such example. There are numerous potential biological applications associated with encapsulating cells in biocompatible polymers. Although researchers have effectively generated mono dispersive polymer beads, they often struggle when attempting to control the number of cells encased by each polymer bead. To circumvent this challenge, Nam *et al.* exploited the increased polymer bead density associated with multiple contained cells. As a particle stream of polymer beads was flowed through a channel, the researchers used a SSAW with pressure nodes at either channel wall to drive the more dense beads to the channel sides while the less dense beads remained in the channel center.[154] In this device, large-cell-quantity alginate beads (LQABs), small-cell-quantity alginate beads (SQABs), and empty beads were collected from their respective outlets (Figure 15.17(b)). The device attained 97% separation efficiency and 98% separation specificity for LQABs. This device demonstrated that SSAWs can be employed to effectively sort particles in a flow stream based on both size and density.

Supplementing the work of previous groups, Jo *et al.* reported a sheathless method to separate equally-sized polystyrene and melamine beads based on their density differential.[155] To achieve sheathless separation, the researchers placed two pairs of IDTs side-by-side along a microfluidic channel. The first IDT pair generated a SSAW field with a pressure node in the channel center in order to focus the particle stream. After passing through the first IDT pair and being focused into a single-file line, the particle stream passed through the second IDT pair. The second IDT pair generated a SSAW field with pressure nodes at the channel walls, pushing the high-density melamine

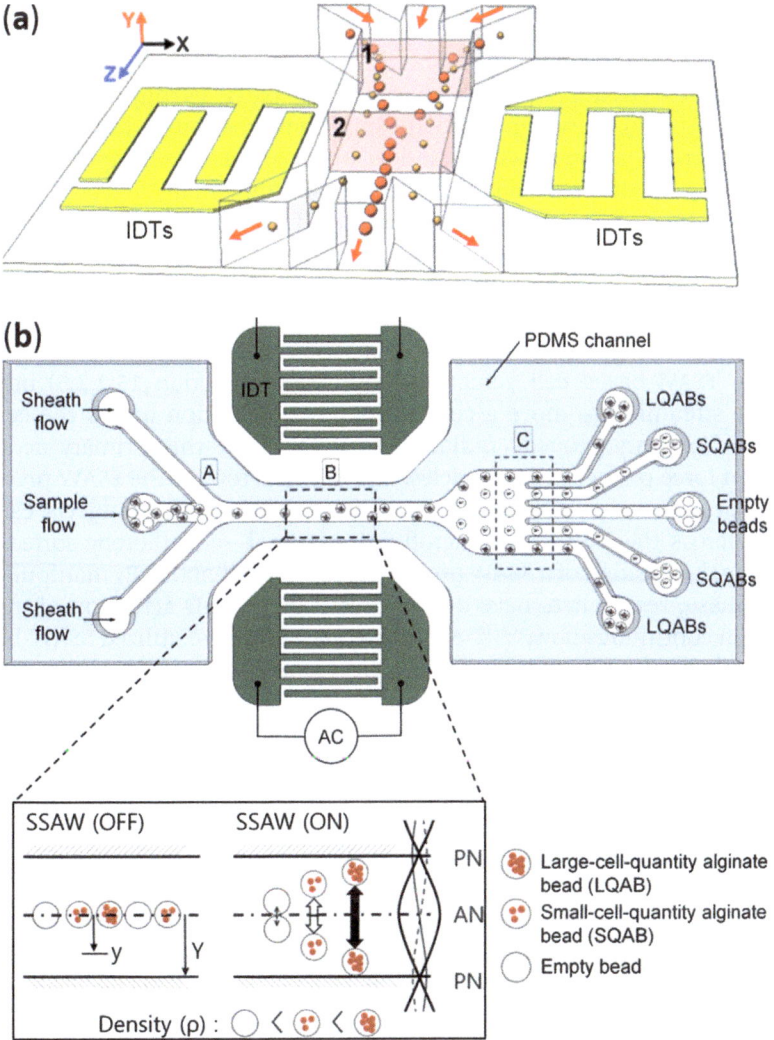

Figure 15.17 (a) Schematic of device for SSAW-based separation of particles in a liquid flow stream, based on particle size. (b) Schematic of device for SSAW-based separation of cell-encapsulating polymer beads in a liquid flow stream, based on bead density. Reprinted with permission from refs. 151 and 154.

beads (1.71 g cm^{-3}) to the channel sides while low-density polystyrene beads (1.05 g cm^{-3}) remained in the center stream. The device achieved 89.4% separation efficiency at a 2 µL min^{-1} flow rate, demonstrating the ability of SSAW to focus and separate particles all in one integrated unit. Guldiken et al. also reported an integration of SSAW-based focusing and separation to achieve sheathless separation of particles in a flow stream.[156]

15.3.2.3 Actuating a Single Particle in a Flow Stream

The previous section described how SSAW could be used to continuously separate a flow stream of particles based on the particles' physical properties (size and density). However, there are situations where it is desirable to sort a single particle moving in a liquid flow stream. A prime example is on-chip flow cytometry. After particles are focused and individually detected they are often sorted based on a property determined during detection (*e.g.*, presence of a cell surface protein as determined by fluorescence). Several mechanisms have been proposed for on-chip particle/cell actuation including electrokinetic, magnetic, optical, and hydrodynamic forces.[157–161] Recently, researchers have also proposed SSAW-based particle/cell actuation.[162–168]

While TSAW-based cell actuation (described in Section 15.2.2.5) utilizes acoustic streaming to move a cell, SSAW-based actuation moves the cell by means of primary acoustic radiation force. Because this primary acoustic radiation force pushes cells, particles, and droplets toward the SSAW pressure nodes (or anti-nodes, depending on their properties), these objects can be moved across the width of a microfluidic channel—and thereby sorted—by changing the position of a SSAW pressure node (or anti-node). By manipulating SSAW phase, researchers have demonstrated precise 1D actuation of microobjects in continuous flow.[162–164] Most recently, Ding *et al.* utilized SSAW-based actuation to construct a tunable device that precisely sorts single cells in a liquid flow stream into as many as five separate outlet channels (Figure 15.18).[165] In their work, the researchers positioned a microfluidic channel between two parallel chirped IDTs; these chirped IDTs operated in a wide frequency range, allowing them to generate a variety of SAW wavelengths. The width of the device channel was carefully selected so that only one pressure node could exist between the channel walls. The researchers tested their sorting device by pumping HL-60 cells through the channel. At frequency f_1 (9.8 MHz), the primary acoustic radiation force pushed the cell towards the SSAW field pressure node positioned at the topmost outlet channel (Figure 15.18(a)). Switching the input frequency to f_5 (10.9 MHz) shifted the pressure node downward and guided the cell to the lowest outlet channel (Figure 15.18(b)). Intermediate input frequencies f_2, f_3, and f_4 were used to guide the cell to the intermediate outlet channels. Figure 15.18(c) shows multichannel cell sorting, in which the device sorted cells contained in a liquid flow stream into five separate channels. This device demonstration shows the feasibility of SSAW-based actuation for the sorting of individual cells in on-chip flow cytometry. Li *et al.* recently applied the same principle to sort water-in-oil droplets into five separate outlets with a throughput of 222 droplets per seconds.[166]

15.3.2.4 Patterning a Group of Particles in Stagnant Fluid

Techniques that can non-invasively arrange particles or cells (contained in a stagnant fluid) into desired patterns are invaluable for many biomedical applications, including microarrays, tissue engineering, and regenerative

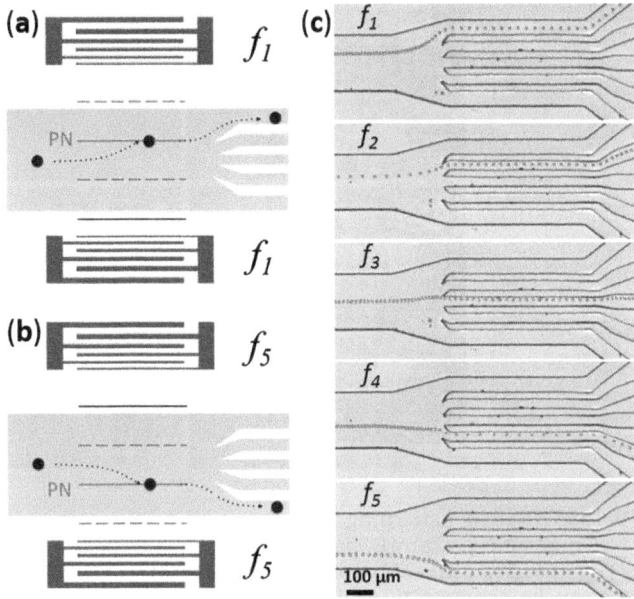

Figure 15.18 Left panel shows the working mechanism of SSAW-based sorting of a single cell contained in a liquid flow stream. (a) At an input signal frequency f_1 (9.8 MHz), the primary acoustic radiation force directs the cell to the pressure node at the upper wall of the channel, and the cell is collected from the topmost outlet. (b) At an input signal frequency f_2 (10.9 MHz), the pressure node is shifted to the lower wall of the channel, and the cell is collected from the lowermost outlet. Right panel (c) shows images of the experiment: HL-60 cells are driven individually to one of five desired outlet channels at five specific frequencies. Reprinted with permission from ref. 165.

medicine. The non-contact, non-invasive primary acoustic radiation forces generated by SSAWs serve as an excellent enabler of such particle/cell patterning techniques. Wood *et al.* exploited SSAWs to demonstrate the linear alignment of particles.[142] Following this work, Shi *et al.* developed "acoustic tweezers" to effectively and non-invasively arrange particles and cells.[141] The acoustic tweezers device consists of a PDMS microfluidic channel and a pair of IDTs deposited on a piezoelectric substrate in a parallel (Figure 15.19(a)) or orthogonal (Figure 15.19(b)) arrangement. After applying a RF signal to both IDTs to generate a SSAW field, particles/cells were patterned in parallel lines or arrays. Figure 15.19(c) and (d) show the distribution of microbeads before and after the 1D and 2D patterning processes. Patterning of massive particles in a large area has also been demonstrated using a similar mechanism and similar devices.[167]

The patterning of the aforementioned devices is static because the working frequency range of the SSAW is very narrow (the regular IDTs used in the devices have a bandwidth of ~0.2 MHz). In order to enable dynamic

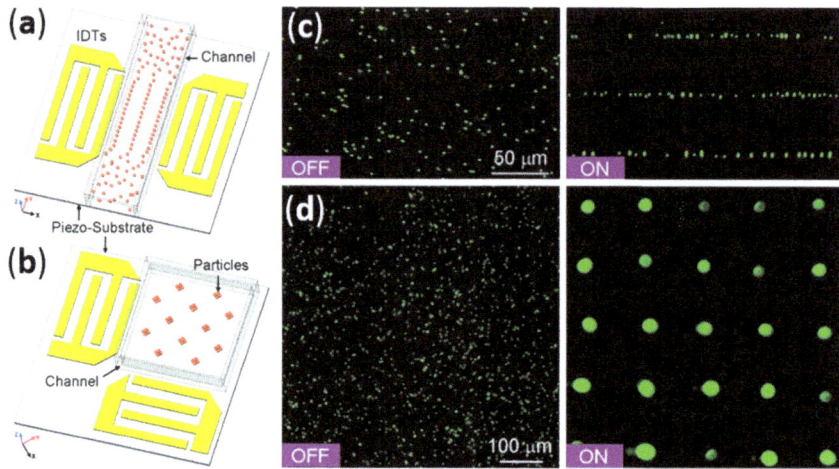

Figure 15.19 SSAW-based patterning of a group of particles in stagnant fluid. Schematic of (a) 1D patterning using two parallel IDTs, (b) 2D patterning using two orthogonal IDTs (the angle between the IDTs can be changed to achieve different patterns). (c) Distribution of fluorescent microbeads before and after the 1D (SAW wavelength was 100 um) patterning process, (d) and the 2D patterning process (SAW wavelength was 200 um). Reprinted with permission from ref. 141.

SSAW-based particle patterning, Ding *et al.* constructed a device with slanted-finger interdigital transducers (SFITs).[41] By tuning the input signal frequency, both the frequency and originating location of the main SAW beam can be tuned, permitting dynamic control of the particle patterning. The authors demonstrate SFIT-produced SSAW-based patterning of HL-60 leukemia cells; the period of the two cell lines was tuned to 60 μm, 78 μm, and 150 μm, using input signal frequencies of 45, 36, and 27 MHz, respectively.

15.3.2.5 Manipulating a Single Particle/Cell/Organism in Stagnant Fluid

In the previous section, we described recent developments in SSAW-based patterning for large numbers of particles/cells in stagnant fluid, with important biomedicine-related applications in microarrays, tissue engineering, and regenerative medicine. In addition to moving multiple particles/cell into a patterned formation, non-invasive manipulation of a single particle or cell contained in stagnant fluid is also invaluable in many biomedical applications, such as the study of cell-to-cell communication and interaction. Currently, many biologists employ optical tweezers for single-cell manipulation due to their excellent precision and versatility.[169] However, optical tweezers depend on complex and expensive optical setups, and the associated focused laser beam can heat the moved object to temperatures that cause physiological damage. To reduce costs and minimize damage, researchers have been developing SSAW-based single-cell manipulation platforms.

When a single particle contained in stagnant fluid is exposed to a SSAW field, the primary acoustic radiation force acts on the particle and pushes it to either a pressure node or antinode (depending on the particle's properties), trapping it at that location. Once the particle has been trapped, it can be moved by changing the frequencies and phase of the constituent SAWs.[162–168] Ding et al. recently demonstrated SSAW-based manipulation on particles, cells, and even millimetre-sized C. elegans worms contained in stagnant fluid inside a microfluidic chamber.[29] Their device consisted of a 2.5 × 2.5 mm^2 PDMS chamber asymmetrically bonded to a LiNbO$_3$ substrate between two orthogonal pairs of chirped IDTs (Figure 15.20(a)). The two pairs of chirped IDTs allowed the device to move the objects trapped in the pressure nodes in the x and y directions independently. Figure 15.20(b) displays the effectiveness of this technique, as the researchers manipulated a single bovine red blood cell through a pre-programmed pattern, taking a photograph at each cell position and compositing the images to trace the cell's path. The researchers examined particle speed and potential for cell damage, finding that a 10 μm polystyrene bead is accelerated to a velocity as high as 1.6 mm s^{-1}, and that HeLa cells exhibited no significant physiological change after being exposed to high power acoustic fields for 10 min. Finally, the researchers placed C. elegans worms inside the chamber to show that they could move an entire organism without damage, as shown in Figure 15.20(c).

The acoustic tweezers device described above is biocompatible, versatile, low-cost, simple in design, and convenient to operate, making it an extremely attractive alternative to conventional optical tweezers for manipulation of single particles or cells. The successful trapping and manipulation of a whole organism also sets the device apart.

15.3.2.6 Protein Manipulation

Supported lipid bilayers (SLBs) are critically important in cellular biology, as they regulate the intracellular and intercellular movement of ions, proteins, and other molecules.[170] Recently, researchers have applied SSAWs to pattern SLBs for use in membrane biology studies. Hennig et al. showed that local concentrations of SLBs can be modulated by applying SSAW to a substrate.[171] In the experiment, standard IDTs generated shear SSAWs on a LiTaO$_3$ substrate to induce lateral reorganization of a lipid bilayer. Higher membrane density was found in the antinodes of the in-plane shear SSAW, while lower membrane density was found in the nodes. The researchers carefully monitored pattern formation and decay by fluorescence microscopy in order to study diffusion times of supported bilayers during the process. Diffusion times matched those attained with conventional methods, confirming the accuracy of SSAW for patterning of SLBs. Hennig et al. further demonstrated the modulation of DNA density on supported lipid bilayers by binding DNA to a catonic SLB.[172]

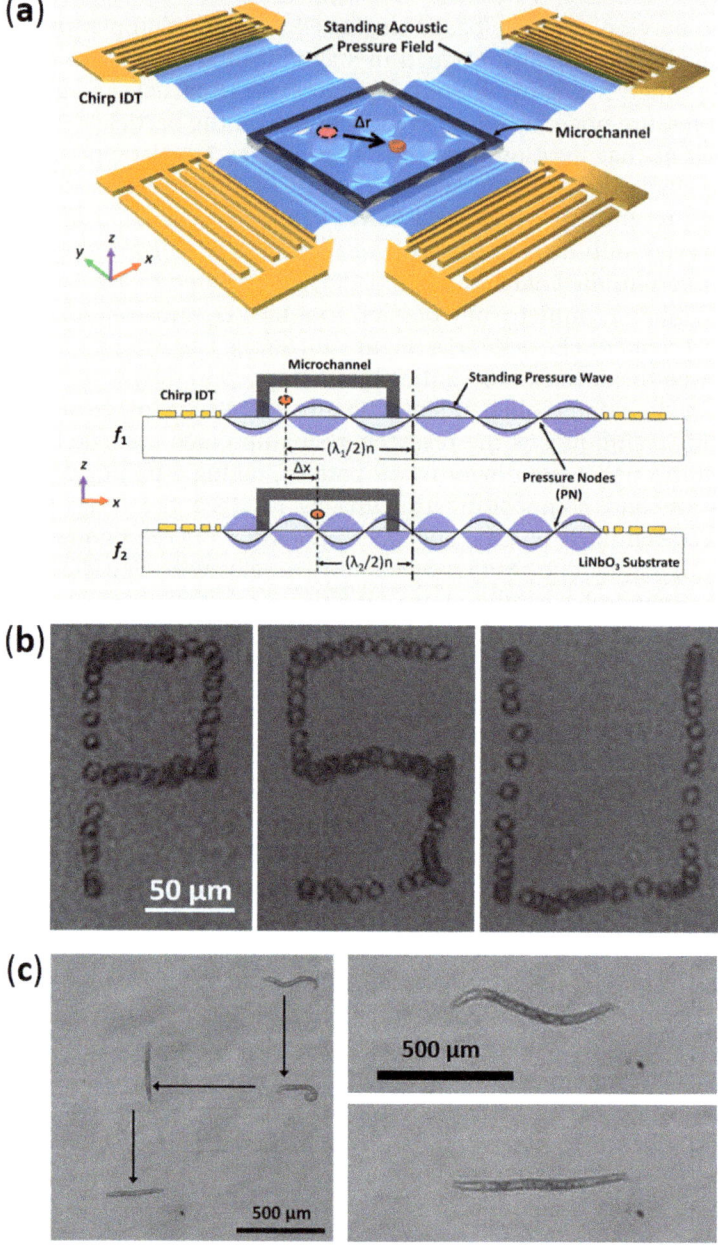

Figure 15.20 (a) Device schematic and working mechanism of SSAW-based manipulation of a single particle contained in stagnant fluid (*i.e.*, acoustic tweezers). (b) Composited image of a single bovine red blood cell translated in two dimensions by changing the frequencies of the constituent SAWs. (c) Optical images showing the manipulation and stretching of a whole *C. elegans* worm. Reprinted with permission from ref. 29.

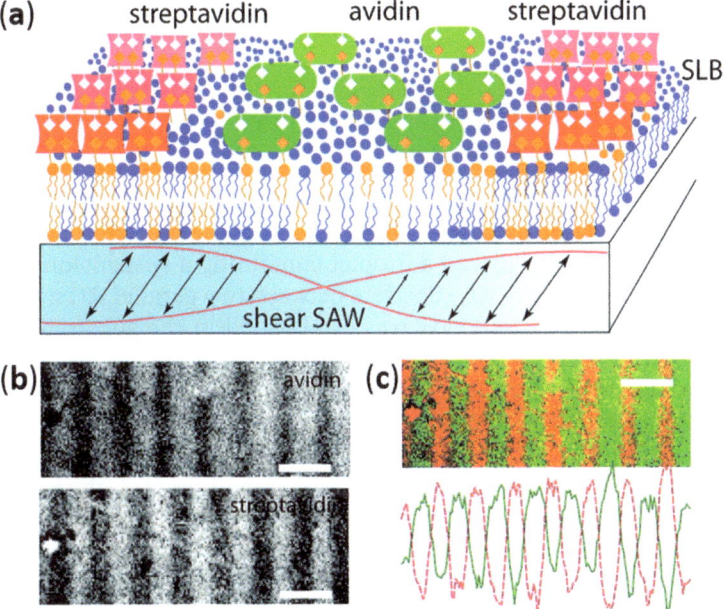

Figure 15.21 (a) Schematic of SSAW-based protein separation mechanism. A LiTaO$_3$ chip with IDTs (153 MHz/wavelength 26.6 μm) on opposite sides is used to generate a shear SSAW. The supported membrane (~4 nm thick) and protein segregation of membrane-bound streptavidin (red, Texas Red label) and avidin (green, Atto 488 label) can be observed by fluorescence microscopy through the optically transparent chip. (b) Fluorescence images of dye-labelled avidin and streptavidin. (c) Overlay images to demonstrate the segregation of avidin and streptavidin. Reprinted with permission from ref. 173.

Based on this lipid-patterning approach, Neumann *et al.* further demonstrated that proteins bound to SLBs can be accumulated, transported, and segregated using shear SSAWs.[173] As shown in recent work, SSAW-induced membrane modulations can achieve accumulation of labeled lipids. When the lipids are labeled with biotin, a biotin-binding protein called neutravidin accumulates alongside the lipids at pressure antinodes. The authors finely controlled the movement of proteins on SLBs by adjusting the phase between the two SAWs that combine to form the SSAWs. Phase adjustments in turn changed the locations of pressure nodes and antinodes, enabling protein transport in the 2D plane of the substrate. Finally, the group used shear SSAWs to separate avidin and streptavidin (two different biotin-binding proteins) based on their sizes (Figure 15.21(a)). Since streptavidin is the smaller of the two, it migrated towards the regions with high lipid bilayer densities at the antinodes. In contrast, SSAWs pushed the larger avidin proteins towards the pressure nodes, resulting in size-based protein segregation

(Figure 15.21(b)–(c)). Thus, SSAW-based microfluidic devices can be used to manipulate proteins on SLBs with operational simplicity and high success rate.

15.3.2.7 Microtube Alignment

In addition to using TSAWs to re-orientate micro/nano objects (see Section 15.2.2.6), researchers have also used SSAWs to align microtubes and nanowires.[174,175] Kong et al.[174] placed a droplet containing a suspension of rolled-up Cr microtubes on a LiNbO$_3$ substrate between two parallel IDTs. When the IDTs were energized and a SSAW was established in the substrate beneath the Cr microtube suspension, the microtubes uniformly aligned parallel to the propagation direction of the SAWs (Figure 15.22). In the SSAW field, these Cr tubes connected together end-to-end, forming tube bridges in a relatively high concentration.

To ascertain the physics underlying the alignment process, the researchers conducted control experiments with two sets of rolled-up microtubes, both conductive (Cr) and non-conductive (SiO/SiO$_2$), and two different piezoelectric surfaces, LiNbO$_3$ with a thin metal cover (to isolate the microtubes from the electric field at the substrate surface) and uncovered LiNbO$_3$ (exposing the microtubes to the electric field at the substrate surface). Experiment results showed that the Cr microtubes did not align on the metal-covered LiNbO$_3$ substrate, but did align on the uncovered LiNbO$_3$ substrate.

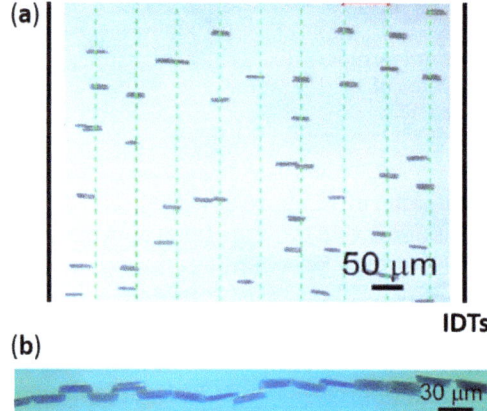

Figure 15.22 (a) Suspension of rolled-up Cr microtubes placed on a LiNbO$_3$ substrate between two parallel IDTs. IDTs are energized to establish a SSAW in the substrate, and the associated electric field uniformly aligned the microtubes parallel to the SAW propagation direction. (b) Optical image of aligned tube chain obtained from 30 μm long Cr tubes confined within the fluidic channel, after application of SAWs (30 MHz, 13 dBm). Reprinted with permission from ref. 174.

Furthermore, the SSAW exerted little influence on the non-conductive microtubes regardless of substrate. This proved that it was the electric field associated with the SSAW, and not the acoustic field, that was responsible for the alignment of the rolled-up metal microtubes.

15.4 Conclusions and Perspectives

As summarized in this book chapter, SAWs have demonstrated tremendous capability in microfluidic applications. TSAWs and SSAWs have, collectively, been used to achieve: rapid and localized fluid mixing, droplet translation, fluid pumping in microchannels, droplet-based rotational micromotors, droplet jetting and atomization, reorientation of nano-objects, as well as particle/cell focusing, sorting, manipulation, and patterning. Some of these functions have been achieved in microfluidics using techniques other than SAW as well. However, SAW has a combination of advantages that competing techniques lack: simple fabrication, high biocompatibility, versatility, compact and inexpensive devices and accessories, fast fluid actuation, contact-free particle manipulation, and compatibility with other microfluidic components. In addition, using SAW microfluidics, one can achieve functions that are unattainable by conventional methods. For example, non-invasive cell manipulation can be performed in a native cell environment, permitting normal cell growth and proliferation subsequent to the manipulation. This would have significant benefits in the study of stem cell differentiation, cell–cell communication, and tissue engineering. We believe that these unique advantages and functions position SAW to be a key component in the use of microfluidics as a tool in medical diagnostics and biological/chemical studies. For this to happen—for SAW microfluidic technologies to progress from research labs to everyday use in real-world applications—a number of hurdles must be overcome. In this section, we identify some of the areas requiring progress from both theorists and technical innovators.

The physics of SAW microfluidics, whether harnessing TSAW or SSAW, encompasses a range of phenomena, including acoustic radiation forces, acoustic streaming, jetting, and atomization. Theoretical work over the last two decades has fleshed out our understanding of this wide-ranging physics. However, work remains to be done if our theoretical understanding is to be complete. For instance, in the case of SAW-induced fluid atomization, there is an order of magnitude difference between the driving SAW frequency and the droplet excitation frequency (leading to atomization). This massive difference between driving and response frequencies remains poorly understood. Similarly, numerical simulations (*e.g.*, computational fluid dynamics modeling) of SAW-induced acoustic streaming have been hampered by the large discrepancy in the time and space domains between the driving SAW and the resulting liquid streaming. More specifically, a very small time step and a very fine mesh are required to capture the SAW actuation of the liquid, but the resulting streaming occurs over a relatively large time scale and large spatial dimensions. Additionally, quantitative estimates of primary acoustic radiation force acting on complexly-shaped biological particles are currently inadequate, impeding the understanding and optimization of SSAW-based microfluidic platforms. Advancements in SAW-based microfluidic devices—and their adoption in real-world applications—will be strongly facilitated by SAW researchers improving the theoretical foundations of the discipline.

While the theorists do their part, there is much room for continued device innovations. As described throughout this chapter, much of SAW microfluidics to date has relied on either acoustic streaming or primary acoustic radiation forces to accomplish useful actions. These two phenomena occur simultaneously for both TSAW and SSAW, but for any given device application, the influence of one effect relative to the other is dictated by factors such as particle size, SAW type, SAW power, SAW frequency, and channel geometry.[176] For particle manipulation techniques that rely on SSAW, primary acoustic radiation forces are the enabler, and acoustic streaming is considered an undesirable concurrence that disrupts otherwise predictable particle movements. To improve the precision of SSAW-based particle manipulation, it would be beneficial for researchers to find ways to minimize acoustic. It should be pointed out that little research has been done on the acoustic streaming induced by SSAW. This is an area in need of attention; perhaps through some clever innovations, researchers may even be able to convert SSAW-induced acoustic streaming from a complicating factor to an enabling mechanism.

Further improvements to the exactness of SSAW-based particle manipulation are also needed. Current acoustic tweezers devices have micrometre-scale resolution. Their resolution could potentially be refined to the nanometre-scale by using high-frequency SAWs. However, increasing SAW frequency results in higher energy attenuation as the acoustic wave transfers from the substrate surface to the contacting fluid. High SAW frequencies also result in stronger acoustic streaming, which compounds the increasing effect of streaming on particle movement as the particle size decreases. Both the increased energy attenuation and augmented acoustic streaming will need to be addressed in order to improve the resolution, and thus the applicability, of SSAW-based particle manipulation.

In addition to the issue of manipulation resolution, researchers have struggled to achieve precise control of particle manipulation in the device out-of-plane (z) direction. Applications such as tissue engineering and cell–cell communications rely on careful control of cells in all three dimensions. For SSAW-based methods to become an all-around particle/cell manipulation solution, researchers will need to gain precise control of the z direction.

Other device innovations may come from the use of new piezoelectric substrates. Among the hundreds of piezoelectric materials, $LiNbO_3$ has found ubiquitous use in SAW microfluidics due to its transparency, which enables easy microscope observation of on-chip events. More effort could be applied to exploring the potential benefits of other substrate materials. For example, the use of phononic crystals could help better manipulate SAW propagation and amplitude, and the use of piezoelectric thin films may lead to reduced device cost. In addition to considering new piezoelectric substrate materials, researchers should expand an existing trend: the use of disposable superstrates. These superstrates, which are bonded to the microfluidic components of the overall device, are acoustically coupled to the substrate *via* a liquid layer.[177,178] Such superstrates serve to separate the expensive piezoelectric substrate and its IDTs from the microfluidic components, making a single SAW-based device reusable by exchanging the superstrates. For example, if a cell patterning and culturing microfluidic device existed on a superstrate, multiple patterned superstrate cultures could be obtained with a single piezoelectric substrate. In combination with investigation of new substrate/superstrate materials, alternate acoustic modes, such as transverse SAW or plate waves, may also offer an opportunity for new innovations.

SAW-based microfluidics has come a long way in the past decade. It has been used to successfully perform a wide range of biomedicine-oriented lab functions, from cell focusing, cell sorting, cell manipulation, cell enrichment, cell lysis, to on-chip PCR, biosensing, and many more. In fact, these functions span the full spectrum of laboratory functions, from sample preparation all the way to final analysis; in other words, SAWs have been used to, on the aggregate, create a lab-on-a-chip. Researchers should now focus on actually creating that lab-on-a-chip. This will require integrating multiple SAW techniques onto a single microfluidic device, with transitions from one stage to another.

In order to make this adoption of SAW-based microfluidics a reality, researchers must build integrated devices capable of performing a complete conventional laboratory process, from start to finish. In this regard, facilitating the application of SAW-based microfluidics to real-world laboratory functions will take more than integrating multiple SAW techniques onto a coordinated device. It will also require creative solutions to a classic problem of microfluidic devices: the mismatch between coin-sized microchips and the large peripheral control and detection systems that still reside off-chip. Researchers will need to integrate function generators, amplifiers, SAW transducers, and signal processors onto a compact system. This integration and miniaturization is doable (SAW devices and accessories have already been integrated into compact, low-cost electronic devices such as cell phones), and it is necessary if SAW-based microfluidic devices are to be adopted for use in biological studies and medical diagnostics that researchers have long predicted.

Acknowledgements

This research was supported by the National Institutes of Health (1DP2OD007209-01) and the Penn State Center for Nanoscale Science (MRSEC) under grant DMR-0820404.

References

1. G. M. Whitesides, *Nature,* 2006, **442**, 368–373.
2. P. Neuzil, S. Giselbrecht, K. Länge, T. J. Huang and A. Manz, *Nat. Rev. Drug Discovery,* 2012, **11**, 620–632.
3. X. Mao and T. J. Huang, *Lab Chip,* 2012, **12**, 1412–1416.
4. M. L. Kovarik, D. M. Ornoff, A. T. Melvin, N. C. Dobes, Y. Wang, A. J. Dickinson, P. C. Gach, P. K. Shah and N. L. Allbritton, *Anal. Chem.,* 2013, **85**, 451–472.
5. T. M. Squires and S. R. Quake, *Rev. Mod. Phys.,* 2005, **77**, 977–1026.
6. A. Manz, N. Graber and H. M. Widmer, *Sens. Actuators, B,* 1990, **1**, 244–248.
7. A. Arora, G. Simone, G. Salieb-Beugelaar, J. T. Kim and A. Manz, *Anal. Chem.,* 2010, **82**, 4830–4847.
8. P. S. Dittrich and A. Manz, *Nat. Rev. Drug Discovery,* 2006, **5**, 210–218.
9. D. J. Beebe, G. A. Mensing and G. M. Walker, *Annu. Rev. Biomed. Eng.,* 2002, **4**, 261–286.
10. A. R. Wheeler, *Science,* 2008, **322**, 539–540.
11. I. A. Eydelnant, U. Uddayasankar, B. Li, M. W. Liao and A. R. Wheeler, *Lab Chip,* 2012, **12**, 750–757.
12. P. Tseng, J. W. Judy and D. D. Carlo, *Nat. Methods,* 2012, **9**, 1113–1119.
13. N. Pamme and A. Manz, *Anal. Chem.,* 2004, **76**, 7250–7256.
14. A. Lenshof and T. Laurell, *Chem. Soc. Rev.,* 2010, **39**, 1203–1217.
15. C. Monat, P. Domachuk and B. J. Eggleton, *Nat. Photonics,* 2007, **1**, 106–114.
16. X. Mao, J. R. Waldeisen, B. K. Juluri and T. J. Huang, *Lab Chip,* 2007, **7**, 1303–1308.
17. H. Schmidt and A. R. Hawkins, *Nat. Photonics,* 2011, **5**, 598–604.
18. Y. Zhao, Z. S. Stratton, F. Guo, M. I. Lapsley, C. Y. Chan, S.-C. S. Lin and T. J. Huang, *Lab Chip,* 2013, **13**, 17–24.
19. D. Psaltis, S. R. Quake and C. Yang, *Nature,* 2006, **442**, 381–386.
20. C. C. W. Ruppel, L. Reindl and R. Weigel, *Microwave Magazine,* 2002, **3**, 65–71.
21. A. Polh, *IEEE Trans. Sonics Ultrason.,* 2000, **47**, 317–332.
22. T. M. Gronewold, *Anal. Chim. Acta,* 2007, **603**, 119–128.
23. K. Länge, G. Blaess, A. Voigt, R. Götzen and M. Rapp, *Biosens. Bioelectron.,* 2006, **22**, 227–232.
24. A. Renaudin, V. Chabot, E. Grondin, V. Aimez and P. G. Charette, *Lab Chip,* 2010, **10**, 111–115.
25. J. Lee, Y.-S. Choi, Y. Lee, H. J. Lee, J. N. Lee, S. K. Kim, K. Y. Han, E. C. Cho, J. C. Park and S. S. Lee, *Anal. Chem.,* 2011, **83**, 8629–8635.
26. S.-C. S. Lin, X. Mao and T. J. Huang, *Lab Chip,* 2012, **12**, 2766–2770.
27. D. L. Miller, N. B. Smith, M. R. Bailey, G. J. Czarnota, K. Hynynen and I. R. S. Makin, *J. Ultrasound Med.,* 2012, **31**, 623–634.
28. M. Wiklund, *Lab Chip,* 2012, **12**, 2018–2028.
29. X. Ding, S.-C. S. Lin, B. Kiraly, H. Yue, S. Li, I. K. Chiang, J. Shi, S. J. Benkovic and T. J. Huang, *Proc. Natl. Acad. Sci. U. S. A.,* 2012, **109**, 11105–11109.

30. H. Li, J. Friend, L. Yeo, A. Dasvarma and K. Traianedes, *Biomicrofluidics*, 2009, **3**, 34102.
31. J. Friend and L. Y. Yeo, *Rev. Mod. Phys.*, 2011, **83**, 647–704.
32. L. Y. Yeo and J. R. Friend, *Biomicrofluidics*, 2009, **3**, 012002.
33. H. Bruus, J. Dual, J. Hawkes, M. Hill, T. Laurell, J. Nilsson, S. Radel, S. Sadhal and M. Wiklund, *Lab Chip*, 2011, **11**, 3579–3580.
34. A. Wixforth, *J. Lab. Automat.*, 2006, **11**, 399–405.
35. D. Beyssen, L. Le Brizoual, O. Elmazria and P. Alnot, *Sens. Actuators, B*, 2006, **118**, 380–385.
36. X. Ding, P. Li, S.-C. S. Lin, Z. S. Stratton, N. Nama, F. Guo, D. Slotcavage, X. Mao, J. Shi, F. Costanzo and T. J. Huang, *Lab Chip*, 2013, **13**, 3626–3649.
37. Z. Wang and J. Zhe, *Lab Chip*, 2011, **11**, 1280–1285.
38. M. Gedge and M. Hill, *Lab Chip*, 2012, **12**, 2998–3007.
39. Y. Q. Fu, J. K. Luo, X. Y. Du, A. J. Flewitt, Y. Li, G. H. Markx, A. J. Walton and W. I. Milne, *Sens. Actuators, B*, 2010, **143**, 606–619.
40. M. K. Tan, J. R. Friend and L. Y. Yeo, *Appl. Phys. Lett.*, 2007, **91**, 224101.
41. X. Ding, J. Shi, S.-C. S. Lin, S. Yazdi, B. Kiraly and T. J. Huang, *Lab Chip*, 2012, **12**, 2491–2497.
42. K. Hashimoto and M. Yamaguchi, *IEEE Trans. Sonics Ultrason.*, 2001, **48**, 1181–1188.
43. C. T. Schröder and W. R. Scott Jr, *J. Acoust. Soc. Am.*, 2001, **110**, 2867–2877.
44. E. Adler, *IEEE Trans. Sonics Ultrason.*, 1994, **41**, 876–882.
45. B. Auld, *Acoustic Fields and Waves in Solids*, Wiley Press, New York, 1973.
46. D. Köster, *SIAM J. Sci. Comput.*, 2007, **29**, 2352–2380.
47. L. Meng, F. Cai, Q. Jin, L. Niu, C. Jiang, Z. Wang, J. Wu and H. Zheng, *Sens. Actuators, B*, 2011, **160**, 1599–1605.
48. M. Wiklund, R. Green and M. Ohlin, *Lab Chip*, 2012, **12**, 2438–2451.
49. A. Wixforth, C. Strobl, C. Gauer, A. Toegl, J. Scriba and Z. V. Guttenberg, *Anal. Bioanal. Chem.*, 2004, **379**, 982–991.
50. Q. Zeng, F. Guo, L. Yao, H. W. Zhu, L. Zheng, Z. X. Guo, W. Liu, Y. Chen, S. S. Guo and X. Z. Zhao, *Sens. Actuators, B*, 2011, **160**, 1552–1556.
51. J. Vanneste and O. Buhler, *Proc. R. Soc. London, Ser. A*, 2010, **467**, 1779–1800.
52. M. Alghane, B. X. Chen, Y. Q. Fu, Y. Li, J. K. Luo and A. J. Walton, *J. Micromech. Microeng.*, 2011, **21**, 015005.
53. X. Y. Du, Y. Q. Fu, J. K. Luo, A. J. Flewitt and W. I. Milne, *J. Appl. Phys.*, 2009, **105**, 024508.
54. W.-K. Tseng, J.-L. Lin, W.-C. Sung, S.-H. Chen and G.-B. Lee, *J. Micromech. Microeng.*, 2006, **16**, 539–548.
55. T. Frommelt, D. Gogel, M. Kostur, P. Talkner, P. Hänggi and A. Wixforth, *IEEE Trans. Sonics Ultrason.*, 2008, **55**, 2298–2305.
56. R. Shilton, M. K. Tan, L. Y. Yeo and J. R. Friend, *J. Appl. Phys.*, 2008, **104**, 014910.
57. D. Köster, PhD Thesis, Augsburg University, 2006.

58. W. L. Nyborg, in *Acoustic Streaming*, ed. W. P. Mason and R. N. Thurston, Academic Press, New York, 1965, vol. 11, pp. 265–329.
59. S. S. Sadhal, *Lab Chip,* 2012, **12**, 2292–2300.
60. C. Bradley, *J. Acoust. Soc. Am.,* 1996, **100**, 1399–1408.
61. H. Bruus, *Lab Chip,* 2012, **12**, 20–28.
62. L. Zarembo, in *Acoustic Streaming*, Plenum Press, New York, 1971, pt. III, pp. 138–199.
63. A. A. Doinikov, *J. Fluid Mech.,* 1994, **267**, 1–21.
64. F. Costanzo, G. L. Gray and P. C. Andia, *Int. J. Eng. Sci.,* 2005, **43**(7), 533–555.
65. P. C. Andia, F. Costanzo and G. L. Gray, *Int. J. Solids Struct.,* 2005, **42**, 6409–6432.
66. L. V. King, *Proc. R. Soc. London, Ser. A,* 1934, **147**, 212–240.
67. K. Yosioka and Y. Kawasima, *Acustica,* 1955, **5**, 167–173.
68. L. P. Gorkov, *Sov. Phys. Dokl.,* 1962, **6**, 773–775.
69. J. Dual, P. Hahn, I. Leibacher, D. Moller, T. Schwarz and J. Wang, *Lab Chip,* 2012, **12**, 4010–4021.
70. H. Bruus, *Lab Chip,* 2012, **12**, 1014–1021.
71. M. Evander and J. Nilsson, *Lab Chip,* 2012, **12**, 4667–4676.
72. R. Barnkob, P. Augustsson, T. Laurell and H. Bruus, *Phys. Rev. E,* 2012, **86**, 056307.
73. T. Frommelt, M. Kostur, M. Wenzel-Schäfer, P. Talkner, P. Hänggi and A. Wixforth, *Phys. Rev. Lett.,* 2008, **100**, 034502.
74. T.-D. Luong, V.-N. Phan and N.-T. Nguyen, *Microfluid. Nanofluid.,* 2011, **10**, 619–625.
75. A. R. Rezk, A. Qi, J. R. Friend, W. H. Li and L. Y. Yeo, *Lab Chip,* 2012, **12**, 773–779.
76. A. Renaudin, P. Tabourier, V. Zhang, J. C. Camart and C. Druon, *Sens. Actuators, B,* 2006, **113**, 389–397.
77. T. A. Franke and A. Wixforth, *ChemPhysChem,* 2008, **9**, 2140–2156.
78. A. Wixforth, *Superlattices Microstruct.,* 2003, **33**, 389–396.
79. Z. Guttenberg, H. Muller, H. Habermüller, A. Geisbauer, J. Pipper, J. Felbel, M. Kielpinski, J. Scriba and A. Wixforth, *Lab Chip,* 2005, **5**, 308–317.
80. M. K. Tan, J. R. Friend and L. Y. Yeo, *Lab Chip,* 2007, 7, 618–625.
81. H. Li, J. R. Friend and L. Y. Yeo, *Biomaterials,* 2007, **28**, 4098–4104.
82. A. R. Rezk, O. Manor, J. R. Friend and L. Y. Yeo, *Nat. Commun.,* 2012, **3**, 1167.
83. L. Schmid, A. Wixforth, D. A. Weitz and T. Franke, *Microfluid. Nanofluid.,* 2011, **12**, 229–235.
84. S. W. Schneider, S. Nuschele, A. Wixforth, C. Gorzelanny, A. Alexander-Katz, R. R. Netz and M. F. Schneider, *Proc. Natl. Acad. Sci. U. S. A.,* 2007, **104**, 7899–7903.
85. L. Masini, M. Cecchini, S. Girardo, R. Cingolani, D. Pisignano and F. Beltram, *Lab Chip,* 2010, **10**, 1997–2000.
86. M. Cecchini, S. Girardo, D. Pisignano, R. Cingolani and F. Beltram, *Appl. Phys. Lett.,* 2008, **92**, 104103.

87. S. Girardo, M. Cecchini, F. Beltram, R. Cingolani and D. Pisignano, *Lab Chip*, 2008, **8**, 1557–1563.
88. M. A. Fallah, V. M. Myles, T. Krüger, K. Sritharan, A. Wixforth, F. Varnik, S. W. Schneider and M. F. Schneider, *Biomicrofluidics*, 2010, **4**, 1–10.
89. C. Fillafer, G. Ratzinger, J. Neumann, Z. Guttenberg, S. Dissauer, I. K. Lichtscheidl, M. Wirth, F. Gabor and M. F. Schneider, *Lab Chip*, 2009, **9**, 2782–2788.
90. R. J. Shilton, N. R. Glass, P. Chan, L. Y. Yeo and J. R. Friend, *Appl. Phys. Lett.*, 2011, **98**, 254103.
91. R. J. Shilton, S. M. Langelier, J. R. Friend and L. Y. Yeo, *Appl. Phys. Lett.*, 2012, **100**, 033503.
92. N. R. Glass, R. J. Shilton, P. Chan, J. R. Friend and L. Y. Yeo, *Small*, 2012, **8**, 1881–1888.
93. J. Eggers, *Rev. Mod. Phys.*, 1997, **69**, 865–929.
94. P. Calvert, *Chem. Mater.*, 2001, **13**, 3299–3305.
95. B.-J. de Gans, P. C. Duineveld and U. S. Schubert, *Adv. Mater.*, 2004, **16**, 203–213.
96. M. K. Tan, J. R. Friend and L. Y. Yeo, *Phys. Rev. Lett.*, 2009, **103**, 024501.
97. A. Lefebvre, *Atomization and Sprays*, CRC Press, New York, 1989.
98. M. Kurosawa, T. Watanabe, A. Futami and T. Higuchi, *Sens. Actuators, A*, 1995, **50**, 69–74.
99. A. Qi, L. Y. Yeo, J. R. Friend and J. Ho, *Lab Chip*, 2010, **10**, 470–476.
100. A. Qi, J. R. Friend, L. Y. Yeo, D. A. V. Morton, M. P. McIntosh and L. Spiccia, *Lab Chip*, 2009, **9**, 2184–2193.
101. S. R. Heron, R. Wilson, S. A. Shaffer, D. R. Goodlett and J. M. Cooper, *Anal. Chem.*, 2010, **82**, 3985–3989.
102. S. H. Yoon, Y. Huang, J. S. Edgar, Y. S. Ting, S. R. Heron, Y. Kao, Y. Li, C. D. Masselon, R. K. Ernst and D. R. Goodlett, *Anal. Chem.*, 2012, **84**, 6530–6537.
103. A. Qi, P. Chan, J. Ho, A. Rajapaksa, J. Friend and L. Y. Yeo, *ACS Nano*, 2011, **5**, 9583–9591.
104. J. Ho, M. K. Tan, D. Go, L. Y. Yeo, J. Friend and H. C. Chang, *Anal. Chem.*, 2011, **83**, 3260–3266.
105. M. Alvarez, J. Friend and L. Y. Yeo, *Nanotechnology*, 2008, **19**, 455103.
106. J. R. Friend, L. Y. Yeo, D. R. Arifin and A. Mechler, *Nanotechnology*, 2008, **19**, 145301.
107. O. V. Abramov, *High-Intensity Ultrasonics*, Gordon and Breach Science Publishers, Amsterdam, 1998.
108. L. Dong, A. Chaudhury and M. K. Chaudhury, *Eur. Phys. J. E: Soft Matter Biol. Phys.*, 2006, **21**, 231–242.
109. M. K. Tan, J. R. Friend, O. K. Mata and L. Y. Yeo, *Phys. Fluids*, 2010, **22**, 112112.
110. D. J. Collins, O. Manor, A. Winkler, H. Schmidt, J. R. Friend and L. Y. Yeo, *Phys. Rev. E*, 2012, **86**, 056312.
111. H. Li, J. R. Friend and L. Y. Yeo, *Biomed. Microdevices*, 2007, **9**, 647–656.
112. P. R. Rogers, J. R. Friend and L. Y. Yeo, *Lab Chip*, 2010, **10**, 2979–2985.

113. M. K. Tan, J. R. Friend and L. Y. Yeo, *Lab Chip,* 2007, **7**, 618–625.
114. T. Franke, A. R. Abate, D. A. Weitz and A. Wixforth, *Lab Chip,* 2009, **9**, 2625–2627.
115. T. Franke, S. Braunmüller, L. Schmid, A. Wixforth and D. A. Weitz, *Lab Chip,* 2010, **10**, 789–794.
116. L. Schmid and T. Franke, *Lab Chip*, 2013, DOI: 10.1039/C3LC41233D.
117. R. H. Baughman, A. A. Zakhidov and W. A. de Heer, *Science,* 2002, **297**, 787–792.
118. C. J. Strobl, C. Schäflein, U. Beierlein, J. Ebbecke and A. Wixforth, *Appl. Phys. Lett.,* 2004, **85**, 1427–1429.
119. T. Smorodin, U. Beierlein, J. Ebbecke and A. Wixforth, *Small,* 2005, **1**, 1188–1190.
120. K. M. Seemann, J. Ebbecke and A. Wixforth, *Nanotechnology,* 2006, **17**, 4529–4532.
121. A. P. Malanoski, V. A. Greanya, B. T. Weslowski, M. S. Spector, J. V. Selinger and R. Shashidhar, *Phys. Rev. E,* 2004, **69**, 021705.
122. Y. J. Liu, X. Ding, S.-C. S. Lin, J. Shi, I.-K. Chiang and T. J. Huang, *Adv. Mater.,* 2011, **23**, 1656–1659.
123. Y. J. Liu, M. Lu, X. Ding, E. S. P. Leong, S.-C. S. Lin, J. Shi, J. H. Teng, L. Wang, T. J. Bunning and T. J. Huang, *J. Lab. Automat.*, 2012, doi: 10.1177/2211068212455632.
124. T.-T. Wu, Z.-G. Huang and S. Lin, *Phys. Rev. B,* 2004, **69**, 1–10.
125. S.-C. S. Lin and T. Huang, *Phys. Rev. B,* 2011, **83**, 174303.
126. J. Shi, S.-C. S. Lin and T. J. Huang, *Appl. Phys. Lett.,* 2008, **92**, 111901.
127. S.-C. Lin, T. Huang, J.-H. Sun and T.-T. Wu, *Phys. Rev. B,* 2009, **79**, 094302.
128. S.-C. S. Lin and T. J. Huang, *J. Appl. Phys.,* 2009, **106**, 053529.
129. S.-C. S. Lin, B. R. Tittmann, J.-H. Sun, T.-T. Wu and T. J. Huang, *J. Phys. D: Appl. Phys.,* 2009, **42**, 185502.
130. T.-T. Wu, Y.-T. Chen, J.-H. Sun, S.-C. S. Lin and T. J. Huang, *Appl. Phys. Lett.,* 2011, **98**, 171911.
131. S.-C. S. Lin, B. R. Tittmann and T. J. Huang, *J. Appl. Phys.,* 2012, **111**, 123510.
132. R. Wilson, J. Reboud, Y. Bourquin, S. L. Neale, Y. Zhang and J. Cooper, *Lab Chip,* 2010, **11**, 323–328.
133. Y. Bourquin, J. Reboud, R. Wilson, Y. Zhang and J. M. Cooper, *Lab Chip,* 2011, **11**, 2725–2730.
134. J. Reboud, Y. Bourquin, R. Wilson, G. S. Pall, M. Jiwaji, A. R. Pitt, A. Graham, A. P. Waters and J. M. Cooper, *Proc. Natl. Acad. Sci. U. S. A.,* 2012, **109**, 15162–15167.
135. Y. Bourquin, R. Wilson, Y. Zhang, J. Reboud and J. M. Cooper, *Adv. Mater.,* 2011, **23**, 1458–1462.
136. J. Reboud, R. Wilson, Y. Zhang, M. H. Ismail, Y. Bourquin and J. M. Cooper, *Lab Chip,* 2012, **12**, 1268–1273.
137. D. Hartono, Y. Liu, P. L. Tan, X. Y. S. Then, L.-Y. L. Yung and K.-M. Lim, *Lab Chip,* 2011, **11**, 4072–4080.

138. J. Shi, S. Yazdi, S.-C. S. Lin, X. Ding, I.-K. Chiang, K. Sharp and T. J. Huang, *Lab Chip,* 2011, **11**, 2319–2324.
139. J. Shi, X. Mao, D. Ahmed, A. Colletti and T. J. Huang, *Lab Chip,* 2008, **8**, 221–223.
140. M. Alvarez, J. R. Friend and L. Y. Yeo, *Langmuir,* 2008, **24**, 10629–10632.
141. J. Shi, D. Ahmed, X. Mao, S.-C. S. Lin, A. Lawit and T. J. Huang, *Lab Chip,* 2009, **9**, 2890–2895.
142. C. D. Wood, S. D. Evans, J. E. Cunningham, R. O'Rorke, C. Wälti and A. G. Davies, *Appl. Phys. Lett.,* 2008, **92**, 044104.
143. L. Johansson, J. Enlund, S. Johansson, I. Katardjiev and V. Yantchev, *Biomed. Microdevices,* 2012, **14**, 1–11.
144. L. Johansson, J. Enlund, S. Johansson, I. Katardjiev, M. Wiklund and V. Yantchev, *J. Micromech. Microeng.,* 2012, **22**, 025018.
145. M. E. Piyasena, P. P. A. Suthanthiraraj, R. W. Applegate, A. M. Goumas, T. A. Woods, G. P. López and S. W. Graves, *Anal. Chem.,* 2012, **84**, 1831–1839.
146. X. Mao, S.-C. S. Lin, C. Dong and T. J. Huang, *Lab Chip,* 2009, **9**, 1583–1589.
147. X. Mao, J. R. Waldeisen and T. J. Huang, *Lab Chip,* 2007, 7, 1260–1262.
148. X. Mao, A. A. Nawaz, S.-C. S. Lin, M. I. Lapsley, Y. Zhao, J. P. McCoy, W. S. El-Deiry and T. J. Huang, *Biomicrofluidics,* 2012, **6**, 024113.
149. M. I. Lapsley, L. Wang and T. J. Huang, *Biomarkers Med.,* 2013, 7, 75–78.
150. Q. Zeng, H. W. L. Chan, X. Z. Zhao and Y. Chen, *Microelectron. Eng.,* 2010, **87**, 1204–1206.
151. J. Shi, H. Huang, Z. Stratton, Y. Huang and T. J. Huang, *Lab Chip,* 2009, **9**, 3354–3359.
152. J. Nam, Y. Lee and S. Shin, *Microfluid. Nanofluid.,* 2011, **11**, 317–326.
153. J. Nam, H. Lim, D. Kim and S. Shin, *Lab Chip,* 2011, **11**, 3361.
154. J. Nam, H. Lim, C. Kim, J. Y. Kang and S. Shin, *Biomicrofluidics,* 2012, **6**, 024120.
155. M. C. Jo and R. Guldiken, *Sens. Actuators, A,* 2012, **187**, 22–28.
156. R. Guldiken, M. C. Jo, N. Gallant, U. Demirci and J. Zhe, *Sensors,* 2012, **12**, 905–922.
157. A. Y. Fu, C. Spence, A. Scherer, F. H. Arnold and S. R. Quake, *Nat. Biotechnol.,* 1999, **17**, 1109–1111.
158. F. Shen, H. Hwang, Y. K. Hahn and J.-K. Park, *Anal. Chem.,* 2012, **84**, 3075–3081.
159. A. J. Mach, O. B. Adeyiga and D. D. Carlo, *Lab Chip,* 2013, **13**, 1011–1026.
160. M. M. Wang, E. Tu, D. E. Raymond, J. M. Yang, H. Zhang, N. Hagen, B. Dees, E. M. Mercer, A. H. Forster, I. Kariv, P. J. Marchand and W. F. Butler, *Nat. Biotechnol.,* 2005, **23**, 83–87.
161. D. D. Carlo, D. Irimia, R. G. Tompkins and M. Toner, *Proc. Natl. Acad. Sci. U. S. A.,* 2007, **104**, 18892–18897.
162. R. D. O'Rorke, C. D. Wood, C. Walti, S. D. Evans, A. G. Davies and J. E. Cunningham, *J. Appl. Phys.,* 2012, **111**, 094911.
163. N. D. Orloff, J. R. Dennis, M. Cecchini, E. Schonbrun, E. Rocas, Y. Wang, D. Novotny, R. W. Simmonds, J. Moreland, I. Takeuchi and J. C. Booth, *Biomicrofluidics,* 2011, **5**, 044107.

164. L. Meng, F. Cai, Z. Zhang, L. Niu, Q. Jin, F. Yan, J. Wu, Z. Wang and H. Zheng, *Biomicrofluidics,* 2011, **5**, 044104.
165. X. Ding, S.-C. S. Lin, M. I. Lapsley, S. Li, X. Guo, C. Y. Chan, I.-K. Chiang, L. Wang, J. P. McCoy and T. J. Huang, *Lab Chip,* 2012, **12**, 4228–4231.
166. S. Li, X. Ding, F. Guo, Y. Chen, M. I. Lapsley, S.-C. S. Lin, L. Wang, J. P. McCoy, C. E. Cameron and T. J. Huang, *Anal. Chem.*, DOI: 10.1021/ac400548d.
167. C. D. Wood, J. E. Cunningham, R. O'Rorke, C. Wälti, E. H. Linfield, A. G. Davies and S. D. Evans, *Appl. Phys. Lett.,* 2009, **94**, 054101.
168. S. B. Q. Tran, P. Marmottant and P. Thibault, *Appl. Phys. Lett.,* 2012, **101**, 114103.
169. D. G. Grier, *Nature,* 2003, **424**, 810–816.
170. R. P. Richter, R. Bérat and A. R. Brisson, *Langmuir,* 2006, **22**, 3497–3505.
171. M. Hennig, J. Neumann, A. Wixforth, J. O. Radler and M. F. Schneider, *Lab Chip,* 2009, **9**, 3050–3053.
172. M. Hennig, M. Wolff, J. Neumann, A. Wixforth, M. F. Schneider and J. O. Radler, *Langmuir,* 2011, **27**, 14721–14725.
173. J. Neumann, M. Hennig, A. Wixforth, S. Manus, J. O. Radler and M. F. Schneider, *Nano Lett.,* 2010, **10**, 2903–2908.
174. X. H. Kong, C. Deneke, H. Schmidt, D. J. Thurmer, H. X. Ji, M. Bauer and O. G. Schmidt, *Appl. Phys. Lett.,* 2010, **96**, 134105.
175. Y. Chen, X. Ding, S.-C. S. Lin, S. Yang, P.-H. Huang, N. Nama, Y. Zhao, A. A. Nawaz, F. Guo, W. Wang, Y. Gu, T. E. Mallouk and T. J. Huang, *ACS Nano,* 2013, **7**, 3306–3314.
176. H. Li, J. Friend and L. Yeo, *Phys. Rev. Lett.,* 2008, **101**, 084502.
177. R. P. Hodgson, M. Tan, L. Y. Yeo and J. Friend, *Appl. Phys. Lett.,* 2009, **94**, 024102.
178. S. M. Langelier, L. Y. Yeo and J. Friend, *Lab Chip,* 2012, **12**, 2970–2976.

CHAPTER 16

Surface Acoustic Wave Based Microfluidics and Droplet Applications

THOMAS FRANKE[*a,b,c,e,f], THOMAS FROMMELT[d], LOTHAR SCHMID[a,e], SUSANNE BRAUNMÜLLER[a,e], TONY JUN HUANG, AND ACHIM WIXFORTH[*a,b,c,e]

[a]University of Augsburg, Augsburg, Germany; [b]Center for NanoScience, CeNS, Munich, Germany; [c]Augsburg Center for Innov. Technologies, ACIT, Augsburg, Germany; [d]SGL Carbon, Meitingen, Germany; [e]Nanosystems Initiative Munich, NIM, 80799 Munich, Germany; [f]School of Engineering, University of Glasgow, G12 8LT; [g]Department of Engineering Science and Mechanics, The Pennsylvania State University, University Park, PA 16802, USA
*E-mail: thomas.franke@glasgow.ac.uk; achim.wixforth@physik.uni-augsburg.de

16.1 Introduction

Ever since the beginning of mankind, the mastering of fluids has played an important role for modern human development. Watering and irrigation of agricultural zones was probably the first application of technological fluid handling. Later on, drinking and service water for industrial use or power generation asked for sophisticated channeling of remote water sources. Also, artificial waterways with locks for transportation of ships and goods to remote locations far from natural streams and rivers developed into an engineering art over the centuries. Complex plumbing schemes, including pumping units, valves and containers, opened a wide area of technological advancements, which finally resulted in the highly complex water systems lancing our megacities nowadays.

On the other hand, the processing of fluids into more valuable goods required sophisticated fluid handling systems. Here, the introduction of bio- or chemo- or physical reactors such as distillation and fermentation played an important role in the development of complex fluid based goods such as wine, beer and gasoline. The development of alchemia and chemistry led to the discovery of even more complex and powerful fluid handling systems without which our modern life would be unthinkable.

In the last centuries, medical and biological applications have entered the field of fluidics, the major developments probably being the invention of dialysis or other life supporting techniques and systems.

In recent years, and probably triggered by the tremendous success of microelectronic 'chips', a new thinking in the field of fluid handling was born. Many laboratories in the world started to develop chip-sized microsystems in which small amounts of fluids can be manipulated, such as the charges on their electronic counterparts. Sophisticated technologies for the creation of microchannels, microreactors and most importantly pumping units were created over the years. Many of them are actually based on the longstanding and proven traditions of semiconductor technology that is still improving to an unprecedented level in modern micro- and nanoelectronics, photonics and other high tech applications.

Apart from standard pumps, such as those based on capillary forces, mechanically actuated pistons, peristaltic pumping or membrane based actuators, all reduced to the microscale, electrically actuated fluid handling systems have proved very successful over the years.[1-3] Also, centrifugal forces[4] or electrical or chemical modulation of the surface energy[5] of solid substrates were shaped into real microfluidic devices. What all these devices have in common is that the fluid behavior is dictated by typically very low Reynold's numbers Re < 1. The Reynold's number basically denotes the ratio of the influence of inertia to viscosity related effects on a fluid. At low Re, typically, the apparent viscosity of the fluid dominates the streaming behavior, making actuation or even mixing a difficult task on such microfluidic chips.

16.2 Acoustic Streaming Effects

This chapter centers around the employment of acoustic fields for fluidic actuation and manipulation purposes. A sound wave propagating within a fluid and having a pressure amplitude $p_{ex}(r,t)$ larger that the static pressure p_0 may lead to a net DC flow in the liquid. It is a result of the time-averaged momentum equation and can be identified with a 'radiation pressure'. This acoustically driven fluid motion is called 'acoustic streaming', a well-known effect that has many applications and implications in hydrodynamics.

Eckart[6] and Nyborg[7] already in the 1950s gave very comprehensive analytical investigations and overviews of this effect.

A special case for the generation of acoustic streaming on a solid chip surface is the use of so-called surface acoustic waves (SAW). A SAW is a mode of elastic energy propagating at the surface of a solid, where all wave-related quantities decay in a quasi-exponential manner with depth below said surface. One of the earliest descriptions of such waves was given by Lord Rayleigh in a seminal paper on earthquakes more than a hundred years ago.[8] SAW are mainly used in high frequency filters[9] and signal processing units for mobile phone applications for example. Also, due to their high sensitivity to disturbances at the surfaces they are propagating at, SAW are widely used as sensors[10] for many different purposes. Not only are they suited to detect minute amounts of mass loading at the surface, in the case of piezoelectric substrates they are also well suited to study changes in the (di-)electric environment at or close to the surface.[11]

On piezoelectrics, SAW are especially simple to excite and detect. Here, so-called interdigital transducers (IDT) are employed to efficiently convert a high frequency electrical input signal into a travelling surface acoustic wave.[12] The geometrical layout of the IDT together with the acoustic properties of the substrate determines the SAW frequency response and wavelength. SAW can also be used to actively and acoustically transport quantities of charge within a layered (semiconductor) substrate.[13,14] Here, the transport mechanism is based on the propagating piezoelectric fields accompanying the SAW at the speed of sound.

Most of the energy of a SAW, however, is stored in the elastic part of the wave, as the piezoelectric effect adds usually only a small contribution in the percentage range. The periodic deformation of the surface—in the case of a pure Rayleigh mode normal to the surface—can very efficiently couple to a fluid. Liquids on the surface are subject to this vibrating force and absorb parts of the energy, leading to acoustic streaming. Here, the acoustic mismatch between the fluid and the substrate leads to a diffraction of the sound wave entering the fluid. The diffraction angle is in first order approximation only determined by the ratio of the sound velocities in either material. For a strongly piezoelectric substrate such as $LiNbO_3$, for example, and water representing the liquid, a typical diffraction angle is of the order of 20°.

In Figure 16.1(a), we depict the situation for water residing on top of a $LiNbO_3$ substrate. The SAW is entering the droplet from the left. The wave fronts indicate the direction of the flow in this case. θ_R represents the diffraction or Rayleigh angle for this situation, which can easily be derived from a phase matching picture.

$$\sin(\theta_R) = \frac{v_{Fluid}}{v_{SAW}} \qquad (16.1)$$

Figure 16.1(b) shows the result of a finite element calculation. Here, the actual pressure amplitudes are calculated that form the SAW induced longitudinal wave in the fluid.[15,16] Here, the SAW is considered as a series of

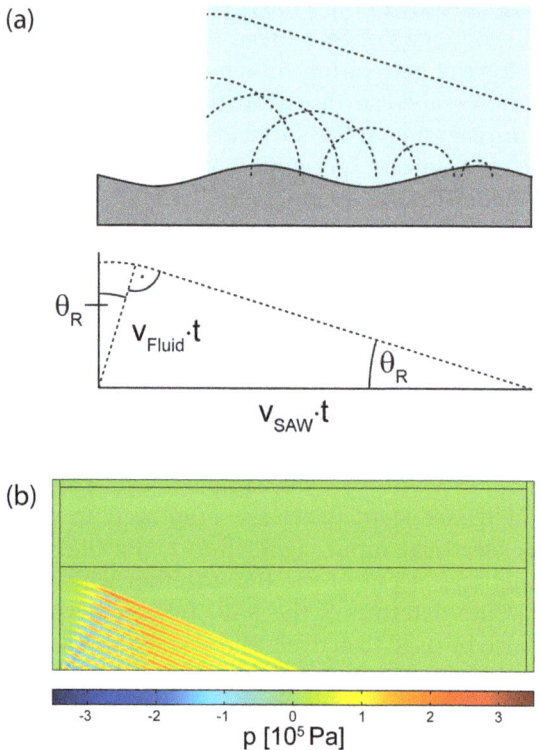

Figure 16.1 (a) Sketch of the SAW induced acoustic streaming effect. A SAW propagating at the surface of a substrate enters a region covered with a fluid. Each point at the surface is considered as a point-like pressure source (Huygens principle), creating longitudinal waves in the fluid. The diffraction angle θ_R can be simply calculated from basic geometrical considerations. (b) Result of a numerical calculation according to eqn (16.1) through eqn (16.3). Here, the SAW induced pressure wave into the fluid is shown.

point sources Q, distributed over the surface of the substrate. The pressure wave equation then reads:

$$\nabla \cdot \left(\frac{1}{\rho_i}\nabla p\right) - \frac{1}{\rho_i v_i^2}\frac{\partial^2 p}{\partial t^2} = Q \quad (16.2)$$

Here, the density and velocity in the regions of the substrate and the water, respectively are denoted by the index i. At the contact plane between water and substrate, the pressure gradient is given by the normal component of the acceleration caused by the SAW:

$$\vec{n}\cdot\nabla p = \rho_{\text{Fluid}}\omega^2 H(v_{\text{SAW}}t - y)A_0\sin(ky - \omega t)\exp(-y/\kappa_{\text{op}}) \quad (16.3)$$

Due to viscous damping, the SAW with an amplitude A_0 becomes exponentially attenuated in the fluid. The coupling constant is assumed to be κ_{op}. The Heaviside function H ensures an abrupt entrance of the SAW into the fluid, as it is assumed in our model. At the interface between substrate and fluid, continuity is ensured by

$$\vec{n} \cdot \left(\frac{1}{\rho_i} \nabla p_i - \frac{1}{\rho_j} \nabla p_j \right) = 0 \qquad (16.4)$$

At low acoustic powers, the pressure gradient drives an internal acoustic streaming in the fluid. This internal flow leads to a complex streaming behavior in most cases, as the exact streaming behavior, although being completely deterministic—if diffusion is neglected for a moment—very strongly depends on the actual geometry of the problem.

16.3 Acoustically Induced Mixing

The complex streaming patterns[17] arising from the interaction between the SAW and a fluid on top of the substrate can be very efficiently used for mixing small quantities of liquid.[18] It should be noted that apart from the SAW described in this article, other acoustic mixers have also been reported.[19] All of them rely on acoustic streaming in one sense or another.

However, as the exact geometry of the flow lines and their temporal development depend on many different parameters such as geometry, viscosity, location of the fluid with respect to the IDT on the chip surface, *etc.*, it is usually not sufficient to simply subject a small quantity of liquid to a SAW on the chip. Instead, one has to make sure that the SAW–fluid interaction leads to a thorough folding of the flow lines during the mixing process. This can be achieved, for example, by either dynamically switching the frequency of the SAW (leading to different flow patterns) or by subjecting the fluid to two or more SAW beams. If those beams are modulated in time with respect to each other, an efficient folding of the flow lines can be achieved, resulting in a quasi-chaotic advection and optimal mixing behaviour.[19] This is demonstrated in Figure 16.2, where we show the flow lines in a circular water disc (confined between two solid plates) subjected to a SAW beam, coming from the right. The difference between the upper two and the lower panels is just the entry point of the SAW. Clearly, one recognizes two different flow patterns. The flow lines have been made visible by superposition of a number of video frames and fluorescent tracer particles within the liquid. The solid lines in the two panels on the right are the result of a numerical, finite element calculation and show an extremely good agreement with the experimentally obtained traces.

The extremely fast and versatile mixing properties of SAW driven microfluidic chips have recently been commercialised in a microarray hybridization chamber for biological purposes.[20,21] Here, a microarray covered with, *e.g.* single stranded DNA oligonucleotides is subjected to a thin layer of fluid

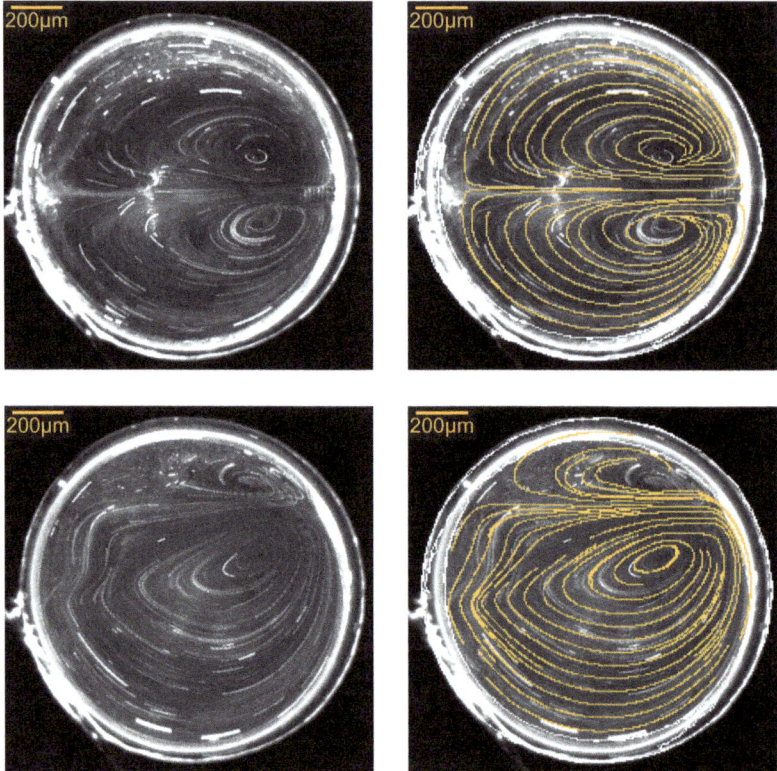

Figure 16.2 Flow lines in a flat circular water disc ($d \approx 500$ μm) confined between the piezoelectric substrate and a glass cover slide. A SAW is impinging from the right. The left panels show the experimentally obtained flow lines employing fluorescent tracer particles and video frame superposition. In the right panels the flow lines as being calculated in a numerical model have been superimposed. The upper two and the lower two panels differ by the entrance point of the SAW beams, resulting in different flow patterns. Switching between these patterns may lead to a quasi-chaotic advection.[19]

containing an unknown sample of evenly single stranded DNA. This sample is usually covered with a glass slide, confining it to a thickness of about 50 μm. The SAW induced actuation of the fluid and the folding of the flow lines induces a dramatic improvement of the hybridization speed and thus a much shorter time to the result.[21]

It is also possible to couple the SAW on the piezoelectric substrate directly through the (solid) bottom of a microfluidic device on top of the SAW chip surface. Here, the SAW first converts into a wave in the container bottom (*e.g.* a cuvette or a microtiter plate), and then enters the fluid within. Again, we observe diffraction and a resulting acoustic streaming in the miniature container. In Figure 16.3, we show such a 'separate media' coupling into the

Figure 16.3 SAW induced fluid actuation in one of the wells in a microtiter plate (left) and in a quartz cuvette (right), visualized by the agitation of some ink initially located at the bottom. In the microwell, the IDT was placed directly under the well and has been coupled through its bottom. In this case, as the IDT is completely symmetric, two jets are excited under the Rayleigh angle θ_R (Advalytix AG, unpublished).

wells of a microtiter plate (left) and into a quartz cuvette (right). Pulsed or multi frequency operation of the SAW leads to perfect mixing behavior within the wells. A detailed analysis of the temporal modulation of the SAW sources with respect to each other is given in the paper by Frommelt et al.[19] One important aspect for this scheme is that it does not require physical contact between the mixant and the mixer. It can be fully integrated into an analysis setup for biological assays. Also, as the SAW chip is separately coupled to the micro wells, in principle each well can be addressed individually. It is also worth mentioning that the SAW can be excited remotely by, e.g. inductively coupling the high frequency radiation into the IDT.

16.4 Acoustic Droplet Actuation

At higher SAW amplitudes and for isolated liquid volumes on the chip surface, an acoustically induced movement of the whole volume can be observed. First described by Shiokawa et al.,[22] the SAW at the surface of a (piezoelectric) substrate can strongly interact with such a droplet.[23–27] Apart from vividly bringing the droplet into rotation, its surface and thus the boundary conditions for the acoustic streaming becomes strongly distorted. At the same time, the center of mass of the droplet becomes pushed into the direction of SAW propagation and its wetting angles at the 'windward' and 'lee' side differ considerably. Both effects lead to a propulsion of the droplet into the direction of the SAW. This effect can be used for microfluidic chips, where droplets act as virtual containers and reactors at a very small scale. Surface tension makes it possible that those virtual test tubes are held together without the need for a real container. Moreover, the surface of the substrate on which the droplets are residing and acoustically actuated can be chemically functionalized in terms of the wettability properties.[24] Then, a whole fluidic track system can be constructed on which the droplets can be moved easily. A manifold of transducers on the chip together with such a fluidic track system can be exploited as electrically addressable microfluidic processors.[28–30]

In Figure 16.4, we show the effect of an impinging SAW on a small droplet ($V \approx 50$ nl) at the surface of a $LiNbO_3$ chip. To demonstrate the strong deformation of the droplet under the influence of the SAW, we superimposed two video frames with and without the SAW induced action. Despite the fact that the back flow effects in the droplet are no longer well defined, one can easily identify the Rayleigh angle in the figure. In Figure 16.4(b), we show the result of a numerical calculation for the droplet deformation together with the internal streaming flow lines, including dynamic boundary conditions caused by the deformation of the droplet shape. This strong deformation, of course, has a back action on the internal (re-)flow, which makes the numerics quite cumbersome.[31]

The possibility to deliberately and controllably actuate small amounts of fluids along the surface of a planar substrate together with the SAW streaming mixing possibilities within the droplet make this technique extremely well suited for laboratories on a chip. The SAW are generated by IDTs, the trajectories of the droplets are predefined either by a chemical surface modification or by the shape of the SAW 'beam' normal to the propagation direction. Due to diffraction effects at the edges of the IDT aperture, the SAW wavefronts usually exhibit typical ripples normal to the propagation direction, providing a self-centering of any droplets within the propagation path. Chemical modification of the surface can be done in many different ways, the most convenient one being a laterally patterned silanization to define hydrophobic and hydrophilic regions on the chip surface. As all the technological steps to produce such SAW driven microfluidic chips

Figure 16.4 Left and center: water droplet on a $LiNbO_3$ substrate under the influence of a pulsed SAW impinging from the left. To visualize the strong deformation of its surface and shape, two video frames have been superimposed. (Advalytix AG, unpublished) Right: example for the streamlines within a droplet as the result of a numerical model. Here, the droplet is deformed only by the viscous stress at the boundary. In this example, the flow is driven by a non-conservative force density alone whereas the conservative part of the force vanishes such that the pressure is constant inside the droplet.[32]

rely on standard techniques from semi-conductor planar technology, and are all carried out at the planar surface, a more complex infrastructure is easily implemented on the surface of the same chip. This could be used to produce heaters, sensors or even micro arrays for an *in situ* control of biological assays. All the active elements on the chip are electrically controllable and can thus be run using computer software. The chips are cheap as they do not involve any complex lithography or doping *etc.* This makes them perfect candidates for single use consumables, which also excludes possible crosscontamination between different batches. All the necessary electronics are combined in an apparatus where the chip is inserted as needed. A bar code on the chip even makes it possible to tell the computer and the electronics the necessary protocol. However, as the chips might differ slightly from batch to batch due to production variances, a 'blind' operation of the protocol might lead to erratic behavior. Therefore, an active feedback of, *e.g.* the droplet position on the chip and the actuation mechanism is needed. For this purpose, a special kind of IDT proves extremely useful. Here, we employ socalled slanted transducers, where the resonance frequency changes across its aperture by varying finger spacing.[33] Thus, a spatial coordinate along the IDT aperture converts directly into a frequency. Hence, sweeping the excitation frequency for the IDT across its operation band and measuring the SAW transmission directly yields the position of a droplet in the propagation path.

In Figure 16.5, we show the result of such position detection on a chip. In the left panel, we show a picture of the chip. On the top and at the bottom of the chip, the slanted IDTs are seen; the two bright lines extending from left to right are the fluidic tracks, made visible by a metallization. On these tracks, a small droplet is moving. The right panel shows the SAW transmission between the IDTs as a function of the excitation frequency, exhibiting a strong absorption over a specific frequency band that can be identified with a specific location on the chip. The same idea has also been followed by other groups[35] and is well established for the active position control of small droplets on acoustically driven microfluidic chips.

16.5 PDMS Microfluidics

Since the versatility of PDMS was demonstrated by Whitesides about a decade ago,[36-39] formation of microfluidic devices in a process called soft lithography has revolutionized fabrication.[40,41] In a rapid prototyping process, this method allows minimization of operating time from concept to experiment. The unique features of PDMS, among which are excellent optical transparency, good biocompatibility enhanced by high permeability to oxygen and carbon dioxide, low toxicity and elastic behavior, have permitted a number of novel approaches and applications. To a large extent, the rapid development of microfluidics was made possible utilizing PDMS technology. Therefore, to take advantage, it is useful to integrate novel designs of experiments and applications into the large number of features that have been realized and developed in PDMS based fluidics.

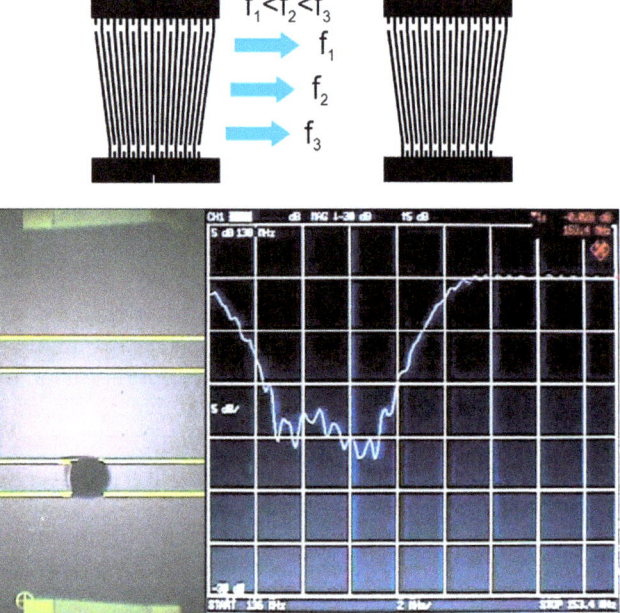

Figure 16.5 Top: slanted IDT for droplet detection on a chip.[33] The transducer fingers have varying spacing along the aperture, leading to a location dependent SAW resonance frequency. Sweeping the excitation frequency over the transmission bandpass leads to the excitation of a narrow ($\approx 10\ \lambda$) SAW beam at a specific location. Bottom left: photograph of the chip with a droplet on it. On the top and at the bottom one can see the slanted transducers whose transmission is monitored while the droplet is moving across the sound path. The horizontal lines indicate the position of fluidic tracks, fabricated by a modulation of the surface energy and made visible by metallization. Right: oscilloscope trace of the SAW transmission between the two slanted transducers as a function of frequency. Over the frequency band corresponding to the position and size of the droplet, one observes a pronounced SAW attenuation that can be used for position feedback control on the chip.[34]

The SAW–PDMS hybrid technique combines the advantages of both technologies to provide a low cost and very versatile platform for manipulation of small amounts of fluids in a closed system, minimizing contamination.

16.5.1 SAW Excitation on a Piezosubstrate and Acoustic Coupling to Standard PDMS Devices

Fabrication and mounting of the acoustic hybrid devices is very simple and involves only a few extra steps in addition to standard PDMS procedures. However, the SAW actuating chip has to be produced in advance. This chip consists of two components. The acoustic part is composed of a piezoelectric

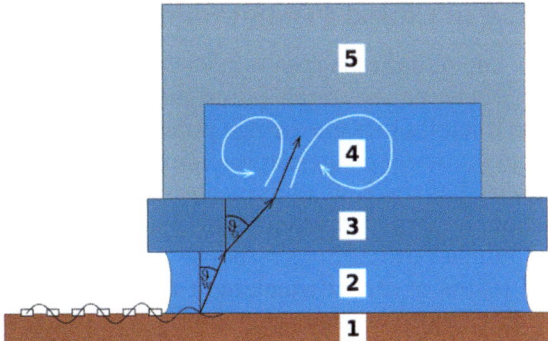

Figure 16.6 Top: side view sketch of the path of the acoustic wave coupling into the channel: a surface acoustic wave is excited by the IDT and travels along the lithium niobate substrate (1). It is refracted into a longitudinal wave in the coupling liquid (water) (2) at a Rayleigh angle $\theta_W = 21.8°$ while the surface wave on the substrate is attenuated on a characteristic length scale of 325 µm. The longitudinal acoustic wave passes through the 150 µm thick water layer and is subsequently refracted into a transversal wave in the glass cover slide at a Rayleigh angle of $\theta_G = 54.4°$. At the top of the glass slide, the wave is refracted again, enters the water-filled channel (4) and transfers momentum to the liquid, causing acoustic streaming as indicated, before it couples into the PDMS layer on top where it is further attenuatued (reproduced from ref. 58 with permission).

substrate with interdigitated electrodes to electrically excite the mechanical surface acoustic wave as described before. The other component is the PDMS channel that can be either directly bonded to the piezo-chip using ozone plasma etching or coupled to the chip after bonding to a glass substrate. Although the former geometry provides a closer coupling by minimization of chip–fluid distance, it requires full assembly of the chip for each experiment. Coupling a standard PDMS–glass device to the acoustic chip enables the reuse of the piezo-chip and saves assembly time. Several PDMS–glass channels may serve as disposables and can be used one after another simply by placing them onto the piezo-chip.

However, the acoustic contact between the PDMS–glass and the chip has to be enabled by a contact liquid and is critical for the robustness and efficiency of the hybrid device. A schematic of the setup is shown in Figure 16.6. When a microfluidic channel is placed on top of the substrate, the Rayleigh wave generates an acoustic wave that couples through the intermediate coupling liquid, for example a water layer, and through the glass bottom of the channel into the fluid, transferring momentum along the direction of propagation and ultimately inducing fluid streaming. Depending on the geometry of the confining PDMS channel in the actuation region, this may create several flow vortices[42] but also may induce a directed flow as is demonstrated in the subsequent paragraph. The flow velocity can be easily controlled electronically by the SAW power.

16.5.2 Pumping in Closed PDMS Channels

Pumping fluid in microfluidic devices is a delicate issue because it often involves an interface between an external pump and the fluid channel. Probably most often used in microfluidics are syringe pumps to drive fluid at a certain volume rate through the device or alternatively, an external pressure applied *via* tubing connection pushes the flow. Both mechanisms suffer from the risk of a leaking connection, in particular when highly complex chip designs involve a large number of pumps. Furthermore, response times in these systems can be as high as several tenths of seconds making it impossible to control a non-stationary flow, for instance to mimic the pulsative flow of a heartbeat.

Although there are several other types of pumps exploiting various physical effects including electrokinetic and electro-osmotic pumps, these pumps have a low pumping efficiency or depend on the fluid properties such as the electrolyte concentration of the liquid. Using multi-layered soft lithography using PDMS to form a peristaltic pump similar to Quake-valves provides faster response times but it needs to be connected to external air pressure generators. Surface acoustic wave actuated pumping avoids these disadvantages by fully integrating the pump and the fluidic chip.

The electrical connections to drive the surface wave can be implemented onto the chip using vapour deposited conductive paths. The actuation principle of acoustic streaming is independent of pH and electrolyte concentration and uses low voltages, which is important when pumping biological samples and to avoid electrochemical effects. Moreover, the method allows the continuous actuation of non-stationary fluid flow with fast response times. A unique feature of the acoustically driven pumps is the opportunity to pump in closed looped channels without the need for fluid inlets or outlets. Issues of cross-contamination can therefore be effectively addressed and looped channel structures for chemical incubation can be envisioned. An example of such a closed loop channel geometry is shown in Figure 16.7. In such geometries, typically, flow velocities in the range of $0 < v_m < \sim 1$ cm s^{-1} can be achieved. Therefore, the flow is laminar all over the microchannel because the Reynolds number is still low (Re < 10) at all positions. We determined the flow velocity in the straight channel of Figure 16.7 using particle tracking techniques at different channel heights and compare in Figure 16.8 the velocities obtained to the theoretical flow profile of a rectangular pressure driven channel.[43]

The plot of both the experimentally measured and the theoretically calculated curve are in excellent agreement demonstrating that the acoustic streaming builds up an effective pressure that drives the fluid flow. In other words, the stationary velocity flow profile can be simply calculated by solving the incompressible Stokes equation with a driving pressure specified by the power of the acoustic actuation and the boundary condition of the constrained channel geometry. This allows the simulation of more complex flow situations and microfluidic channel designs with standard techniques such as commercially available finite element methods.

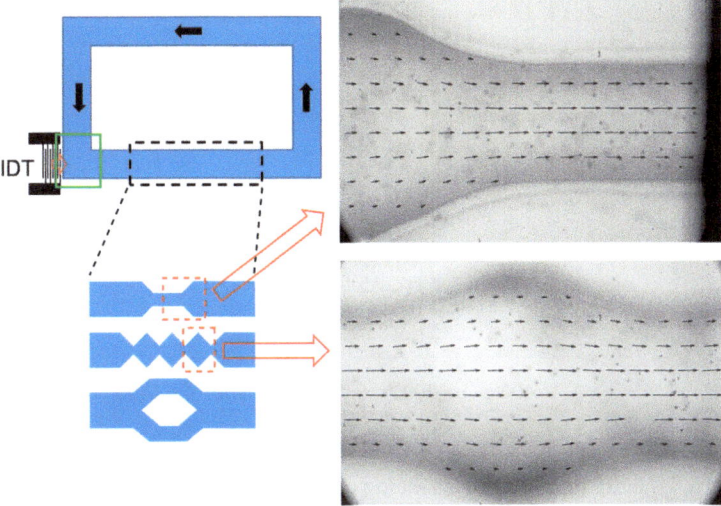

Figure 16.7 Left: schematic of the hybrid PDMS–SAW chip as seen from above. The basic channel (width = 1 mm, height = 0.75 mm, loop circumference = 42 mm) consists of a closed rectangular PDMS channel and is bonded onto a glass substrate. The IDT is carefully positioned below the channel and pumps the fluid all around the closed channel without inlets or outlets (black arrows indicate flow direction). The IDT consists of 42 fingers and has an aperture of 624 μm and a length of 1233 μm. It is positioned in parallel to the fluid channel exactly at the PDMS–fluid boundary. Enclosed by a green line is the coupling region of the acoustics to the fluid. We have used different channel designs as indicated by the channel sections in the schematic below: a narrowing nozzle-like channel, an oscillating zig-zag channel and a bifurcated channel. Right: micrographs of the different channels are superposed with the experimentally measured flow velocity vectors (black arrows) as obtained from particle tracking with focus in the symmetry plane, *i.e.* middle of the channel. (reproduced from ref. 58 with permission).

For non-stationary flows, the response time of the system has to be taken into account. As pointed out by Stone[44] for syringe pumps, compressibility of the fluid and small air bubbles in the fluidic system can seriously affect the response time. Even without bubbles this transient time can reach values of a minute before a stationary flow has been established, a fact many experimentalists are aware of. The highest oscillation frequency for a non-stationary flow that is accessible to a system is given by the inverse of the response time η. We have determined the response time for the aforementioned straight channel geometry by applying a square wave modulated HF signal to the IDT electrodes and observed the corresponding flow relaxation for different values of the channel dimensions (Figure 16.9). The measured response times are well below 100 ms that correspond to a maximal oscillation frequency of 10 Hz. For physiologically relevant situations such as the pulsative flow of the heartbeat, a typical frequency of ∼1 Hz is observed.

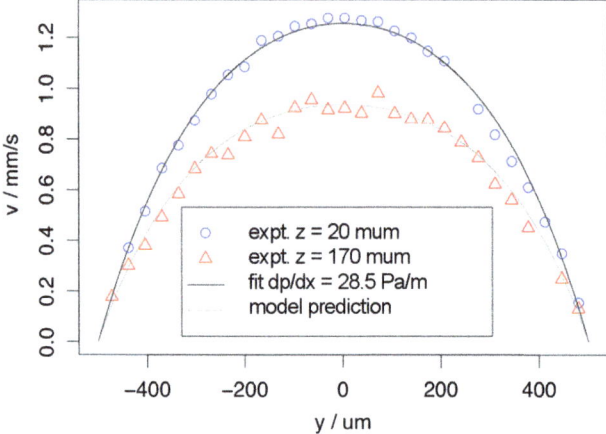

Figure 16.8 Left: flow profile as being measured in the PDMS channel (basic rectangular channel as shown in Figure 16.7; width = 1000 μm, height = 750 μm). The fluid moves in the x-direction, y is the horizontal and z is the vertical direction. The origin of the coordinate system is at the center of the channel. Blue circles represent the flow velocity measured near the channel center ($z = 20$ μm), red triangles show the velocity measured at $z = 170$ μm. The black line is a fit of the analytical model to the measurement at $z = 20$ μm (blue circles) and yields a pressure gradient $G = \mathrm{d}p/\mathrm{d}x = 28.5$ Pa m^{-1} as a free fitting parameter. The dashed line shows the prediction of the analytical model for off-center flow ($z = 170$ μm) using G as determined above. This is in good agreement with measurements (red triangles). Reproduced from ref. 58 with permission.

Hence, with the acoustic pump, a simple platform to mimic blood flow in small capillaries can be envisioned. Together with the biocompatibility of PDMS and the opportunity of cell growth on its surface, a simplified model of a blood vessel under realistic flow conditions can be obtained.

16.5.3 Droplet Based Fluidics

Although favorable in some cases, one-phase microfluidics as demonstrated in the previous paragraph inherently suffers from defining and manipulating distinct volumes of fluids,[45] which is a considerable prerequisite for many applications. Firstly, the integrity of a defined liquid volume disperses simply because diffusion becomes significant on the micron length scale. Secondly, in microchannels, which are widely used, the parabolic flow velocity profile of a pressure driven flow enhances mixing by Taylor dispersion.[46,47] Even though the latter can be circumvented using electro-osmotically driven flows generating a plug flow profile, these problems are still an important issue. Both problems can be avoided using droplet fluidics.[48] In droplet fluidics, emulsion droplets are employed to define liquid containers of picolitre volume.

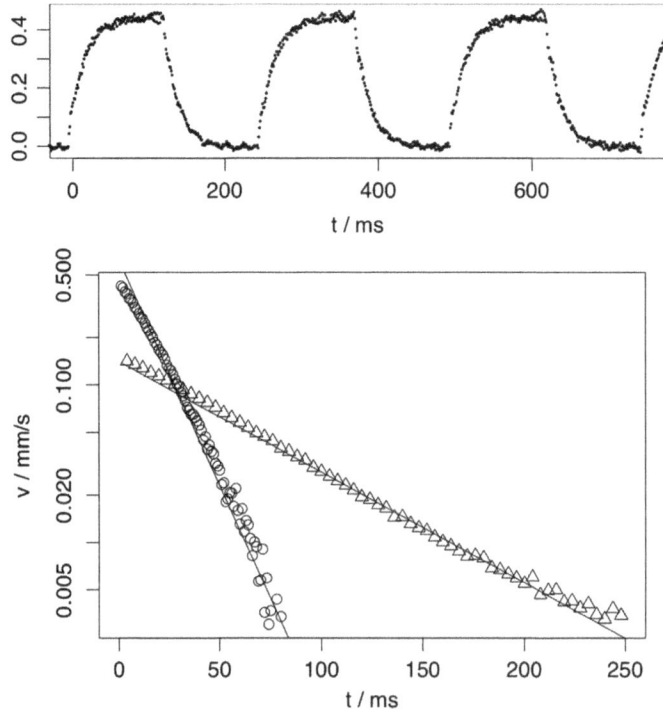

Figure 16.9 Response of the flow velocity to a 2 Hz square-wave modulation of the SAW (20.5 mW, 13.1 dBm). The top right inset shows the time-dependent flow velocity of a 4 Hz modulation in a smaller channel (63.5 mW, 18.0 dBm). After repeated voltage cut-off, the flow velocity decays exponentially. The plot below shows on a logarithmic scale the decay in two different channels both of height $h = 0.75$ mm and a width of 1 mm and 0.5 mm, respectively. The exponential fits result in a decay time $\tau = 61.7 \pm 0.4$ ms for the wider channel and a decay time $\tau = 15.1 \pm 0.4$ ms for the smaller channel (reproduced from ref. 58 with permission).

These containers can be utilized to encapsulate reactants, cells, bacteria or polymers such as DNA or proteins. They provide a unique microenvironment and inhibit diffusive as well as convective mixing. Precise control of droplet volume in combination with coalescence and splitting of drops enables many analytical applications and can be regarded as digital microfluidics, like the single phase droplets described earlier. Droplet formation is driven by interfacial tension and can occur right at the tip of the capillary orifice (dripping) or further downstream by disintegration of the fluid jet (jetting). The two regimes can be characterized by the capillary number of the outer fluid and the Weber number of the inner fluid.[49,50] An alternative method for droplet production in glass capillaries is the formation in a microchannel formed by the elastomer PDMS.[51] Both of these techniques allow for serial production of multiple emulsions of higher order and hierarchy. In multiple

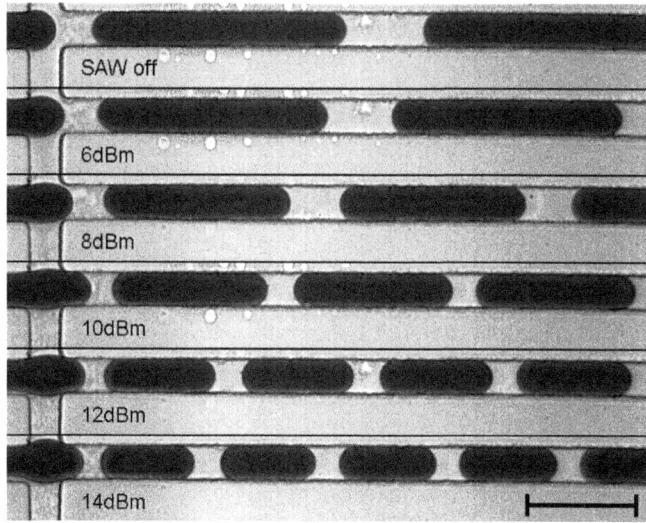

Figure 16.10 Drop formation in a flow focusing geometry. Stack of a dropmaker cross-junction driven at different acoustic powers. Upper image: in the absence of an acoustic field, drops break up at the junction forming extended fluid drops. Lower images: depending on the power, the time for drop formation can be decreased, causing drops of shorter length (reproduced from ref. 51 with permission).

emulsions, the emulsion drops themselves contain a controlled number of drops. Double emulsions in particular are useful in vesicle preparation[48] that can be applied to encapsulate reagents or magnetic beads.

Although drops can be produced at rates as high as several kHz by changing the flow rates of the dispersed and continuous phase, transients can be extreme.[44] For a syringe driven system, equilibration times can already reach several minutes. This seriously limits the controllability of the dropmaker and the fast production and processing of well-defined drop volumes.

Very recently, we have demonstrated that this limitation can be overcome by using SAWs. In a flow-focusing device we actively control the volume of each drop in real time.[51] The intensity of the acoustic field regulates the dimensions of the drops as shown in Figure 16.10.

Pinch-off occurs in the so-called squeezing-regime[52,53] of drop formation. In this regime we have demonstrated that the length of drops can be described using a linear model with a constant slope $\partial l/\partial P_{SAW}$ of drop length l on SAW power P_{SAW} The fast response times and precise control allow a fully electronically controlled way of drop formation in real time. A similar approach can be followed using a T-junction geometry; see Figure 16.11.

The rapid development of biomedicine and diagnostics creates a great demand for high throughput systems to analyze tiny amounts of biofluids, a goal that can be achieved by applying droplet based fluidics in PDMS designed devices. However, this requires ultrafast processing techniques to

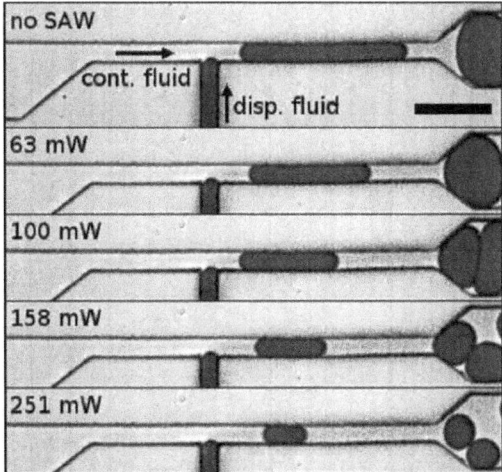

Figure 16.11 Dropmaker T-junction. The continuous fluid (oil) enters from the left, the dispersed aqueous fluid from the bottom, where it breaks up into droplets that flow to the right. The scale bar denotes 100 µm. The images show the droplet length at different SAW powers while the external pressure of the inlets was held constant at 5.7 kPa (continuous fluid) and 11.7 kPa (dispersed fluid) (reproduced from ref. 59 with permission).

manipulate drops and cells on the chip. The available state of the art techniques suffer from low processing rates or a dependence of sorting efficiency on the properties of the objects (droplets, cells) to be processed. For example, switching valves are limited to frequencies of ~10 Hz and techniques based on dielectrophoresis clearly depend on the contrast of permittivity between continuous fluid and objects. Both restrictions can be overcome using a surface acoustic wave based technique. SAW actuated droplet and cell direction allow us to sort cells to rates of up to ~10 kHz. To demonstrate the superior control of droplets, we integrated a microfluidic drop maker and the acoustic chip on a PDMS device. The acoustic part consists of a channel that splits into two outlet channels. Adjacent to the sorting region and just before the branching, an IDT is positioned that creates acoustic streaming to laterally deflect the droplets, as shown in Figure 16.12, and direct them into either one of the outlet channels. The droplets are aqueous drops suspended in an immiscible continuous environment that consists of HFE-7500 fluorocarbon oil with 5% (vol/vol) $1H,1H,2H,2H$-perfluoro-1-octanol, stabilized by 1.8 wt% of the fluorosurfactant ammonium carboxylate of DuPont Krytox 157. The same principle can be applied to hydrogel particles that provide a significantly lower contrast when suspended in water. Therefore, we have synthesized PAM particles using microfluidic droplet polymerization in a separate PDMS device. We added 10% of the acrylamide monomer as well as 10% of the cross-linker BIS-acrylamide to the middle channel, while oil flowed through the two side channels. The particles were then suspended in

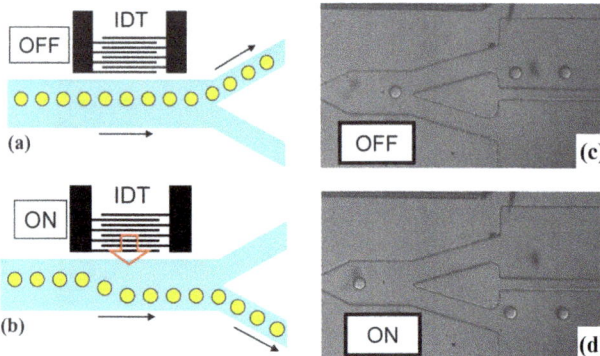

Figure 16.12 (a) and (b): Schematic of the hybrid PDMS–SAW chip as seen from above: a branched PDMS channel is coupled to a SAW device. (a) If the SAW power is switched off, all drops flow along the upper channel because of its lower flow resistance. (b) If the SAW power is switched on, the acoustic streaming induced by the IDT (red arrow) drives the droplets in the lower channel of the branch. (c), (d): Top view of the hybrid device showing a branched PDMS channel. Monodisperse water droplets are entering at constant separation from the left and are produced further upstream in a nozzle combining oil and water inlets (not shown here). The IDT is positioned below the PDMS and is carefully aligned parallel to the channel and can be partly seen at the top of the images (dark horizontal lines). The dark tip on the right side of the IDT is a feature to facilitate alignment and should roughly aim at the upper left corner of the upper outlet reservoir. However, a misalignment of approximately 100 mm did not affect the ability to direct the drops in our experiments. (c) When the SAW is switched off, all the drops take the upper outlet channel because its cross-section is designed to be slightly larger. (d) When a RF signal of approximately 10 dBm is applied to the IDT, all drops are pushed into the lower channel. The device is 50 mm high and 100 mm in width right before the branch. Flow rates were 100 ml h^{-1} for the dispersed phase and 1000 ml h^{-1} for the continuous phase. (Reproduced from ref. 57 by permission).

water and introduced into the sorting device. Again, we were able to control the gel particle flow by application of the SAW.

Cell sorting at high rates is a problem that has been addressed often in microfluidic systems because cell populations typically possess a natural distribution of single cell properties or dimensions. Since most physical actuation mechanisms used for cell deflection and sorting directly exploit the cell's properties, the deflection effect varies within a cell population. This problem can be avoided using highly monodisperse droplets as templates for cell encapsulation.[55] However, loading drops with discrete objects such as cells or particles obeys Poisson statistics and yields many unloaded and multiply-loaded droplets. This disadvantage can be avoided to some extent using close-packed deformable particles.[52] Moreover, SAW based flow direction steering enables cell sorting that operates in continuous flow without the need for encapsulation prior to sorting. Using the acoustic streaming technique, we are able to deflect cells directly from the bulk solution such as cell buffer or blood plasma. The sorting principle is independent of the cell's size or properties and can operate at sorting rates as high as several kHz.[53] To minimize accidental sorting of multiple cells at the same time (Poisson statistics) and to ensure a serial flow of one at a time, we align cells by hydrodynamic focusing using a sheath flow of the cell buffer solution. When entering the focusing region, cells are accelerated and accordingly their cell to cell distance increases reducing erroneous collective sorting. Sheath and cell flow rates can be controlled in a running experiment.

Using this acoustic device, we have successfully directed various cell types including MV3 melanoma cells, murine fibroblasts and HaCaT cells. Because of the inherently small shear stress of this method, the viability of sorted cells was excellent as has been proven by viability tests.[53,54]

16.6 Conclusions

We have shown that acoustic streaming induced by surface acoustic waves is a versatile tool for microfluidics enabling superior control of small fluid volumes. It can enhance mixing in tiny volumes such as nl drops or the actuation of a droplet container on a chip substrate surface. Moreover, the technology is compatible with the widely used PDMS soft lithography. This combination allows for high throughput applications such as cell and droplet sorting at rates of $\sim 10^4$ drops per second. Its ability to pump fluid in a closed loop channel without external tubing connections is useful for cell assays that need repeated incubation periods.

References

1. N.-T. Nguyen, S. T. Wereley, *Fundamentals and applications of microfluidics*, Artech House, 2002.
2. G. Fuhr, T. Schnelle and B. Wagner, *J. Micromech. Microeng.*, 1994, **4**, 217.
3. A. Hatch, A. E. Kamholz, G. Holman, P. Yager and K. F. Böhringer, *J. Microelectromech. Syst.*, 2001, **10**, 2.
4. M. Inganas, H. Derand, A. Eckersten, G. Ekstrand, A. K. Honerud, G. Jesson, G. Thorsen, T. Soderman and P. Andersson, *Clin. Chem.*, 2005, **51**, 1985.
5. W. J. J. Welters and L. G. J. Fokkink, *Langmuir*, 1998, **14**, 1535.
6. C. Eckart, *Phys. Rev. Lett.*, 1948, **73**, 68.
7. W. L. Nyborg, Acoustic streaming, in *Physical Acoustics*, ed. W. P. Mason, Academic Press, 1965, vol. 2B, pp. 265–330.
8. L. J. W. S. Rayleigh, *Proc. London Math. Soc.*, 1885, **17**, 4.
9. (a) A. A. Oliner, *Acoustic Surface Waves*, Springer, 1973; (b) H. Matthews, *Surface Wave Filters*, Wiley, 1977.

10. D. S. Ballantine Jr, R. M. White, S. J. Martin, A. J. Ricco, G. C. Frye, E. T. Zellars and H. Wohltjen, *Acoustic Wave Sensors—Theory, Design, and Physico-Chemical Applications*, Elsevier, 1997.
11. A. Wixforth, M. Wassermeier, J. Scriba, J. P. Kotthaus, G. Weimann and W. Schlapp, *Phys. Rev. B,* 1989, **40**, 7874.
12. (a) R. M. White and F. W. Voltmer, *Appl. Phys. Lett.,* 1965, 7, 12; (b) D. P. Morgan, Rayleigh Wave Transducers, in *Rayleigh-Wave Theory and Application*, ed. E. A. Ash and E. G. S. Paige, Springer, 1985.
13. C. Rocke, S. Zimmermann, J. P. Kotthaus, G. Böhm and G. Weimann, *Phys. Rev. Lett.,* 1997, **78**, 4099.
14. C. Wiele, F. Haake and A. Wixforth, *Phys. Rev. A,* 1998, **58**, R2680.
15. T. Frommelt, D. Gogel, M. Kostur, P. Talkner, P. Hänggi and A. Wixforth, *IEEE Trans. Ultrason. Ferroelectr. Freq. Control,* 2008, **55**, 2298.
16. H. Antil, R. Glowinski and R. H. W. Hoppe, *et al., J. Comput. Math.,* 2010, **23**, 149.
17. Z. v. Guttenberg, A. Rathgeber, S. Keller, J. O. Rädler, A. Wixforth, M. Kostur, M. Schindler and P. Talkner, *Phys. Rev. E,* 2004, **70**, 056311.
18. K. Sritharan, C. J. Strobl, M. F. Schneider, A. Wixforth and Z. v. Guttenberg, *Appl. Phys. Lett.,* 2006, **88**, 054102.
19. T. Frommelt, M. Kostur, M. Wenzel-Schäfer, P. Talkner, P. Hänggi and A. Wixforth, *Phys. Rev. Lett.,* 2008, **100**, 034502.
20. A. Tögl, R. Kirchner, C. Gauer and A. Wixforth, *J. Biomol. Tech.,* 2003, **14**, 197.
21. Z. v. Guttenberg, H. Mueller, H. Habermueller, A. Geisbauer, J. Pipper, J. Felbel, M. Kielpinski, J. Scriba and A. Wixforth, *Lab Chip,* 2005, **5**, 308.
22. S. Shiokawa, Y. Matsui and T. Ueda, *Proc. - IEEE Ultrason. Symp.,* 1989, 643.
23. A. Wixforth, *Superlattices Microstruct.,* 2003, **33**, 389.
24. A. Wixforth, C. J. Strobl, C. Gauer, A. Tögl, J. Scriba and Z. v. Guttenberg, *Anal. Bioanal. Chem.,* 2004, **379**, 982.
25. T. D. Luong and N. T. Nguyen, *Micro Nanosyst.,* 2010, **2**, 239.
26. L. Y. Yeo and J. R. Friend, *Biomicrofluidics,* 2009, **3**, 012002.
27. A. Renaudin, P. Tabourier, V. Zhang, J. C. Camart and C. Druon, *Sens. Actuators, B,* 2006, **113**, 389.
28. C. J. Strobl, Z. v. Guttenberg and A. Wixforth, *IEEE Trans. Ultrason. Ferroelectr. Freq. Control,* 1432, **51**.
29. Z. v. Guttenberg, H. Müller, H. Habermüller, A. Geisbauer, J. Pipper, J. Felbel, M. Kielpinski, J. Scriba and A. Wixforth, *Lab Chip,* 2005, **5**, 308.
30. S. Thalhammer, in SPIE Microtechnologies for the New Millennium, *Proc. SPIE,* 2009, **7364B**, doi: 10.1117/12.820942.
31. M. Schindler, P. Talkner and P. Hanggi, *Phys. Fluids,* 2006, **18**, 103303.
32. M. Schindler, PhD Thesis (Dissertation), University of Augsburg, 2006.
33. M. Streibl, A. Wixforth, J. P. Kotthaus, A. O. Govorov, C. Kadow and A. C. Gossard, *Appl. Phys. Lett.,* 1999, **75**, 4139.
34. T. Frommelt, PhD thesis (dissertation), University of Augsburg, 2007.
35. T. T. Wu and I. H. Chang, *J. Appl. Phys.,* 2005, **98**, 1.

36. Y. Xia and G. M. Whitesides, *Angew. Chem. Int. Ed.*, 1998, **37**, 550–575.
37. G. M. Whitesides, E. Ostuni, S. Takayama, X. Jiang and D. E. Ingber, *Annu. Rev. Biomed. Eng.*, 2001, **3**, 335–373.
38. Y. Xia and G. M. Whitesides, *Annu. Rev. Mater. Sci.*, 1998, **28**, 153–184.
39. G. M. Whitesides, *Nature*, 2006, **442**, 368–373.
40. J. C. McDonald and G. M. Whitesides, *Acc. Chem. Res.*, 2002, **35**, 491–499.
41. Dong Qin, Younan Xia and G. M. Whitesides, Soft lithography for micro- and nanoscale patterning, *Nat. Protoc.*, 2010, **5**, 491–502.
42. T. Franke, S. Braunmüller, T. Frommelt and A. Wixforth, *Proc. SPIE*, 2009, **7365**, 73650O.
43. P. Tabeling, *Introduction to Microfluidics*, Oxford University Press, USA, 2005.
44. H. A. Stone, A. D. Stroock and A. Ajdari, *Annu. Rev. Fluid Mech.*, 2004, **36**, 381–411.
45. R. F. Ismagilov, A. D. Stroock, P. J. A. Kenis, G. Whitesides and H. A. Stone, *Appl. Phys. Lett.*, 2000, **76**, 2376.
46. R. F. Ismagilov, *Angew. Chem. Int. Ed.*, 2003, **42**, 4130–4132.
47. H. Song, J. D. Tice and R. F. Ismagilov, *Angew. Chem.*, 2003, **115**, 792–796.
48. R. K. Shah, H. C. Shum, A. C. Rowat, D. Lee, J. J. Agresti, A. S. Utada, L.-Y. Chu, J.-W. Kim, A. Fernandez-Nieves, C. J. Martinez and D. A. Weitz, *Mater. Today*, 2008, **11**, 18–27.
49. A. S. Utada, L. Y. Chu, A. Fernandez-Nieves, D. R. Link, C. Holtze and D. A. Weitz, *MRS Bull.*, 2007, **32**, 702–708.
50. A. S. Utada, A. Fernandez-Nieves, J. M. Gordillo and D. A. Weitz, *Phys. Rev. Lett.*, 2008, **100**, 014502–014504.
51. L. Schmid and T. Franke, *Lab Chip*, 2013, **13**, 1691–1694.
52. P. Garstecki, H. A. Stone and G. M. Whitesides, *Phys. Rev. Lett.*, 2005, **94**, 164501.
53. A. R. Abate, A. Rotem, J. Thiele and D. A. Weitz, *Phys. Rev. E*, 2011, **84**, 1–5.
54. L. Schmid, D. A. Weitz and T. Franke, *Lab Chip*, 2014, **14**, 3710–3718.
55. A. R. Abate, C. H. Chen, J. J. Agresti and D. A. Weitz, *Lab Chip*, 2009, **9**, 2628–2631.
56. T. Franke, S. Braunmüller, L. Schmid, A. Wixforth and D. A. Weitz, *Lab Chip*, 2010, **10**, 789–794.
57. T. Franke, A. R. Abate, D. A. Weitz and A. Wixforth, *Lab Chip*, 2009, **9**(18), 2625.
58. L. Schmid, A. Wixforth, D. A. Weitz and T. Franke, *Microfluid. Nanofluid.*, 2011, **12**(1), 229.
59. L. Schmid and T. Franke, *Appl. Phys. Lett.*, **104**(13), 133501.

CHAPTER 17

Ultrasound-Enhanced Immunoassays and Particle Sensors

MARTIN WIKLUND[*a], STEFAN RADEL AND JEREMY HAWKES[c]

[a]Dept. of Applied Physics, Royal Institute of Technology, Stockholm, Sweden; [b]Institute of Chemical Technologies and Analytics, Institute of Applied Physics, Vienna University of Technology, Vienna, Austria; [c]Manchester Interdisciplinary Biocentre, The University of Manchester, Manchester, UK
*E-mail: martin.wiklund@biox.kth.se

17.1 Introduction

Many diagnostic methods are based on detecting the binding or immobilization of molecules, cells or other bio-particles onto a substrate. Examples include immunoassays where antigens bind to capture antibodies immobilized on beads in suspension or on the surface of a microplate, and biosensors based on immobilized antibodies against cell membrane-bound antigens for cell or bacteria detection. For clinical diagnostics, it is important to identify antigens or other ligands where the existence or concentration level of the specific antigen can be correlated with a certain disease or health state.[1]

This chapter discusses the utilization of ultrasonic standing waves in order to enhance the performance of immunoassays and cell per particle sensors. In these applications, the function of the ultrasound is to induce particle–particle contact, to concentrate particles or cells, or to deposit particles or cells onto or near a sensor surface. The chapter includes ultrasound-enhanced bead-based immunoassays (Part 17.2), ultrasound-enhanced cell per particle sensors (Part 17.3) and ultrasound-enhanced vibrational-spectroscopy sensing (Part 17.4). The chapter focuses on the bioapplications of the technology, while the fundamental principles and device designs of ultrasonic standing wave particle manipulation are found in Chapters 4 and 6.

17.2 Ultrasound-Enhanced Bead-Based Immunoassays

Enhancement of immunoassays by the use of ultrasonic standing wave manipulation has been demonstrated for two different assay formats: agglutination assays and fluorescence assays, see Figure 17.1. Both these assay formats utilize antibody-functionalized polystyrene beads (Figure 17.1(b)) as a solid substrate for capturing and concentrating the target analyte (Figure 17.1(a)). The binding events are indirectly detected by either measuring the degree of clumping of beads (Figure 17.1(d)), or by measuring the fluorescence light from secondary tracer antibodies (Figures 17.1(c) and (e)). Both assays discussed here are of a sandwich-type, and have been operated in the homogeneous (wash-free) format when enhanced by ultrasound, which makes them simpler and less labour-intensive.

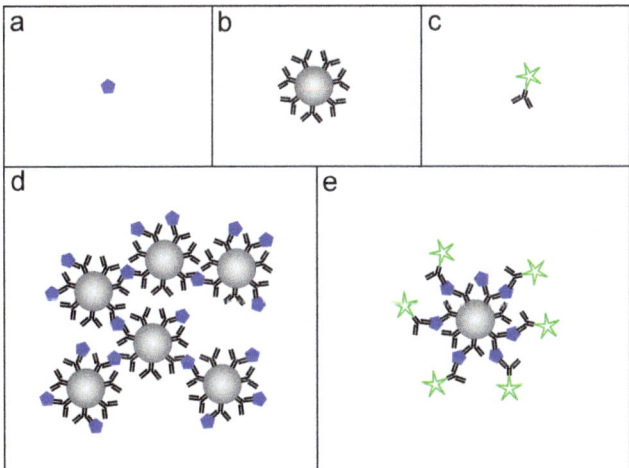

Figure 17.1 The building blocks of a bead-based immunoassay: the analyte (a) bead-immobilized capture antibodies (b) and fluorophore-labeled tracer antibodies (c). The two bead-based assay formats that have been enhanced by ultrasound: the agglutination assay (d) and the fluorescence assay (e). Reproduced from ref. 67.

In this section we will discuss the basic principles of bead-based immunoassays and briefly review the field of ultrasound-enhancement of such assays. In this context, enhancement means to improve the speed and sensitivity of the assay. More detailed reviews of this particular topic are found in refs. 2 and 3.

17.2.1 Agglutination Assays

The agglutination assay was invented in the 1950s by Singer and Plotz.[4] They used 0.8 µm polystyrene beads (latex) coated with human gamma globuline (HGG), and noticed the beads were clumped when mixed with serum from patients with rheumatoid arthritis.[5] Since then, a large number of latex agglutination tests (LATs) have been developed for a range of different diagnostic purposes.[6]

In its simplest format, a sample containing the analyte (Figure 17.1(a)) is mixed with a solution of capture beads functionalized with a suitable analyte-specific receptor molecule, *e.g.* an antibody (Figure 17.1(b)). Initially, this analyte–bead mixture has a uniform color and texture. Following a gentle agitation of the mixture, the interaction between the analyte and the capture beads causes the formation of multi-bead aggregates (Figure 17.1(d)). These aggregates can be detected qualitatively by the naked eye, by observing the change in texture from the uniform to a more turbid appearance of the mixture. Although this simple "yes/no"-format is the most common for commercially available LAT kits, it is also possible to perform quantitative LATs. Methods include turbidimetry,[7] nephelometry,[8] particle counting detectors[9] and flow cytometry.[10]

The major limitation of LATs is the sensitivity. Typically, the detection limit is in the nanomolar range. The reason for this is the relatively small beads (<1 µm) and high bead concentrations needed for obtaining sufficiently high diffusion and bead collision rates, respectively. Furthermore, smaller beads have a higher degree of non-specific agglutination, which increases the background noise.

Ultrasound can be used to enhance latex agglutination tests.[3] The main function of the ultrasound is to increase the probability of bead collisions by pushing and concentrating the beads into the pressure nodes of the standing wave. The most obvious enhancement caused by the increased collision rate is improved speed of the assay (about one to two orders of magnitude).[11] However, if the concentration of capture beads is decreased, an increase in sensitivity is also achieved (between one to three orders of magnitude).[12] Without ultrasound, decreased concentration of capture beads would directly lead to increased incubation times. Thus, there is a mutual benefit of the ultrasound for both the assay speed and the sensitivity.

The device used in ref. 11 and 12 was based on a 2 mm inner-diameter glass capillary inserted into a cylindrical ultrasonic resonator operated at 1.97 MHz or 4.59 MHz. This device, described in more detail in ref. 3, was later commercialized under the name "Immunosonic", Electro-Medical

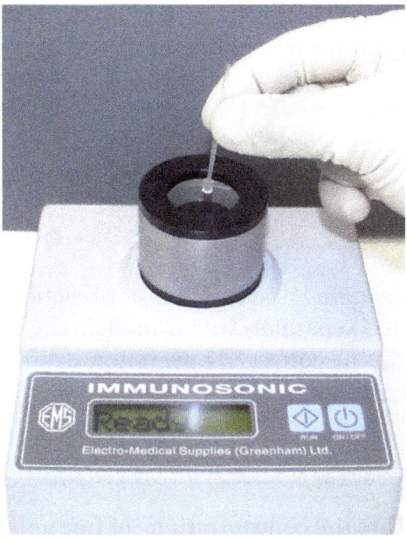

Figure 17.2 The Immunosonic instrument from Electro-Medical Supplies, based on ultrasonic standing wave action on an immuno-agglutination assay. The sample was inserted into a glass capillary which was placed along the symmetry axis of a cylindrical ultrasonic resonator. The ultrasound was used to boost the bead–bead interaction by concentrating the beads into the pressure nodes of the ultrasonic standing wave. More detailed descriptions of the device design and operation are found in ref. 11 and 12.

Suppliers, Wantage, UK, see Figure 17.2. However, despite the benefits of ultrasound-enhanced LATs, the commercial device was outcompeted by standard laboratory culture and PCR-based methods,[13] and is no longer available for sale. A possible reason for this was the choice of target disease (which was meningococcal meningitis) when launching the Immunosonic instrument.[14] On the other hand, ultrasound-enhanced agglutination assays still have great potential and would also be very suitable to implement in the lab-on-a-chip format. Thus, micro-scaled ultrasound-enhanced LATs are yet to be demonstrated.

It is relevant to ask why a decrease in capture bead concentration leads to higher sensitivity in an agglutination assay. This has been investigated both theoretically and experimentally by Wiklund *et al.*[10] They studied the initial stage of immuno-agglutination and developed a model for predicting the probability of both specific and non-specific agglutination per one collision between individual beads and/or small bead clusters. The model is based on combining two reactions: (I) the binding of antigens, A, (which here is considered as the analyte) to the bead-immobilized capture antibodies, Y, and (II) linking of beads or bead aggregates, L_i, to other beads or bead aggregates, L_j (where i and j are the number of beads in each aggregate).

Each of the two reactions can be described by the rate constants, k_{on} and k_{off} (for the antigen–antibody interaction) and k_{ij} (for the bead–bead interaction):

$$A + Y \underset{k_{off}}{\overset{k_{on}}{\rightleftarrows}} = AY \qquad (17.1)$$

$$L_i + L_j \overset{k_{ij}}{\rightarrow} L_{i+j} \qquad (17.2)$$

Since eqn (17.1) is considered as a much faster reaction than eqn (17.2), the two reactions are treated separately in our model. The antigen–antibody rate constants (k_{on} and k_{off}) in eqn (17.1) are often expressed in terms of the dissociation constant, K_D, as:

$$K_D = \frac{k_{off}}{k_{on}} = \frac{[A] \cdot [Y]}{[AY]} \qquad (17.3)$$

where $[A]$, $[Y]$ and $[AY]$ are the concentrations of free antigens, free antibodies and bound antigen–antibody complexes at equilibrium, respectively. The bead–bead rate constant (k_{ij}) in eqn (17.2) is based on von Smoluchowski kinetics[15] and is in our model treated as an irreversible interaction, and can be expressed as:[10]

$$k_{ij}(\alpha, \beta) = \left(\frac{1}{\alpha + ijc_{ij}\beta} + 1\right) \frac{8k_B T}{3\eta} \qquad (17.4)$$

Here, k_{ij} is a function of the probabilities of non-specific and specific agglutination at each particle collision, α and β, respectively. Other parameters in eqn (17.4) are the Boltzmann constant (k_B), the temperature (T), the viscosity of the medium (η) and the cluster-surface steric hindrance coefficient (c_{ij}). The latter coefficient is defined as $0 < c_{ij} < 1$, and is introduced in order to take into account that not all surface of the beads in the cluster is available for contact with another cluster. This coefficient must be numerically determined by averaging all possible geometrical configurations of beads in a cluster.[10]

The non-specific agglutination probability (α) is assumed to be constant for a given assay protocol. The specific agglutination probability (β), on the other hand, can, for a given antigen–antibody interaction (*cf.* Eqn (17.1) and (17.3)), be treated purely geometrically as:[16]

$$\beta = \frac{b^3}{8R^3} \cdot \frac{K_D/Y}{(1 + K_D/Y)^2} N_{max}^2 \qquad (17.5)$$

where b is the radius of the binding site of the antibodies (assumed to be circular), R is the radius of the beads and N_{max} is the binding capacity of the beads. If the size of the analyte is small compared to the average distance between the binding sites, then N_{max} is the same as the number of antibodies on each bead.

Let us now use eqn (17.1)–(17.5) in order to define a theoretical detection limit (*i.e.* sensitivity) of the agglutination assay. This limit can be defined as the analyte concentration when $\beta = \alpha$, *i.e.*, when the amounts of specific and non-specific and agglutinations are equal. If we assume a high-affinity immunoassay ($K_D \ll A_0$ and $K_D \ll Y_0$), this minimum analyte concentration, $A_{\min}$, can be shown to be:[10]

$$A_{\min} = \frac{1}{2} n_0 N_{\max} \left(1 - \sqrt{1 - \frac{4\alpha(Y_0 - A_0)A_0}{\beta Y_0^2}} \right) \quad (17.6)$$

where n_0 is the initial bead concentration, and Y_0 and A_0 are the initial concentrations of bead-bound antibodies and analyte, respectively. Interestingly, eqn (17.6) shows that the detection limit, $A_{\min}$, is proportional to the bead concentration, n_0. Furthermore, we see that the practical limit is defined by the ratio between the probabilities of non-specific and specific agglutinations (α/β). In summary, ultrasound-enhanced agglutination assays are simple and straightforward, but are in practice limited to assays with analyte concentrations not lower than about 100 pM.[3]

17.2.2 Fluorescence Assays

An alternative to the immuno-agglutination assay is the immuno-fluorescence assay, see Figure 17.1(e). This assay also uses antibody-functionalized polystyrene beads (Figure 17.1(b)) for capturing and concentrating the analyte (Figure 17.1(a)). However, instead of relying on bead–bead interactions and clumping, the fluorescence assay uses fluorescently labelled secondary antibodies, or tracer antibodies (Figure 17.1(c)), that bind to another epitope of the analyte. The analyte is then quantified by measuring the fluorescence intensity from each bead. The bead-based immuno-fluorescence assay is particularly suitable for multiplexing. This can be done by encoding the bead with various amounts of two or more fluorophores, a technology called suspension array technology (SAT).[17] The read-out instrument for SAT is typically a flow cytometer, and this method has also been commercialized (Luminex Corporation, Austin, Texas, USA). There also exists read-out technology performed in bulk solution, *e.g.* by the use of confocal fluorescence microscopy[18] or two-photon excitation (TPX) technology.[19]

Just as for the immuno-agglutination assay, the sensitivity of a fluorescence assay is dependent on the concentrations of beads in the suspension. This is a consequence of the concentration of capture antibodies attached to the beads, relative to the analyte concentration in the sample. In 1989, Ekins proposed the concept of ambient analyte immunoassay,[20] for which he demonstrated an increase in sensitivity by using very low concentrations of capture antibodies (<0.01 K_D). Ekins applied this concept to microspot array technology,[21] where the total concentration of capture antibodies was very small, but locally concentrated to small spots on a solid substrate. The main principle of ambient analyte condition is to increase the fractional occupancy of the capture antibody binding sites by using an excess of analyte and tracer antibodies, resulting in a fluorescence intensity dependent on analyte concentration only.

Ultrasound can be used for improving the sensitivity of bead-based fluorescence assays. However, the improved sensitivity is not a direct consequence of concentrating beads by ultrasound, but rather a consequence of applying ambient analyte condition to the bead-based immunoassay. Since the ambient analyte condition results in very low concentrations of capture beads (typically a few beads per µL), a method for concentrating the beads and delivering them to the detection site is needed. This has been demonstrated by Wiklund et al.,[18] who used ultrasonic standing wave manipulation on a 96-well plate platform combined with confocal fluorescence detection for enhancing the sensitivity in a bead-based immunoassay, see Figure 17.3. In their set-up, the beads where positioned into a set of planar, compact monolayers matching the horizontal x–y scanning plane of the laser focus of the confocal microscope. These monolayers were positioned into the pressure nodes of the ultrasonic standing wave. Thus, a benefit of this method is the possibility to predict the vertical position of the monolayers (each separated by half the acoustic wavelength), and use a time-efficient scanning procedure. Typically, the device used in ref. 20 used ~2500 beads of size 3 µm in a 100 µL sample volume in one of the wells in the 96-well plate (corresponding to a volume fraction of 10^{-7}), which were rearranged into 6 monolayer aggregates (each separated by half the acoustic wavelength = 190 µm at 4 MHz). The bead enrichment efficiency (i.e. the ratio between trapped and total number of beads) was about 10%, corresponding to ~50 trapped beads in each aggregate. Since the laser focus of the confocal microscope could be tuned to match the size of the beads (3 µm), the local bead concentration in the scanning area was 10^6–10^7 times the initial concentration. This is 10^3–10^4 times better enrichment than would be achieved with simple centrifugation or sedimentation to the bottom of the 96-well plate.

The assay used to test the potential of ultrasound-enhanced fluorescence assays was the human thyroid stimulating hormone (hTSH) assay.[18] This assay used bead-immobilized capture antibodies and fluorescent tracer antibodies with affinity constants ($K_A = 1/K_D$, cf. Eqn (17.1)) of 2×10^{10} M^{-1} and 1×10^{10} M^{-1}, respectively. These antibodies were directed against two different epitopes of the TSH molecule. The estimated limit of detection was as low as 20 fM when the assay was enhanced by ultrasound. It is interesting to compare this limit with the expected number of analyte molecules bound to each bead. The experiment was carried out with ~2500 beads in the 100 µL sample, which means that there are ~480 TSH molecules per bead in the sample at the analyte concentration 20 fM. This number can be compared with the theoretical detection limit[10] of the TSH assay, which for the affinity constant ($K_A = 10^{10}$ M^{-1}) and bead capacity (10^6 capture antibodies per bead) corresponds to ~130 TSH molecules bound to each bead. This means that about 30% of the analyte is expected to be bound at this limit. This is in good

Ultrasound-Enhanced Immunoassays and Particle Sensors 427

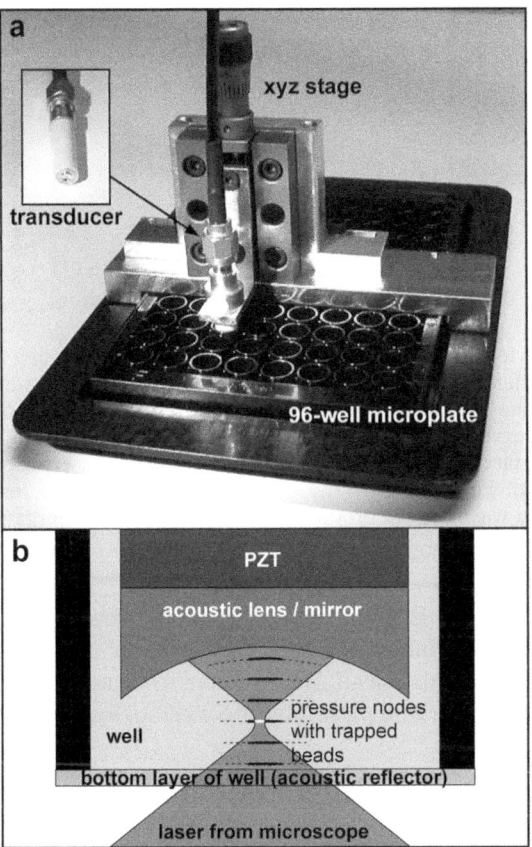

Figure 17.3 The 96-well microplate platform for ultrasonic enhancement of the bead-based human thyroid stimulating hormone assay developed by Wiklund et al.[18] (a) The 4 MHz transducer (upper left corner) was submerged into one of the wells of the 96-well plate by a *xyz* precision translation stage. (b) The beads were concentrated into the pressure nodes of the standing wave (dotted lines) formed by reflections in the acoustic lens/mirror (front part of the transducer) and the bottom layer of the well. Each pressure node was then scanned by an inverted confocal microscope. The figure is taken from ref. 18.

agreement with the detection limit of the confocal fluorescence system used, which was about 100 fluorophores per bead.

17.3 Ultrasound-Enhanced Particle Sensors

This section describes in general terms the phenomena of attracting particles onto a wall by the use of ultrasound. A specific example is described in Section 17.4. This attraction wall, like the particles described in Section 17.2,

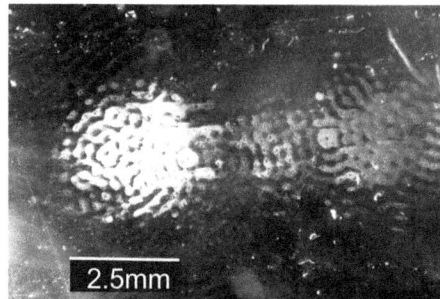

Figure 17.4 Device B in Table 17.1. Clumps of *Bacillus subtilis var. niger* (BG) spores adhered to a microscope slide coated with an anti-BG antibody. The white areas are regions where spores have adhered to the surface. The slide formed a half wavelength thick (at 2.9 MHz) attraction surface over a 10^8 ml^{-1} suspension of spores, exposed to sound for 5 min. Pictured after the glass slide was removed from the chamber and washed.[22] Reproduced with permission of Pergamon.

may be coated with an immuno-selective agent. The terms "pushing" and "attraction" are interchangeable for the action of a standing wave force carrying particles towards a wall. Here the term "attraction" is used because we are interested in the role played by the wall surface.

17.3.1 The Distinctive Pattern of Clumps in Contact with a Surface

When particles are attracted to a wall by ultrasound, they form clumps in a quite distinctive pattern, which is seen in Figure 17.4. In this example,[22] spores were pushed onto an antibody coated glass microscope slide. The same close arrangement of clumps is seen with yeast cells on a 50 μm thick polystyrene film in Figure 17.5(a); this chamber had no sides so that the water depth could be adjusted. When the water depth was significantly greater than ¼ wavelength, another pattern of clumps appeared. These clumps were more widely spread, they were suspended in the fluid and were not in contact with any surface. The systems in Figures 17.4 and 17.5 are very different, yet, when the water layer thickness was ¼ wavelength or less, both produced the distinctive close pattern of clumps against the wall surface.

17.3.2 The Need for a Cell Attracting Wall in Microsystems

For continuous on-line environmental monitoring for the unlikely appearance of pathogen cells, the addition of highly specific immuno-coated particles for detection as described in Section 17.2 is not a reasonable option since those particles would need to be selectively retained over long periods, requiring continuous extraction from inhomogeneous environmental

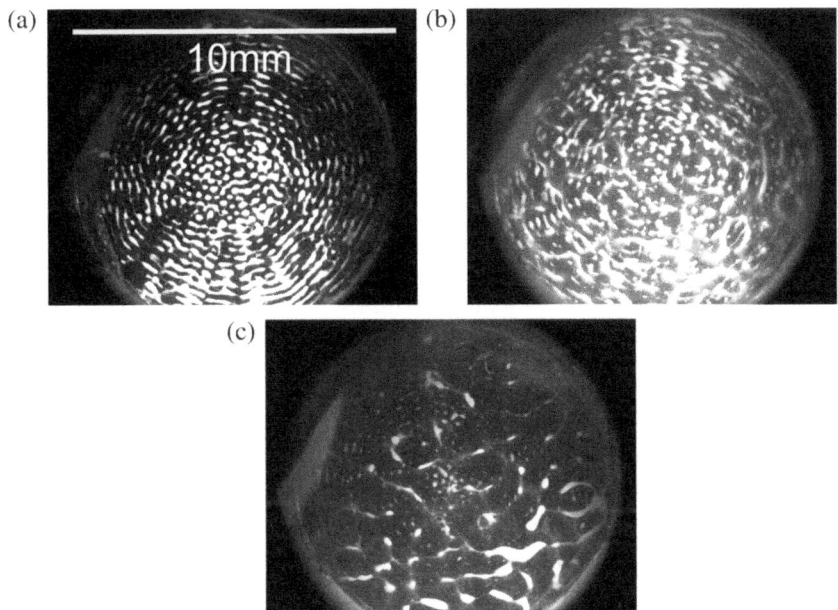

Figure 17.5 Device G in Table 17.1. Yeast cells (at 10^7 ml^{-1}) in a thin layer of water between a vibrating (1 MHz) microscope slide and a 50 μm thick polystyrene film, (white areas are cell clumps). The water layer thickness was adjusted with a micrometer stage to move the film. The acoustic wavelength (λ) in water at 1 MHz was 0.75 mm. (a) Water thickness <λ/4, clumps in contact with the surface. (b) Water thickness >λ/4, some clumps in contact with the surface and some free in the suspension. (c) Water thickness ≫ λ/4, most clumps free in the suspension. Reproduced from ref. 67.

samples. A solution for longer term monitoring is to place the immunoselective element on the chamber wall. However, antibody coated, cell specific, "biosensor" surfaces rely on the cells landing on the surface and since, under gravity alone, the terminal velocity of 2 μm diameter cells is in the region of 2×10^{-7} ms^{-1}, small cells often take minutes or hours to make contact with the surface. The lack of sedimentation is exacerbated by both Brownian motion, which has a significant influence on particles <2 μm diameter, and a shear-force-driven migration towards the central axis of microfluidic channels, which has the greatest influence on particles >2 μm diameter especially at high particle concentration.[23] It is because small cells in passive systems do not quickly make contact with walls that antibody coated surfaces are rarely used for rapid cell detection. Ultrasound provides a solution, typically attracting cells to a surface in less than a second in sub-mm channels. At the surface, antibodies can adhere to specific cells and after the sound is turned off or switched to a repelling frequency, unattached cells flow away and the remaining cells can be counted.

The method has some limitations: when the attraction surface is more than a few millimetres across, cells always collect into clumps. The clump separation is near to half a wavelength in the fluid; the pattern is empirically predictable but its physical origin has not been fully investigated. Systems developed to use the phenomenon of attracting particles to a surface must accept this limitation of non-uniform cell coverage. Another limitation which must be appreciated is that currently, the smallest cells that can be reliably manipulated with ultrasound standing waves are just over 1 µm diameter due to competing acoustic steaming drag forces. Therefore many, but not all, bacteria can be manipulated by this method.

17.3.3 Device Design

Possibly the most important principle of cell manipulation in ultrasound standing waves is that most biological cells suspended in aqueous media move to pressure nodes. For particle sensors where cells are attracted to a wall, a pressure node must be placed on the wall, but at the interface between a fluid and a sound-reflector (a non-moving wall), standing waves always form a pressure antinode. The requirement for a pressure node (velocity antinode) at the interface conflicts with this usual situation at an interface. Despite this difficulty, the feat of allowing the wave's velocity antinode to straddle an interface between two materials has been achieved by several different methods; successful solutions to the problem are illustrated in Table 17.1 and the properties of each attractor wall is briefly described below (further construction details are given in Chapter 7 and in the paper by Glynne-Jones *et al.*[24]). With the exception of the first method, all have used a water thickness of less than a ½ wavelength to allow only one pressure node in the system.

17.3.4 The Cell Attractor Wall

A. *A flexural plate wave* on a 3 µm thick attractor plate,[25,26] which creates evanescent displacements because the sound velocity in the thin plate is less than the velocity in the water. Displacements do not propagate through the fluid, they simply cancel with adjacent opposite phase regions on the plate.[35] The cells are attracted to the displacement antinodes on the surface.

B. *A half wavelength attractor wall.*[22,24,28,29] Driving the chamber at the resonance for the attractor wall produces a pressure node at the liquid interface to attract cells. This active pressure release method was demonstrated 10 years ago and has been developed further but it should be noted that the operating frequency is very narrow because of the need for simultaneous resonances in the attractor wall, the PZT and their combination with the water. This has encouraged the search for less demanding alternatives.

C. *A thick polymer attractor wall.*[32] Making use of the sound absorbing properties of polymers, a thick absorbing layer rather than a resonant attractor wall allows some travelling wave components that displace cells towards the liquid–attractor wall interface. To date, this system has been designed for manipulating cells a short distance from the wall; some design alterations would be needed to adapt it to become an attractor wall. The system is transparent and the viewing direction is conveniently parallel to the pressure node plane.

   D. *A thin attractor wall*.[34] An air interface displacement occurs at the surface. The result is the same as with a resonating half wavelength wall, however, without the need for resonance, it operates over a wider frequency range than the half wavelength wall.

   E. Sound *introduced from the side*[27] brings new engineering possibilities. The figure also shows that both symmetric (right hand side) and antisymmetric (left hand side) waves can be present in one system.

   F. *A rigid horizontal base,* vibrations introduced to the fluid by a non-evanescent flexural plate.[30,31] Vertical pressure nodes form and cells sediment down to the base.

   G. *A thin membrane driven by a thicker flexural plate wave*.[33] This is a combination of the thin wall (D) and the sideways introduction of sound (E). The end-on-drive creates a system where the PZT sound source has only minimal coupling to the chamber avoiding many interference problems, which always occur when two surfaces are bonded.

1D models have been used to develop some of the systems A–G. These, of course, cannot be used to explain the clumping patterns observed across the surface (Figures 17.4 and 17.5) and further model development is needed if we are to control the clump locations. This author believes that evanescent acoustic waves may be involved in the attraction of particles towards the walls and the formation of the clump patterns; evanescent waves are certainly possible for the membrane walls: Oberti[31] has calculated that at MHz frequencies, wavelengths suitable for evanescent fields to occur can be produced when the wall thickness is less than 300 µm, which is thicker than the membrane walls currently used.

17.3.5 Immuno-Based Selective Cell Capture and Detection by Light

When the attracting wall of the chamber is coated with an antibody, cells or particles with the antigen on their surface become attached when they make contact. After switching the sound off, cells without the antigen are easily washed away. To quantify the number of attached cells, the chamber can be taken apart and the attraction plate examined under a microscope.[22] This method is cumbersome and the mechanical disturbance risks losing many positive binding events. Here we describe a method based on scattered evanescent light to continuously detect the number of cells at the surface.

   The arrival of cells at a surface can be continuously observed with a microscope, but this does not distinguish between cells at a surface and those near a surface. The distinction becomes important when determining whether a cell is bound to an antibody on the surface or is floating just above

Table 17.1 Attractor wall types: red shading/curves indicates the acoustic velocity in the attractor wall. PZT poling direction indicated by grey shading. Light blue indicates the fluid layer. Flexural waves are shown where described in the publications.

Particle attraction walls in experimentally proven systems (simplified cross section views)

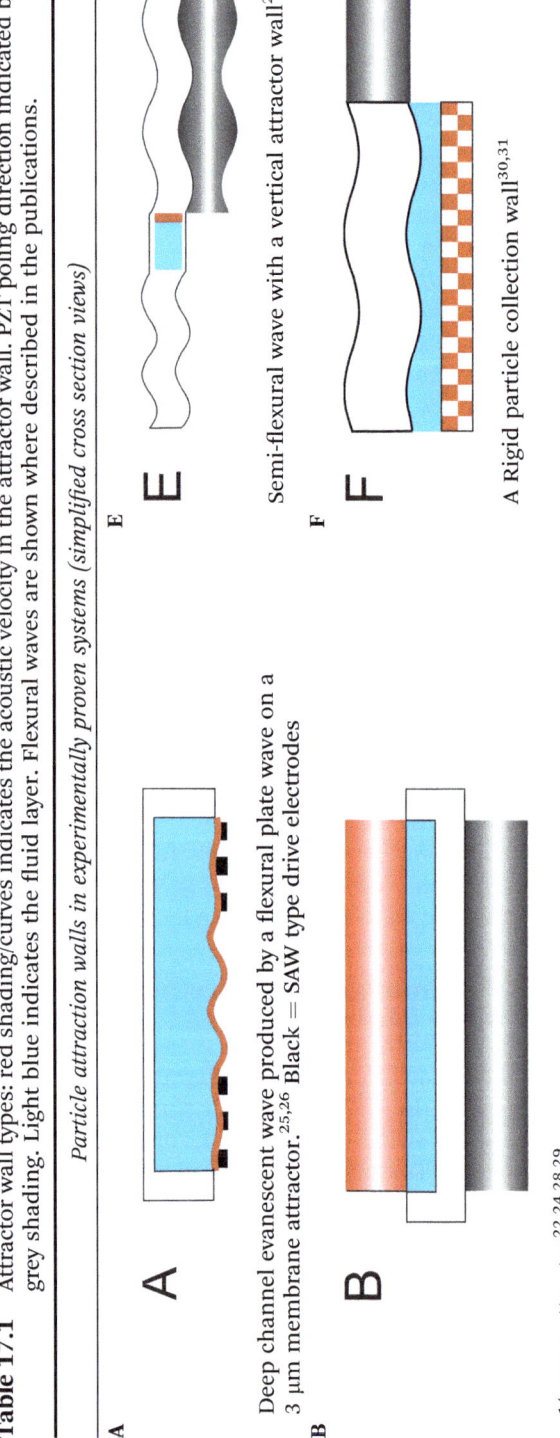

A

B

½ wave attractor[22,24,28,29]

Deep channel evanescent wave produced by a flexural plate wave on a 3 μm membrane attractor.[25,26] Black = SAW type drive electrodes

E

Semi-flexural wave with a vertical attractor wall[27]

F

A Rigid particle collection wall[30,31]

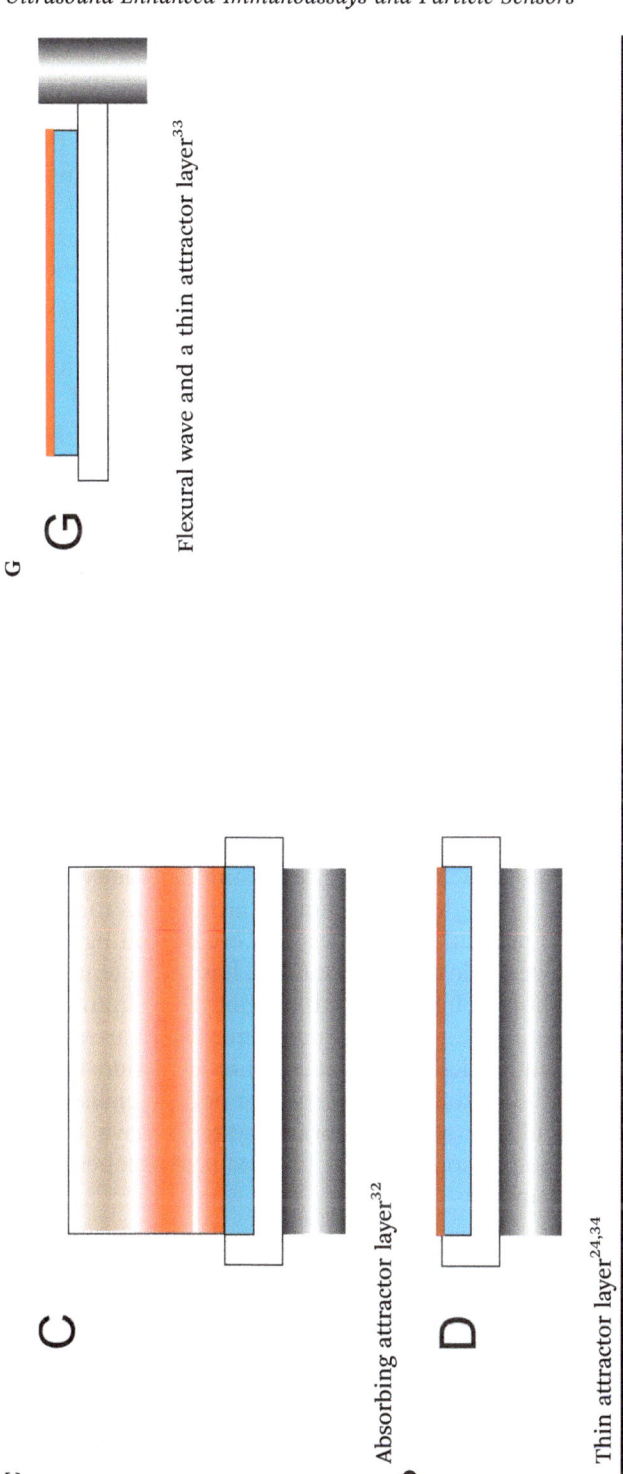

it. Evanescent light has a decay length of $1/2\pi$ light wavelengths except when light is introduced near the critical angle for total internal reflection (TIR) where the decay length extends slightly further. Evanescent light can therefore be used to selectively illuminate only surface features. This thin evanescent field is used to illuminate molecules in surface plasmon resonance but it is not suitable for cells the size of bacteria (~1 μm diameter) and above; for these large particles, a slightly thicker field is created using a metal-clad-leaky-waveguide (MCLW).[36] In these systems, the metal acts to push the wave further into the fluid. The metal also reduces light loss to the glass substrate producing a sharper waveguide mode.[37]

The evanescent light guide has been integrated with the acoustics to form a cell detector where ultrasound brings the cells to the surface of the detector.[29,34] Evanescent light skims along the interface between water and a waveguide; the magnetic component remains in the guide and part of the electric component moves in the fluid, without propagating away.[36] The waveguide is a thin layer, usually a solgel spin coated onto a semi-transparent metal coating on a support substrate, usually glass. Laser light is introduced to the waveguide through a prism and index matching fluid or grating at an angle, which achieves total internal reflection. Particles on the surface disturb the propagation, destroying the evanescence and displacing light out of the waveguide. The scattered light is easily detectable against a dark background. BG spores attracted to a surface have been detected with the MCLW guide on a glass microscope slide surface at a concentration down to 10^3 ml^{-1}.[29] Similar results have been obtained by Glynne-Jones *et al.*[28] using potassium ion waveguides whose field penetrates only 100 nm into the fluid. They have also had success by introducing light simply through the roughened edge of a microscope slide, which is clearly not a single mode system.

The combination of three technologies: ultrasound to attract cells to walls; antibodies to selectively retain particular cell types; and evanescent light for quantifying the attachments even at low concentrations, has the potential to provide a very rapid cell detection tool. However, this is not yet a commercial system primarily because when a half wavelength attractor wall is used, the narrow band of operating frequencies for the ultrasonic attraction is too highly dependent on temperature and fluid properties for non-specialist use. Recent developments[34,38] of plastic and membrane systems have much broader operating frequency ranges and these are likely to be developed into easily used cell detection systems.

17.3.6 The Next Stage of Developments

The three way combination of attracting particles to a surface with ultrasonic standing waves, antibody coated surfaces and evanescent light detection, is a strong contender for the development of label free rapid cell characterization to monitor environmental or medical samples. However, since antibodies provide the selectivity and they have a short life, this has taken the developments towards very low cost disposable units;[24,28,38] an example of the

Figure 17.6 Very low cost chambers have been made[33] using vacuum forming techniques (also used for low cost blister packaging). Held with a clip against a transducer, as shown in Table 17.1 G. Reproduced from ref. 67.

disposable section is shown in Figure 17.6. In the future we expect to see this extended to multi-antibody arrays and controlled cell movement between regions.

17.4 Ultrasound-Enhanced Vibrational Spectroscopy

The techniques of ultrasonic particle manipulation were exploited in the regime of vibrational spectroscopy sensing applications. Manipulation in the given context means to have control over the spatial distribution of suspended particles in respect to where particles are driven to (and concentrated) when the suspension is exposed to an ultrasonic plane standing wave and on the other hand to generate regions depopulated of particles, that is locations where no particles will be found when radiation forces are exerted. Particles used for the various examples were all particles as such, hence were solid and therefore denser and harder than the liquid. As a result the position of agglomeration of the beads, crystals and yeast cells was always the pressure nodal region of the respective ultrasonic standing wave.[24,39] The set-ups used throughout this section were layered resonators, *i.e.* stacks of parallel sheets of different materials (PZT, glass or aluminium carrier, suspension liquid, reflector). The excitation frequency always was around 2 MHz.

Vibrational spectroscopy[40] comprises a group of optical measurement techniques increasingly popular in process analytical chemistry because of the ability to directly provide molecular-specific information about a given sample under investigation. The term "vibrational" refers to the measurement of frequencies of periodic motions of atoms in a molecule. The excitement of these modes is governed by quantum mechanics; on a less sophisticated level we see the interaction of the molecule with photons of certain wavenumbers. The wavenumber is the reciprocal of the light frequency; it is specific for a given molecule due to its dependency on atoms or atom groups and their chemical bonds. In a nutshell, the spectrum of light collected after a sample has been exposed to electro-magnetic excitation carries specific information about the sample's chemical composition.

Such spectra can be recorded from solids, liquids and gases, providing qualitative as well as quantitative information on the chemical composition of the sample. In the context of real-time process analytical chemistry, advances in the field are of interest as a great part of the desired (bio) chemical information can be extracted from the recorded spectra. Spectra can be measured in various ways. Methods such as absorption spectroscopy (in the near or mid-infrared) or Raman spectroscopy are exploited.

Both ultrasonic and vibrational spectroscopy techniques are at a stage of development at which crucial improvements can be expected in the near future, while at the same time mature enough to be endorsed in industrial environments. The presented examples of incorporating an ultrasonic standing wave field in different vibrational spectroscopy sensing applications have been successful and will be shown to be advantageous in various respects. The potential of influences due to the combination of the two techniques, such as cross-sensitivities or cross-talking caused by driving electronics, was given close attention, however, no adverse effects were observed throughout the experiments.

17.4.1 Agglomeration of Crystals for Raman Spectroscopy

In Raman spectroscopy,[41] a sample is irradiated with a focused laser beam and the rare inelastically scattered photons are recorded. The measured intensities at different wavenumbers are termed Raman spectrum, providing information on vibrational transitions characteristic for the chemical composition of the sample under study. The spatial resolution of Raman micro-spectroscopy in the low-micrometre scale and its ability to probe samples under *in vivo* conditions allow for new insights into living single cells without the need for fixatives, markers, or stains.[42,43]

However, the quality of measurements is highly dependent on the number of molecules exposed to the incoming light, which is *e.g.* of special relevance when monitoring reacting suspensions, where Raman spectroscopy holds great promise due to possible non-invasive measurement strategies. It turns out to be difficult for standard on-line Raman spectroscopy to discriminate between Raman photons from the solid matter and signals originating from the pure liquid phase. This problem is of special relevance in the case of low concentrations of suspended particles here.

Means to concentrate samples in the light path are regularly used. Acoustic levitation in air for monitoring containerless chemical reactions has been used traditionally in Raman micro-spectroscopy.[44] Recently, the investigation of red blood cells and micro-organisms with this technique was reported.[45]

The combination of ultrasonic particle manipulation and confocal Raman micro-spectroscopy is a novel approach to increase selectivity and sensitivity of on-line Raman measurements of suspensions. Aggregating particles in suspension by an ultrasonic standing wave can provide an experimentally simpler approach to immobilize and manipulate micro-particles. They may be deliberately concentrated or removed from the Raman measurement spot, thus allowing the selective measurement of the liquid and solid phases, respectively. In addition to the improved selectivity, an increase in sensitivity may be expected due to the local enrichment of the particles.

A resonator comprising a PZT ceramic glued to the side of a small glass cuvette (2 mL) was used to control the spatial distribution of suspended theophylline particles relative to the light path of the Raman microscope, *i.e.* agglomerates of particles were deliberately positioned within and out of the focus of the instrument.[46]

The set-up was placed under the Raman microscope with sound propagation direction perpendicular to the light path (Figure 17.7, top). Thus, the pressure nodal planes where oriented parallel to the incident beam, allowing the control of their locations relative to the light path by slightly changing the excitation frequency, for instance to the next resonance frequency. Illustrating the influence of the radiation forces on the spatial distribution of particles, a light micrograph of the agglomeration of theophylline crystals is shown in the enlargement at the bottom of Figure 17.7.

The results of investigations of dissolved as well as suspended theophylline are depicted in Figure 17.8. The aim was to compare the Raman signal of homogeneously suspended theophylline crystals (black line) with measurements of agglomerates brought about by the ultrasonic standing wave (black with dots). Figure 17.8 shows a significant increase (3 to 6-fold) of scatter intensity when the ultrasonic field was applied. Moreover, the data suggest better resolution, *e.g.* between wavenumbers 1600 and 1700 cm^{-1} when the theophylline crystals were concentrated by the ultrasonic standing wave.

In contrast, no significant differences were found for regions where no particles are present (in the velocity nodes). The Raman spectrum of a crystal-free theophylline solution (Figure 17.8, grey line) was not different from a measurement taken with the optical focus positioned within this depleted region (Figure 17.8, grey with dots), hence the very low concentration of particles between the nodal planes enables one to specifically take measurements of the host liquid's composition.

When comparing sediment—hence supposed to be packed tightly—theophylline crystals with agglomerates brought about by the ultrasonic radiation forces, a slight increase of scatter intensity was measured when the ultrasonic field was present (data not shown). Very similar results were obtained using yeast cells as a model for bio-suspensions; again, the Raman measurements indicated that the agglomeration of particles in suspension by the ultrasonic field was a suitable means to achieve higher Raman scatter intensities.

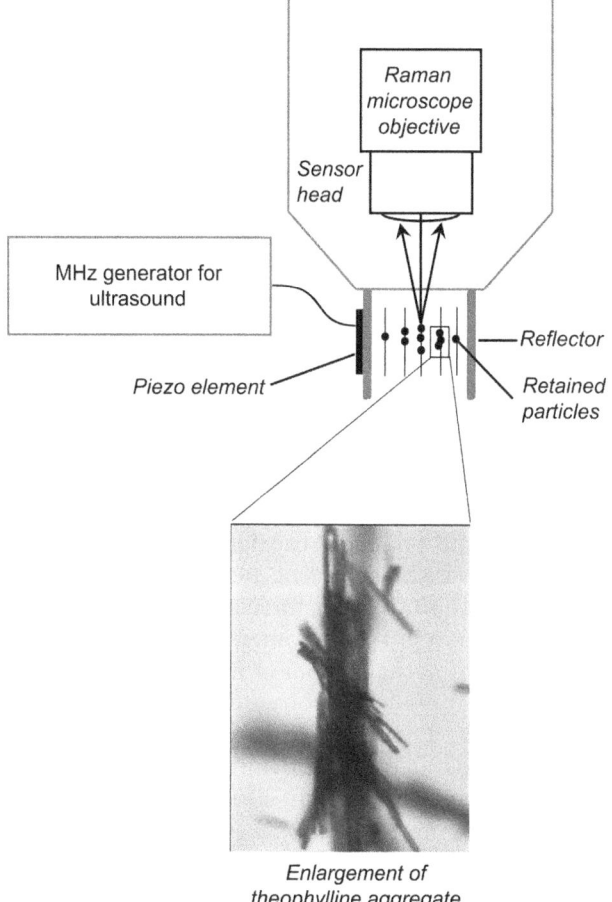

Figure 17.7 Raman microscope at the top with light path into cuvette resonator (right). The picture at the bottom shows the theophylline crystals agglomerated by the ultrasonic field. The figure is taken from Radel et al.[46]

Another study aimed at the strategy of trapping particles by radiation forces in a flow cell allowed for exposure to a variety of reactants and thus the execution of a given set of chemical reactions while being continuously monitored by Raman micro-spectroscopy. The potential of this ultrasonic trapping method to monitor on-bead chemical reactions was investigated with an example of automated on-line generation of a surface-enhanced Raman (SER)-active layer on ultrasonically trapped beads.[47] The process included steps of preparation and subsequent recording of SER spectra of different analytes during which the beads were retained in the focus of the Raman microscope. Furthermore, the discharge of the silver-loaded beads by switching off the ultrasonic field is advantageous in comparison to previously

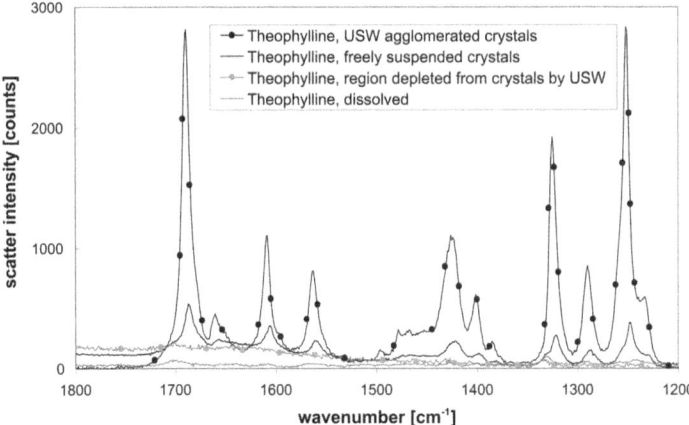

Figure 17.8 Raman spectra of theophylline solution (grey) and freely suspended theophylline crystals (black) in comparison to theophylline crystals agglomerated by ultrasound (black with dots) and the theophylline solution in a region where the crystals were depleted by the ultrasonic standing wave (grey with dots). The figure is taken from Radel et al.[46]

described procedures that also employed beads as carriers for SER substrates.[48]

A broad variety of assays can be performed with this application in many fields, among which solid-phase synthesis and bead-based bioassays can be envisioned. The robust but gentle handling also allows for prolonged studies on beads as well as on living cells.

17.4.2 Enhancement of Stopped Flow FT–IR ATR Spectroscopy

The growing use of bioprocesses as a manufacturing route for, *e.g.*, antibiotics and other medical compounds, enforces the development of reliable, automated sensors for bioprocess monitoring. Such sensors are a key for optimal system performance[49] as continued analysis is needed in order to control the bioprocess in question. State-of-the-art sensor systems provide information on physical parameters (pressure, temperature) but only a few chemical parameters such as *e.g.* pH, oxygen and carbon dioxide concentration in liquid and gas. Real-time information on the chemical composition and on the physiological status of the employed microorganism would be of high diagnostic value. Fast response times of at least one order of magnitude faster compared to the generation time of the observed microorganism are necessary.[50] Such sensors are not available so far due to experimental difficulties, however, the development of new sensor designs is being triggered by the rapid progress in biotechnology.

Infrared (IR) absorption spectroscopy is a well-developed method in chemical analysis.[51] The incident IR radiation excites parts of the molecules in the sample, therefore, a certain amount of light energy at a given light wavelength is converted into vibrational energy, hence is absorbed. The acquisition of an IR absorption spectrum can be conducted in minutes to seconds by Fourier transform infrared (FT–IR) spectroscopy, delivering specific molecular information about the sample in the optical pathway.[40] Constantly, new devices and concepts for advanced chemical analysis have been developed during recent years.[52]

FT–IR spectroscopy in combination with attenuated total reflection (ATR) sensing elements is a currently developing, very promising means for process and bioprocess monitoring. The ATR scheme exploits the occurrence of total reflection of light at the interface of two media with different optical densities. FT-IR ATR spectroscopy is a surface sensitive technique, the detection range is only a few μm. Any substance covering the said interface influences the incoming light at certain wavenumbers and thus specific information about its chemical composition can be obtained from the absorption spectrum. Due to the exponential decay within the evanescent field, the closer the sample is located to the ATR surface the higher its contribution to the recorded spectrum, whereas almost no absorption takes place at greater distances. Therefore additional measures are necessary to bring a sufficient amount of sample, *e.g.* suspended particles, into this region.

The limited detection range is especially advantageous when evaluating aqueous samples in the information rich mid-IR spectral range. Due to the strong infrared absorption of water, the optical path must be short (<10 μm) to keep the detection limit low in common FT-IR spectrometers using thermal radiation emitters. Short optical path lengths like this are realized by the ATR technique without putting geometrical constraints on the sample volume.[53]

The surface sensitivity opens the possibility to measure the spectra of suspended particles (cells) and the supernatant (*i.e.* the liquid component). In the case of the stopped flow technique[54]—see Figure 17.9 for a sketch of the flow cell used—the basic idea is to keep cells (or other particles) from the horizontal ATR surface by pumping the suspension through a detection cell in bypass while the spectrum of the suspending liquid is measured (Figure 17.9(a)). When the throughput is stopped, the cells settle on the surface into the evanescent field of the ATR. Then a spectrum can be taken, representing the chemical composition of the cells constituting this layer of sediment (Figure 17.9(b)). In case of cells such as *E. coli,* the cleanness and thus the sensitivity of the ATR element could be maintained by periodic rinsing with $NaHCO_3$ (Figure 17.9(c)).

However, when trying to measure glucose and ethanol in a baker's yeast fermentation with the stopped flow technique, the formation of a bio-film on the horizontal ATR element was observed.[55]

The formation of biofilms is a widely known problem in medically, biochemically and industrially used sensors and filters.[56] In the regime of the ATR technique it poses a serious problem due to its surface sensitivity. The

Ultrasound-Enhanced Immunoassays and Particle Sensors 441

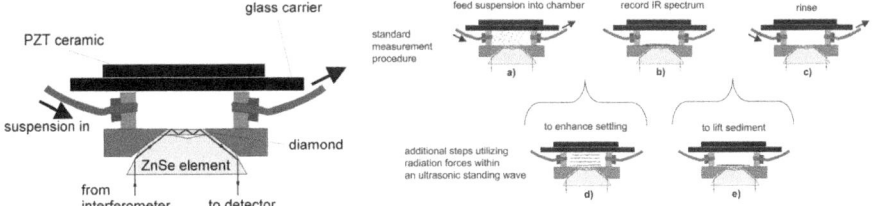

Figure 17.9 *left*: Flow cell comprising the ATR element at the bottom and the PZT-sandwich transducer at the top. *right*: Stopped flow technique to specifically measure the IR absorbance of suspended particles: the suspension is pumped into the detection volume (a). When the flow is switched off, particles settle onto the ATR surface and the spectrum is recorded (b). After the measurement, the cell is rinsed (c). An ultrasonic standing wave was applied to accelerate the measurement time by agglomerating yeast cells prior to the settling (d) and to improve the cleaning by actively lifting the sediment from the ATR prior to the rinse (e). The figure is taken from Radel et al.[57]

upper left graph in Figure 17.10 shows the influence on selectivity and sensitivity of a bio-film on measurements recorded in a model yeast suspension every 30 min. The upper right hand side graph in Figure 17.10 represents the growth of the bio-film at the respective times (details in ref. 57). Reliability and robustness are key necessities when industrial applications for on-line fermentation monitoring are envisaged. Therefore the findings are detrimental; a thorough cleaning protocol is needed to prevent or reduce biofilm formation to a minimum.

To overcome this detrimental contamination of the sensor, an ultrasonic transducer was used as the lid of the set-up, the ATR element at the bottom represented the reflector. Consequently, a standing wave with horizontal nodal planes was built up, enabling the manipulation of particles (cells) within the flow cell. The radiation forces exerted on particles were used to actively lift the material covering the horizontal ATR sensor surface during rinsing as shown in Figure 17.9(e) and thus improving its cleanness.

The application of ultrasound performed well (see spectra in Figure 17.10, lower graphs). More than 70% of the protein and 90% of carbohydrates could be removed. The reason for this gap in cleaning performance was interpreted to be a result of the different particle sizes. High carbohydrate contents seemed to be associated with large particles like cells or cell debris, while protein molecules might exist in solution or be contained by smaller particles, which are significantly less effected by radiation forces.

An obvious measure to remove a biofilm is the application of liquid cleaning agents.[58,59] The effects of acids, surfactants and oxidizing agents on biofilm removal in this flow cell have been investigated in comparison to the effect of the ultrasonic field.[60] The detailed comparison of data for protein and carbohydrate removal respectively delivered the best results for sodium hypochlorite (NaOCl), which was able to remove 100% of the protein and

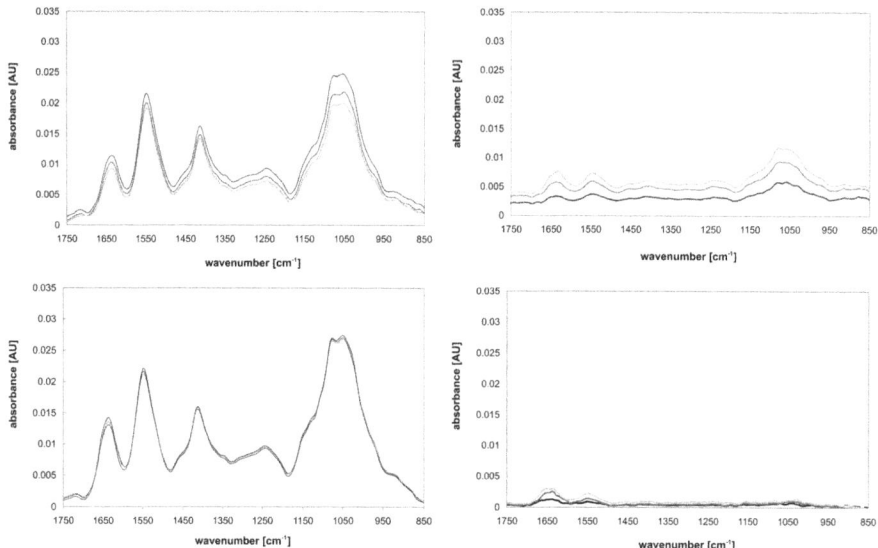

Figure 17.10   IR spectra recorded to observe bio-film formation while the ATR was exposed to a yeast suspension. Measurements were taken at 30 (black), 60 (dark grey), 90 (lighter grey) and 120 min (lightest grey). *Top left*: result when the basic procedure (*cf.* Figure 17.9(a)–(c)) was used. *Top right*: measurements representing the bio-film causing a decrease in resolution and sensitivity of the measurements. *Bottom left*: result when an ultrasonic standing wave was applied during the rinse (*cf.* Figure 17.9(e)). *Bottom right*: spectra when the ultrasonic standing wave was applied, resulting in significantly lower bio-film. The figure is adopted from figures in Radel *et al.*[57]

95% of carbohydrates in the biofilm. However, using this chemical (a.k.a. bleach) places constraints, due to its aggressive nature, on the range of materials that can be used as a sensor. Sodium dodecyl sulfate (SDS), one of the standard surfactants used for cleaning, showed removal abilities of 93% for the protein and 94% for carbohydrates respectively. The downside here was a prolonged removal time; it took 30 minutes of rinsing until the agent had completely left the chamber.

When developing a sensor for fermentation monitoring, time resolution is an important issue. In order to increase the settling rate and therefore decrease the measurement time, agglomeration of the yeast cells was induced by having an ultrasonic standing wave present when the flow cell was filled with suspension (see Figure 17.9(d)). The method was evaluated by comparing the carbohydrate value of absorption as a measure for the settling speed.[57]

In the absence of the ultrasonic standing wave, the suspended yeast cells settled slowly on the ATR surface at a constant increase. A maximum was reached after approximately 185 s, signalling a complete coverage of the sensitive surface.

When the ultrasonic field was present during the filling and subsequently switched off with stopping the flow, a different behaviour occurred. For 15 s, no cells were detected by the ATR as no material was present at the sensing surface. Following that, a strong increase of absorbance was detected; it took some 70 s to reach the same maximum as above. The acceleration due to ultrasound-induced aggregation was therefore increasing the settling rate by a factor of more than two.

17.4.3 ATR Probe for Inline Bioprocess Monitoring

Sensors taking measurements inside a reactor are the preferred option for bioprocess monitoring due to sterility and practicability issues. However, as no recalibration or verification of sensor response is possible, such inline sensing approaches are especially demanding in regard to long-term robustness and calibration stability.

Inline ATR sensors connected to the spectrometer by mid-IR fibre optics or optical conduits for infrared absorption spectroscopy have previously been exploited in bioprocess monitoring. A variety of small molecules such as sugars, alcohols, organic and amino acids as well as phosphate at concentrations of a few g L^{-1} were successfully determined in the fermentation broth.[61,62] Only chemical information of the supernatant could be assessed with these inline configurations, as the cells in culture do not reach the ATR's evanescent field. However, from a bioprocess control point of view, it would be very desirable to access chemical information about the culture as well.

This demand triggered investigations to combine an ultrasonic standing wave with an inline ATR fibre probe. The ability of ultrasonic standing wave fields to deposit particles on a surface was investigated with functionalised surfaces[22] and by optical means.[28] The target here was to independently measure mid-IR spectra of the supernatant and the suspended cells by deliberately populating or de-populating the ATR's evanescent field of cells. The concept of pushing particles towards a surface described in Section 17.3 has been investigated.

To combine the optical and the ultrasonic techniques, it was necessary to implement an ATR element in the proximity of an ultrasonic standing wave, permitting particle manipulation within the evanescent field.[63] The way to accomplish this was to use the ATR element as the reflector of an ultrasonic resonator (see Figure 17.11). Sound was propagating in the direction of the axis of the ATR fibre probe, hence suspended particles were agglomerated in the nodal planes which were oriented parallel to the evanescent field. The precise location of the agglomerate could therefore be controlled by the ultrasonic frequency.

Figure 17.12 shows four images obtained when polystyrene beads suspended in methanol were used to study the manipulation of particles in proximity to the ATR fibre probe. A set-up such as that sketched in Figure 17.11 was used in a vessel containing a model yeast suspension agitated by a magnetic stirrer; bead concentration was kept low to maintain good visual access.

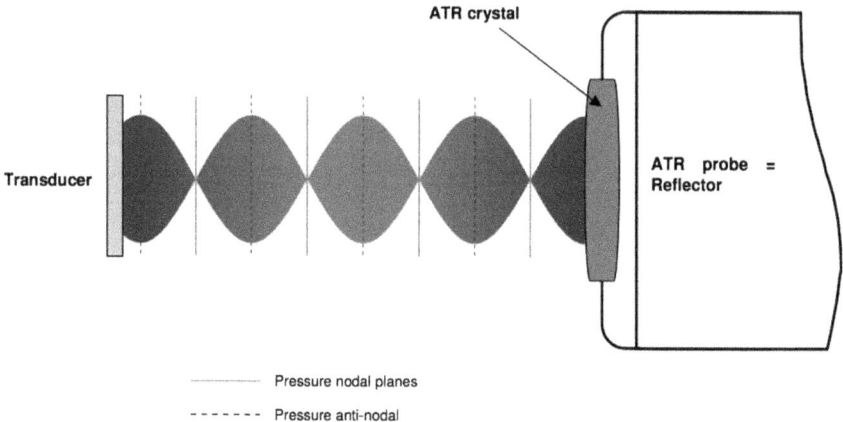

Figure 17.11 Sketch of the ultrasonic resonator where the ATR probe acts as reflector for the incident acoustic waves. The envelope of the standing wave represents the resonant state as in Figure 17.12(a). Changing the frequency will alter the position of the pressure nodes and therefore the location of particle aggregation. The figure is taken from Radel et al.[64]

Figure 17.12(a) shows a strong alignment of the particles in the gap between the ATR probe head (left) and transducer (right), indicating high acoustic energy densities at the indicated frequency of 1.85 MHz. The visible arrangement reflects the nodal planes half a wavelength apart (sound wavelength approx. 590 μm for methanol). The arrangement in the pressure nodal planes was established within seconds from the activation of the ultrasound signal and maintained as long as an ultrasonic field was present.

In Figure 17.12(b), the operating frequency was changed to 1.854 MHz. The result was a slightly less pronounced pattern in respect to the particles' alignment. A few of the beads in the leftmost nodal plane seemed to touch the diamond ATR surface, but still the majority of the beads remained a short distance from the cone tip. An increase of the frequency to 1.860 MHz loosened the aggregates further (Figure 17.12(c)). The top plateau was covered more substantially with beads, but still a vast number of beads remained in the nodal plane in front of the ATR without touching it.

Figure 17.12(d) shows the result when the frequency was changed to 1.869 MHz and eventually the major part of the first aggregate came in contact with the diamond. However, the strong alignment observed at 1.850 MHz was gone, indicating that the energy density was not very high in the fluid layer at this stage.

The analysis of these images suggests the possibility of manipulating suspended particles onto a smooth, rigid surface orientated perpendicular to the sound propagation direction. A broad hint that particles were actually pushed into the evanescent field was achieved with the results in Figure 17.13 obtained with a yeast suspension.[63] While the respective pushing frequency

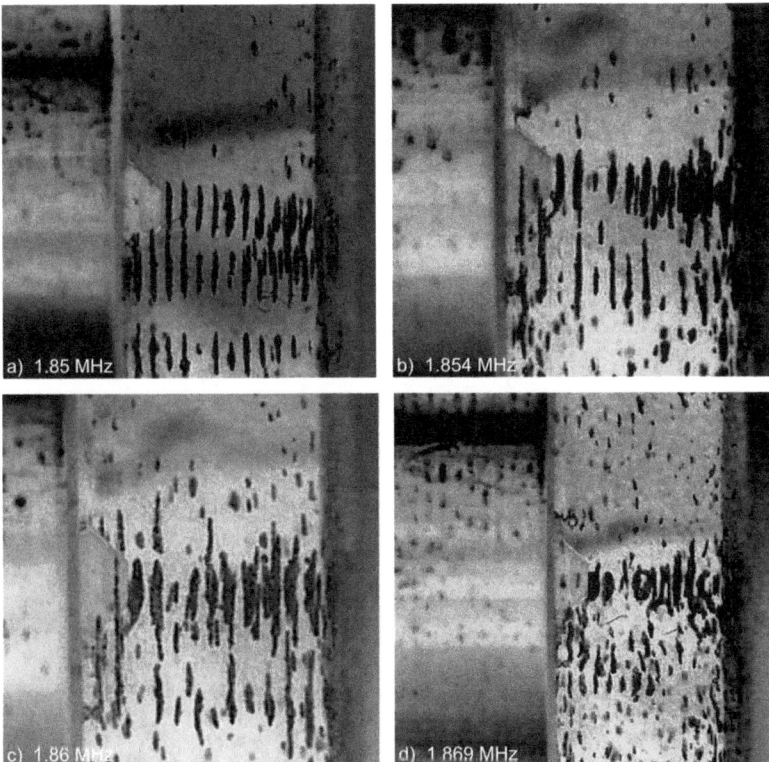

Figure 17.12 Images of polystyrene resin beads (100 μm) suspended in methanol and agglomerated in the pressure nodal planes of an ultrasonic standing wave between a transducer (right) and the head of the ATR fibre probe (left). The acoustic path length was adjusted to 3.18 mm. Each image was taken at a distinct ultrasonic field frequency stated below the respective image (a)–(d). Due to an increase of frequency from 1.850 MHz to 1.869 MHz, the suspended beads continuously approached the ATR surface. The figure is adopted from Radel et al.[64]

f_2—indicated by the grey bar at the bottom of Figure 17.13—was applied, a strong increase of the infrared absorption at 1047 cm^{-1} was observed. This wavenumber is associated with carbohydrates, which are contained by the yeast. The data therefore were interpreted as cells entering the sensitive zone of the ATR. Switching to the retracting frequency f_1 (blue sections of bar) resulted in a steep decrease back to the base level. Obviously the cells were pulled out of the evanescent field. This data were interpreted as proof of concept; the ultrasonic standing wave could be used to control the spatial distribution of various particles in close proximity to an ATR sensor and infrared spectra of particles and host liquid could be obtained independently.

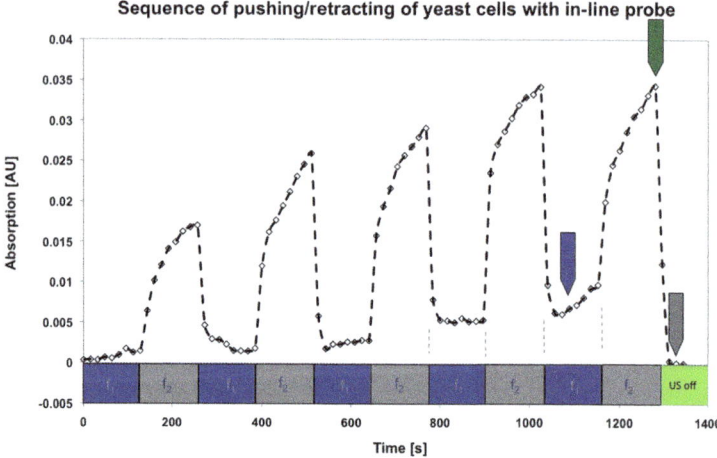

Figure 17.13 Sequence of repeated applications of the pushing and the retracting frequency recorded with the inline probe. The ATR's population/depopulation is indicated by the temporal development of an absorption band (carbohydrates at 1047 cm^{-1}) of the yeast cells. The figure is taken from Radel et al.[65]

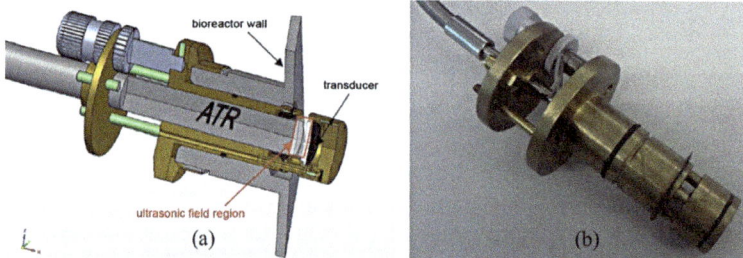

Figure 17.14 Left: Workshop drawing of the inline probe connected to the bioreactor. The position of the ATR probe can be adjusted with the micrometre screw gauge along the axis, thus changing the resonator length between ATR probe and transducer. Right: Photograph of the inline probe. The figure is taken from Koch et al.[63]

A gradual increase of the absorptions peak value over several cycles of switching between the retracting and pushing frequencies was found, possibly caused by the agglomerate slowly growing during the experiment. Additionally, the base line was not reached completely after the third cycle, supposed for the same reason: the aggregate was too big to be completely pulled away from the ATR. When finally switching off the field (green bar), the absorption signal fell back to the initial level.

The successful experiences led to the construction of a prototype inline probe for use inside a bioreactor.[63] The probe (see Figure 17.14) was designed in accordance with FDA regulations, hence only biocompatible materials

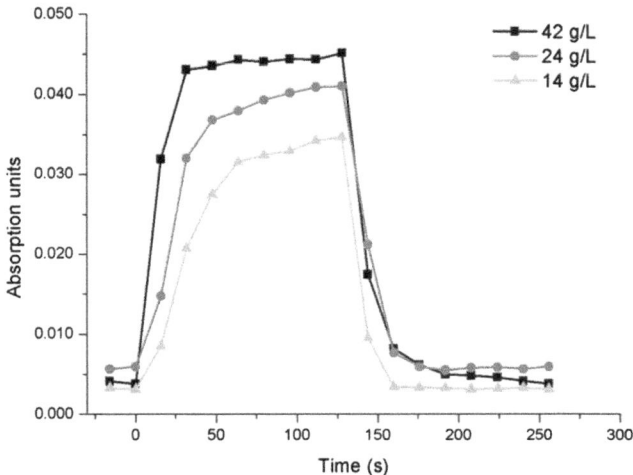

Figure 17.15 Change in absorption of the carbohydrate band with yeast cell concentration when the pushing frequency is applied (recorded with the inline probe). Each data point was calculated as the average absorption measured at the time elapsed (Δt) after sonication at pushing frequency: $t = 0$ s corresponds to the last spectrum before application of the pushing frequency, $\Delta t = 16$ s is the average of the first spectrum after pushing frequency was switched on, and so on. The figure is taken from Koch et al.[63]

(brass, Macor®, Viton®) are used. The sound source is a 10 mm diameter, 1 mm thick PZT transducer glued to a Macor® carrier. It is situated exactly opposite the ATR probe by three rods, one of which acts as a guide for the cable connecting the transducer with the frequency power synthesizer. The distance between the ATR probe and the transducer can be adjusted with a micrometre screw gauge moving the ATR probe along its axis. Results obtained with this probe resemble Figure 17.13 and are therefore not shown.

As an important influence, especially from an application point of view, the cell concentration was identified. Firstly the maximum absorption was found to increase with cell concentration again in a yeast suspension. The average absorptions recorded for yeast concentrations of 14 g L^{-1}, 24 g L^{-1} and 42 g L^{-1} are shown in Figure 17.15. Moreover, a change in the rate as the absorption increases during a pushing cycle with the yeast cell concentration is noticeable. Values close to maximum absorption are reached faster for a 42 g L^{-1} cell concentration than for 14 g L^{-1}.

A comparison of the absorption achieved when cells were pushed into the evanescent field by the ultrasonic standing wave with the absorption reached by sedimentation of yeast cells on the ATR (data not shown) was performed. It revealed that approximately 25–40% of the achievable absorption of the sediment was reached by the inline probe; most probably this is caused by different inclinations up to 135° where cells have to be pushed upwards.

17.5 Conclusions

Ultrasonic standing wave technology for the manipulation of particles and cells has been shown to be suitable for enhancing immunoassays and particle per cell sensors. For immunoassays, the method has been applied to bead-based agglutination assays and bead-based fluorescence assays. Both these assays have been investigated in the homogeneous (wash-free) format. The simplest and most straightforward method is to apply ultrasonic enhancement to latex agglutination tests. Such tests can be improved in speed (readout within a few minutes) and the sensitivity can be improved from the >nM-range to the 100-pM-range. If higher sensitivity is needed, the bead-based fluorescence assay is more suitable. The reason for this is that the unenhanced format has a typical detection limit in the low pM-range. With ultrasonic enhancement, the sensitivity of the assay was demonstrated to be in the 10–100 fM-range. This is an impressive sensitivity if we take into account that the assay was operated in the homogeneous (wash-free) format. In future, a full implementation of the methods into the lab-on-a-chip format would make them very attractive. It would also be interesting to investigate whether combined acoustic streaming[66] and acoustic trapping could be used for enhancing the binding reaction of the fluorescence assay.

Another field where ultrasound has been used is to attach (and detach) particles or cells onto a sensor surface. Here, the most mature application is to use ultrasound for enhancing vibrational spectroscopy sensing. This has been demonstrated by using mid-infrared absorption spectroscopy for retrieving molecular-specific information of a certain sample, *e.g.* for on-line/inline and real-time monitoring of an *in vivo* cell suspension. In particular, the use of surface-sensitive optical detection methods based on, *e.g.*, evanescent field and attenuated total reflection technologies show great promise and would be very suitable for integration into the lab-on-a-chip format. Still, as shown in Table 17.1, designing devices for driving cells or particles denser than the fluid onto a solid surface by ultrasound is not trivial and is also more complicated than designing devices for standard acoustophoretic operation (where particles are driven to pressure nodes in the bulk fluid). Thus, more fundamental investigations are needed and should be combined with modeling in order to verify experimental results and to push forward the development of the promising technique of using ultrasound for enhancing immunoassays and surface-based particle and cell sensors.

References

1. M. I. Mohammed and M. P. Y. Desmulliez, *Lab Chip*, 2011, **11**, 569–595.
2. R. W. Ellis and M. A. Sobanski, *J. Med. Microbiol.*, 2000, **49**, 853–859.
3. M. Wiklund and H. M. Hertz, *Lab Chip*, 2006, **6**, 1279–1292.
4. J. M. Singer and C. M. Plotz, *Am. J. Med.*, 1956, **21**, 888–892.
5. D. Stollar, *Can. Med. Assoc. J.*, 1960, **83**, 950–954.

6. J. A. Molina-Bolívar and F. Galisteo-González, *J. Macromol. Sci. C*, 2005, **45**, 59–98.
7. W. J. Litchfield, A. R. Craig, W. A. Frey, C. C. Leflar, C. E. Looney and M. A. Luddy, *Clin. Chem.*, 1984, **30**, 1489–1493.
8. W. H. Kapmeyer, W. H. Pauly and P. Tuengler, *J. Clin. Lab. Anal.*, 1988, **2**, 76–83.
9. P. L. Masson, *J. Pharm. Biomed. Anal.*, 1987, **5**, 113–117.
10. M. Wiklund, O. Nord, R. Gothäll, A. V. Chernyshev, P.-Å. Nygren and H. M. Hertz, *Anal. Biochem.*, 2005, **338**, 90–101.
11. M. A. Grundy, W. E. Bolek, W. T. Coakley and E. Benes, *J. Immunol. Methods*, 1993, **165**, 47–57.
12. M. A. Grundy, K. Moore and W. T. Coakley, *J. Immunol. Methods*, 1994, **176**, 169–177.
13. R. J. Porritt, J. L. Mercer and R. Munro, *Pathology*, 2003, **35**, 61–64.
14. M. A. Sobanski, R. A. Barnes and W. T. Coakley, *Methods Mol. Med.*, 2001, **67**, 41–59.
15. M. Z. V. Smoluchowski, *J. Phys. Chem.*, 1917, **92**, 129–168.
16. I. V. Surovtsev, M. A. Yurkin, A. N. Shvalov, V. M. Nekrasov, G. F. Sivolobov, A. A. Grazhdantseva, V. P. Maltsev and A. V. Chernyshev, *Colloids Surf. B*, 2003, **32**, 245–255.
17. J. P. Nolan and L. A. Sklar, *Trends Biotechnol.*, 2002, **20**, 9–12.
18. M. Wiklund, M. Tirri, J. Toivonen, P. Hänninen and H. M. Hertz, *J. Appl. Phys.*, 2004, **96**, 1242–1248.
19. P. Hänninen, A. Soini, N. Meltola, J. Soini, J. Soukka and E. Soini, *Nat. Biotechnol.*, 2000, **18**, 548–550.
20. R. P. Ekins, *J. Pharm. Biomed. Anal.*, 1989, 7, 155–168.
21. R. Ekins and F. Chu, *Ann. Biol. Clin.*, 1992, **50**, 337–353.
22. J. J. Hawkes, M. J. Long, W. T. Coakley and M. B. McDonnell, *Biosens. Bioelectron.*, 2004, **19**, 1021–1028.
23. M. Frank, D. Anderson, E. R. Weeks and J. F. Morris, *J. Fluid Mech.*, 2003, **493**, 363–378.
24. P. Glynne-Jones, R. J. Boltryk and M. Hill, *Lab Chip*, 2012, **12**, 1417–1426.
25. J. P. Black, R. M. White and J. W. Grate, *Proc. - IEEE Ultrason. Symp.*, 2002, 475–479.
26. R. M. White, *Faraday Discuss.*, 1997, **107**, 1–13.
27. P. Glynne-Jones, M. Hill, N. R. Harris, R. J. Townsend and S. Ravula, in *Proceedings of Acoustics 08, Paris*, Société Francaise d'Acoustique, 2008, pp. 5989–5993.
28. P. Glynne-Jones, R. J. Boltryk, M. Hill, F. Zhang, L. Q. Dong, J. S. Wilkinson, T. Melvin, N. R. Harris and T. Brown, *Anal. Sci.*, 2009, **25**, 285–291.
29. M. Zourob, J. J. Hawkes, W. T. Coakley, B. J. T. Brown, P. Fielden, M. B. McDonnell and N. J. Goddard, *Anal. Chem.*, 2005, **77**, 6163–6168.
30. S. Oberti, *Micromanipulation of small particles within micromachined fluidic systems using ultrasound*, PhD Thesis, ETH Zürich, 2009.
31. S. Oberti, A. Neild and J. Dual, *J. Acoust. Soc. Am.*, 2007, **121**, 778–785.

32. I. Gonzalez, L. J. Fernandez, T. E. Gomez, J. Berganzo, J. L. Soto and A. Carrato, *Sens. Actuators, B,* 2010, **144**, 310–317.
33. J. J. Hawkes, S. Mohr, B. Bastini, N. J. Goddard and M. McDonnell, in preparation.
34. P. Glynne-Jones, R. J. Boltryk, N. R. Harris, P. Baclet and M. Hill, *J. Acoust. Soc. Am. Express Lett.,* 2009, **126**, EL75–EL79.
35. E. G. Williams, *Fourier Acoustics: Sound radiation and nearfield acoustical holography*, Academic Press, Cambridge, United Kingdom, 1999.
36. N. Skivesen, *Metal-Clad Waveguide Sensors*, PhD Thesis, Risø National Laboratory, Roskilde, Denmark, 2005.
37. M. Zourob, S. Mohr, B. J. T. Brown, P. R. Fielden, M. McDonnell and N. J. Goddard, *Sens. Actuators, B,* 2003, **90**, 296–307.
38. A. C. Sorando, J. J. Hawkes, P. R. Fielden and I. González, in *International Congress on Ultrasonics, AIP Conf. Proc., Gdansk*, AIP Publishing, 2012, pp. 757–760.
39. A. Lenshof, M. Evander, T. Laurell and J. Nilsson, *Lab Chip,* 2012, **12**, 684–695.
40. J. Chalmers and P. Griffiths, *Handbook of Vibrational Spectroscopy*, John Wiley & Sons Inc., New York, 2002.
41. L. A. Lyon, C. D. Keating, A. P. Fox, B. E. Baker, L. He, S. R. Nicewarner, S. P. Mulvaney and M. J. Natan, *Anal. Chem.,* 1998, **70**, 341R–361R.
42. J. R. Baena and B. Lendl, *Curr. Opin. Chem. Biol.,* 2004, **8**, 534–539.
43. R. Swain and M. Stevens, *Biochem. Soc. Trans.,* 2007, **35**, 544–549.
44. N. Leopold, M. Haberkorn, T. Laurell, J. Nilsson, J. R. Baena, J. Frank and B. Lendl, *Anal. Chem.,* 2003, **75**, 2166–2171.
45. L. Puskar, R. Tuckermann, T. Frosch, J. Popp, V. Ly, D. McNaughton and B. R. Wood, *Lab Chip,* 2007, **7**, 1125–1131.
46. S. Radel, J. Schnöller, A. Dominguez, B. Lendl, M. Gröschl and E. Benes, *Elektrotechnik & Informationstechnik,* 2008, **125**, 82–85.
47. M. Ruedas-Rama, A. Dominguez-Vidal, S. Radel and B. Lendl, *Anal. Chem.,* 2007, **79**, 7853–7857.
48. M. J. A. Cañada, A. R. Medina, J. Frank and B. Lendl, *Analyst,* 2002, **127**, 1365–1369.
49. P. Harms, Y. Kostov and G. Rao, *Curr. Opin. Biotechnol.,* 2002, **13**, 124–127.
50. L. Olsson, U. Schulze and J. Nielsen, *Trends Anal. Chem.,* 1998, **17**, 88–95.
51. Aldrich Library of FT-IR Spectra, Sigma-Aldrich, Milwaukee, WI, 1997.
52. N. Harrick, *Internal Reflection Spectroscopy*, John Wiley & Sons Inc., New York, 1967.
53. J. Bertie and Z. Lan, *Appl. Spectrosc.,* 1996, **50**, 1047–1057.
54. G. Jarute, A. Kainz, G. Schroll, J. Baena and B. Lendl, *Anal. Chem.,* 2004, **76**, 6353–6358.
55. J. Schnöller and B. Lendl, *Proc.IEEE Sensors,* 2004, **2**, 742–745.
56. K. Merritt, V. Hitchins and S. Brown, *J. Biomed. Mater. Res. B,* 2000, **53**, 131–136.
57. S. Radel, J. Schnöller, B. Lendl, M. Gröschl and E. Benes, *Elektrotechnik & Informationstechnik,* 2008, **125**, 76–81.

58. X. Chen and P. Stewart, *Water Res.*, 2000, **34**, 4229–4233.
59. B. Meyer, *Int. Biodeterior. Biodegrad.*, 2003, **51**, 249–253.
60. S. Radel, J. Schnöller, M. Gröschl, E. Benes and B. Lendl, *IEEE Sens. J.*, 2010, **10**, 1615–1622.
61. D. J. Pollard, R. Buccino, N. C. Connors, T. F. Kirschner, R. C. Olewinski, K. Saini and P. M. Salmon, *Bioprocess Biosyst. Eng.*, 2001, **24**, 13–24.
62. D. L. Doak and J. A. Phillips, *Biotechnol. Prog.*, 1999, **15**, 529–539.
63. C. Koch, M. Brandstetter, B. Lendl and S. Radel, *Ultrasound Med. Biol.*, 2013, **39**, 1094–1101.
64. S. Radel, M. Brandstetter and B. Lendl, *Ultrasonics,* 2010, **50**, 240–246.
65. S. Radel, M. Brandstetter, C. Koch and L. Bernhard, in *Proceedings of 1st EAA - EuroRegio, Ljubljana*, European Acoustics Association, 2010, p. 8.
66. M. Wiklund, R. Green and M. Ohlina, *Lab Chip,* 2012, **12**, 2438–2451.
67. M. Wiklund, S. Radel and J. J. Hawkes, *Lab Chip,* 2013, **13**, 25–39.

CHAPTER 18

Multi-Wavelength Resonators, Applications and Considerations

JEREMY J. HAWKES[*a] AND STEFAN RADEL[b]

[a]Manchester Interdisciplinary Biocentre, The University of Manchester, Manchester, UK; [b]Institute of Chemical Technologies and Analytics, Institute of Applied Physics, Vienna University of Technology, Vienna, Austria
*E-mail: JeremyjHawkes@gmail.com

18.1 Introduction

Many resonant systems, for example musical instruments, operate at their fundamental frequency which, depending on the boundaries, is often either a half or quarter wavelength. Particle manipulation chambers, however, operate additionally very effectively at higher harmonics, *i.e.* when many wavelengths fit into the resonator. Therefore, multi-wavelength chambers can be used as these have more internal space, which reduces blockages while retaining higher frequencies. There are two reasons for preferring high frequencies: (1) the acoustic force on the particles increases in proportion to frequency as described in Chapter 4.[1] (2) Above 1 MHz, cavitation and its destructive effects are almost absent—the reasons are explained fully in Chapter 21[2]—and therefore frequencies below 1 MHz are often avoided when working with aqueous media; this limits the sound path-length in half-wavelength chambers to 0.75 mm or less.

Descriptions given here will be limited to ultrasonic frequencies in the range of a few ten kHz (for aerosol-filled chambers) up to 10 MHz (for aqueous suspension-filled chambers). The devices described have internal dimensions typically in the range 1–100 mm; this size range is suitable for supporting *ultrasonic standing waves* (USW) in liquids and gases. This chapter describes the use of aqueous media for most of the examples, but details are given so that designs can be created for other fluids including gases. Multi-wavelength devices are generally larger than quarter- or half-wavelength resonators, however, when very high frequencies are used, even a 50 μm channel can be a multi-wavelength device.[3]

The most successful multi-wavelength system is the ultrasound enhanced sedimentation filter (also known as ultrasonically enhanced settling), UES. This use of sedimentation is a different filtration principle to the filter described in Chapters 6, 7 and 8.[4-7] In brief, the latter filter (which we will identify as the hydrodynamic acoustic filter, HAF) splits lines of acoustically-concentrated cells from the clear fluid by careful flow adjustment to outlet channels, which are usually outside the standing wave region. The HAF is better suited to low cell concentrations. UES requires the formation of cell clumps and so is best suited for samples with high cell numbers, such as whole blood (40% haematocrit) and for fermentation broth filtration (10^8 cells per mL and above). At high particle concentrations where the nodes become overloaded, the performance of UES degrades more slowly than the HAF. For whole blood filtration, the result is that clearer plasma and a greater volume of plasma is produced (see Table 18.1). A second advantage of UES is that the dimensions are less critical, even to the extent that the UES chamber used in Table 18.1 and Figure 18.1(b) and (d) was simply the Vacutainer (BD, Franklin Lakes) used to draw the blood.

Vacutainers are an example of a batch system, however, UES is often used in flow-through systems, and we will use the flow-through system as the main example for the rest of this chapter.

Table 18.1 Separation efficiency of plasma from whole blood comparing a centrifuge, ultrasound enhanced sedimentation (UES) and hydrodynamic acoustic filtration (HAF). UES has a clear advantage, although the final column shows a very high flow rate for a hydrodynamic acoustic filter, achieved when an additional buffer is used to overcome node overloading.

	Centrifuge	UES		HAF
		Whole blood alone		Blood + receiving buffer[8]
		1 stage[9]	4 stages	
Cells left in plasma (%)	0.01	0.3	3.5	<1	5
Recovered plasma (%)	60	40	10	10	—
Blood volume processed (ml hr^{-1})	—	60	4.8	4.8	1000

With an emphasis more on construction than applications, this chapter aims to give the reader the understanding and sources needed to design and construct a multi-wavelength chamber. We also attempt to indicate areas where new development approaches may be most valuable. Earlier chapters have covered flow dynamics (Chapter 1),[10,11] conditions needed for resonance

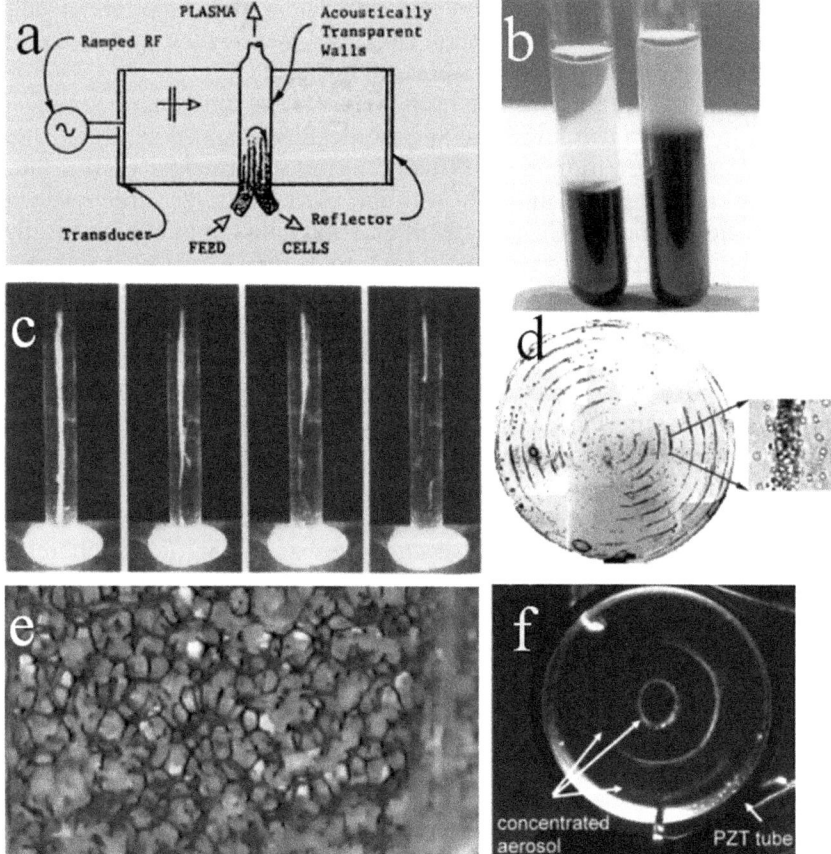

Figure 18.1 Multi-wavelength ultrasound manipulation systems and applications in blood, water, gel, mesh and air. (a) Pseudo-standing wave/sedimentation blood separator[19] © IEEE. Reprinted, with permission. (b) Left, centrifuged blood, right, UES separation where the tube is the chamber.[20] Reproduced with permission of Elsevier Inc. (c) Pseudo-standing wave progressively lifting a column of 9 μm latex particle clumps in a 50 mm high chamber.[21] Reproduced with permission of Elsevier BV. (d) Node lines from (b) fixed in position in an agar gel, set while the sound was applied.[22] Reproduced with permission of Elsevier Inc. (e) Cells (black) adhered to a porous mesh. (f) Water droplet aerosol driven to nodes.[23] Reproduced with permission of American Institute of Physics for the Acoustical Society of America.

(Chapters 2 and 3),[12,13] devices for creating standing waves (Chapters 5, 6 and 20),[4,14,15] the primary and secondary acoustic standing wave forces acting on cells (Chapter 4)[1] and acoustic streaming (Chapter 13).[16-18] This chapter revisits some of these concepts to show how they have been applied to larger dimension systems and provides further details where additional considerations are needed.

The chapter is divided into four parts:

Part 18.2 Acoustic filters: gives examples of multi-wavelength resonators.

Part 18.3 Resonators: gives an overview of the processes that have been considered in the design of multi-wavelength systems, and describes some of the observations of complex effects.

Part 18.4 Flow changes produced by scale-up: provides tools to determine when flow becomes turbulent and also defines other limits to the flow velocity.

Part 18.5 New design approaches: outlines how models commonly used for non-destructive testing can be introduced to resonant chamber designs.

18.2 Acoustic Filters

This section is called acoustic filters because almost all the larger scale applications of particle manipulation are solely for filtration. Filtration is a vital (and often a limiting) step for many industrial and laboratory processes. The following brief history of ultrasound manipulation chambers shows their potential to fit a very diverse range of filtration processing applications.

Peterson et al.[19] created one of the first clearly application-driven systems (Figure 18.1(a)), for separating blood cells from plasma. A pseudo-standing wave was used to drive the cells across a rising flow towards an outlet that received the concentrated cell sediment. This achieved haematocrits of 80% and 1.5% on the outflows from a 40% haematocrit inflow. Cousins[20] used a fixed frequency circular wave at 1.6 MHz to enhance sedimentation directly in the Vacutainers used to draw blood, leaving only 0.3% of the cells in the plasma after <6 min (Figure 18.1(b)). Whitworth[21] also used a pseudo-standing wave to gain impressive control of a column of 9 µm latex beads moving up and down a 5 cm high tube at up to 20 mm s^{-1} (Figure 18.1(c)). A similar but more sophisticated node movement in harmonic steps is known as drifting field;[24] recently, this transverse movement has been used to move particles out of a flow.[25] Using the Vacutainer again (Figure 18.1(b)), the node pattern was fixed for subsequent observation by suspending cells in a viscous agar gel and allowing it to set while the sound was applied (Figure 18.1(d)).[22] Feke[26] has used a polyester mesh, 1.25 mm pore size, to attract 99% of 30 µm diameter particles and 95% of hybridoma cells to the fibres with ~1.12 MHz. This attraction is the result of non-uniformities in the field produced by the mesh. Particle manipulation is not limited to water or even liquids; Kaduchak (Figure 18.1(f))[23] and Anderson[27] have both brought aerosols of 5 µm water droplets to the nodal planes.

There are other multi-wavelength processes that do not require standing waves. The three main systems use frequencies below 40 kHz. The system most similar to the standing wave systems is for agglomeration of smoke, developed for large scale filtration.[28] In this approach the sound increases collisions between large and small particles because of their different entrainment in the oscillating air,[29] leading to increased clump formation and sedimentation. The other two multi-wavelength processes are sonochemistry[30] and cell sonication.[31] These use the extreme pressures created by cavitation[32] to produce unusual chemical reactions and to destroy cells. These three systems will not be further described.

To allow more detailed descriptions the rest of the section concentrates on two systems, an UES system and a type of HAF system known as an h-shape separator. Both are flow-through systems with one inlet and two outlets: one for the clarified liquid and one for the enriched suspension. These systems with centimetre scale ducts are even more tolerant to bubbles and oversized particles than the hydrodynamic acoustic filters (themselves more tolerant than many other filters).

18.2.1 Ultrasonically Enhanced Sedimentation

In brief: in ultrasonically enhanced sedimentation filters, ultrasound creates clumps as it draws single cells into vertical nodal planes. The clumps sediment against the upward moving incoming flow and highly clarified fluid emerges from the upper outlet. This approach has been developed over the past two decades mainly in commercial/industrial environments. The principle utilizes the ultrasonic agglomeration of particles bringing about a locally decreased surface to volume ratio. Due to the increased diameter of this "super-particle", the ratio of Stokes drag force and gravitational force falls leading to higher terminal settling velocity (sedimentation rates) of the loose aggregates.[33–36] A quantified description is given in Section 18.2.2.

Figure 18.2 shows the stages of the UES process: in the beginning the particles are freely distributed in the liquid (a). After the ultrasonic field has built up, the axial primary radiation force drives them into nodal planes (b). Typically, it does not take more than 2–3 s until such a spatial distribution is reached; at this stage bands of particles become apparent as shown in Figure 18.3. It depends on the acoustic contrast between particle and liquid; if this force points towards the pressure nodes or towards the displacement nodes for a given suspension, dense particles such as cells are driven into the pressure nodes.

The transverse primary radiation force concentrates the particles further within these planes (c). This force perpendicular to the sound propagation direction is less well understood but is at least partly a result of an uneven distribution displacement amplitude over the chamber wall. This transverse primary radiation force is weaker; it takes typically 10–30 s until the concentration within the planes is finished (d).

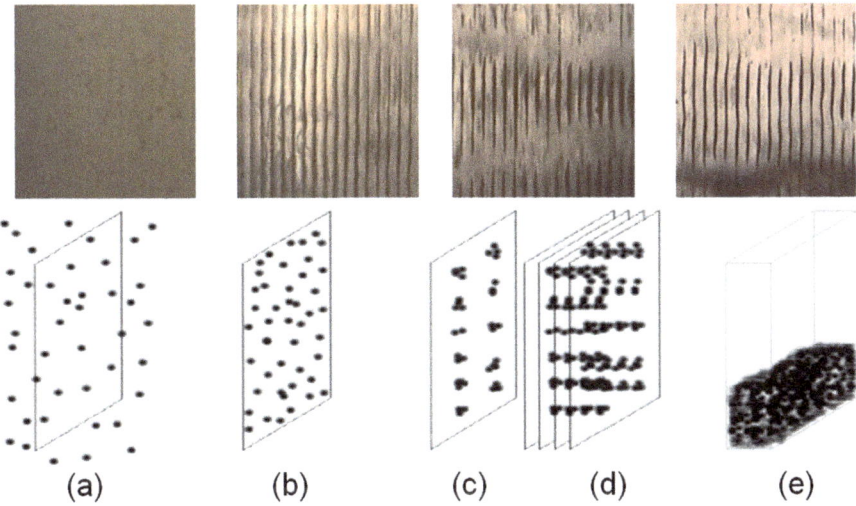

Figure 18.2 Stages of ultrasonically enhanced settling: homogeneously dispersed particles (a), accumulation in planes (b), and further concentrated within the planes (c), viewed as a multi-wavelength resonator, aggregates are stacked in rods or horizontal columns (d). The aggregates finally settle at the bottom of the vessel (e). Adapted from ref. 37, reproduced with permission of Kluwer Academic Publishers.

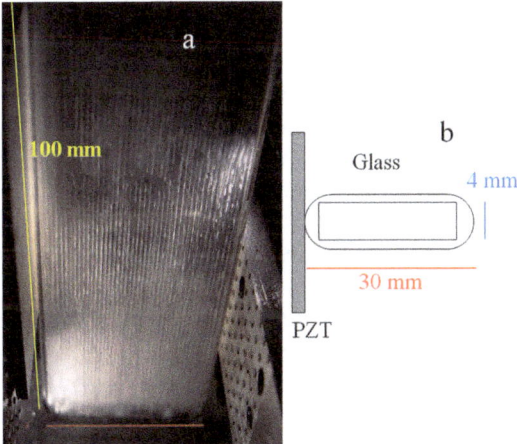

Figure 18.3 (a) Enhanced sedimentation of yeast, near 1 MHz, with upward flow in a vertical glass duct, viewed from above at an angle of ~30°. Showing vertical banding pattern and some accumulation at the base. (b) Cross section showing piezoelectric plate, pressed on the edge of the left side, without glue.

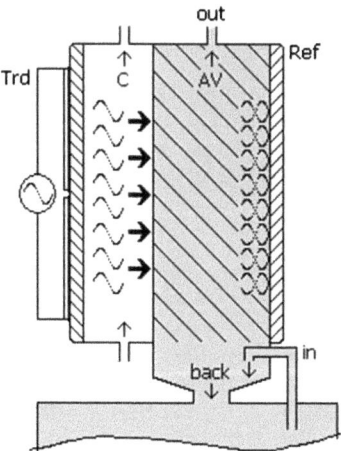

Figure 18.4 UES separation on top of a reservoir, *e.g.* a bio-reactor. The ultrasound is emitted from the transducer (Trd, 25 6 25 mm^2), passing a cooling volume (C) and the active volume (AV) holding the suspension and finally reflected at a reflector (Ref). The thickness of C and AV is 12.3 and 22.25 mm, respectively. From ref. 37, reproduced with permission of Kluwer Academic Publishers.

It was shown in Chapter 4[11] that the axial force is typically greater than gravity. The transverse radiation force generally is lower and therefore these aggregates finally settle at the bottom of the vessel (e) due to gravitational force (see Section 18.2.2).

Figure 18.4 shows a pilot series UES system (USSD-05, Anton Paar GmbH, Austria); unless stated otherwise this is the design we have in mind throughout this chapter when discussing UES systems. The device sits at the top of a reservoir holding the suspension. Currently, acoustic filters are mainly used in bio-technology, hence the vessel will commonly be a bio-reactor. In applications like this, the temperature of the cell suspension is of utmost importance. Therefore the liquid layer comprises two separate compartments: water circulating in the cooling layer (C) inhibits thermal stress transferred from the transducer into the adjacent active volume (AV) filled with the suspension.

In operation, *e.g.* as a filter for perfusion reactors, the clarified liquid is harvested at the top (out). The suspension is pumped into the system from the side at the bottom (in) and together with the bottom outlet (back), this builds up a re-circulation loop by which the settled particles are immediately fed back into the bio-reactor. The vertical nodal planes allow an upward streaming of the clarified liquid between the settling aggregates.

The main advantage of cell filters based on this technique is the complete absence of moving parts. The systems can be hot-steam sterilized *in situ*, the materials used, such as stainless steel and glass, are bio-compatible. The scale-up of the technology has progressed; perfusion filtering systems

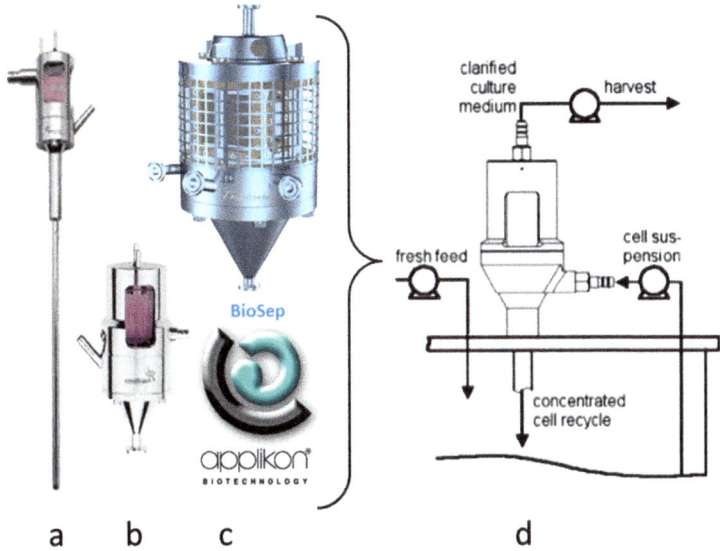

Figure 18.5 Commercial Biosep filters supplied by Applikon, The Netherlands. (a) 10 L d^{-1}, (b) 50 L d^{-1}, (c) 200 L d^{-1}, (d) sketch shows the UES on top of the bio-reactor; clarified medium is harvested at the top. Reproduced with permission of Applikon Biotechnology.

capable of 200 L d^{-1} and more are commercially available. Figure 18.5 shows a range available from BioSep by Applikon (Schiedam, The Netherlands).

18.2.2 Enhanced Sedimentation

A general calculation of sedimentation from Chapter 4[11] can be used to calculate the filtration efficiency of the UES filter. Good agreement with experiments is obtained by assuming that the clumps are spherical and reach diameters of one quarter of a wavelength[36] (at 1 MHz: a clump of tightly packed cells could contain >400 000 5 µm diameter cells). The gravitational force acting on cell clumps of radius r_c, density ρc, is given by:

$$F_g = \left(\frac{4}{3}\right)\pi r_c^3 (\rho_c - \rho_f) \cdot g \quad (18.1)$$

Viscous (Stokes') drag on the clump moving at velocity v_c in a fluid moving at velocity v_f is given by $6\pi\eta r_c (v_f - v_c)$. At its terminal velocity v_c, the gravitational force is fully opposed by the viscous drag force and when the sedimentation velocity is equal to the rising fluid velocity then $v_c = 0$ and:

$$6\pi\eta r_c (v_f - v_c) - \left(\frac{4}{3}\right)\pi r_c^3 (\rho_c - \rho_f) \cdot g = 0 \quad (18.2)$$

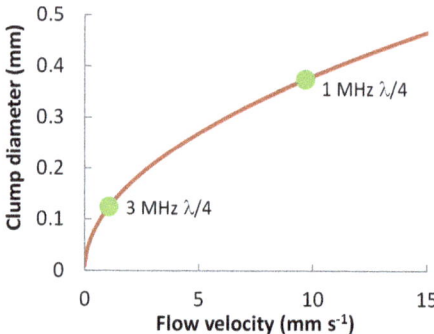

Figure 18.6 Diameters at which cell clumps no longer sediment against an upward flow. Markers show clumps with diameters equal to a quarter wavelength at 1 MHz and 3 MHz, which indicate the maximum flow velocities for filters operating at these frequencies. Calculated for an aqueous suspension of yeast ρ_c 1.114 × 10^3 kg m^{-3}. Adapted from ref. 36. Reproduced with permission of Blackwell Publishing Ltd.

We find:

$$v_f = \frac{2r_c^2(\rho_c - \rho_f)g}{9\eta} \qquad (18.3)$$

This upper limit for sedimentation against a rising flow is shown by the red line in Figure 18.6. Since the maximum clump diameter is ∼¼ wavelength, $\lambda/8$ can be substituted for r_c to find the frequency-dependent maximum velocity of clarified fluid through the system, plotted as two data points in Figure 18.6.

The calculation of maximum velocity assumes that the ultrasound field is long enough for most particles to arrive at the nodal plane[11] and also that the input concentration is sufficient to form large clumps.

18.2.3 Influences on Separation Efficiency (UES)

The operating parameters of UES systems have been extensively characterised.[33,36–38] The more recent study employed a commercial UES system as described here to investigate the influence of process parameters, such as flow-through rate, cell concentration and true electric power input, on the separation efficiency for the case of yeast/saline suspensions. It was found that up to 99.6% yeast cells can be retained by the acoustic filter. Cell concentrations of 5–50 g L^{-1} wet weight and throughput of 5–20 L d^{-1}, in certain cases 46 L d^{-1}, were found to be the favourable operation conditions in a 50 L d^{-1} system. In respect to the electrical energy an upper threshold was found, at which an increase of signal amplitude, *i.e.* loudness from the acoustical point of view, did not improve the separation efficiency any further. On the contrary, even higher levels of sound pressure led to unwanted introduction of heat into the system.

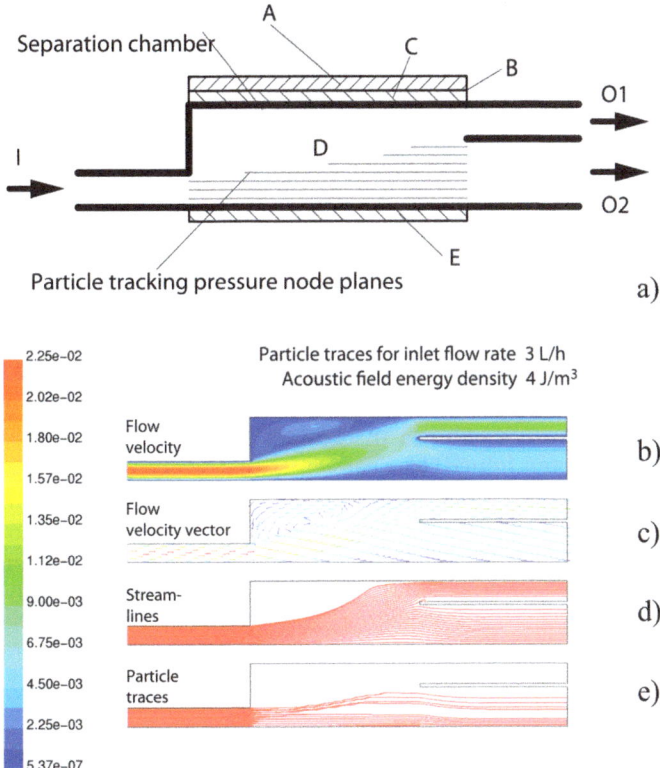

Figure 18.7 Schemes of the h-shape separator. (a) On the left side the suspension is fed into inlet I of the separator, on the right side the upper outlet O_1 is for the separated clean liquid and the lower outlet O_2 for the particle enriched suspension. The upper wall of the active volume D (approx. 10 ml) of the system is a PZT-glass-compound-transducer (A, B, C) 50 × 19 mm^2 emitting sound towards the reflector (E) 10 mm away. (b–d) show the flow within the device when flows to outlets O_1 and O_2 are equal. Flow velocity (b) and flow velocity vectors (c) were determined by solving the Navier–Stokes equation for the given geometry. In (d) the velocities and vectors have been transformed to streamlines using Netwon's equations of motion. Finally (e) the radiation force is introduced, delivering the trajectories of the particles and showing their removal from the upper outlet (calculated for 20 μm polystyrene particles). © IEEE. Reprinted with permission from ref. 41.

18.2.4 Flow Splitters in the h-Shape Separator

In contrast to the UES principle, the ordering of the ultrasonic radiation forces is utilized directly in an h-shape separator; this uses a flow splitter and is therefore a type of HAF. The layout of the device is shown in Figure 18.7(a).[39,40] The ultrasound standing wave (USW) is exploited for separating the liquid, some of which follows the flow lines into the cleaned outlet O_1, from the particles, which are diverted by the sound into the particle enriched outlet O_2.

The concept was investigated thoroughly in a combined simulation of flow and force field, see Figure 18.7(b)–(e). The FEM simulation resulted in a maximum throughput of 3 L h^{-1} for 20 μm polystyrene particles. This direct ultrasonic separation concept is not relying on gravity. The h-shape separator has therefore been tested successfully under microgravity conditions in the European Space Agency 29th zero g parabolic flight campaign within the frame of the ESA Melissa project.[39]

18.3 Resonators

This section deals with various aspects to be considered when designing multi-wavelength ultrasonic resonators (Chapter 6[4] is more general although focused on sub-wavelength systems). The description assumes a simple stack of plane parallel layers as shown in Figure 18.8, comprising a piezoelectric element (P) connected (glued) to a carrier layer (C) very similar to the device under test in Chapter 7.[6] This composite transducer (T), *i.e.* the combination of piezoelectric element and carrier, emits the sound signal into a cavity holding the suspension liquid (L), which is terminated by a reflector (R) at the opposite side. In the literature, the liquid layer (L) is occasionally termed as active volume (AV).

18.3.1 Construction

A vast choice of materials for the construction of an ultrasonic resonator is available. Every application will pose its own demands, nevertheless some general remarks shall be given here along with reports about failure where it has occurred. It has to be mentioned that the following remarks are limited to the construction of multi-wavelength ultrasonic resonators.

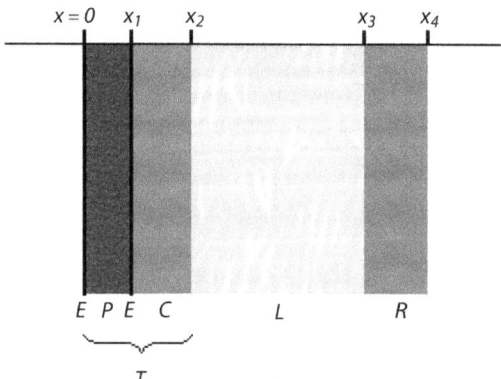

Figure 18.8 Layered resonator comprising a transducer T, *i.e.* PZT ceramic P with two electrodes E glued to a glass carrier C, a liquid layer L and a reflector R. From ref. 42.

Regarding the piezoelectric element, most applications envisaged in this work use PZT (lead–zirconate–titanate) ceramics, which are more suitable for high power applications than the well-known quartz resonators. Devices are available in different shapes and sizes and at low cost, and more importantly for almost every frequency range in the kHz to ten MHz range. A strong ultrasonic field will be excited by elements made from "hard" PZT types, which show high electro-mechanical coupling factors. Furthermore these materials can be driven in resonance mode, as mechanical losses are low; a detailed discussion can be found in Chapter 5.[14] Apart from PZT, special materials like barium titanate (lead free, low mass density) or lithium niobate (high Curie temperature) exist. In addition, new materials such as single crystals from lead magnesium niobate/lead titanate (PMN-PT) or lead zirconate niobate/lead titanate (PZN-PT) are being investigated.

To employ the piezoelectric effect an electrical field is necessary, and therefore a PZT is equipped with (at least) two electrodes. Usually "thick printed" silver electrodes (approx. 10 μm) or thin film electrodes from gold or CuNi are available; the shape can be tailored to any needs. The excited mode is in first order governed by the geometry of the electrodes in relation to the polarisation direction; in the example used here a sheet of PZT carries electrodes at the top and the bottom. As the polarisation is oriented perpendicular to those surfaces the application of a voltage at the electrodes leads to a thickness-extension mode, and the PZT emits a plane longitudinal wave.

There are different approaches to coupling the sound from the PZT into the rest of the device, or in other words, how to build up the resonator. Some groups use a very flexible suspension of the PZT.[43] A successful method used in the past was the use of very thin layers of thermally cured glue[34] in a pressure node of the transducer (dimensions will be discussed later). Then, as the thickness of the glue layer is far below the acoustic wavelength it will be mainly displaced ("shaken") by the USW, but not stressed. The latter would result in losses, as acoustical energy is converted into heat where a material is compressed and expanded by the excitement. One important point here is that for high performance systems it is necessary to match the thermal expansion coefficients of the PZT and the carrier, otherwise the strong glue will lead to lateral stresses and terminal disruption of one of the two layers when in operation.

When choosing materials it is also important to remember that carrier and reflector will be in touch with the suspension, and they must therefore be adequately chemically inert. This is needed for two reasons: on the one hand, *e.g.* bio-technical applications will have to be at least food-safe, but moreover might have to withstand further chemical/mechanical/temperature stresses, *e.g.* during cleaning procedures employing high temperatures, pressure and/or chemically aggressive solvents.

It is important that materials in the acoustic pathway should show low acoustic absorption. Losses are expressed by the absorption coefficient α; when considering resonances this can be used to define the resonance quality factor.

$$Q_{\text{eff}} = \frac{\pi}{\lambda \alpha_{\text{eff}}} \tag{18.4a}$$

The expression is valid for high Q-factors ($Q \gg 1$). The Q-factor is derived from the absorption coefficient α, which is governed by the viscous losses of the fluid (or any single component) only; it does not describe the resonance of the whole system. For this reason the *effective Q-factor*, Q_{eff} representing all mechanisms attenuating the signal is denoted in eqn (18.4a). Examples are losses by scattering from particles and sidewalls, and the initial loss in transduction of electrical to mechanical oscillations.

The Q-factor also serves as a measure for the ratio of energy stored, ε, by a system under investigation over the energy dissipated (see ref. 44 and Chapters 2 and 3[12,13]). The latter can be expressed as time-averaged power loss $\langle P^{\text{loss}} \rangle$, i.e.

$$Q_{\text{eff}} = \frac{2\pi f \cdot \varepsilon}{\langle P^{\text{loss}} \rangle} \tag{18.4b}$$

Hence Q_{eff} is of utmost importance for filtration describing the maximum acoustic energy stored in the liquid layer for a given electric energy supplied.

Several methods to measure the Q-factor are discussed in Chapter 20.[15,45] Intuitively the simplest (although error prone as discussed in Chapter 20) is to apply frequency bursts and measure the time, τ to rise 99.8% or "ring-down" to 0.2% power. The expression

$$Q_{\text{eff}} = \pi f \tau \tag{18.4c}$$

is then used to obtain the Q-factor[15,45] for the given frequency. In other words the Q-factor $= \pi$ multiplied by the number of cycles in time τ.

Precision of mechanical components is important; an analysis of a 2 MHz resonator with a liquid layer of 32 mm[35] delivered a decrease of 10–20% with respect to the resonance quality factor Q_{eff} when transducer and reflector were inclined by 1 mrad. Interestingly, this fall was greater for liquid layers with a higher quality factor. It was concluded that for the system in question ($L = 32$ mm, transducer surface 25 × 25 mm), manufacturing tolerances should not exceed 30 µm. Therefore materials commonly found in devices are either supplied in sheet form or easily machined with necessary precision, *e.g.* metals such as brass, aluminium and stainless steel. To meet the precision requirement, flat flush-mounted gaskets as spacers/sealings are used with success. Often Perspex is found in bench top set-ups, as it is widely available and easy to machine. However, Perspex should not be used for parts within the sound path, since high absorption induces high temperatures which can potentially induce melting.

As mentioned, chemical requirements have to be met by all parts touching the suspension. In this context interestingly Borosilicate glass (Tempax, Schott, Germany) was reported early as suitable for carrier layers.[44] There are a couple of reasons for this: firstly for its chemical inertness. Also its acoustic impedance matches the needs of a transfer layer between the PZT and the suspension, which is almost always an aqueous suspension. On the down side one finds its fragility and the limited possibilities to shape glass, *e.g.* it cannot be milled. In respect to this a substitute was found in Corning Macor®, which is described as glass ceramics by the manufacturer. It shows similar chemical and acoustical properties as glass, while being machineable, and thus allowing for more complex shapes of transducer carrier layers. Glass is also used as a reflector adding the possibility for visual inspection, which is often needed for reasons of control. The identification of glass as a suitable material led to the use of glass cuvettes in the industrially available acoustic filters.[46]

18.3.2 Tools for Development

There are principal challenges connected with the design and evaluation of ultrasonic resonators. One is that sound fields spread throughout a device—especially if a set-up comprises materials with low acoustic attenuation—and often one ends up with a multimode vibration state much more complicated than foreseen. The number of modes increases with dimension, simply because more harmonics can be excited. This tendency is therefore intensified in multi-wavelength devices.

Another challenge is that USWs and the resulting radiation forces are hard to measure *in situ* because any sensing device, for instance a hydrophone, will alter the field by its mere presence. A method for high-precision observation of the particle movement to evaluate the acoustic pressures and radiation forces has been presented in ref. 47 and 48 and Chapter 4.[1] In multi-wavelength devices, however, the view onto the particles is often obstructed, hence another helpful method to freeze the spatial distribution, *i.e.* the effect of the forces for further examination, has been developed[22] (see Section 18.3.4).

18.3.3 Models and Measurements

When designing a resonator for a defined purpose it is valuable to evaluate different approaches prior to construction of a set-up as the latter step is time consuming and expensive. Very helpful at this stage is a model, *i.e.* a mathematical description of the sound field to be expected based on the properties of the various parts of the set-up (dimension, shape, materials). Moreover a model is also useful to test a given proposition for having the desired characteristics and, to a certain degree, opens the possibility for improvements during construction. It is of low significance if an analytical model[49-51] or a numerical method such as FEM[14] is applied, as long as the geometrical configuration can be described properly.

A second pre-requisite is a measurement device. Electrical measurements are commonly made over a frequency range at the PZT electrodes. Depending on the measurement apparatus applied (simple oscillator, impedance analyser, phase locked loop, *etc.*), the absolute or complex values for impedance/admittance over frequency become available. For the latter case, measurements with dedicated systems designed in-house[52,53] or network analysers[54] have been carried out. Recently, a commercial device (Z-Check, SinePhase) for this purpose has become available.

When performing high-precision measurements, one has to be aware that the device under test seen by the measurement and driving electronics comprises the piezoelectric element and an additional static capacitance C_0 caused by connectors and cable, but foremost by the device's electrodes. These are facing each other and therefore represent a simple electric capacity. The well-known Butterworth-van-Dyke equivalent circuit for piezoelectric elements models these influences as parallel capacity; when evaluating the measurements this capacitance has to be compensated for.[35,55]

As mentioned, the process of evaluating a resonator in question is iterative. The natural starting point is the PZT itself. Often, the material properties are available from the manufacturer, hence the properties of the explicit specimen can be tested by a measurement/simulation comparison. To be more precise, usually mass density and speed of sound values in data sheets are of high quality, however, electromechanical coupling factors given tend towards the theoretical maximum. Thus using a measurement, the best specimen out of a group can be chosen.

The next step is the fabrication of the composite transducer, *i.e.* the gluing process. Now already some material parameters of the carrier (and the glue, if taken into account) might be missing, and can be estimated iteratively. If possible, it is of some use to compare measurements of the isolated transducer (*e.g.* lying on cotton wool) with measurements when the transducer is connected to the empty resonator cavity. Strong differences in such measurement results indicate an excitement of the device itself, possibly interfering with the main target to carry acoustic energy into the suspension.

Already here and in the next step, a common phenomenon is the occurrence of resonances that are not backed by the model. This is, as mentioned, a consequence of multi-wavelength devices being bigger and therefore exhibiting more possibilities to excite other modes, *e.g.* flexural waves. Sometimes it is difficult to match the measured spectrum and the prediction of the model. In such cases it can help to change the temperature of the device and measure again, as the change in frequency is usually different for longitudinal and transversal waves, hence the former can be identified easily.

The final step is the evaluation of the resonator filled with a well-characterised liquid, *e.g.* distilled, degassed water or alternatively with the pure host liquid used in the desired application. A resonator characterised in this way can be very valuable when dealing with suspensions of unknown properties (as will be the most common scenario). As only the parameters regarding the sonicated volume are influencing the measurement, just a handful of parameters are to be fitted.

Simulation-supported measurements were used to estimate the dependency of the acoustic quality factor on the particle load.[42] A simple cuvette resonator was filled with suspensions of yeast cells at various concentrations c_y. Measurements of the admittance were fitted with an analytical model,[50] the presence of 5 g L[21] bakery yeast did lower the original quality factor of 6000 moderately to 5300, an addition of 50 g L,[21] however, led to a considerably decreased value of 3000. The comparison delivered a relation of

$$Q_{\text{susp}} = Q^l_{\text{eff}} * e^{-0.1*c_y} \qquad (18.5)$$

between the resonance quality factor Q_{susp} of the resonator filled with suspension at a cell concentration c_y and the value for Q^l_{eff} when the liquid layer consisted of pure host liquid, in this case water. The coefficient of c_y in the exponent is specific for the respective resonator; for a different device, a value of −0.08 was found.

One successful application of this process of comparing measurements and modelled data was the assessment of the driving frequencies of a multi-wavelength resonator. Typically, these devices show spectra as in Figure 18.8(a). The admittance over frequency is dominated by the transducer resonance, i.e. the main trend in Figure 18.9(a), which is superimposed by a large number of "water resonances" (see also Chapter 5[14]).

Two characteristic frequencies are marked in Figure 18.9(a): f_1 coinciding with the fundamental resonance frequency of the piezo ceramic and therefore electrically most strongly pronounced, and f_2, which seems to be less optimal due to its lower admittance.

Figure 18.9(b) shows the acoustic energy density along the axial direction of the resonator calculated by an analytical model.[50] The graphs represent the results for the two selected frequencies f_1 (black) and f_2 (red). Surprisingly, the maximum energy density in the suspension layer was found at the electrically much less pronounced system's resonance f_2. Thus, the optimum

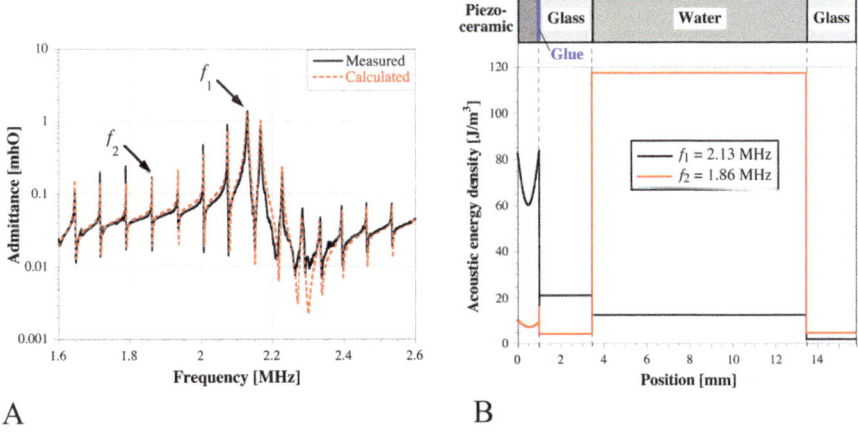

Figure 18.9 Investigation of two resonance frequencies: (a) electrical admittance, (b) spatial distribution of energy density at 4 W_{rms}. © IEEE. Reprinted with permission from ref. 41.

system performance is achieved at resonance frequencies not coinciding with the Eigenfrequencies of the piezoelectric ceramic. For practical reasons even the opposite is true: at f_1, one would face heating of the transducer due to the high energy density in the piezoelectric element.

18.3.4 Gel Technique

For the investigation of the sound-generated spatial arrangement of particles within a multi-wavelength device, a novel technique was developed overcoming the reconfiguration of ultrasonic fields upon any change of environment, *e.g.* the use of a hydrophone. The liquid fraction of the suspension was replaced by a gel initiated polymerization chemically or thermally entrapping the particle arrangement during the application of the ultrasonic field. Thus the particles were "frozen" at their positions caused by the ultrasonic forces. The resulting gel-block could be retrieved for further examination with a range of microscopy methods.[22,56]

In Figure 18.10, a set gel formed from a yeast/polyacrylamide suspension was sliced and examined by light microscopy. With this method it was possible to evaluate the influence of the axial and transverse primary radiation force separately. The periodical variation of the axial primary radiation force in the direction of sound propagation was reflected by the cells' organization in lines corresponding to the USW's pressure nodal planes (see Figure 18.10(a)).

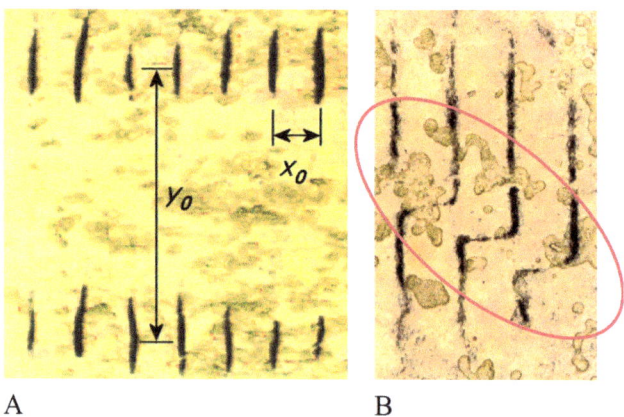

Figure 18.10 Light micrographs of a gel-block cut parallel to the direction of sound propagation. (a) Representation of the influence of the primary radiation forces. The half-wavelength was measured to be $x_0 = 379.6 \pm$ s.d. 21.7 μm, the distance between the columns was $y_0 = 3427.7 \pm$ s.d. 88.9 μm. (b) The "dancing of the cells", *i.e.* cells streaming from one pressure nodal plane to another (see red mark) is not explained theoretically yet. Adapted from ref. 56, reproduced with permission of Kluwer Academic Publishers.

An aberration of the predicted behaviour was detected (see Figure 18.10(b)) where cells were concentrated in regions along the direction of sound propagation. This phenomenon is dynamic, *i.e.* the cells stream from one nodal plane to the next one in the direction of sound propagation.

The gel-method allows for more sophisticated assessments; Figure 18.11 shows the results of a laser confocal microscopy investigation of the ultrasound field within a cylindrical resonator used for increasing the sensitivity of agglutination tests for clinically relevant compounds such as meningitis-causing bacteria.[57] The results reveal a structure of three cylindrical shells in which particles are concentrated.

Of interest for the construction of microfluidic applications might be the possibility to apply electron microscopy assessments to a gel specimen. Figure 18.12 shows a scanning electron microscopy image of the inner structure of an aggregate of yeast cells. The picture suggests that the cells are not touching each other and therefore the supply with nutrients and oxygen can be assumed for cells within an USW.

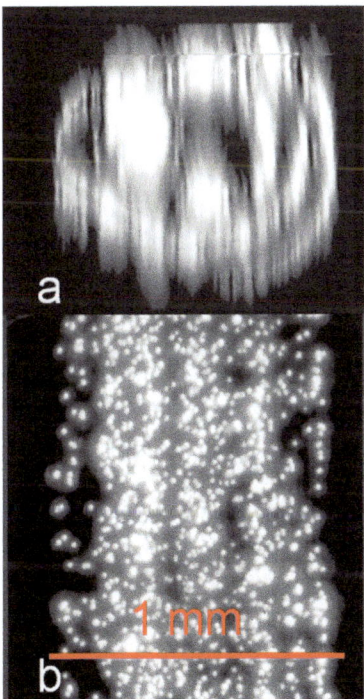

Figure 18.11 Confocal microscopy of the gel rod retrieved from a cylindrical resonator driven at 3 MHz. Transverse (a) and longitudinal (b) sections of fluorescent particles in a rod of alginate gel. The particles are concentrated along three surfaces of vanishing pressure.[22] Reproduced with permission of Elsevier Inc.

From the experiments in these studies, an additional important observation was made, when yeast cells were concentrated in the nodal plane in a nutrient gel (data not shown). After a four day incubation the amount of biomass at the location of the agglomerate significantly increased, indicating that the yeast cells were able to reproduce.[22] Thus the exposure to an USW did not harm yeast cells. It has previously been shown that megahertz sonication does not significantly affect yeast cell viability.[58] Only when the cells left the pressure nodes of the USW were significant alterations of viability and other similar parameters observed.[37] Detailed information about cell viability in microfluidic acoustic resonators can also be found in Chapter 21.[2]

The use of the gel technique was further employed to confirm a report[59] about the change of resonance frequency in multi-wavelength devices while the familiar arrangement of particles was brought about by the USW. Measurements of the true electrical power input vs. the frequency of a layered resonator (shown in Figure 18.8) were conducted (data not shown) with the "suspension" layer consisting of a gel-block with suspended yeast.[42] Data from randomly distributed cells were compared to data acquired when an USW was used to arrange the cells prior to gelation. The result was a significant influence on the resonance frequency by the ordering of cells in the pressure nodal planes; the re-arrangement by the USW influences the properties of the resonator as a whole.

18.3.5 Control of Acoustic Signal

It is necessary to be able to control the driving frequency for a stable operation for several reasons: one is the mentioned change in frequency as cells come into the nodal planes and agglomerate, hence it is necessary to

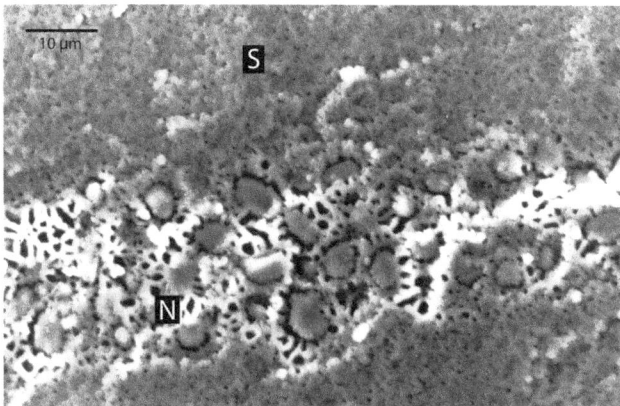

Figure 18.12 SEM image of a single band as shown in Figure 18.9. It shows the yeast cell distribution within the band and the structure of the treated gel. A rare net (N) of gel formation is visible within the band around the single cells. A more compact structure (S) is formed in the internodal space. The fracture was performed along the band. From ref. 56, reproduced with permission of Kluwer Academic Publishers.

tune the signal when this happens. Moreover, the speed of sound is affected by the temperature[60] and so is the wavelength. Consequently, resonance conditions are not met when the system's temperature changes, *i.e.* due to dissipation. Correct tuning of the driving frequency was accomplished by controlling and/or logging the temperature change[8,48] or measuring the excitement.[33,61]

Additionally, of course, electronic measurements are available. In this context it is important that the chosen electrical criterion meets the acoustic behavior. It was shown[35] that for resonances between the transducer Eigenfrequencies *e.g.* f_2 in Figure 18.9(a), high acoustic energy density in the liquid layer coincides with the maxima of the true power consumption, which could easily be utilized by the control electronics to tune the system properly. This resulted in high performance numbers,[44] *i.e.* optimal usage of a given electrical power consumption.

18.3.6 Dimensions

Transducer (T). The composite transducer (T) consists of the PZT (P) covered with electrodes (E) on both sides glued to a carrier layer (C), see Figure 18.8. Experience shows that for frequencies in proximity of the piezoelectric element's resonance frequency, the glue layer can be neglected. For frequencies away from the PZT Eigen frequency, however, some influence can occur.[54]

The reason is that around the PZT's resonance, *i.e.* when an odd multiple of half wavelengths (usually one) fits into the piezo layer, the glue layer is situated in or close to a pressure node, hence no forces are exerted. An analysis showed that some applications benefit from carrier plates close to odd multiples of a quarter wavelength. This, however, means that stress on the glue layer will not vanish and the type of glue is of importance; thin layers of low-viscosity types were used successfully.

Reflector (R). The perfect reflector, *i.e.* showing a reflection coefficient of 100%, would be the termination of the resonator by a liquid–vacuum interface. However, such a "free end", *i.e.* a force/pressure-free boundary condition, is obviously not practical. For construction purposes the termination by a layer showing a high acoustic impedance (hard and heavy) has turned out to be the best solution. To minimize the power loss in this reflector, the material should have low mechanical absorption; the thickness should be a (small) odd multiple of a quarter of the wavelength. This dimension leads to a standing wave with small amplitude (destructive interference) within the reflector expressing a pressure antinode at the interface with the liquid layer.[6,62]

For certain applications the use of microscopy cover slips or other thin reflectors was reported.[33,63] These are thinner than ¼ wavelength yet they are very efficient reflectors, probably by producing a condition similar to a liquid terminated by a vacuum.

18.3.7 Thickness Limits and Scale Considerations

Multi-wavelength resonators usually work well with acoustic path lengths up to a few tens of millimetres. They are often used with high particle concentrations, hence attenuation poses a limit. A strong resonance occurs when a wave is reflected many times along the same path. Attenuation by the suspension will lead to the situation in Figure 18.13: due to the damping the amplitude of the standing wave depicted by the grey envelope is strongest at the reflector and decreases across the cavity towards the transducer. This decrease is dependent on the thickness of the liquid layer. When attenuation reaches a point where only one significant reflection is expressed, the filtration breaks down because the amplitude of the returning wave becomes too low and particles are not held in place by the radiation forces. In addition, a residual propagating wave always exists (red line in Figure 18.13), as the reflected wave's amplitude interfering with the incoming wave (and thus building up the standing wave) has decreased. This travelling share will induce turbulent streaming, and particles will be dragged from the transducer towards the reflector. The phenomenon is known as Eckart streaming.

The preceding description about dimensioning is rather straightforward; the reader is referred to ref. 64 if interested in high performance applications from an industrial point of view, *i.e.* consideration of efficient use of electrical power, precise prediction in respect to heat production *etc.*

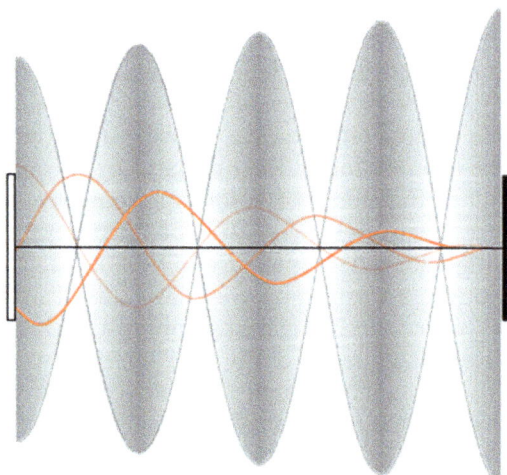

Figure 18.13 A standing wave (indicated by the amplitude envelope fixed in space) and a residual propagating wave (depicted by the red lines showing the displacement at different times). Adapted from ref. 42.

18.3.8 Acoustic Contrast

The observation of a steady movement of particles and liquid from the transducer towards the reflector can also be due to the lack of acoustic contrast between particles and liquid. The acoustic contrast is a coefficient depending on the speeds and sound and the mass densities of particle and liquid, respectively, in the calculation of the primary radiation force. In the case of a standing wave it is determining whether the force points towards the pressure or the displacement nodes for a given suspension. For a travelling wave the acoustic contrast can only be positive; the respective force always points in the direction of wave propagation.[65]

Figure 18.14 shows the ratio of the acoustic contrasts in the plane of mass density ratio Λ and speed of sound ratio σ of particle and liquid. In other words, the material properties of a given suspension correspond to one data point in Figure 18.14. The important fact is that acoustic contrasts for

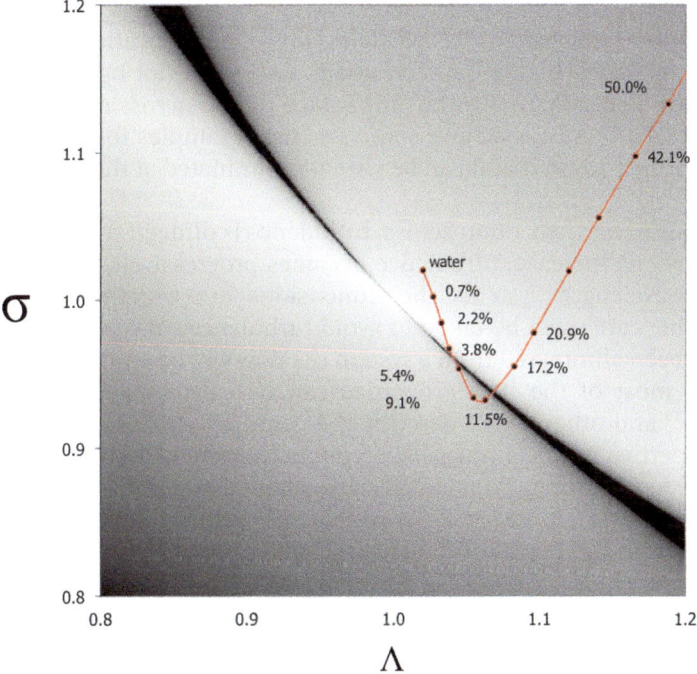

Figure 18.14 The acoustic contrast factor ratio K_p/K_s of a progressive and a standing wave over mass density ratio Λ and sound speed ratio σ of particle and liquid. Dark regions depict high values where K_p is exceeding K_s. As both coefficients are zero for different values, the ratio reaches infinity (within the black regions). The red line refers to specific values of yeast cells suspended in water–ethanol mixtures; the numbers give the ethanol concentration. Taken from ref. 42.

propagating and standing wave reach zero for different combinations of Λ and σ. That means that for suspensions with certain characteristics—in the black region in Figure 18.14—the force of the standing wave becomes very low or zero while the force of the propagating wave—although usually being much smaller—is still exerted on the particles. As a residual propagating wave is always present (see Figure 18.13), particles are driven through the resonator.

For yeast cells in water–ethanol mixtures—the red line in Figure 18.14 indicates the material properties of mixtures with increasing ethanol concentration—this was observed to cause decreased viability and changes in the morphology.[37] It should be emphasized that the standing wave was present; however, due to the lack of acoustic contrast the respective radiation force was not exerted on the particles.[42]

18.4 Flow Changes Produced by Scale-up

18.4.1 Non-Turbulent Flow Required

Well organised movement of particles into nodal locations is severely disturbed by flow crossing between nodal planes but not by flow along the planes. Turbulence is usually the only source of flow across nodal planes. In Sections 18.4.2 to 18.4.6 we give some practical examples for predicting the onset of turbulence so that it can be virtually eliminated at the system design stage.

In channels less than 1 mm across, turbulence is difficult to produce but as the dimensions increase, turbulence becomes progressively more probable. In multi-wavelength systems the dimensions are usually large and so considerable care must be taken to avoid turbulence. However, once turbulence has been eliminated from a system the flow will be laminar, and in this condition most of the flow properties can be calculated very accurately. Chapter 1[10] and other texts[66] describe the concepts required for calculating many properties in the laminar flow regime. Here we will give one example, simplified to two dimensions: in straight channels the flow is parabolic (as seen in Figure 18.15, width $w = 8$ mm); the velocity v_z of each band at distance z from the centre of the channel is described by:[67]

$$v_z = \frac{6Q'}{w^3} \cdot \left(\frac{w^2}{4} - z^2\right) \tag{18.6}$$

where Q' is the flow rate through a channel of unit depth (note we are using (') to distinguish flow rate Q' from the resonance Q-factor).

The certainty of velocity at each position in a laminar flow, together with the certainty of acoustic node positions, in principle allows complex hydrodynamic manipulation downstream from the sound field as described for sub-wavelength systems in Chapters 1, 2, 4, 6, 7 and 8.[1,4-7,11,13] However, surprisingly there are as yet no examples where hydrodynamic manipulation is used in combination with systems of more than two wavelengths.

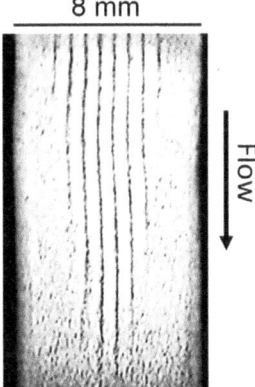

Figure 18.15 Parabolic flow profile visualised by bands of yeast cells a few seconds after switching on 9.9 MHz sound in a 1.1 mm wide resonating channel, which has been expanded to 8 mm downstream from the sound field, without disruption to the parabolic flow. Adapted from ref. 68. Reproduced with permission of Elsevier BV.

18.4.2 Calculating the Onset of Turbulence

There are no high-precision methods to calculate the exact condition where laminar flow will become turbulent, and so to prevent turbulence our designs should allow a large margin for error and err on the side of caution. The Reynolds number is a good indicator of turbulence; the probability of flow changing from laminar to turbulent increases rapidly when the Reynolds number is large, usually between 500 and 2000.[69] The lower value is used in examples here, since any flow detachment from the walls and not just full vortex shedding can reduce particle manipulation efficiency.

The design process for a new system should start with the analytical calculations set out in the next section. If this leaves any uncertainty about the possibility of turbulence then, although many more equations are available, the next step is usually to carry out initial tests either using 3-dimensional models or physical experiments.

18.4.2.1 Reynolds Number Dependence on Duct Profiles

For an infinite length parallel sided tube, the Reynolds number is given by:

$$\text{Re} = \frac{\rho_f v_{av} L_o}{\eta} \tag{18.7}$$

where ρ_f is the fluid density, v_{av} is the *average* velocity through the tube obtained from cross section area per flow rate Q', and η is the dynamic viscosity of the fluid.

L_0 the length scale,[10] also known as D_h hydraulic diameter, is the tube diameter for circular cross section tubes. When considering non-circular ducts of area A, the length scale is obtained from:[70]

$$L_0 = \frac{4A}{\text{wetted perimeter}} \qquad (18.8\text{a})$$

For rectangular ducts this is

$$L_0 = \frac{4wd}{2(w+d)} = \frac{2wd}{(w+d)} \qquad (18.8\text{b})$$

where w is the width and d is the depth of the channel.

18.4.3 Effect of Scale-up on the Initiation of Turbulence

Here we define the relationship between scale-up and the switch to turbulent flow, for long parallel-sided ducts, such as the main channel of the UES filter.[36,71]

The upper flow rate Q'_{max} for laminar conditions (we are defining this condition as Re = 500) is obtained from eqn (18.7) and (18.8b) as:

$$Q'_{max} = \frac{500\eta A}{\rho_f L_0} \qquad (18.9)$$

In water filled square ducts:

1×1 mm, $Q'_{max} = 0.45$ ml s^{-1} $v_{av} = 450$ mm s^{-1}.

10×10 mm, $Q'_{max} = 4.5$ ml s^{-1}, $v_{av} = 45$ mm s^{-1}.

This small advantage gained by increased channel cross section is shown as the lower (blue) line in Figure 18.16. Two better solutions for scale-up are also shown in Figure 18.16; these are:

(1) Increase only one dimension.
(2) Use many parallel channels.

The compromises from these two low dimension, high velocity scale-up solutions are: increased blockage potential from oversized particles, an increase in inter-channel wall area and a need to supply equalised ultrasound levels to all channels.

For enhanced sedimentation systems, slightly larger chambers are possible since the maximum sedimentation velocity at 1 MHz is 10 mm s^{-1} (calculated in Section 18.2.2), whereas the observed upper velocity in hydrodynamic separation systems is ~50 mm s^{-1} (5 µm diameter particles in a 10 mm field length[67]). At these flow velocities, in square ducts, flow becomes turbulent at 45 and 9 mm widths respectively.

18.4.4 Heating

Convection from heating introduces a further mechanism for fluid movement which may destroy particle alignment. In vertical chambers, when one wall is at a higher temperature than the opposite wall (e.g. heated by the

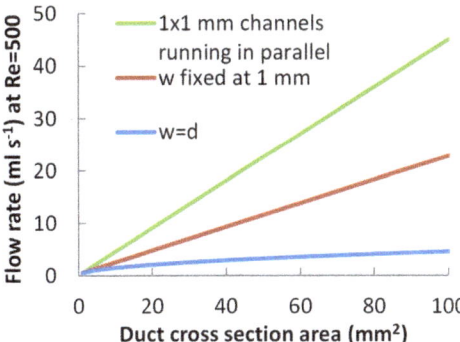

Figure 18.16 Maximum non-turbulent flow rates for three parallel-sided channel geometries: square, aspect ratio increased with area and multiple square 1 mm² channels. Calculated for water filled ducts $\rho_f = 997$ kg m^{-3}, $\eta = 9 \times 10^{-4}$ kg m^{-1} s^{-1}, $g = 9.8$ m s^2. Flow rates at the onset of turbulence (identified as Re = 500).

presence of an ultrasound transducer[5,74]), fluid against the heated wall rises and falls near the opposing wall, forming a chamber-high circulation as shown in Figure 18.17(a).[72,73] However, one circulation is not a significant problem when the nodal planes are vertical since it causes little cross-over between nodal planes in the central region of the chamber. Further heating increases the circulation velocity and a new regime occurs where the flow takes short cuts to the opposite wall forming smaller vortices, and this is disruptive for vertical alignment of particles. In water the new flow pattern may be much more complex than the stable pattern shown in Figure 18.17(b). There is no precise definition for when the disruptive circulation will begin but convection becomes significant when the Nusselt number Nu (convective heat transfer/conductive heat transfer) is above 1, and in water disruptive vortices have been observed for values above 10.[75]

A calculation of the Nusselt number depends on conditions; in the stable conditions in which we are interested, then for a chamber height h and width w, the Nusselt number is given empirically as:[72]

$$\text{Nu} = 0.36 \text{Pr}^{0.051} (h/w)^{-0.11} \text{Ra}_h^{0.25} \qquad (18.10)$$

where Pr (Prandtl number) = v/κ, v = kinematic viscosity (η/ρ_f) and κ is the thermal diffusivity $k/\rho Cp$, where k is the thermal conductivity of the fluid and Cp is the specific heat capacity.

Ra$_H$ (Rayleigh number for the chamber height) = Gr Pr where Gr (Grashof number) = $g\alpha_v(T_h - T_c)h^3/v^2$. g is acceleration due to gravity, α_v is volumetric thermal expansion coefficient, T_h and T_c are the temperatures of the hot and cold walls.

Figure 18.18 shows Nusselt numbers for chambers of differing heights where $h/w = 3$. The figure shows that Nusselt numbers near 10 are obtained from 1 °C heating in a 33 mm high chamber or 10 °C heating in a 15 mm high

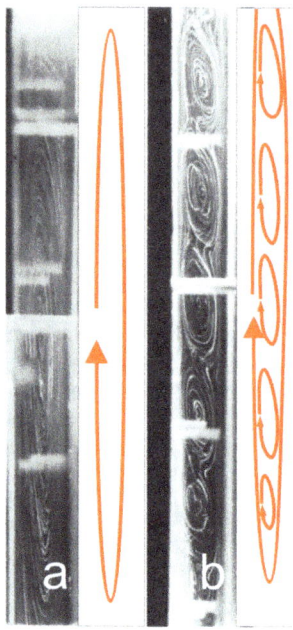

Figure 18.17 Flow lines in a vertical duct heated on the left wall to a constant temperature. The streak photographs show 0.2 mm polystyrene spheres (acting as flow tracers) moving through transformer oil. $h/w = 15$ ($w = 20$ mm). (a) Whole channel circulation Nu = 53. (b) Circulation breaks into multiple vortices Nu = 65. Photographs.[72,73] Reproduced with permission of Cambridge University Press.

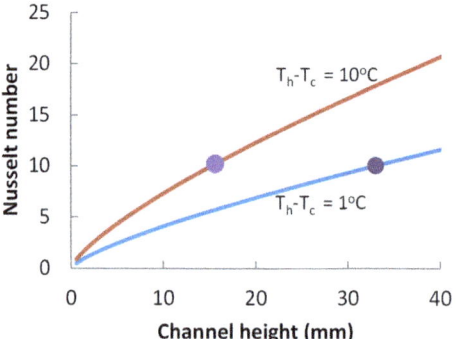

Figure 18.18 Dependence of thermal convection, indicated by the Nusselt number, on chamber height with 1 and 10 °C gradients across a parallel plate chamber. Disruptive convection (multiple vortices) in water occurs with Nusselt numbers above 10 in water. The markers show Nu = 10 at 1 °C and 10 °C temperature gradients indicating the maximum acoustic chamber heights for these temperature gradients. Calculated for water filled channels with aspect ratios $h/w = 3$. $\alpha_v = 0.256 \times 10^{-3}$ kg m^{-1} s^{-1}, $C_p = 4180$ J kg^{-1} °C^{-1}, $k = 0.607$ Wm^{-1} °C^{-1}, $g = 9.8$ m s^2.

chamber. Since the chamber width has little effect on the Nusselt number, a general rule is that the chamber height should not greatly exceed 30 mm and if high temperature gradients cannot be avoided smaller chambers should be used.

In 1 mm high chambers up to nearly 20 mm wide, the Nusselt number remains below 2 with a 10 °C temperature gradient. For chambers heated from below, the Rayleigh number can be used to determine the onset of convection; the critical value in this case is 1700.[73] In a 1 mm high chamber this is reached with a 90 °C temperature gradient while in a 50 mm high chamber (see Figure 18.1(c)) convection will start with a 7×10^{-4} °C temperature difference. When convection starts in chambers where h/w is small, it breaks into small Rayleigh-Bénard convection cells, and when particles are present, the cells organise the particles into patterns that could easily be confused with acoustic particle clump formations.

We can conclude that for chambers where $h < 1$ mm thermal convection is unlikely. In tall systems with horizontal acoustic nodes, convection crossing the nodes will cause some disruption as shown by two observations in a 50 mm high $\times$ 10 mm diameter tube (extensively used for observing horizontal nodes, where sedimentation is absent). On the lab bench without heat sources, convective flow was typically 100 µm s^{-1}, a temperature increase of 2 °C was observed after applying sound for 1 min and this increased the convection to 200 µm s^{-1}.[76] 1.3 µm diameter particles could not be held in 1 MHz nodal planes with this level of flow across the nodes. However, in a microgravity experiment, convection stopped within 1 s and the particles were brought to the nodal planes until collapsing when gravity returned.[77]

18.4.5 Entrance Condition

When a turbulent fluid enters a channel with a Reynolds number low enough for laminar flow to develop, it travels some distance before all turbulence ceases and a fully developed parabolic flow profile forms. The maximum entrance length L_e for fully developed laminar flow has been refined by a number of authors.[78–81] A straightforward approximation for Reynolds numbers in the region of 100 is given in Chapter 1.[11]

$$L_e = \frac{l_0 \text{Re}}{24} \tag{18.11}$$

For square section ducts with a flow rate at Re = 500, the increase in entrance length with width is shown in Figure 18.19. In a 1×1 mm duct, L_e is 21 mm; in a 10×10 mm duct it extends to 210 mm. Since large systems are usually operated at higher Reynolds numbers, the entrance length becomes increasingly important but curved parts as described in the next section can be used to reduce the length.

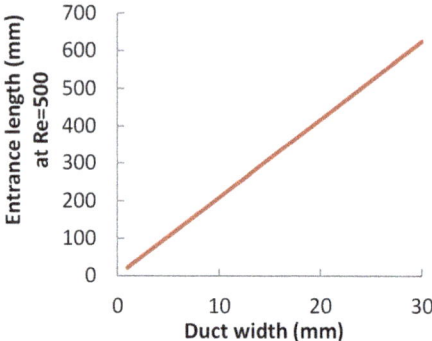

Figure 18.19 Relationship between the duct width and the entrance length L_e required for fully developed laminar flow. Calculated for square cross-section ducts Re = 500, *i.e.* the flow rate increases with duct width but remains just below the level for transition to full turbulence.

In systems with adjacent flow,[82] which is a technique used extensively in sub-wavelength systems[83] and described in Chapter 8,[5] a failure to observe the entrance length criteria can twist the flow so that side by side flows become above and below; only in extreme cases would detached vortices be seen.

At constant flow rate, the entrance length falls with increasing fluid viscosity.[22]

18.4.6 Flow Expansion and Contraction without Disruption or Dead Volumes

When flow in a channel is laminar, expanding the channel can give three possible outcomes, dependent on the angle of expansion, as shown in Figure 18.20(A). Full turbulence with vortices shedding from the discontinuity; these can be detrimental further downstream in the system. (B) Attached vortices, and flow line separation from the wall (see Figure 18.7(b), top left corner for example); when this happens the streamlines become concentrated at the centre (the pressure drop at expansion can actually become a pressure increase as the flow becomes squeezed to the centre lines). Alternatively, if streamlines follow the expansion with no separation from the wall then no turbulence develops (C and D).

The detachment criteria have been identified to a reasonable level of precision for aerofoils[84] but not for liquids. Where precision is needed, chemical engineering texts recommend resorting to numerical modelling.[70] However, some early experimental work with channels (~40–100 mm wide) provide good indicative values which are similar to computational fluid dynamic model values.[85] In large water-filled ducts no flow separation (Figure 18.20(C)) was observed when the angles between the diverging walls

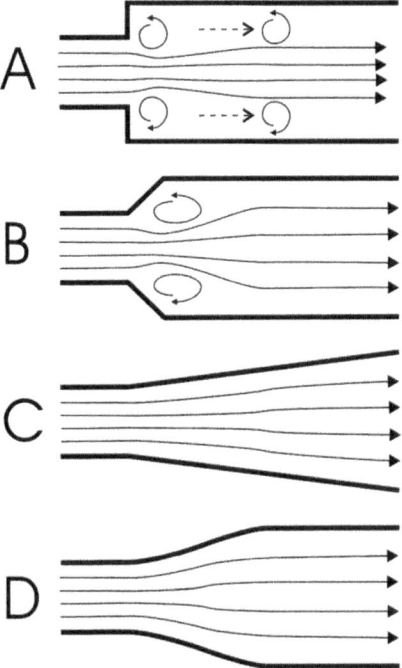

Figure 18.20 Channel expansion. (A) At high flow rates, an abrupt expansion produces vortex shedding. (B) Less abrupt expansion produces stable attached vortices and separation of streamlines from the walls. (C) Low angle expansion maintains fully formed laminar flow. (D) Curves decrease the length required for expansion.

was <5° for circular channels, and 11° for rectangular ducts expanding in one-dimension[86] (14° for air[87]). In contracting ducts, detachment is less likely and not usually seen with angles <45°. The flow patterns A–D appear to be almost independent of flow rate[86] and scale in Chapter 12,[17] for example, flow streaming (similar to Figure 18.20(B)) is described around spheres down to the scale of cells.[16,17] Scale independence allows very large expansions, such as upstream from Figure 18.15, where the channel expanded seven-fold (1.1–8 mm) at up to 20° without measurable disruption to the particle formations created with ultrasound.[67] This steep angle is probably partly achieved by the use of curved walls designed to follow the flow lines as shown in (D).

In some systems, separation from the wall may be unavoidable; this creates dead-volumes, however, provided the vortices remain stable, their presence may be acceptable. Instability of the vortices leads to fluid oscillations throughout the system and should be avoided. Detachment, *i.e.* full turbulence (A) (indicated by pressure drop), should of course always be avoided; it occurs with expansion angles >60°.

18.5 New Design Approaches

18.5.1 Heating Mitigation

At low power, mechanical losses in the transducer and at interfaces are the main source of heat, while at high power levels, electrical losses in the transducer overtake mechanical losses and the piezoelectric ceramics can become very hot (sufficient to depole the transducer *i.e.* ~200 °C).[88] Mechanical losses are often controllable and it is possible to run transducers at 5–10 °C above ambient, although of course temperature differentials make particle manipulation increasingly unstable (due to loss of resonance and convection). The best results are usually obtained with no part heating by more than 1 °C. When maximum power is needed, outputs have been pushed upwards 2–3 fold by adding some form of cooling; in sub-wavelength devices the piezoelectric element has been separated from the chamber wall with a metal block heat sink cooled with a Peltier element.[8,48] Separation is also used in larger devices where forced water cooling replaces the metal block;[33,71] or an alternative is forced air cooling, which has also been used very successfully for cell sonication.[31] However, heat removal becomes increasingly difficult with increasing system volume. Designing efficient acoustics is essential to achieve a high ratio of pressure amplitude to heat. 10 years ago most acoustic chambers were designed as bonded multilayer structures operating in the compression mode to form a single span coherently resonating stack of the type shown in Figure 18.8. Several one-dimensional model approaches[6,40,51] were developed[17,25,39] for this structure, which were experimentally verified[6,50,54] as seen in Figure 18.9. However, over the past 10 years, the design of most micro-fluidic acoustic systems has moved to applying the sound from one side. In these systems the mode carrying the vibrations into the fluid chamber is not a controlled compression wave. A multi-wavelength system which operates in a similar manner is shown in Figure 18.3. In this case the piezoelectric element is pressed to one side of a glass duct. This approach has minimal heat transfer from the drive element to the duct and since all parts are relatively unconstrained, interface losses are also minimised. The next section describes some of the requirements for developing this type of system.

18.5.2 Creating a Standing Wave: Selection of a Particular Wall Mode Is Not Always Required

Compression waves entering a liquid from a solid experience very little attenuation over long distances. Travelling shear waves, in contrast, decay in water within 0.5 μm of the surface at 5 MHz.[89] A small propagation length (an evanescent wave) is also produced when the wavelength in the solid is less than in the liquid, as described in Chapters 3 and 17.[12,90] However, when the water dimension is resonant, there is very little difference between a wave produced by a compression or a shear device.[89] In both cases the vibration energy converts to a compression mode in the water. Based on the assumption that it does not matter which mode produces the vibrations, then the critical factor is delivering vibration energy at the resonant frequency of the fluid cavity. For an in-depth description of waveforms at fluid interfaces, Chapter 14[91] is recommended.

There is also freedom to switch between mode types in the solid components. This is demonstrated by the example of the multimode system in Figure 18.3 where a compression mode in the PZT converts to a flexural mode in the glass duct and then back to a compression mode in the liquid. The conversion between the two solids is achieved by maintaining minimal contact between them, ensuring that variation in mode shape between these parts does not cause interference (or transfer heat).

Minimal contact also allows each element to be separately modelled and optimised for maximum vibrations. Although mode-conversion is widely used with sub-wavelength systems only one other minimal contact particle manipulation system has been reported.[61] However, the modelling approach is well established within the field of non-destructive testing applications and an outline description is given in the next section.[92–94] Since this minimal contact approach is at an early stage of development for particle manipulation, researchers choosing this method should proceed with some caution: although minimal contact potentially allows high Q and therefore increased efficiency, the weak coupling can lead to complex "beating" variations in power levels between the components. Also systems requiring fast switching between modes (currently used to enhance separation in multi-wavelength systems,[82] and to control position in sub-wavelength systems) will have their speed limited by the "ring down" and rise as Q is increased (the relationship is given in Section 18.3.1).

In air-based systems the loading and coupling back from the air to the structure can be ignored and so even the fluid (air) and its container can be modelled separately. This is not possible for liquids that will strongly affect the resonance frequency of the driving structure. Therefore, although the PZT drive can be modelled separately, a liquid and its chamber should always be modelled together.

18.5.3   Modelling Resonant Parts

The Modelling Path

(1) Choose the operating frequency: in water-filled chambers, this should be above 1 MHz to avoid cavitation effects; in air, much lower frequencies can be used but forces increase with frequency.
(2) Define the internal resonant dimensions: for a parallel wall chamber, the distance between the walls must be a multiple of half wavelengths, $n.c/2f$ where n is an integer. For tubular chambers (see Figure 18.1(B), (D), (F) and 18.10), the diameter is $Xn.c/\pi f$ where Xn is from the Bessel series 3.813, 7.015, 10.173 ($J1(X) = 0$). When $n > 20$, the size is less critical since a 5% change in operating frequency is usually possible.

(3) Match the wall mode shape to the fluid mode shape at the operating frequency. This is not usually carried out for MHz chambers with substantial walls because at these frequencies there are so many resonances that selection of one mode is difficult. In lower frequency air fill chambers, matching mode shape of the fluid and the wall is used[95] while for higher frequencies this is under development.

18.5.4 Modelling the Chamber Wall

For simplicity a rectangular bar of aluminium is described here. The symmetric (compression) modes (S) occur when the bar length, l, is an integer multiple, n, of the ½ wavelengths (the first two harmonics are shown in Figure 18.21). At low frequencies the phase velocity v_s and resonant frequency, f_r are accurately predicted by the approximation:[97]

$$v_s = \sqrt{\frac{E}{\rho_p(1-v^2)}} \tag{18.12a}$$

$$f_r = \frac{v_s}{2nl} \tag{18.12b}$$

where E is the elastic (Young's) modulus (70.8 GPa), ρ is the density 2700 kg m^{-3} and v is Poisson's ratio (0.3375). The wavelength depends on the phase velocity not the bulk velocity but phase velocity varies with frequency as shown in the dispersion plot in Figure 18.22 (obtained with the modelling program Disperse[98]) and therefore the approximation eqn (18.12a) is not valid at higher frequencies (as indicated by the divergence of the open red symbols in Figure 18.22). Despite this, analytical solutions remain a very useful first calculation step and the tables compiled by Blevins[99] are highly recommended; these give solutions for many shapes and the anti-symmetric (flexural plate) mode (A), which is also shown in Figure 18.21. The analytical solution for the A mode is more difficult but the agreement with the disperse curves is good at all frequencies. Numerical (3D) solutions for both the compression (S) and the flexural plate (A) modes for the 100 × 50 × 20 mm aluminium bar are also plotted in Figure 18.22 (including those presented in Figure 18.21). Agreement between numerical solutions and dispersion curves is very high (velocity values for the 3D solutions have been calculated on the basis of $n0.5\lambda$ for resonances in the compression mode and $(n0.5 + 0.25)\lambda$ for the flexural wave mode).

It is convenient to use all three modelling methods together. Starting with an analytical calculation such as eqn (18.12) to identify the frequency range, which can be confirmed using a mode tracking program such as Disperse, this program also provides useful representations of the mode shape in single, multi-layer stacks and tube systems, including fluid layers but with only one finite dimension. Finally, 3D stress analysis models produced by programs such as Comsol[100] and Abaqus[96] can in some cases provide results close enough to experimental systems for dimensions to be transposed directly to

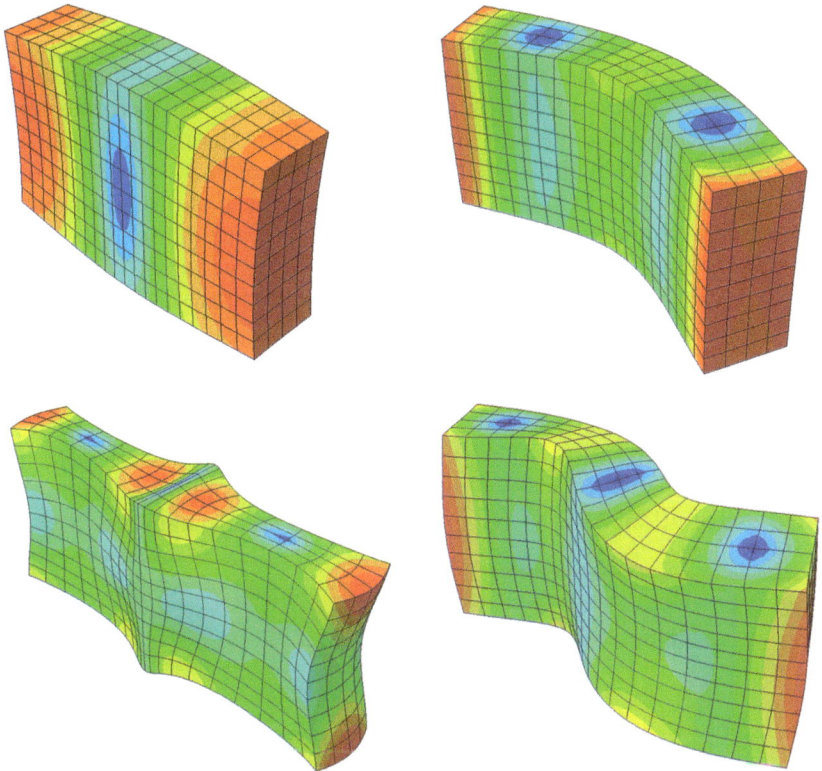

Figure 18.21 Aluminium bar 100 × 50 × 20 mm showing displacement (not to scale) predicted by Abaqus.[96] First harmonics (upper line) and second harmonics (lower line) are shown for zero order antisymmetric and symmetric waves in the length direction. Colours indicate displacement magnitude (three dimensions combined), blue low, red high.

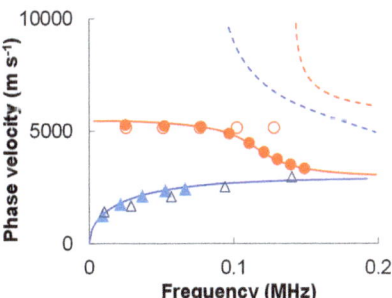

Figure 18.22 Resonant modes in an aluminium bar 100 × 50 × 20 mm, red = compression S0, blue = flexural A0 correspond with Figure 18.21 (dotted lines top right are higher modes). Lines are predicted by Disperse; open symbols are numerical predictions, filled symbols are 3D models from Abaqus.[96]

engineering sketches (and require significant amounts of specialist knowledge). Although multi-wave resonators have many advantages, the maximum number of waves is limited: identifying modes becomes more difficult as the number of wavelengths increases; stability of operation also becomes more difficult, therefore, as a rule, the number of wavelengths in one dimension should not exceed 15 in either the fluid or one element of the structure.

It is likely that in the future, 3D modelling will be used to design chambers where the wave shape in the wall matches the wave shape in the fluid. This is expected to increase efficiency and control streaming patterns.

The next step is to test the system and confirm that the wave-shapes have been created.

18.5.5 Experimental Confirmation

The ideal method to confirm that model predictions of the acoustic fields are correct is to test the chamber with a particle suspension and look at the particle distribution. This is not possible using small chambers that are

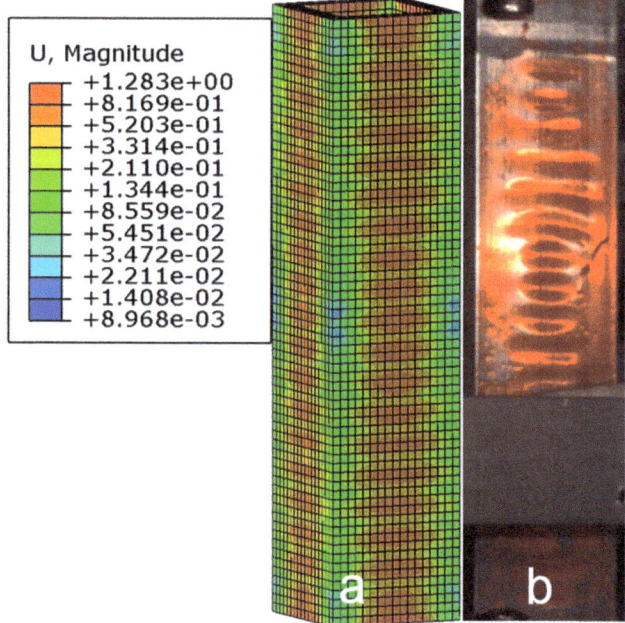

Figure 18.23 Experimental confirmation of modelled mode shape predictions in an aluminium duct (20 × 20 × 100 mm OD). (a) A stress analysis (Abaqus) model of the duct showing transverse velocity antinodes nodes at 206 kHz. (b) Confirmation of the model with a liquid Chladni figure. Water with red food colouring moves to the same pattern of antinodes on the surface of the duct when driven at 203 kHz. (The horizontal bar across the lower section is holding the piezoelectric transducer in place.)

either opaque or manifest acoustic streaming. However, in large chambers, there are several options that can be used:

(1) A high frequency microphone is useful for confirming the sound output level; a nozzle placed on the front can help monitor small regions.
(2) Sand, glass beads or liquid placed on the surface will identify the mode shape by forming Chladni figures,[101,102] see Figure 18.23.
(3) A scanning laser vibrometer (if available) provides non-invasive confirmation of all the mode shapes across frequency range.

Agreement between model and modes on the outside of a chamber gives confidence in the model predictions for the inside.

Since it is possible to build a system without knowledge of the wave shape in the walls, these considerations in Sections 18.5.4 and 18.5.5 are not usually taken into account. However, as the need for efficiencies of power and particle positioning increase, the techniques described in this section are likely to become more important.

18.6 Appendix

Nomenclature

C_p	Specific heat capacity at constant pressure. 4180 (J kg^{-1} °C^{-1})
c_0	Speed of sound
c_y	Cell concentration, yeast ml^{-1}
g	Acceleration, due to gravity 9.8 (m s^{-2}).
D	Channel depth (m)
e	Base of the natural logarithm
E	Elastic modulus (Pa)
F	Frequency (Hz)
F_r	Resonant frequency (Hz)
Gr	Grashof number, $g\alpha(T_h - T_c)H^3/\nu^2$
h	Internal channel height (m)
k	Thermal conductivity 0.607 (Wm^{-1} °C^{-1})
L_0	The length scale (m)
L_e	Entrance length (m)
Nu	Nusselt number (convective heat transfer/conductive heat transfer)
Pr	Prandtl number, ν/κ
Q'	Flow rate (m^3 s^{-1})
Q'_{max}	Maximum non turbulent flow rate
Q_{eff}	Effective Q-factor of multiple parts $2\pi\varepsilon$/power loss
Q_{susp}	Q-factor of the suspension
r_c	Cell clump of radius $\lambda/8$ (m)
Ra	Rayleigh number, $Gr\,Pr$
Re	Reynolds number, $\rho_f v_{av} L_0/\eta$

T_c T_h	Temperature of walls (°C)
v_{av}	Average velocity across tube (m s^{-1})
v_f	Bulk fluid velocity (m s^{-1})
v_c	Clump velocity (m s^{-1})
v_z	Velocity at z from the channel centre (m s^{-1})
w	Internal chamber width (m)
Xn	Bessel series
z	Distance from centre of channel (m)

Greek letters

A	Acoustic Absorption coefficient
α_v	Volume (or cubic) thermal expansion dynamic viscosity of the fluid 0.256 × 10^{-3} (kg m^{-1} s^{-1}).
ε	Stored energy (J)
κ	Thermal diffusivity, $k/\rho Cp$ 1.45 × 10^{-7} (J^2 °C^{-2} m^{-4} s^{-1})
λ	Wavelength of sound (m)
Λ	Mass density ratio particle fluid
P	Density (kg m^{-3})
ρ_c	Cell density—yeast 1.114 × 10^3 (kg m^{-3})
ρ_f	Fluid density 997 (kg m^{-3})
H	Dynamic viscosity of the fluid 9 × 10^{-4} (kg m^{-1} s^{-1} or Pa s^{-1})
Σ	Speed of sound ratio particle/fluid
Y	Kinematic viscosity, η/ρ_f (m^2 s^{-1})

Values given for water are at 25 °C[103] where not stated otherwise.

References

1. H. Bruus, *Lab Chip*, 2012, **12**, 1014–1021.
2. M. Wiklund, *Lab Chip*, 2012, **12**, 2018–2028.
3. G. G. Yaralioglu, I. O. Wygant, T. C. Marentis and B. T. Khuri-Yakub, *Anal. Chem.*, 2004, **76**, 3694–3698.
4. A. Lenshof, M. Evander, T. Laurell and J. Nilsson, *Lab Chip*, 2012, **12**, 684–695.
5. A. Lenshof, C. Magnusson and T. Laurell, *Lab Chip*, 2012, **12**, 1210–1223.
6. P. Glynne-Jones, R. J. Boltryk and M. Hill, *Lab Chip*, 2012, **12**, 1417–1426.
7. P. Augustsson and T. Laurell, *Lab Chip*, 2012, **12**, 1742–1752.
8. J. D. Adams, C. L. Ebbesen, R. Barnkob, A. H. J. Yang, H. T. Soh and H. Bruus, *J. Micromech. Microeng.*, 2012, **22**, 1–8.
9. A. Lenshof, A. Ahmad-Tajudin, K. J. rås, A.-M. S. rd-Nilsson, L. Åberg, G. Marko-Varga, J. Malm, H. Lilja and T. Laurell, *Anal. Chem.*, 2009, **81**, 6030–6037.

10. H. Bruus, *Lab Chip,* 2011, **11**, 3742–3751.
11. H. Bruus, *Lab Chip,* 2012, **12**, 1578–1586.
12. J. Dual and T. Schwarz, *Lab Chip,* 2012, **12**, 244–252.
13. H. Bruus, *Lab Chip,* 2012, **12**, 20–28.
14. J. Dual and D. Möller, *Lab Chip,* 2012, **12**, 506–514.
15. J. Dual, P. Hahn, I. Leibacher, D. Möller and T. Schwarz, *Lab Chip,* 2012, **12**, 852–862.
16. S. S. Sadhal, *Lab Chip,* 2012, **12**, 2292–2300.
17. S. S. Sadhal, *Lab Chip,* 2012, **12**, 2600–2611.
18. M. Wiklund, R. Green and M. Ohlina, *Lab Chip,* 2012, **12**, 2438–2451.
19. S. Peterson, G. Perkins and C. Baker, in *IEEE 8th Annual Conference of the Engineering in Medicine and Biology Society*, Institute of Electrical and Electronics Engineers, IEEE, New York City, USA, 1986, pp. 154–156.
20. C. M. Cousins, P. Holownia, J. J. Hawkes, M. S. Limaye, C. P. Price, P. J. Keay and W. T. Coakley, *Ultrasound Med. Biol.,* 2000, **26**, 881–888.
21. G. Whitworth, M. A. Grundy and W. T. Coakley, *Ultrasonics,* 1991, **29**, 439–444.
22. L. Gherardini, C. Cousins, J. J. Hawkes, J. Spengler, S. Radel, H. Lawler, B. Devcic-Kuhar, M. Gröschl, W. T. Coakley and A. J. McLoughlin, *Ultrasound Med. Biol.,* 2005, **31**, 261–272.
23. S. Kogan, G. Kaduchak and D. N. Sinha, *J. Acoust. Soc. Am.,* 2004, **116**, 1967–1974.
24. E. Benes, F. Hager, W. Bolek and M. Gröschl, in *Ultrasonics International '91*, Butterworth-Heinemann, Oxford, UK, 1992, pp. 167–170.
25. B. Lipkens, J. Dionne, A. Trask, B. Szczur, A. Stevens and E. Rietman, *Phys. Procedia,* 2010, **3**, 263–268.
26. Z. Wang, P. Grabenstetter, D. L. Feke and J. M. Belovich, *Biotechnol. Prog.,* 2004, **20**, 384–387.
27. M. J. Anderson, R. S. Budwig, K. S. Line and J. G. Frankel, *IEEE Ultrason. Symp.,* 2002, 464–467.
28. E. Riera, J. A. Gallego-Juarez and T. J. Mason, *Ultrason. Sonochem.,* 2006, **13**, 107–116.
29. I. Gonzalez, T. L. Hoffmann and J. A. Gallego, *J. Aerosol Sci.,* 2000, **31**, 1461–1468.
30. T. J. Mason and J. P. Lorimer, *Applied sonochemistry: uses of power ultrasound in chemistry and processing*, John Wiley & Sons, Chichester, 2002.
31. K. A. J. Borthwick, W. T. Coakley, M. B. McDonnell, H. Nowotnyc, E. Benes and M. Gröschl, *J. Microbiol. Methods,* 2005, **60**, 207–216.
32. E. A. Brujan, T. Ikeda and Y. Matsumoto, *Exp. Therm. Fluid Sci.,* 2008, **32**, 1188–1191.
33. J. J. Hawkes and W. T. Coakley, *Enzyme Microb. Technol.,* 1996, **19**, 57–62.
34. M. Gröschl, *Acta Acust. Acust.,* 1998, **84**, 815–822.
35. M. Gröschl, *Acta Acust. Acust.,* 1998, **84**, 632–642.
36. J. J. Hawkes, M. S. Limaye and W. T. Coakley, *J. Appl. Microbiol.,* 1997, **82**, 39–47.

37. S. Radel, L. Gherardini, A. J. McLoughlin, O. Doblhoff-Dier and E. Benes, *Bioseparation,* 2000, **9**, 369–377.
38. S. Radel, *Elektrotech. Inftech,* 2009, **126**, 51–57.
39. H. Böhm, L. G. Briarty, K. C. Lowe, J. B. Power, E. Benes and M. R. Davey, *Biotechnol. Bioeng.,* 2003, **82**, 74–85.
40. M. Hill and R. J. K. Wood, *Ultrasonics,* 2000, **38**, 662–665.
41. E. Benes, M. Gröschl, H. Nowotny, F. Trampler, T. Keijzer, H. Böhm, S. Radel, L. Gherardini, J. J. Hawkes, R. König and C. Delouvroy, in *IEEE ultrason. symp., Atlanta,* 2001, pp. 649–659.
42. S. Radel, in *Institut für allgemeine physik,* TU Wien, Vienna, 2003.
43. S. Gupta and D. L. Feke, *AIChE J.,* 1998, **44**, 1005–1014.
44. M. Gröschl, *Acta Acust. Acust.,* 1998, **84**, 432–447.
45. R. Nawrodt, A. Zimmer, T. Koettig, C. Schwarz, D. Heinert, M. Hudl, R. Neubert, M. Thürk, S. Nietzsche, W. Vodel, P. Seidel and A. Tünnermann, *J. Phys.: Conf. Ser.*, 2008, **122**, 012008.
46. T. M. P. Keijzer, F. Trampler, A. Oudshoorn, O. Doblhoff-Dier and H. v. d. Berg, in *Cell culture engineering VII, Snowmass village, Colorado, USA,* 2002.
47. R. Barnkob, P. Augustsson, T. Laurell and H. Bruus, *Lab Chip,* 2010, **10**, 563–570.
48. P. Augustsson, R. Barnkob, S. T. Wereley and H. Bruus, *Lab Chip,* 2011, **11**, 4152–4164.
49. M. Hill, Y. Shen and J. J. Hawkes, *Ultrasonics,* 2002, **40**, 385–392.
50. M. Hill, R. J. Townsend and N. R. Harris, *Ultrasonics,* 2008, **48**, 521–528.
51. H. Nowotny and E. Benes, *J. Acoust. Soc. Am.,* 1987, **82**, 513–521.
52. M. Schmid, E. Benes and R. Sedlaczek, *Meas. Sci. Technol.,* 1990, **1**, 970–975.
53. R. Schnitzer, C. Reiter, K.-C. Harms, E. Benes and M. Gröschl, *IEEE Sens. J.,* 2006, **6**, 1314–1322.
54. J. J. Hawkes, W. T. Coakley, M. Gröschl, E. Benes, S. Armstrong, P. J. Tasker and H. Nowotny, *J. Acoust. Soc. Am.,* 2002, **111**, 1259–1266.
55. E. Benes, M. Gröschl, F. Seifert and A. Pohl, *IEEE Trans. Ultrason., Ferroelectr., Freq. Control.,* 1997, **45**, 1314–1330.
56. L. Gherardini, S. Radel, S. Sielemann, O. Doblhoff-Dier, M. Gröschl, E. Benes and A. J. McLoughlin, *Bioseparation,* 2001, **10**, 153–162.
57. M. A. Sobanski, C. R. Tucker, N. E. Thomas and W. T. Coakley, *Bioseparation,* 2001, **9**, 351–357.
58. S. Radel, A. McLoughlin, L. Gherardini, O. Doblhoff-Dier and E. Benes, *Ultrasonics,* 2000, **38**, 633–637.
59. C. S. Kwiatkowski and P. L. Marston, *J. Acoust. Soc. Am.,* 1998, **103**, 3290–3300.
60. M. Greenspan and C. E. Tschiegg, *J. Acoust. Soc. Am.,* 1959, **31**, 75–76.
61. G. Goddard and G. Kaduchak, *J. Acoust. Soc. Am.,* 2005, **117**, 3440–3447.
62. E. Benes and F. Hager, *Austrian Patent* PCT/AT/1989/000098, 1990.

63. P. Glynne-Jones, R. J. Boltryk, N. R. Harris, P. Baclet and M. Hill, *J. Acoust. Soc. Am. Express Lett.*, 2009, **126**, EL75–EL79.
64. F. Trampler, *Accumulation of Acoustic Energy for Fluid-Particle Separation in Industrial Processes*, PhD Thesis, nst. für Allgemeine Physik, Vienna University of Technology, 2000.
65. K. Yosioka and Y. Kawasima, *Acustica*, 1955, **5**, 167–172.
66. H. Bruus, *Theoretical microfluidics*, Oxford University Press, Oxford, 2008.
67. J. J. Hawkes and W. T. Coakley, *Sens. Actuators, B*, 2001, **75**, 213–222.
68. J. J. Hawkes, D. Barrow, J. Cefai and W. T. Coakley, *Ultrasonics*, 1998, **36**, 901–903.
69. J. F. Douglas, J. M. Gsiorek, J. A. Swaffield and L. B. Jack, *Fluid mechanics*, Prentice Hall, Harlow, England, New York, 2011.
70. F. M. White, *Viscous fluid flow*, McGraw-Hill, New York, 1991.
71. F. Trampler, S. A. Sonderhoff, P. W. S. Pui, D. G. Kilburn and J. M. Piret, *Biotechnology*, 1994, **12**, 218–284.
72. N. Seki, S. Fukusako and H. Inaba, *J. Fluid Mech.*, 1978, **84**, 695–704.
73. D. J. Tritton, *Physical fluid dynamics*, Clarendon Press, Oxford, 1988.
74. M. Wiklund, *Lab Chip*, 2012, **12**, 2018–2028.
75. N. Seki, S. Fukusako, H. Inaba and Sapporo, *Warme Stoffubertrag*, 1978, **11**, 145–156.
76. M. A. Grundy, in *Pure and applied biology*, University of Wales Cardiff, Cardiff, 1994.
77. J. J. Hawkes, J. J. Cefai, D. A. Barrow, W. T. Coakley and L. G. Briarty, *J. Phys. D: Appl. Phys.*, 1998, **31**, 1673–1680.
78. B. Atkinson, M. P. Brocklebank, C. C. H. Card and J. M. Smith, *AIChE J.*, 1969, **15**, 548–553.
79. R. Y. Chen, *J. Fluids Eng.*, 1973, **95**, 153–158.
80. N. Dombrowski, E. A. Foumeny, S. Ookawara and A. Riza, *Can. J. Chem. Eng.*, 1993, **71**, 472–476.
81. M. Friedmann, J. Gillis and N. Liron, *Appl. Sci. Res.*, 1968, **19**, 426–438.
82. Y. Liu and K.-M. Lim, *Lab Chip*, 2011, **11**, 3167–3173.
83. J. J. Hawkes, R. W. Barber, D. R. Emerson and W. T. Coakley, *Lab Chip*, 2004, **4**, 446–452.
84. L. Castillo, X. Wang and W. K. George, *J. Fluids Eng.*, 2004, **126**, 297–304.
85. E. M. Sparrow, J. P. Abraham and W. J. Minkowy, *Int. J. Heat Mass Transfer*, 2009, **52**, 3079–3083.
86. A. H. Gibson, *Philos. Trans. R. Soc. London*, 1910, **83**, 366–378.
87. D. L. Cochran and S. J. Kline, in *National advisory committee for aeronautics*, Stanford University, 1958.
88. K. Uchino, *Smart Mater. Struct.*, 1998, 7, 273–285.
89. T. W. Schneider and S. J. Martin, *Anal. Chem.*, 1995, **67**, 3324–3335.
90. M. Wiklund, S. Radel and J. J. Hawkes, *Lab Chip*, 2012, **13**, 25–39.
91. M. Gedge and M. Hill, *Lab Chip*, 2012, **12**, 2998–3007.
92. M. J. S. Lowe and P. Crawley, *J. Nondestruct. Eval.*, 1994, **13**, 185–200.
93. M. Castaings and M. J. S. Lowe, *J. Acoust. Soc. Am.*, 2008, **123**, 696–708.

94. W. J. Jacobi, *J. Acoust. Soc. Am.*, 1949, **21**, 120–127.
95. G. Kaduchak, D. N. Sinha and D. C. Lizon, *Rev. Sci. Instrum.*, 2002, **73**, 1332–1336.
96. S. Abaqus Unified FEA software, Dassault systemes.
97. S. W. Wenzel and R. M. White, *IEEE Trans. Electron Devices,* 1988, **15**, 735–743.
98. B. Pavlakovic and M. Lowe, *Disperse, software for generating dispersion curves*, Mechanical Engineering, Imperial College, London, 2011, 2.0.16i.
99. R. D. Blevins, *Formulas for natural frequency and mode shape*, Krieger Publishing Company Malbar, Florida, 2001.
100. COMSOL, Finite Element Analysis and Engineering Simulation Software.
101. E. F. F. Chladni, *Entdeckungen über die Theorie des Klanges, discoveries concerning the theory of music*, Weidmanns Erben und Reich, Leipzig, 1787.
102. M. Faraday, *Philos. Trans. R. Soc. London,* 1831, **121**, 299–340.
103. CRC handbook of chemistry and physics, ed. D. R. Lide, CRC Press, 2002–2003.

CHAPTER 19

Microscopy for Acoustofluidic Micro-Devices

MARTIN WIKLUND*, HJALMAR BRISMAR AND
BJÖRN ÖNFELT

Department of Applied Physics, Royal Institute of Technology, SE-10691 Stockholm, Sweden
*E-mail: martin.wiklund@biox.kth.se

19.1 Introduction

One of the most frequently used tools for the investigation of a lab-on-a-chip device is the optical microscope. The purpose of using microscopy ranges from the basic visual inspection of microchannels using a standard microscope (*e.g.*, in acoustophoresis[1] and dielectrophoresis[2]), to advanced optical imaging,[3–5] and imaging methods integrated within the micro-device itself.[6–8] Even if it is obvious that many researchers in the lab-on-a-chip community routinely use microscopy, it is clear that the specific requirements high quality microscopy can pose on the device design are not always a top priority.

In this chapter we focus on the use and implementation of microscopy for lab-on-a-chip devices, with a special emphasis on micro-devices for the ultrasonic manipulation of cells. The chapter includes the basic principles and practice of microscopy (Section 19.2), a brief overview of different modes of microscopy (Section 19.3), and finally some practical guidelines and recommendations to consider when using microscopy for studies in a micro-device along with examples of such implementations in studies of acoustofluidic micro-devices (Section 19.4).

The first question to ask when identifying the microscope to use is: what is the size of the object of interest and how do fine details need to be resolved in the images? Typical objects studied with optical microscopy range from a few millimetres all the way down to the resolution limit, approximately 200 nm in optical microscopy (Figure 19.1). The resolution needed in an experiment will define which type of microscope to use. Often one must make a compromise between resolution and ease of use. High resolution comes with short working distance and short depth of field, factors that should be considered when designing the micro-device. The resolution limit of the selected microscope and its lenses is defined by the natural law of diffraction, discussed in more detail in Section 19.2.2.

The second question to ask is: how do I get contrast between the imaged object and its background/surroundings? This question is particularly important when imaging cells. A biological cell suspended in water is a highly transparent, weakly refracting object which may appear invisible unless a contrast-enhancing method is used. Examples of such methods are phase contrast, dark field, differential interference contrast (DIC) and fluorescence. These methods are described in more detail in Section 19.3, with examples shown in Section 19.4.

19.2 Basic Principles of Optical Microscopy

Generally, optical microscopy is classified as either wide-field microscopy or different variations of point-illumination/detection techniques, *e.g.* confocal microscopy. We will start by discussing wide-field microscopy, which means standard optical microscopy where the entire field of view is uniformly illuminated with light and the image is observed by the operator through the eyepiece or by a camera.

A schematic illustration is shown in Figure 19.2. The main parts of the microscope can be divided into: (1) the illumination system (either epi- and/or trans-illumination), (2) the objective, (3) optional filters, and (4) the imaging detector (either an eye-piece and the eye, or a camera). The most basic implementation of a microscope has only an objective and imaging detector; while more advanced microscopes also have an integrated illumination system. A filter cube (*i.e.* a set of optical filters) is typically used in fluorescence microscopy for separation of the fluorescence light from the excitation light. It is important to keep in mind that the objective is the single most important component of the microscope. Often, only the manufacturer and model of the microscope is specified in scientific papers using microscope images. Unless the objective used is specified, this information is of limited value. Finally, Figure 19.3 shows how the different optical components are assembled in practice for both an upright (looking from above) and inverted microscope (looking from below). The upright microscope corresponds to the schematic arrangement in Figure 19.2, while the inverted microscope corresponds to flipping Figure 19.2 vertically.

Microscopy for Acoustofluidic Micro-Devices 495

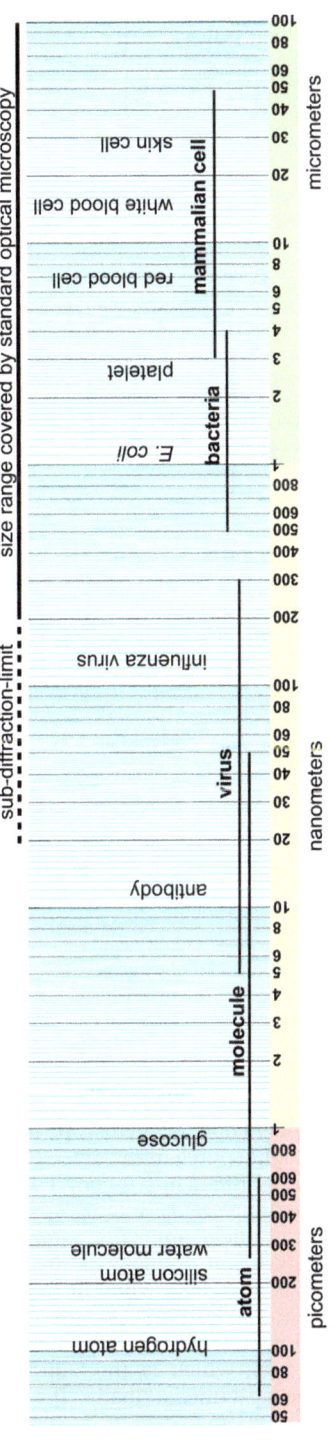

Figure 19.1 Schematic diagram of the relative sizes of different biological objects of interest and the coverage of optical microscopy operating in the visible light regime.

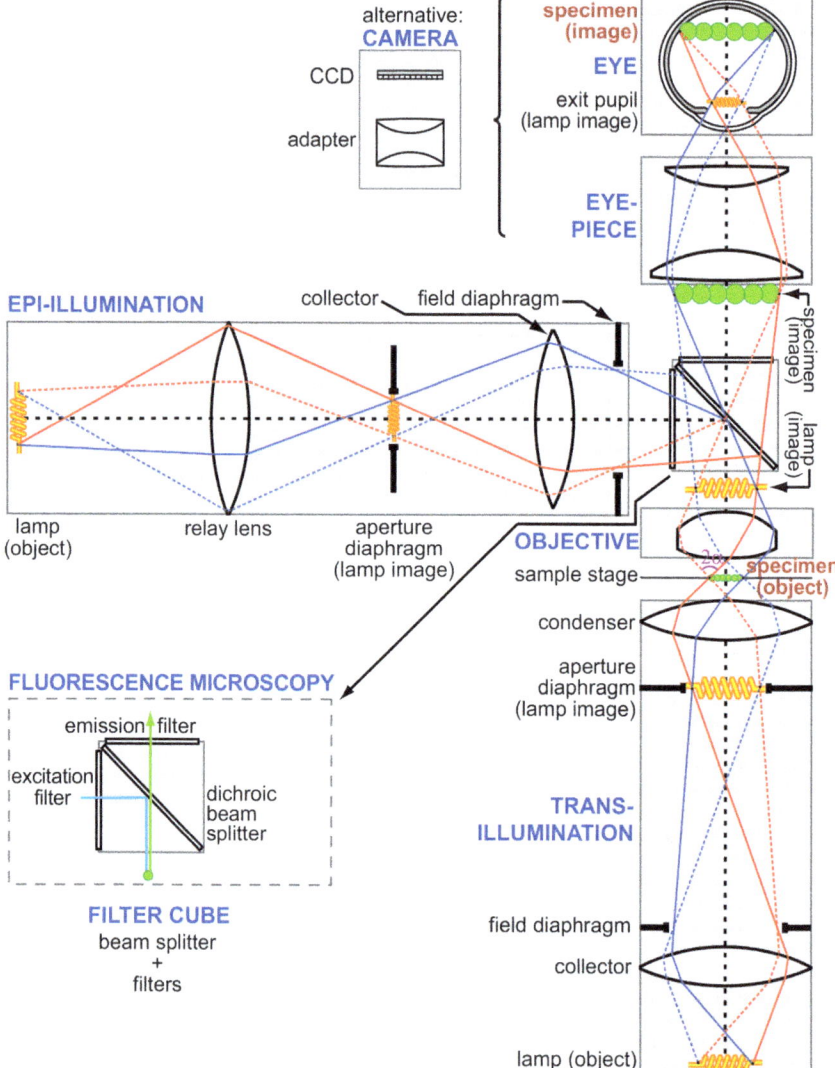

Figure 19.2 Schematic of a wide-field optical microscope including an illustration of light rays from the light source to the detector. The diagram illustrates trans-illumination bright-field microscopy, epi-illumination bright-field microscopy and epi-fluorescence microscopy. The difference between epi-illumination bright-field and epi-fluorescence microscopy is the choice of filter cube/beam splitter: in the former, a simple semi-transparent mirror is sufficient, while in the latter, the beam splitter is wavelength-sensitive (short wavelengths are reflected while long wavelengths are transmitted). The light rays illustrate the Köhler illumination technique, which is the standard method of uniform illumination of the specimen. Finally, the two standard detection options in wide-field microscopy are illustrated: manual observation through an eyepiece, or electronic detection with a camera. The figure is based on illustrations from ref. 12.

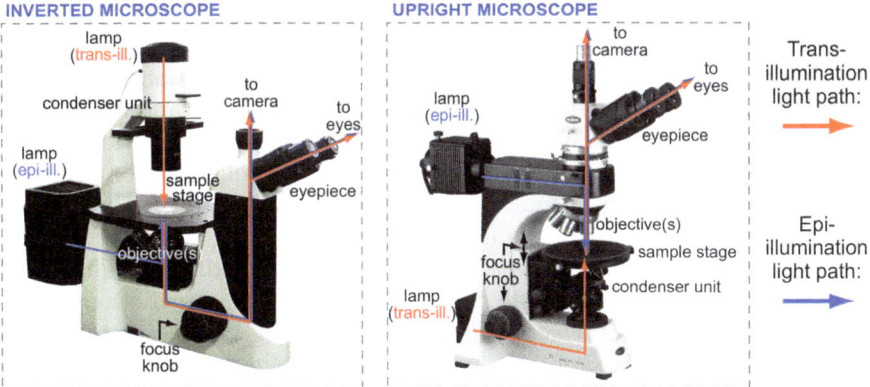

Figure 19.3 Photos of two standard compound microscopes: an inverted microscope (left) and an upright microscope (right) compatible with both bright-field and fluorescence microscopy. The red and blue light rays show the path for trans-illumination and epi-illumination microscopy, respectively.

19.2.1 Illumination System

We can start by looking at the trans-illumination, wide-field microscope with a "human detector" (right part of Figure 19.2) by following the light rays from the light source *via* the specimen (object) and to the detector where the magnified image of the object should be located. Naturally, most specimens of interest are not inherently luminous and, therefore, need to be illuminated. The standard illumination technique in wide-field microscopy is called Köhler illumination.[9] The basic idea of Köhler illumination is to provide a uniform illumination of the specimen. This is achieved by a series of optical elements and a collector–condenser system which are aligned so that the image of the light source (*e.g.*, the filament of a halogen lamp) is maximally defocused in the specimen plane. In Figure 19.2, this can be seen by following the light rays from two different points in the light source (here, a spiral-shaped lamp filament). At any point where the pair of dotted or solid lines converges, an image of the light source is formed. In the trans-illumination microscope, three such images ("lamp images") are formed between the light source and the detector. The specimen itself is located in between two such lamp images, causing rays from a single point in the lamp filament to fill out the whole field of view in the specimen plane. Simply speaking, we have achieved a maximally blurred lamp image in the specimen plane. However, illumination uniformity is not the only important parameter in wide-field microscopy; the Köhler arrangement provides the operator with additional degrees of freedom for control. In the specimen plane, it is possible to control the illuminated area, the light intensity and the solid angle under which the specimen is illuminated (*i.e.*, the "diffusivity" of the illumination). This is achieved by adjusting the apertures of two diaphragms in the collector–condenser system and is of particular importance for different contrast-enhancing techniques discussed later (Section 19.2.3).

An alternative to trans-illumination is epi-illumination (*cf.* mid-left part of Figure 19.2). In this mode, the light from the specimen is collected and detected from the same side as where the sample is illuminated. This is a common method when studying non-transparent (scattering/reflecting) objects such as a printed circuit board or when using fluorescence. In microfluidics, many microchips are not transparent, making this illumination mode the only possible alternative. In fluorescence microscopy, epi-illumination facilitates the separation of excitation light from fluorescence emission. To accomplish epi-illumination, a beam splitter is used to separate light from the light source and from the specimen. The beam splitter is either a semi-transparent mirror (in bright-field epi-illumination microscopy) or a filter cube comprising a dichroic mirror and a set of optical filters (in epi-fluorescence microscopy). The dichroic mirror is designed to reflect short wavelengths (the excitation light) and transmit long wavelengths (the emission light). For this reason, it is important to use the correct filter cube for the fluorophore used.

It is possible to combine trans-illumination and epi-fluorescence in real time by operating the two illumination systems simultaneously. An example is shown in Figure 19.4, where a cluster of cells trapped in an ultrasonic "cage"[10,11] is viewed in epi-fluorescence mode (Figure 19.4(a)), trans-illumination bright-field mode (Figure 19.4(b)) and finally in combined mode (Figure 19.4(c)). The combined mode clearly reveals which of the cells in the cluster are labeled with green fluorescence. In this case the green cell is a human embryonic kidney (HEK) cell labeled with calcein-AM, which is trapped in a cluster of natural killer (NK) cells. This mode of microscopy is particularly useful, for example, to simultaneously image the microfluidic structure (*e.g.* the channel walls) and objects (*e.g.* labelled cells) within the

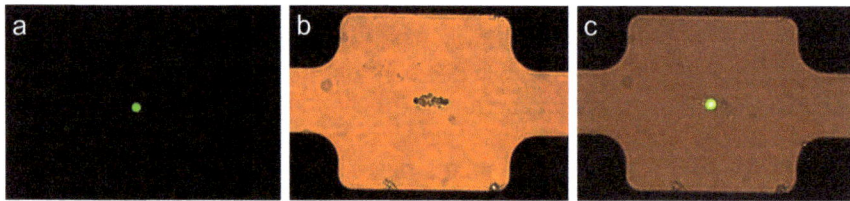

Figure 19.4 Practical examples of different illumination techniques in optical microscopy. (a) Epi-fluorescence microscopy. (b) Trans-illumination bright-field microscopy. (c) Combined epi-fluorescence and trans-illumination bright-field microscopy. The sample is an aggregate consisting of one human embryonic kidney (HEK) cell labeled with the green-fluorescent viability assay calcein-AM and approximately eight natural killer (NK) cells. The cell cluster is trapped and retained in a flow in an ultrasonic cage, which is described in more detail in ref. 11.

device. However, this only works for strongly fluorescent samples as the fluorescence must be clearly distinguishable from the bright background. In high-sensitivity, quantitative fluorescence applications, any background light must be carefully suppressed.

19.2.1.1 Magnification

A modern research microscope is built up with several modules and is called a compound microscope. The first and most important choice when setting up the microscope is the choice of objective. Often a number of different objectives are mounted in a rotatable objective turret. Objectives are made with different properties. The most obvious to a user is the magnifying power (typically between 1× and 100×). The magnified image is viewed through an eyepiece which typically magnifies the image another 10×, or is detected by a camera by projecting the image onto a CMOS or CCD chip with the help of an adapter lens. The adapter lens adjusts the size of the projected image to cover the full size of the camera chip, normally magnifying between 0.5× and 2.5×. The total magnification is given by the product of the objective's magnification and the eyepiece's or camera adapter's magnification.

When the specimen is viewed directly through the eyepiece, magnification refers to the viewing angle. Thus, the standard definition of the total magnification of a microscope is to compare the apparent angle the specimen is covering viewed through the eyepiece compared to the corresponding angle when viewed by the naked eye at a distance of 25 cm. The distance 25 cm is chosen since it corresponds approximately to the preferred distance between a nearby object and the accommodated eye (such as when reading in a book). In the case of camera detection, the concept of total magnification is simpler since it can be directly related to the size of the CCD chip in the camera (*i.e.*, viewing angle is not relevant in this case). When choosing a camera, the pixel number is not the most important for the resolution. Instead, it is the pixel size that needs to be matched with the optical resolution of the microscope, *cf.* Section 19.2.2. In particular, note that higher magnification does not automatically lead to higher resolution. Instead, it is the light collecting angle and the aberrations of the objective that determine the resolution. This is discussed in Section 19.2.2. Another important matter when using a camera is to choose a suitable camera adapter. In microscopy, the camera adapter corresponds to the camera lens used for normal photography. Most microscope objectives are designed to provide a field of view in the image plane between 20 and 25 mm. However, the CCD chip in the camera is often smaller, typically in the range 5–10 mm wide. Therefore, camera adapters often have 0.5× magnification to reduce this mismatch. If the CCD chip size is known, the field of view in the camera image can easily be calculated by dividing the CCD chip size with the total magnification, see example in Table 19.1.

Table 19.1 Magnification and numerical aperture of objectives commonly used in microfluidics.

Magn.	Typical numerical aperture (N.A.)	Approx. resolution (D)	Approx. depth of field	Typical working distance	Approx. field of view (in camera image)[a]	Application example in microfluidics
1×–1.25×	0.025–0.03 (dry)	>10 μm	600–900 μm	3–4 mm	11–14 mm	Imaging a whole micro-fluidic chip of size >1 cm
2.5×	0.075 (dry)	4 μm	100 μm	6–15 mm	5.6 mm	Imaging sharply in a 100 μm deep channel
5×	0.12–0.16 (dry)	2–3 μm	20–40 μm	10–20 mm	2.8 mm	Intermediate resolution with good field of view, long working distances available
10×	0.20–0.50 (dry)	0.7–1.7 μm	2–14 μm	2–16 mm	1.4 mm	Very good trade-off between resolution and field of view
20×–25×	0.30–0.80 (dry) 1.0 (water)	0.3–0.8 μm	0.5–3.4 μm	0.2–12 mm	500–700 μm	Often sufficient resolution (sub-μm) in many applications
40×	0.65–0.9 (dry) 1.0–1.2 (water)1.3–1.4 (oil)	0.2–0.5 μm	0.3–1.3 μm	0.16–2 mm	350 μm	Best resolution and performance, good for single-cell imaging
100×	0.65–0.95 (dry) 1.0–1.25 (water) 1.25–1.46 (oil)	0.2–0.5 μm	0.3–1.3 μm	0.1–1 mm	140 μm	Similar performance as for 40× objectives

[a]Estimation of the scale of the longest width in the camera image based on the following assumptions: 7 mm wide CCD chip and a 0.5 × camera adapter. This corresponds to a 14 mm wide image before the camera adapter, which must be smaller than the specified field of view of the objective (typically between 2025 mm).

19.2.1.2 Resolution

After magnification, the next important property of the objective is the numerical aperture (N.A.). This is a measure of the objective's maximum resolving power and is defined as:

$$\text{N.A.} = n \cdot \sin\alpha \qquad (19.19.1)$$

where n is the refractive index of the immersion medium between the specimen and the objective, and α is half the maximum light-collecting angle of the objective (cf. Figure 19.2).

For a given N.A., the diffraction-limited resolution can be estimated from the Rayleigh criterion as:[12]

$$D = \frac{0.61 \cdot \lambda}{\text{N.A.}} \qquad (19.19.2)$$

where D is the smallest resolvable distance between two closely lying point sources (or details) in the object, and λ is the wavelength of the emitted light from the object. Thus, the resolution of the microscope is dependent on n, α and λ. From eqn (19.1) and (19.2), we see that the resolution can be increased by using a high-index immersion medium, n (e.g. water or oil instead of air), by increasing the light-collecting angle of the objective, α, and by using short wavelength light, λ. The N.A. (i.e., n and α) is normally defined by the design of the objective where the size of the front lens, the working distance and the immersion medium are fixed parameters. The working distance is often marked with "WD" on the objective. Typically, high-N.A. objectives have short working distances in order to increase the light-collecting angle. When imaging a closed microfluidic device, it is important to use an objective with a sufficiently long working distance so that the full depth of the device (e.g. its microchannel) can be placed in focus, i.e. imaged sharply. It is important to note here that objectives are often designed to work with a certain coverslip thickness, i.e. the lid of the microfluidic device. The coverslip can actually be considered as a component of the objective and the wrong coverslip thickness will degrade performance significantly. The majority of all objectives are made for borosilicate coverslips #1.5, i.e. a thickness of 150–190 µm. This is discussed in more detail in Section 19.4. Objectives with different magnification and N.A. are listed in Table 19.1, together with microfluidic application examples.

As a practical example of diffraction-limited resolution (cf. Eqn (19.2)), an object viewed with blue light (e.g. the blue line of an argon-ion laser, $\lambda = 458$ nm) through a high-index oil-immersion objective (e.g. N.A. $= 1.4$) results in $D = 200$ nm. This is roughly the maximum resolution that can be achieved with a standard optical microscope and visual light. An example of the Rayleigh resolution criterion is shown in Figure 19.5, which illustrates the predicted (Figure 19.5(a)–(c)) and real (Figure 19.5(d)) diffraction-limited images of point-like objects (e.g. sub-µm fluorescent beads). However, note that the definition in eqn (19.2) provides no information about the smallest

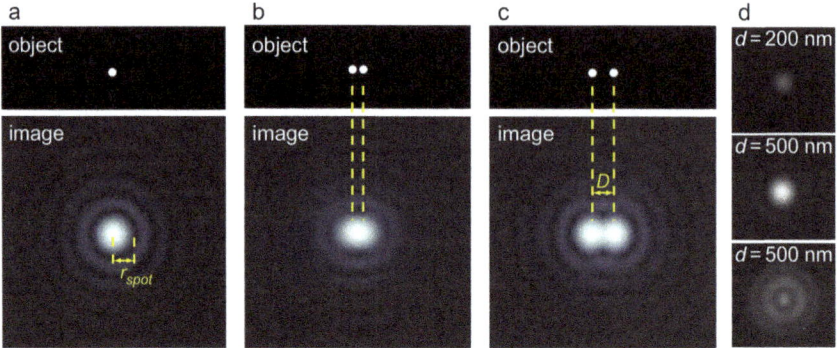

Figure 19.5 (a)–(c) Conceptual-model illustration of the Rayleigh criterion for the minimum resolvable distance, D, between two small point-like objects. (a) The diffraction-limited predicted image (lower panel) of a small object (upper panel) consisting of a bright sphere. (b) The corresponding image of two such small objects, unresolved case. (c) Same as in (b) but the limit where the spacing, D, is just large enough for resolving the two small spheres. The radius of the imaged spot, r_{spot}, in (a) is equal to the minimum resolvable distance, D, in (c). The situation in (c) illustrates the Rayleigh criterion. This criterion is defined as when the first intensity minima from the center of the imaged spot from one of the objects coincides with the intensity maxima of the other object, i.e., when $D = r_{spot}$. (d) Real microscopy images of a 200 nm fluorescent bead (upper panel) and a 500 nm fluorescent bead (middle and lower panels) acquired with a 100 × /0.95 (magnification/numerical aperture) objective. The upper and middle panels show the images of the beads in the focal plane, while the lower panel shows the effect of imaging a bead slightly out-of-focus.

detectable object. For example, a fluorescent bead with size much smaller than D can still be visible in a standard optical microscope given that it is bright enough relative to its background and that it does not move too much. This results in an imaged spot (called Airy disc) having a radius $r_{spot} = D$ (cf. Figure 19.5(a) and (c)). In Figure 19.5(d), real microscope images of green-fluorescent 200 nm and 500 nm beads are shown to verify the predicted images in Figure 19.5(a)–(c). The images in Figure 19.5(d) are acquired with a 100 × dry-immersion objective with N.A. = 0.95. Using eqn (19.2), the diameter ($d = 2r_{spot} = 2D$) of the imaged spot from a point source is equal to 660 nm (for $\lambda = 515$ nm and N.A. = 0.95). Since this lower limit is larger than both 200 nm and 500 nm, we can expect that the imaged spots of 200 nm beads and 500 nm beads will not differ much in size. This is verified in Figure 19.5(d), where the image of the 500 nm bead is only about 20% wider than the image of the 200 nm bead. If diffraction was the only limiting factor, theory predicts about 30% difference in imaged widths under these circumstances.

Another important effect of high-resolution imaging is the short depth of field for high-N.A. objectives. Depth of field is defined as the thickness of the layer where the specimen is imaged sharply. This thickness is proportional to $\lambda/\text{N.A.}$,[2,12] which means that for the objective and fluorescence color in Figure 19.5(d), the depth of field is approximately equal to the bead size (500 nm). As a result, a 500 nm bead vertically displaced on the order of 1 μm is imaged as in Figure 19.5(d), lower panel. In microfluidics, objectives with high N.A. should only be used if the sample is positioned with μm-precision, preferably on the bottom of the channel. This is further discussed in Section 19.4.

The Rayleigh criterion can be used to estimate the maximum resolution that can be achieved with a chosen objective. This is a fundamental limit originating from the wave nature of light. However, in reality other engineering-related limitations often define the image quality and resolution, in particular for less expensive objectives. These limitations are called aberrations, and are summarized in Table 19.2. In general, an optical aberration is any type of deviation in the imaging performance of an ideal lens or lens system. Such a deviation originates from rays passing through the periphery of a lens and/or at a large angle relative to the optical axis. Thus, while the definition of N.A. suggests the use of large-diameter lenses and wide-angle light cones, aberrations typically increase with lens diameter and light angle. Still, even small lenses are not aberration-free. The solution is to correct as many aberrations as possible by using complex lens systems containing combinations of several lenses and lens compounds (doublets and triplets) with up to 15–20 lens elements in total for advanced objectives. Table 19.2 lists different aberration-corrected objective classes, together with their performance and relative cost. All modern objectives are corrected for spherical and chromatic aberration to some extent, but more expensive objectives are better corrected for chromatic aberrations and also for field curvature. The latter is important when imaging at high resolution over a broad spectral range and in a wide field of view, *e.g.*, in fluorescence microscopy and when recording images with a sensor. A final note about aberrations is that a glass layer between the objective and the specimen can act as a source of spherical aberration. As previously discussed, objectives are often designed for 150–190 μm borosilicate coverslip use. Thus, this is the optimal material and thickness of a glass layer for sealing of a microfluidic channel. If another thickness and/or material are used, an objective with a correction collar should be used for optimal microscopy performance.

When using microscopy in an acoustofluidic micro-device, the choice of objective should be matched with the illumination and detection systems, with the properties of the specimen/sample, and with the micro-device. An example of this is shown in Figure 19.6, where a 375×110 μm^2 (width × height) transparent acoustofluidic channel filled with a suspension of 4.5 ± 0.7 μm polyamide beads ("blood-mimicking fluid", Danish Phantom Design, Denmark) and actuated at 5.68 MHz is imaged in trans-illumination mode with two different objectives: one with N.A. = 0.075 and one with N.A. = 0.30. First, we notice that the pattern of aggregated beads in the pressure nodes of

Table 19.2 Objective classes (corrections for aberrations).

Name	Correction	Typical use	Relative price (ranging from about 100 Euro to 10 000 Euro)
Achromat	Color-corrected, narrow band	Basic visual inspection	$
Plan Achromat	Color-corrected, narrow band, field curvature correction	Basic visual inspection, need for sharp images within the whole field of view, and photographic recording	$$
Fluorite	Color-corrected, intermediate band	Suitable for fluorescence applications with short-wavelength excitation	$$$
Plan Fluorite	Color-corrected, intermediate band, field curvature correction	Suitable for fluorescence applications with short-wavelength excitation, need for sharp images within the whole field of view, and photographic recording	$$$$
Apochromat	Color-corrected, wide band	Suitable for demanding applications, typically high-resolution multiple-probes fluorescence	$$$$$
Plan Apochromat	Color-corrected, wide band, field curvature correction	Suitable for demanding applications, typically high-resolution multiple-probes fluorescence, need for sharp images within the whole field of view, and photographic recording	$$$$$$

the acoustic standing wave is clearly resolved in both cases. However, single-bead resolution and identification is only possible in Figure 19.6(b). Note that the field of view relative to the resolution and camera pixel size is well matched in both cases: the theoretical radius of the imaged spot from a point source, r_{spot}, (cf. eqn (19.2)) is approximately equal to the pixel size in both Figure 19.6(a) ($r_{spot} \approx$ pixel size $\approx$ 4 µm) and in Figure 19.6(b) ($r_{spot} \approx$ pixel size $\approx$ 1 µm). Thus, in this experiment, the digital resolution (pixel size) is matched with the optical resolution (r_{spot}). As expected, the ratios of these resolutions ($r_{spot,a}/r_{spot,b} = 4$ and pixel size$_a$/pixel size$_b \approx 4$) both correspond to the ratio of the numerical apertures (N.A.$_b$/N.A.$_a = 4$). Furthermore, if we need to image and identify single beads of size 4–5 µm, the imaging performance in Figure 19.6(a) is not sufficient; we need at least ~5 times better optical and digital resolution relative to the bead diameter to correctly identify single beads (cf. Figure 19.6(c)). This is provided that all beads are in the focal plane of the objective, i.e., that the depth of field is large enough. However, if we take into account the sampling theorem[13] and compare with eqn (19.2), the pixel-to-pixel distance should not be more than $0.4 \times D$ in order to avoid subsampling. For the objective used in Figure 19.6(b) (N.A. = 0.30), the depth of field is roughly about the same size as the bead diameter (cf. Table 19.1). This indicates that most beads in Figure 19.6(b) are located on the channel bottom. If beads are uniformly distributed in a ~100 µm high channel, an objective with slightly lower magnification and N.A. than in Figure 19.6(b) is a better choice, e.g. a 10×/0.25 objective as used in ref. 14.

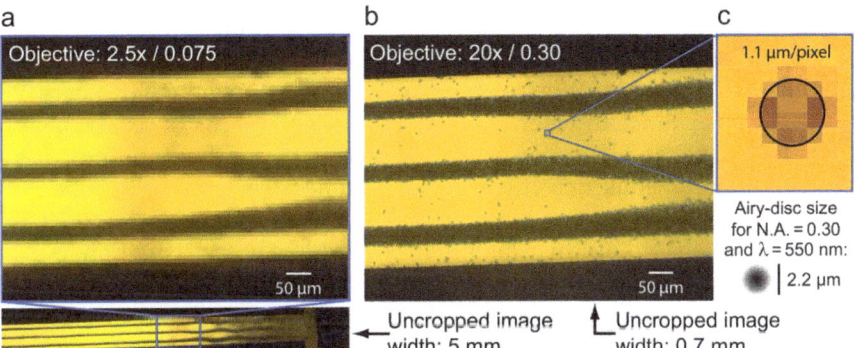

Figure 19.6 Comparison between images acquired with (a) a 2.5×/0.075 (magnification/numerical aperture) objective and (b) a 20×/0.30 objective. The image in (a) is cropped to match the scale bar in the image in (b). (c) Zoom-in displaying the image of a single 4.5 µm diameter polyamide bead, including the pixel size and the theoretical diffraction-limited image of a point source (the Airy disc) for N.A. = 0.30 and λ = 550 nm. The circle indicates the size of a 4.5 µm diameter circle scaled by the use of the channel width (375 µm) as a reference. The images show the pattern of beads undergoing acoustophoresis at the ultrasonic frequency 5.68 MHz in a 375 × 110 µm² (width × height) microchannel.

19.2.1.3 Contrast

Contrast is an important parameter to consider when imaging a microfluidic channel, in particular when the specimen is a water-based cell suspension. Cells are difficult to distinguish from water because they have high water-content and are highly transparent. Optically, this means that cells have a refractive index, n_{cell}, close to the refractive index of water, n_{water}. As a result, light rays are neither deflected nor reflected by the cell to any significant extent. Thus, the image of the cell suspension may appear as pure water. Two different strategies can be used to overcome this problem: either to stain the cells with a color substance or fluorescent probe, or to use an optical method to enhance the natural contrast between cells and suspension medium. Fluorescence microscopy is discussed in Section 19.3.2 while bright-field-based techniques for contrast enhancement are discussed in Section 19.3.1. The latter includes phase contrast, dark field and differential interference contrast (DIC).

19.2.1.4 Light Sources and Detectors/Cameras

The standard light sources in a compound microscope are a halogen lamp for bright-field microscopy, a mercury-vapor lamp or LED for (wide-field) fluorescence microscopy, and lasers for (laser-scanning) confocal fluorescence microscopy. In Figure 19.3, a mercury lamp is used for epi-illumination, and a halogen lamp is used for trans-illumination. Mercury lamps have higher efficiency than lamps of incandescent filament-type such as the halogen lamp. However, they are not as flexible in start-up time and intensity adjustments as the halogen lamp. In addition to their high intensity, mercury lamps have a strong emission in the lower wavelength regime (*i.e.*, near-UV, blue and green). Since the mercury lamp spectrum consists of several peaks and valleys, optical filters are often matched with the fluorophores used enabling simultaneous detection of fluorescence (*cf.* the "filter cube" in Figure 19.2). In recent years, LEDs have become an attractive alternative to mercury lamps based on their long lifetime and low heat dissipation.

The standard detectors are a CCD or CMOS camera in wide-field microscopy, and a photo-multiplier tube (PMT) in confocal microscopy. Note that a PMT is a single-element detector that requires a serial scanning procedure for generating an image, rendering confocal microscopy a rather slow technique. On the other hand, a PMT can be more sensitive than a CCD although today, CCDs are sufficiently sensitive for most applications.

19.3 Modes of Optical Microscopy

In this section we discuss different modes of optical microscopy that are used when imaging acoustofluidic micro-devices. The typical specimen in these devices is either cells or polymer beads suspended in a water-based fluid inside a micro-channel or micro-chamber. Although the examples are chosen from acoustofluidic devices, the discussions are relevant for any type of micro-device of similar design.

19.3.1 Bright-Field Microscopy

Bright-field is the basic mode of microscopy where the specimen is viewed with white-light illumination and contrast is provided naturally by a combination of transmission, reflection, refraction, scattering and absorption mechanisms. The name refers to the bright background seen around a shadowing object. However, due to the low contrast when imaging cells, bright-field microscopy is often operated in a contrast-enhancing mode. The three most important modes are phase contrast, dark field and differential interference contrast (DIC). These contrast methods are typically performed in trans-illumination mode.

19.3.1.1 Phase Contrast

This method utilizes the phase difference introduced by a ring-shaped phase plate between scattered and unscattered rays. The phase plate is combined with a ring aperture in the condenser that gives cone-shaped illumination of the specimen. In this way, rays that are unscattered pass through the ring part of the phase plate, while scattered rays pass through other parts of the phase plate (having a different thickness to the ring part). Interference between these rays increases the contrast between the image background and the specimen (*e.g.* the structures of a cell). In practice, the phase plate is integrated in the objective, which is marked with "*Ph*" or similar. Thus, an objective specifically designed for phase contrast is needed. Furthermore, a ring aperture needs to be inserted and aligned in the condenser system. Another remark is that optimal contrast is achieved for the phase difference $\pi/2$ between the scattered and unscattered rays. Exactly this shift can only be generated for one wavelength. Hence, the optimal effect is achieved with monochromatic light (typically green light around 550 nm). However, phase contrast can also be performed with good results with white light. It should be noted that a common artifact in phase contrast images is a halo around objects such as a cell. Furthermore, the best effect is achieved for thin and transparent samples such as a monolayer of cells. An example of phase contrast imaging of a monolayer of cells aggregated and trapped by ultrasound in a half-wave microfluidic channel is seen in Figure 19.7(b) and compared with standard bright-field imaging in Figure 19.7(a).

19.3.1.2 Dark Field

This is a rather straightforward and useful method to discriminate between direct (unscattered) rays and rays scattered by the object, and is particularly suitable for weakly scattering objects and for low-concentration samples of small particles. Just as with phase contrast, this method utilizes a ring

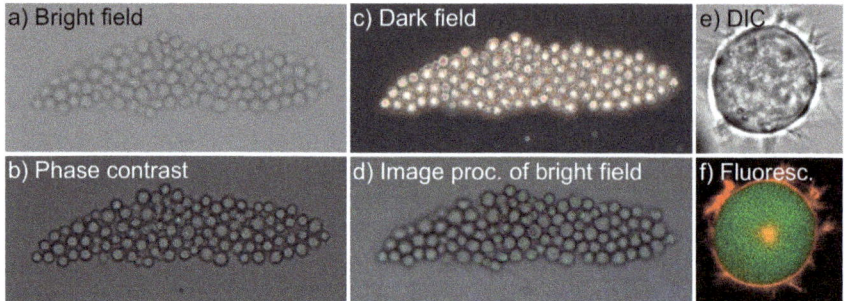

Figure 19.7 Ultrasonically trapped cells imaged with different contrast modes: bright field (a), phase contrast (b), dark field (c), digital image processing of the bright field (d), differential interference contrast (e) and dual-color fluorescence (f). The cells in (a)–(d) are COS-7 cells, and in (e)–(f) a human B cell.

aperture in the condenser system to produce cone-shaped illumination. However, instead of a phase plate, dark field imaging uses a condenser with higher N.A. than of the objective. This, together with a suitable ring aperture alignment, causes unscattered rays to completely miss the objective. For cell imaging, the result is bright cell structures on a dark background which may reveal slightly different image information compared with, *e.g.*, phase contrast images. No particular objective is needed for dark field imaging, as long as its N.A. is lower than the N.A. of the condenser. An example of dark field imaging is shown in Figure 19.7(c) (same object as in Figure 19.7(a) and (b)). When using a stereo microscope with a flexible-arm illumination, a dark-field effect can be achieved by adjusting the arm into a suitable oblique illumination angle relative to the specimen plane.

19.3.1.3 DIC

Differential interference contrast (DIC) microscopy, or Nomarski microscopy, is based on imaging local gradients in optical path lengths in a sample by the use of polarized light. A simplified description of the method is as follows: linearly polarized light is split by a so-called Wollaston prism into two differently polarized and laterally displaced beams. These beams pass through the sample. A second Wollaston prism then recombines the beams and they are filtered by a second polarizer (analyzer). Interference between the recombined beams after the second polarizer produces an image where small local variations in refractive index of the sample are enhanced, see example in Figure 19.7(e). The variations are in the displacement direction defined by the orientation of the Wollaston prisms, which can be adjusted by the operator. It should be noted that the DIC images of a cell, for example, often appear as relief-type images which should not be mistaken for topological details. Although useful, DIC images are not always easy to interpret

In practice, DIC microscopy is useful for thicker samples and doesn't suffer from the halo effect seen in phase contrast imaging. In microfluidics, DIC cannot be used with plastic devices since the birefringence of plastics destroys the DIC effect. Thus, plastics are perfect depolarizers. The optimal device material for DIC is glass. Care should also be taken when clamping a glass chip in a chip holder, since induced stresses also cause depolarization.

19.3.2 Fluorescence Microscopy

Fluorescence microscopy is a sensitive and useful technique allowing quantitative analysis with high contrast and selectivity.[15] In this section, we primarily focus the discussion to cellular applications of fluorescence microscopy. The light source in the microscope system, typically a mercury arc lamp or a laser (cf. Figures 19.2 and 19.3), is used to transfer fluorophores to the excited state and fluorescence can then be collected by the microscope detector system. The fluorescence light is most often of a longer wavelength (red-shifted) compared to the excitation light and is emitted incoherently in all directions independently of the direction of the excitation light. From an optical perspective, this makes fluorescence microscopy simpler than the bright-field-based contrast methods. On the other hand, in order to minimize contaminating light from the excitation source, fluorescence is typically detected in the opposite direction from the excitation (i.e. in epi-illumination mode) and through a filter only allowing passage of red-shifted photons (cf. filter cube in Figure 19.2) and not back-scattered excitation light.

19.3.2.1 Fluorescent Probes

The chemical structure largely determines how efficiently and at what wavelength the molecule absorbs and emits light. The molecular extinction coefficient is a measure of the absorption and the quantum yield specifies what fraction of the relaxation back to the ground state takes place through fluorescence. Thus, both of these parameters should be high for efficient fluorescence. Some fluorophores have a fluorescence quantum yield that is sensitive to the local micro-environment and can be used as sensors for e.g. pH or Ca^{2+} concentration. Several molecules occurring naturally in cells absorb UV light, but the amount of cellular molecules absorbing visible light (400–700 nm) efficiently is relatively low. This provides a spectral window where cells are largely transparent and also non-fluorescent. This window can be utilized for fluorescence microscopy by using fluorophores that are excited by visible light. Typically, such molecular probes are rather small polyaromatic hydrocarbons or heterocycles and, thus, much smaller than many biomolecules, e.g. proteins. By chemical modification of the fluorophores, both their fluorescent properties, e.g. wavelength or excited-state lifetime, and intracellular localization can be tuned.

19.3.2.2   Cellular Labeling with Fluorescent Probes

It is important to consider what property of the cell needs to be analyzed when selecting a fluorescent probe. If the purpose is to detect cells or distinguish different populations of cells, there are several tracking dyes available. Such dyes typically distribute in the cellular cytoplasm (*e.g.* calcein-AM), bind covalently to intracellular proteins (*e.g.* carboxyfluorescein succinimidyl ester—CFSE), or accumulate in the plasma and intracellular membrane of cells (*e.g.* dicarbocyanine—DiO). Such dyes often stain cells brightly and are available in a variety of colors.

A widely used technique for fluorescent detection of specific proteins is to use antibodies. Antibodies can either be directly labeled with the fluorophore of choice or combined with secondary fluorescently-conjugated antibodies that bind to the first (primary) antibody. The advantage of using secondary antibodies is that the library of primary antibodies can be kept lower as there is no need to maintain the same antibody directly labeled with different fluorophores. When using antibodies it is important to keep track of which species the antibody was raised in and care should be taken to use appropriate controls with non-specific antibodies of the same isotype. If the antibody binds non-specifically to cells, erroneous conclusions can be drawn about the protein distribution and quantity.

Molecular biology can be used to introduce fluorescent proteins, *e.g.* green fluorescent protein (GFP), into cells and is an excellent strategy for many imaging applications, in particular for live cell imaging. Some drawbacks are that it can be time consuming or difficult to generate cells expressing the protein of interest, especially if one wants to use primary cells. Other limitations are that the GFP-tag may affect the natural distribution and behavior of the protein of interest, and the fact that the vectors carry different promoter regions (*i.e.* specialized DNA sequences where the transcription process starts) compared to the wild-type DNA. This often leads to protein overexpression, which facilitates detection but may affect the natural function and localization of the protein. Despite these limitations and concerns, the use of fluorescent protein, such as GFP, is now the predominant approach for introducing fluorescence in live cell applications.

19.3.2.3   Live Cell Fluorescence Microscopy

Live cell imaging poses some additional challenges since the cells generally need to be maintained in a culture medium and at a physiological atmosphere (37 °C and humidified air complemented with 5% CO_2) if they are to be imaged for long periods of time (of the order of one hour or longer). There are different approaches to create chambers allowing control of the environment around the cells while imaging. Two common ways are to either place the cells in a rather small heating block that is placed on top of the microscope stage or to encapsulate the whole microscope stage in a larger chamber where the atmosphere is kept at the desired level. For both approaches it is common to have a supply of humidified air supplemented with 5% CO_2 that purges the cells. Cells can survive at lower temperatures in a buffered environment for a limited time but it should be remembered that dynamic cellular processes are temperature dependent. One cannot expect that the rate of enzymatic activity, molecular transport or diffusion will be the same at lower temperatures.

Another challenge in live cell imaging is that many fluorescent probes that stain fixed cells cannot be used to efficiently stain live cells. Major reasons for this are that many probes do not easily cross the plasma membrane of live cells or that they are toxic to cells, sometimes increasingly so during imaging due to the formation of reactive oxygen species. Despite these limitations, there are several probes that are useful for live cell imaging in addition to fluorescent proteins (discussed in the previous section). Examples are found among cytoplasmic dyes, lipid dyes and some nuclear dyes. As a general rule, one should strive to use as low a dye concentration as possible for live cell imaging and, if possible, use experimental controls that probe if normal cellular function is maintained also when the cells are stained.

The fact that the plasma membrane provides a natural boundary that is difficult for some molecules to cross can be used to design different labeling strategies. For example, dead cells can be distinguished since some dyes can only leak through the compromised plasma membrane of a dead cell and not the intact membrane of live cells.[16] This can be exploited by the use of dyes that leak out from the cell or into the nucleus as it dies. Examples of viability probes suitable for acoustofluidic cell handling are provided in Chapter 21.

19.3.2.4 Multiple Fluorescent Probes

When selecting probes for cellular experiments, the choice of fluorophores needs careful consideration, taking into account their individual absorption and emission spectra as well as the properties of the imaging system used. Even if modern high-end confocal microscopes are equipped with multiple laser lines for excitation and several detectors combined with tunable filters, it is generally difficult to image more than 3–4 dyes in the same sample without getting unwanted fluorescence bleed-through or cross talk between the channels. To limit the cross talk it is desirable to use dyes with narrow fluorescence spectra. An example of a dual-labeled cell trapped by ultrasound is seen in Figure 19.7(f).

19.3.2.5 Optical Sectioning Techniques

It is sometimes desirable to get three-dimensional (3D) information about the sample studied and the quality of that information depends heavily on the imaging system. With a conventional fluorescence microscope, the resolution along the optical axis is generally quite poor. However, when using a technique allowing optical sectioning (examples include laser-scanning confocal microscopy, spinning-disc confocal microscopy, structured-illumination microscopy and two-photon microscopy),[17] the quality of the 3D information can be much higher, especially if a 3D reconstruction is made using specialized software. However, this requires multiple acquisitions, which obviously leads to longer imaging times. Furthermore, if the imaged objects move between sequential images, significant distortions in the 3D reconstructed image can occur.[18] Distortions due to cell drift can be corrected, but morphological changes of a cell during acquisition are more difficult.

19.4 Implementation of Optical Microscopy in an Acoustofluidic Micro-Device

In this section we discuss how to design an acoustofluidic micro-device intended for use in an optical microscope (Section 19.4.1). The discussion includes choices of materials and dimensions, how to match the device with a certain objective and illumination method, and how to match the acoustic performance with the optical performance. Finally, we provide examples of applications of acoustic manipulation of cells in micro-devices characterized by optical microscopy (Section 19.4.2).

19.4.1 Design Criteria

A microfluidic chip used for acoustophoresis has carefully selected thicknesses and supporting structure materials, in order to host and control acoustic resonances. The typical chip design is a half-wave fluid channel surrounded by an acoustically hard material such as silicon or glass.[19] Glass has optimal material properties for both hosting acoustic resonances, as well as for high-resolution optical microscopy. A polymer material is a less suitable choice for two reasons: (1) its high acoustic absorbance, and (2) its optical birefringence (which makes DIC microscopy impossible). Therefore, most acoustophoresis chips use a combination of glass and silicon: glass for optimal visual access and silicon for optimal channel processing. If the supporting structures around the fluid channel have thicknesses corresponding to an odd multiple of a quarter wave in that material, the design is a layered resonator.[15]

A suitable starting point when building a layered acoustic resonator compatible with high-resolution microscopy is to use the same materials and thicknesses as established as a microscopy standard: a 170 ± 20 μm thick borosilicate layer (*i.e.*, a standard coverslip) and a 0.8–1.0 mm thick borosilicate layer (*i.e.*, a standard microscope slide, typical dimension 25 × 75 mm^2 or 1 × 3 inches^2). If other glass materials than borosilicate are used, the refractive index, n, should preferably be as close as possible to the index of borosilicate (*i.e.*, n between 1.51 and 1.54). By proper acoustic frequency selection, the thinner glass layer can act as an acoustic quarter-wave reflector on the side of the chip facing the microscope objective. For example, a 170 μm thick Pyrex layer (one type of borosilicate glass) has a quarter-wave resonance frequency of about 8.3 MHz (given a speed of sound = 5661 m s^{-1}).[19] This resonance frequency defines the driving frequency of the transducer, which, in turn, defines the half-wave thickness of the water-filled fluid channel to about 90 μm (given a speed of sound = 1497 m s^{-1}).[19] The other reflecting layer on the opposite side of the channel should preferably be thicker than a quarter-wave, otherwise the total stack of layers will become fragile. Two options are available: either to use a two-layer chip with the channel etched in silicon or glass and closed by a glass layer, or to use a three-layer chip with the channel etched throughout the mid-layer.[20] The latter is the most flexible since the channel can be etched in any material and then closed by two glass layers. Furthermore, this option also makes it possible to use the standard coverslip/microscope slide combination for the two glass layers building up the micro-device. This is important since all microscope suppliers have designed their microscope components to be compatible with these materials and dimensions: a coverslip facing the objective and a glass slide facing the condenser system. Thus, using the 170 ± 20 μm and 0.8–1.0 mm combination for the glass layers will result in a chip compatible with both high-Q acoustic resonances and any type of optical microscopy (both epi- and trans-illumination). In theory, the optimal choice is 170 μm and 850 μm. This is exactly equivalent to $\lambda/4$ and $5\lambda/4$ acoustic reflectors at 8.3 MHz, which perfectly matches the microscopy coverslip/glass slide standards.

If the mid-layer in between the two glass layers is silicon, acoustic resonances can also be built up in the transversal direction perpendicular to the layered resonator enabling three-dimensional acoustic particle manipulation.[10] This design is a hybrid between the layered resonator and the transversal resonator described in ref. 20. An example of such a layered/transversal resonator hybrid is seen in Figure 19.8. A similar example is described in ref. 10 and uses a 200 μm bottom layer in Pyrex, which is near but not fully optimized for high-resolution microscopy. Thus, the device is a three-layer chip compatible with both three-dimensional acoustic particle manipulation and trans-illumination microscopy. Here, a 110 μm thick silicon layer is sandwiched between two acoustic reflectors: a 200 μm ($\lambda/4$-wave) Pyrex layer and a 1 mm (5 × $\lambda/4$-wave) Pyrex layer. Thus, the axial resonance (layered resonator) is around 7 MHz causing "levitation"[21] of particles in the vertical direction, while the transversal resonances in the horizontal directions are defined by the channel width and length, respectively. High spatial control of the three-dimensional manipulation is achieved by implementing ultrasonic "cages" with half-wave dimensions along all the three dimensions width, height and length.[10] Imaging of levitated particles and cells should be performed with an objective having a working distance of at least 200μm + 55 μm (corresponding to the thickness of the bottom glass layer plus half the channel height).

The inverted microscope is the most convenient to use for imaging in a microfluidic chip, see Figure 19.3. The inverted microscope has a beam line arrangement corresponding to flipping the schematic Figure 19.2 vertically: the objective, epi-illumination source and the detector are below the sample

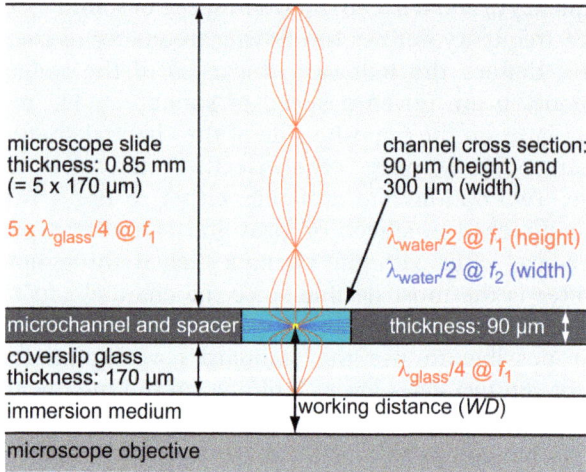

Figure 19.8 An optimized design of a layered acoustic resonator for 2D ultrasonic manipulation in the fluid channel, compatible with high-resolution optical microscopy. The chip is a three-layer structure based on a coverslip glass (bottom) and a microscope slide glass (top). A spacer layer is sandwiched in between the two glass layers defining the channel height. The chosen thicknesses correspond to a vertical resonance frequency f_1 of 8.3 MHz, and a horizontal resonance frequency of another lower frequency defined by the channel width (e.g. 2.5 MHz for the width 300 µm).

stage, while the condenser system (i.e., the trans-illumination source) is above the sample stage. This setup makes it possible to arrange all tubing, connectors, wires and ultrasonic transducers on the upper side of the microfluidic chip. This is particularly useful in high-resolution microscopy where high-N.A. objectives with short working distances are used. As previously mentioned, a closed microfluidic chip should preferably have a glass bottom layer (facing the objective in the inverted microscopy) of a thickness near 170 µm. Furthermore, the objective used should be coverslip-corrected. This information is found in the second row below the magnification and N.A. labels on the objective. For example, the label "∞/0.17" means infinity-corrected objective to be used with a glass coverslip of thickness 0.17 mm. This thickness is the most common standard in microscopy, sometimes called No. 1½. Examples of other common labels are "∞/0" (no coverslip), "∞/-" (insensitive for coverslip) and "∞/0.14–0.19" (adjustable for different coverslip thicknesses). The latter type of objective with a correction collar is strongly recommended for use with microfluidic chips, since the chip layers may differ from the standard thickness of microscopy coverslips, or the glass material may have a different refractive index to the index of standard coverslips. The other label ("∞") stands for infinity-corrected objectives, or infinity corrected system (ICS). This is the present standard for objectives, which means that the objective produces an intermediate image placed in infinity (in contrast to the older standard of 160 mm fixed tube length). The advantage with infinity-corrected objectives is the possibility to insert various optical components (such as filters, polarizers, beam splitters, *etc.*) in the optical path after the objective without causing image artifacts due to refraction. The ICS objectives are combined with a tube lens placed in between the objective and the eyepiece to relocate the intermediate image to a finite distance from the objective.

For high-resolution objectives with N.A. ≥ 1, an immersion medium is always needed (*cf.* eqn (19.1)). This medium is oil, water, or, less commonly, glycerol. Oil-immersion is the best choice for thin microfluidic channels or in applications where only the channel bottom is imaged. Oil has the same refractive index as glass, which means that no refraction takes place between the cover slip and the objective. In principle, oil immersion objectives are insensitive to changes in the thickness of the glass layer. This means that any glass thickness shorter than the working distance of the objective can be used. On the other hand, refraction takes place between the coverslip and the fluid channel (given a water-based fluid inside the channel). This means that in applications where the object is located at a distance from the channel bottom, a water-immersion objective is a better choice. Thus, if water is the medium on both sides of the glass layer, image distortion is minimized. In acoustofluidics, a vertically oriented half-wave channel used for trapping or aligning particles in the center of the channel should be imaged with a water-immersion objective if high-resolution microscopy is needed. However, it should be noted that the immersion medium will introduce acoustic transmission losses, which in turn may decrease the acoustophoresis performance. If a vertical resonance is used (*cf.* Figure 19.8), dry objectives are the best choice from an acoustic point of view.

19.4.2 Applications of Microscopy to Acoustofluidics

In this section we provide a few examples of acoustofluidic applications where optical microscopy is a central tool of the method. Fluorescence microscopy was used by Radel *et al.*[22] to measure alterations in the vacuole structure of yeast cells exposed to ultrasound. Spengler *et al.*[23] performed quantitative measurements of the fractal perimeter dimension to investigate the morphology and stability of bead aggregates trapped by ultrasound in different solutions. The measurements were based on epi-illumination microscopy images. Quantitative fluorescence measurements were performed by Wiklund *et al.*[24] by the use of confocal laser-scanning microscopy (CLSM). They demonstrated a detection limit in the low femtomolar range for a bead-based human thyroid stimulating hormone (hTSH) assay by combining ultrasonic bead enrichment with high-sensitivity confocal fluorescence microscopy.[25] This work was performed in a standard 96-well plate. Bazou *et al.*[26] used both bright field and fluorescence microscopy to measure the rate of membrane–membrane spreading and the distribution of different trans-membrane proteins responsible for cell–cell adhesion in ultrasound-mediated aggregates of C6 neural cells. Similar measurements were also performed on prostate cell lines,[27] chondrocytes[28] and a liver cell line.[29] Viability of cells trapped by ultrasound has been measured by fluorescence microscopy, *e.g.* by the use of the viability probes calcein-AM[30,31] and acridine orange.[32] The immune synapse between a natural killer cell and a target cell trapped by ultrasound was investigated with high-resolution confocal fluorescence microscopy by Christakou *et al.*[18]

The position of acoustophoretically focused particles in a microchannel can be measured with optical microscopy. Manneberg *et al.*[10] used confocal fluorescence microscopy to measure the 3D position of trapped beads in a sono-cage. Similar measurements were performed by Evander *et al.*[33] using a wet-etched glass chip. These experiments show that image quality and resolution do not need to be very advanced to estimate the position with sufficient accuracy.

Particle image velocimetry (PIV) has been used by many researchers for measuring either fluid velocities in acoustic streaming[34,35] or the acoustic radiation forces acting on suspended particles in acoustofluidic devices.[36–38] Although there are advanced and expensive PIV systems available, it is possible to use free-source-code software applied to basic microscopy video clips. Barnkob *et al.*[14] and Iranmanesh *et al.*[39] used a simpler optical approach than PIV to quantify acoustic energy density based on measuring the transmitted light through a microchannel. This approach has the advantage of not requiring advanced or high-resolution microscopy.

19.5 Conclusions

We conclude this chapter by summarizing a few practical recommendations when implementing optical microscopy in a lab-on-a-chip device.

Case 1: How to design a micro-device compatible with optical microscopy.

- Use the materials and thicknesses that are established as a standard[40,41] in optical microscopy: borosilicate glass (*e.g.* Pyrex), coverslip thickness = 170 µm and glass slide thickness = 0.8–1.0 mm. This combination makes it possible to use the device with any mode of optical microscopy, both trans- and epi-illumination based.
- The combination 170 µm and 850 µm (for the bottom and top glass layers) and 90 µm for the fluid channel is optimal for both optical microscopy and for a layered acoustic resonator.
- In general, use as shallow a micro-channel as possible, preferably a few times the size of objects studied (*e.g.* cells or beads).
- Avoid polymer materials in the optical path when performing differential interference contrast (DIC) microscopy. Polymers are also less suitable for hosting acoustic resonances.
- Optimal high-resolution imaging in the center of a fluid channel is done with a water immersion objective.

- Optimal high-resolution imaging on the bottom of a fluid channel is done with an oil immersion objective.
- Water and oil immersion mediums may cause decreased acoustophoretic performance due to acoustic transmission losses.

Case 2: How to select and operate an optical microscope for optimal performance with an existing micro-device.

- The microscope objective should not have higher magnification and numerical aperture (N.A.) than needed. 20 × magnification and 0.3–0.7 N.A. is sufficient in many applications such as identification and characterization of individual cells. Such an objective is not very sensitive to deviations in material and thicknesses of the chip layers.
- If high-resolution microscopy is needed, an objective with a correction collar is often beneficial, in particular if the glass layer facing the objective deviates from the optimal thickness of 170 μm.
- If the glass layer facing the objective is significantly thicker than 170 μm, long-working distance objectives should be used, and high-N.A. objectives (around or above 1) should be avoided. The optimal choice is to use a long-working distance objective with a correction collar and a moderate N.A. For example, there exists 40 × /0.5 N.A. objectives which can be corrected for a cover glass ranging from 0 to more than 1 mm.
- An inverted microscope is the best choice for most micro-devices. This allows for assembling all fluid connections, tubing, transducers and other bulky equipment on the upper side of the chip, while imaging from below.
- If the micro-device is non-transparent (*e.g.* a two-layer silicon-glass chip), epi-fluorescence microscopy is the most useful imaging method.
- A simple contrast-enhancing bright-field method is to insert paper to block a part of the illumination light, which results in oblique illumination causing a dark-field like effect.

Acknowledgements

The authors are grateful for financial support from the Swedish Research Council and the EU FP-7 RAPP-ID project.

References

1. A. Lenshof, C. Magnusson and T. Laurell, *Lab Chip,* 2012, **12**, 1210–1223.
2. M. Wiklund, C. Günther, R. Lemor, M. Jäger, G. Fuhr and H. M. Hertz, *Lab Chip,* 2006, **6**, 1537–1544.
3. C. Muñoz-Pinedo, D. R. Green and A. van der Berg, *Lab Chip,* 2005, **5**, 628–633.
4. M. Ochsner, M. R. Dusseiller, H. M. Grandin, S. Luna-Morris, M. Textor, V. Vogel and M. L. Smith, *Lab Chip,* 2007, 7, 1074–1077.

5. B. Wang, J. Ho, J. Fei, R. L. Gonzalez Jr and Q. Lin, *Lab Chip,* 2011, **11**, 274–281.
6. E. McLeod, W. Luo, O. Mudanyali, A. Greenbaum and A. Ozcan, *Lab Chip,* 2013, **13**, 2028–2035.
7. X. Mao, J. R. Waldeisen, B. K. Juluri and T. Jun Huang, *Lab Chip,* 2007, **7**, 1303–1308.
8. J. Wu, G. Zheng and L. M. Lee, *Lab Chip,* 2012, **12**, 3566–3575.
9. A. Köhler and Z. Wiss. Mikrosk, *Mikrosk. Tech.,* 1893, **10**, 433–440.
10. O. Manneberg, B. Vanherberghen, J. Svennebring, H. M. Hertz, B. Önfelt and M. Wiklund, *Appl. Phys. Lett.,* 2008, **93**, 063901.
11. O. Manneberg, B. Vanberberghen, B. Önfelt and M. Wiklund, *Lab Chip,* 2009, **9**, 833–837.
12. K. Carlsson, *Light microscopy*, KTH Physics, Stockholm, 2007, available online at http://www.biox.kth.se/kjellinternet/Compendium.Light.Microscopy.pdf: http://www.biox.kth.se/kjellinternet/Compendium.Light.Microscopy.pdf.
13. K. Carlsson, *Imaging Physics*, KTH Applied Physics, Stockholm, 2009, available online at: http://www.biox.kth.se/kjellinternet/Compendium.Imaging.Physics.pdf.
14. R. Barnkob, I. Iranmanesh, M. Wiklund and H. Bruus, *Lab Chip,* 2012, **12**, 2337–2344.
15. B. Herman, *Fluorescence microscopy*, Springer, USA, 1998.
16. M. Wiklund, *Lab Chip,* 2012, **12**, 2018–2028.
17. J. P. Pawley, *Handbook of biological confocal microscopy*, Springer, USA, 3rd edn, 2006.
18. A. E. Christakou, M. Ohlin, B. Vanherberghen, M. Khorshidi, N. Kadri, T. Frisk, M. Wiklund and B. Önfelt, *Integr. Biol.,* 2013, **5**, 712–719.
19. H. Bruus, *Lab Chip,* 2012, **12**, 20–28.
20. A. Lenshof, M. Evander, T. Laurell and J. Nilsson, *Lab Chip,* 2012, **12**, 684–695.
21. O. Manneberg, J. Svennebring, H. M. Hertz and M. Wiklund, *J. Micromech. Microeng.,* 2008, **18**, 095025.
22. S. Radel, L. Gherardini, A. J. McLoughlin, O. Doblhoff-Dier and E. Benes, *Bioseparation,* 2001, **9**, 369–377.
23. J. F. Spengler and W. T. Coakley, *Langmuir,* 2003, **19**, 3635–3642.
24. M. Wiklund, J. Toivonen, M. Tirri, P. Hänninen and H. M. Hertz, *J. Appl. Phys.,* 2004, **96**, 1242–1248.
25. M. Wiklund and H. M. Hertz, *Lab Chip,* 2006, **6**, 1279–1292.
26. D. Bazou, G. A. Foster, J. R. Ralphs and W. T. Coakley, *Mol. Membr. Biol.,* 2005, **22**, 229–240.
27. D. Bazou, G. Davies, W. G. Jiang and W. T. Coakley, *Cell Commun. Adhes.,* 2006, **13**, 279–294.
28. D. Bazou, G. P. Dowthwaite, I. M. Khan, C. W. Archer, J. R. Ralphs and W. T. Coakley, *Mol. Membr. Biol.,* 2006, **23**, 195–205.
29. D. Bazou, W. T. Coakley, A. J. Hayes and S. K. Jackson, *Toxicol. In Vitro,* 2008, **22**, 1321–1331.

30. J. Hultström, O. Manneberg, K. Dopf, H. M. Hertz, H. Brismar and M. Wiklund, *Ultrasound Med. Biol.*, 2007, **33**, 145–151.
31. B. Vanherberghen, O. Manneberg, A. Christakou, T. Frisk, M. Ohlin, H. M. Hertz, B. Önfelt and M. Wiklund, *Lab Chip,* 2010, **10**, 2727–2732.
32. M. Evander, L. Johansson, T. Lilliehorn, J. Piskur, M. Lindvall, S. Johansson, M. Almqvist, T. Laurell and J. Nilsson, *Anal. Chem.*, 2007, **79**, 2984–2991.
33. M. Evander, A. Lenshof, T. Laurell and J. Nilsson, *Anal. Chem.*, 2008, **80**, 5178–5185.
34. L. A. Kuznetsova and W. T. Coakley, *J. Acoust. Soc. Am.*, 2004, **116**, 1956–1966.
35. M. Ohlin, A. E. Christakou, T. Frisk, B. Önfelt and M. Wiklund, *J. Micromech. Microeng.*, 2013, **23**, 035008.
36. S. M. Hagsäter, T. G. Jensen, H. Bruus and J. P. Kutter, *Lab Chip,* 2007, 7, 1336–1344.
37. O. Manneberg, S. M. Hagsäter, J. Svennebring, H. M. Hertz, J. P. Kutter, H. Bruus and M. Wiklund, *Ultrasonics,* 2009, **49**, 112–119.
38. P. Augustsson, R. Barnkob, S. T. Wereley, H. Bruus and T. Laurell, *Lab Chip,* 2011, **11**, 4152–4164.
39. I. Iranmanesh, R. Barnkob, H. Bruus and M. Wiklund, *J. Micromech. Microeng.,* 2013, **23**, 105002.
40. H. Becker, *Lab Chip,* 2010, **10**, 1894–1897.
41. H. van Heeren, *Lab Chip,* 2012, **12**, 1022–1025.

CHAPTER 20

Experimental Characterization of Ultrasonic Particle Manipulation Devices

JÜRG DUAL*, PHILIPP HAHN, IVO LEIBACHER, DIRK MÖLLER, AND THOMAS SCHWARZ

Institute of Mechanical Systems, Department of Mechanical and Process Engineering, ETH Zentrum, CH-8092 Zurich, Switzerland
*E-mail: dual@imes.mavt.ethz.ch

20.1 Introduction

Ultrasonic particle manipulation devices in lab on chip systems typically work at frequencies from 0.1 to 10 MHz given their typical size. They are usually quite complicated systems, consisting of a fluid volume containing particles, a solid bounding the fluid volume and transducers attached to the device either permanently or in a removable way. The transducers are piezoelectric devices[1–4] or have interdigitated transducers (IDT).[5] Due to the complexity of the devices and scatter in their parameters (*e.g.* material properties or geometrical parameters including the thickness of glue layers), as well as the temperature dependence of their properties, the modeling has its limitations, resulting in the need to characterize the devices non-destructively after assembly. When characterizing devices used for ultrasonic manipulation, a number of techniques are valuable, among others:

- admittance curves
- laser interferometry and schlieren images

- modal analysis
- microscopy
- particle image velocimetry (PIV).

In this tutorial the focus is put on laser interferometry, schlieren images, admittance curves and modal analysis exemplified by several examples.

When a device is in use for manipulation, microscopy, possibly in combination with particle image velocimetry, is an additional useful tool.[6,7]

20.2 Laser Interferometry

Laser interferometry is a powerful tool when displacements or velocities on surfaces need to be measured on small devices, because it is a contactless measurement technique. Therefore, the measurement does not disturb the motion to be measured.

The propagation of light is described by Maxwell's equations.[8] Light propagates with the wave speed $c = 2.998 \times 10^8$ m s^{-1} in a vacuum. The frequency f of the visible light ranges from about 4.0×10^{14} Hz to 7.9×10^{14} Hz, resulting in a wavelength $\lambda = cf^{-1}$ from about 380 nm to 750 nm.

A linearly polarized plane harmonic light wave (*e.g.* from a laser with sufficiently long coherence length) propagating in the direction of $\boldsymbol{e}_z$ is described with respect to an orthonormal *xyz* coordinate system by

$$\boldsymbol{E} = E_y \boldsymbol{e}_y = A\cos(\omega t - kz)\boldsymbol{e}_y, \qquad (20.1)$$

where $\boldsymbol{E}$ is the electrical field, A the amplitude, $\omega = 2\pi f$ the angular frequency and $k = \omega/c$ is the wavenumber.

The frequency of the visible light is too high to be detected directly, therefore detectors are used that measure the time averaged energy flux, the intensity I, which is given by

$$I = c\varepsilon_0 \frac{1}{T} \int_T \boldsymbol{E} \cdot \boldsymbol{E} \mathrm{d}t = c\varepsilon_0 \langle \boldsymbol{E} \cdot \boldsymbol{E} \rangle \qquad (20.2)$$

where ε_0 is the vacuum permittivity, T the period of the wave and $\langle \rangle$ denotes the time average. The factor $c\varepsilon_0$ will be omitted in the following.

Various types of interferometers exist.[9] In the most common setup used for particle manipulation, the light energy from a laser is split into two beams using a beam splitter, one of which is called the reference beam, the other is called the object beam.

The reference beam is reflected from a fixed mirror, while the object beam is reflected from the surface moving with displacement $u(t)$ in the direction of the object beam. If the two beams $\boldsymbol{E}_1$ and $\boldsymbol{E}_2$ are recombined and superposed, *i.e.* interfere with each other, the resulting intensity is

$$I = \langle \boldsymbol{E}_1^2 \rangle + \langle \boldsymbol{E}_2^2 \rangle + 2 \langle \boldsymbol{E}_1 \cdot \boldsymbol{E}_2 \rangle, \qquad (20.3)$$

where the last term is the phase-dependent interference term. At the detector, assuming the two fields have the same polarization and propagation directions one obtains for the total intensity

$$I = I_1 + I_2 + \sqrt{I_1 I_2} \cos \delta, \tag{20.4}$$

where $\delta = k(\Delta z - 2u(t))$, Δz is the static path length difference of the two beams and I_i is the intensity of beam i.

The intensity varies between a minimum ($\delta = (2N-1)\pi$) and a maximum ($\delta = 2N\pi$), depending on the phase difference δ between the two waves. N is an integer number. If u changes with time, the intensity varies with time. This intensity variation is detected with a photodiode.

In order to discriminate between motion towards and motion away from the laser as well as to improve the signal to noise ratio, often a process called heterodyning is used, where one of the beams is shifted in frequency with respect to the other e.g. using an acoustooptic modulator. Upon interference the voltage signal at the photodetector then has a carrier frequency f_C (typically about 100 MHz) equal to the frequency difference $f_2 - f_1$ of the two beams.

$$I(t) = \frac{1}{T} \int_T A_1^2 \cos^2[2\pi f_1 t] + A_2^2 \cos^2[2\pi f_2 t + \delta] \tag{20.5}$$
$$+ A_1 A_2 \{ \cos[2\pi(f_2 - f_1)t + \delta] + \cos[2\pi(f_2 + f_1)t + \delta] \} dt.$$

Because of the low pass behavior of the photodetector, the first two terms are averaged to yield a DC voltage V_0. The term with the sum of the frequencies will average to zero, while the term with the frequency difference will yield the signal to be demodulated.

$$V(t) = V_0 + K A_1 A_2 \cos\left[2\pi(f_2 - f_1)t + \frac{2\pi}{\lambda}(\Delta z - 2u(t))\right], \tag{20.6}$$

where K is a constant. Eqn (20.6) describes a phase modulated signal with carrier frequency $f_C = f_2 - f_1$.

Alternatively, this can be interpreted as a frequency modulated signal according to

$$\dot{\phi}(t) = 2\pi f^*(t), \tag{20.7}$$

where ϕ and f^* are instantaneous phase and frequency of the signal.

Combined with eqn (20.6) one obtains

$$\dot{\phi} = 2\pi f_C - \frac{4\pi}{\lambda} \dot{u} = 2\pi(f_C - f_D), \tag{20.8}$$

where $f_D(t) = 2\dot{u}/\lambda$ is the Doppler frequency. The frequency modulated signal is then written as

$$V(t) = V_0 + K A_1 A_2 \cos[2\pi(f_C - f_D)t]. \tag{20.9}$$

Accordingly, eqn (20.6) and (20.9) are demodulated using

- fringe counting or phase demodulation to yield displacement
- frequency demodulation to yield velocity, respectively.

Displacement amplitudes in ultrasonic manipulation devices are often much less than the wavelength, typically of the order of 10 nanometres. Fringe counting is therefore less suitable than phase demodulation. However, low frequency motions (caused by thermal drift, shocks, *etc.*) must be eliminated. For this purpose the reference mirror can be mounted on a piezoelectric element (for low frequency path length stabilization) or its influence on the signal is eliminated in the demodulation process.[10] Commercial phase demodulators have a frequency range up to about 20 MHz, with a maximum displacement of about $\lambda/8$, which is about 80 nm for a typical HeNe laser. Corresponding velocity decoders have slightly lower frequency range and are suitable up to similar amplitudes.

When the structure to be investigated is very small, the object beam needs to be precisely focused. This is best done with a fiber optic interferometer, where a spot size of <5 µm can be obtained. When both arms of such an interferometer are reflected at different points on the object, the difference of the displacements or velocity is measured.

In-plane displacements can also be measured, *e.g.* using a Bauernfeind prism rotating about the optical axis. Because here the light needs to be retro reflected, a special tape or paint is used on the surface. However, this might not be possible without affecting the motion, if the structures are very small, because of the additional mass loading.

20.2.1 Obtainable Resolution in Interferometry

When comparing different interferometer systems, the SNR (signal to noise ratio) and the frequency range are crucial. The theoretical maximum resolution is limited by shot noise at the detector.

As a measure for the resolution one can take the displacement for SNR = 1. In general, it is given by,[11]

$$u_{min} = k\sqrt{\frac{\Delta \tilde{f}}{\eta P_0}} \qquad (20.10)$$

where η detector efficiency (*e.g.* 10%)
 P_0 laser power (*e.g.* 1 mW)
 $\Delta \tilde{f}$ bandwidth used in the measurement
 k a constant, depending on the configuration of the optical setup

A typical value is

$$u_{min}/\sqrt{\Delta \tilde{f}} = 10^{-14} \text{m}/\sqrt{\text{Hz}}. \qquad (20.11)$$

For a bandwidth $\Delta \tilde{f} = 1$ MHz one can therefore expect a maximum resolution of 10^{-11} m. By limiting the bandwidth or averaging n times, the resolution can be further improved according to the $1/\sqrt{n}$ law, where n is the number of averages.

20.3 Frequency Analysis, Admittance Curves and Modal Analysis

The ultrasonic manipulation system without the particles is usually assumed to be linear and time invariant. Therefore, the tools available from linear systems analysis are applicable. In particular, the system excited with one frequency will have a stationary solution with a motion of the same frequency (Figure 20.1).

In order to determine the system properties, one can excite the system harmonically with an excitation $x(t)$ and scan the excitation frequency through the range of interest. Alternatively, repetitive broad band spectra can be used as excitation, for example a linear sweep signal or band limited white noise. Then the Fourier transform[12] can be applied to both excitation and measured signal, resulting in the corresponding spectra. For measurement data, the signals first need to be digitized, then the discrete Fourier transform is applied using the FFT algorithm to yield the desired spectra. A detailed analysis of the discrete Fourier transform and its comparison with the continuous Fourier transform is given in ref. 13.

A general signal $x(t)$ discretized with sample frequency $f_s = 1/T$ (Figure 20.2) will yield meaningful spectra up to the Nyquist frequency $f_N = \frac{1}{2} f_s$. This is the result of the sampling theorem. The maximum frequency in the

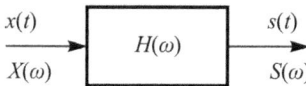

Figure 20.1 Linear system, characterized by the transfer function $H(\omega)$ with input $x(t)$ and output $s(t)$ and corresponding spectra $X(\omega)$ and $S(\omega)$. Reproduced from ref. 34.

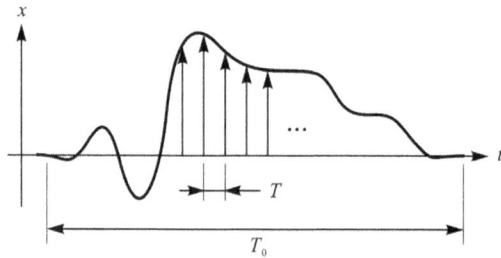

Figure 20.2 A general voltage signal $x(t)$ is discretized with sampling frequency $f_s = 1/T$ for a sample length T_0. The arrows (not all are shown) represent the sampling points, which are equally distributed across the whole sample length T_0. Reproduced from ref. 34.

signal therefore needs to be smaller than f_N, otherwise aliasing errors occur.[13] This is assured by using anti-aliasing analog filters before the sampling. Also, one should be aware that suitable windows need to be applied before the FFT in order to avoid leakage.[13] The frequency resolution of the result of the FFT will be given by $\Delta f = 1/T_0$, where T_0 is the length of the sample.

Referring to Figure 20.1, in the frequency domain, the concept of the transfer function $H(\omega)$ is used, which is defined as

$$H(\omega) = \frac{S(\omega)}{X(\omega)}. \qquad (20.12)$$

$H(\omega)$ is the quotient of the Fourier transforms of output $S(\omega)$ and input $X(\omega)$. The accuracy of the result can be assessed using the coherence function γ^2,[14] which is used to obtain confidence intervals. An estimate $\hat{\gamma}^2$ of the coherence function γ^2 can be gained from

$$\hat{\gamma}^2(\omega) = \frac{\overline{|S(\omega)X^*(\omega)|}}{\overline{X(\omega)X^*(\omega)}\ \overline{S(\omega)S^*(\omega)}}, \qquad (20.13)$$

where the superposed bar denotes averaging over the number of sampled signals and the superscript* means the complex conjugate. Without noise the coherence function equals 1. As soon as there is noise $\hat{\gamma}^2 < 1$. Bad coherence function values can be the result of many effects, e.g. bad signal to noise ratios at a particular frequency, undefined trigger levels, scatter in the excitation and others.

In order to understand the behavior of an ultrasonic manipulation system, modal analysis is a widely used method. Using linear system analysis a structure is characterized dynamically in terms of its resonance frequencies, damping and modes of vibrations. For every device an infinite number of resonances exist. In many cases the lowest resonances are most important. For every mode, in principle the whole structure including the excitation is involved.

For ultrasonic manipulation devices, typically, piezoelectric elements are used to excite the structure. Laser interferometers are used to determine the mechanical response. If the mode shapes are also important, then the structure can be scanned in 1D or 2D with a laser interferometer, which might be a challenge if the structure is small, however, it can give detailed information about displacement distributions.

If only the resonance frequencies and their damping are needed, in principle it is sufficient to measure the transfer function between excitation at one position and response at another position. If neither of the two positions is a node of one of the resonance modes, all the frequencies and damping values can be found.

When the damping of the mode is not too large and the electromechanical coupling is strong (e.g. near a transducer resonance), admittance curves of the transducer can also be used to determine the resonance frequencies and damping.[15] The electrical admittance Y is a measure of how easily an element will allow an electrical current to flow and is defined as

$$Y = \frac{1}{Z} = G + iB \qquad (20.14)$$

where Z is the impedance, G the conductance and B the susceptance. All are functions of frequency.

If the resonance modes are separated enough (depending on damping), one can decouple the modes and consider them as single degree of freedom oscillators with modal coordinate s_j and modal excitation $\bar{f}_j$. This also corresponds to the modal decomposition in a FEM analysis.[16,17] The differential equation describing mode j is

$$m_j\ddot{s}_j + \delta_j\dot{s}_j + k_js_j = \bar{f}_j, \tag{20.15}$$

where m_j, δ_j and k_j are the modal mass, damping and spring constant of mode j, respectively. The resonance frequency is given by

$$\omega_{0j} = \sqrt{\frac{k_j}{m_j}}, \tag{20.16}$$

and the transfer function H_j for harmonic excitation by

$$H_j = \frac{s_j}{\bar{f}_j} = \frac{1}{m_j} \frac{1}{\omega_{0j}^2 - \omega^2 + i\omega\delta_j/m_j}. \tag{20.17}$$

Using partial fractions one obtains

$$H_j = \frac{1}{2i\omega_{0j}m_j}\left(\frac{1}{i\omega + \lambda_j} - \frac{1}{i\omega + \lambda_j^*}\right), \tag{20.18}$$

where

$$\begin{aligned}\lambda_j &= \frac{\delta_j}{2m_j} - i\omega_{0j}, \\ \lambda_j^* &= \frac{\delta_j}{2m_j} + i\omega_{0j}.\end{aligned} \tag{20.19}$$

For positive ω and damping which is not too large, the first term in eqn (20.18) dominates in the vicinity of ω_0. If one neglects the second term and plots the resulting $\text{Re}(H_j)$ and $\text{Im}(H_j)$ in the complex plane (Figure 20.3), we obtain the equation of a circle, which can be used to fit the damping δ_j.[17]

$$\text{Re}(H_j)^2 + \left(\text{Im}(H_j) + \frac{1}{2\omega_{0j}\delta_j}\right)^2 = \left(\frac{1}{2\omega_{0j}\delta_j}\right)^2 \tag{20.20}$$

Also the quality factor Q_j can be defined for the respective mode j

$$Q_j = \frac{m_j\omega_{0j}}{\delta_j} = \frac{\omega_{0j}}{\omega_{uj} - \omega_{lj}} = \frac{\omega_{0j}}{\Delta\omega_j}. \tag{20.21}$$

ω_{uj} and ω_{lj} are the angular frequencies, which correspond to phase values of the phase at resonance $+/- \pi/4$. For $Q > 10$, these are also about equal to the

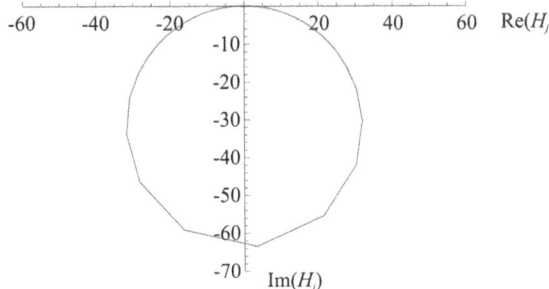

Figure 20.3 Complex representation of the approximation of the transfer function $H_j(\omega)$ of mode j in the complex plane near a resonance frequency ω_{0j}. Please note the insufficient frequency resolution that might result from the application of the discrete Fourier transform when the damping is low. Reproduced from ref. 34.

frequencies, for which the amplitude is $1/\sqrt{2}$ of the maximum amplitude. Q is therefore given by the resonance frequency divided by the bandwidth.

The transfer function of the whole system can then be composed of the sum

$$H = H_1 + H_2 + H_3 + \ldots, \quad (20.22)$$

where each mode can have different parameters.

20.4 Schlieren Imaging of 2D Pressure Fields in Cavities

The schlieren method is an optical method that makes visible spatial variations in the refractive index of the liquid. Hence any quantity related to the refractive index can be visualized, these are *e.g.* temperature variations or density variations caused by fluid flow or by acoustic waves. A detailed account of the schlieren technique in air is given by Settles.[18]

The schlieren method is a proven tool for the experimental characterization of optically transparent liquids and for the visualization of ultrasonic fields.[19] The optical nature of the method results in a number of advantages over methods such as particle tracking or hydrophones. The method is non-invasive, that is no seeding particles are required which might influence the fluid properties or the acoustic field. The method can cover a large image area and track changes over a long time frame and in real time. The complete pressure field within the image area can be recorded. The resolution and frequencies of the ultrasonic field that can be imaged are mainly limited by the optical wavelength and the resolution of the optical components used. Reports show schlieren images of ultrasound in water of over 100 MHz.[20] The structure containing the fluid, like any obstacle in the optical light path, causes interference fringes. This can be a limiting factor

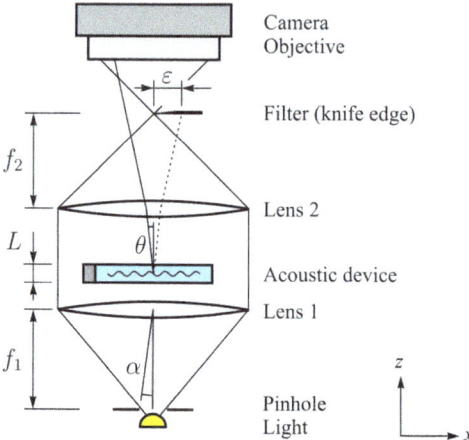

Figure 20.4 Schematic of a vertically mounted three lens schlieren setup. Reproduced from ref. 19.

for microfluidic systems or if one is interested in features close to a wall. The same problem can arise when using the schlieren method in combination with particles.

Reports suggest a large variety of different setups based on mirrors and transparent lenses including bidirectional setups with a mirror as background.[21] Specialized methods are *e.g.* the background oriented schlieren method,[22] the colour schlieren method[18] or rainbow schlieren.[23] An example of quantitative imaging of ultrasonic waves with real-time pulsed schlieren imaging is given in ref. 24. The schlieren method has similarities to a number of microscopy imaging techniques presented in Chapter 19.[25] These are *e.g.* phase contrast and interferometry methods such as the differential interference contrast method.

The layout of a simple three lens schlieren setup is presented in combination with general design criteria for a schlieren setup suitable to visualize ultrasonic fields. A more detailed account of the setup, including experimental results, can be found in the paper by Möller *et al.*[19]

Figure 20.4 shows a conventional (Töpler) three lens schlieren system which has been mounted vertically. The basic components are a point light source, two field lenses (L1, L2), a filter and a camera with objective.

The field lenses are a collimating lens (L1) and a focusing lens (L2). The focal plane of the objective is at the position of the planar acoustic device with waves travelling in the x- or y-direction, resulting in an image where the spacing of the interference fringes is equal to the ultrasonic wavelength multiplied by the lateral magnification of real images.[26] In the cut-off or Fourier plane, different filters can be placed. Examples are a knife edge filter which emphasizes directional effects dependent on its rotational position, a dark field filter, or colour filters.

Given is a system with optical wavelength λ, acoustic wavelength Λ and acoustic beam thickness L. The dominating type of diffraction can be determined with the Klein–Cook parameter $Q_{KC} = 2\pi\lambda L/\Lambda^2$. With $Q_{KC} \gg 1$ Bragg diffraction is dominating, while the case with $Q_{KC} \ll 1$ is called Raman–Nath diffraction.[27] In the latter case, the assumption holds that the beam thickness L, or length that the optical waves cross the acoustic waves, is sufficiently small for the effect of multiple diffraction to be neglected. Most ultrasonic systems are in the Raman–Nath regime, where the diffraction angle θ is given by $\sin\theta = \lambda/\Lambda$ for the first intensity peak. For ultrasonic systems the Bragg diffraction angle for one main peak provides similar results, since both angles are approximately the same. The diffracted beam separation distance $\varepsilon = \theta f_2$ is a measure for the sensitivity of the schlieren setup. For a large ratio of focal length f_1 to pinhole diameter d_p, the relation $\alpha \propto d_p/f_1$ holds, where α is the beam deviation angle at the collimation lens L1 which should be as small as possible. For d_p there is a lower limit in the order of the Airy disc and Rayleigh criterion $d_p = 2\sqrt{2f_1\lambda}$.

For the imaging of standing wave fields, high power LEDs are well suited. For travelling waves or other applications sensitive to illumination time, xenon stroboscopic lamps or lasers are other possibilities. The schlieren method shows the flaws of the optical components, particularly of the focusing optics (L2). Typically, a correction of spherical aberration and coma is particularly important, while the importance of further corrections, such as chromatic aberration and astigmatism, is more individually dependent on the application and light source.

20.5 Measurements

In the following section, measurements of the electrical and mechanical behavior of piezoelectrically driven systems are outlined. The understanding of the measured results is supported by comparison to simulation results. First, a simple piezoelectric element is characterized by admittance and interferometer measurements, then these methods are applied to several manipulation devices.

20.5.1 Characterization of a Transducer

20.5.1.1 Experimental Setup

The experimental setup for interferometer measurements is illustrated in Figure 20.5. For the excitation of a piezoelectric element, a function generator (Stanford research systems, model DS345) is used in connection with a broadband power amplifier (ENI ®2100L). For laser vibrometry, a Polytec OFV 303 sensor head is connected to a Polytec OFV 3001 vibrometer controller. The velocity measurement signal of the vibrometer is then connected to a LeCroy WaveSurfer™ 424 oscilloscope for further data analysis. The measurement and excitation signals are filtered by a low-pass filter

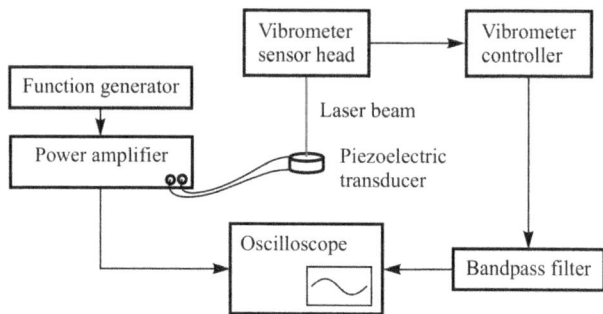

Figure 20.5 Schematic illustration of the experimental setup for laser interferometry measurements of a piezoelectric transducer. The path of the analog excitation and measurement signals is marked by arrows. Reproduced from ref. 34.

(Krohn-Hite model 3945). For admittance measurements of piezo elements, an impedance analyzer (Sine Phase Z-Check 16777k) is used, which has a range of 1 kHz to 16.7 MHz. Another method for the admittance measurement has been reported by Neild:[28] a current probe was connected to an oscilloscope and the current signal together with the voltage signal were recorded and processed.

20.5.1.2 Correlation between Electrical and Mechanical Measurements

As a simple first example, the electrical properties of a piezoelectric transducer are compared to its mechanical behavior over a certain frequency range. A cylindrical piezoelectric element of diameter 6.3 mm and thickness 2.9 mm (PZ 27, Ferroperm Piezoceramics) with polarization direction along the cylindrical axis is chosen. The two electrodes at the top and bottom of the cylinder are electrically connected to wires, which also allow for a free mounting of the piezoelectric element.

The admittance curve of the described transducer is measured in a range from 50 kHz to 1.5 MHz. Its absolute value is plotted in Figure 20.6(a). Five major maxima peaks represent the transducer's resonance frequencies. In the following, the admittance and transfer function are studied.

For the mechanical characterization of the piezoelectric element, its transfer function $H(\omega) = S(\omega)/X(\omega)$ is calculated between the excitation voltage $X(\omega)$ of the piezoelectric transducer and the velocity amplitude $S(\omega)$ as measured by laser vibrometry. The laser beam is pointed at an electrode surface of the cylindrical piezoelectric element. In order to measure the vibration along the cylindrical axis of the piezoelectric element, the laser beam is adjusted to be parallel to this axis. An excitation over a large frequency range is desired, hence for $X(\omega)$, white noise is used. The chosen white noise signal has a root mean square (RMS) value of 23.6 V, which is

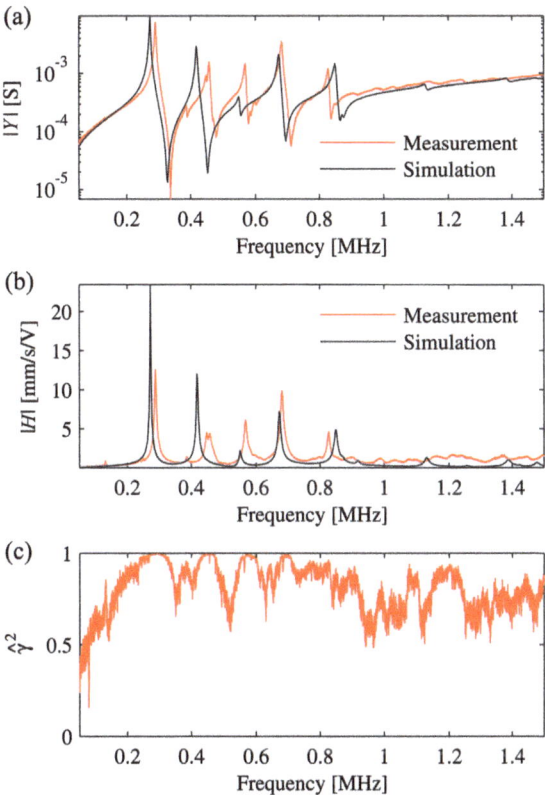

Figure 20.6 Characterization of a piezoelectric element: (a) the absolute value of its admittance $|Y|$, (b) the absolute value of the transfer function $|H|$ as well as (c) the estimate of the coherence function $\hat{\gamma}^2$ are plotted against frequency. The comparison of admittance and transfer function reveals the coupling between these quantities, especially at the five clearly visible resonance peaks. For (b) and (c), a driving voltage with a RMS value of 23.6 V is applied in the experiment. Simulation results are also plotted in (a) and (b). In (c), a coherence close to 1 ascertains the quality of the measurement. Reproduced from ref. 34.

high enough to obtain a sufficient coherence. Alternatively, a frequency sweep over a certain frequency range could also be applied.

A 10 ms time signal of both the excitation and the velocity measurement is recorded at 14 well-distributed random points on the top surface of the cylindrical piezoelectric element. The measured maximum velocity, averaged over all measurement points, accounts for 149.2 mm s^{-1}, and the RMS value of the measured velocity is 41.0 mm s^{-1}, also averaged over all measured points. By means of a fast Fourier transformation (FFT), the measurement signal is transformed into the frequency domain with a resolution of 100 Hz. Low-pass filters with a cut-off frequency of 2 MHz are applied to avoid aliasing errors. In order to reduce the spectral leakage, the time signals are enveloped by Hamming windows before the FFT.[13]

In Figure 20.6(b), the resulting transfer function is plotted. Five major peaks can be observed in the plotted frequency range, which correspond to different resonant modes of the piezoelectric element. For verification of the measurement quality, the estimate of the coherence $\hat{\gamma}^2$ between $S(\omega)$ and $X(\omega)$ is plotted in Figure 20.6(c). Especially around the resonance frequencies, a coherence close to 1 is observed, which underlines the validity of the measurement. Between the resonance peaks the coherence is below one, reflecting the reduced SNR at these frequencies.

The five major peaks of the transfer function occur at the same resonance frequency as the five positive peaks in the transducer's admittance curve, revealing the electromechanical resonances. Whereas the five positive peaks in the admittance curve at resonance frequency result in a minimal impedance (series resonance), the corresponding proximate negative peaks with maximal impedance are known as the antiresonance (parallel resonance). The physical interpretation of the demonstrated relation between the admittance curve and the transfer function lies in the interplay between the inverse and the direct piezoelectric effect. The electromechanical coupling of the piezoelectric material can be seen as the conversion from electrical to mechanical energy and *vice versa*.

For a better understanding of the resonances, the piezoelectric element is also simulated in COMSOL Multiphysics® 4.1 with a finite element model. The piezoelectric material PZ27 is modeled with a density of $\rho = 7700$ kg m^{-3} and a stiffness matrix with the components[29]

$c_{E11} = c_{E22} = 147$ GPa, $c_{E33} = 113$ GPa, $c_{E44} = c_{E55} = 23.0$ GPa, $c_{E66} = 21.2$ GPa, $c_{E12} = c_{E21} = 105$ GPa and $c_{E13} = c_{E23} = c_{E31} = c_{E32} = 93.7$ GPa, a coupling matrix with the components $e_{15} = e_{24} = 11.64$ C m^{-2}, $e_{31} = e_{32} = -3.09$ C m^{-2}, $e_{33} = 16.0$ Cm^{-2} and a complex relative permittivity $(1 - i \cdot 0.017)$ with the components $\varepsilon_{rS11} = \varepsilon_{rS22} = 1130$ and $\varepsilon_{rS33} = 914$.

Complex material parameters are set to yield a quality factor of $Q = 100$. The model allows the evaluation of the current across electrodes for a given voltage, so the transducer's admittance can be calculated. The simulation results for the velocity amplitude and the admittance are also plotted in Figure 20.6. However, in contrast to the measurements, the velocity amplitude was averaged over the whole top surface of the cylindrical piezoelectric element. Differences between the measurement and the simulation can be caused by material variations, the electrical wiring, noise in the electrical signal, as well as the different averaging.

The advantage of a simulation model lies in the more profound understanding of the resonant modes. For example, the resonance at about 0.68 MHz in the measurement corresponds to the resonance at 0.6725 MHz in the simulation. Its resonant mode can be identified as a thickness mode of the piezoelectric element as illustrated in Figure 20.7. The three dimensional plot of the cylindrical transducer shows the z-component of the displacement field, whereas the z-axis is parallel to the axis of the cylinder. The

Experimental Characterization of Ultrasonic Particle Manipulation Devices 533

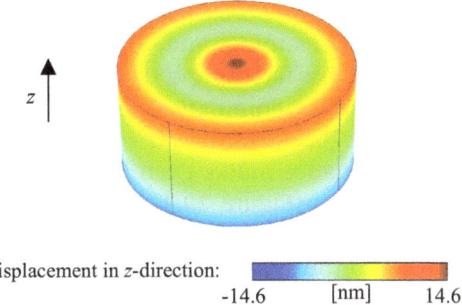

Figure 20.7 In the simulated resonant mode, the displacement field in the z-direction of the cylindrical piezoelectric transducer is shown. A harmonic excitation with an amplitude of 10 V was applied between the top and bottom electrode at a resonance frequency of 672.5 kHz.

displacement is half a wavelength along the axis of the cylinder, where the highest amplitudes occur.

For the individual piezoelectric element as presented here, the electromechanical resonance frequencies were studied. When a piezoelectric element is attached to an ultrasonic particle manipulation device, the transfer function as well as the admittance curve will show many more resonance frequencies of the system, which are caused by additional structural resonance modes, as will be shown later.

20.5.1.3 Q-Factor

Different types of resonances are utilized in the field of ultrasonic particle manipulation. Resonances of the fluid cavity cause the radiation forces that are used for particle manipulation, while transducer resonances are leveraged to enhance the amplitudes. Structural resonances of the enclosure are sometimes desired to couple with the fluid, but they can also disturb the standing pressure field in the fluid. However, due to the coupling between the transducer, the structure and the fluid, most resonances are affected by all components and cannot be associated with only one element of the device. All resonances combined lead to the complex patterns that can be observed in Figures 20.14 and 20.16. The damping associated with each material can be expressed by a material loss factor.[16] In experiments, however, the energy transferred into the support of the device is another important factor that affects each resonance mode. This effect is frequency dependent, since the rate at which the energy is transferred into the support strongly depends on the specific vibration pattern of the device. The electromechanical coupling of piezoelectric elements shows significant frequency dependent damping as well. Typically, low frequencies show higher damping.[30] Additionally, there exist a number of more complex damping mechanisms within the fluid. They can be traced back to viscous boundary layers, acoustic streaming and to the radiation forces. The combination of the material loss factors and the other damping mechanisms attenuates each resonance of a manipulation device, leading to well defined peaks, each with a specific bandwidth and magnitude. Both parameters depend on the specific resonance mode. Even if the system consists of only one material, as is the case in Figure 20.4, the resonance peaks will still display individual bandwidths and magnitudes due to resonance mode dependent energy flow into the support and frequency dependent material behavior. In order to characterize each resonance peak, the Q-factor is used.[16]

Material loss factors are of primary interest for use in simulations. Here, the attainable pressure amplitudes in the fluid strongly depend on them. Furthermore, the relation between the material loss factors for the fluid and the solid device components defines how strongly structural resonances can affect the standing pressure field in the fluid. In simulations it is difficult to directly simulate loss mechanisms other than material losses. For this reason the additional loss is often accounted for by artificially increasing the material losses in simulations.

In contrast to material loss factors, the Q-factors defined for each resonance are of practical importance. Particle manipulation devices that use one specific resonance have an operating frequency range that strongly depends on the Q-factor of the resonance. Moreover, strong standing pressure fields can be expected only for high resonance peaks, typically the ones with a high Q-factor. As compared to material loss factors, which are widely available in the literature, the Q-factors corresponding to a specific resonance mode need to be determined either in simulations or in experiments.

In order to demonstrate different methods, the Q-factor of one resonance is determined on the basis of the admittance curve and the transfer function, measured in Section 20.5.1.2. As an example, the thickness mode of the piezoelectric element at a frequency of around 0.68 MHz is chosen. It is important to note that the demonstrated techniques only work for strong and distinct resonance peaks that can clearly be separated from each other. This becomes a problem for complex devices with a vast number of resonances (compare Figure 20.16).

The Q-factor of a resonance can be deduced from measurements by means of various methods. Here, only the most common techniques are demonstrated, starting with those for the admittance curve. In order to minimize the error, all of them require resonance peaks that display a resonance–antiresonance attenuation difference of at least 25 dB. If this ratio is smaller, the error steeply increases up to invalid values at around 10 dB. More sophisticated methods exist that can be used for weak resonances; however, these methods are beyond the scope of this chapter.[21]

The resonance–antiresonance attenuation difference is illustrated in Figure 20.8 between the maximum and the minimum of the absolute admittance curve $|Y|$. The chosen peak leads to a ratio of $A = 36$ dB which is well in the range at which the classical techniques provide accurate results. Based on the derivation presented in section 20.3, the Q-factor can be calculated by dividing the resonance frequency f_0 by the bandwidth Δf of the resonance peak. The bandwidth is generally given as the difference between an upper frequency f_u and a lower frequency f_l.

$$Q = \frac{f_0}{\Delta f} = \frac{f_0}{f_u - f_l} \qquad (20.23)$$

The resonance frequency $f_0 = 681.5$ kHz can be found at the point of maximum conductance G in Figure 20.8. Also illustrated in Figures 20.8 and 20.9 are three different ways that f_u and f_l can be found. The first option is to choose the frequencies at which the conductance reaches a level ½ or −6 dB of its maximum value ($f_{u(-6\ \text{dB})}$ and $f_{l(-6\ \text{dB})}$). The second option is to choose the frequencies corresponding to the susceptance extremes ($f_{u(s)}$ and $f_{l(s)}$). Finally, the third option is to choose the frequencies in the phase plot (see Figure 20.9), which have a phase shift of ±45° relative to the resonance frequency ($f_{u(\pm 45°)}$ and $f_{l(\pm 45°)}$). All specific frequency values are shown in Table 20.1 together with the corresponding Q-factors.

For a physical interpretation of the different components in the admittance plot in Figure 20.8, it is useful to consider two basic relations. First, the current is the product of the applied voltage and the admittance. This leads to the conclusion that the part of the current that is in phase with the applied voltage at a certain frequency is proportional to the conductance G. Second, the converted power is the product of the applied voltage and the part of the

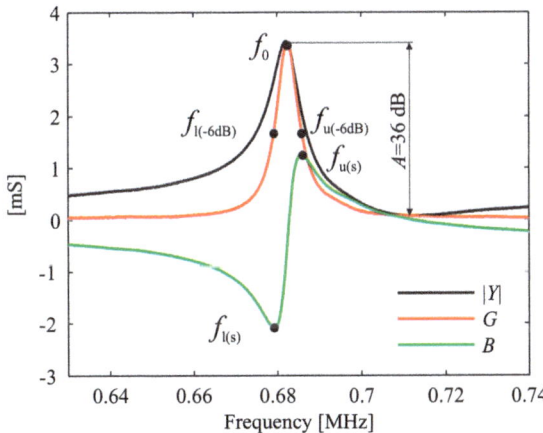

Figure 20.8 Q-factor determination using the admittance curve. As an example, a thickness mode of a piezoelectric element without mechanical load is measured. The absolute value of the admittance Y has already been shown in Figure 20.6(a). The upper and lower frequencies f_u and f_l can be evaluated using the conductance G and the susceptance B. Reproduced from ref. 34.

current that is in phase with the latter. In this sense, the conductance is a measure of the converted power. Of course, not all of the electrical power is converted into mechanical power because a part is transferred into heat. A similar interpretation could follow for the susceptance B. However, the information of both conductance and susceptance can also be visualized more intuitively in the form of an absolute value $|Y|$ and a phase as demonstrated in Figures 20.8 and 20.9. The absolute value is then a measure of the absolute current while the angle between the applied voltage and the current is given by the phase.

The Q-factor of the resonance can also be deduced from the velocity transfer function. This might be the more intuitive choice, however, compared to admittance measurements it requires more expensive equipment. Additionally, in some cases measurements might not even be possible

Table 20.1 The different upper and lower frequencies f_u and f_l, obtained from the admittance measurement in Figures 20.8 and 20.9 and the resulting Q-factor. Since the three different ways of measuring the Q-factor are based on the same assumptions, the differences are due to measurement errors.

f_0	681.5 kHz	
$f_{l(-6dB)}$	678.2 kHz	$Q = 103$
$f_{u(-6dB)}$	684.8 kHz	
$f_{l(s)}$	678.2 kHz	$Q = 105$
$f_{u(s)}$	684.7 kHz	
$f_{l(\pm 45°)}$	678.3 kHz	$Q = 99$
$f_{u(\pm 45°)}$	685.2 kHz	

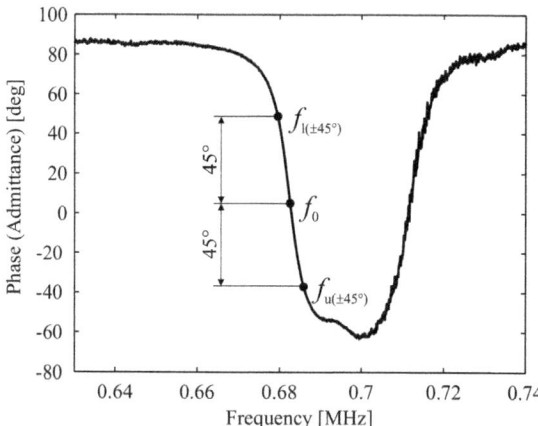

Figure 20.9 Q-factor determination at the example of a thickness mode of a piezoelectric element using the phase plot corresponding to the admittance measurement. The upper and lower frequencies f_u and f_l can be found at the phases of $\pm 45°$ relative to the resonance frequency f_0. Reproduced from ref. 34.

if the frequencies exceed the frequency range of the measurement equipment or if the velocity amplitudes are simply too low.

The resonance frequency $f_0 = 681.2$ kHz is obtained from Figure 20.10 at the maximum of the magnitude of the transfer function $|H|$. The frequency is slightly different from the one obtained using the admittance curve. However, the difference is usually very small. Typically, two methods can be used to deduce f_u and f_l from the transfer function. The points at which the absolute value of the transfer function reaches a level of $1/\sqrt{2}$ or -3 dB of its maximum value are marked in Figure 20.10. The points with a phase shift of $\pm 45°$ relative to the resonance frequency can be seen in Figure 20.11.

All specific frequency values are shown in Table 20.2 together with the corresponding Q-factors.

The physical interpretation of the real and complex components of the transfer function H in Figure 20.10 is not an easy task since the physical effects involved in the transfer of the electrical input voltage to the mechanical output velocity are quite complicated. For this reason it is more useful to consider only the absolute value of the transfer function, which relates to the sensitivity of the system and the phase, which gives the angular relation between the input voltage and the output velocity. In comparison to the admittance measurements, the Q-factors determined using transfer function are decreased by about 20%. This is due to the higher excitation voltage necessary to obtain a good signal to noise ratio in the laser vibrometry measurements. Non-linear material behaviour of the piezoelectric transducer at high amplitudes causes the energy to leak out of the resonance and effectively reduces the Q-factor. Measurements of the Q-factor for resonances

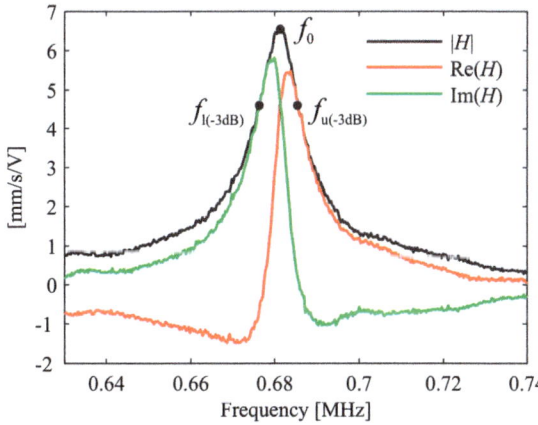

Figure 20.10 Q-factor determination at the example of a thickness mode of a piezoelectric element without mechanical load using the transfer function H. The magnitude of the transfer function $|H|$ has already been shown in Figure 20.6(b). The upper and lower frequencies f_u and f_l can be found from the magnitude at a level of -3 dB relative to the resonance frequency f_0. Reproduced from ref. 34.

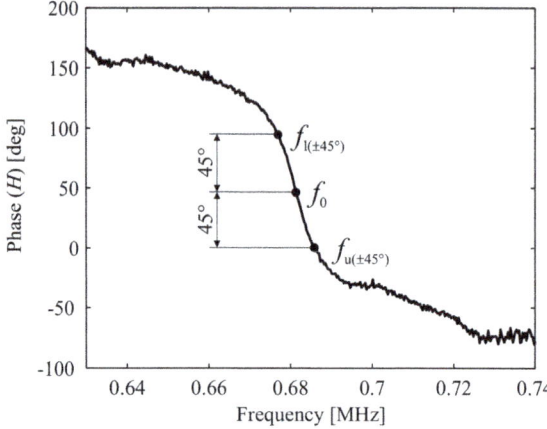

Figure 20.11 Q-factor determination at the example of a thickness mode of a piezoelectric element using the phase plot of the transfer function H. The upper and lower frequencies f_u and f_l can be found at the phases of $\pm 45°$ relative to the resonance frequency f_0. Reproduced from ref. 34.

Table 20.2 The different upper and lower frequencies f_u and f_l, obtained from the measured transfer function in Figures 20.10 and 20.11 and the resulting Q-factor.

f_0	681.2 kHz	
$f_{l(-3\ dB)}$	676.6 kHz	$Q = 77$
$f_{u(-3\ dB)}$	685.4 kHz	
$f_{l(\pm 45°)}$	677.3 kHz	$Q = 85$
$f_{u(\pm 45°)}$	685.3 kHz	

that are used for the manipulation of particles should therefore be done in the same amplitude range as the actual working excitation. Low-amplitude measurements can lead to an overestimation.

The Q-factor can also be calculated from the time constant τ obtained by measuring the decaying velocity amplitude after a monochromatic resonant excitation is turned off. However, for high-Q systems, this method is very error prone because a slight deviation from the exact resonant frequency results in drastically reduced Q-factors. Additionally, abruptly switching off the excitation might excite other resonances of the system, which also increases the error. Nevertheless, this method provides an intuitive way of calculating the Q-factor of a system. Figure 20.12 shows the velocity amplitude of the piezoelectric element over time.

After the vibration has reached a quasi-stationary state, the excitation frequency of 681.2 kHz is switched off and the velocity amplitude decays. The left point in Figure 20.12 is selected to be at a position shortly after the excitation was switched off. The right point is defined at the position where

Experimental Characterization of Ultrasonic Particle Manipulation Devices 539

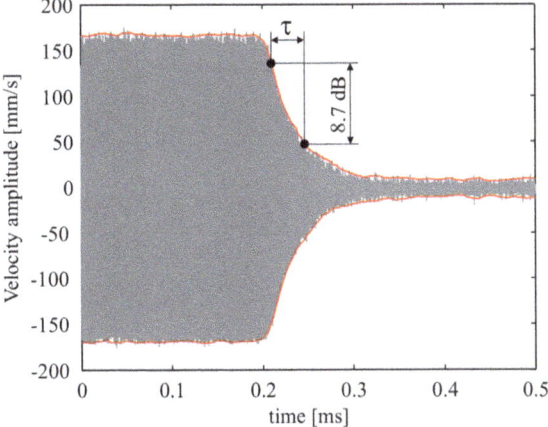

Figure 20.12 Q-factor determination at the example of a thickness mode of a piezoelectric element utilizing the characteristic time of the decaying velocity amplitude. The envelope of the velocity measurement shows an exponential decay after the resonant excitation has stopped. The velocity measurement was done by laser interferometry. Reproduced from ref. 34.

the velocity amplitude has decayed to $1/e$ or -8.7 dB of the amplitude at the left point. Calculating the time difference between both points leads to the characteristic time τ, which is 37.2 µs in the measurement above. The Q-factor can then be calculated using the relation: $Q = \pi f_0 \tau$, leading to a Q-factor of 81. Again, a relatively high excitation voltage is necessary to obtain a good signal to noise ratio. As explained above, non-linear effects at high deformation amplitudes reduce the Q-factor.

20.5.2 Characterization of Manipulation Devices

Here a planar resonator is used to show the relation between admittance measurements and resonances in the fluid cavity. The planar resonator consists of the following five layers as shown in Figure 20.13: a piezoelectric transducer (Ferroperm PZ 26, 0.5 mm), adhesive layer (Epotek H20E, 8 µm), carrier (glass, 1 mm), fluid (water, 1 mm) and reflector (glass, 0.56 mm).

The resonator is simulated with a one-dimensional transfer matrix model reported by Gröschl.[31] In Figure 20.14, the admittance measurement and simulation is plotted *vs.* frequency. The fundamental resonance frequency of the piezoelectric transducer alone would be at 4.2 MHz. The peaks marked with F represent the resonances dominated by the fluid. For example, the first peaks at 2.26 MHz and 2.97 MHz correspond to 1.5 and 2 wavelengths in the fluid layer and therefore 3 and 4 nodal pressure planes, respectively. The resonant layers are determined with the transfer matrix model by plotting the pressure distribution for the different layers or by

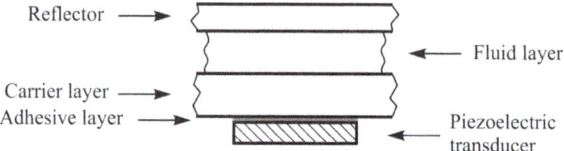

Figure 20.13 Schematic view of a planar resonator consisting of the following five layers: piezoelectric transducer, adhesive layer, carrier, fluid and reflector, with thicknesses of 0.5 mm, 8 μm, 1 mm, 1 mm and 0.56 mm, respectively. Reproduced from ref. 34.

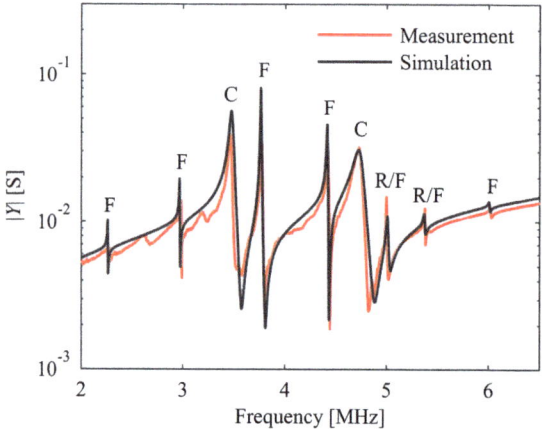

Figure 20.14 Admittance measurement and simulation for a planar resonator. The model allows the assignment of each peak to the corresponding layer that dominates the resonance. The peaks marked with F correspond to fluid resonances, C to carrier resonances and R/F to a combination of reflector and fluid resonances. Reproduced from ref. 34.

varying the layer thickness of the resonant layer in the model and observing the frequency shift of the peak in the admittance curve. The peaks C are dominated by a resonance in the carrier and R/F is a combination of a reflector and fluid resonance. The knowledge of which peak corresponds to which layer is important for the operation of the device. It is possible to drive the system exactly at one of the fluid resonances. When conditions such as temperature are changing, the resonance frequency will also change and so will the peak in the admittance plot. As reported by Hammarström et al.,[32] an online tracking of the resonance frequency can be implemented using a real-time analysis of the impedance curve around the operating frequency. In this way, optimal device functionality can be kept over longer periods of time. The excitation automatically follows resonance frequency shifts due to e.g. temperature changes, particle agglomeration or washing steps in flow-through devices.

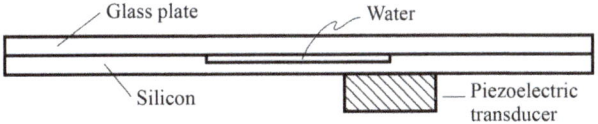

Figure 20.15 Schematic view of a microfluidic device excited with a piezoelectric transducer. Reproduced from ref. 34.

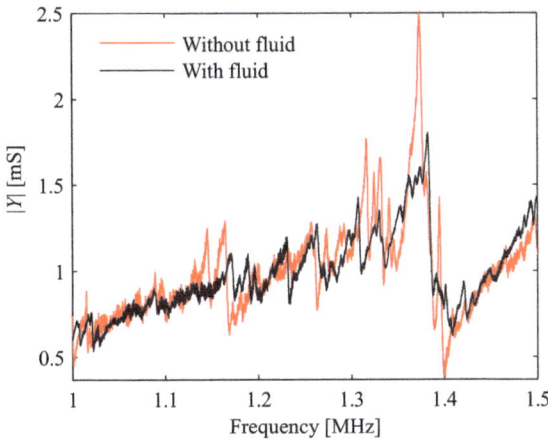

Figure 20.16 Admittance plot of a piezoelectric element mounted on a microfabricated device. The red curve represents the measurement without fluid (air-filled cavity) and the black curve the measurement with a fluid (water) filled cavity. Reproduced from ref. 34.

As a further example, the admittance curve of a typical microfabricated device (see Figure 20.15) is measured. The main part is a 16 mm × 16 mm silicon substrate (thickness 500 μm) where a 5 mm × 6 mm cavity is etched 200 μm deep inside. The cavity is covered with a glass plate (thickness 500 μm) by anodic bonding. The actuation is done with a 6 mm × 2.5 mm piezoelectric transducer (thickness 1 mm) glued to the bottom of the device.

The admittance curve of the mounted piezoelectric element is measured for an air-filled as well as a water-filled chamber and is shown in Figure 20.16. Contrary to the previous simple examples, the complex structure results in a large number of peaks in the admittance curve, which are caused by additional structural resonance modes. The peaks in the admittance curve are generally less pronounced for the water-filled device, as the water causes additional damping. The determination of resonance modes by means of the admittance curve is less obvious in these complex systems, however, *e.g.*, the peak in the admittance curve at 1.42 MHz was found to result in a resonance at which 10 lines of particles are formed in the chamber. Due to the high number of nearby resonance peaks, an online frequency tracking as explained by Hammarström *et al.*[32] is not an option for these setups.

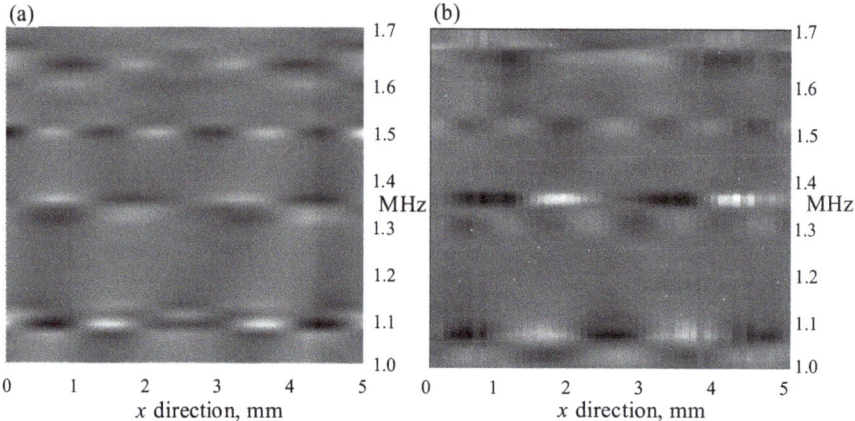

Figure 20.17 Vertical displacement along a line on the underside of a piezo-electric element mounted on a microfabricated device. The simulation results (a) can be compared to measurements by laser interferometry (b). Reprinted from ref. 28 with permission from Elsevier.

In addition to the admittance measurements shown, the dynamic behaviour of manipulation devices can also be measured by laser vibrometry. This tool is especially useful for the validation of FEM simulations. An example for a 1D measurement is reported by Neild et al.[28] as shown in Figure 20.17. The measurement (b) shows the vertical displacement on the underside of a piezoelectric element, which was mounted on a microfabricated device. The measurement along a line on the piezoelectric element is plotted over a frequency range of 1.0–1.7 MHz after a FFT. The frequencies at which the device is operated in resonance can be observed as well as the number of wavelengths along the measured line. The measurement is in good agreement with the simulation results shown in (a). Finally, an example of scanning laser vibrometry is outlined by Oberti et al.,[33] where the mode shape of a piezoelectric element was measured in 2D.

20.6 Conclusions

In this paper, a number of techniques have been shown and applied, which can be used to characterize devices used for ultrasonic manipulation. The mechanical behaviour is best studied using laser interferometers with either displacement or velocity decoders. The 2D pressure fields in fluid-filled cavities can be investigated using a schlieren set up. The electrical characteristics of the piezoelectric elements also give information about resonance frequencies where strong electromechanical coupling occurs. The Q-factor is a measure of how efficiently energy can be used for the manipulation. The higher the Q-factor, the lower the required voltages. On the other hand, a high Q results in narrow bandwidth, which in turn might pose some difficulties in stabilizing the operation when the temperature and therefore the resonance frequency changes. Several methods have therefore been demonstrated to determine the Q based on various measured quantities. When the device is filled with a fluid, Q will be decreased, as additional damping mechanisms are introduced. This will lower the amplitudes generated and therefore also the forces acting on particles in the fluid.

References

1. W. T. Coakley, D. W. Bardsley, M. A. Grundy, F. Zamani and D. J. Clarke, *J. Chem. Technol. Biotechnol.,* 1989, **44**, 43.
2. N. R. Harris, M. Hill, S. Beeby, Y. Shen, N. M. White, J. J. Hawkes and W. T. Coakley, *Sens. Actuators, B,* 2003, **95**, 425.
3. A. Nilsson, F. Petersson, H. Jonsson and T. Laurell, *Lab Chip,* 2004, **4**, 131.
4. A. Haake and J. Dual, *Ultrasonics,* 2004, **42**, 75.
5. J. Friend and L. Y. Yeo, *Rev. Mod. Phys.,* 2011, **83**, 647.
6. R. Barnkob, P. Augustsson, T. Laurell and H. Bruus, *Lab Chip,* 2010, **10**, 563.
7. J. F. Spengler, M. Jekel, K. T. Christensen, R. J. Adrian, J. J. Hawkes and W. T. Coakley, *Bioseparation,* 2000, **9**, 329.
8. M. Bass, Handbook of Optics, McGraw-Hill, 1995, vol. 1.
9. M. Bass, Handbook of Optics, McGraw-Hill, 1995, vol. 2.
10. J. Dual, M. Hageli, M. R. Pfaffinger and J. Vollmann, *Ultrasonics,* 1996, **34**, 291.
11. J. W. Wagner, Optical Detection of Ultrasound, *Physical Acoustics* XIX, Academic Press, 1990, 201.
12. I. N. Sneddon, *Fourier Transforms*, McGraw-Hill, 1951.
13. E. O. Brigham, *The Fast Fourier Transform*, Prentice Hall, 1974.
14. J. S. Bendat and A. G. Piersol, *Random Data: Analysis and Measurement Procedures*, Wiley-Interscience, New York, 1971.
15. J. Dual and D. Möller, *Lab Chip,* 2012, **12**, 506.
16. J. Dual and T. Schwarz, *Lab Chip,* 2012, **12**, 244.
17. D. J. Ewins, *Modal Testing: Theory and Practice*, Research Studies Press, John Wiley, 1995.
18. G. Settles, *Experimental Fluid Mechanics*, Springer, 2001.
19. D. Möller, N. Degen and J. Dual, *J. Nanobiotechnol.,* 2013, **11**, 21.
20. C. I. Zanelli and S. M. Howard, *Ultrasonics,* 2006, **44**, e105.
21. G. Speak and D. Walters, *Rep. Memoranda,* 1950, **2859**.
22. H. Richard and M. Raffel, *Meas. Sci. Technol.,* 2001, **12**, 1576–1585.
23. W. L. Howes, *Appl. Opt.,* 1984, **23**, 2449.
24. A. Hanafy and C. Zanelli, *IEEE 1991 Ultrason. Symp.,* 1991, 1223–1227.
25. M. Wiklund, H. Brismar and B. Önfelt, *Lab Chip,* 2012, **12**, 3221.
26. M. Ohno, N. Tanaka and Y. Matsuzaki, *Jpn J. Appl. Phys.,* 2003, **42**, 3067–3071.
27. C. V. Raman and N. S. Nagendra Nath, *Proc. Indian Acad. Sci.,* 1935, **A2**, 406–412.

28. A. Neild, S. Oberti and J. Dual, *Sens. Actuators, B,* 2007, **121**, 452.
29. Ferroperm Piezoceramics A/S, http://www.ferroperm-piezo.com/.
30. A. V. Mezheritsky, *IEEE Trans. Ultrason., Ferroelectr., Freq.Control,* 2005, **52**, 2120.
31. M. Gröschl, *Acustica,* 1998, **84**, 432.
32. B. Hammarström, M. Evander, J. Walström, T. Laurell and J. Nilson, *Proceedings of the Acoustofluidics* 2013, 12–13. Sept., Southampton, pp. 49–50.
33. S. Oberti, A. Neild and J. Dual, *J. Acoust. Soc. Am.,* 2007, **121**, 778.
34. J. Dual, P. Hahn, I. Leibacher, D. Möller and T. Schwarz, *Lab Chip,* 2012, **12**, 852.

CHAPTER 21

Biocompatibility and Cell Viability in Acoustofluidic Resonators

MARTIN WIKLUND

Department of Applied Physics, Royal Institute of Technology, SE-10691 Stockholm, Sweden
E-mail: martin@biox.kth.se

21.1 Introduction

Microdevices for acoustic manipulation of particles and cells typically utilize the acoustic radiation force[1] acting on the suspended objects. This force is stronger if the acoustic field is of standing wave type and of high frequency (>MHz), thus in the ultrasound regime. When handling biological cells it is important to know if the acoustic field is capable of causing any stress or damage to them. Generally, acoustic energy is of a mechanical nature in the form of vibrations ("shaking") and pressure fluctuations ("squeezing"), *i.e.* it consists of kinetic and potential energy. In a plane propagating wave, the shaking and squeezing effects are equally large anywhere in the wave. In a standing wave, however, there is a phase difference between the maximum vibration and maximum pressure leading to a spatial separation between the velocity antinodes and the pressure antinodes, respectively. For that reason, it is important to consider not only the energies or pressure/velocity amplitudes, but also the types of acoustic fields (*e.g.*, standing or propagating waves) and the position of cells in that field when investigating the impact on cell viability. Furthermore, the time constant of the fluctuations, *i.e.* the frequency, is important; typically, cavitation is more likely to occur in the low ultrasound frequency regime (<1 MHz) while absorption leading to heating increases with frequency.

Many past studies of ultrasonic bioeffects (*i.e.*, any observable biological effect that is likely caused by ultrasound) are related to ultrasonography (*i.e.*, diagnostic ultrasound). This is of interest since the acoustic energies and frequencies employed in diagnostic ultrasound are comparable in magnitude to the ones used in acoustic manipulation of cells in micro-devices. Diagnostic ultrasound has been used for about half a century[2] and is the most widely used diagnostic imaging tool used at clinics in the world. The general conclusion is that no epidemiologic evidence exists of any adverse effects in humans caused by routine use of diagnostic ultrasound.[3–5] However, as pointed out by Barnett *et al.*[5] any past safety record should not be mistaken for a guarantee that harm can never occur. This general advice is also applicable when using micro-devices for ultrasonic manipulation of cells. However, it is important to note that there are different bioeffects in the human body compared to an *in vitro* cell culture in a micro-device when using similar acoustic field parameters, and these effects may also vary between different cell types. Therefore, a general guideline is that the cell viability is difficult to model and predict; it needs to be experimentally investigated for each device and for each cell type used.

Several ultrasound applications exist where an adverse effect on cells or tissue is intended. For example, shock-wave lithotripsy and high-intensity focused ultrasound (HIFU) are ultrasonic methods used for, *e.g.*, destroying kidney stones, gall stones and similar. This therapeutic ultrasound technology has also been used for creating localized haemostasis to prevent internal bleeding as well as tissue necrosis for cancer therapy without damaging nearby tissue.[6] The mechanisms involved are complicated but typically involve heating combined with bubble activity (cavitation) at acoustic intensities in the range 10^3–10^4 W cm^{-2}. Furthermore, a common laboratory method is to lyse cells in suspension by "sonication" in an acoustic field in the low-frequency ultrasound regime, typically at 30–50 kHz. This method is well-known among cell biologists, but should not be mistaken for ultrasound activity in general since the lysing effect is based on cavitation which is more likely to occur at lower ultrasound frequencies.

In this chapter the physical mechanisms of ultrasound capable of causing various bioeffects on cells are discussed. The discussion is primarily limited to devices for ultrasonic standing wave manipulation of mammalian cells. The physical mechanisms of interest can be divided into thermal and non-thermal effects. Here, cavitation is the most important non-thermal effect, but other non-thermal effects such as acoustic radiation forces and acoustic streaming are also discussed. Furthermore, different observed bioeffects on cells are discussed as well as different available methods to quantify the impact of ultrasound on cell viability.

21.2 Physical Mechanisms of Ultrasound Causing a Bioeffect

This section discusses the origin of different physical mechanisms of ultrasound affecting cells and tissue, and provides a few design criteria important for controlling or improving the biocompatibility of an ultrasonic manipulation device.

21.2.1 Thermal Effects

When ultrasound is absorbed by a material the mechanical energy is primarily converted into heat. However, the high-Q resonators used for ultrasonic manipulation of particles and cells are typically made of low-loss materials such as silicon, glass, steel and a water-based suspension. Therefore the absorption of ultrasound is in most cases very small, in particular for frequencies in the range ~1–10 MHz. On the other hand, temperature elevation remains a problem to be dealt with in an ultrasonic manipulation device. There are two reasons for this: (1) heat generated from electromechanical losses in the piezoelectric layer in the transducer and (2) losses in thin glue layers in between the different supporting solid layers in the resonator. The localized heat generated primarily in the transducer and glue layers may then be conducted to the fluid and cells in the channel/chamber of the resonator causing a thermal bioeffect. Care should also be taken if polymers are used as supporting layers, although they are not commonly used materials in ultrasonic particle manipulation devices due to their high losses.

In vitro cell lines generally have a wider range of tolerable temperatures compared to whole organisms. For example, cryopreservation is a widely used technique for cell storage at very low temperatures. Thus, a lethal temperature for a whole organism may be tolerable for an individual cell.[5] However, moderate temperature increases (a few degrees) may still lead to molecular changes in cells, such as misfolding, entanglement and or unspecific aggregation of proteins. These effects may cause a variety of different cellular responses, of which the most apparent is cell cycle arrest and a stagnation of the growth rate.[7] Other reported effects are cytoskeletal defects due to actin filament reorganization and misplacement or fragmentation of organelles.[8] Thus, although the cell may be technically viable (*e.g.* having an intact cell membrane), its functions and internal structure may be severely affected by a moderate increased temperature.

Temperature effects and recommendations for human cells are summarized schematically in Figure 21.1. A rough guideline for most mammalian cells is that cell growth is optimal at 37 °C but often tolerable between 33–39 °C. If the temperature exceeds ~43 °C, cell death is most likely and above 45 °C, proteins become denaturated. In between ~39–43 °C, the bioeffect is highly diverse depending on, *e.g.* the cell type, heating rate and time of retained and elevated temperature. In addition to the production of heat shock

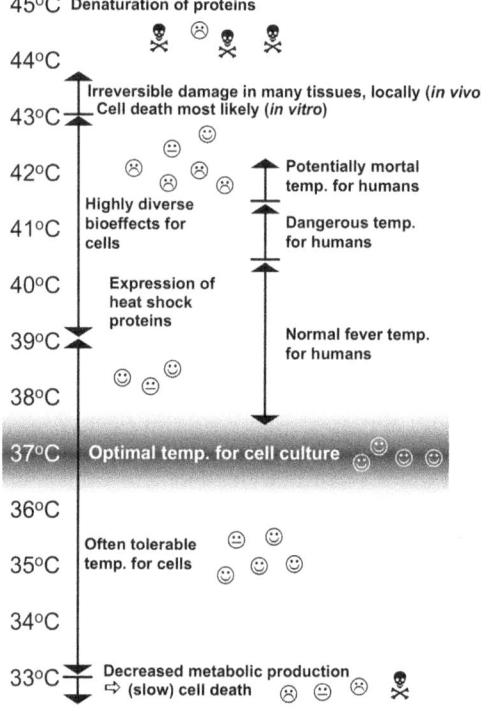

Figure 21.1 Overview of the temperature dependence of different observed bioeffects in humans and for mammalian cells.

proteins, typical effects in this temperature interval include increased metabolic activity and cell cycling rates. On the other hand, at sub-physiological temperatures (~33–37 °C), the cellular response is also diverse depending on cell type and system. Al-Fageeh et al.[9] reviewed the "cold shock" response in cultured mammalian cells, which typically leads to a modulation of the cell cycle, metabolism, transcription translation and the cell cytoskeleton. He pointed out that the bioeffects at sub-physiological temperatures do not necessarily have to be inhibitive or detrimental. For example, reports exist where the production of recombinant proteins is increased at lower temperatures.[9] However, for most mammalian cell lines the proliferation ceases although the viability is prolonged at lower temperatures.[9,10] This eventually results in slow cell death if the temperature is kept below approx. 33 °C for prolonged times. On the other hand, since the metabolic rate typically decreases with decreasing temperature, cells are more sensitive to changes in nutrient and oxygen supply at 37 °C than at lower temperatures.

In summary, what is the best recommended temperature when using ultrasound as a manipulation tool for mammalian cells? Naturally, it is to keep the system at the physiological temperature 37 °C with a tolerance of

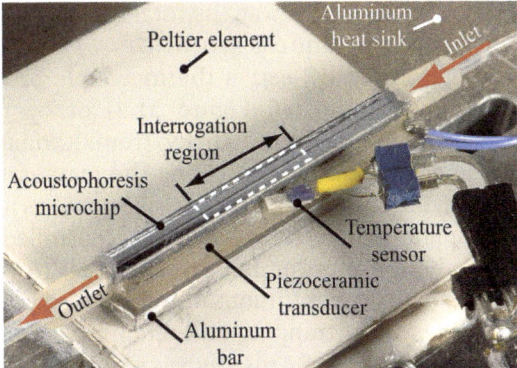

Figure 21.2 The temperature-controlled acoustophoretic chip by Augustsson et al.[13] based on a Peltier element. The chip, piezo transducer (PZT), aluminium bar and Peltier element were sandwiched and glued together on a microscope stage. A Pt100 thermoresistive element was glued onto the PZT for feedback to the temperature control loop.

about ±1 °C. This will reduce the risk that ultrasound-generated heat variations interfere with any measured cellular process of interest. The recommended tolerance of ±1 °C is also in line with the safety definition used for ultrasonic medical imaging. In medical imaging, the *thermal index* (*TI*) is used as an international standard, and is defined as the power produced by the transducer divided by the power needed to raise the temperature by 1 °C. Thus, *TI* = 1 means that there is a risk of heating the examined tissue by approximately 1 °C. If *TI* exceeds 1, the operator should carefully consider the benefits of the examination relative to the risks. Concerning resonators for ultrasonic particle manipulation, there is an additional and very important advantage of controlling the temperature. A stable temperature will also improve the resonance stability by keeping the temperature-dependent acoustic wavelength constant. Thus, both the ultrasonic manipulation performance as well as the cellular environment benefit from a controlled and constant temperature.

A few different strategies have been reported for temperature-regulation in ultrasonic standing wave manipulation devices. One method is to cool the system by a water loop,[11] a fan[12] or a Peltier element[13] close to the piezoelectric layer, see Figure 21.2. A water loop is suitable as a heat sink for large-scale devices, while a Peltier element is better suited to miniaturized devices. Another method is based on utilizing the heat generated by the ultrasonic device, typically in the transducer, in an active temperature regulation scheme.[14,15] For example, provided that the ultrasound-generated heat does not lead to an increase of more than 10–15 °C, the temperature will not exceed 37 °C if the device is operated at room temperature. For smaller temperature increases, the ultrasound-generated heat can be combined with an external heating system such as a heatable chip holder or a hood with a controlled air environment. In this way the total temperature increase can be tuned so that the final temperature is kept at 37 °C. The monitoring of the temperature can be performed by, *e.g.*, a thermocouple probe attached to a suitable location on the chip (*cf.* Figure 21.2) or by measuring the temperature-dependent fluorescence intensity from Rhodamine B.[16] The latter method has the advantage of monitoring the temperature directly at the location of the cells, but the method is not as accurate as a thermocouple probe. In a fully temperature-controlled and CO_2-controlled micro-device, Vanherberghen *et al.* showed that human immune cells (B cells) were kept viable for up to three days of continuous ultrasonic exposure at relevant pressure amplitudes needed for trapping and retaining cells in well-defined clusters in a multi-well microplate.[15] This study is of interest since it confirms the high biocompatibility reported in commercially available macro-scaled ultrasonic manipulation devices, *e.g.* the SonoSep™ and BioSep™ systems by Applikon Biotechnology (Schiedam, The Netherlands).[17]

21.2.2 Cavitation-Based Effects

Apart from thermally induced bioeffects, the other important effect to consider is cavitation-based effects. Cavitation is a physical phenomenon that can be induced by different means, of which ultrasound is one.[18] It can be defined as the formation and or activity of gas/vapour filled cavities, *i.e.* bubbles, in a fluid medium. Such bubbles are efficiently formed and driven by low-frequency ultrasound, typically in the 20–200 kHz range. For example, a commercial ultrasonic cleaner used for cleaning, degassing, cell lysis *etc.* operates at around 20–50 kHz. However, many reports show clear cavitation effects for frequencies up to approximately one or a few MHz.[5] Thus, there is no guarantee that cavitation will not occur even in the standard frequency range used for ultrasonic manipulation devices, *i.e.* 1–10 MHz. The stimulated bubble activity is classified as (1) *stable* cavitation and (2) *inertial* or *transient* cavitation. The first class, stable cavitation, is defined as the continuous oscillation of bubbles over a large number of cycles driven by the pressure fluctuations of the applied acoustic field. This type of cavitation typically produces highly localized acoustic streaming which may cause shear stresses on a nearby cell. The bioeffects are usually minimal but the presence of vibrating bubbles often completely destroys attempts to aggregate cells in a standing wave. The other class, inertial cavitation, is defined as a transient type of cavitation where the bubble oscillates heavily over a single or a few cycles eventually leading to a violent collapse/implosion. The collapse is associated with localized high pressures and temperatures and high-velocity liquid jets. Typical scales of the parameters involved are >1000 °C temperature formed during 1 μs in a region of size ~10 μm leading to internal pressures of 10–100 MPa (100–1000 atmospheres). Naturally, this is not a good environment for cells. The motion of the bubble surface is most often asymmetric, in particular for bubbles close to a rigid boundary, which may lead to directed liquid jets towards the surface. As a rough measure, bubbles that grow to more than twice their initial size within one or two cycles undergo inertial cavitation; otherwise stable cavitation is more likely.

The threshold for obtaining cavitation is very important for cell viability considerations. This threshold is dependent on both frequency and amplitude of the acoustic wave. However, even more important for the threshold is the presence of potential cavitation nuclei in the fluid in the form of dissolved or undissolved gases and helping solid surfaces. This is similar to the process of heating water to its boiling point in a pan or pouring beer into a glass; bubbles are typically formed in the irregularities/crevices on the inside surface of the pan or glass. Thus, a general guideline to reduce the risk of bubbles in microfluidics is to use smooth, clean surfaces and degassed fluids (and non-permeable chip and tubing materials). If no helping surface or dissolved gasses are present, bubble creation and growth starting in the sub-μm domain is difficult. The reason for this is the high tension associated with the curved surface of a very small bubble; a tension of 100 MPa (1000 atmospheres) corresponds to the spontaneous creation of a bubble of radius 1.4 nm at room temperature in water.[19]

On the other hand, this theoretical calculation does not reflect reality since such pure fluids are difficult to produce. Still, it is important to reduce the amount of potential cavitation nuclei in the sample, *e.g.* by degassing the medium buffer with an ultrasonic cleaner before adding the cells. Apfel and Holland theoretically investigated the threshold acoustic pressure amplitude for obtaining inertial cavitation as a function of the size of pre-existing gaseous nuclei in water and other fluids for different ultrasonic frequencies.[20] Such pre-existing nuclei are typically microscopic pockets of undissolved gases and/or vapour, which can easily be stabilized in the fluid on solid helping surfaces. Their results are shown in Figure 21.3. As predicted by theory, the threshold pressure is high for nuclei-free or nm-sized-nuclei fluid media. In such media, cavitation would practically never occur. On the other hand, if gaseous nuclei with sizes in the 100 nm–1 μm range are present in the fluid, the threshold pressure is within realistic levels often employed in acoustofluidic applications. This pressure level is typically 0.1–1 MPa (corresponding to 1–10 atmospheres). We also note that for these pressures it makes a significant difference if the device is operated close to 1 or 10 MHz. From a cavitation perspective, 10 MHz is safer as near this frequency there is both a higher level of minimum threshold pressure as well as a steeper curve for larger nuclei sizes compared to ~1 MHz. The latter means that for a pressure exceeding the minimum threshold, a wider range of nuclei sizes will cavitate at 1 MHz than at 10 MHz.

In analogy to the thermal index (*TI*) used to estimate the risk of sample heating in ultrasound imaging, an international standard exists for estimating the risk of sample cavitation. This index is called *mechanical index* (*MI*) and is defined as the ratio of the (negative) peak pressure of the acoustic wave (in MPa) and the square root of the acoustic frequency (in MHz), *i.e.*, $MI = P_{\text{neg.peak}}/\sqrt{f}$. A general guideline is to avoid $MI > 1$, which, if we compare with the predicted threshold pressure in Figure 21.3, assumes

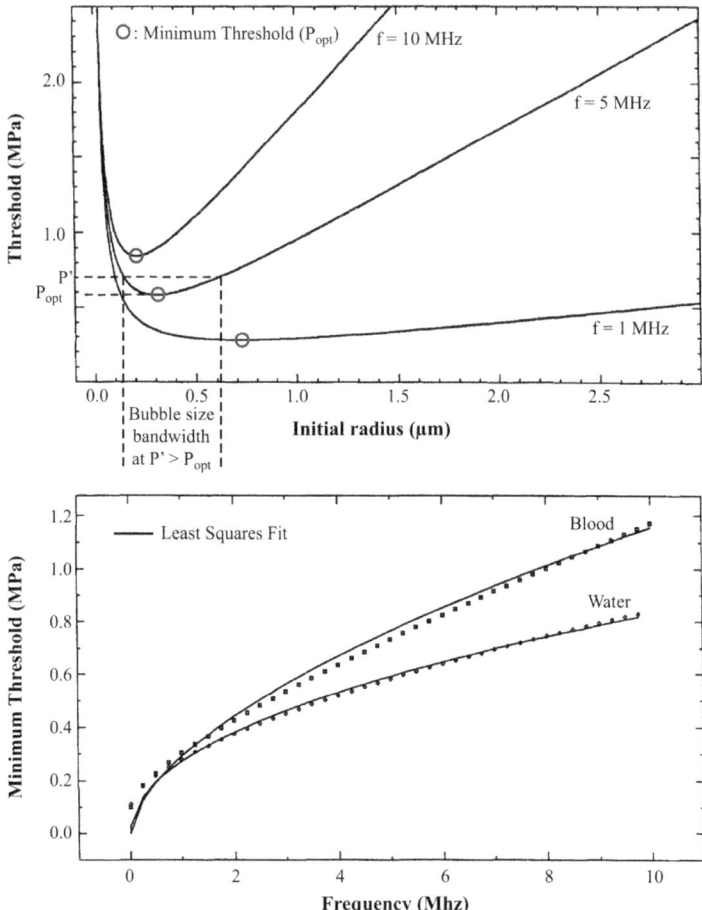

Figure 21.3 Prediction of cavitation threshold levels. Upper panel: calculation of the cavitation threshold in water as a function of the size of pre-existing inertial cavitation nuclei (modeled as air bubbles in room temperature) for three different driving frequencies. P_{opt} is the lowest threshold for the optimal bubble size, and P' marks the bubble size bandwidth for a given pressure $>P_{opt}$. Lower panel: the calculated minimum threshold pressure, P_{opt}, for the optimal bubble size in the upper panel is plotted as a function of frequency for two different fluids: blood and water. Least squares fits are made to the theoretical data, indicating that the threshold pressure is $\alpha f^{-0.60}$ for blood and $\alpha f^{-0.48}$ for water. The figure is taken from Apfel and Holland.[20]

a relatively nuclei-free sample (no bubbles larger than 100 nm present). Furthermore, from the MI definition we expect the risk of cavitation to be $\alpha f^{-0.5}$, where f is the frequency. This is a good approximation for most relevant biological fluids and soft tissue. A more detailed theoretical study was performed by Apfel and Holland[20] who predicted the cavitation risk as $\alpha f^{-0.60}$ for blood and $\alpha f^{-0.48}$ for water, cf. the lower panel in Figure 21.3.

If cavitation occurs in a cell suspension, severe physical and biological damage may occur. For stable cavitation, the most likely effect is cell membrane rupture caused by shear stresses from microstreaming.[21,22] The effects of inertial cavitation can be more violent including extreme temperatures, stresses and liquid jets leading to direct cell destruction and death. Because of the high temperatures associated with inertial cavitation it is often difficult to distinguish pure cavitation effects from thermal effects. Anyway, a number of biological consequences of cavitation have been reported, *e.g.* necrosis[23] and strand breaks in DNA.[24] The reason for the latter is the high local energies associated with inertial cavitation comparable in magnitude with ionizing radiation. A related field is called sonochemistry, where inertial cavitation is used for enhancing chemical reactions, *e.g.* by the production of free radicals which may catalyze reactions.[25] On the other hand, there also exist reports on beneficial bioeffects of cavitation. From stable cavitation, controlled microstreaming can lead to enhanced fluid stirring, possibly beneficial for transport of nutrients or enhanced blood perfusion.[21] Another field is the use of drug-loaded microbubbles for targeted imaging and therapeutics.[26] Furthermore, there is a cavitation-based technique for controlled and reversible membrane rupture called sonoporation.[27] This is a technique where pulsed-mode cavitating microbubbles added to a cell sample temporarily create cell membrane pores, which increases the transmembrane permeability, followed by membrane recovery/repair.[28] Sonoporation has been suggested as an attractive method for drug delivery and cellular uptake of, *e.g.* large molecules that otherwise cannot penetrate the cell membrane.[29] However, the method should be used with caution; although sonoporated cells normally recover and remain viable,[30] a recent study shows that sonoporation can lead to an arrest in the cell cycle and/or apoptosis, possibly due to a disturbance of the intracellular calcium ion concentration.[31] In a recent study, Carugo *et al.* demonstrated a microfluidic device for microbubble-free sonoporation.[32] The device was operated at a resonance around 2 MHz and voltages between 0 and 40 V_{pp}, and was suggested to provide a simpler and more gentle methodology for drug delivery and uptake.

21.2.3 Effects of Acoustic Radiation Forces

Steady-state radiation forces are central for this tutorial topic since they are the basis for any acoustophoretic action on particles and cells. In standing wave fields where the size of the cell is much smaller than the acoustic wavelength, the direct effect is clumping/aggregation of cells in bands at half wavelength intervals. This phenomenon is covered in more detail in other parts of this tutorial series.

Already in the early 70s, Dyson *et al.* observed the reversible arrest of red blood cells *in vivo* in blood vessels of a living chick's embryonic tissue due to ultrasonic standing wave exposure.[33] Actually, this work can be considered to be the first demonstration of acoustic trapping in a microfluidic system, approximately three decades before the technology was implemented in lab on a chip devices. The bioeffect of *in vivo* red cell arrest is blood stasis, which may cause similar effects to thrombosis.[34] The authors noticed a strong dependence of blood cell arrest on the blood vessel orientation relative to the orientation of the utilized ultrasonic resonator, and suggested that for imaging applications, potential stasis could be minimized by continuously moving the ultrasonic probe during an examination. They also observed occasionally that some blood vessel endothelial cells were damaged in the plasma membranes on the luminal surface, resulting in cell debris and membrane fragments leaking out into the vessel cavity. Red cells were observed to sometimes adhere to these damaged spots in the endothelial membrane and could lead to sites of irreversible stasis and thrombus formation. However, it should be noted that the latter finding was rare and had unknown causes, most probably not a direct effect of the radiation force but rather due to cavitation. Another study that has received some attention shows a correlation between ultrasonic exposure and dislocation of neurons in the developing brain in fetal mice.[35] These authors showed that upon >30 min of ultrasound exposure at clinically relevant levels, some cortical neurons failed to reach their final correct position in the superficial gray matter of the brain when migrating from their site of development deeper inside the brain. The authors argued that it was unlikely that heating or cavitation was involved; instead they suggested that acoustic radiation forces or microstreaming could be the reason for the disturbed neuron migration. However, it should be mentioned that the conclusions from this study are not applicable to humans due to their larger brain mass and brain developing times compared to mice.

21.2.4 Effects of Acoustic Streaming

Since it is well-known that shear stresses from a fluid flow can cause membrane rupture and cell lysis,[36] it is not surprising that acoustic streaming could be a potential source of cell damage. Hughes and Nyborg verified this already in the 60s in work where bacteria, protozoa and erythrocytes were damaged by acoustic microstreaming from an 85-kHz vibrating tip.[37] However, the bioeffect of streaming depends largely on the magnitude of flow velocity; a moderate streaming may cause beneficial stirring and fluid exchange effects[23] or lead to mechanically activated cell signal pathways.[38] Good examples of the benefits of fluid exchange are microfluidic perfusion cultures, where controlled delivery and removal of soluble factors are used for defining microenvironments at the level of single cells.[39] On the other hand, Ankrett *et al.* suggested in a recent study that Rayleigh-type acoustic streaming may reduce viability at sufficiently high actuation voltages (corresponding to 20–30 V_{pp} in their device).[40] Thus, acoustic streaming may damage cells even if the temperature is controlled and the pressure is below the cavitation threshold.

21.2.5 Effects Not Caused by Ultrasound

When handling cells in a micro-scaled device, effects other than ultrasound may be of interest for biocompatibility. Most important is the choice of material facing the microchannel/microchamber, in particular if cells are in contact with this material. Many of the standard materials used in microfabrication technology have been successfully implemented in micro-scaled cell culture systems. Examples include polydimethylsiloxane (PDMS), silicon, glass and polymethylmethacrylate (PMMA), or combinations of such materials.[39] Since glass is a reliable material in standard bulk cell cultures it is straightforward to use for microscaled devices. However, in microfluidic perfusion cultures PDMS is the most popular material. One reason for this is its gas permeability, a property not valid for silicon or glass. On the other hand, PDMS is a lossy and weakly reflecting material and is therefore not suitable in an acoustic resonator unless used only as a passive spacer.[17,41] Additional challenges with PDMS are its high hydrophobicity and permeability for organic solvents. Thus, glass and silicon are recommended as the first choice for biocompatible ultrasonic cell handling in chips.[15] Other material aspects to consider when handling cells for extended periods of time include the use of bio-coated materials for facilitating cell adhesion (e.g., immobilization of collagen, laminin and fibronectin onto the surface),[39] and different surface treatment methods to enhance the biocompatibility of, e.g., silicon.[42] Finally, it should be noted that it is important to establish a similar environment in the chip as in a standard cell incubator. In addition to the temperature-control around 37 °C that was treated in Section 21.2.1, a 5% CO_2-level is appropriate for maintaining the correct pH level of the culture medium. Here, an open microdevice for acoustic manipulation of cells has the advantage of providing a simple method for temperature and CO_2-control directly at the liquid–gas interface.[15] In a closed system, CO_2-control can be implemented either by driving the system in perfusion-mode or using a gas-permeable material such as PDMS. In a continuous-mode acoustophoretic microdevice, surface treatment and CO_2-control are of less importance since cells are only in the chip environment for orders of a few seconds. However, material choice and temperature stability are always important since they define the quality and stability of the acoustic resonance.

21.3 Observed Bioeffects on Cells in Ultrasonic Standing Wave Manipulation Devices

In an ultrasonic standing wave device, the bioeffects may be different from the effects of ultrasound in general. For example, cells in a water-based medium move by the acoustic radiation force[1] to the pressure nodes while small (of a few µm in diameter or less) bubbles—if present or generated—typically move to the pressure antinodes at frequencies of a few MHz and at moderate amplitudes.[43] This means that potential cavitation nuclei are physically separated from the cells in a standing wave. Furthermore, since the cavitation threshold depends on the pressure amplitude, it is more likely that cavitation will occur in the pressure antinodes than in the pressure nodes. Thus, the radiation force in an ultrasonic standing wave actually provides a protective effect on cells. This implies that cells may be unaffected even if cavitation is present given that the field is of standing wave type and that the size range of cavitation effects is less than the antinode-to-node distance $\lambda/4$ (approximately 0.4 mm at 1 MHz in water). Church commented on this protective effect on cells in standing wave fields already in 1982 in a work on ultrasound-induced cell lysis.[44] He noticed that at moderate intensities (1 W cm^{-2}) and a frequency of 1 MHz, cell lysis was only efficient if the cell sample was rotated. Sample rotation led to a mixing of cells and bubbles that otherwise would have been separated into the node and antinodes, respectively, in the standing wave. However, it is well known that the motion of a bubble in a standing wave field of a given frequency is dependent on the bubble size; bubbles smaller than a cut-off size (corresponding to the resonance frequency of the bubble) move to the pressure antinodes while bubbles larger than this cut-off size move to the pressure nodes.[45] For example, at 1 MHz, bubbles with diameter <7 μm move to the pressure antinodes while bubbles with diameter >7 μm move to the pressure node where the cells are trapped. Church predicted that the bubbles responsible for cavitation were those with sizes just below the resonance size, *i.e.*, with diameters 4–7 μm at 1 MHz driving frequency. Even if larger bubbles may be present at the location of cells in the pressure nodes, they will most likely produce little or no damage to cells because (1) the pressure amplitude is very weak in the pressure node, and (2) the bubbles do not respond as strongly as they would if driven below resonance.[44] It should, however, be pointed out that bubble motion in a sound field is very complex. For example, at sufficiently high pressure amplitudes, active sub-resonant-size bubbles can behave like larger bubbles in a standing wave field and accumulate in the pressure nodes and cause cell damage.[46–48]

A comparison of the impact on cellular viability between propagating fields and standing wave fields was performed by Böhm *et al.*, who studied plant cells (*Petunia hybrida*) exposed to different energy densities, times and wave types of ultrasound.[49] The major finding from this study is seen in Figure 21.4. For the stored energy density 8.5 J m^{-3} and frequency around 2 MHz, practically no reduction in viability was observed for standing wave exposure relative to the no-exposure control, while a dramatic reduction in viability was found during 20 minutes of propagating wave exposure of the same magnitude. For standing wave exposure, a reduction in viability was found for the stored energy densities 44 J m^{-3} and 70 J m^{-3} but still at a lower rate than for a propagating wave at a stored energy density of 8.5 J m^{-3}.[49] The authors concluded that the locations of cells in the pressure nodes of a standing wave field correspond to positions of zero or low displacement gradient, where the latter is associated with mechanical stress. Thus,

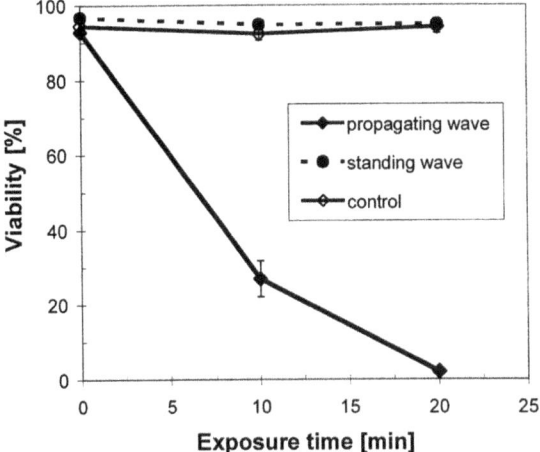

Figure 21.4 Comparison of the cell viability of *Petunia hybrida* exposed to propagating-wave ultrasound and standing-wave ultrasound of equal stored energy densities (8.5 J m^{-3}, $n = 2$). The figure is taken from Böhm et al.[49]

according to Böhm, for both displacement and pressure it is the field gradient and not magnitude that has a negative impact on the cells. Interestingly, Böhm also noticed that cells in mitosis were more susceptible to mechanical stress and showed a lower viability than cells in other stages of the cell cycle.

Similar studies to Böhm's[49] were performed on yeast cells by Radel et al.[50,51] In agreement with Böhm's findings, Radel et al. noticed that cells agglomerated in the pressure nodal planes appeared to be less damaged by ultrasound than cells located elsewhere in a standing wave field or than cells located in a propagating wave field. However, although the cell membranes were intact and the cells were still viable, electron microscopy[50] and fluorescence microscopy[51] studies revealed morphological changes to intracellular structures when the yeast cells were exposed to <1 MPa and ~2 MHz standing wave ultrasound. For example, the ultrasound seemed to disrupt the integrity of the vacuole membrane, but not the nucleus or plasma membranes. In agreement with Böhm's study, Radel et al. also noticed that the growth rate seemed to be hampered by ultrasound exposure, but only for cells that escaped from the pressure nodes. However, they suggested that this was due to intracellular damage of elements responsible for mitosis.[50] Furthermore, the ability to collect the yeast cells in the pressure nodal planes of the standing wave was dependent on the suspension properties: yeast cells suspended in water were more strongly retained in the pressure nodes (and therefore less damaged) than yeast cells suspended in a water–ethanol mixture.[50] This result is of interest for systems using buffers that concentrate cells in the pressure antinodes; such systems may be less biocompatible than systems that concentrate cells in the pressure nodes.

A thorough study of the physical environment of neural cell aggregates trapped by standing wave ultrasound was performed by Bazou et al.[52] They concluded that in their device the cavitation threshold pressure amplitude was approx. 2 MPa, which is more than twice the pressure amplitude needed in most ultrasonic particle and cell manipulation devices. Furthermore, at 0.54 MPa pressure amplitude, an acoustic streaming velocity of 70 µm s^{-1} was measured, which corresponds to orders of magnitude lower levels of hydrodynamic stress on cells compared to standard centrifugation at ∼100–1000 g used routinely for cell culture preparation.

21.4 Methods for Measuring the Impact of Ultrasound on Cell Viability

The most simple and straightforward method for quantifying the impact of ultrasound on cell viability in microscaled devices is to measure the integrity of the cell membrane by light microscopy. The thin cell membrane is one of the most fragile parts of the cell and also the boundary to the extracellular environment. Many of the physical mechanisms discussed in Section 21.2 can rupture the cell membrane and it is therefore highly relevant to choose a membrane-integrity-based method. The most commonly used dyes for viability studies in ultrasonic standing wave devices are trypan blue and propidium iodide (PI).[53–59] Both these methods are based on dye exclusion, meaning that the dye is prevented from passing through the membrane in living cells. If the membrane is ruptured, the dye can traverse the plasma membrane and stain the cell. Trypan blue appears blue when illuminated with white light, while PI is a red-fluorescent dye used together with fluorescence microscopy settings. An additional advantage with PI is that its fluorescence increases by a factor of ∼30 when bound to a nucleic acid, such as DNA, in the nucleus. This means that there is practically no background fluorescence from PI outside the cells yielding a strong signal to noise ratio. An alternative to the PI assay is the ethidium homodimer-1 (EthD-1) assay, which has been used for measuring the impact of ultrasonic standing waves on, e.g. a liver cell line.[60] The opposite of the exclusion method is to use a dye that stains living but not dead cells, e.g. using the green-fluorescent (or orange-fluorescent) dye calcein AM.[15,41,57] Calcein AM (which initially is non-fluorescent), penetrates the membrane of living cells and the fluorescence is activated by an esterase-based enzymatic reaction present only in living cells. Thus, the green (or orange) fluorescence indicates two properties of a living cell: (1) that the cell has esterase activity and (2) that the esterase product (the fluorescent by-product of calcein AM) is retained within the cell, which means that the cell has an intact membrane. The latter is because the fluorescent form of calcein has much lower membrane permeability than the non-fluorescent form. When a cell dies, the fluorescence disappears primarily due to leakage of fluorescent calcein through the damaged membrane out from the cell. A similar method to the calcein AM assay is the acridine orange assay, which is based on a dye whose emission spectrum changes upon interaction with DNA and RNA in the cell. Evander et al. used this assay in a perfusion setting to monitor the viability of acoustically trapped neural stem cells in a microfluidic chip.[16] Recently, Augustsson et al. used the XTT assay to measure the mitochondrial dehydrogenase activity present in living cells from prostate cancer cells[61] and BV2 microglial cells[62] with very promising results. The latter study included several different assays with the purpose of studying the effect of ultrasound on cell viability, inflammatory status, functional phenotype and mitochondrial respiration. In summary, a selection of different dyes that have been used for measuring the viability of cells exposed to ultrasound is shown in Table 21.1.

In fluorescence microscopy it is practical to use a combination of dyes, e.g. green-fluorescent calcein AM together with red-fluorescent PI. This way, all cells are clearly visible at all time points, either in green (live cells) or in red (dead cells), see Figure 21.5. However, in this figure, the red dye is not PI but far-red DDAO-SE. This is not a viability dye, but a cytoplasm dye which fluoresces independently on the viability state. During cell death the cells often go from green via yellow (both colors visible) to red, which may reveal some information about the "death dynamics". It should be noted, however, that in viability assays based on PI or calcein AM, the definition of cell death is given by the assay principle. For example, in PI assays the definition of cell death is when PI binds to the nucleus while in calcein AM assays the

Table 21.1 Selected dyes used for measuring viability in ultrasonic cell manipulation devices.

Dye	Exclusion/inclusion method	Indicator of:
Trypan blue	Stains dead cells	Damaged plasma membrane
Propidium iodide (PI)	Stains dead cells	Damaged plasma membrane Note: toxic for long incubation times
Ethidium homodimer-1 (EthD-1)	Stains dead cells	Damaged plasma membrane
Calcein-AM	Stains living cells	Intact plasma membrane Esterase activity
Acridine orange (AO)	Stains living cells (color-specific)	Cell cycle (emission spectrum depends on interaction with DNA and RNA)
XTT assay	Stains living cells	Mitochondrial dehydrogenase activity
Annexin V	Stains apoptotic cells	Apoptosis (translocation of phosphatidylserine in the cell membrane)
Far-red DDAO-SE	Stains the cytoplasm (of both living and dead cells)	Must be used in combination with other dyes, e.g. calcein-AM Note: compatible with very long term incubation

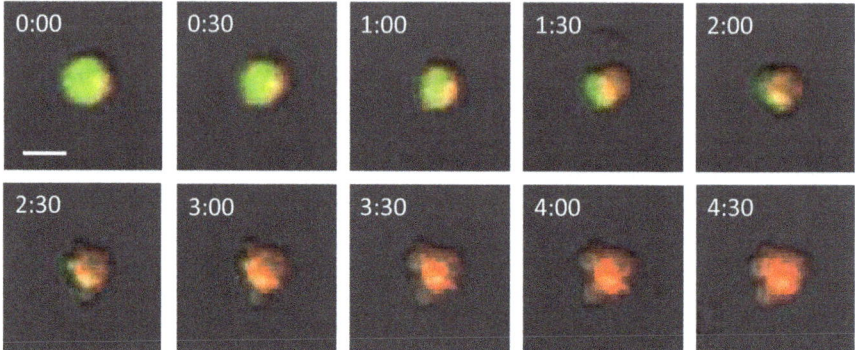

Figure 21.5 A 4.5 hour long time lapse study of the dynamics of a human embryonic kidney cell (293T) undergoing apoptosis. The cell is labeled with calcein AM (green fluorescent) and far-red DDAO-SE (red-fluorescent). Both these dyes are compatible with long-term studies (approx. 12 h). The images show early (1:00–2:00 hours) and late (2:30–4:30 hours) apoptosis. During early apoptosis, the cell shrinks and the blebs start forming. During late apoptosis, the blebs become larger and the cell finally breaks up into fragments. The scale bar in the first image is 10 μm. The images are courtesy of Dr Bruno Vanherberghen.

definition is either when the esterase-activated form of calcein leaks out through the damaged cell membrane, or (for already dead cells before the staining) when it has no esterase activity and therefore cannot produce the fluorescent form of calcein.

It is also possible to classify a cell as live or dead based on cell morphology. This can be seen in Figure 21.5 where the "blebbing" of the cell membrane during programmed cell death, *i.e.* apoptosis, is shown. Here, blebbing means that the cell shape turns into a more irregular form with clear bulges or "blebs" in the membrane (*cf.* Figure 21.5). These blebs, typical for early stage apoptosis, can later detach from the cell, which finally breaks up into fragments (late stage apoptosis), see Figure 21.5. Apoptosis may be of interest in studies of delayed or long-term bioeffects of ultrasound on cells. For example, Bazou *et al.* studied the effect of early and late stage apoptosis of cells exposed to standing wave ultrasound for up to 1 h.[57] For early stage apoptosis, the fluorescence assay Annexin V was used, and for the late stage, PI was used. The Annexin V assay is based on the translocation of membrane phospholipid phosphatidylserine (PS) from the inner face of the plasma membrane to the cell surface, a process specific for early stage apoptosis.[63] Morphology (blebbing) classification is a simple and straightforward alternative for measuring apoptosis, although it is more difficult to automate than fluorescence-based methods.

In addition to physical and visual attributes, such as membrane integrity and blebbing, viability can be measured indirectly by different means. In ultrasonic standing wave devices, examples include measuring the release/leakage of naturally occurring intracellular components such as potassium ions, haemoglobin and lactate dehydrogenase from erythrocytes.[64–66] Thus, if the membrane is damaged, the concentration of intracellular components outside the cells increases.[50] This method could be a better choice than using the different dyes discussed earlier, since the dyes are not natural and may negatively affect the cells, particularly in long-term studies (>12 h). Another method was reported by Zhang *et al.*, who measured protein expression and virus production in a baculovirus/insect cell expression system using an ultrasonic standing wave cell retention device compared to production levels without using ultrasound.[67] Similarly, in a standing wave device of considerable power used for manipulation of a hybridoma culture, Chisti reported a method where the rates of glucose uptake and antibody production were used as a viability measure.[68] Furthermore, the proliferation rate of different cell types exposed to standing wave ultrasound has been used as a measure of delayed or long-term effects of ultrasonic exposure, both on macro-scale[55,60,69] and micro-scale systems.[16,41,59] One interesting observation noticed by both Hultström[41] and Bazou[60] is the importance of cell–cell interactions for the maintenance of cell viability and growth rate. This seems to be important for adherent cells forming tissues such as kidney[41] and liver[60] cells. For example, the growth rate of ultrasonically aggregated kidney cells was higher than non-aggregated control cells in low-concentration samples, possibly due to the lack of cell–cell contact in the control sample not treated with ultrasound.[41]

The final work to be mentioned here is a paper by Dykes *et al.*[70] Instead of just measuring the direct viability, Dykes *et al.* also studied the functional impact of ultrasound on different blood cells after acoustophoretic separation in a microfluidic device. The functional tests included the colony-forming ability of peripheral blood progenitor cells (PBPC) and activation levels of platelets. They concluded that the investigated cellular functions were preserved after the ultrasound exposure in their device. A similar study with corresponding conclusions was published by Burguillos *et al.*,[62] valid for short-term but high-amplitude acoustophoresis.

21.5 Conclusions

A general guideline when handling cells by ultrasound is to follow the same recommendation as James Bond's; a cell prefers to be shaken, not stirred. In a pressure node where cells are typically located in ultrasonic manipulation devices, the velocity of the fluid medium elements has a maximum while both the pressure and the pressure gradient have a minimum. Therefore, the cells are more shaken than squeezed, and most cavitation bubbles with the potential to cause shear stresses from fluid stirring or direct damage to the cells are physically separated from the cells. Furthermore, any temperature increase from ultrasound can be limited and controlled. A viability-optimized ultrasonic manipulation device can manipulate cells for hours and days without any apparent negative impact on cellular state and growth.[15] However, few studies exist where the cellular functions under the effect of long-term ultrasound exposure are investigated. One such example is a (several-hour-exposure) study by Christakou *et al.*, measuring the killing efficiency of natural killer cells aggregated with target cells using standing wave ultrasound in a multi-well microplate.[71] In future it should be possible to use ultrasound either as a minimally influencing cell handling tool, or for triggering a cellular response, depending on the system design, operation and intended bio-application.

Acknowledgements

The author is grateful for financial support from the Swedish Research Council and the EU FP-7 RAPP-ID project, and for advice and feedback from Dr Bruno Vanherberghen and Dr Björn Önfelt concerning cell biology.

References

1. M. Settnes and H. Bruus, *Phys. Rev. E,* 2011, **85**(016327), 1–12.
2. W. L. Nyborg, *Ultrasound Med. Biol.,* 2000, **26**, 911–964.
3. M. C. Ziskin and D. B. Petitti, *Ultrasound Med. Biol.,* 1988, **14**, 91–96.
4. M. W. Miller, W. L. Nyborg, W. C. Dewey, M. J. Edwards, J. S. Abramowicz and A. A. Brayman, *Int. J. Hyperthermia,* 2002, **18**, 361–384.
5. S. B. Barnett, G. R. Ter Haar, M. C. Ziskin, W. T. Nyborg, K. Maeda and J. Bang, *Ultrasound Med. Biol.,* 1994, **20**, 205–218.
6. M. R. Bailey, V. A. Khokhlova, O. A. Sapozhnikov, S. G. Kargl and L. A. Crum, *Acoust. Phys.,* 2003, **49**, 437–464.
7. K. Richter, M. Haslbeck and J. Buchner, *Mol. Cell,* 2010, **40**, 253–266.
8. D. M. Toivola, P. Strnad, A. Habtezion and M. B. Omary, *Trends Cell Biol.,* 2010, **20**, 79–91.
9. M. B. Al-Fageeh, R. J. Marchant, M. J. Carden and C. M. Smales, *Biotechnol. Bioeng.,* 2006, **93**, 829–835.
10. H. Kaufmann, X. Mazur, M. Fussenegger and J. E. Bailey, *Biotechnol. Bioeng.,* 1999, **63**, 573–582.
11. J. J. Hawkes and W. T. Coakley, *Enzyme Microb. Technol.,* 1996, **19**, 57–62.
12. K. A. J. Borthwick, W. T. Coakley, M. B. McDonnell, H. Nowotny, E. Benes and M. Gröschl, *J. Microbiol. Methods,* 2005, **60**, 207–216.
13. P. Augustsson, R. Barnkob, S. T. Wereley, H. Bruus and T. Laurell, *Lab Chip,* 2011, **11**, 4152–4164.
14. J. Svennebring, O. Manneberg and M. Wiklund, *J. Micromech. Microeng.,* 2007, **17**, 2469–2474.
15. B. Vanherberghen, O. Manneberg, A. Christakou, T. Frisk, M. Ohlin, H. M. Hertz, B. Önfelt and M. Wiklund, *Lab Chip,* 2010, **10**, 2727–2732.
16. M. Evander, L. Johansson, T. Lilliehorn, J. Piskur, M. Lindvall, S. Johansson, M. Almqvist, T. Laurell and J. Nilsson, *Anal. Chem.,* 2007, **79**, 2984–2991.

17. I. Z. Shirgaonkar, S. Lanthier and A. Kamen, *Biotechnol. Adv.,* 2004, **22**, 433–444.
18. R. E. Apfel, *J. Acoust. Soc. Am.,* 1997, **101**, 1227–1237.
19. M. Voltmer and A. Weber, *Z. Phys. Chem.,* 1926, **119**, 277–301.
20. R. E. Apfel and C. K. Holland, *Ultrasound Med. Biol.,* 1991, **17**, 179–185.
21. W. L. Nyborg, *Ultrasound Med. Biol.,* 2001, **3**, 301–333.
22. P. Marmottant and S. Hilgenfeldt, *Nature,* 2003, **423**, 153–156.
23. E. L. Carstensen, M. W. Miller and C. A. Linke, *J. Biol. Phys.,* 1974, **2**, 173–192.
24. D. L. Miller, R. M. Thomas and M. E. Frazier, *Ultrasound Med. Biol.,* 1991, **17**, 729–735.
25. K. S. Suslick, *Science,* 1990, **247**, 1439–1445.
26. J. R. Lindner, *Nat. Rev. Drug Discovery,* 2004, **3**, 527–532.
27. C. X. Deng, F. Sieling, H. Pan and J. Cui, *Ultrasound Med. Biol.,* 2004, **30**, 519–526.
28. Y. Z. Zhao, Y. K. Luo, C. T. Lu, J. F. Xu, J. Tang, M. Zhang, Y. Zhang and H. D. Liang, *J. Drug Targeting,* 2008, **16**, 18–25.
29. S. Mitragotri, *Nat. Rev. Drug Discovery,* 2005, **4**, 255–260.
30. R. Karshafian, P. D. Bevan, R. Williams, S. Samac and P. N. Burns, *Ultrasound Med. Biol.,* 2009, **35**, 847–860.
31. W. Zhong, W. H. Sit, J. M. F. Wan and A. C. H. Yu, *Ultrasound Med. Biol.,* 2011, **37**, 2149–2159.
32. D. Carugo, D. N. Ankrett, P. Glynne-Jones, L. Capretto, R. J. Boltryk, P. A. Townsend, X. Zhang and M. Hill, Proc. of μTAS 2011, Seattle, USA, ed. J. Landers, CBMS, 2011, pp. 106–108.
33. M. Dyson, J. Pond and B. Woodward, *Nature,* 1971, **232**, 572–573.
34. M. Dyson, J. B. Pond, B. Woodward and J. Broadbent, *Ultrasound Med. Biol.,* 1974, **1**, 133–148.
35. E. S. B. C. Ang, V. Gluncic, A. Duque, M. E. Schafer and P. Rakic, *Proc. Natl. Acad. Sci. U. S. A.,* 2006, **103**, 12903–12910.
36. A. McQueen, E. Meilhoc and J. E. Bailey, *Biotechnol. Lett.,* 1987, **9**, 831–836.
37. D. E. Hughes and W. L. Nyborg, *Science,* 1962, **138**, 108–114.
38. S.-F. Chang, C. A. Chang, D.-Y. Lee, P.-L. Lee, Y.-M. Yeh, C.-R. Yeh, C.-K. Cheng, S. Chien and J.-J. Chiu, *Proc. Natl. Acad. Sci. U. S. A.,* 2008, **105**, 3922–3927.
39. L. Kim, Y.-C. Toh, J. Voldman and H. Yu, *Lab Chip,* 2007, 7, 681–694.
40. D. N. Ankrett, D. Carugo, J. Lei, P. Glynne-Jones, P. A. Townsend, X. Zhang and M. Hill, *J. Nanobiotechnol.,* 2013, **11**, 20.
41. J. Hultström, O. Manneberg, K. Dopf, H. M. Hertz, H. Brismar and M. Wiklund, *Ultrasound Med. Biol.,* 2007, **33**, 145–151.
42. T. W. Frisk, M. A. Khorshidi, K. Guldevall, B. Vanherberghen and B. Önfelt, *Biomed. Microdevices,* 2011, **4**, 683–693.
43. L. A. Crum and A. I. Eller, *J. Acoust. Soc. Am.,* 1970, **48**, 181–189.
44. C. C. Church, H. G. Flynn, M. W. Miller and P. G. Sacks, *Ultrasound Med. Biol.,* 1982, **8**, 299–309.

45. F. G. Blake, *J. Acoust. Soc. Am.*, 1949, **21**, 551.
46. T. Watanabe and Y. Kukita, *Phys. Fluids A*, 1993, **5**, 2682–2688.
47. A. A. Doinikov, *Phys. Fluids*, 2002, **14**, 1420–1425.
48. S. Khanna, N. N. Amso, S. J. Paynter and W. T. Coakley, *Ultrasound Med. Biol.*, 2003, **29**, 1463–1470.
49. H. Böhm, P. Anthony, M. R. Davey, L. G. Briarty, J. B. Power, K. C. Lowe, E. Benes and M. Gröschl, *Ultrasonics*, 2000, **38**, 629–632.
50. S. Radel, A. J. McLoughlin, L. Gherardini, O. Doblhoff-Dier and E. Benes, *Ultrasonics*, 2000, **38**, 633–637.
51. S. Radel, L. Gherardini, A. J. McLoughlin, O. Doblhoff-Dier and E. Benes, *Bioseparation*, 2001, **9**, 369–377.
52. D. Bazou, L. A. Kuznetsova and W. T. Coakley, *Ultrasound Med. Biol.*, 2005, **31**, 423–430.
53. D. G. Kilburn, D. J. Clarke, W. T. Coakley and D. W. Bardslay, *Biotechnol. Bioeng.*, 1989, **34**, 559–562.
54. O. Doblhoff-Dier, T. Gaida, H. Katinger, W. Burger, M. Gröschl and E. Benes, *Biotechnol. Prog.*, 1994, **10**, 428–432.
55. P. W. S. Pui, F. Trampler, S. A. Sonderhoff, M. Groeschl, D. G. Kilburn and J. M. Piret, *Biotechnol. Prog.*, 1995, **11**, 146–152.
56. Z. Wang, P. Grabenstetter, D. L. Feke and J. M. Belovich, *Biotechnol. Prog.*, 2004, **20**, 384–387.
57. D. Bazou, G. A. Foster, J. R. Ralphs and W. T. Coakley, *Mol. Membr. Biol.*, 2005, **22**, 229–240.
58. S. Khanna, B. Hudson, C. J. Pepper, N. N. Amso and W. T. Coakley, *Ultrasound Med. Biol.*, 2006, **32**, 289–295.
59. P. Thévoz, J. A. Adams, H. Shea, H. Bruus and T. Soh, *Anal. Chem.*, 2010, **82**, 3094–3098.
60. D. Bazou, W. T. Coakley, A. J. Hayes and S. K. Jackson, *Toxicol. In Vitro*, 2008, **22**, 1321–1331.
61. P. Augustsson, C. Magnusson, M. Nordin, H. Lilja and T. Laurell, *Anal. Chem.*, 2012, **84**, 7954–7962.
62. M. A. Burguillos, C. Magnusson, M. Nordin, A. Lenshof, P. Augustsson, M. J. Hansson, E. Elmér, H. Lilja, P. Brundin, T. Laurell and T. Deierborg, *PLoS One*, 2013, **8**, e64233.
63. S. J. Martin, C. P. M. Reutelingsperger, A. J. McGahon, J. A. Rader, R. C. A. A. van Schie, D. M. LaFace and D. R. Green, *J. Exp. Med.*, 1995, **182**, 1545–1556.
64. K. Yasuda, S. S. Haupt and S. Umemura, *J. Acoust. Soc. Am.*, 1997, **102**, 642–645.
65. K. Yasuda, *Sens. Actuators, B*, 2000, **64**, 128–135.
66. C. M. Cousins, P. Holownia, J. J. Hawkes, M. S. Limaye, C. P. Price, P. J. Keay and W. T. Coakley, *Ultrasound Med. Biol.*, 2000, **26**, 881–888.
67. J. Zhang, A. Collins, M. Chen, I. Knyazev and R. Gentz, *Biotechnol. Bioeng.*, 1998, **59**, 351–359.
68. Y. Chisti, *Trends Biotechnol.*, 2003, **21**, 89–93.

69. L. Gherardini, C. M. Cousins, J. J. Hawkes, J. Spengler, S. Radel, H. Lawler, B. Devcic-Kuhar, M. Gröschl, W. T. Coakley and A. J. McLoughlin, *Ultrasound Med. Biol.*, 2005, **31**, 261–272.
70. J. Dykes, A. Lenshof, I.-B. Åstrand-Grundström, T. Laurell and S. Scheding, *PloS One*, 2011, **6**, e23074.
71. A. E. Christakou, M. Ohlin, B. Vanherberghen, M. Ali Khorshidi, N. Kadri, T. Frisk, M. Wiklund and B. Önfelt, *Integr. Biol.*, 2013, **5**, 712–719.

Subject Index

Acoustically induced mixing, 403–405
Acoustic droplet actuation, 405–407
Acoustic eigenmodes, 38–42
    boundary conditions, 38–39
    viscous and radiative losses, 40–42
    water-filled rectangular channel, 39–40
Acoustic filters, 455–462
    enhanced sedimentation, 459–460
    h-shape separator, flow splitters, 461–462
    separation efficiency influence, 460–461
    ultrasonically enhanced sedimentation, 456–459
Acoustic levitators, 278–279
Acoustic manipulation technology, 242–254
    electrical fields, dielectrophoresis, 248–250
    gravitational forces, 243–245
    hydrodynamic forces, 245–248
    magnetic forces, 250–252
    optical forces, 252–253
Acoustic particle manipulation
    acoustic radiation forces, 127–128
    acoustic wave field, generation, 128–129
    planar resonators, modelling, 128–131
    position control, resonators, 138–141
    resonator congurations, 131–138
Acoustic particle sorting, 154–157
Acoustic radiation, 56–59
Acoustic radiation forces, 65–79, 127–128
    acoustophoretic particle tracks, 74–77
    applied piezo voltage, 77–78
    cylinder, inviscid fluid, 234–235
    effects of, 553–554
    energy density, 77–78
    particle near a wall, 236–237
    plate vibrating harmonically, 58–59
    scattering theory, 67–74
    stratified liquids, 163, 179–181
    viscous corrections, 78–79
Acoustic radiation torque, non-spherical particles, 220–222, 237–238
Acoustic resonances, 35–38
Acoustic resonators, 100–123
    actuation, 116–123
    design configurations, 106–116
    material, choice, 103–106
    symmetry breaking in, 43–45
Acoustic streaming effects, 32–35, 400–403
    acoustic levitators, 278–279
    analysis of, 256–308
    applications of, 312–333

Subject Index

bubbles, acoustic fields, 296–308
cavitation microstreaming, 317–318, 327–331
continuous-phase Stokes layer, 283–284
Eckart streaming, 316–317, 326–327
effects of, 554
first-order solution $[O(\varepsilon)]$, 294–296
incompressible flow approximation, 263–266
inner solutions, 273–274
leading-order solutions, 282–284, 294
microfluidic applications of, 318–333
oscillatory flows, 259–263
$O(\varepsilon)$ solutions, 284–286
outer solution, 272–273
problem statement, 272
qualitative description of, 313–318
quartz wind, 266–267
radial oscillations, 299–301
Rayleigh streaming, 268–270, 318–325
semi-cylindrical bubble, 302–305
solid particle/liquid drop, velocity antinode, 290–294
solid particles, interaction, 276–296
solid sphere, nodes, 289–290, 294–296
steady streaming, 274–276
Stokes drift, 270–271
surface acoustic wave, 331–333
transverse oscillations, 297–301
between two plates, 271–276
Acoustic trapping
applications in, 189–208
bioassays, microparticles, 199–201
cell population studies, 201–208
methods for, 193–196
particle studies, 196–199
theory, 190–192
Acoustic wave equation, 30–32
Acoustofluidic micro-devices
microscopy for, 493–517
optical microscopy, 494–516
Acoustofluidic resonators
biocompatibility, 545–562
cavitation-based effects, 550–553
cell viability in, 545–562
glass, ultrasonic cell handling, 555
silicon, ultrasonic cell handling, 555
thermal bioeffect, 547–550
Acoustophoretic particle tracks, 74–77
Actuation, 116–123
coupled resonance modes, 118–119
electrode and transducer modifications, 119–121
focused transducers, 121–123
transducer coupling, 116–118
Admittance curves, 524–527
Agglutination assays, 422–425
Anisotropic media, 349–350
Applied piezo voltage, 77–78

Basic continuum fields, 2–3
Bi-directional fractionation, 174–175
Biocompatibility, 545–562
Boundary layer acoustic streaming, 314–316
Bright-field microscopy, 507–509
dark field, 507–508
differential interference contrast microscopy, 508–509
phase contrast, 507

Carbon nanotubes (CNTs), 368–370
Cavitation microstreaming, 317–318, 327–331
Cell attractor wall, 430–431
    flexural plate wave, 430
    half wavelength attractor wall, 430
    rigid horizontal base, 431
    sound, 431
    thick polymer attractor wall, 430–431
    thin attractor wall, 431
    thin membrane, flexural plate wave, 431
Cell population studies, 201–208
    enrichment/washing of cells, 204–207
    size sorting and separations, 208
Cellular labeling, 510
Cell viability, ultrasound impact, 558–561
Chemiluminescence, 203
CNTs. *See* Carbon nanotubes
COMSOL multiphysics, 62–63, 95
Concentration factor (f_c), 150
Continuity equation, 5–6
Continuous flow acoustophoresis
    acoustic particle sorting, 154–157
    applications in, 148–185
    buffer exchange, devices, 163–167
    cell analysis, 182–183
    cells bound to beads, 176–177
    cells, medium exchange, 157–167
    clarification, particles, 152–153
    cytometry applications, 154–157
    frequency switching, 177–179
    microparticles, medium exchange, 157–167
    negative contrast particles, 177
    particle concentration, 150–152
    particles separation, acoustophysical properties, 167–176
    positive contrast particles, 176–177
    separation performance, measuring, 183–184
    size limitations, particles, 184–185
    stratified liquids, acoustic forces, 163
    temperature aspects, 182
    transport mechanisms, 158–163
    variations, acoustic field, 181
Continuous-phase Stokes layer, 283–284
Continuum mechanics, 46–64
Coupled resonance modes, 118–119
Cylindrical capillaries, 151
Cytometry, 154–157

DEP. *See* Dielectrophoresis
Design approaches, resonators
    chamber wall, modelling, 484–486
    experimental confirmation, 486–487
    heating mitigation, 482
    modelling resonant parts, 483–484
    standing wave creation, 482–483
Design configurations, acoustic resonators, 106–116
    acoustic traps, 111–114
    capillaries, 114–116
    1D and 2D acoustic focusing, continuous flow, 110–111
    flow splitter design, 109–110
    layered resonators, 106–107
    surface acoustic wave devices, 108–109
    transversal resonators, 107–108

Subject Index

DIC microscopy. *See* Differential interference contrast (DIC) microscopy
Dielectrophoresis (DEP), 248–250
Differential interference contrast (DIC) microscopy, 508–509
Diffusion-limited colloid aggregation (DLCA), 197
Dimensionless numbers, 22–24
DLCA. *See* Diffusion-limited colloid aggregation

Eckart streaming, 256, 316–317, 326–327
Effective Q-factor (Q_{eff}), 464
Eigenmodes, acoustic modes, 38–42
    boundary conditions, 38–39
    viscous and radiative losses, 40–42
    water-filled rectangular channel, 39–40
Electrical admittance, 92, 94
Electrical fields, dielectrophoresis, 248–250
Electrical impedance, 92, 93
Electric displacement, 83
Energy density, 77–78
Entrance length, microfluidics, 24–25
Equivalent circuit modeling, 16–20
    hydraulic compliance, 18–20
    hydraulic inductance, 20
    hydraulic resistance, 17–18
Euler–Bernoulli beam theory, 52
Eulerian picture, continuum fields, 2

FACS. *See* Fluorescence activated cell sorters
FEA. *See* Finite element analysis (FEA) model
FEM. *See* Finite element method (FEM) simulations
FFA. *See* Free flow acoustophoresis
FFF. *See* Field flow fractionation (FFF) scheme

Field flow fractionation (FFF) scheme, 247
Finite element analysis (FEA) model, 130, 131
Finite element method (FEM) simulations, 42–45, 61, 95–98
Finite volume method (FVM), 233–234
First-order perturbation theory, 30–32
Flow rate, 16, 21–22
Flow solutions, 11–16
    flow rate, 16
    hydrostatic pressure, 11–12
    Poiseuille flow, 13–16
    Reynolds number, 12–13
    Stokes equation, 12–13
Flow splitter design, 109–110
Flow stream
    continuous separation of particles, 379–381
    particles, focusing of, 378–379
    single particle, actuating, 382
Fluid structure interaction, 56–61
    acoustic radiation, 56–59
    device vibration, particle manipulation, 59–61
Fluorescence activated cell sorters (FACS), 155–156
Fluorescence assays, 425–427
Fluorescence microscopy, 509–512
    cellular labeling, 510
    live cell fluorescence microscopy, 510–511
    multiple fluorescent probes, 511
    optical sectioning techniques, 511–512
    probes, 509
Focused transducers, 121–123
Fourier transform infrared (FT-IR) attenuated total reflection (ATR) spectroscopy, 439–443
Fractionation, particle separation, 172–174

Free flow acoustophoresis (FFA), 168–172
Frequency analysis, 524–527
Frequency sweeping, 131
Frequency switching, 177–179
FVM. *See* Finite volume method

Geometrical effects, 42–45
Gravitational forces, 243–245

Hagen–Poiseuille law, 17–18
Half-wave devices, 133–137
Half wavelength acoustic resonator, 101, 102
Heat-transfer equation, 9–11
Hydraulic compliance, 18–20
Hydraulic inductance, 20
Hydraulic resistance, 17–18
Hydrodynamic forces, 245–248
Hydrostatic pressure, 11–12

Incompressible flow approximation, 263–266
    three-dimensional axisymmetric flows, 265–266
    two-dimensional flows, 265
Inertial time scale, 25–27
Inline bioprocess monitoring, 443–447
Integrated selective enrichment target (ISET), 206
Interface waves, 347–349
ISET. *See* Integrated selective enrichment target

Kirchhoff plate theory, 52, 54
KLM model, 129, 130

Lab-on-a-chip technologies, 354–391
Laser interferometry, 521–524
Layered resonators, 106–107
Leading-order solutions, 282–284
Lead zirconate titanate (PZT), 82
Leaky Rayleigh waves, 344–347
Liquid crystal reorientation, 370

Live cell fluorescence microscopy, 510–511
Longitudinal-Rayleigh combination, 341

Magnetic forces, 250–252
Mason model, 129–130
Mathematical notation, microfluidics, 3–5
Mechanical index (MI), 551, 552
Mechanical model, ultrasonic cavity, 61–63
Mechanical strain, 84
MI. *See* Mechanical index
Microchambers, pattern formation, 214–219
    device description, 214
    standing waves, superposition, 214–219
Microfluidic acoustic resonators. *See* Acoustic resonators
Microfluidics
    basic continuum fields, 2–3
    continuity equation, 5–6
    dimensionless numbers, 22–24
    entrance length, 24–25
    equations in, 1–27
    equivalent circuit modeling, 16–20
    Eulerian picture, continuum fields, 2
    flow rate, 16, 21–22
    flow solutions, 11–16
    heat-transfer equation, 9–11
    hydraulic compliance, 18–20
    hydraulic inductance, 20
    hydraulic resistance, 17–18
    hydrostatic pressure, 11–12
    inertial time scale, 25–27
    mathematical notation, 3–5
    Navier–Stokes equation, 7–9
    Poiseuille flow, 13–16
    Reynolds number, 12–13
    scaling laws, 20–27
    Stokes equation, 12–13

Microfluidic technologies, TSAW
  droplet/cell sorting, 368
  fluid mixing, 359–361
  fluid translation, 361–365
  jetting and atomization, 366–367
  microfluidic pumping, 363–364
  microfluidic rotational motor, 364–365
  nano-objects, reorientation, 368–370
  particle/cell concentration, 367–368
Microgripper, particle positioning, 227–229
Micro-particles, rotation, 220–223
  non-spherical particles, acoustic radiation torque, 220–222
  spherical particles, acoustic viscous torque, 222–223
Microtube alignment, 388–389
Mindlin plate theory, 54
Modal analysis, 524–527
Multiple fluorescent probes, 511
Multi-wavelength resonators
  acoustic contrast, 473–474
  acoustic filters, 455–462
  acoustic signal control, 470–471
  applications and considerations, 452–487
  construction of, 462–465
  design approaches, 482–487
  development, tools for, 465
  dimensions, 471
  entrance condition, 479–480
  flow expansion and contraction, channel, 480–481
  gel technique, 468–470
  heating, 476–479
  models and measurements, 465–468
  non-turbulent flow, 474–475
  scale considerations, 472
  thickness limits, 472
  turbulence initiation, scale-up, 476
  turbulence, onset, 475–476

Navier–Stokes (N–S) equation, 7–9, 233–234
Negative contrast particles, 177
N–S equation. *See* Navier–Stokes (N–S) equation
Nusselt number, 477, 478

Optical forces, 252–253
Optical microscopy
  bright-field microscopy, 507–509
  contrast, 506
  design criteria, 512–515
  fluorescence microscopy, 509–512
  illumination system, 497–506
  implementation of, 512–516
  light sources and detectors/cameras, 506
  magnification, 499–500
  modes of, 506–512
  principles of, 494–506
  resolution, 501–505
Optical sectioning techniques, 511–512
Optical trapping, 190
Oscillatory flows, 259–263
$O(\varepsilon)$ solutions, 284–286

Particle image velocimetry (PIV), 516
Particle sensors, 420–448
Particles separation, acoustophysical properties
  bi-directional fractionation, 174–175
  fractionation, 172–174
  free flow acoustophoresis, 168–172
  medium, altering, 176
  pre-alignment, 172–174

Particle transport
    device description, 225–226
    experimental results, 227
    numerical simulation, 226
PDMS microfluidics. *See* Polydimethylsiloxane (PDMS) microfluidics
Perturbation methods, 256–308
Perturbation theory, 29–45
    first-order, 30–32
    second-order, 32–35
Phononic crystal-assisted TSAWs, 370–372
    jetting and nebulization, 373–374
    particle/cell concentration, 372–373
Piezoelectricity, 81–98
    basic equations, 82–85
    element vibration, applied electrical voltage, 85–87
    transducers, mechanical vibrations, 87–94
Piezoelectric transducers, mechanical vibrations, 87–94
PIV. *See* Particle image velocimetry
Planar resonant devices
    cell-interaction studies, 143–144
    filtration, washing and separation, 141–142
    modelling and applications of, 127–144
    sensing and detection, 142–143
    sonoporation, 143–144
Poiseuille flow, 13–16
Polydimethylsiloxane (PDMS) microfluidics, 407–408
    acoustic coupling, 408–409
    droplet based fluidics, 412–417
    pumping fluid in, 410–412
    SAW excitation, 408–409
Positive contrast particles, 176–177
Pre-alignment, particle separation, 172–174
Protein manipulation, 385–388
Pz26 piezoelectric ceramics, 84

Quality factor (Q-factor), 60, 464, 533–539
Quarter-wave resonators, 137–138
Quartz wind, 266–267

Raman spectroscopy, crystal agglomeration, 436–439
Rayleigh streaming, 268–270, 318–325
Rayleigh waves, 338–344
Reaction-limited colloid aggregation (RLCA), 196
Resonator congurations
    half-wave devices, 133–137
    quarter-wave resonators, 137–138
    thin-reflector resonators, 138
Reynolds number, 12–13, 475–476
RLCA. *See* Reaction-limited colloid aggregation
Runge–Kutta algorithm, 234

SAW. *See* Surface acoustic waves
Scaling laws
    dimensionless numbers, 22–24
    entrance length, 24–25
    flow rate, 21–22
    inertial time scale, 25–27
Scattering theory, 67–74
    dipole coeffcient f_2, 71–72
    monopole coefficient f_1, 71
    near-field potential, 70–71
    and radiation force, 68–70
    standing plane wave, radiation force, 72–74
Schlieren imaging, pressure fields, 527–529
Scholte waves, 347
Second-order perturbation theory, 32–35
Sonoporation, 143–144
SSAW. *See* Standing surface acoustic wave

Stagnant fluid
  particle/cell/organism, manipulating, 384–385
  particles patterning, 382–384
Standing surface acoustic wave (SSAW)
  microfluidic technologies, 377–389
  theory, 375–377
State-of-the-art sensor systems, 439
Stokes drift, 270–271
Stokes equation, 12–13
Stoneley waves, 349
Surface acoustic waves (SAW), 108–109, 401
  acoustically induced mixing, 403–405
  acoustic droplet actuation, 405–407
  acoustic streaming effects, 400–403
  anisotropic media, 349–350
  interface waves, 347–349
  leaky Rayleigh waves, 344–347
  PDMS microfluidics, 407–417
  piezoelectric considerations, 349–350
  Rayleigh waves, 338–344
  Scholte waves, 347
  Stoneley waves, 349
  streaming, 331–333
  surface waves generation, 350–352
  theory of, 337–352
Symmetry breaking, 43–45

Thermal bioeffect, 547–550
Thin-reflector resonators, 138
Time-averaged acoustic effects
  cylinder, viscous fluid, 235–232
  Navier-Stokes (N-S) equation, 233–234
  non-spherical particles, acoustic radiation torque, 237–238
  numerical techniques for, 229–239
  perturbation method, 231–232
Transducer characterization, 529–539
  electrical and mechanical measurements, 530–533
  experimental setup, 529–530
  Q-factor, 533–539
Transducer coupling, 116–118
Transport mechanisms, acoustophoresis, 158–163
  acoustic streaming, 160–161
  diffusion, 160
  flow perturbations, 162–163
  hydrodynamic interactions, 161–162
Transversal resonators, 107–108
Transverse-Rayleigh combination, 341
Travelling surface acoustic wave (TSAW)
  microfluidic technologies, 358–374
  theory, 357–358
TSAW. See Travelling surface acoustic wave

Ultrasonic microrobotics, cavities, 212–239
  microchambers, pattern formation, 214–219
  micro-particles, rotation, 220–223
  particle manipulation, 214–229
  particle positioning, microgripper, 227–229
  particle transport, 223–227
  time-averaged acoustic effects, 229–239

Ultrasonic particle manipulation
    acoustic fields, excitation, 81–98
    admittance curves, 524–527
    characterization of, 520–543
    continuum mechanics, 46–64
    FEM model example of, 95–98
    fluid structure interaction, 56–61
    frequency analysis, 524–527
    laser interferometry, 521–524
    linear elastodynamics, 47–51
    manipulation devices characterization, 539–542
    mechanical model, example, 61–63
    modal analysis, 524–527
    Schlieren imaging, pressure fields, 527–529
    structure deformations, 51–56
    transducer characterization, 529–539
Ultrasonic standing wave manipulation devices, cell bioeffects, 555–558
Ultrasound-enhanced bead-based immunoassays, 421–422
    agglutination assays, 422–425
    fluorescence assays, 425–427

Ultrasound-enhanced particle sensors, 427–428
    cell attracting wall, microsystems, 428–430
    cell attractor wall, 430–431
    cell capture and detection, selective, 431–434
    clumps, distinctive pattern of, 428
    device design, 430
    stage of developments, 434–435
Ultrasound-enhanced vibrational spectroscopy, 435–436
    FT–IR ATR spectroscopy, 439–443
    inline bioprocess monitoring, ATR probe, 443–447
    Raman spectroscopy, 436–439
Ultrasound resonances, 29–45

Viscous corrections, radiation force, 78–79
Viscous liquids, acoustic resonances, 35–38
Volumetric flow rate, 16

Womersley number, 257